Electromechanics and MEMS

Offering a consistent, systematic approach to capacitive, piezoelectric, and magnetic MEMS, this textbook equips students to design and develop practical, system-level MEMS models.

- Includes a concise yet thorough treatment of the underlying principles of electromechanical transduction.
- Makes extensive use of easy-to-interpret electrical and mechanical analogs, such as electrical circuits, electromechanical two-port models, and the cascade paradigm.
- Each chapter features extensive worked examples, and numerous homework problems.

Thomas B. Jones is Professor of Electrical Engineering at the University of Rochester. An experienced educator involved in teaching for over 40 years, his research has focused on electric field-mediated manipulation and transport of particles and liquids. He holds a Ph.D. from MIT, is the author of *Electromechanics of Particles* (Cambridge University Press, 1995) and is a Fellow of the IEEE.

Nenad G. Nenadic is a Research Associate Professor at the Rochester Institute of Technology. His career, spanning both industry and academia, has involved him in many aspects of MEMS, including design and analysis, system-level simulation, test development, and marketing. He holds a Ph.D. from the University of Rochester, where he assisted in the teaching of graduate-level MEMS courses.

"This is an excellent textbook presenting the fundamentals of electromechanics required by every practicing MEMS engineer. The authors treat the arduous concepts of coupled electrical and mechanical systems simultaneously with lucidity and a thorough pedagogical rigor that comes from deep appreciation of the field and the love to impart that knowledge as a teacher. The book elucidates the concepts with very topical examples of microelectromechanical systems such as MEMS microphones, comb drive actuators, gyroscopes, energy harvesters, and piezoelectric and magnetic devices, including MATLAB models and a comprehensive set of problems at the end of each chapter."

Srinivas Tadigadapa, The Pennsylvania State University

"A fantastic book for the student seeking a solid foundation in electromechanical device design and an essential reference for the expert MEMS engineer. Jones and Nenadic present the fundamental theory behind electromechanical transduction, with a focus on capacitive drive and sense microsystems. The authors systematically frame the device fundamentals into real world micro scale device applications that provide relevance to the underlying physics. This book captures and dutifully explains the foundational physics at work in the MEMS devices we often unknowingly use daily in our automobiles, mobile phones and electronic devices."

Chris Keimel, GE Global Research

"*Electromechanics and MEMS* is a thorough treatment of fundamental MEMS analysis for both the student and the practitioner. The readers are presented with the tools to methodically build system models that are comprehensive yet manageable."

Eric Chojnacki, MEMSIC, Inc.

Electromechanics and MEMS

THOMAS B. JONES

University of Rochester, New York

NENAD G. NENADIC

Rochester Institute of Technology, New York

CAMBRIDGE
UNIVERSITY PRESS

CAMBRIDGE
UNIVERSITY PRESS

Shaftesbury Road, Cambridge CB2 8EA, United Kingdom

One Liberty Plaza, 20th Floor, New York, NY 10006, USA

477 Williamstown Road, Port Melbourne, VIC 3207, Australia

314–321, 3rd Floor, Plot 3, Splendor Forum, Jasola District Centre, New Delhi – 110025, India

103 Penang Road, #05–06/07, Visioncrest Commercial, Singapore 238467

Cambridge University Press is part of Cambridge University Press & Assessment, a department of the University of Cambridge.

We share the University's mission to contribute to society through the pursuit of education, learning and research at the highest international levels of excellence.

www.cambridge.org
Information on this title: www.cambridge.org/9780521764834

First published 2013

A catalogue record for this publication is available from the British Library

Library of Congress Cataloging-in-Publication data
Jones, T. B. (Thomas Byron), 1944–
Electromechanics and MEMS / Thomas B. Jones, University of Rochester, New York, Nenad G. Nenadic, Rochester Institute of Technology.
pages cm
Includes bibliographical references.
ISBN 978-0-521-76483-4 (hardback)
1. Microelectromechanical systems. I. Nenadic, Nenad G. II. Title.
TK7875.J66 2012
621.381 - dc23 2012021830

ISBN 978-0-521-76483-4 Hardback

Additional resources for this publication at www.cambridge.org/mems

Contents

Preface

The growing interest in microsystems, and particularly in MEMS technology, has reasserted electromechanics as a key discipline. This book fills the need for a textbook that presents the fundamentals of electromechanics, classifies structures according to their functional capabilities, develops systematic modeling methods for the design of MEMS devices integrated into electronic systems, and provides practical examples derived from selected microdevice technologies. It is written for engineering students and physical science majors who want to learn about such systems. A further ambition is that the book will find its way slowly onto the shelves of practicing engineers involved in MEMS design and development.

Organization

The organization proceeds from basics to systems-oriented applications. The first three chapters focus on fundamentals of circuits and lumped parameter electromechanics. Chapter 1 provides some historical context, introducing key terminology and then offering a general description of electromechanical transducers based on power and energy considerations. Chapter 2 introduces the crucial concept of circuit-based modeling. Because the vast majority of MEMS devices are capacitive, this chapter focuses on circuits with capacitors and resistors. Chapter 3, drawing heavily on Part 1 of H. H. Woodson and J. R. Melcher's text, *Electromechanical Dynamics*, presents the classic, energy-based formulation for electromechanical interactions. The treatment here differs from their text by concentrating on capacitive microelectromechanical devices and introducing the geometries and dimensions characteristic of MEMS technology.

Next, we present in detail a systematic method for modeling and then analyzing practical MEMS. Chapter 4 introduces the general small-signal transducer formalism found in H. K. P. Neubert's classic text, *Instrument Transducers*. This formalism serves as the robust backbone for the chapters that follow. Representing the coupled small-signal behavior – mechanical and electrical – of a MEMS device is the key to developing a systematic, integrated coverage of the transient dynamics, frequency response, and mechanical stability of transducers and sensors.[1] The critical design and performance

[1] The formal basis of this integrated treatment was first published by H. A. Tilmans (*Journal of Micromechanics and Microengineering* 6, 1996, 157–176; *Journal of Micromechanics and Microengineering* 7, 1997, 285–309).

issues can be addressed in terms of a mechanical variable-dependent capacitance, $C(x)$, and its first two derivatives, i.e., $\mathrm{d}C/\mathrm{d}x$, and $\mathrm{d}^2C/\mathrm{d}x^2$. Small-signal analysis also facilitates the modeling of virtually any cascaded system-on-a-chip comprising both the MEMS device and the electronics that controls or monitors it.

The next three chapters deal in some depth with small-signal modeling of practical MEMS geometries, such as cantilevered beams, membranes, and diaphragms, and practical systems, such as pressure sensors, actuators, and gyroscopes. Chapter 5 introduces some operational amplifier-based electronics topologies for capacitive sensing devices. While the treatment is elementary, some prior exposure to electronic circuits may be helpful. These circuits are employed with variable-gap and variable-area capacitive sensors configured in single-ended and half-bridge configurations, with both DC and AC excitation. The chapter includes a brief summary of amplitude- and double-sideband suppressed carrier modulated schemes for capacitive sensors. Then, after a very concise presentation of noise, the advantages of modulated schemes for circumventing the often-prevalent $1/f$ noise are revealed. The chapter concludes with a consideration of phase-locked loop drives for resonant actuators.

Chapter 6 is devoted to mechanical modeling of deformable continua, such as cantilevered beams and plates. It is shown by analytical approximation that beams and plates can be reasonably well represented by linear, single-degree-of-freedom models and formulated as electromechanical two-ports, as introduced in Chapter 4. The more general, multiple-degree-of-freedom modeling approach is then presented. With proper attention given to the existence of multiple resonant modes, reduced order modeling is again restored. Practical MEMS devices amenable to this approach include pressure sensors, microphones, mirror arrays, and energy harvesters. Chapter 7 focuses on a few important examples of MEMS devices that utilize identifiable mechanical continua, including pressure transducers, accelerometers, gyroscopes, and energy harvesters. Specific geometries for each of these devices are considered in turn, with small-signal lumped parameter models as outcomes of the modeling exercise. This approach provides an opportunity to introduce important design considerations and higher-order effects that influence practical MEMS.

Because piezoelectric technology is important in certain microdevice applications, and will probably remain so, Chapter 8 offers a brief presentation of this subject. First, the standard phenomenological models, expressed in terms of mechanical and electrical field variables, are presented for the three important piezoelectric effects, that is, longitudinal, transverse, and shear modes. Then, adhering to the strategy of the previous chapters, these models are reduced to small-signal, two-port circuits useful in analysis of actuator and sensor devices.

Chapter 9, the last, offers a concise, largely self-contained coverage of the electromechanics of microscale magnetic transducers and linear, small-signal models for them. This chapter was prepared to anticipate likely breakthroughs in materials processing and fabrication methods for high-aspect ratio magnetic MEMS devices.[2] The

[2] See, for example, Section 2.02 of the new three-volume reference edited by Y. B. Gianchandani, O. Tabata, and H. Zappe, *Comprehensive Microsystems*, vols. 1, 2, and 3, Elsevier, 2008.

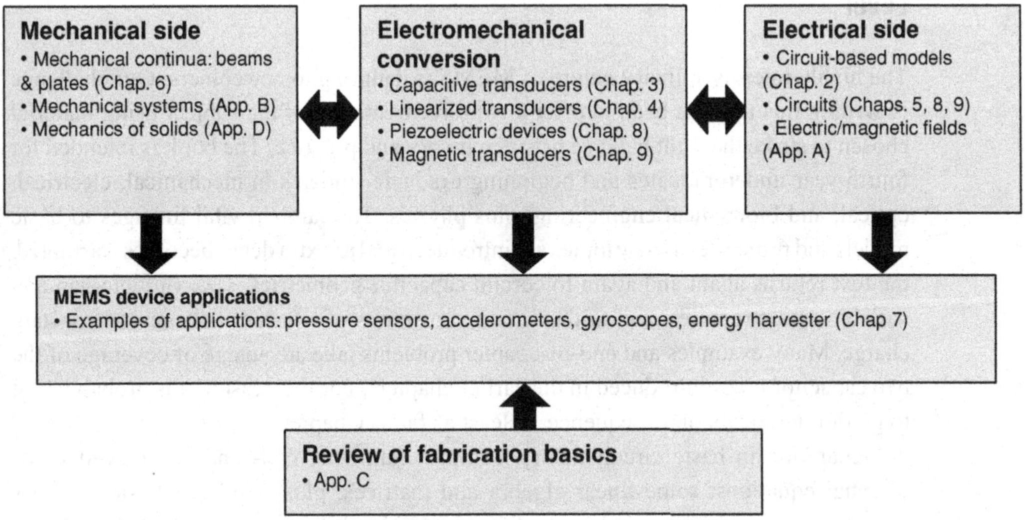

A block diagram representation of the organization of this textbook.

organization largely parallels the far more extensive treatment of capacitive devices found in Chapters 3, 4, and 5. A brief presentation of the basics of lumped parameter magnetic transducers comes first. Then, Neubert's small-signal formulation and the accompanying two-port circuit models are presented. The basic types of magnetic transducer are exemplified, as much as possible by introducing geometries that have already been or could be fabricated on the MEMS scale. The chapter concludes by presenting analyses of current bridge and linear variable differential transformer based sensors connected to operational amplifiers.

The diagram above reveals the organization of the text in a format mimicking standard block diagrams used throughout the book to represent general electromechanical systems. Thus, sections of the text that address the electromechanical conversion, the mechanical system, the electrical system, and MEMS device applications are separated into individual blocks.

The appendices found at the end of the text provide essential background and summaries of topics that intersect MEMS technology. Appendix A contains a very brief review of certain essentials from electromagnetic field theory. Appendix B covers mechanical systems, principally resonance of single- and multiple-degree-of-freedom systems. Appendix C offers a concise review of MEMS fabrication technology. The solid mechanics of typical MEMS structures, such as beams and plates, is covered in Appendix D.

Numerous examples and end-of-chapter problems reinforce and extend the important principles. A few more challenging design-oriented exercises have been included in some of the later chapters, and these are specially marked. Some of these exercises are good choices for assignment to teams composed of students from different disciplines.

Level

The highly interdisciplinary nature of MEMS is daunting to newcomers. Our challenge in writing this text has been to offer a cohesive treatment of the subject using material chosen to strike the right balance between theory and practice. The book is intended for fourth-year undergraduates and beginning graduate students in mechanical, electrical, optical, and biomedical engineering, plus physics. To maintain vital linkages to basic models and principles as new topics are introduced and as extra detail becomes warranted, the text returns again and again to certain capacitor geometries, viz., variable-gap and variable-area structures, and to the circuit constraints of constant voltage and constant charge. Many examples and end-of-chapter problems take advantage of coverage of the two capacitor types introduced in the earlier chapters. For this reason, it is probably best to go through the book in sequence, at least as far as Chapter 7.

Preparation in basic circuit theory, complex numbers, first- and second-order differential equations, some linear algebra and matrices, plus standard physics courses on mechanics, electricity, and magnetism are essential. In a typical setting, one might expect electrical and mechanical engineering students to dominate a class population, the electrical engineers having more understanding of circuits and linear systems and the mechanical engineers bringing their knowledge of mechanics and materials. Such an interdisciplinary mix is fertile. We are confident that industrious students will be able to overcome their respective "deficiencies" and gain deep understanding of the important basics of microelectromechanical interactions and how real MEMS devices are integrated into real systems. Our experience in teaching this material is that motivated students have little difficulty in overcoming any initial unfamiliarity with topic areas falling outside their undergraduate preparation. In fact, engineering and mechanical engineers and other students quickly form effective study teams and assist each other quite effectively in learning the material.

The sections with headings flagged by an asterisk provide details on related but peripheral topics. These sections are optional and may be passed over safely in a one-semester course.

Some limitations

While the acronym MEMS now seems to cover virtually all micromechanical devices, whether or not their actuation or transduction mechanism is really electromechanical in nature, this book is limited to devices with a true electromechanical mechanism. Thus, there is no coverage of magnetoresistive, bimetallic, or other thermally actuated devices. There is clearly a need for such a text, but we are not the ones to provide it.

Our coverage of mechanical continua is highly focused. A special effort has been made to demonstrate that the performance of the most common mechanical continua, e.g., the cantilevered beam and the circular plate diaphragm, can be reasonably approximated by reduced-order, lumped parameter models. Such models will prove useful to the systems engineer whose ambition is to integrate a MEMS device with the drive electronics in a real system.

The fabrication of MEMS is a diverse and ever-evolving enterprise. Because of the pace of change and development in the field, this text limits its coverage to the very concise summary of microfabrication found in Appendix C. Critical terminology and basic processing methods are introduced but MEMS packaging is not covered. We believe that reliance on the fine reference books already available is the best strategy for those ready to learn how to build MEMS devices.

Guide to the use of this text

The study of MEMS is a very broad and highly interdisciplinary subject. That being the case, the length of this text is more a testament to the breadth of the subject than to authorial diligence or ambitions for thoroughness. Indeed, we had great difficulty – and some vigorous arguments – about what to include and what to leave out. The instructor trying to decide whether or not to adopt this book faces a related dilemma; namely, does this text adequately cover the material for an established MEMS course with given objectives. The table on page xviii, which identifies the sections, examples, and end-of-chapter problems relevant to many of the important MEMS technologies, provides guidance in making the right decision. The information is organized by the set of technologies appearing in the leftmost column. This set is no doubt incomplete but, we think, reasonably illustrative.

Selected references on MEMS

Advanced students, design engineers, and researchers might not find adequate coverage of certain topics relevant to their specific interests. These individuals can refer to the many fine texts, monographs, and reference volumes available for assistance. These books, listed chronologically by date of publication and accompanied by very brief synopses of their contents, should be of help. Students are urged to familiarize themselves with some of these MEMS resources.

An early MEMS compendium is *Micromechanics and MEMS, Classic and Seminal Papers to 1990* (IEEE Press, 1997), edited by W. D. Trimmer and containing seminal MEMS papers. This is required reading for students entering the field.

A general MEMS reference volume, Gregory T. Kovacs' *Micromachined Transducers Sourcebook* (McGraw-Hill, 1998), offers a very broad survey of sensors and actuators, including some MEMS devices.

On the subject of MEMS microwave systems, *Microelectromechanical (MEM) Microwave Systems* (Artech House, 1999), by Héctor de los Santos, offers a concise treatment of MEMS devices and mechanisms applied in microwave systems.

Microsystem Design (Kluwer Academic, 2001) by Stephen D. Senturia provides broad coverage of a large amount of material. Since its publication, practicing engineers in

Guide to the use of this textbook organized by the more well-recognized MEMS technologies. Relevant sections found in the text itself, plus examples and end-of-chapter problems, are tabulated

Technology	Relevant sections & material	Relevant examples & end-of-chapter problems
Pressure sensors & microphones: Section 7.2	• Simple microphone model: 4.7.2 • Deformation of plates & diaphragms: 6.6, 6.7, App. D.10, D.11 • Distributed capacitive modeling: 6.7, App. A • Three-plate (differential) sensors: 4.7.5 • Modulation: 5.5, 5.6 • Noise: 5.8	Examples: 5.7, 6.9 Problems: 6.15, 6.16, 6.17, 7.1, 7.2, 7.3, 7.4, 7.5, 7.6, 7.7, 7.8
MEMS switches	• Vibrating beam mechanics: 6.2 to 6.5, App. D.5, D.7 • SDF resonance & transients: App. B • Magnetic MEMS: Chapter 9, App. A.4	Examples: 6.2
Accelerometers: Section 7.3	• Variable-gap & variable-area capacitors: 3.4, 4.3.3 • Three-plate (differential) sensors: 4.7.5 • Half-bridge amplification: 5.3, 5.4 • Switched capacitance: 5.7 • Modulation: 5.5, 5.6 • Noise: 5.8	Example 2.3, 5.6, 7.1, 7.2 Problems: 4.11, 7.9, 7.10, 7.11, 7.12, 7.13, 7.14, 7.15, 7.16
Gyroscopes: Section 7.4	• Variable-gap & variable-area capacitors: 3.4 • Multiport couplings: 3.3, 4.3.5 • Three-plate (differential) sensors: 4.7.5 • Half-bridge amplification: 5.3, 5.4 • Switched capacitance: 5.7 • Modulation: 5.5, 5.6 • Noise: 5.8 • Resonant drives: 5.9 • Vibration isolation: 6.8.2, 6.8.3	Examples: 3.4, 7.3 Problems: 5.16, 7.17, 7.18, 7.19, 7.20, 7.21
Energy harvesters: Section 7.5	• Mechanical resonance: App. B • Vibrating beam mechanics: 6.2 to 6.5, App. D.5, D.7 • Piezoelectric MEMS: Chapter 8 • Electret-based MEMS: 3.6	Examples: 4.2, 7.4 Problems: 5.17, 6.2, 6.8, 6.10, 7.21, 7.22, 7.23, 7.24, 7.25, 9.23
Rotating mirror displays	• Rotational capacitive transducers: 3.5 • Rotational MEMS devices: 4.3.4	Examples: 3.5

the field have relied on this volume as a standard reference for the design of MEMS systems.

Fundamentals of Microfabrication: The Science of Miniaturization (CRC, 2nd edition, 2002), written by Marc Madou, is a recently updated and very complete resource for those interested in learning about MEMS fabrication.

For mechanical modeling of MEMS devices, John A. Pelesko and David H. Berstein's *Modeling MEMS and NEMS* (Chapman and Hall/CRC, 2003) chiefly concerns modeling of beams, diaphragms, and other continua on the microscale. A particularly valuable feature is its coverage of numerical analysis methods relevant to MEMS devices.

On the subject of piezoelectric MEMS devices, *Micromechatronics* (Marcel Dekker, 2003), written by Kenji Uchino and Jayne R. Giniewicz, is the best modern reference available on ferroelectric phenomena. It provides treatment of the relevant solid mechanics and examples of piezoelectric devices.

Chang Liu's book, *Foundations of MEMS* (Pearson Education, 2006), offers a very general treatment of microsystems and MEMS topics, although with somewhat limited coverage of electromechanics.

Another general MEMS reference, *Comprehensive Microsystems* by Y. B. Gianchandani, O. Tabata, and H. Zappe, in three volumes (Elsevier, 2008), is exhaustively complete and up-to-date.

Finally, V. Kaajakari's *Practical MEMS* (Small Gear Publishing, 2009) is a new textbook featuring coverage of many areas of MEMS with excellent practical examples distributed throughout the text.

Special acknowledgments

To provide the student with concrete examples of working MEMS devices, we have incorporated images of MEMS devices throughout the text. These inclusions were made possible through permissions granted by the engineers, students, and faculty researchers who created the images. We are humbly grateful for this generosity. Further, Weiqiang Wang obtained for us the SEM of the pyramidal etched pit shown that appears in Fig. C.13. We acknowledge James Moon, who thoroughly reviewed Appendix C, and Erica MacArthur, who helped us by preparing some of the SEM images. Additional assistance from Scott Adams, Zeljko Ignjatovic, Kelly Lee, Christopher Keimel, and Paul H. Jones is gratefully acknowledged.

Final note

The sources and inspirations for this text are many, and we can rightly claim full credit only for the errors. More than anything else, it was excellent undergraduate-level teaching that fostered our appreciation of electromechanics. The lead author (TBJ) was introduced to the subject in the Fall Semester of 1968 at MIT by Herman Schneider, who delivered crystal-clear, virtually error-free lectures without resort to any notes. A few years later, though not quite having mastered the ability to lecture without notes, TBJ got the chance to teach this same course. Anyone teaching the class in those days relied upon a thick, unwieldy binder of mimeographed notes, which were destined to become the textbook entitled *Electromechanical Interactions* and written by H. H. Woodson and J. R. Melcher.

In the first year of his electrical engineering undergraduate studies at the University of Novi Sad in Serbia, the second author (NN) was confronted by the requirement to take a course titled *Introduction to Mechanics*. Any doubts harbored about the value of this course were rapidly dispelled by the inspiring lectures of Đorđe Đukić and Teodor Atanacković. In 1996, during the first year of his graduate studies at the University of Rochester, NN enrolled in a course entitled *Transducers and Actuators*. This course revealed MEMS technology to be a tightly woven fabric of mechanics, electricity and magnetism, circuit theory, electronics, and beam mechanics. The lecture notes and problems prepared for this course by TBJ served as the foundation of the present text.

1 Introduction

1.1 Background

The trend toward device miniaturization and large-scale integration, which has already revolutionized electronics, now promises a profound transformation of engineered mechanical systems, reducing their size by orders of magnitude while vastly increasing their capabilities. Microelectromechanical systems (MEMS) are now found in automotive airbag systems, computer projectors, digital cameras, gyroscopic sensors, and many other devices. Their small size invites a high degree of on-chip integration with essential drive, detection, and signal conditioning circuitry. These collective advances have spawned another new term, the *system-on-a-chip*, with its own inevitable acronym, SOC. Consider as an example the digital camera. Nowadays, even rather inexpensive models have MEMS chips installed to sense the camera's orientation with respect to gravity and to detect and compensate for the inadvertent jolts and motions of the picture taker. Such features would have been well beyond the expectations of the owner of even the most expensive SLR camera of 20 years ago. The capabilities mentioned above are made possible by mechanical devices with dimensions less than a millimeter or so, fabricated on a chip side by side with all the required control and drive electronics.

Additional evidence for the vitality of this new technology is that the microfabrication industry has appropriated the term *foundry* to describe their facilities. This word dates from sixteenth-century French. One of the authors (TBJ) has a vivid childhood recollection from the 1950s of the nightly spectacle of fumes and fire belching impressively from the venting chimneys atop a metal casting foundry in the small Midwestern city where he grew up. This plant produced manhole covers and other essential yet mundane components of the urban infrastructure. In this new century, however, the products manufactured in what we call a foundry range from heavy metal castings, with dimensions of meters, down to very intricate parts made of silicon and having dimensions of the order of microns.

As technology advances, etymological take-overs like this are sometimes accompanied by resurrections of words that fall out of favor but then find a path back to prominence. The term *electromechanics*, embedded in the acronym MEMS, provides a fine example of this phenomenon. Fifty years ago, an engineer specializing in electromechanics would certainly have anticipated a career devoted to relays, solenoids, motors, servos, or AC synchronous alternators. These critical components were inevitably magnetic devices and ranged in size from centimeters to meters. They still do. While the education of this engineer certainly would have involved perfunctory attention to capacitive devices, the greatest share of his time would have been spent on inductance, windings, large currents, and magnetic flux. A half-century ago, capacitive microphones and speakers were only starting to make inroads into established markets. There were no large, capacitively based electromechanical converters. There never will be, because the physics governing electromechanics unambiguously awards the advantage for power conversion applications to magnetic devices, at least in devices larger than approximately one centimeter.

Consider the modern microfabrication facility that calls itself a foundry but engages in the manufacture of tiny components with feature sizes as small as a few microns. The same laws of physical scaling that award the prize to magnetic devices of large dimensions strongly favor electrostatic, i.e., capacitive, devices for physical dimensions below approximately 10^2 microns. Refer to Section A.5 of Appendix A for a brief consideration of these scaling considerations. Indeed, the vast bulk of MEMS systems are capacitive. The flexibility of modern microfabrication technology in combination with designer ingenuity leads to MEMS actuator and sensor geometries that go far beyond the familiar parallel-plate capacitor to very novel structures, some of which could hardly have been manufactured at all 50 years ago. So, the twenty-first century still finds us making the components needed for the modern world in *foundries*, but this facility belches no smoke and fumes, the workers wear white suits, masks, and funny hats, the devices themselves are tiny, and they are apt to be actuated by electrostatic forces.

1.2 Some terminology

The growing interest in microsystems has reinvigorated *electromechanics* as a key engineering discipline. The term electromechanics is, in general, reserved to describe devices and systems that use electrical (or magnetic) forces to cause mechanical motion or sense motion and then induce measurable electrical signals. Electromechanics is subdivided into (i) electrical systems, where charge, voltage, capacitance, and small electrical current are important, and (ii) magnetic systems, where magnetic flux, induced voltage, inductance, and large currents are the critical quantities.

It is helpful to introduce some additional terminology to aid in categorizing intended applications of MEMS technology. We start with the word *transducer*, a surprisingly recent term dating from the 1920s. A transducer takes power in one form and converts it into another form. There are of course many types of transduction mechanisms. For example, a bimetallic strip converts thermal energy to mechanical form. Of more interest to us, an *electromechanical transducer* changes electrical power into mechanical form or mechanical to electrical.

Another important word, dating back to *c.* 1864, is *actuator*. This term is in fact derived from the verb *actuate*, which was first used in the seventeenth century and derives from Latin. An actuator is a mechanical device that creates or controls mechanical motion. Thus, an *electromechanical actuator* is a type of transducer that uses electrical input to create mechanical motion. Motors and electromechanical relays are traditional examples.

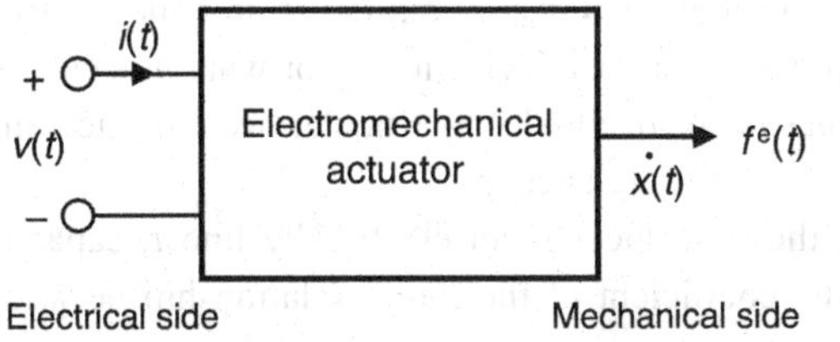

Figure 1.1 Basic electromechanical transducer with one electrical terminal pair and one mechanical degree of freedom. The system is assumed to be lossless.

A third term of importance to our subject is *sensor*. This word is also of early twentieth-century origin, though it springs from Latin. A sensor is any device that responds to some physical stimulus by emitting a signal. There are many, many examples of sensors. A particularly intriguing new one, coming from biochemistry, is the *chemical sensor*, where the output "signal" is actually a chemical response. Within our subject, MEMS, a microphone exemplifies an *electromechanical sensor* because it detects mechanical sound energy and converts it to voltage or current.

Another term may be introduced here, the *brake*. Braking action slows or dampens mechanical motion, and sometimes electromechanical systems are employed to achieve it. Such a brake may be thought of as a type of actuator.

1.3 Electromechanical systems

An electromechanical transducer is a device possessing mass and physical dimensions. It has electrical leads and at least one movable element. The heart of any such transducer is a mechanism for converting power between electrical and mechanical forms. Let us assume for expediency that the device has a single electrical terminal pair and a single mechanical degree of freedom. Although such restrictions will be abandoned later, we can learn much by considering this special case. We may represent this mechanism by the deceptively simple diagram of Fig. 1.1. For now, the electrical leads (terminals) are located on the left side and the mechanical system on the right. On the electrical side, there is a time-dependent voltage $v(t)$ and a time-dependent current $i(t)$. To represent the mechanical side, we need two more variables. For the present, it proves convenient to use the mechanical variable $x(t)$ along with the electrical force $f^e(t)$ that the transducer exerts on the external mechanical system. Later, $x(t)$ will be replaced by velocity, that is, $\dot{x} = dx/dt$.

A useful physical interpretation of the nature of the force $f^e(t)$ emerges from a consideration of power flow. The instantaneous electrical power input to the device is

$$p_e(t) = v(t)\,i(t) \tag{1.1}$$

and the instantaneous mechanical power output is

$$p_m(t) = f^e(t)\,\dot{x}. \tag{1.2}$$

One might be tempted here to set $p_e(t) = p_m(t)$, but doing so is not justified because neither the nature of the electromechanical device itself nor what it is capable of doing with input electrical power has been specified. In particular, we must accommodate the possibility that the device can itself store energy.

To proceed, assume that the transducer is an electrically linear, capacitive device. Capacitance is defined as the coefficient of the *linear* relationship between electrical charge $q(t)$ and voltage $v(t)$,

$$q(t) = C[x(t)]\, v(t). \tag{1.3}$$

Equation (1.3) recognizes that the capacitance will be a function of time through its dependence on the mechanical variable $x(t)$. If charge conservation is now employed to express the current $i(t)$ using Eq. (1.3), then

$$i(t) = \frac{\mathrm{d}q}{\mathrm{d}t} = C(x)\frac{\mathrm{d}v}{\mathrm{d}t} + \frac{\mathrm{d}C}{\mathrm{d}x}v\frac{\mathrm{d}x}{\mathrm{d}t}. \tag{1.4}$$

This current expression contains the familiar *capacitive current*, $C(x)\mathrm{d}v/\mathrm{d}t$, plus another term that depends on the velocity, $\mathrm{d}x/\mathrm{d}t$. This new term, called the *motional current*, should be a tip-off that the device is more than a simple capacitor. Combining Eq. (1.4) with Eq. (1.1) yields

$$p_e(t) = C(x)\frac{\mathrm{d}v}{\mathrm{d}t}v + \frac{\mathrm{d}C}{\mathrm{d}x}v^2\frac{\mathrm{d}x}{\mathrm{d}t}, \tag{1.5}$$

which can be manipulated to obtain

$$p_e(t) = \frac{\mathrm{d}}{\mathrm{d}t}\left[\frac{C(x)}{2}v^2\right] + \frac{1}{2}\frac{\mathrm{d}C}{\mathrm{d}x}v^2\frac{\mathrm{d}x}{\mathrm{d}t}. \tag{1.6}$$

The first term on the right-hand side of Eq. (1.6) is the time derivative of a familiar quantity, $Cv^2/2$, the electrostatic energy storage associated with the capacitance. Evidently, some of the electrical power flowing into the transducer can be intercepted and stored in electrostatic form within the device rather than being converted directly to mechanical power. So, if we assume that there is no loss inherent in the transduction mechanism and that the electromechanical coupling depicted in Fig. 1.1 has no other mechanism of energy storage, then it is permissible to equate the remaining term in Eq. (1.6) to the mechanical output power, $p_m(t)$, defined by Eq. (1.2). Thus,

$$p_e(t) = \frac{\mathrm{d}}{\mathrm{d}t}\left[\frac{C(x)}{2}v^2\right] + p_m(t). \tag{1.7}$$

Combining Eqs. (1.2), (1.6), and (1.7) yields an expression for the force f^e:

$$f^e = \frac{v^2}{2}\frac{\mathrm{d}C}{\mathrm{d}x}. \tag{1.8}$$

A more rigorous approach to evaluating the force, f^e, based on a thermodynamic argument, is detailed in Chapter 3. Nevertheless, Eq. (1.8) is a correct and very general result. While the exclusion of loss from the coupling mechanism precludes consideration of certain types of transducer, such as bimetallic devices and shape memory alloys, it presents no restriction concerning true electromechanical transducers. Further, we will find in a later chapter that restricting the energy storage to the capacitive terms facilitates a systematic approach to modeling real transducers with both electrical and mechanical constraints.

Note that the force f^e, henceforth to be referred to as the *force of electrical origin*, is proportional to the product of the square of the voltage and the first derivative of the capacitance. It takes its sign from the derivative $\mathrm{d}C/\mathrm{d}x$. Furthermore, the mechanical power flow will be positive (out) or negative (in), depending on the sign of the product: $(\mathrm{d}C/\mathrm{d}x)(\mathrm{d}x/\mathrm{d}t)$.

The sign of the electrical power can be investigated conveniently by considering the special case of constant voltage, that is, $\mathrm{d}v/\mathrm{d}t = 0$. Then, we have

$$p_e(t)|_{v=\text{constant}} = v^2 \frac{\mathrm{d}C}{\mathrm{d}x}\frac{\mathrm{d}x}{\mathrm{d}t}. \tag{1.9}$$

The sign-determining product, $(\mathrm{d}C/\mathrm{d}x)(\mathrm{d}x/\mathrm{d}t)$, reappears in Eq. (1.9). Furthermore, $p_e(t)|_{v=\text{constant}} = 2\,p_m(t)$, showing that when $(\mathrm{d}C/\mathrm{d}x)(\mathrm{d}x/\mathrm{d}t) > 0$, half of the electrical input energy is converted directly to mechanical form and the other half is stored in the capacitor. On the other hand, if $(\mathrm{d}C/\mathrm{d}x)(\mathrm{d}x/\mathrm{d}t) < 0$, the system delivers electrical power output, half from the mechanical side and the other half provided by the capacitor.

Though derived for the special case of constant voltage, the conclusion that the transducer can convert energy from electrical to mechanical form or vice versa is general. Transducer operation is controlled by the external electrical and mechanical constraints, topics to be addressed in Chapter 3.

One systematic way to distinguish between the various operational modes of a transducer is to keep track of the signs of power flows, as defined for the lossless coupling shown in Fig. 1.1. First, consider time-average power, denoted by

$$\langle p(t)\rangle = \frac{1}{T}\int_0^T p(t)\,\mathrm{d}t, \tag{1.10}$$

where T is some appropriate time scale for averaging. Then, for an actuator operating in steady-state, the device delivers mechanical power output from electrical power input, that is, $\langle p_e(t)\rangle > 0$ and $\langle p_m(t)\rangle > 0$. A sensor operating in the steady-state is characterized by the conditions $\langle p_e(t)\rangle < 0$ and $\langle p_m(t)\rangle < 0$, converting mechanical to electrical power. Thinking of a sensor as an energy converter may seem strange but, after all, *all* measurements require conversion of energy from one form to another. In particular, the detection of a weak signal requires efficient conversion and then amplification.

To consider other modes, it is better not to restrict attention to time-average power. In one form of *dynamic braking*, the conditions $p_e(t) > 0$ and $p_m(t) < 0$ are maintained for a transient interval of time until some moving element has stopped. Both electrical and mechanical power are supplied to the transducer and this energy is stored in the device.

A rather different example of device operation is the *force-balance* sensor, which employs feedback to balance mechanical and electrical forces so that the mechanical element does not move at all. Sensor information is extracted by monitoring the feedback system. Because $p_m(t) \approx 0$, there is no conversion of energy. One final case might be hypothesized, where both electrical and mechanical power are extracted from the transducer, that is, $p_e(t) < 0$ and $p_m(t) > 0$. This situation can exist only during a transient period as energy stored in the transducer is drained away on the electrical and mechanical sides until depleted.

Example 1.1 **An interdigitated MEMS transducer**

Figure 1.2 shows a basic version of the interdigitated electrode geometry of the well-known *comb-drive* capacitive transducer. This device is found in automotive airbag sensors, gyroscopes, and many other commercially successful devices. Usually, one electrode is fixed and the other is constrained to move only in one direction, here x. In a real comb drive, the constraint to x-directed motion is only approximately achieved and then only over a limited frequency range. The device capacitance is

$$C(x) = 2N\varepsilon_0 \frac{h(L_{\text{eff}} + x)}{g} + C_s.$$

Here, $2N$ = number of air gaps, L_{eff} = overlap between the fixed and the moving electrodes when $x = 0$, g = air gap between the electrodes, h = height, and $\varepsilon_0 = 8.854 \cdot 10^{-12}$ F/m, the permittivity of free space. C_s, the parasitic capacitance, is likely to be large compared with the x-dependent term if the electrodes are connected on the substrate via long conductive traces. This problem is common to most MEMS devices.

This capacitive transducer exemplifies *variable-area* capacitive transducers, which will be treated in subsequent chapters. For the device shown, the moving electrode bank has $N = 6$ active movable plates. Using $g = 2$ μm, $h = 10$ μm, and $L_{\text{eff}} = 15$ μm, the air capacitance is calculated to be $C(x = 0) = C_0 \approx 8 \cdot 10^{-15}$ F = 8 fF, an exceedingly small value. The capacitance gradient is

$$\frac{\mathrm{d}C}{\mathrm{d}x} = 2N\varepsilon_0 \frac{h}{g} \approx 0.5 \text{ fF}/\mu\text{m}.$$

This parameter can be increased by adding more electrode pairs or by increasing the electrode height-to-gap ratio. Note that the electrode length, L_{eff}, does not affect dC/dx. To conserve chip area, L_{eff} should be as small as possible, but large enough for sufficient overlap to accommodate the expected operational range of motion.

The electrostatic force in a capacitive transducer can only be attractive, pulling the electrodes together. The necessary restoring force in the negative x direction is achieved with a spring.

Assume a square pulse voltage excitation of amplitude 5 V with a 50% duty cycle. If T is the period of the periodic excitation, then, from Eq. (1.8),

$$f^e = \frac{v^2}{2}\frac{\mathrm{d}C(x)}{\mathrm{d}x} = \begin{cases} 6.25 \cdot 10^{-9}\,N, & 0 \le t \le T/2, \\ 0, & T/2 < t < T, \end{cases}$$

the time average of which is

$$\langle f^e \rangle = \int_0^T f^e(t)\,\mathrm{d}t \approx 3.12 \cdot 10^{-9} \text{ N}.$$

To appreciate the scale of this force, consider terrestrial gravity acting on a 1 cm length of human hair. Average human hair has a mass per unit length of ~30 μg/cm, so in a gravitational field of 9.81 m/s^2, this hair experiences a force of ~290 nN, almost 100 times larger than the calculated f^e. It is also instructive to compute the electrostatic energy stored in the capacitor at $v = 5$ V and $x = 0$ m:

$$\frac{1}{2} C_0 v^2 = \frac{1}{2}(8 \cdot 10^{-15}) \times 25 \approx 10^{-13}\ \text{J}.$$

For comparison, the gravitational potential energy of this 1 cm length of hair lifted 1.8 m above the floor is $290 \cdot 10^{-9} \times 1.8 \approx 5.2 \cdot 10^{-7}$ J, a value many orders of magnitude larger than the capacitive energy!

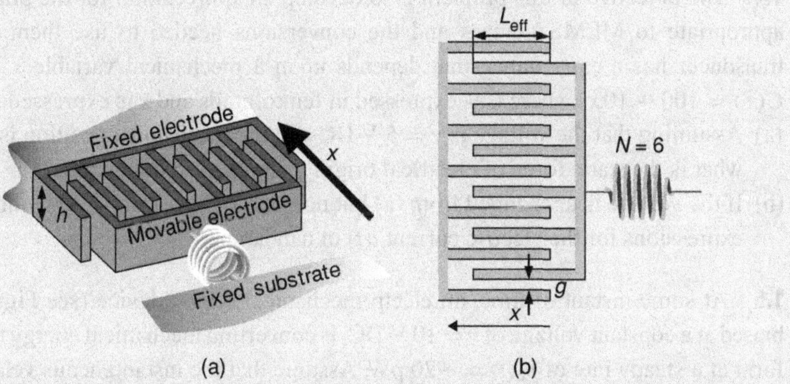

Figure 1.2 Interdigitated electrode geometry of a basic comb-drive capacitive transducer consisting of fixed and movable electrodes. The movable electrode bank is attached to a mechanical restoring spring, which constrains it to move only in the x direction. (a) Perspective view of the structure. (b) Top view with definitions for the principal geometric dimensions.

1.4 Conclusion

The values calculated in Example 1.1 for capacitance, force, and electrostatic energy of the comb-drive structure are tiny. In this book, we will show that, despite such seemingly small numbers, these and other MEMS devices can be put to work effectively in broad classes of engineered microsystems.

Problems

1.1 For mechanical variable $x < x_0$, the capacitance of a MEMS sensor is $C(x) = C_0(1 - x/x_0)$. Find the current $i(t)$ for these sets of conditions:

(a) $v(t) = V_0$ and $x(t) = 0$.
(b) $v(t) = V_0$ and $x(t) = u_0 t$.

(c) $v(t) = V_0\,[1 - \exp(-\alpha t)]$ and $x(t) = 0.5x_0$.
(d) $v(t) = V_0 \cos(\omega t)$ and $x(t) = u_0 t$

1.2 For the capacitor described in Problem 1.1 and the conditions (a) through (d), find time-dependent expressions for the force of electrical origin $f^e(t)$.

1.3 Assume that the capacitor described Problem 1.1 is charged to voltage V_0 with $x = 0$, then open-circuited to fix the electric charge. Under this condition with the voltage v no longer constrained externally, what is the algebraic relationship between v and the mechanical variable x?

1.4 The objective of this problem is to develop an appreciation for the practical units appropriate to MEMS devices and the conversions needed to use them. A MEMS transducer has a capacitance that depends upon a mechanical variable x as follows: $C(x) = 100 + 10x$, where C is expressed in femtofarads and x is expressed in microns.
(a) Assuming that the voltage is $v = 5$ V-DC and the mechanical position is $x = 5$ μm, what is the static force of electrical origin f^e expressed in newtons?
(b) If the voltage is unchanged from (a) but now $x(t) = 5\cos(2\pi \cdot 10^4 t)$, find predictive expressions for the electric current $i(t)$ in nanoamps.

1.5 At some instant of time, an electromechanical sensor device (see Figure below), biased at a constant voltage of $v = 10$ V-DC, is converting mechanical energy to electrical form at a steady rate of $p_e(t) = -20$ pW. Assume that the instantaneous velocity of the moving element in this device is a constant 1.0 mm/s and the acceleration is zero.

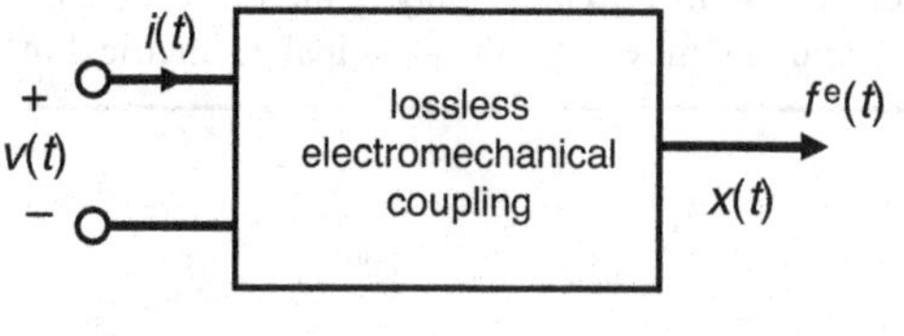

(a) Find the value of the derivative of the capacitance dC/dx for this device in pF/μm.
(b) What are the values of the supplied instantaneous mechanical power $p_m(t)$ in picowatts?
(c) What is the rate of change of the stored energy in picowatts?

1.6 An initially uncharged capacitive MEMS device is charged to $v = 5$ V-DC. With the voltage fixed at this value by a battery, the capacitance is then changed at a uniform rate over a time interval of $\Delta t = 1$ ms from $C = 5$ fF to 10 fF.
(a) Account for all energy flows occurring in this process, quantifying and identifying where power is coming from and where it is flowing to.
(b) With the device now mechanically fixed so that the capacitance $C = 10$ fF, it is connected to a resistor $R = 10$ kΩ starting at $t = 0$ and discharged. Determine the transient decay of voltage and then account quantitatively for all energy flows.

1.7 Example 1.1 described an interdigitated capacitive transducer with fixed electrode gaps and variable overlap. The same interdigitated geometry can be configured as a variable-gap device, as shown below. Obtain expressions for the capacitance $C(x)$ and the force of electrical origin f^e of this device, assuming it has $2N$ variable gaps and fixed overlap of the electrodes.

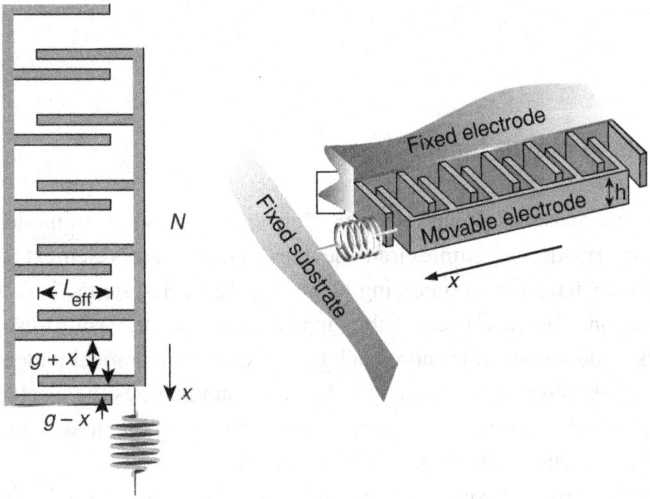

1.8 The capacitance of a MEMS transducer depends upon a single mechanical variable x; this dependence is $C(x) = C_0 \exp(-\alpha x)$. Assume that the device is initially charged up to a DC voltage V_0 with the movable element held fixed at $x = x_0$, and then open-circuited so that the charge is constant. Then, at time $t = 0$, the movable element is released and the voltage is monitored with an oscilloscope as a function of time. It is found that the voltage is $v(t) = V_0 \exp(-\beta t)$ for $t \geq 0$. Obtain an expression for the time dependence of the mechanical variable, $x(t)$.

Reference

1. V. H. Price and E. Menefee, Quantitative estimation of hair growth I. Androgenetic alopecia in women: effect of minoxidil. *Journal of Investigative Dermatology*, 95 (1990), 683–687.

2 Circuit-based modeling

2.1 Fundamentals of circuit theory

This chapter summarizes basic circuit theory concepts, shows how to model capacitive MEMS actuator structures as simple circuit devices, and introduces elements of multiport network theory for later use in modeling electromechanical systems. The approach is a very practical one, based on reasonable approximations and easy-to-use inspection methodologies. Students seeking more background on the theoretical underpinnings of circuit modeling should refer to Appendix A, which summarizes the electroquasistatic and magnetoquasistatic approximations, reveals the origins of the models for capacitors and inductors, and relates them to basic circuit theory.

Our emphasis on circuit-based representations for MEMS devices is motivated by the fact that they can be embedded directly into the electronic system models for the control and sensing circuitry. With this groundwork, we will later investigate conventional implementations of capacitive sensors and actuators, including inverting operational-amplifier circuits, two-plate and three-plate topologies, and the half-bridge differential scheme. For sufficiently complex systems, software tools such as PSPICE or CADENCE might be used, but in this text on fundamentals, we restrict the focus to systems that can be treated analytically.

2.1.1 Motivation

Electromechanical energy transduction occurs in consequence of the interaction of electric fields with charge, or magnetic fields with electric current. For this reason, reliable prediction of the performance of microelectromechanical devices would seem to hinge upon detailed knowledge of the electric field (or magnetic field) inside the device. Such an impression can be intimidating when one starts to consider the huge variety and complexity of MEMS geometries. "Where to begin?" the student might reasonably ask. Solutions for the electric or magnetic fields in these geometries often necessitate the use of numerical computation tools. Fortunately, the situation for the newcomer to MEMS is not nearly so bad. One can uncover and then study virtually all the important behavior of capacitive MEMS devices – including frequency-dependent response, sensitivity, dynamic response, and stability – from knowing the capacitance $C(x)$ and its first two derivatives.

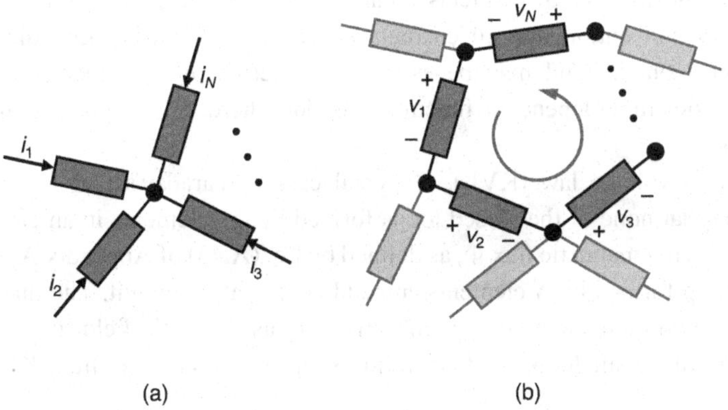

Figure 2.1 Fundamental electric circuit relations defined in terms of interconnected circuit elements. (a) General definition of circuit node for Kirchhoff's current law. (b) General definition of circuit loop for Kirchhoff's voltage law.

At some point in the design and optimization process, numerical field solutions may be needed to improve modeling accuracy; however, the often considerable effort to perform such analyses can be made separately and in parallel with the system-level modeling that is the subject of this text. Therefore, because the focus of this book is MEMS, finite-element analysis (FEA) is not covered. The reader interested in numerical modeling is encouraged to find reference volumes that deal with this vast topic.

Equation (1.8) revealed that the electrical force in a capacitive device is a function of two lumped parameter quantities, namely, capacitance – actually, its derivative – and voltage. This comes as good news. Chapter 4 shows that a fully developed predictive model for a real MEMS device can be formulated from accurate knowledge of the capacitance $C(x)$, plus descriptions of the external electrical and mechanical constraints. The external constraints are accounted for, respectively, via conventional circuit theory and the laws of mechanics. Likewise, in Chapter 9, it will be revealed that lumped parameter inductance $L(x)$, its first two derivatives, and the current are entirely sufficient for the purpose of developing integrated models for magnetic MEMS devices.

2.1.2 Kirchhoff's current and voltage laws

The analysis of circuits is based on two approximations derived from physical laws: Kirchhoff's current law (KCL) and Kirchhoff's voltage law (KVL). Kirchhoff's current law is a special case of charge conservation. It is applicable in any electric circuit as long as the net electric charge accumulation at the circuit nodes is either negligible or can be accounted for by incorporation of a lumped, parasitic capacitance. Kirchhoff's current law for the node depicted in Fig. 2.1a may be written

$$\sum_{n=1}^{N} i_n = 0, \tag{2.1}$$

where N is the number of elements connected to the node. Equation (2.1) employs the arbitrary convention that all currents i_n are defined flowing into the node. The law is quite general, and may be expressed in terms of any electric signal representation, from time-dependent functions, as done here, to AC phasors and Laplace transforms.

Kirchhoff's voltage law (KVL), a special case of Faraday's law, is based on the assumption that none of the closed loops formed by the elements in an electric circuit link time-varying magnetic flux Φ, as defined by Eq. (A.43) of Appendix A. Figure 2.1b depicts a loop formed by N elements embedded in a larger circuit. The flux restriction does not exclude inductive components. As long as magnetic field lines do not leak out and link a circuit loop, KVL is valid. Subject to that restriction, KVL requires that

$$\sum_{n=1}^{N} v_n = 0. \tag{2.2}$$

The direction taken around the loop is arbitrary. A counterclockwise convention is used here for the summation of voltages v_n in Eq. (2.2). Just as for KCL, we may express KVL with time-dependent, phasor, or Laplace transform voltage quantities.

2.1.3 Circuit elements

The solution for the voltages and currents in a circuit is always based on KCL and KVL, but it also requires knowledge of the elements themselves. The generalized elements comprising the networks in Figs. 2.1 and 2.2 may constrain (i) voltage or (ii) current, or (iii) they may impose a relationship between the two. Such relationships, which may be linear or non-linear, are sometimes referred to as *v–i characteristics*. The most familiar circuit elements, resistors, capacitors, and inductors, all passive and linear, are supplemented by voltage and current sources. Refer to Table 2.1.

Electromechanical transducers are also circuit elements, usually capacitive or inductive, but they differ from their conventional counterparts because they depend upon at least one mechanical variable. Refer back to Chapter 1, specifically to Eq. (1.4) for the current flowing into a capacitor $C(x)$, where $x(t)$ is a mechanical variable, and to Appendix A, Eq. (A.38), for the voltage across an inductor $L(x)$.

Note the distinction made in Table 2.1 between *independent* and *dependent* sources. In general, independent sources depend only on time. For dependent sources, v_s or i_s will depend (usually linearly) on a voltage or current somewhere in the circuit. Starting in Chapter 4, we will introduce dependent sources that depend on a mechanical variable to create models for electromechanical two-port devices.

2.1.4 Tellegen's theorem: power and energy

In Chapter 1, we invoked power conservation for a capacitive electromechanical transducer to show that electrostatic energy storage within the device is essential to electrical $\leftrightarrow$ mechanical power conversion. This observation leads naturally to a more general

Table 2.1 Common electric circuit elements and sources

Component type		Circuits	Transducers
Resistors		$v(t) = Ri(t)$	$v(t) = R(x)i(t)$
Capacitors	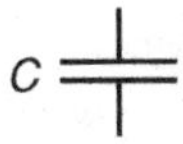	$i(\mathrm{t}) = C\dfrac{\mathrm{d}v}{\mathrm{d}t}$	$i(t) = C(x)\dfrac{\mathrm{d}v}{\mathrm{d}t} + v\dfrac{\mathrm{d}C}{\mathrm{d}x}\dfrac{\mathrm{d}x}{\mathrm{d}t}$ [a]
Inductors		$v = L\dfrac{\mathrm{d}i}{\mathrm{d}t}$	$v(t) = L(x)\dfrac{\mathrm{d}i}{\mathrm{d}t} + i\dfrac{\mathrm{d}L}{\mathrm{d}x}\dfrac{\mathrm{d}x}{\mathrm{d}t}$ [b]
Independent voltage sources	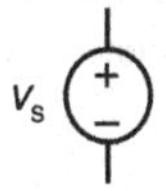	$v = v_s(t)$	$v = v_s(t)$
Dependent voltage sources	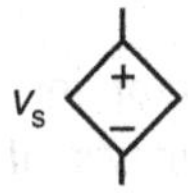	$v_s = f(v)$ or $f(i)$	$v_s = f(v), f(i),$ or $f(x)$
Independent current sources		$i = i_s(t)$	$i = i_s(t)$
Dependent current sources		$i_s = f(v)$ or $f(i)$	$i_s = f(v), f(i),$ or $f(x)$

[a] Refer to Section A.3.1 of Appendix A.
[b] Refer to Section A.5.1 of Appendix A.

consideration of power flow in electric circuits. Consider the network shown in Fig. 2.2, which consists of N internal circuit elements and one external terminal pair. We place no constraints whatsoever on the circuit elements: they may be active or passive, linear or non-linear. Now imagine performing two independent experiments with this circuit. In the first experiment, some arbitrary, time-dependent voltage $v_e(t)$ is applied at the external terminal pair. For this particular excitation, we can determine all internal voltages $(v_1, v_2, \ldots, v_N)$, all internal currents $(i_1, i_2, \ldots, i_N)$, and the external terminal current $i_e(t)$, by invoking KVL, KCL, and the v–i characteristics of the elements.

Next, we perform a second, completely independent, experiment by applying a different voltage $v'_e(t)$ at the external terminal. For this new excitation, one can again solve for all the internal voltages $(v'_1, v'_2, \ldots, v'_N)$, internal currents $(i'_1, i'_2, \ldots, i'_N)$, and $i'_e(t)$. Keep in mind that the unprimed and primed quantities are associated, respectively, with the first and second experiments. For the network shown in Fig. 2.2, Tellegen's theorem [1] states that

$$v_e i'_e = \sum_{n=1}^{N} v_n i'_n \tag{2.3}$$

Equation (2.3) provides a relationship between the sums of all $v_n i'_n$ products from two, *completely independent* experiments performed on the electric circuit. Tellegen's

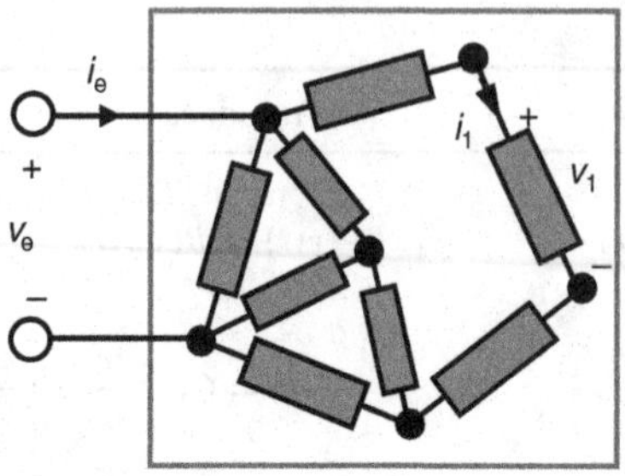

Figure 2.2 An electrical network featuring one external electrical port with N elements located inside. The elements can be passive or active, linear or non-linear. In addition to the external port voltage v_e and current i_e, there are the voltages v_n and currents i_n associated with each internal element. Note that the sign convention employed for current flow associated with the internal elements is opposite that for the external port.

theorem is in fact derived from very general assumptions about the voltages and currents, leading to an alternate and entirely equivalent way of looking at Eq. (2.3), namely, that (i) the set of voltages $(v_1, v_2, \ldots, v_N; v_e)$ satisfies KVL and (ii) the set of currents $\left(i_1', i_2', \ldots, i_N'; i_e'\right)$ independently satisfies KCL. According to this interpretation, Tellegen's theorem is indeed very general.

For the special case of $v_e(t) = v_e'(t)$, Eq. (2.3) reduces to

$$v_e i_e = \sum_{n=1}^{N} v_n i_n, \tag{2.4}$$

which is a statement of instantaneous power conservation, namely, that the electrical power consumed inside the box in Fig. 2.2 is equal to the power delivered to the input terminal pair. It is reassuring to learn that Tellegen's theorem, as general as it is, is consistent with power conservation.

2.1.5 AC circuits, impedance, and admittance

Crucial to circuit modeling are the concepts of electrical impedance and admittance. These quantities make it possible to determine the distribution of voltages and currents in actuators and sensors.

The definition of electrical impedance is based on an assumption that the voltages and currents in a circuit are harmonic functions, that is, $v(t) = V_0 \cos(\omega t + \phi)$ and $i(t) = I_0 \cos(\omega t + \varphi)$, where $\omega = 2\pi f$ is the angular electric frequency in radians per second, f is the frequency in Hz, and ϕ and φ are electrical phase quantities. The most convenient way to proceed is to employ exponential time dependence $e^{j\omega t}$ for the voltage and current:

$$v(t) = \mathrm{Re}[\underline{V} e^{j\omega t}] \quad \text{and} \quad i(t) = \mathrm{Re}[\underline{I} e^{j\omega t}], \tag{2.5}$$

Table 2.2 Impedances and admittances of basic passive elements

Element	Impedance $\underline{Z}$ [Ω]	Admittance $\underline{Y}$ [S]
Capacitor C (F)	$1/\mathrm{j}\omega C$	$\mathrm{j}\omega C$
Inductor L (H)	$\mathrm{j}\omega L$	$1/\mathrm{j}\omega L$
Resistor R (Ω)	R	$G = 1/R$

where $\mathrm{j} = \sqrt{-1}$. The coefficients of $\mathrm{e}^{\mathrm{j}\omega t}$ in Eq. (2.5) are the voltage and current phasors, respectively, $\underline{V}$ and $\underline{I}$.[1] Phasors are complex numbers conveying magnitude and phase information about electrical signals. For the given time-dependent voltage and current functions, $\underline{V} = V_0\angle\phi = V_0\mathrm{e}^{\mathrm{j}\phi}$ and $\underline{I} = I_0\angle\varphi = I_0\mathrm{e}^{\mathrm{j}\varphi}$. In this definition, Re[$\underline{V}$] and Re[$\underline{I}$] are the peak rather than RMS (root-mean-square) magnitudes. Phasors will be used to represent mechanical variables as well, starting in Chapter 4.

Impedance ($\underline{Z}$) and admittance ($\underline{Y}$), examples of linear v–i characteristics, are always defined in terms of voltage and current phasors:

$$\underline{Z} \equiv \underline{V}/\underline{I} \quad \text{and} \quad \underline{Y} \equiv \underline{I}/\underline{V}. \tag{2.6}$$

Note that $\underline{Z} = 1/\underline{Y}$. It is never correct to define either $\underline{Z}$ or $\underline{Y}$ in terms of time-varying voltages and currents. The impedance and admittance expressions for capacitors, inductors, and resistors are summarized in Table 2.2.

Expressed in terms of phasors, the time-average electrical power is [2]

$$\langle p_\mathrm{e}(t)\rangle = \frac{1}{2}\mathrm{Re}[\underline{V}\,\underline{I}^*], \tag{2.7}$$

where $\underline{I}^*$ designates the complex conjugate of $\underline{I}$. Combining Eq. (2.7) with the definitions for impedance and admittance, the time-average power may be expressed in terms of the real part of $\underline{Z}$ or $\underline{Y}$:

$$\langle p_\mathrm{e}(t)\rangle = \frac{1}{2}\,|\underline{I}|^2\,\mathrm{Re}\,[\underline{Z}] = \frac{1}{2}\,|\underline{V}|^2\,\mathrm{Re}[\underline{Y}^*]. \tag{2.8}$$

Thus, for any network containing only ideal (lossless) capacitors or inductors with purely imaginary $\underline{Z}$ or $\underline{Y}$, $\langle p_\mathrm{e}(t)\rangle = 0$. In other words, while capacitors and inductors can store energy, they do not consume it.

2.2 Circuit models for capacitive devices

The geometries of many capacitive MEMS devices can be described as *parallel-plate* structures, where the electrodes are planar and parallel. Further, the spacing between the electrodes in these devices is often quite small compared with the areal dimensions. In such cases, it is justified to ignore the fringing electric fields that exist around the edges. Refer to Section A.3 of Appendix A, where the *parallel-plate approximation* is examined in detail.

[1] The convention of using upper-case variables for phasors is restricted to this chapter. Starting in Chapter 4, where linearization and small-signal AC variables will be introduced, lower-case will be used for all phasor variables.

Example 2.1 **Capacitance of a simple MEMS geometry**

Surface micromachining lends itself to MEMS designs featuring air gaps that are very small compared with the areal dimensions. Figure 2.3 depicts a cantilevered beam formed from polysilicon upon a crystalline silicon substrate. Both materials are sufficiently conductive to serve as electrodes. Focusing here on the device capacitance, we defer to Chapter 6 any explanation of the nature of beams as movable electrode elements.

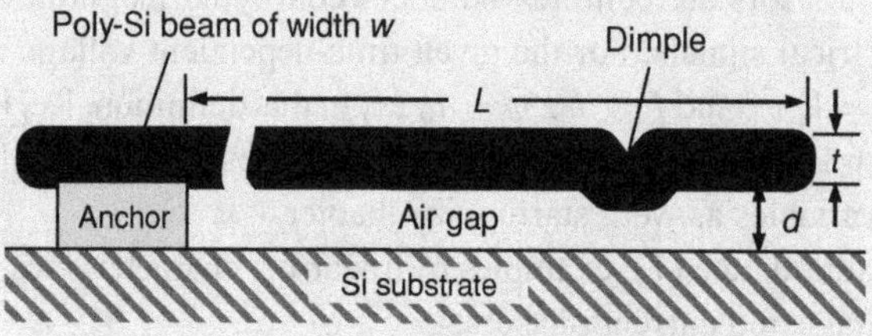

Figure 2.3 Side view of MEMS cantilevered beam modeled as parallel-plate capacitor. Length is not shown to scale because $d \ll L$. The anchor, made of SiO_2, is an electrical insulator. The dimple, also not to scale, serves as a mechanical stop and is ignored in the capacitance model.

According to the parallel-plate approximation, the capacitance is

$$C \approx \varepsilon_0 A/d,$$

where $A = wL$ is the electrode area. Representative dimensions for the deformable polysilicon beam are: $w = 5$ μm, $L = 50$ μm, and $t = 1$ μm. In the undeflected state, the air gap is $d = 1$ μm. For the dimensions given, the capacitance is $C \approx 2.21$ fF, where 1 fF (femtofarad) is 10^{-15} F. Such small values create a major electronic design challenge in practical MEMS because parasitics and stray contributions are usually much larger than the transducer capacitance.

2.2.1 Basic RC circuit building block

The simple geometry shown in Fig. 2.4a is the basic building block to be used to create circuit models for MEMS devices. It is a plane, parallel structure consisting of a planar slab of dielectric material of dielectric constant κ and electrical conductivity σ situated between parallel electrodes. Conductivity is usually not significant in electromechanical devices, but we account for it here to preserve some generality for its future utility. Appendix A treats this problem and uses the integral laws of electroquasistatics to derive the following differential equation relating current and voltage:

$$i = C\frac{\mathrm{d}v}{\mathrm{d}t} + Gv. \tag{2.9}$$

The physical origins of the two terms on the right-hand side of Eq. (2.9) are, respectively, capacitive charging of the electrodes and ohmic current flow through the medium. The expressions $C \approx \kappa\varepsilon_0 A/d$ and $G \approx \sigma A/d$, obtained in Sections A.3.3 and A.3.4 of

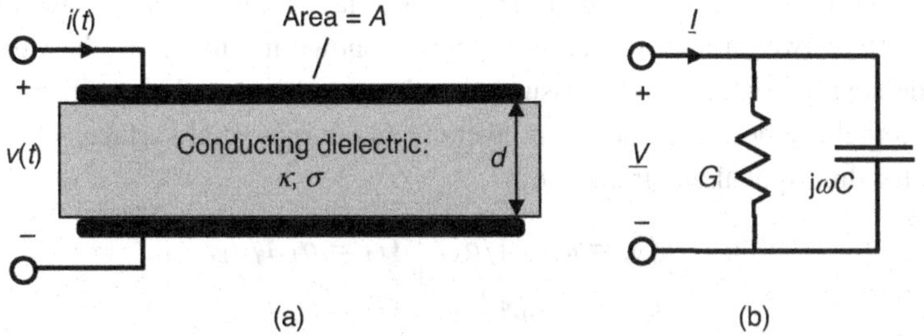

Figure 2.4 Basic parallel-plate geometry. (a) Electrodes are separated by a slab of ohmically conducting dielectric material. (b) The equivalent electric circuit model consists of a capacitor C in parallel with a linear conductance G.

Appendix A, reflect the parallel-plate approximations for capacitance and conductance, respectively. The equivalent circuit for the device described by Eq. (2.9) is the parallel RC network of Fig. 2.4b.

Assuming sinusoidal time dependence for the applied voltage, then the current must also be a sinusoid. Plugging exponential forms for $v(t)$ and $i(t)$ from Eq. (2.5) into Eq. (2.9), and then solving for $\underline{I}$ in terms of $\underline{V}$, yields

$$\underline{Y} = \frac{\underline{I}}{\underline{V}} = \mathrm{j}\omega C + G. \tag{2.10}$$

The simple, parallel RC network depicted in Fig. 2.4b will serve as a basic building block. In fact, we will use it to construct models for far more complex geometries by simple inspection. Two important examples, the so-called series and parallel configurations, provide adequate opportunity to demonstrate this method.

2.2.2 The series capacitive circuit

Consider first the geometry of Fig. 2.5a, sometimes referred to as the Maxwell capacitor, which consists of two, homogeneous, conductive, dielectric slabs of uniform thicknesses d_1 and d_2, dielectric constants κ_1 and κ_2, and electrical conductivities σ_1 and σ_2. By

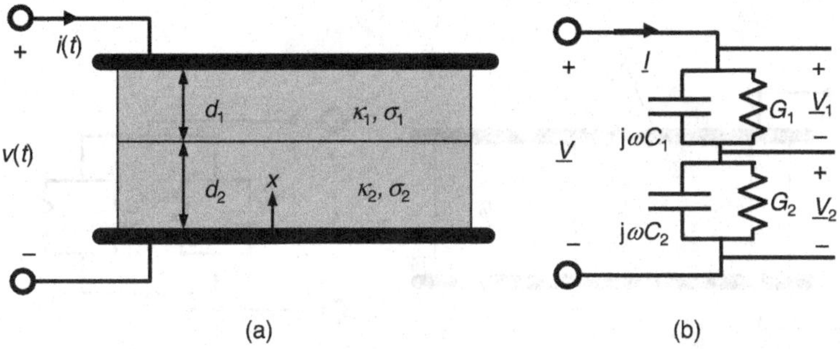

Figure 2.5 The Maxwell capacitor. (a) The two-layer capacitor geometry shown in side view. (b) AC circuit model as the series connection of two parallel RC networks.

inspection, one may recognize that these stacked layers can be represented by the series connection of two parallel RC building blocks shown in Fig. 2.5b. The validity of this equivalence is predicated on the assumption that the electric field is uniform within each layer, so that the boundary between layers is an equipotential surface. The component values for this equivalent circuit are

$$\begin{aligned} C_1 &= \kappa_1 \varepsilon_0 A/d_1, \quad G_1 = \sigma_1 A/d_1, \\ C_2 &= \kappa_2 \varepsilon_0 A/d_2, \quad G_2 = \sigma_2 A/d_2. \end{aligned} \tag{2.11}$$

Section A.3.5 of Appendix A derives the identical result directly from the integral laws of electroquasistatics and the definition of voltage.

Because $\underline{I}_1 = \underline{I}_2 \equiv \underline{I}$, it should be evident that the impedance of this network is

$$\underline{Z}_{\text{series}} = \frac{\underline{V}_1 + \underline{V}_2}{\underline{I}} = \frac{1}{j\omega C_1 + G_1} + \frac{1}{j\omega C_2 + G_2}. \tag{2.12}$$

Extension of this method to models for stacks of any number of such parallel laminae will be obvious. The distribution of voltage and electric field within each layer of a capacitive MEMS device is easily predicted using complex impedances and the voltage divider relation. For the Maxwell capacitor,

$$\underline{V}_1 = \frac{\underline{Z}_1}{\underline{Z}_1 + \underline{Z}_2}\underline{V} \quad \text{and} \quad \underline{V}_2 = \frac{\underline{Z}_2}{\underline{Z}_1 + \underline{Z}_2}\underline{V}, \tag{2.13}$$

where $\underline{Z}_1 = \underline{V}_1/\underline{I}_1$ and $\underline{Z}_2 = \underline{V}_2/\underline{I}_2$ are the AC impedances of the two layers. The x-directed phasor components of the electric field are then $\underline{E}_{x1} = -\underline{V}_1/d_1$ and $\underline{E}_{x2} = -\underline{V}_2/d_2$ The minus signs here signify that the positive terminal is the upper electrode, and so the electric field vector points downward. The practical utility of being able to predict electric fields becomes apparent if one of the layers in the transducer is either an air gap or a dielectric material susceptible to electrical breakdown.

2.2.3 The parallel capacitive circuit

Next, consider the parallel geometry of Fig. 2.6a, with two homogeneous, ohmic, dielectric media arranged side by side to fill the space between the plates. Once again, we employ the inspection method, this time arriving at the parallel RC circuit shown in

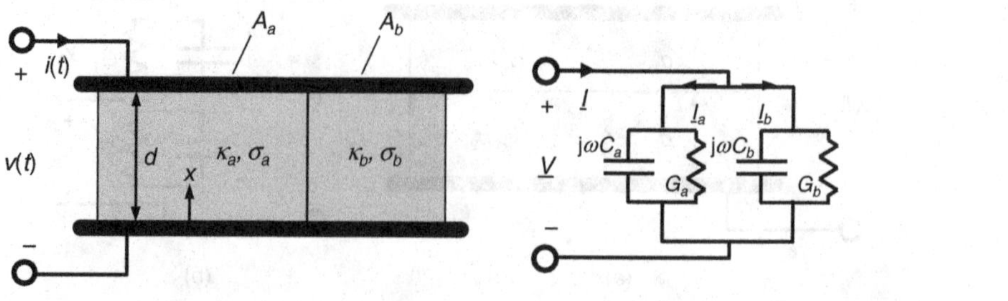

Figure 2.6 Parallel ohmic dielectric geometry. (a) Two ohmic dielectric slabs shown in side view. (b) AC circuit model as the shunt (parallel) connection of two parallel RC networks.

Fig. 2.6b. The expressions for the component values are

$$C_a = \kappa_a \varepsilon_0 A_a/d, \quad G_a = \sigma_a A_a/d,$$
$$C_b = \kappa_b \varepsilon_0 A_b/d, \quad G_b = \sigma_b A_b/d. \tag{2.14}$$

The admittance of this structure is

$$\underline{Y}_{\text{parallel}} = \text{j}\omega(C_a + C_b) + G_a + G_b. \tag{2.15}$$

If the structure contains more materials, one merely incorporates additional, parallel RC sections into the circuit model.

Using the current divider relation expressed in terms of admittances, the current distribution for the circuit in Fig. 2.6b is

$$\underline{I}_a = \frac{\underline{Y}_a}{\underline{Y}_a + \underline{Y}_b}\underline{I} \quad \text{and} \quad \underline{I}_b = \frac{\underline{Y}_b}{\underline{Y}_a + \underline{Y}_b}\underline{I}, \tag{2.16}$$

where $\underline{I}$, the total current, consists of $\underline{I}_a$ and $\underline{I}_b$, the current flowing to materials "a" and "b," respectively. For convenience here, admittances quantities $\underline{Y}_a = \text{j}\omega C_a + G_a$ and $\underline{Y}_b = \text{j}\omega C_b + G_b$ have been defined.

Example 2.2 **Interconnection trace capacitance**

A MEMS device is usually interconnected to the drive and detection circuitry using narrow metal traces patterned on a thin dielectric layer. See Fig. 2.7.

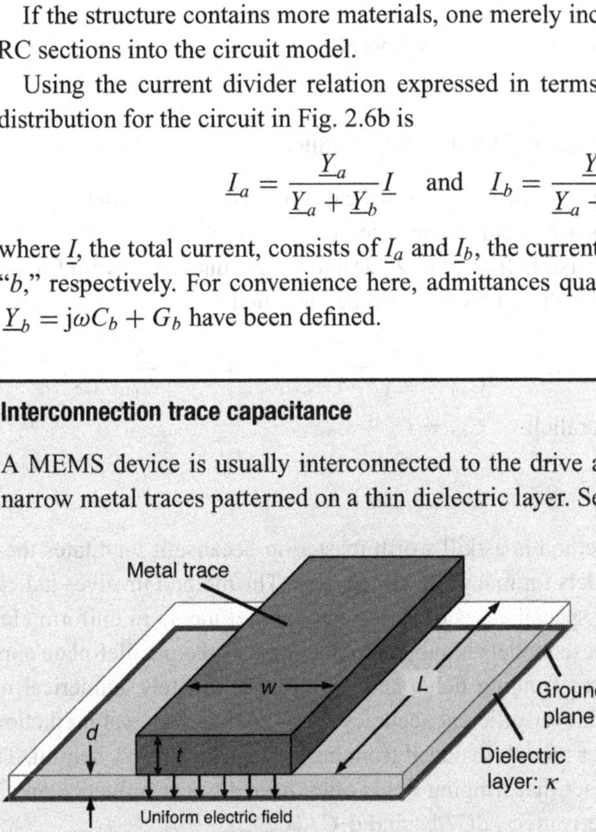

Figure 2.7 Metallic interconnection trace of width w and thickness t patterned on a dielectric-coated substrate of thickness d and dielectric constant κ. If $w \gg d$, the uniform field approximation can be used for $\overline{E}$ beneath the trace.

Estimating trace capacitance is important because parasitics degrade sensor sensitivity. Typically, $w \gg d$, so the electric field is uniform between the trace. Then, the capacitance per unit length is

$$C/L \approx \kappa\varepsilon_0 w/d, \quad t \sim d \ll w.$$

Using $w = 5$ μm, $d = 1$ μm, and $\kappa = 3.9$, the dielectric constant of SiO_2, one obtains $C/L \approx 0.17$ fF/μm. The cable capacitance of a trace 200 μm long having these dimensions is $C_{\text{trace}} \approx 35$ fF, which may be much larger than the transducer capacitance itself.

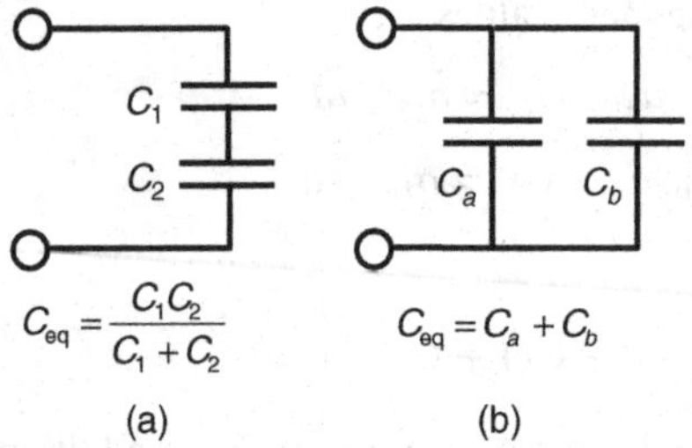

Figure 2.8 Special cases for capacitive connections. (a) Series capacitance. (b) Parallel capacitance.

2.2.4 Special cases: series and parallel capacitance

For modeling MEMS devices, the special cases of series and parallel capacitance, shown in Figs. 2.8a and b, are worth special attention.

Removing the resistors from Eq. (2.12) and the conductances from Eq. (2.15), equivalent capacitance for these networks can be identified:

$$\begin{aligned} \text{series:} \quad & \frac{1}{C_{\text{eq}}} = \frac{1}{C_1} + \frac{1}{C_2} \quad \text{or} \quad C_{\text{eq}} = \frac{C_1 C_2}{C_1 + C_2}, \\ \text{parallel:} \quad & C_{\text{eq}} = C_a + C_b. \end{aligned} \tag{2.17}$$

2.2.5 Summary

This inspection method is a skill worth mastering because it facilitates the creation of reduced-order models for many MEMS devices. The method involves judicious combinations of the two special cases of series and parallel media in uniform electric fields. The accuracy of these models is subject to the limits of the parallel-plate approximation itself, which ignores fringing fields and coupling. Ultimately, numerical methods are often needed to achieve sufficient accuracy for MEMS design, yet predictions obtained using reduced-order models deduced from inspection can be very helpful. This claim is supported by the fact that fringing fields often have a larger influence on $C(x)$ than on its all-important derivatives, $\mathrm{d}C/\mathrm{d}x$ and $\mathrm{d}^2C/\mathrm{d}x^2$.

Series and parallel equivalents are not restricted in their application to planar geometries. They can be employed in other geometries, for example, concentric cylindrical structures, which are not relevant to MEMS, but important in technologies including millimeter microwave engineering.

Example 2.3 **A mechanically complex capacitive MEMS geometry**

The SEM in Fig. 2.9 shows an accelerometer fabricated from (100) crystalline Si using a dry reactive ion etch (DRIE) process. Despite appearances, this high-aspect ratio structure is rather simple, consisting of two variable-gap shunt-connected banks of parallel-plate capacitors. The truss structures provide essential mechanical rigidity. Compared with the widths of the trusses, the air gaps between the elements are quite small.

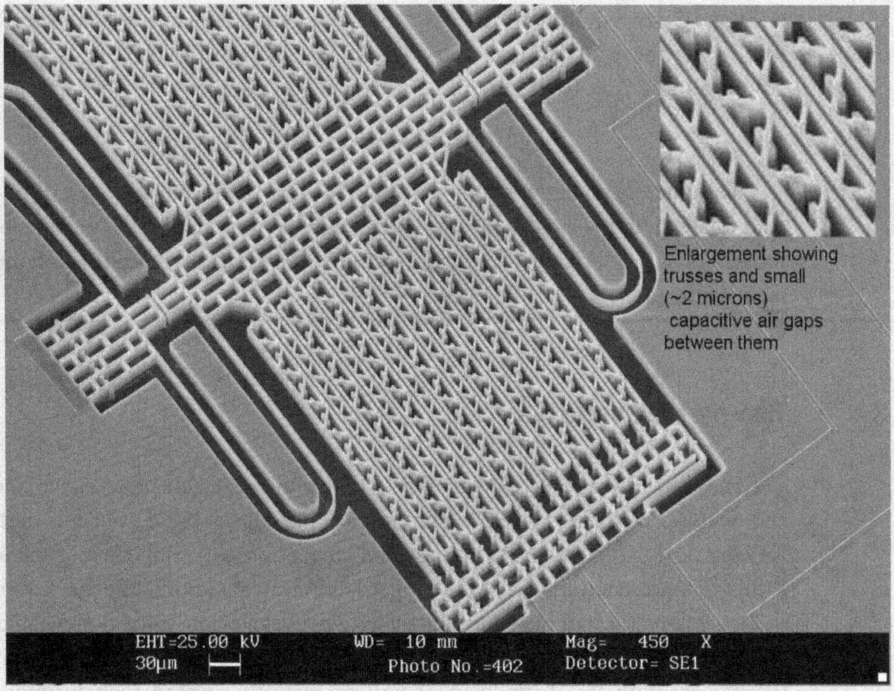

Figure 2.9 SEM image of a high-aspect-ratio accelerometer fabricated using DRIE process on a (100) crystalline Si wafer. There are 56 air gaps, with 28 on each side of the central support truss. The moving element (*proof mass*) is suspended and held in place by the sets of curved beams visible at the lower left and upper right. Image provided courtesy of Kionix, Inc.

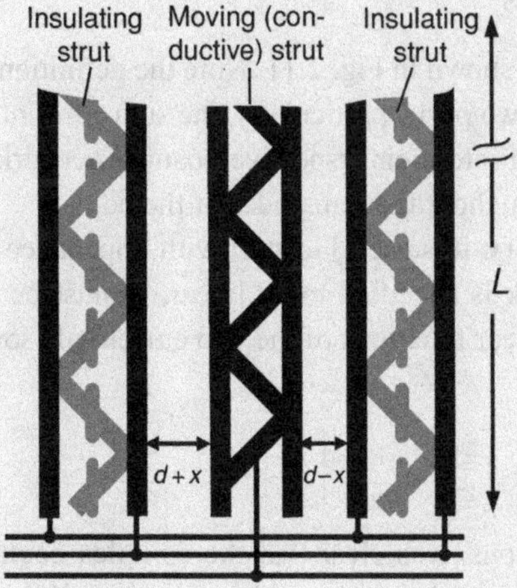

Figure 2.10 Detail of high-aspect-ratio capacitor bank from accelerometer device of Fig. 2.9 showing trusses and air gaps (considerably exaggerated). The structure width and length are w and L, respectively.

Figure 2.10 shows that the trusses of the moving elements are electrically conductive, while those of the stationary electrodes insulate the electrodes on either side. If $w = 50\ \mu\text{m}$, $L = 1000\ \mu\text{m}$, $d = 2\ \mu\text{m}$, and $N = 28$, then

$$C(x = 0) \approx N\varepsilon_0 wL/d \approx 6.2\ \text{pF}.$$

This value, much larger than that for the surface-fabricated structure of Example 2.1, suggests that high-aspect-ratio structures might be less susceptible to parasitics. Unfortunately, DRIE-fabricated designs need much longer traces, so ultimately little advantage is accrued.

2.3 Two-port networks

The two-port network formalism offers a powerful, systematic approach to the modeling and analysis of linear systems. This section introduces basic concepts and principles for electrical two-ports using voltage and current phasors, impedance, and admittance. While this presentation focuses on electric circuits, its ramifications are more general. In fact, Chapter 4 will introduce an analogy between electrical and mechanical systems that permits us to exploit two-port theory for the modeling of any electromechanical transducer. With this formalism, it will be easy to develop realistic models for the electrically and mechanically constrained behavior of MEMS sensors and actuators. Because this approach lays the foundation for the later chapters, the treatment of electrical two-ports presented here should be studied closely.

2.3.1 Impedance and admittance matrices

Consider the linear electrical network shown in Fig. 2.11. Note the definitions of phasor voltages and currents for each of the two ports, particularly, the sign convention that the currents $\underline{I}_1$ and $\underline{I}_2$ are positive flowing into their respective positive network terminals. This convention will be followed throughout the remainder of the book.

As an introduction to two-port theory, it is useful to start with impedance and admittance quantities. Because the network is specified to be linear, it must be possible to express the two voltage phasors as linear functions of the two current phasors. Thus, in matrix form,

$$\begin{bmatrix} \underline{V}_1 \\ \underline{V}_2 \end{bmatrix} = \begin{bmatrix} \underline{z}_{11} & \underline{z}_{12} \\ \underline{z}_{21} & \underline{z}_{22} \end{bmatrix} \begin{bmatrix} \underline{I}_1 \\ \underline{I}_2 \end{bmatrix}. \tag{2.18}$$

From simple dimensional considerations, it is clear that the complex coefficients $\underline{z}_{11}$, $\underline{z}_{12}$, $\underline{z}_{21}$, and $\underline{z}_{22}$ have units of ohms; $\underline{z}_{11}$ and $\underline{z}_{22}$ are self-impedances, while $\underline{z}_{12}$ and $\underline{z}_{21}$ are mutual impedances. The 2×2 matrix on the right-hand side of Eq. (2.18) is called the impedance matrix:

$$\overline{\overline{\underline{z}}} = \begin{bmatrix} \underline{z}_{11} & \underline{z}_{12} \\ \underline{z}_{21} & \underline{z}_{22} \end{bmatrix}. \tag{2.19}$$

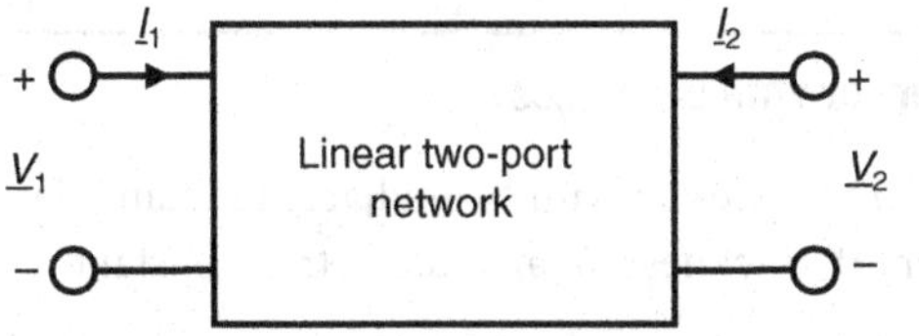

Figure 2.11 Definitions of phasor voltages and currents for a linear, two-port electrical network. Note the sign convention used for the currents.

Formal definitions for the coefficients making up the impedance matrix may be written in terms of the phasors:

$$\underline{z}_{11} = \left.\frac{\underline{V}_1}{\underline{I}_1}\right|_{\underline{I}_2=0}, \quad \underline{z}_{12} = \left.\frac{\underline{V}_1}{\underline{I}_2}\right|_{\underline{I}_1=0}, \quad \underline{z}_{21} = \left.\frac{\underline{V}_2}{\underline{I}_1}\right|_{\underline{I}_2=0}, \quad \text{and} \quad \underline{z}_{22} = \left.\frac{\underline{V}_2}{\underline{I}_2}\right|_{\underline{I}_1=0}. \tag{2.20}$$

These definitions can be used with KVL and KCL to derive the individual matrix elements for any network.

Alternatively, the currents can be expressed as linear functions of the voltages,

$$\begin{bmatrix} \underline{I}_1 \\ \underline{I}_2 \end{bmatrix} = \begin{bmatrix} \underline{y}_{11} & \underline{y}_{12} \\ \underline{y}_{21} & \underline{y}_{22} \end{bmatrix} \begin{bmatrix} \underline{V}_1 \\ \underline{V}_2 \end{bmatrix}, \tag{2.21}$$

where the coefficients $\underline{y}_{11}, \underline{y}_{12}, \underline{y}_{21}$, and $\underline{y}_{22}$ are complex admittance quantities, and

$$\overline{\overline{\underline{y}}} = \begin{bmatrix} \underline{y}_{11} & \underline{y}_{12} \\ \underline{y}_{21} & \underline{y}_{22} \end{bmatrix} \tag{2.22}$$

is the admittance matrix. The formal coefficient definitions are

$$\underline{y}_{11} = \left.\frac{\underline{I}_1}{\underline{V}_1}\right|_{\underline{V}_2=0}, \quad \underline{y}_{12} = \left.\frac{\underline{I}_1}{\underline{V}_2}\right|_{\underline{V}_1=0}, \quad \underline{y}_{21} = \left.\frac{\underline{I}_2}{\underline{V}_1}\right|_{\underline{V}_2=0}, \quad \text{and} \quad \underline{y}_{22} = \left.\frac{\underline{I}_2}{\underline{V}_2}\right|_{\underline{V}_1=0}. \tag{2.23}$$

The impedance and admittance matrices are related via the inverse matrix operation:

$$\overline{\overline{\underline{y}}} = (\overline{\overline{\underline{z}}})^{-1} \tag{2.24}$$

If the network contains only passive elements, that is, resistors, capacitors, and inductors, but no sources, then reciprocity applies, $\underline{z}_{12} = \underline{z}_{21}$ and $\underline{y}_{12} = \underline{y}_{21}$.[2] When applicable, reciprocity is a vital tool for analyzing two-port networks.

2 Refer to pages 312–315 of Bose and Stevens [2] for a definition and concise discussion of two-port network reciprocity.

Example 2.4 **Circuit model for three-plate sensor with parasitics**

Figure 2.12a depicts a *three-plate capacitor* with two adjacent rectangular electrodes patterned on a dielectric layer of thickness d and dielectric constant κ. The self-capacitances are

$$C_1 \approx \kappa\varepsilon_0 w L_1/d \quad \text{and} \quad C_2 \approx \kappa\varepsilon_0 w L_2/d.$$

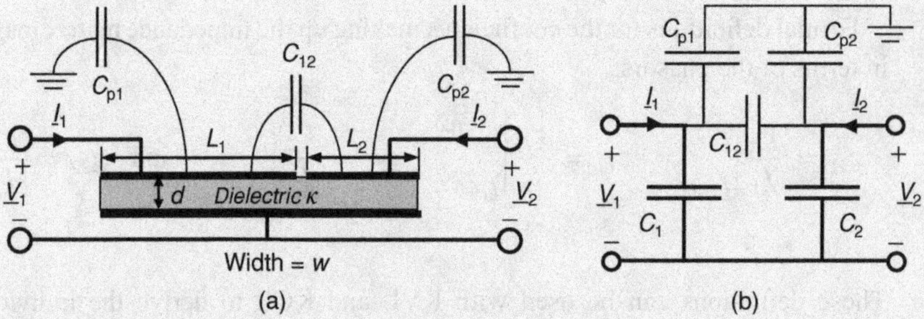

Figure 2.12 Three-plate capacitive structure with parasitic capacitances due to interconnection traces. (a) Side view of parallel-plate structure. (b) Equivalent circuit model for the system. These parasitics can reduce the sensitivity and precision of MEMS sensors.

C_{12} accounts for capacitive coupling between the electrodes. It is likely that the reference ground will be separate from the common terminal of the two-port. The resulting parasitic capacitances, C_{p1} and C_{p2}, which may be significant, can be computed numerically or sometimes estimated from published tables or charts.

The equivalent circuit is shown in Fig. 2.12b. In general, parasitics such as C_{p1} and C_{p2} degrade the resolution and accuracy of capacitive MEMS sensors.

2.3.2 The transmission matrix

Impedance and admittance matrices, while intuitive and familiar, are not particularly useful in the analysis of linear systems. Another form, the transmission matrix, has distinct advantages for many practical configurations; in particular, cascaded networks. This matrix relates the voltage and current variables at one set of ports to those at the remaining ports. For the case of the two-port network,

$$\begin{bmatrix} \underline{V}_1 \\ \underline{I}_1 \end{bmatrix} = \begin{bmatrix} \underline{A} & \underline{B} \\ \underline{C} & \underline{D} \end{bmatrix} \begin{bmatrix} \underline{V}_2 \\ -\underline{I}_2 \end{bmatrix}. \tag{2.25}$$

The 2×2 transmission matrix, sometimes called the ABCD matrix, is

$$\underline{\underline{t}} = \begin{bmatrix} \underline{A} & \underline{B} \\ \underline{C} & \underline{D} \end{bmatrix}. \tag{2.26}$$

The formal circuit definitions for the coefficients in $\underline{\bar{\bar{t}}}$ are

$$\underline{A} = \left.\frac{\underline{V}_1}{\underline{V}_2}\right|_{\underline{I}_2=0}, \quad \underline{B} = \left.\frac{\underline{V}_1}{-\underline{I}_2}\right|_{\underline{V}_2=0}, \quad \underline{C} = \left.\frac{\underline{I}_1}{\underline{V}_2}\right|_{\underline{I}_2=0}, \quad \text{and} \quad \underline{D} = \left.\frac{\underline{I}_1}{-\underline{I}_2}\right|_{\underline{V}_2=0}. \tag{2.27}$$

In the case of passive networks, the reciprocity condition takes the form

$$\underline{A}\,\underline{D} - \underline{B}\,\underline{C} = 1. \tag{2.28}$$

For a reciprocal and lossless two-port, $\underline{A}$ and $\underline{D}$ are pure real numbers, while $\underline{B}$ and $\underline{C}$ are pure imaginary. Another useful transmission matrix form for this two-port network expresses the right-side variables of the network in terms of those on the left side, that is,

$$\begin{bmatrix} \underline{V}_2 \\ -\underline{I}_2 \end{bmatrix} = \underbrace{\begin{bmatrix} \underline{A}' & \underline{B}' \\ \underline{C}' & \underline{D}' \end{bmatrix}}_{\underline{\bar{\bar{t}}'}} \begin{bmatrix} \underline{V}_1 \\ \underline{I}_1 \end{bmatrix}. \tag{2.29}$$

Note the minus sign that still attends the current $\underline{I}_2$ on the left-hand side. The coefficients are

$$\underline{A}' = \left.\frac{\underline{V}_2}{\underline{V}_1}\right|_{\underline{I}_1=0}, \quad \underline{B}' = \left.\frac{\underline{V}_2}{\underline{I}_1}\right|_{\underline{V}_1=0}, \quad \underline{C}' = \left.\frac{-\underline{I}_2}{\underline{V}_1}\right|_{\underline{I}_1=0}, \quad \text{and} \quad \underline{D}' = \left.\frac{-\underline{I}_2}{\underline{I}_1}\right|_{\underline{V}_1=0}. \tag{2.30}$$

This new matrix is of course the inverse of $\underline{\bar{\bar{t}}}$, that is,

$$\underline{\bar{\bar{t}}'} = \begin{bmatrix} \underline{A} & \underline{B} \\ \underline{C} & \underline{D} \end{bmatrix}^{-1} = \frac{1}{\underline{A}\,\underline{D} - \underline{B}\,\underline{C}} \begin{bmatrix} \underline{D} & -\underline{B} \\ -\underline{C} & \underline{A} \end{bmatrix}. \tag{2.31}$$

The reciprocity condition for $\underline{\bar{\bar{t}}'}$ is

$$\underline{A}'\underline{D}' - \underline{B}'\underline{C}' = 1. \tag{2.32}$$

Using the definitions from Eq. (2.30), it may be shown that this equation is identical to Eq. (2.28).

2.3.3 Cascaded two-port networks

The most persuasive advantage of the transmission matrix over other forms is its ease of application for cascaded systems. For example, the relations amongst the port voltages and currents of the system in Fig. 2.13 are

$$\begin{bmatrix} \underline{V}_1 \\ \underline{I}_1 \end{bmatrix} = \begin{bmatrix} \underline{A} & \underline{B} \\ \underline{C} & \underline{D} \end{bmatrix} \begin{bmatrix} \underline{V}_2 \\ -\underline{I}_2 \end{bmatrix} \quad \text{and} \quad \begin{bmatrix} \underline{V}_3 \\ \underline{I}_3 \end{bmatrix} = \begin{bmatrix} \underline{a} & \underline{b} \\ \underline{c} & \underline{d} \end{bmatrix} \begin{bmatrix} \underline{V}_4 \\ -\underline{I}_4 \end{bmatrix}. \tag{2.33}$$

Recognizing that $\underline{V}_2 = \underline{V}_3$ and $\underline{I}_2 = -\underline{I}_3$, matrix multiplication can be used to relate the left-hand-side variables, $\underline{V}_1$ and $\underline{I}_1$, to the right-hand-side variables, $\underline{V}_4$ and $\underline{I}_4$:

$$\begin{bmatrix} \underline{V}_1 \\ \underline{I}_1 \end{bmatrix} = \begin{bmatrix} \underline{A} & \underline{B} \\ \underline{C} & \underline{D} \end{bmatrix} \begin{bmatrix} \underline{a} & \underline{b} \\ \underline{c} & \underline{d} \end{bmatrix} \begin{bmatrix} \underline{V}_4 \\ -\underline{I}_4 \end{bmatrix} = \underbrace{\begin{bmatrix} \underline{a}\,\underline{A} + \underline{c}\,\underline{B} & \underline{b}\,\underline{A} + \underline{d}\,\underline{B} \\ \underline{a}\,\underline{C} + \underline{c}\,\underline{D} & \underline{d}\,\underline{D} + \underline{b}\,\underline{C} \end{bmatrix}}_{\text{cascaded transmission matrix}} \begin{bmatrix} \underline{V}_4 \\ -\underline{I}_4 \end{bmatrix}. \tag{2.34}$$

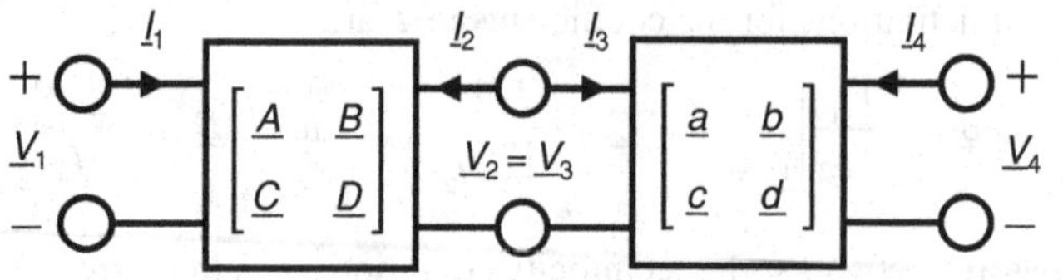

Figure 2.13 Cascaded system consisting of a pair of linear, two-port, electrical networks.

The noteworthy thing about this exercise is that the left-hand-side variables $(\underline{V}_1, \underline{I}_1)$ are so easily expressed in terms of the right-hand-side variables $(\underline{V}_4, \underline{I}_4)$ through matrix multiplication. The need for matrix inversions and other tedious operations is obviated. One can use matrix multiplication to cascade any number of two-port networks. We will exploit this property of transmission matrices in later chapters, once the electromechanical two-port network has been introduced. If care is taken to ensure that the numbers of rows and columns are matched correctly, general N-port networks can be cascaded in the same fashion by matrix multiplication.

Example 2.5 **Linear response of three-plate sensor**

Three-plate sensors featuring geometric antisymmetry provide a highly linear relation between displacement and capacitance change. For the basic coplanar structure of Fig. 2.14a, the middle electrode moves back and forth in the $\pm x$ directions about $x = 0$.

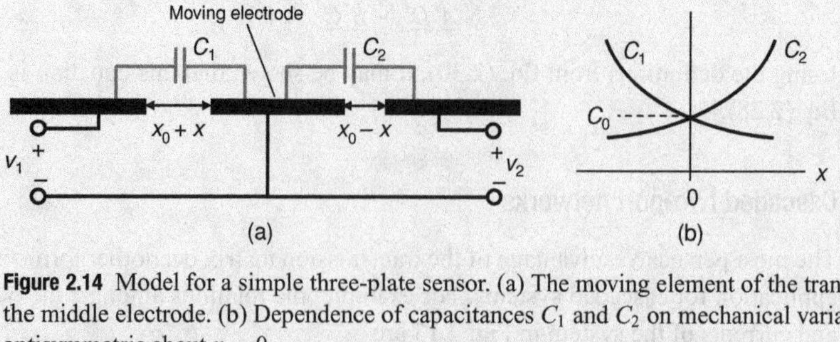

Figure 2.14 Model for a simple three-plate sensor. (a) The moving element of the transducer is the middle electrode. (b) Dependence of capacitances C_1 and C_2 on mechanical variable x is antisymmetric about $x = 0$.

The defined capacitances exhibit antisymmetry with respect to their x dependence:

$$C_1(x) = C_0 f(x) \quad \text{and} \quad C_2(x) = C_0 f(-x),$$

where $f(x)$ is a monotonically increasing function of x and $f(0) = 1$; see Fig. 2.14b. The critical performance measure of three-electrode sensors is the difference between C_1 and C_2, that is,

$$\Delta C = C_1(x) - C_2(x).$$

Using the definitions of C_1 and C_2 to expand ΔC in a Taylor series about x_0, we obtain

$$\Delta C = C_0 \left[2\frac{\mathrm{d}f}{\mathrm{d}x} x + O_3 \right],$$

where $O_3 \propto x^3$. Dramatic improvement in the linear dynamic range of such sensors results from cancellation of the second-order terms arising from C_1 and C_2.

2.3.4 Some important two-port networks

There is a small set of two-port networks and transmission matrices to go with them that will prove useful later in modeling electromechanical transducers. Two of these are the series impedance and shunt admittance circuits shown in Figs. 2.15a and b, respectively.

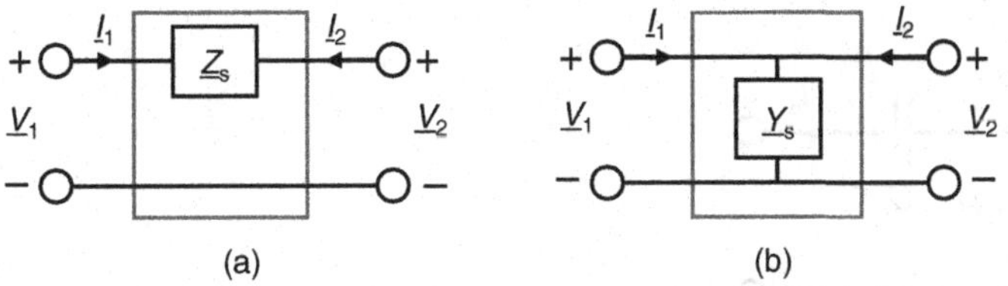

Figure 2.15 Two basic two-port networks. (a) Series impedance network: $\underline{Z}_s$. (b) Shunt admittance network: $\underline{Y}_s$.

We can use the definitions for ($\underline{A}, \underline{B}, \underline{C}, \underline{D}$) from Eq. (2.27) and ($\underline{A}', \underline{B}', \underline{C}', \underline{D}'$) from Eq. (2.30), or matrix inversion, that is, Eq. (2.31), to determine the transmission coefficients:

$$\text{series impedance (Fig. 2.15a):}\quad \underline{\bar{\bar{t}}} = \begin{bmatrix} 1 & \underline{Z}_s \\ 0 & 1 \end{bmatrix} \quad \text{or} \quad \underline{\bar{\bar{t}}}' = \begin{bmatrix} 1 & -\underline{Z}_s \\ 0 & 1 \end{bmatrix}, \tag{2.35}$$

$$\text{shunt admittance (Fig. 2.15b):}\quad \underline{\bar{\bar{t}}} = \begin{bmatrix} 1 & 0 \\ \underline{Y}_s & 1 \end{bmatrix} \quad \text{or} \quad \underline{\bar{\bar{t}}}' = \begin{bmatrix} 1 & 0 \\ -\underline{Y}_s & 1 \end{bmatrix}. \tag{2.36}$$

These matrices satisfy the reciprocity condition of Eq. (2.28). The transmission matrices for certain combinations of these simple networks will be used extensively later on. As a first example, consider the cascade connection of two series impedance networks shown in Fig. 2.16a.

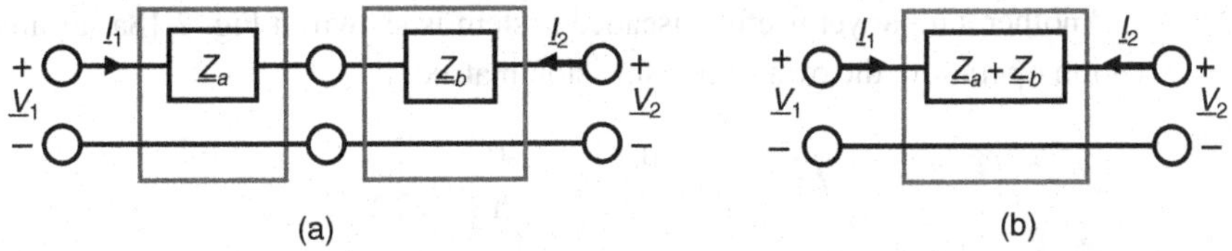

Figure 2.16 Simplification of basic, two-port electrical networks. (a) Cascade connection of two series impedances, $\underline{Z}_a$ and $\underline{Z}_b$. (b) Equivalent two-port realization sums the two impedances.

We may obtain the overall transmission matrix for this linear system from a matrix multiplication operation:

$$\bar{\bar{t}} = \begin{bmatrix} 1 & \underline{Z}_a \\ 0 & 1 \end{bmatrix} \begin{bmatrix} 1 & \underline{Z}_b \\ 0 & 1 \end{bmatrix} = \begin{bmatrix} 1 & \underline{Z}_a + \underline{Z}_b \\ 0 & 1 \end{bmatrix}. \tag{2.37}$$

Equation (2.37) reveals that the network of Fig. 2.16a reduces to the series connection of $\underline{Z}_a$ and $\underline{Z}_b$ shown in Fig. 2.16b, a result already obvious from inspection of the network. Of course, this new transmission matrix is reciprocal.

Example 2.6 **Cascaded circuit matrix operations**

Two-port matrix methods make it easy to perform certain operations upon cascaded circuits. The network in Fig. 2.17 consists of the cascade of a shunt resistor R and series capacitor C.

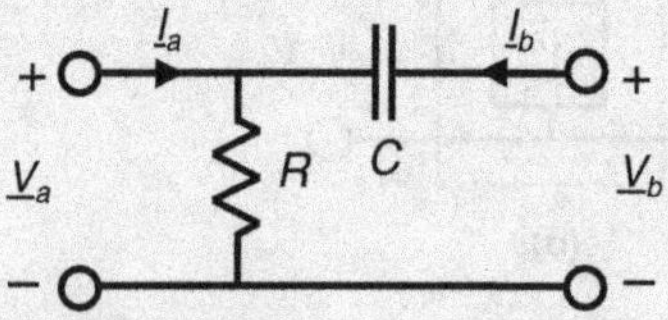

Figure 2.17 Simple example of a cascaded network consisting of a shunt resistor R and series capacitor C.

The transmission matrix of this network is the product of the transmission matrices of the two elements found in Table 2.3. Assuming $R = 1\ \text{M}\Omega$ and $C = 1$ pF,

$$\begin{bmatrix} \underline{V}_a \\ \underline{I}_a \end{bmatrix} = \underbrace{\begin{bmatrix} 1 & 0 \\ 10^{-6} & 1 \end{bmatrix} \begin{bmatrix} 1 & 10^{12}/(\mathrm{j}\omega) \\ 0 & 1 \end{bmatrix}}_{\bar{\bar{t}}} \begin{bmatrix} \underline{V}_b \\ -\underline{I}_b \end{bmatrix}.$$

Matrix multiplication yields

$$\bar{\bar{t}} = \begin{bmatrix} 1 & 10^{12}/(\mathrm{j}\omega) \\ 10^{-6} & 1 + 10^{6}/(\mathrm{j}\omega) \end{bmatrix}$$

This multiplication is not commutative; order is important in such matrix operations.

Another simple yet useful cascaded system is shown in Fig. 2.18a. From a multiplication operation, the overall transmission matrix is

$$\bar{\bar{t}} = \begin{bmatrix} 1 & 0 \\ \underline{Y}_a & 1 \end{bmatrix} \begin{bmatrix} 1 & 0 \\ \underline{Y}_b & 1 \end{bmatrix} = \begin{bmatrix} 1 & 0 \\ \underline{Y}_a + \underline{Y}_b & 1 \end{bmatrix}. \tag{2.38}$$

Matrix multiplication yields a result that is obvious from inspection, and of course, this matrix is reciprocal. The equivalent two-port network is shown in Fig. 2.18b.

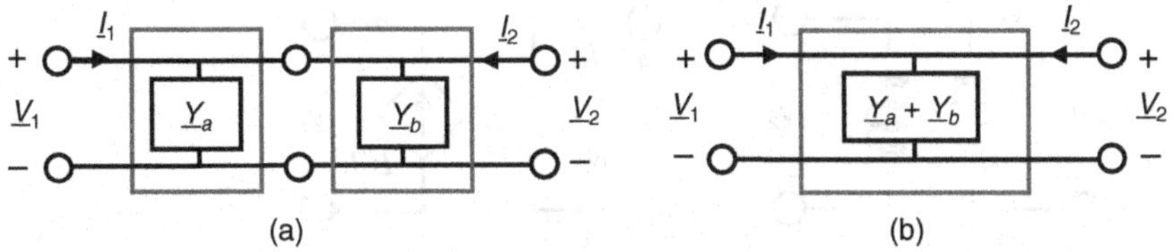

Figure 2.18 Simplification of basic, two-port electrical networks. (a) Cascade connection of two shunt admittances, $\underline{Y}_a$ and $\underline{Y}_b$. (b) Equivalent two-port realization sums the two admittances.

2.3.5 The gyrator and the transformer

Two other two-port networks of use to us are the *gyrator* and the *transformer*. Along with the previously introduced series impedance and shunt admittance networks, they are all we will need in Chapter 4 to construct practical models for electromechanical devices.

Even to electrical engineering students, the *gyrator*, the standard circuit symbol for which is introduced in Fig. 2.19a, might be unfamiliar. First described by Tellegen [3], the gyrator transmission matrix is

$$\underline{\bar{\bar{t}}} = \begin{bmatrix} 0 & -\underline{M} \\ -1/\underline{M}^* & 0 \end{bmatrix} \quad \text{and} \quad \underline{\bar{\bar{t}}}' = \begin{bmatrix} 0 & -\underline{M}^* \\ -1/\underline{M} & 0 \end{bmatrix}, \tag{2.39}$$

where $\underline{M}^*$ is the complex conjugate of $\underline{M}$. Formal definitions for the transimpedance coefficient $\underline{M}$, written in terms of the phasor voltages and currents, enforce the reciprocity constraint on the network. Equation (2.28) may be used to confirm that $\underline{M}$ must be a purely imaginary quantity:

$$\underline{M} = \left.\frac{\underline{V}_1}{\underline{I}_2}\right|_{\text{any } \underline{I}_1, \underline{V}_2} \quad \text{and} \quad \frac{1}{\underline{M}^*} = -\left.\frac{\underline{I}_1}{\underline{V}_2}\right|_{\text{any } \underline{V}_1, \underline{I}_2}. \tag{2.40}$$

A commonly invoked circuit realization for the gyrator, shown in Fig. 2.19b, uses two linear, current-dependent voltage sources.

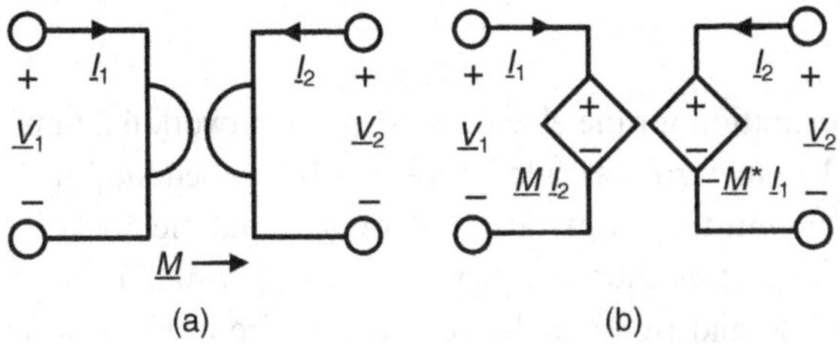

Figure 2.19 The gyrator, a linear, symmetric, two-port network. (a) The standard symbolic form for the gyrator with transimpedance $\underline{M}$. (b) Dependent source-based realization for the gyrator.

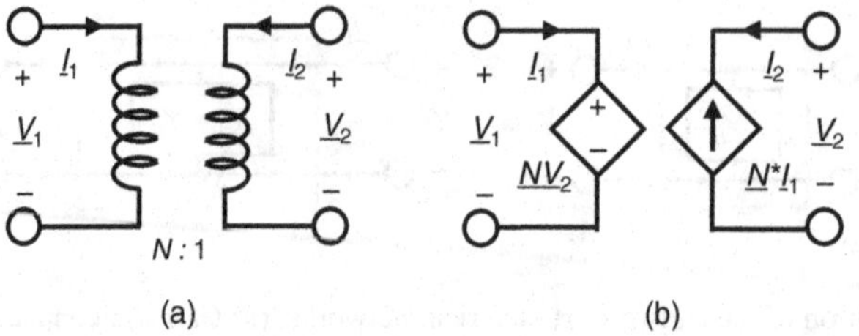

Figure 2.20 The transformer, a linear, symmetric, two-port network. (a) The familiar symbolic form used for a magnetic transformer having turns ratio $N = N_1/N_2$. (b) A dependent source realization for the transformer. The interpretation of this circuit as a transformer is not directly useful for the electromechanical actuator models of Chapter 4.

The other element needed is the *transformer*. The schematic shown in Fig. 2.20a is familiar from the usual first course in circuit theory. The transmission matrix of the transformer is

$$\underline{\bar{\bar{t}}} = \begin{bmatrix} \underline{N} & 0 \\ 0 & 1/\underline{N}^* \end{bmatrix} \quad \text{and} \quad \underline{\bar{\bar{t}}}' = \begin{bmatrix} 1/\underline{N} & 0 \\ 0 & \underline{N}^* \end{bmatrix}, \tag{2.41}$$

where

$$\underline{N} = \left.\frac{\underline{V}_1}{\underline{V}_2}\right|_{\text{any } \underline{I}} \quad \text{and} \quad \frac{1}{\underline{N}^*} = \left.\frac{\underline{I}_1}{-\underline{I}_2}\right|_{\text{any } \underline{V}}. \tag{2.42}$$

The transformer is a passive lossless device that, like the gyrator, cannot store energy. From Eq. (2.28), it is easy to show that the matrices of Eq. (2.41) are reciprocal as long as $\underline{N}$ is a purely real number. The linear, dependent source network realization for the transformer drawn in Fig. 2.20b will be used extensively in Chapter 4. It is only for notational consistency that we use $\underline{N}$ instead of N.

Though not directly relevant for electromechanical two-port theory, an instructive interpretation of the transformer is to identify the coefficient $\underline{N}$ appearing in Eq. (2.41) as the ratio of the number of windings in the primary (N_1) and secondary (N_2) of a real transformer, that is, $\underline{N} \to N = N_1/N_2$.

While not unique, the dependent source realizations for the gyrator and transformer of Figs. 2.19b and 2.20b will be employed in Chapter 4. For convenience, the matrices from Eqs. (2.39) and (2.41) are collected in Table 2.3.

2.3.6 Embedded networks

A useful exercise in preparation for the electromechanical network manipulations to be employed in Chapter 4 is to form cascaded networks by embedding either the gyrator or the transformer within the series- and shunt-connected elements introduced in Section 2.3.4. In conducting this exercise, there are many permutations and combinations for the left-hand-side and right-hand-side variables from which to choose. One can, for example, exchange the order of voltage and current in the column matrices and change the signs of the currents. Figures 2.21a and b present two important embedded networks to be used in Chapter 4.

Table 2.3 Different forms of the transmission matrices for the four building blocks of cascaded, two-port networks: the series impedance, the shunt admittance, the gyrator, and the transformer

	Standard matrix forms		Inverse matrix forms	
Basic linear two-port networks	$\begin{bmatrix}\underline{V}_1\\ \underline{I}_1\end{bmatrix} = \bar{\bar{\underline{t}}}\begin{bmatrix}\underline{V}_2\\ -\underline{I}_2\end{bmatrix}$	$\begin{bmatrix}\underline{I}_1\\ \underline{V}_1\end{bmatrix} = (\bar{\bar{\underline{t}}})_{\text{alt}}\begin{bmatrix}-\underline{I}_2\\ \underline{V}_2\end{bmatrix}$	$\begin{bmatrix}\underline{V}_2\\ -\underline{I}_2\end{bmatrix} = \bar{\bar{\underline{t}}}'\begin{bmatrix}\underline{V}_1\\ \underline{I}_1\end{bmatrix}$	$\begin{bmatrix}-\underline{I}_2\\ \underline{V}_2\end{bmatrix} = (\bar{\bar{\underline{t}}}')_{\text{alt}}\begin{bmatrix}\underline{I}_1\\ \underline{V}_1\end{bmatrix}$
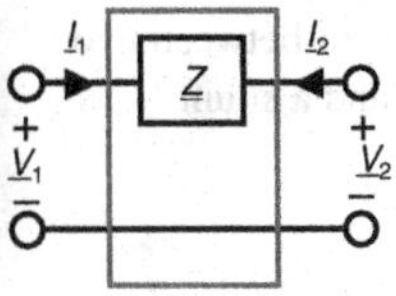	$\begin{bmatrix}1 & \underline{Z}\\ 0 & 1\end{bmatrix}$	$\begin{bmatrix}1 & 0\\ \underline{Z} & 1\end{bmatrix}$	$\begin{bmatrix}1 & -\underline{Z}\\ 0 & 1\end{bmatrix}$	$\begin{bmatrix}1 & 0\\ -\underline{Z} & 1\end{bmatrix}$
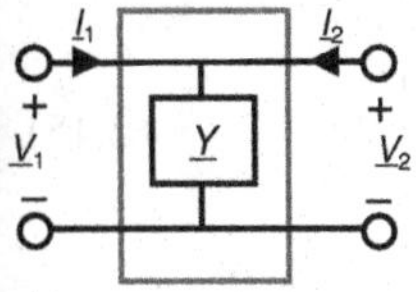	$\begin{bmatrix}1 & 0\\ \underline{Y} & 1\end{bmatrix}$	$\begin{bmatrix}1 & \underline{Y}\\ 0 & 1\end{bmatrix}$	$\begin{bmatrix}1 & 0\\ -\underline{Y} & 1\end{bmatrix}$	$\begin{bmatrix}1 & -\underline{Y}\\ 0 & 1\end{bmatrix}$
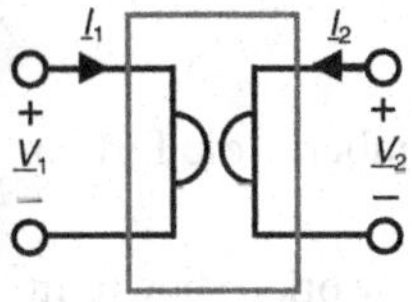	$\begin{bmatrix}0 & -\underline{M}\\ -1/\underline{M}^* & 0\end{bmatrix}$	$\begin{bmatrix}0 & -1/\underline{M}^*\\ -\underline{M} & 0\end{bmatrix}$	$\begin{bmatrix}0 & -\underline{M}^*\\ -1/\underline{M} & 0\end{bmatrix}$	$\begin{bmatrix}0 & -1/\underline{M}\\ -\underline{M}^* & 0\end{bmatrix}$
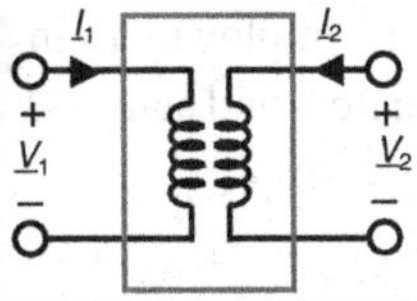	$\begin{bmatrix}\underline{N} & 0\\ 0 & 1/\underline{N}^*\end{bmatrix}$	$\begin{bmatrix}1/\underline{N}^* & 0\\ 0 & \underline{N}\end{bmatrix}$	$\begin{bmatrix}1/\underline{N} & 0\\ 0 & \underline{N}^*\end{bmatrix}$	$\begin{bmatrix}\underline{N}^* & 0\\ 0 & 1/\underline{N}\end{bmatrix}$

Consider first a gyrator element embedded between two series impedances, shown in Fig. 2.21a. The networks are cascaded so the overall system matrix is the multiplication of three matrices:

$$\begin{bmatrix}\underline{V}_a\\ \underline{I}_a\end{bmatrix} = \underbrace{\begin{bmatrix}1 & \underline{Z}_a\\ 0 & 1\end{bmatrix}\begin{bmatrix}0 & -\underline{M}\\ -1/\underline{M}^* & 0\end{bmatrix}\begin{bmatrix}1 & \underline{Z}_b\\ 0 & 1\end{bmatrix}}_{-\frac{1}{\underline{M}^*}\begin{bmatrix}\underline{Z}_a & \left|\underline{M}\right|^2+\underline{Z}_a\underline{Z}_b\\ 1 & \underline{Z}_b\end{bmatrix}}\begin{bmatrix}\underline{V}_b\\ -\underline{I}_b\end{bmatrix}. \tag{2.43}$$

Remember that the reciprocity condition imposed upon the gyrator requires the coefficient $\underline{M}$ to be purely imaginary. A useful alternate form for this matrix relationship results from reversing the order of the current and voltage variables on the right-hand side:

$$\begin{bmatrix}\underline{V}_a\\ \underline{I}_a\end{bmatrix} = -\frac{1}{\underline{M}^*}\begin{bmatrix}|\underline{M}|^2+\underline{Z}_a\underline{Z}_b & \underline{Z}_a\\ \underline{Z}_b & 1\end{bmatrix}\begin{bmatrix}-\underline{I}_b\\ \underline{V}_b\end{bmatrix}. \tag{2.44}$$

For the transformer embedded between an impedance on the left and admittance on the

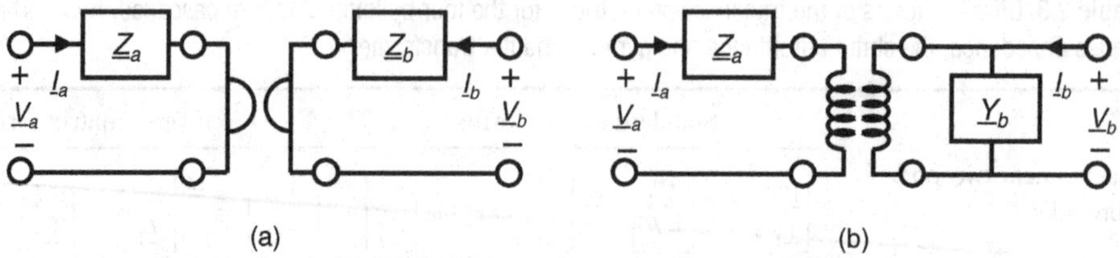

Figure 2.21 Two important cascaded reciprocal networks. (a) A gyrator embedded between two series impedances. (b) A transformer embedded between a series impedance and a shunt admittance.

right, as depicted in Fig. 2.21b, we have

$$\begin{bmatrix} \underline{V}_a \\ \underline{I}_a \end{bmatrix} = \underbrace{\begin{bmatrix} 1 & \underline{Z}_a \\ 0 & 1 \end{bmatrix} \begin{bmatrix} \underline{N} & 0 \\ 0 & 1/\underline{N}^* \end{bmatrix} \begin{bmatrix} 1 & 0 \\ \underline{Y}_b & 1 \end{bmatrix}}_{\frac{1}{\underline{N}^*} \begin{bmatrix} \left|\underline{N}\right|^2 + \underline{Z}_a \underline{Y}_b & \underline{Z}_a \\ \underline{Y}_b & 1 \end{bmatrix}} \begin{bmatrix} \underline{V}_b \\ -\underline{I}_b \end{bmatrix}. \tag{2.45}$$

It is important to recall that, for reciprocity, $\underline{N}$ must be a pure real number. Note further that this matrix is canonically identical to that appearing in Eq. (2.44).

In later chapters, we will extensively employ the cascaded networks shown in Fig. 2.21 and their associated matrix forms. In particular, Chapter 4 will employ them in circuit models for the small-signal behavior of electromechanical actuators and sensors.

2.3.7 Source and impedance reflection

Sometimes it is useful to transform a circuit containing a gyrator or transformer into an equivalent circuit where all impedances, admittances, and sources are collected on one side of an equivalent network. This operation is called *impedance reflection*. These transformed circuits are usually easier to analyze and often provide insights into the behavior of electromechanical devices.

Consider the example in Fig. 2.22a with impedance $\underline{Z}_s$ and voltage source $\underline{V}_s$ on the right-hand side of a gyrator.

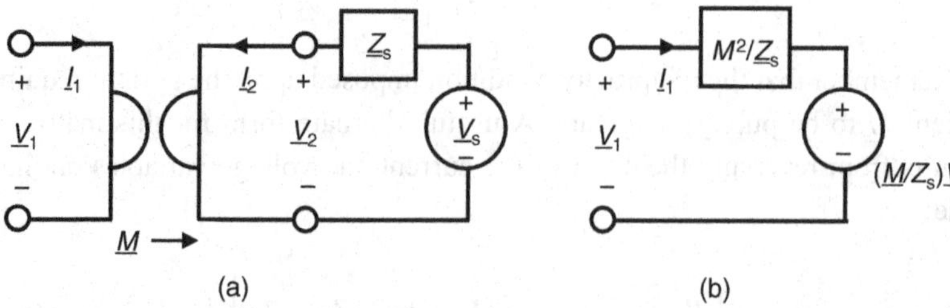

Figure 2.22 Example of impedance and source reflection across a gyrator. (a) Original circuit with source $\underline{V}_s$ and impedance $\underline{Z}_s$ on the right side of the gyrator. (b) Equivalent circuit with source and impedance reflected across to the left side.

Using Table 2.3, the system can be described by the following cascade:

$$\begin{bmatrix} \underline{V}_1 \\ \underline{I}_1 \end{bmatrix} = \begin{bmatrix} 0 & -\underline{M} \\ -1/\underline{M}^* & 0 \end{bmatrix} \begin{bmatrix} 1 & \underline{Z}_s \\ 0 & 1 \end{bmatrix} \begin{bmatrix} \underline{V}_s \\ -\underline{I}_2 \end{bmatrix}, \tag{2.46}$$

which reduces to

$$\begin{bmatrix} \underline{V}_1 \\ \underline{I}_1 \end{bmatrix} = \begin{bmatrix} 0 & -\underline{M} \\ -1/\underline{M}^* & -\underline{Z}_s/\underline{M}^* \end{bmatrix} \begin{bmatrix} \underline{V}_s \\ -\underline{I}_2 \end{bmatrix}. \tag{2.47}$$

Utilizing the pair of linear equations in Eq. (2.47), $\underline{I}_2$ may be eliminated to obtain

$$\underline{V}_1 = \frac{|\underline{M}|^2}{\underline{Z}_s}\underline{I}_1 + \frac{\underline{M}}{\underline{Z}_s}\underline{V}_s. \tag{2.48}$$

A *Thevenin source* equivalent circuit for Eq. (2.48) is shown in Fig. 2.22b. Here, both the impedance $\underline{Z}_s$ and the voltage source $\underline{V}_s$ have been reflected across to the left side, but it should be evident that either one could be mapped across, leaving the other component in its original location. This observation reveals the non-uniqueness of circuit models.

Similarly, for the network with the embedded transformer of Fig. 2.23a, one may write

$$\begin{bmatrix} \underline{V}_1 \\ \underline{I}_1 \end{bmatrix} = \begin{bmatrix} \underline{N} & 0 \\ 0 & 1/\underline{N}^* \end{bmatrix} \begin{bmatrix} 1 & 0 \\ \underline{Y}_s & 1 \end{bmatrix} \begin{bmatrix} \underline{V}_2 \\ -\underline{I}_s \end{bmatrix}, \tag{2.49}$$

which reduces to

$$\begin{bmatrix} \underline{V}_1 \\ \underline{I}_1 \end{bmatrix} = \begin{bmatrix} \underline{N} & 0 \\ \underline{Y}_s/\underline{N}^* & 1/\underline{N}^* \end{bmatrix} \begin{bmatrix} \underline{V}_2 \\ -\underline{I}_s \end{bmatrix}, \tag{2.50}$$

and finally

$$\underline{I}_1 = \frac{\underline{Y}_s}{|\underline{N}|^2}\underline{V}_1 - \frac{1}{\underline{N}^*}\underline{I}_s. \tag{2.51}$$

Equation (2.51) is readily cast in the form of the *Norton equivalent* shown in Fig. 2.23b. Either or both $\underline{Y}_s$ and $\underline{I}_s$ can be reflected across to the left side of the

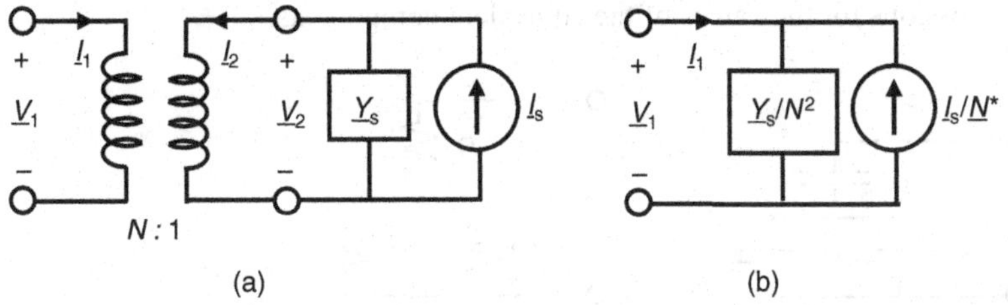

Figure 2.23 Example of admittance and source reflection across a transformer. (a) Original circuit with source $\underline{V}_s$ and admittance $\underline{Y}_s$ on the right side of the transformer. (b) Equivalent circuit with source and impedance reflected across to left side.

two-port, so we again encounter the non-unique nature of circuit equivalents. In the interpretation of these results, bear in mind that $\underline{N}$ is a pure real number and that its designation as complex is only for notational consistency with Chapter 4.

2.4 Summary

This chapter started with a brief review of some circuit theory basics, principally KCL and KVL, AC phasors, impedance, and admittance, and expressions for time-average electric power. The next topic was circuit models for simple, parallel-plate, RC elements. We presented an inspection technique for reducing complicated capacitor and resistor geometries to simpler RC networks, which will be used extensively in Chapter 3 and later. The student wishing to gain a deeper understanding of these circuit concepts and modeling methods is encouraged to refer regularly back to this chapter and also to Appendix A, which reviews the electroquasistatic approximation and shows that KVL and KCL are special cases of the more general field-based integral laws. Appendix A also shows how to use the integral formulation of electroquasistatics to derive the familiar models for capacitors and resistors.

Section 2.3 focused on the topic of two-port networks. Of all the choices for two-port network representations, the transmission matrix turns out to be the most natural because of the ease with which it can be used to construct cascade-based systems. To facilitate later efforts, we presented a set of four linear reciprocal networks: the series impedance, the shunt admittance, the transformer, and the gyrator. Chapter 4 will reveal that, as long as its operation is restricted to the linear range, any electromechanical actuator or sensor can be represented in a very general form – requiring only capacitance and its first two derivatives – using transmission matrices and appropriate selections from these four basic network types. These two-port models are readily integrated with the electronic and mechanical subsystems.

Problems

2.1 Simplify the purely capacitive and purely resistive circuits in (a) and (b) and find expressions for the values of the equivalent components.

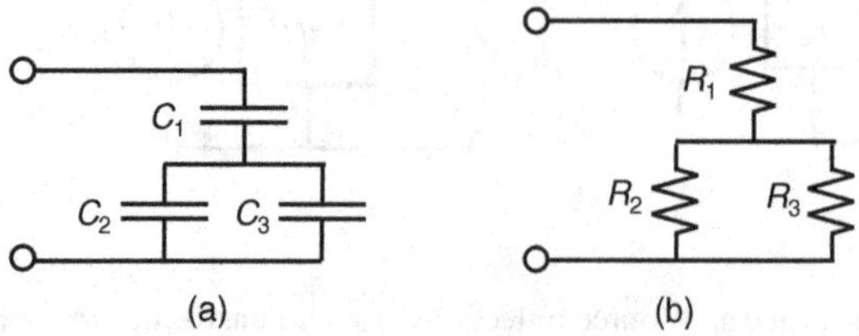

2.2 The transducer below is subjected to a force, causing the dielectric slab to move back and forth according to the equation:

$$x(t) = a + b\cos(\omega t).$$

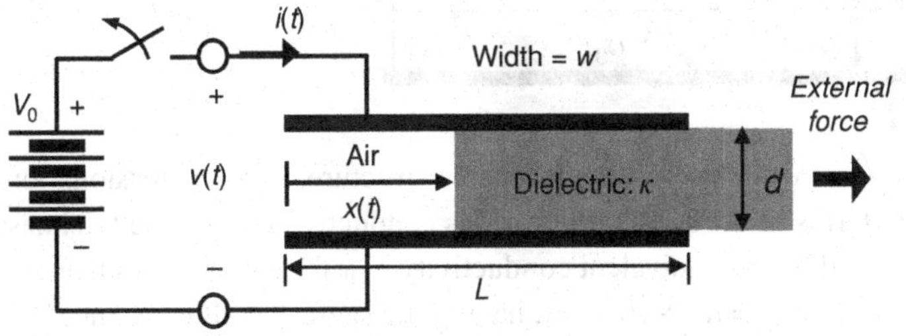

Assume that the amplitude of the motion is sufficiently small so that neither end of the slab moves close to the edges of the parallel electrodes.

(a) If the voltage of the device is fixed at $v = V_0$, what is the current $i(t)$?
(b) In a second test the voltage is set first to V_0 with the slab stationary at $x = a$. The device is then *open-circuited* so that $i(t) = 0$. If the previously specified sinusoidal motion of the slab is once again imposed, what is the time dependence of the voltage $v(t)$?

2.3 A layered device is constructed from four parallel laminae of dielectric material, as shown below.

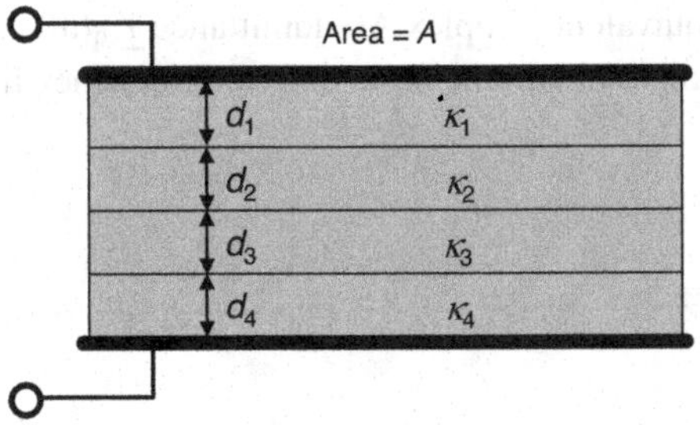

(a) What is the equivalent circuit model for the device? Provide expressions for the individual elements of your model.
(b) If the net voltage across the device is V, find expressions for the voltages across each of the four laminae.
(c) Find the equivalent dielectric constant κ_{eff} for a homogeneous capacitor of the same area A, same total thickness $d = d_1 + d_2 + d_3 + d_4$, and same overall capacitance.

2.4 A layered device is constructed from three parallel laminae of electrically conductive material, as shown.

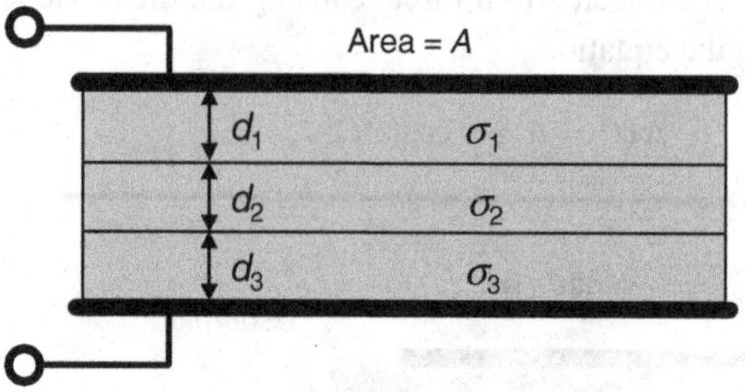

(a) Obtain an equivalent DC circuit for this structure. Find expressions for each resistance in terms of the area A, the various conductivities, and the thicknesses.
(b) What would be the equivalent conductivity σ_{eff} for a device of identical area A and total thickness $d = d_1 + d_2 + d_3$, having the same overall resistance?

2.5 Find an expression for the equivalent complex AC impedance $\underline{Z}_{eq}(\omega)$ of the circuit shown below.

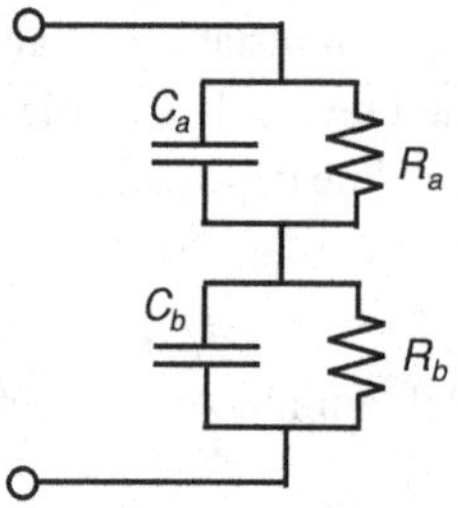

2.6 Find an expression for the equivalent complex AC admittance $\underline{Y}_{eq}(\omega)$ of the circuit shown, and then use this expression to obtain the high- and low-frequency limits.

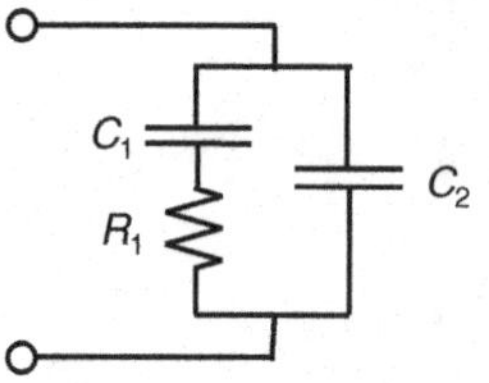

2.7 Find an expression for the equivalent complex AC impedance $\underline{Z}_{eq}(\omega)$ of the circuit shown below.

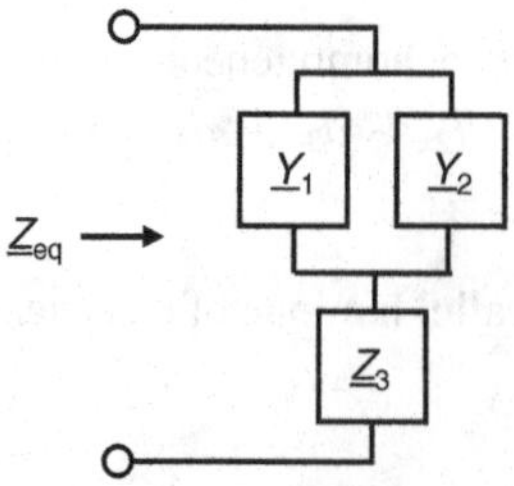

2.8 The parallel-plate device below consists of three sections with different conductivities and dielectric constants.

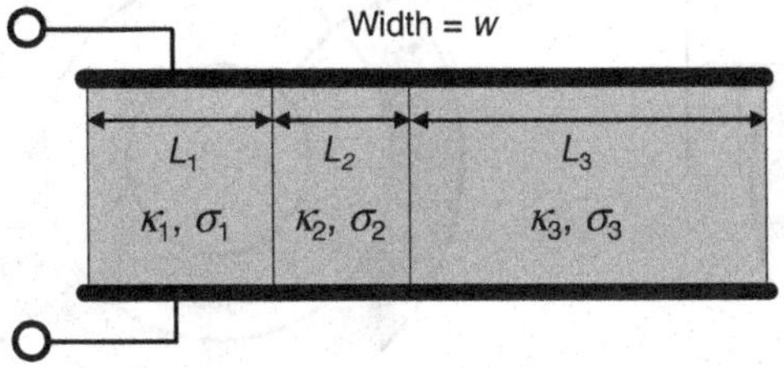

(a) Use inspection to find the simplest equivalent circuit for this geometry, obtaining expressions for all circuit components needed in terms of the various conductivities, dielectric constants, and dimensions of the slabs. Explain the assumptions inherent in the inspection method.
(b) What are the equivalent conductivity σ_{eff} and dielectric constant κ_{eff} for a homogeneous slab having the same admittance and the same total overall dimensions?

2.9 Use the integral form of Gauss's law from Appendix A to obtain an approximate expression for the capacitance of the device shown below. To work this problem, assume that the gap dimensions, d_r and d_l, are much smaller than the areal dimensions, L and w, so that the x-dependent electric field magnitude in the dielectric can be approximated by the applied voltage divided by the local gap spacing. Neglect fringing.

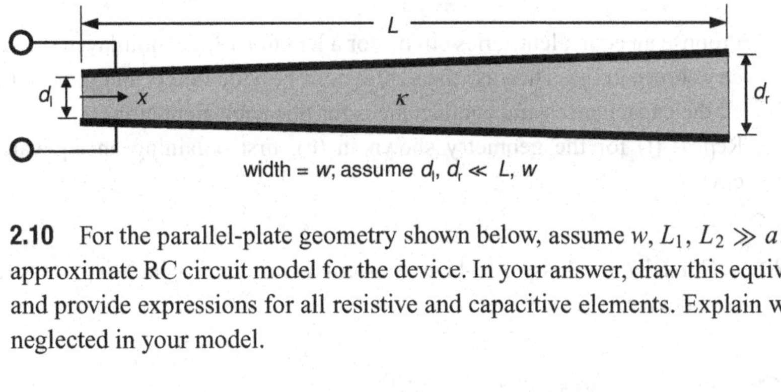

2.10 For the parallel-plate geometry shown below, assume $w, L_1, L_2 \gg a$. Propose an approximate RC circuit model for the device. In your answer, draw this equivalent circuit and provide expressions for all resistive and capacitive elements. Explain what is being neglected in your model.

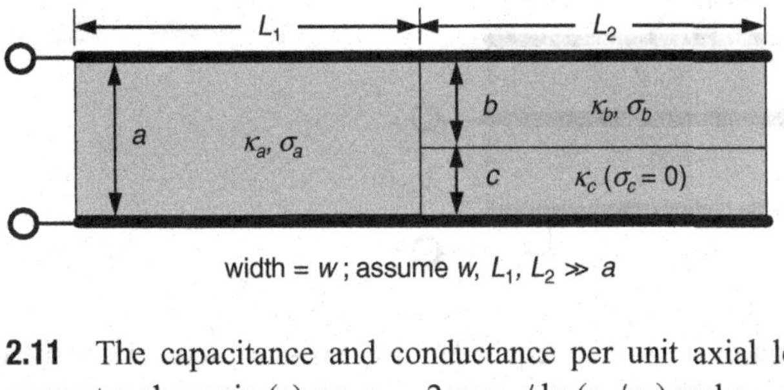

2.11 The capacitance and conductance per unit axial length for the basic coaxial geometry shown in (a) are $c_l = 2\pi\kappa\varepsilon_0 / \ln(r_2/r_1)$ and $g_l = 2\pi\sigma / \ln(r_2/r_1)$.

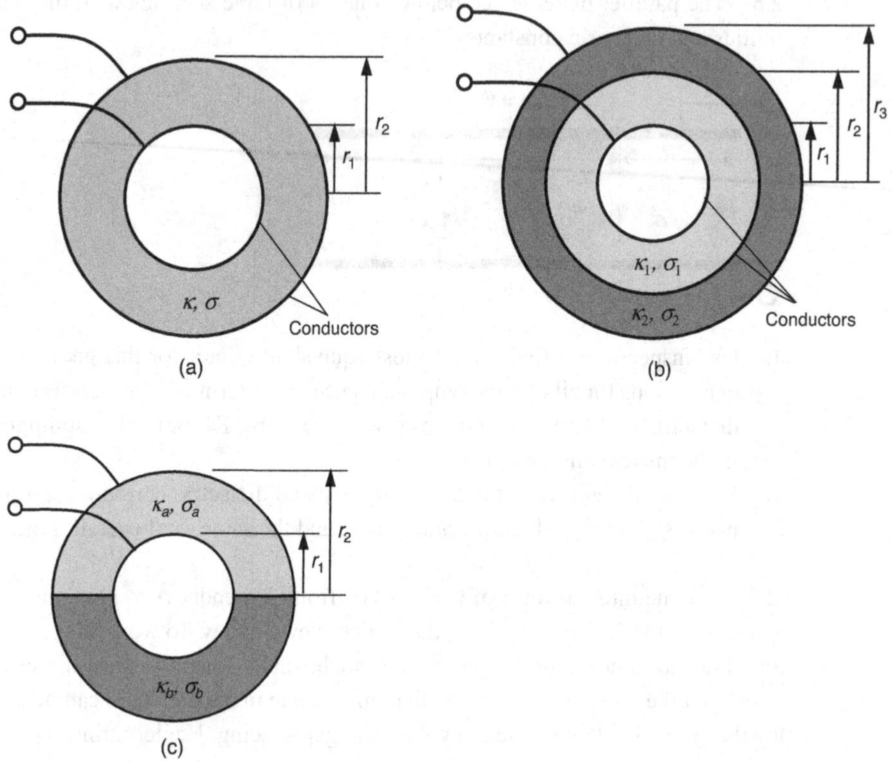

(i) Propose an equivalent series circuit for a length L of the multilayered coaxial geometry shown in (b). Then use the expressions provided for c_l and g_l from (a) to obtain all the capacitances and conductances for this equivalent circuit.

(ii) Repeat (i) for the geometry shown in (c), first obtaining an equivalent parallel circuit.

2.12 Consider the lossy two-layer capacitive structure, which is connected up as a two-port device.

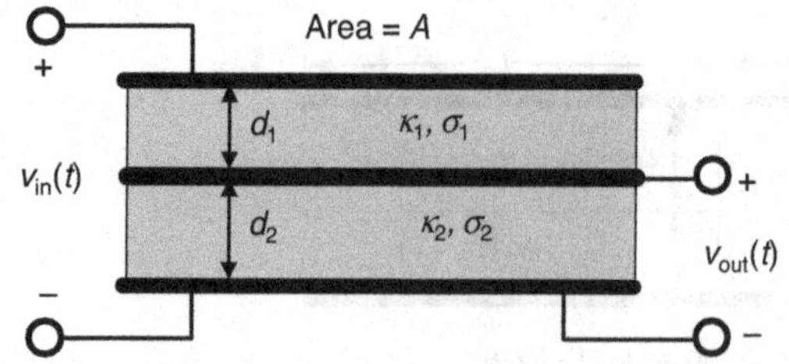

(a) Use inspection to propose an equivalent RC two-port network for the two-layered device.

(b) Assuming sinusoidal electrical excitation at frequency ω, find an expression for the impedance matrix $\bar{\bar{z}}$ for this device.
(c) What is the transmission matrix $\bar{\bar{t}}$ for this device?

2.13 For the two-port device shown, it is observed that the instantaneous power into port 1 is not always equal to the instantaneous power coming out of port 2, that is, $v_1(t) \cdot i_1(t) \neq -v_2(t) \cdot i_2(t)$. However, the time averages of these two values of power are equal, that is, $\langle v_1(t) \cdot i_1(t) \rangle = -\langle v_2(t) \cdot i_2(t) \rangle$ for any excitation. What can be said about this network? Note: here $\langle f(t) \rangle$ represents the time average of any time-dependent function $f(t)$.

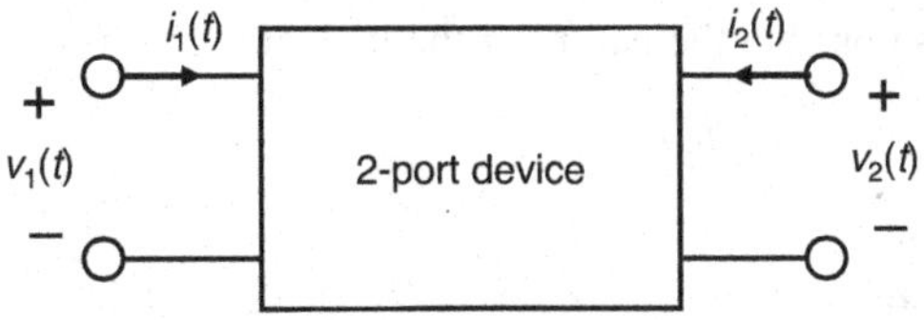

2.14 Find the impedance and transmission matrices, $\bar{\bar{z}}$ and $\bar{\bar{t}}$, for the resistive two-port network shown below. Verify reciprocity for each matrix.

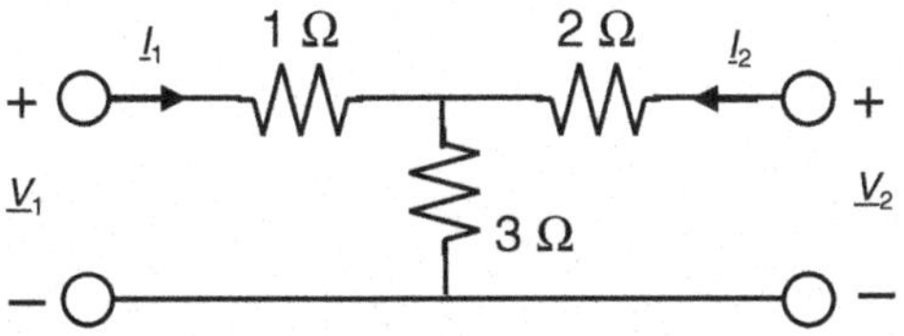

2.15 Employ the formal methods of Section 2.3.1, specifically, Eqs. (2.20) and (2.23), to obtain all elements of the impedance and admittance matrices for the capacitive two-port network shown below. Verify reciprocity.

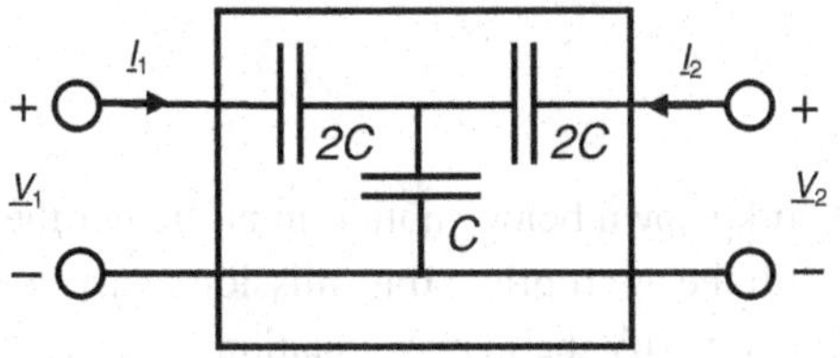

2.16 Find the transmission matrix $\bar{\bar{t}}$ for the capacitive two-port network shown below. Pay close attention to units here. Verify reciprocity for the matrix.

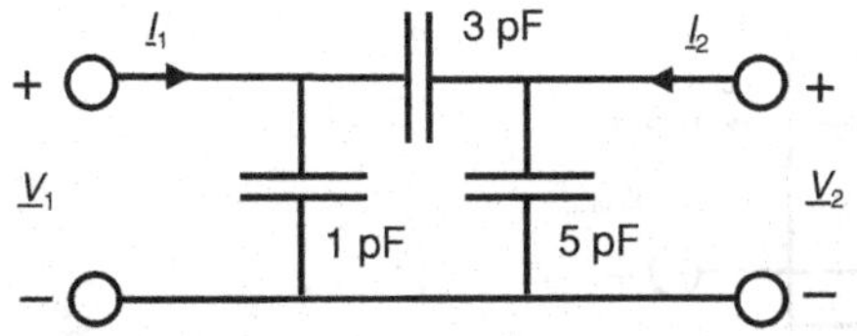

2.17 Employ the formal methods of Eq. (2.27) to obtain all elements of the transmission matrix for the capacitive two-port network shown. Verify that the network is reciprocal.

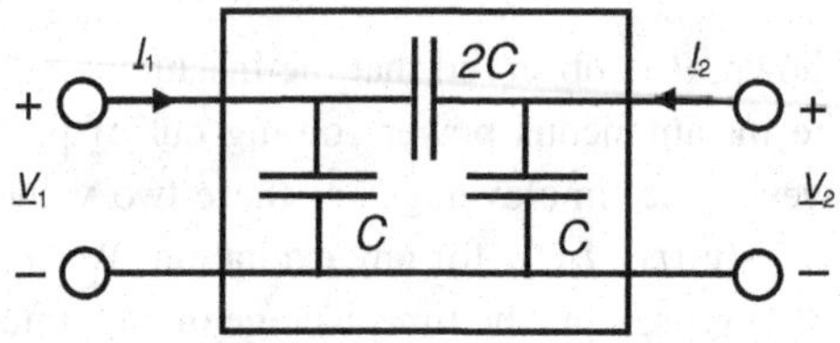

2.18 Consider the two-port network shown below. For each of the transmission matrices listed, characterize as fully as possible the nature of the network.

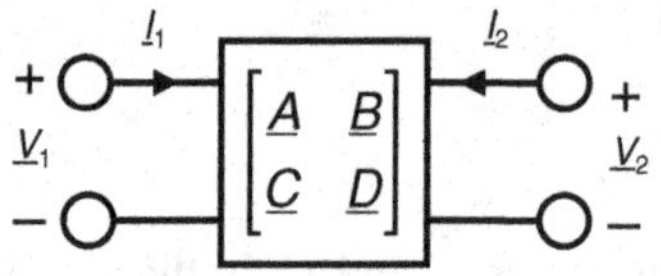

(a) $\begin{bmatrix} 1 & 3 \\ 0 & 1 \end{bmatrix}$;

(b) $\begin{bmatrix} 1 & 0 \\ 3 & 1 \end{bmatrix}$;

(c) $\begin{bmatrix} 1 & 0 \\ 0 & 2 \end{bmatrix}$;

(d) $\begin{bmatrix} 1 & j \\ -j & 1 \end{bmatrix}$;

(e) $\begin{bmatrix} 1 & -1 \\ -1 & 1 \end{bmatrix}$;

(f) $\begin{bmatrix} 0 & j \\ j2 & 0 \end{bmatrix}$;

(g) $\begin{bmatrix} 0 & j \\ -j & 0 \end{bmatrix}$.

2.19 Consider the cascaded network shown below, noting in particular the definitions of positive current flow. Starting with the appropriate transmission matrices provided in Table 2.3, use matrix multiplication to verify the matrix equation

$$\begin{bmatrix} i_{\text{in}} \\ v_{\text{in}} \end{bmatrix} = \begin{bmatrix} 1 & G_b \\ R_a & 1 + R_a G_b \end{bmatrix} \begin{bmatrix} i_{\text{out}} \\ v_{\text{out}} \end{bmatrix}.$$

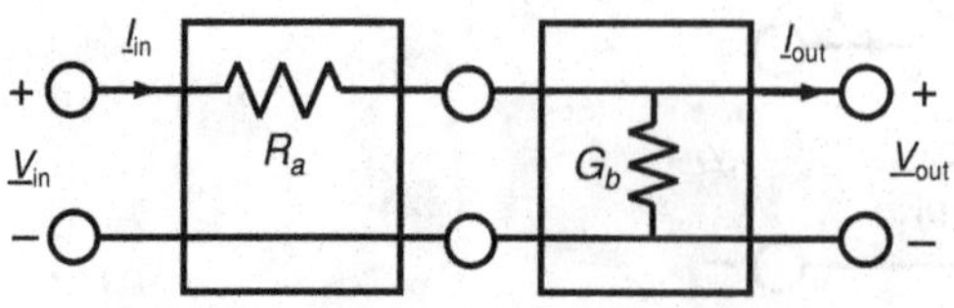

2.20 Making no assumptions whatever about the ABCD two-port network, find the transmission matrix for the cascaded network shown in (a). Repeat for the network shown in (b). Can these two matrices be reconciled if it is assumed that the ABCD two-port is reciprocal?

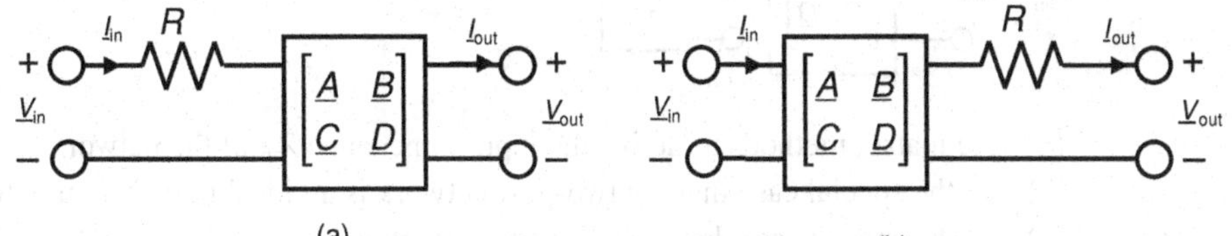

(a) (b)

2.21 Use phasors to obtain expressions for the complex coefficients of the transmission matrices of the three networks shown below.

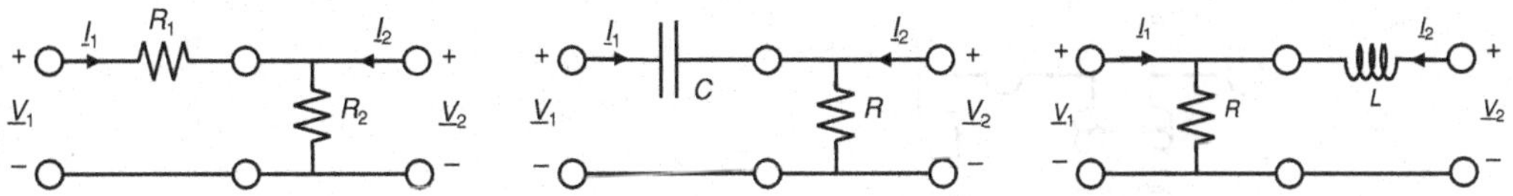

(a) (b) (c)

2.22 Find the transmission matrices for the cascaded networks shown in (a) and (b). Obtain the first of these by inspection of the equivalent transformer network using the circuit shown in Fig. 2.20b and then invoking inspection. Check your results using matrix methods. Assume that N_1 and N_2 are pure real numbers.

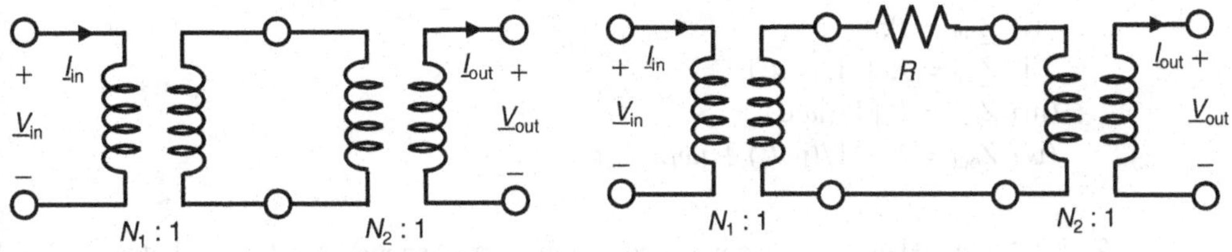

(a) (b)

2.23 Find the transmission matrices for the two cascaded networks shown. Obtain the first of these by inspection of the equivalent transformer network using the circuit shown in Fig. 2.19b and then invoking inspection. Check your result using matrix methods. Assume that $\underline{M}_1$ and $\underline{M}_2$ are pure imaginary numbers.

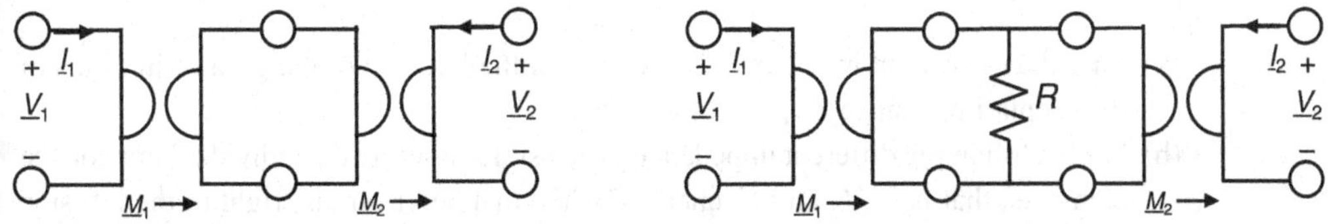

(a) (b)

2.24 An impedance $\underline{Z}_L$ is cascade-connected to a general two-port network having coefficients $\underline{A}$, $\underline{B}$, $\underline{C}$, and $\underline{D}$, as shown below.

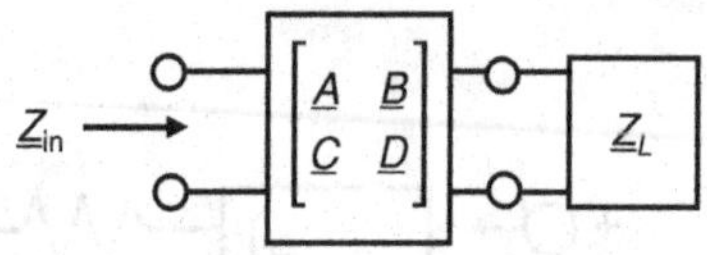

(a) Use matrix methods to derive the input impedance $\underline{Z}_{\text{in}}$ of the network.
(b) For the special case that the two-port network is an ideal transformer, what is $\underline{Z}_{\text{in}}$?
(c) For the special case that the two-port network is an ideal gyrator, what is $\underline{Z}_{\text{in}}$? Verify units for the coefficients in your result.

2.25 An impedance $\underline{Z}_{\text{out}}$ is connected to a transformer, as shown below.

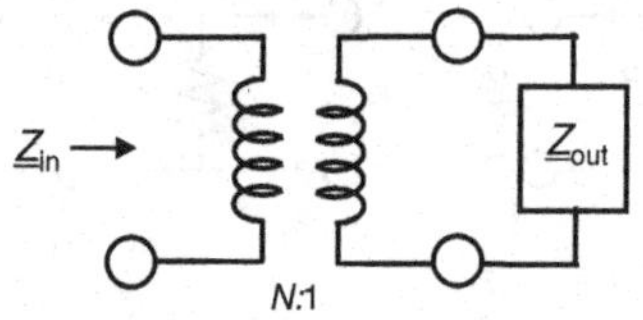

(a) Find the equivalent impedance $\underline{Z}_{\text{in}}$ seen from the primary of the transformer in terms of the output impedance $\underline{Z}_{\text{out}}$.
(b) For the following different impedance expressions, how are the individual impedance elements, that is, R, L, and C, changed when mapped from the secondary to the primary side of the transformer?
 (i) $\underline{Z}_{\text{out}} = R$;
 (ii) $\underline{Z}_{\text{out}} = R + 1/(\text{j}\omega C)$;
 (iii) $\underline{Z}_{\text{out}} = R \,\|\, 1/(\text{j}\omega C)$;
 (iv) $\underline{Z}_{\text{out}} = R + 1/(\text{j}\omega C) + \text{j}\omega L$.

2.26 An impedance $\underline{Z}_{\text{out}}$ is connected to a gyrator as shown in the figure below.

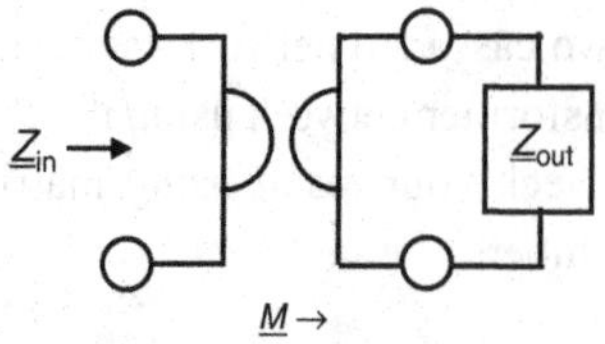

(a) Find the equivalent impedance $\underline{Z}_{\text{in}}$ seen from the left side of the gyrator in terms of the output impedance $\underline{Z}_{\text{out}}$.
(b) For the following different impedance expressions, how are the individual impedance elements, that is, R, L, and C, changed when mapped from the right to the left side of the gyrator?

(i) $\underline{Z}_{\text{out}} = R$;
(ii) $\underline{Z}_{\text{out}} = R + 1/(\text{j}\omega C)$;
(iii) $\underline{Z}_{\text{out}} = R \,\|\, 1/(\text{j}\omega C)$;
(iv) $\underline{Z}_{\text{out}} = R + 1/(\text{j}\omega C) + \text{j}\omega L$.

2.27 The equivalent circuit for the transformer shown in Fig. 2.20b is not unique. Propose an alternative circuit model using dependent sources.

2.28 The equivalent circuit for the gyrator shown in Fig. 2.19b is not unique. Propose an alternative circuit model.

References

1. B. D. H. Tellegen, A general network theorem. *Philips Research Reports*, 7 (1952), 259–269.
2. A. G. Bose and K. N. Stevens, *Introductory Network Theory* (New York: Harper and Row, 1965), Chapter 7.
3. B. D. H. Tellegen, The gyrator, a new electric network element. *Philips Research Reports*, 3 (1948), 81–101.

3 Capacitive lumped parameter electromechanics

3.1 Basic assumptions and concepts

In Chapter 1, we introduced the notion of the lossless electromechanical coupling. Then, in Chapter 2, the principles and approximations of circuit-based device modeling were reviewed. The primary goal of Chapter 3 is to introduce the energy-based technique for determining the electrical force operative in a capacitive transducer. This force effects the electromechanical transduction of energy between electrical and mechanical forms, and we cannot predict the behavior of a MEMS device without it. Electromechanical interactions in capacitive devices arise from either of two physical origins. First and far more familiar is the Coulombic interaction of electric charges at a distance. The force exerted on an electrostatic charge q is $q\overline{E}$, where $\overline{E}$, the vector electric field, is the superposition of the *force fields* created by all the other charges. The other, less well-known force mechanism originates from the interactions of an electric field with the dipoles that constitute liquids and solids. These dipoles can be either induced or permanent. The essential requirement for an observable force is a non-uniform electric field. All dipoles have zero *net* charge, but if the positive and negative charge centers experience slightly different electric field vectors, there will be a net force. For a small dipole having moment $\overline{p}$, this force may be approximately expressed by $\overline{p} \cdot \nabla \overline{E}$. An ensemble of dipoles in any solid (or liquid) can experience a net body force, called the *ponderomotive* effect. The classic book by Landau and Lifshitz presents a general electroquasistatic formulation for the volume density of the ponderomotive force [1].

In a text on lumped parameter electromechanics, there is no need to delve into Coulombic or polarization forces because, as we will discover, the capacitance-based energy method introduced here automatically accounts for both of them. It does so while making no distinction between the two mechanisms and without any need for direct reference to either $q\overline{E}$ or $\overline{p} \cdot \nabla \overline{E}$. This fortunate circumstance is a direct consequence of the mutual consistency of the laws of electromagnetism, the principles of circuit theory, and energy conservation.

The principal assumption inherent in the theory of electromechanics is that the device is lossless. Because no real MEMS device is truly lossless, this assumption might seem unrealistic and overly restrictive. It turns out, however, that for the vast majority of capacitive transducers, the electrical and mechanical loss mechanisms can be separated from

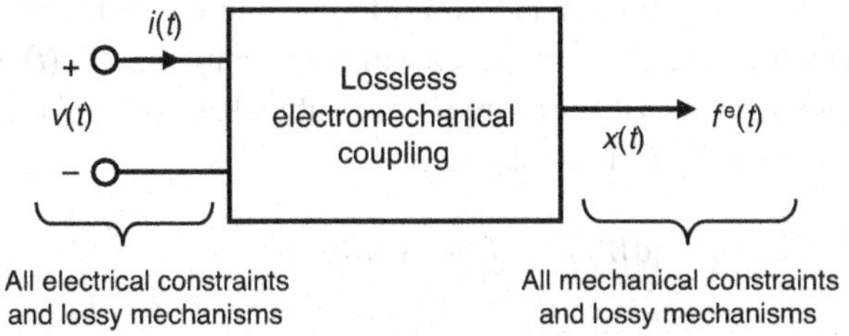

Figure 3.1 The lossless electromechanical coupling. All lossy components, such as electrical resistors and mechanical dampers, as well as other constraints, including mass, mechanical springs, and electrical circuit connections, are external to the coupling.

the energy transduction mechanism and accounted for separately as *external constraints*. We will rely heavily on this fact here and in subsequent chapters.

Because of their predominance in MEMS technology, attention in this chapter and in most of those to follow is limited to capacitive devices. The reader may refer to Appendix A, Section A.2, for a review of the essentials of electroquasistatics and field-based definitions for voltage and capacitance. A concise, parallel coverage of magnetic transducers is found in Chapter 9.

3.1.1 The lossless electromechanical coupling

For clarity and consistency, and in deference to the intellectual origin of this book, we adopt the terminology and conventions of Woodson and Melcher [2], referring to the ideal (lossless) transducer as an *electromechanical coupling*. Figure 3.1 provides a representation of a simple electromechanical coupling with one electrical and one mechanical port. The electrical side is quantified by voltage $v(t)$ and current $i(t)$, while the mechanical side has one mechanical (translational) degree of freedom $x(t)$ and an associated *force of electrical origin* $f^{\mathrm{e}}(t)$.

We assume that the coupling is lossless, but can store energy W_{e} and then later return it unattenuated. The origin of this stored energy can be either electrical or mechanical. As stated in Chapter 1, the term "force of electrical origin" reminds us that $f^{\mathrm{e}} = 0$ if there is no electrical excitation. For most capacitive transducers, it is only necessary to set charge or voltage to zero to achieve this condition; however, in devices such as the electret, greater care must be taken.

It should be borne in mind that all external constraints, including mass, restoring spring forces, and voltage or current sources, as well as all contributions to loss, such as mechanical damping and electrical resistance, are excluded from the electromechanical coupling.

Paralleling the argument of Section 1.3, we may express a power conservation law for the coupling as follows:

$$vi = f^{\mathrm{e}} \frac{\mathrm{d}x}{\mathrm{d}t} + \frac{\mathrm{d}W_{\mathrm{e}}}{\mathrm{d}t}. \tag{3.1}$$

Equation (3.1) equates the electrical input power to the sum of the mechanical output power and the rate at which energy W_e is stored within the coupling. If $q(t)$ is the charge in the capacitive device, we may use the charge conservation law, $i(t) = dq/dt$, to rewrite Eq. (3.1) in a more convenient, differential, form:

$$dW_e = -f^e dx + v dq. \tag{3.2}$$

3.1.2 State variables and conservative systems

The key to determining f^e is to obtain the energy function W_e. Setting aside for now the external electrical and mechanical constraints, we must choose a set of independent *state variables* for the lossless coupling. One evident choice would be x and q; then, the voltage and force of electrical origin are expressed in terms of them, that is, $v = v(x, q)$ and $f^e = f^e(x, q)$. For a *conservative system*, the energy function is also fully specified by these state variables: $W_e = W_e(x, q)$. The total differential of the energy function can be expressed in terms of partial derivatives of W_e with respect to the state variables:

$$dW_e = \left.\frac{\partial W_e}{\partial x}\right|_q dx + \left.\frac{\partial W_e}{\partial q}\right|_x dq. \tag{3.3}$$

Because q and x are independent until the external constraints are imposed, Eqs. (3.2) and (3.3) combine to yield

$$v(x, q) = \left.\frac{\partial W_e}{\partial q}\right|_x \quad \text{and} \quad f^e(x, q) = -\left.\frac{\partial W_e}{\partial x}\right|_q . \tag{3.4}$$

For a linear capacitive transducer, we know that $v(x, q) = q/C(x)$. What we do not have is an expression for the force of electrical origin, but Eq. (3.4) teaches that we will be able to obtain it if we can determine the energy function.

In a conservative system, the change in energy depends only on the initial and final conditions. In integral form, based on Eq. (3.2),

$$W_e(x_b, q_b) - W_e(x_a, q_a) = \int_{(x_a, q_a)}^{(x_b, q_b)} [-f^e dx + v dq], \tag{3.5}$$

where the initial and final states are (x_a, q_a) and (x_b, q_b), respectively. As suggested by Fig. 3.2, the integral on the left-hand side of Eq. (3.5) may be performed along any path between the two points.

3.1.3 Evaluation of energy function

For the electromechanical coupling depicted in Fig. 3.1, evaluation of W_e involves integration over the two state variables (x, q) using Eq. (3.2). The conservative nature of

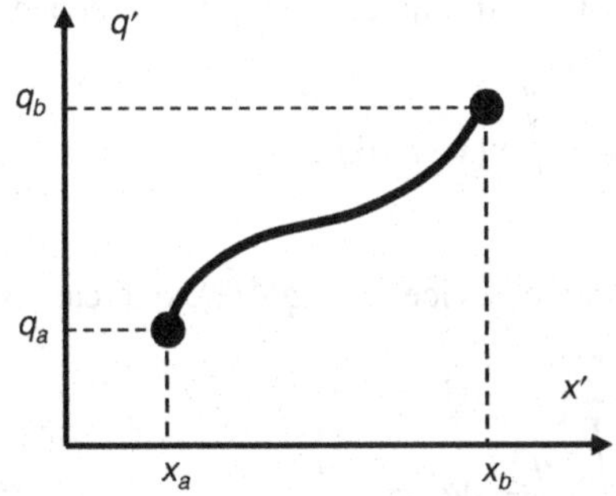

Figure 3.2 Integral path in state space for Eq. (3.5) to determine the change in the energy function between two states: (x_a, q_a) and (x_b, q_b). If the system is conservative, then the result of the integration is independent of the path.

the system permits us to perform this integral along any path that might be convenient. With that property in mind, consider the path shown in Fig. 3.3, where the mechanical variable x' is integrated first holding $q' = 0$, followed by the electrical variable, q', holding $x' = x$. Thus,

$$W_e = \int_{(1)} dW_e + \int_{(2)} dW_e. \tag{3.6}$$

With q' and x' fixed along $\int_{(1)}$ and $\int_{(2)}$, respectively, Eq. (3.6) reduces to a pair of definite integrals:

$$W_e(x, q) = -\int_0^x f^e(x', q' = 0)\,dx' + \int_0^q v(x, q')Fq', \tag{3.7}$$

where x' and q' denote variables of integration, while (x, q) is the final state of the system. Now recall that the definition of f^e demands that it go to zero in the absence of any

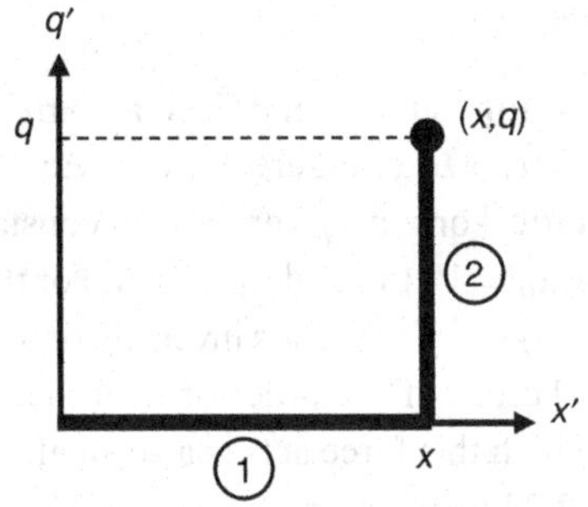

Figure 3.3 Convenient integration path for the energy function $W_e(x, q)$ that takes advantage of the property $f^e(x', q' = 0) = 0$ for a capacitive transducer.

electrical excitation, that is, $f^e(x', q' = 0) = 0$. Therefore, Eq. (3.7) reduces to

$$W_e(x, q) = \int_0^q v(x, q')\, dq'. \tag{3.8}$$

Using the terminal relation for a capacitive device, $v = q/C(x)$, one can evaluate this integral:

$$W_e(x, q) = \int_0^q \frac{q'}{C(x)}\, dq' = \frac{q^2}{2\,C(x)}. \tag{3.9}$$

3.1.4 Force of electrical origin

The force of electrical origin is determined using W_e from Eq. (3.9) in Eq. (3.4):

$$f^e(x, q) = -\left.\frac{\partial W_e}{\partial x}\right|_q = \frac{q^2}{2\,C^2(x)} \frac{d\,C(x)}{d\,x}. \tag{3.10}$$

After the partial derivative is performed on the energy function with charge q held constant, one can use the terminal relation $v = q/C(x)$ to rewrite f^e in the form

$$f^e(x, v) = \frac{v^2}{2} \frac{dC(x)}{dx}, \tag{3.11}$$

which is identical to Eq. (1.8).

It is important not to use the capacitive terminal relationship in the energy function until the derivative of W_e has been taken, because doing so would make it impossible to maintain the constant charge constraint dictated by the energy argument that gave us Eq. (3.4). The next section introduces an alternative energy function called the coenergy, which uses voltage v as the electrical state variable instead of charge q.

Example 3.1 **Force magnitude of a variable-gap capacitor**

A variable-gap capacitive device is 250 μm square and has a variable gap x. If $w \approx L \ll x_0$, the capacitance is $C(x) \approx \varepsilon_0 w L/x$. Figure 3.4 uses Eqs. (3.33) and (3.34), respectively, to plot the force of electrical origin f_x^e versus x for constant voltage and constant charge about an equilibrium at $x_0 = 10$ μm and $v_0 = 10$ V. For the constant charge curve, the charge is fixed at $q_0 = C(x_0)\, v_0$. f_x^e depends inversely on x at constant voltage, but is constant if charge is fixed. Forces of the order of micronewtons seem small, but the mass of the electrode upon which this force acts is also small. For a plate of 2 μm thick polysilicon of mass density 2.33 g/cm³,

$$m = 2.33 \cdot 10^3 (250 \cdot 10^{-6})^2 (2 \cdot 10^{-6}) = 2.9 \cdot 10^{-10}\ \text{kg}, \quad \text{or} \quad 0.29\ \mu\text{g}.$$

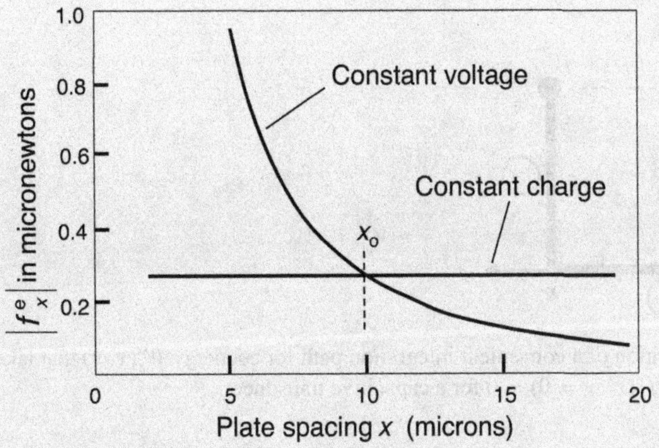

Figure 3.4 Magnitude of force of electrical origin $|f^e|$ for simple variable-gap capacitor with areal dimensions $w = L = 250$ μm and air gap x. Fringing fields are ignored. Equilibrium is at $x_0 = 10$ μm when $v_0 = 10$ V for both constant voltage and current constraints.

3.2 Coenergy – an alternate energy function

The function $W_e(x, q)$ is not unique as a measure of the electrostatic energy of a conservative, capacitive, electromechanical actuator. An alternative is the *coenergy*, $W'_e(x, v)$, which uses voltage v and displacement x as state variables.

3.2.1 Definition of coenergy

In its origin, coenergy is formally similar to the Helmholtz free energy commonly used in thermodynamics. It is defined using a *Legendre transform*, that is,

$$W_e + W'_e = qv. \tag{3.12}$$

Taking the total differential of this equation

$$dW_e + dW'_e = q dv + v dq, \tag{3.13}$$

and combining it with Eq. (3.2) to eliminate dW_e yields

$$dW'_e = f^e dx + q dv. \tag{3.14}$$

This differential expression may be interpreted as a statement of "coenergy conservation." The state variables are (x, v) and the complete differential is

$$dW'_e = \left.\frac{\partial W'_e}{\partial x}\right|_v dx + \left.\frac{\partial W'_e}{\partial v}\right|_x dv. \tag{3.15}$$

Comparing Eq. (3.15) with (3.14) yields

$$q(x, v) = \left.\frac{\partial W'_e}{\partial v}\right|_x \quad \text{and} \quad f^e(x, v) = \left.\frac{\partial W'_e}{\partial x}\right|_v. \tag{3.16}$$

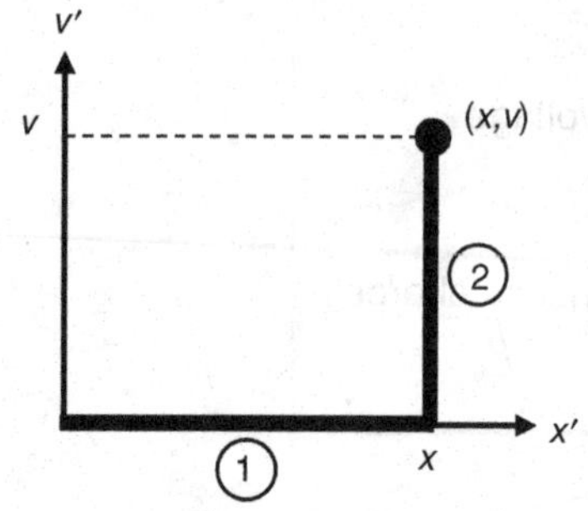

Figure 3.5 Definition of a convenient integration path for coenergy $W'_e(x, v)$ that takes advantage of the property $f^e(x', v = 0) = 0$ for a capacitive transducer.

Note the absence of the minus sign when f^e is defined in terms of the derivative of the coenergy.

3.2.2 Integral evaluation of coenergy

The coenergy is determined by integrating Eq. (3.14) in (x, v) state space. Just as for the energy, the coenergy integral is path-independent. If the mechanical variable is integrated first holding $v' = 0$, as shown in Fig. 3.5, then

$$W'_e(x, v) = \int_0^x f^e(x', v' = 0)\, dx' + \int_0^v q(x, v')\, dv'. \tag{3.17}$$

This integral path takes advantage of the definition of the force of electric origin, namely that $f^e = 0$ when $v = 0$.

Therefore, the first integral term in Eq. (3.17) is zero. For the second term, the mechanical variable is now fixed, that is, $x' = x$. Employing the linear electrical terminal relation, $q = C(x)v$, the resulting expression for coenergy is

$$W'_e(x, v) = \int_0^v q(x, v')\, dv' = \tfrac{1}{2}\, C(x)\, v^2. \tag{3.18}$$

This is the *coenergy function*, which takes the familiar form commonly used to quantify the electrostatic (capacitive) energy.

For linear capacitive transducers, W_e and W'_e are numerically equal. The one distinction between them is the choice for the electrical state variable: electrical charge q for W_e and voltage v for W'_e. Figure 3.6 provides a way to interpret and compare these functions in terms of their integral definitions, Eqs. (3.8) and (3.18). If the terminal relation relating charge q and voltage v is electrically linear, that is, $q = C(x)v$, then the two triangular areas are equal, as shown in Fig. 3.6a. On the other hand, if the relation is non-linear, then $W_e \neq W'_e$, as depicted in Fig. 3.6b. Irrespective of such inequality, nevertheless, Eqs. (3.4) and (3.16) yield numerically equal expressions for the force of electrical origin f^e.

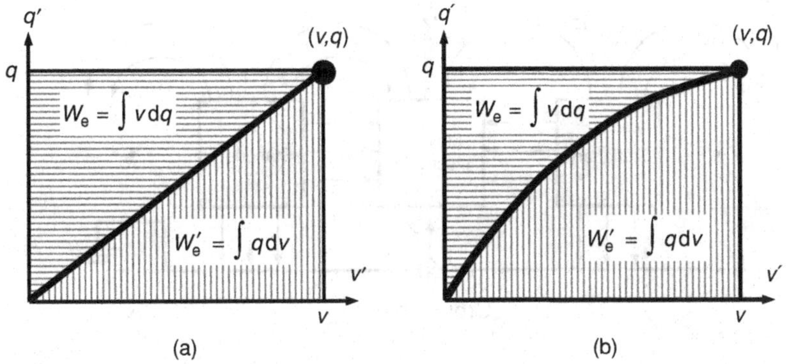

Figure 3.6 Integrals for energy W_e and coenergy functions W'_e. (a) When the terminal relation is linear, that is, $q = C(x)v$, $W_e = W'_e$. (b) When the terminal relation is non-linear, as might be the case for certain dielectric materials, $W_e \neq W'_e$.

3.2.3 Evaluation of force of electrical origin

Using Eq. (3.18) for the coenergy in Eq. (3.16), we obtain

$$f^e(x, v) = \frac{v^2}{2}\frac{dC(x)}{dx}. \tag{3.19}$$

When the voltage is fixed, f^e depends only on the derivative of capacitance, not on $C(x)$ itself. This attribute of the coenergy formulation will later prove advantageous for certain capacitive geometries.

The energy and coenergy formulations using q or v as the electrical variable, respectively, yield different algebraic expressions for f^e, but they are entirely equivalent. Once a force expression has been obtained, then the electrical terminal relation, be it linear or non-linear, can be used in principle to convert the expression into the form most convenient for a given circumstance.

Example 3.2 **Vertical force on comb-drive elements**

Comb-drive electrodes, common in capacitive MEMS, pack a maximum of useable capacitance into a small chip area, but a problem with the geometry is an unwanted electrostatic force component that pulls the moving electrodes out of alignment, thereby degrading performance [3]. Figure 3.7a shows the cross-section of the interdigitated electrodes, with stationary electrodes connected to a voltage source and grounded elements between them that move in the y direction, i.e., normal to the page. The sketched field lines reveal that the moving electrodes experience a z-directed force, $f_z^e > 0$, in consequence to ground plane asymmetry. Without counterbalancing by mechanical stiffening, this force pulls the moving electrodes out of alignment with the fixed electrodes, thereby reducing sensitivity.

To reduce this effect, the device may be operated as a three-plate system using bipolar voltage excitation. Figure 3.7b shows that this configuration greatly reduces f_z^e.

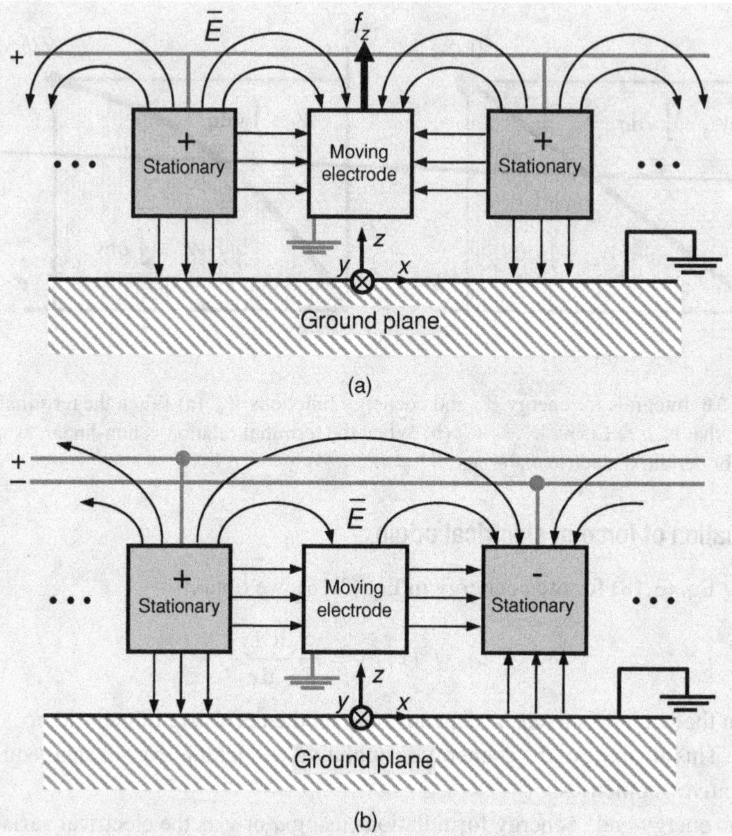

Figure 3.7 Cross-section of comb-drive structure showing influence of voltage connections on the z-directed electrostatic force f_z^e. (a) Comb-drive configuration with single polarity excitation. (b) Comb drive operated as a three-plate capacitive system with bipolar (±) excitation.

3.3 Couplings with multiple ports

Many MEMS devices have more than one electrical terminal pair and, in some cases, more than one mechanical degree of freedom. Good examples of such complexity are the MEMS-scale vibrating gyroscopes now available commercially. The typical gyroscope has two mechanical degrees of freedom and three electrical ports.[1] Also, we will later show that many mechanical continua, such as cantilevered beams and diaphragms, can be modeled accurately and efficiently as dynamic systems with finite numbers of degrees of freedom. These reasons provide the impetus to consider the generalized, multiple port system depicted in Fig. 3.8, which has N_e electrical and N_m mechanical ports.

[1] Refer to Example 3.4.

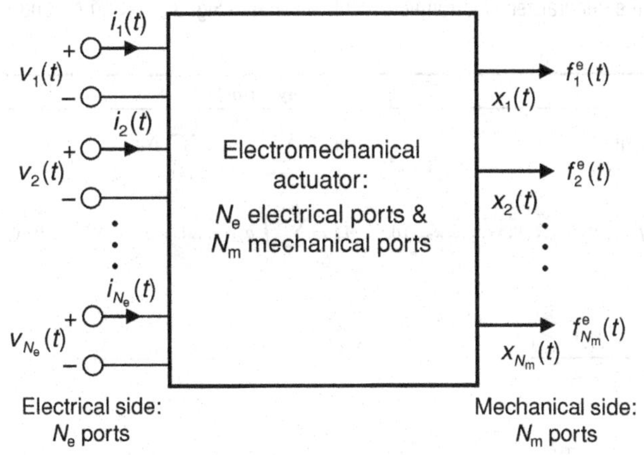

Figure 3.8 Generalized lossless electromechanical coupling featuring N_e electrical terminal pairs and N_m mechanical ports. The system energy W_e depends on the mechanical variables $(x_1, x_2, \ldots, x_{N_m})$ and the charge variables $(q_1, q_2, \ldots, q_{N_e})$.

3.3.1 Energy conservation relation

The starting point for modeling the multiport coupling is the same power conservation principle used for the simpler system of Fig. 3.1. No new concepts are involved, but care must be taken with the algebraic bookkeeping to keep track of all the input and output variables:

$$dW_e = -\sum_{k=1}^{N_m} f_k^e \, dx_k + \sum_{k=1}^{N_e} v_k \, dq_k. \tag{3.20}$$

The energy W_e is a function of all the electrical charge and mechanical variables, that is, $W_e = W_e(x_1, x_2, \ldots, x_{N_m}; q_1, q_2, \ldots, q_{N_e})$.

Writing out the total differential of energy W_e and comparing it with Eq. (3.20) provides a set of relations for the voltages,

$$v_k = \left.\frac{\partial W_e}{\partial q_k}\right|_{\substack{x_1, x_2, \ldots, x_{N_m}; \\ q_1, q_2, \ldots, q_{k-1}, q_{k+1}, q_{k+2}, \ldots, q_{N_e}}}, \tag{3.21}$$

and for the electrical forces,

$$f_k^e = -\left.\frac{\partial W_e}{\partial x_k}\right|_{\substack{x_1, x_2, \ldots, x_{k-1}, x_{k+1}, x_{k+2}, \ldots, x_{N_m}; \\ q_1, q_2, \ldots, q_{N_e}}}. \tag{3.22}$$

As evident in Eq. (3.20), evaluating the energy function involves $N_m + N_e$ integrations in state space. The system is still conservative and the energy function retains its analytical property. Because the change in W_e between any two states is independent of the path taken, one can choose the order of integration based on convenience. Making obvious accommodation for the additional electrical terminals and mechanical degrees

Table 3.1 Summary of results for the generalized electromechanical coupling of Fig. 3.8 having N_m and N_e mechanical and electrical terminals, respectively.

	Energy formulation	Coenergy formulation
Energy conservation	$dW_e = \sum_{j=1}^{Ne} v_j dq_j - \sum_{j=1}^{Nm} f_j^e dx_j$	$dW_e' = \sum_{j=1}^{Ne} q_j dv_j + \sum_{j=1}^{Nm} f_j^e dx_j$
State space integral for energy function	$W_e = \sum_{j=1}^{Ne} \int_0^{q_j} v_j(q_1, \ldots, q_{j-1}, q_j', 0, \ldots, 0; x_1, \ldots, x_{Nm})\, dq_j'$	$W_e' = \sum_{j=1}^{Ne} \int_0^{v_j} q_j(v_1, \ldots, v_{j-1}, v_j', 0, \ldots, 0; x_1, \ldots, x_{Nm})\, dv_j'$
Force of electrical origin	$f_j^e = -\dfrac{\partial W_e(q_1, \ldots, q_{Ne}; x_1, \ldots, x_{Nm})}{\partial x_j}$	$f_j^e = \dfrac{\partial W_e'(v_1, \ldots, v_{Ne}; x_1, \ldots, x_{Nm})}{\partial x_j}$
Electrical terminal relations	$v_j = \dfrac{\partial W_e(q_1, \ldots, q_{Ne}; x_1, \ldots, x_{Nm})}{\partial q_j}$	$q_j = \dfrac{\partial W_e'(v_1, \ldots, v_{Ne}; x_1, \ldots, x_{Nm})}{\partial v_j}$

For linear capacitor network: $q_k = \sum_{j=1}^{Ne} C_{kj}(x_1, x_2, \ldots, x_{Nm}) v_j$, where $k = 1, 2, \ldots, N_e$.

Electric currents defined by: $i_k = dq_k/dt$.

Legendre transform relating energy and coenergy: $W_e + W_e' \equiv \sum_{j=1}^{Ne} v_j q_j$.

of freedom, the same trick employed before and illustrated in Fig. 3.3 can be used again to determine the energy function. In other words, we integrate all the mechanical variables first to take advantage of the original definition for the force of electrical origin in generalized form, that is,

$$f_k^e = 0, \quad \text{for all } k \text{ if } q_1 = q_2 = \ldots = q_{N_e} = 0. \tag{3.23}$$

Integrations over the charge variables $(q_1, q_2, \ldots, q_{Ne})$ are then performed using the electrical terminal relations. As already pointed out, no new conceptual hurdle exists here. Correctly executing the required integrations is primarily a bookkeeping matter. An entirely analogous set of relations arises when coenergy is used. Refer to Table 3.1. The next section provides an instructive example of a coupling with multiple ports.

3.3.2 System with two electrical and two mechanical ports

To exemplify an electromechanical device with multiple electrical and mechanical ports, consider the system shown in Fig. 3.9 with $N_e = 2$ electrical and $N_m = 2$ mechanical ports. From Eq. (3.20), the energy conservation equation is

$$dW_e = -f_a^e dx_a - f_b^e dx_b + v_1 dq_1 + v_2 dq_2. \tag{3.24}$$

Equation (3.24) defines a four-dimensional state space: (x_a, x_b, q_1, q_2). The energy function for this conservative system depends on these variables, that is,

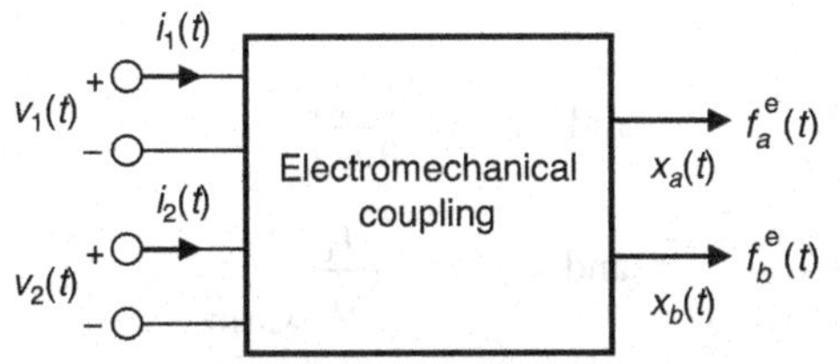

Figure 3.9 Conservative capacitive electromechanical coupling featuring two electrical and two mechanical ports.

$W_e = W_e(x_a, x_b; q_1, q_2)$. Taking the total differential of the energy function and using now familiar arguments, we obtain

$$v_1 = \left.\frac{\partial W_e}{\partial q_1}\right|_{x_a, x_b, q_2} \quad \text{and} \quad v_2 = \left.\frac{\partial W_e}{\partial q_2}\right|_{x_a, x_b, q_1}, \tag{3.25a}$$

$$f_a^e = -\left.\frac{\partial W_e}{\partial x_a}\right|_{x_b, q_1, q_2} \quad \text{and} \quad f_b^e = -\left.\frac{\partial W_e}{\partial x_b}\right|_{x_a, q_1, q_2}. \tag{3.25b}$$

To integrate the energy function in state space, the terminal relations of the system are needed. For a linear, capacitive system with two electrical ports, the most commonly used matrix form is

$$\begin{bmatrix} q_1 \\ q_2 \end{bmatrix} = \begin{bmatrix} C_{11} & C_{12} \\ C_{21} & C_{22} \end{bmatrix} \begin{bmatrix} v_1 \\ v_2 \end{bmatrix}. \tag{3.26}$$

Coefficients C_{11} and C_{22} are self-capacitances, while C_{12} and C_{21} are mutual capacitances. In general, each term may depend on both mechanical variables, that is, $C_{11} = C_{11}(x_a, x_b)$, etc. As discussed in Section 2.3.1, the passive nature of the capacitive network guarantees reciprocity, that is, $C_{12} = C_{21}$.

To evaluate the integrals for W_e, we need to express the voltages in terms of the charges but extracting these from Eq. (3.26) involves a matrix inversion. Furthermore, after we have done this, the result will be forces expressed, rather inconveniently, in terms of the charges. For these reasons, it is a better tactic to use coenergy instead of energy.

As for the derivation in Section 3.2, we use a Legendre transform, this time employing the sum of the products of the two lumped parameter electrical variables:

$$W_e + W_e' = q_1 v_1 + q_2 v_2. \tag{3.27}$$

After some straightforward manipulation to combine Eqs. (3.24) and (3.27), the *coenergy conservation relation* may be written as

$$dW_e' = f_a^e dx_a + f_b^e dx_b + q_1 dv_1 + q_2 dv_2. \tag{3.28}$$

Then, because $W'_e = W'_e(x_a, x_b; v_1, v_2)$,

$$q_1 = \left.\frac{\partial W'_e}{\partial v_1}\right|_{x_a, x_b, v_2} \quad \text{and} \quad q_2 = \left.\frac{\partial W'_e}{\partial v_2}\right|_{x_a, x_b, v_1}, \tag{3.29a}$$

$$f_a^e = \left.\frac{\partial W'_e}{\partial x_a}\right|_{x_b, v_1, v_2} \quad \text{and} \quad f_b^e = \left.\frac{\partial W'_e}{\partial x_b}\right|_{x_a, v_1, v_2}. \tag{3.29b}$$

The final step before determining the force is to integrate Eq. (3.28) in $(x_a, x_b; v_1, v_2)$ state space. As before, the mechanical variables are integrated first to take advantage of the fact that $f_a^e = f_b^e = 0$ when $v_1 = v_2 = 0$. Because the integrations over x_a and x_b are zero, the integral reduces to

$$W'_e(x_a, x_b; v_1, v_2) = \int_0^{v_1} q_1(x_a, x_b; v'_1, v'_2 = 0)\, dv'_1 + \int_0^{v_2} q_2(x_a, x_b; v_1, v'_2)\, dv'_2. \tag{3.30}$$

To understand this operation, it is worthwhile examining the arguments of the charge functions q_1 and q_2 very closely, paying particular heed to distinguish the variables of integration, all of which are primed, from the (unprimed) state variables. Because the coupling is conservative, the order of integration – v_1 first, then v_2, or vice versa – should not matter; and of course it does not, because $C_{12} = C_{21}$. Using the terminal relations from Eq. (3.26),

$$W'_e = \tfrac{1}{2} C_{11}(x_a, x_b) v_1^2 + \tfrac{1}{2} C_{22}(x_a, x_b) v_2^2 + C_{12}(x_a, x_b) v_1 v_2. \tag{3.31}$$

Equation (3.31) has two familiar-looking terms that depend on the self-capacitances and squared voltages, but also a mutual term that depends on the product $v_1 v_2$. f_a^e and f_b^e can be obtained by taking partial derivatives of W'_e according to Eq. (3.29b).

3.4 Basic capacitive transducer types

Despite great diversity in the designs and geometries for capacitive MEMS, the overwhelming majority of them can be categorized as *variable-gap* or *variable-area* transducers. This section presents simple examples of these two types and examines their lumped parameter electromechanical behavior. The student is advised to become familiar with them now because they will be encountered repeatedly throughout the rest of the book.

3.4.1 Variable-gap capacitors

Figure 3.10 shows the side view of a simple, variable air gap capacitor with fixed area A and variable gap $x(t)$. Under the assumption that the gap x remains small, the capacitance for this geometry is[2]

$$C(x) = \varepsilon_0 A / x. \tag{3.32}$$

[2] See Section A.3 of Appendix A for a derivation of this capacitance function from the laws of electroquasistatics.

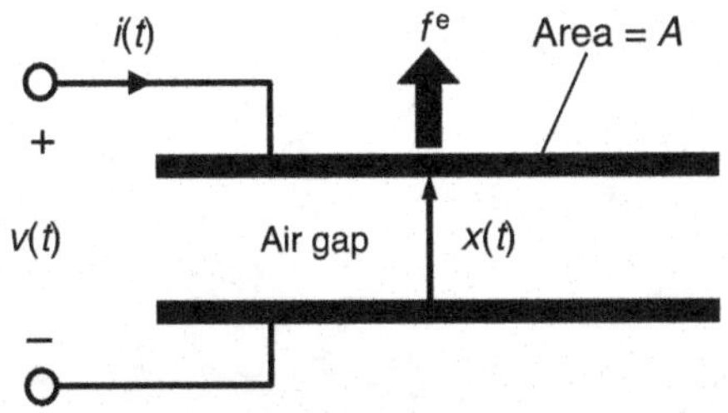

Figure 3.10 Side view of the variable-gap capacitor with parallel electrodes of fixed area. If the gap $x(t)$ remains small with respect to all areal dimensions, fringing fields can be neglected. In that case, we may employ the simple parallel-plate model introduced in Appendix A, Section A.3.3.

Assuming the lower electrode to be fixed and immobile, then the upper plate is the moving element and the force of electrical origin f^e is exerted upon it. Using Eq. (3.32) in Eq. (3.19) yields

$$f^e = -\frac{\varepsilon_0 A}{2x^2} v^2. \tag{3.33}$$

f^e is negative, that is, it acts in the $-x$ direction, because the two electrodes have opposite charge and attract each other. A second observation is that, at least with the voltage fixed, the force depends inversely on x^2. This strong dependence explains the excellent sensitivity of variable-gap geometries in MEMS applications, such as pressure sensors and microphones.

The above force expression can be rewritten in terms of the electric charge using the terminal relation, $v = q/C(x)$, in Eq. (3.33):

$$f^e = -\frac{1}{2\varepsilon_0 A} q^2. \tag{3.34}$$

Equation (3.34) reveals the important prediction that f^e does not depend on x if the electrical charge q is fixed. If fringing field effects are taken into account, either by numerical analysis or by employing an empirical expression of improved accuracy for $C(x)$, the calculated force exhibits a weak, decreasing dependence as the spacing x increases at fixed charge. Of course, when the spacing approaches the areal electrode dimensions, this dependence becomes stronger.

Comparison of Eqs. (3.33) and (3.34) proves that the electrical constraint imposed on a variable-gap capacitive transducer has a profound influence on its behavior. A good way to illustrate the influence of electrical constraints is to plot the force f^e versus displacement x for the cases of constant voltage and constant charge, as done in Fig. 3.11. This plot employs the condition $q_0 = C(x_0)\, v_0$ as a normalization. In Chapter 4, there will be more to learn about the influence of electrical constraints on the behavior of capacitive transducers.

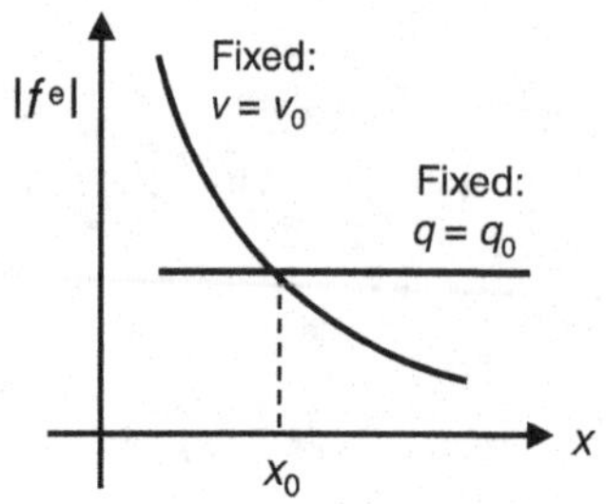

Figure 3.11 Dependence of force magnitude $|f^e|$ of a variable-gap capacitive transducer upon displacement x for constant voltage and constant charge constraints. The crossover at x_0 is defined for convenience by the condition $q_0 = C(x_0)\,v_0$. Note that any fringing field, which would influence the fixed charge case, is ignored here.

Example 3.3 **AC-excited capacitive actuator**

Energy methods used to analyze capacitive transducers treat either charge or voltage as a state variable. Once the force expression f^e is obtained, there is no restriction on the time-dependent nature of the electrical variable. For example, a sinusoidal voltage input may be assumed: $v(t) = V_0 \cos(\omega\, t)$, where V_0 is the peak magnitude and ω is the frequency in rad/s. Using Eq. (3.33) for the force on a variable-gap transducer,

$$f^e(t) = -\frac{\varepsilon_0 A}{2x^2} V_0^2 \,[\cos(\omega\, t)]^2\,.$$

Using a trigonometric identity to replace the squared cosine yields

$$f^e(t) = \underbrace{-\frac{\varepsilon_0 A}{4\,x^2} V_0^2}_{\text{time average}} \underbrace{- \frac{\varepsilon_0 A}{4\,x^2} V_0^2 \cos(2\omega\, t)}_{\text{double-frequency term}}\,.$$

This force consists of a constant term and a double-frequency sinusoidal component that can lead to undesirable vibrations. One way to eliminate such vibration is to use square-wave excitation. This solution is attractive because pulsed voltage is readily available in digital circuitry. The square-wave signal applied to the capacitor does present another potential problem, namely, large current transients.

3.4.2 Variable-area capacitors

A basic variable-area capacitive transducer is illustrated in Fig. 3.12. The electrode gap d is fixed and the area depends on the mechanical variable x. If we assume that the fixed spacing d is small compared with the areal dimensions of the electrodes, w and L, then the capacitance is proportional to the area of overlap of the two electrodes, that is,

$$C(x) = \frac{\varepsilon_0 w(L - x)}{d} \quad \text{for} \quad \sim d < x < L - d. \tag{3.35}$$

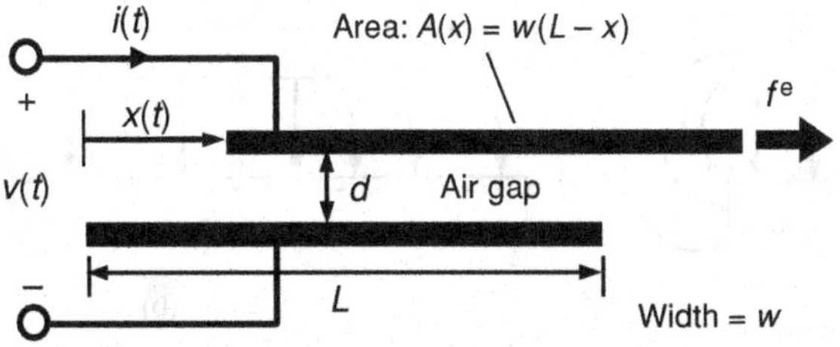

Figure 3.12 The variable-area capacitor. The air gap is fixed and the area is variable with respect to one degree of freedom. If the gap is small with respect to areal dimensions, fringing can be neglected and the simple model introduced in Appendix A, Section A.3.3, may be employed.

The limits on x specified for Eq. (3.35) justify the neglect of fringing fields. Over this range, the capacitance decreases linearly with x. Using this capacitance expression in Eq. (3.19), the voltage-dependent force is

$$f^{e} = -\frac{\varepsilon_0 w}{2d} v^2. \tag{3.36}$$

The force is negative and independent of x. On the other hand, if we substitute the electrical terminal relation, $v = q/C(x)$, back into Eq. (3.36), the force expression becomes

$$f^{e} = -\frac{d}{2\varepsilon_0 w (L-x)^2} q^2. \tag{3.37}$$

Under the constant electric charge constraint, f^e is strongly dependent on x, as shown in Fig. 3.13. These force-versus-displacement dependencies are simply the opposite of those for the variable-gap capacitive transducer. For the variable-area capacitor, it is the constant-charge condition that leads to a strong dependence of the force on the displacement. This dependence results from the redistribution of the charge as the plate moves.

For both transducer types, the force of electromechanical origin f^e always acts to increase system capacitance. This very general rule of thumb applies to all capacitive electromechanical couplings, irrespective of geometry. The only way to overcome the tendency of the force to raise $C(x)$ is to introduce an external constraint such as feedback.

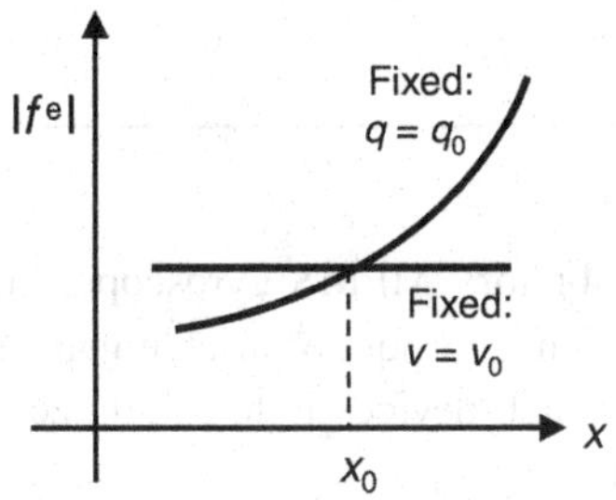

Figure 3.13 Dependence of force magnitude $|f^e|$ of a variable-area capacitive transducer on displacement x for constant voltage and constant charge constraints. The crossover at x_0 is defined for convenience by the condition $q_0 = C(x_0)\, v_0$.

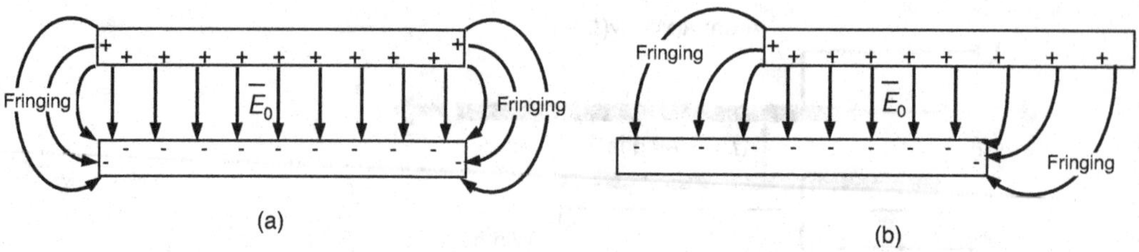

Figure 3.14 Electric fields and surface charge distribution for basic capacitive transducers. (a) For the variable-gap device, the force drawing the electrodes together is due to the uniform electric field acting on the uniform surface charge. The fringing fields have a negligible influence on the force if the gap is small compared with the areal dimensions of the electrodes. (b) For the variable-area device, the fringing fields acting on charge concentrated at the edges of the electrodes are actually responsible for the force component parallel to the electrodes.

3.4.3 Comparison of variable-gap and variable-area actuators

One way to understand why variable-gap and variable-area capacitive transducers behave differently under constant charge and constant voltage conditions is to contrast the distribution of the surface charge on the electrodes and the electric fields for the two geometries. For the variable-gap transducer of Fig. 3.14a, the electric field, which is uniform except near the edges, acts on the induced surface charge, also uniform, to pull the electrodes together. As long as the electrode spacing is small compared with the areal dimensions of the electrodes, the non-uniform field in the fringing region has negligible effect. The uniform field approximation introduced in Appendix A ignores fringing altogether.

On the other hand, Fig. 3.14b reveals that it is the fringing field acting on the concentrated surface charges along the edges of the electrodes that creates the force in the variable-area capacitor. This force pulls the electrodes into alignment. That it is possible to neglect the fringing field in the capacitive model, as in Eq. (3.35), and still get an accurate result for the force testifies to the robustness of the lumped parameter energy method. Some important implications of fringing-field-based forces for electromechanics will be revealed in Section 4.8.4.

Example 3.4 **Model for a MEMS gyroscope**

Figure 3.15a shows an SEM image of a vibratory MEMS gyroscope. Figure 3.15b highlights the drive and sense electrodes, which incorporate variable-gap and variable-area structures. A big design challenge for such devices is to minimize mechanical coupling between x- and y-directed motions.

The x suspension spring is designed to be very rigid in the y direction and the y suspension very rigid in the x direction. Thus, the effective mass $(m_x + m_y)$ moves only in the x direction, while the inner mass m_y moves only in the y direction.

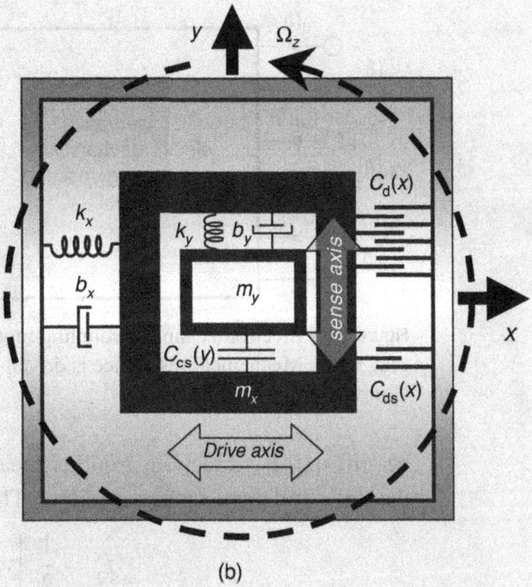

Figure 3.15 Vibratory MEMS gyroscope designed to detect rotational motions in x–y plane. (a) SEM of surface micromachined device [4] ©IEEE. (b) Simplified electromechanical model of vibratory gyroscope with three electrical ports: drive actuator C_d; drive sensor C_{ds}; Coriolis sensor C_{cs}; and two mechanical degrees of freedom: driven x motion of effective mass $m_x + m_y$, spring constant k_x, damping b_x; Coriolis-induced y motion of mass m_y, spring constant k_y, and damping b_y.

For large-amplitude, x-directed resonant motion, C_d is a variable-area capacitor. To accommodate this large stroke with uniform sensitivity, C_{ds} is also a variable-area device. To achieve good detection sensitivity for the very small Coriolis-induced, y-directed motion of mass m_y, C_{cs} is a variable-gap capacitor.

The electrical ports of the lumped parameter model of Fig. 3.16 are identified by their subscripts: "d" for the external periodic drive, "ds" for sensing the driven motion, and "cs" for sensing the Coriolis motion. The mechanical ports are $x(t)$, for the large-amplitude driven resonant motion, and $y(t)$, for the much smaller Coriolis-force-induced displacement.

A 3×3 capacitance matrix specifies the electrical terminal relations:

$$\begin{bmatrix} q_d \\ q_{ds} \\ q_{cs} \end{bmatrix} = \begin{bmatrix} C_d(x,y) & 0 & 0 \\ 0 & C_{ds}(x,y) & 0 \\ 0 & 0 & C_{cs}(x,y) \end{bmatrix} \begin{bmatrix} v_d \\ v_{ds} \\ v_{cs} \end{bmatrix}.$$

Bear in mind that $|v_d| > |v_{ds}| \sim |v_{cs}|$ and $|x| \gg |y|$.

The mutual capacitive coupling must be negligible. This requirement, evident in the capacitance matrix, is designed into the device by physical separation.

Any y-directed motion of m_x caused by the drive is undesirable because it induces m_y motion that can overwhelm the Coriolis response. Thus, the x suspension spring must

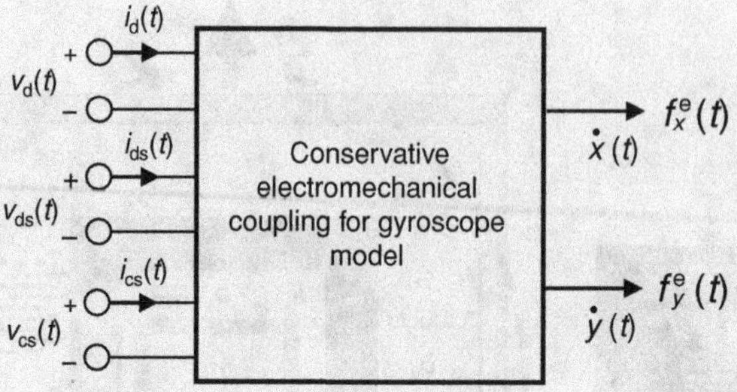

Figure 3.16 Electromechanical coupling model of the vibratory MEMS gyroscope shown in Fig. 3.15b. Mechanical resonance is driven in the x direction and the y motion is induced by the Coriolis force.

be stiff in the y direction. Another measure is to minimize any y-directed force due to the drive and drive sense capacitors. The y-directed force is

$$f_y^e = \frac{1}{2}\left[v_d^2\frac{\partial C_d}{\partial y} + v_{ds}^2\frac{\partial C_{ds}}{\partial y}\right].$$

To minimize motion due to the first term, the y dependence of C_d must be very weak. The second term may be ignored as long as the drive sense circuit operates at high frequency compared with the mechanical resonance. Another problem arises if the Coriolis sense capacitor has significant x dependence. From the matrix expression above,

$$q_{cs} = C_{cs}(x, y)\, v_{cs}.$$

Because $|x(t)| \gg |y(t)|$, any x dependence of C_{cs} must be very weak so that the Coriolis signal is not overwhelmed by current induced due to the x motion. Care is required in the electromechanical design to achieve these requirements imposed upon the respective y and x dependences of C_d and C_{cs}. MEMS-based gyroscopes are considered in detail in Section 7.4.

3.4.4 Transducer stroke

Force versus displacement curves facilitate comparisons among conservative lumped parameter actuators, ranging from mechanical springs to electromechanical transducers. They make the task of choosing the right device for a given system easier. Refer to the generalized force versus displacement curve shown in Fig. 3.17, noting the definitions for *stroke* and *force magnitude*. For a transducer, stroke quantifies the range of useable motion. If the mechanical element is conservative and the voltage fixed, then the work done against the element depends only on the initial and final values of the energy.

$$\int_{x_a}^{x_b} -f\,\mathrm{d}x = W_b - W_a \equiv \Delta W. \tag{3.38}$$

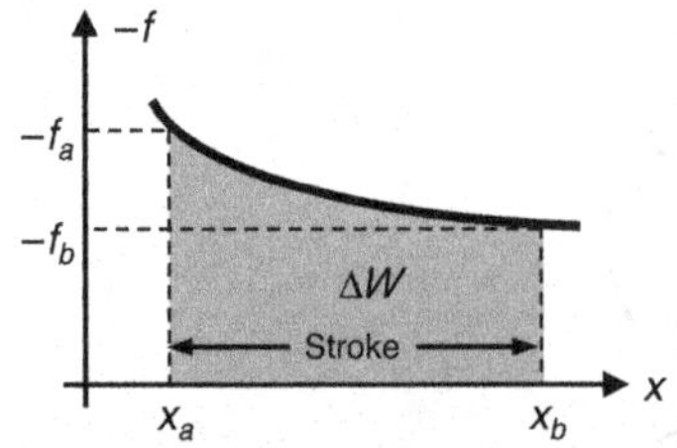

Figure 3.17 Force versus displacement plot of a transducer. Here, x_a and x_b are lower and upper physical limits imposed upon x-directed displacements by the transducer design. The stroke is $x_b - x_a$. The shaded area under the curve and between these limits is ΔW.

The shaded area in the figure is ΔW, the work done. Comparing transducer types based on fixed ΔW facilitates a general discussion of how to make the proper design choice for a given application.

Depending on the application, a transducer with a long or a short stroke might be favored, but there is an important trade-off to be recognized. For two transducers with identical ΔW, the mechanism with the longer stroke exerts a weaker force while the one with the shorter stroke provides a stronger force.

Refer back to Figs. 3.11 and 3.13, showing force versus displacement curves for variable-gap and variable-area capacitive transducers, respectively. The constant-charge constraint applied to a variable-gap device or, similarly, constant-voltage constraint applied to a variable-area device maximizes the transducer stroke. On the other hand, a constant-voltage constraint applied to the variable-gap device or a constant-charge constraint for the variable-area device maximizes the force. Table 3.2 summarizes the most important operational distinctions between variable-gap and variable-area capacitors when used as actuators.

Table 3.2 Comparison of certain operational attributes of variable-gap and variable-area capacitive transducers

Basis of comparison	Variable-gap capacitive transducers	Variable-area capacitive transducers[a]
Stroke length	Shorter	Longer
Force magnitude	Stronger	Weaker
Constant charge constraint	$f^e \sim$ constant	$f^e \sim x^{-2}$
Constant voltage constraint	$f^e \sim x^{-2}$	$f^e \sim$ constant
Sensitivity for sensor applications	Higher	Lower
Attributes for actuator applications	Favored in short-stroke or strong-force applications	Favored in long-stroke or weak-force applications
Failure modes	Pull-in	Out-of-plane shocks and accelerations
Typical MEMS applications	Accelerometers, pressure transducers, microphones, environmental sensors	Drives for gyroscopes, position control actuators, wherever highly linear response is needed

[a] Coplanar comb-drive structures share all essential attributes with variable-area capacitive transducers.

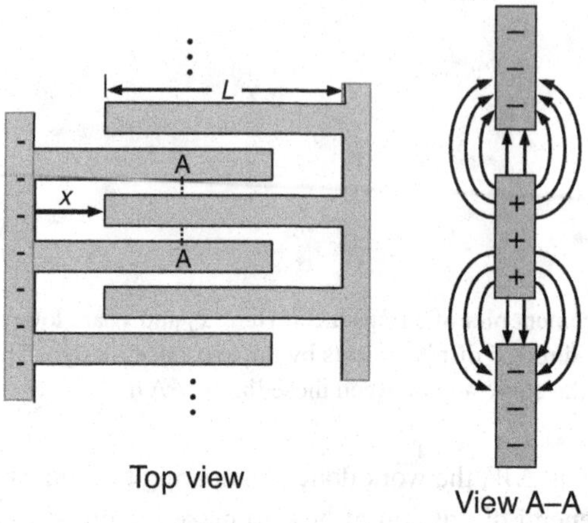

Figure 3.18 Top view and cross-section of the interdigitated comb drive. Actual devices are likely to have hundreds of interdigitated fingers. View A–A shows electric field lines between adjacent electrodes. Note that the capacitance is determined largely by fringing fields. Because the electric field is fixed along most of the overlapping section, the capacitance per unit length c_e may be used in a capacitance model.

3.4.5 The comb-drive geometry

Comb drives, interdigitated electrode structures used in a variety of MEMS actuators, are designed to fit the maximum amount of useable working gap into a confined chip area. Their characteristic behavior is usually that of the variable-area capacitor. They can take either of two distinct forms: high-aspect ratio geometries, such as the device described in Example 1.1, or the more common, coplanar surface structures depicted in Fig. 3.18. The high-aspect ratio structures are often fabricated using the DRIE (dry reactive ion etch) process. Coplanar structures feature thin electrodes fabricated using surface micromachining. Refer to Appendix C for brief descriptions of these processes.

The electric field in coplanar comb-drive structures is highly non-uniform, meaning that the uniform field approximation cannot be used. Nevertheless, by defining c_e as capacitance per unit length along two adjacent electrode edges, one can write a general form for $C(x)$:

$$C(x) = 2N_p c_e (L - x) + C_0, \tag{3.39}$$

where N_p is the number of opposed pairs of electrode edges and C_0 is the constant capacitance quantity associated with the ends of the electrode fingers. The edge capacitance per unit length c_e, a function of the electrode thickness and the interelectrode spacing, can be determined using finite-element analysis.

Using Eq. (3.39) in (3.19), the force of electrical origin is

$$f^e = -N_p c_e v^2. \tag{3.40}$$

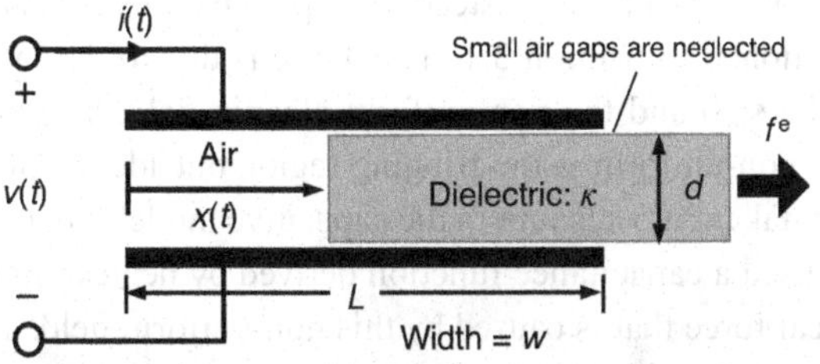

Figure 3.19 Another variable-area capacitor with fixed gap and a moving dielectric element. The dielectric slab is acted upon by the ponderomotive force. If d is small with respect to L and w, fringing can be neglected in the capacitor model so the parallel-media model introduced in Appendix A may be employed.

Comparing Eqs. (3.39) and (3.35) shows that the capacitances of the comb-drive and variable-area geometries have the same canonical form. Thus, comb-drive structures provide a force of electrical origin that is independent of x for fixed voltage.

3.4.6 Another variable-area capacitor

Figure 3.19 illustrates, in simplified form, a very different type of transducer. The electrodes are parallel and immobile, and the moving element between them is a solid slab of homogeneous dielectric material that is free to slide back and forth in the x direction. The parallel-media configuration illustrated in Fig. A.5 of Appendix A suggests a circuit model for this structure. Neglecting the fringing fields and the very small clearance gaps above and below the slab, the total capacitance is the sum of contributions from the open gap to the left and dielectric slab to the right:

$$C(x) = \frac{\varepsilon_0 w x}{d} + \frac{\kappa \varepsilon_0 w (L - x)}{d}. \tag{3.41}$$

For convenience, $C(x)$ may be rewritten as

$$C(x) = \frac{\kappa \varepsilon_0 w\, L}{d} - \frac{(\kappa - 1)\, \varepsilon_0 w x}{d}. \tag{3.42}$$

The latter form is similar to Eq. (3.35) for the capacitance of the variable-area capacitor shown in Fig. 3.12. This similarity means that the force of electrical origin is independent of x when voltage v is fixed:

$$f^{\mathrm{e}} = -\frac{(\kappa - 1)\, \varepsilon_0 w}{2d} v^2. \tag{3.43}$$

Expressed in terms of charge q, the force is inversely related to x:

$$f^{\mathrm{e}} = -\frac{(\kappa - 1)\, d}{2\varepsilon_0 w [\kappa L - (\kappa - 1)\, x]^2} q^2. \tag{3.44}$$

The sign of f^{e} is negative, so the electromechanical interaction pulls the dielectric slab into the plates, thus tending to increase capacitance.

The reader may well be puzzled about the physical origin of the force derived in Eq. (3.44). Because there is no free electrical charge induced upon it, the slab does not experience the qE (Coulombic) force. Instead, the operative mechanism here is the *ponderomotive force* mentioned in Section 3.1. This force is due to the interactions of *induced dipoles* within the slab and the non-uniform electric field. The only location where the electric field is non-uniform is the fringing region outside the plates, but the fringing field has not been taken into account in the capacitive model, Eq. (3.41). This is the second time we have used a capacitance function derived by neglecting the fringing field to predict an electrical force that is caused by this non-uniform field. The first was the variable-area device considered in Section 3.4.2. The apparent paradox is really no paradox at all; the force depends on the *change* of the energy (or the coenergy), not its absolute value. Therefore, the derivative of capacitance, $\mathrm{d}C/\mathrm{d}x$, which depends only very weakly on the fringing field, is all that is needed.

Figure 3.20 Scanning electron micrograph of a MEMS electrostatic motor first fabricated in the Berkeley Sensor & Actuator Center at the University of California, Berkeley [5] © IEEE. The structure, made of polysilicon with SiO_2 as sacrificial scaffolding material, was fabricated using the MUMPS foundry process [6]. Note the ten-micron scale bar at the bottom right.

3.5 Rotational transducers

While the majority of MEMS devices are designed for translational motion, there are some important examples of rotational devices. The rotation may be continuous, as in a motor, or it may be produced by actuators operative over a limited range of angles. Figure 3.20 shows a very early example, dating from 1990, of an electrostatic micromotor having diameter $\sim$0.2 mm [5]. This device was fabricated using the MUMPS process [6]. While this particular device had the problem that its bearings tended to wear out quickly, nevertheless it quite deservedly has achieved iconic status as the exemplar of MEMS potentialities. Interest in electrostatic micromotors has not burgeoned. On the other hand, some important MEMS devices operate over a limited range of angular displacements. The best example is the digital micromirror array used in optical projection systems. Texas Instruments pioneered these devices and they are in common use today [7].

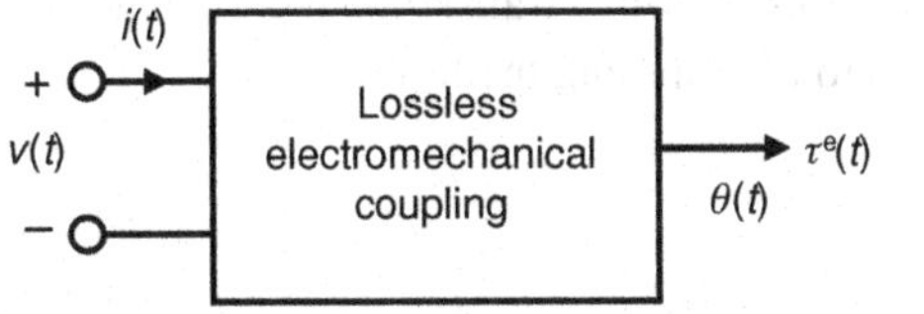

Figure 3.21 A lossless conservative, rotational electromechanical coupling. The chosen mechanical variables are angle, θ, and torque of electrical origin, τ^e. The capacitance depends on the angle, that is, $C = C(\theta)$.

A different sort of rotational electromechanical actuation shows up in certain MEMS designs that exhibit unanticipated rotations or flexural distortions. Such effects are usually the consequence of mechanical design flaws that reveal themselves only when a device is subjected to large accelerations or mechanical shocks. When this happens, MEMS designers find themselves troubleshooting unexpected rotational motion to establish performance limits, search for possible failure modes, and consider design remedies.

3.5.1 Modeling rotational electromechanics

Consider the conservative electromechanical coupling of Fig. 3.21. On the mechanical side, a rotational displacement angle, θ, and a *torque of electrical origin*, τ^e, replace the translational displacement variable and the force, respectively. The electrical terminal relation has its usual form, $q = Cv$, though now capacitance is a function of θ, that is, $C = C(\theta)$. Following the same method of Section 3.1.1 and recalling that rotational mechanical power is equal to the product of torque and angular velocity, that is, $p_{\text{rot}}(t) = \tau^e \, d\theta/dt$, power conservation requires

$$vi = \tau^e \frac{d\theta}{dt} + \frac{dW_e}{dt}, \tag{3.45}$$

where W_e is the electrostatic energy stored within the coupling.

Using $i(t) = dq/dt$, the differential form of Eq. (3.45) is

$$dW_e = -\tau^e d\theta + v dq. \tag{3.46}$$

The independent state variables are θ and q; v and τ^e are expressed in terms of them, that is, $v = v(\theta, q)$ and $\tau^e = \tau^e(\theta, q)$. The total differential of the energy function $W_e = W_e(\theta, q)$ is defined in terms of partial derivatives of W_e with respect to these variables:

$$dW_e = \left.\frac{\partial W_e}{\partial \theta}\right|_q d\theta + \left.\frac{\partial W_e}{\partial q}\right|_\theta dq. \tag{3.47}$$

Until the external constraints are imposed, q and θ are independent. Thus, Eqs. (3.46) and (3.47) combine to give

$$v(\theta, q) = \left.\frac{\partial W_e}{\partial q}\right|_\theta \quad \text{and} \quad \tau^e(\theta, q) = -\left.\frac{\partial W_e}{\partial \theta}\right|_q. \tag{3.48}$$

Just as before, when we sought the force f^{e} for a translational electromechanical device, the energy function W_{e} is the key to determining the torque τ^{e}.

3.5.2 Torque of electrical origin

State-space integration can be performed to obtain the energy, based on the postulate that the torque of electrical origin is zero in the absence of electrical excitation, that is, $\tau^{\mathrm{e}}(\theta', q' = 0) = 0$. Paying attention to the order of integration, we have

$$W_{\mathrm{e}}(\theta, q) = \underbrace{-\int_0^{\theta} \tau^{\mathrm{e}}(\theta', q' = 0)\,\mathrm{d}\theta'}_{=0} + \int_0^{q} v(\theta, q')\,\mathrm{d}q'. \tag{3.49}$$

In Eq. (3.49), θ' and q' denote variables of integration, while (θ, q) represents the final state of the system. Employing the terminal relation for a linear capacitor, $q = C(\theta)v$,

$$W_{\mathrm{e}}(x, \theta) = \int_0^{q} \frac{q'}{C(\theta)}\,\mathrm{d}q' = \frac{q^2}{2\,C(\theta)}. \tag{3.50}$$

The torque of electrical origin is then

$$\tau^{\mathrm{e}} = \frac{q^2}{2\,C^2}\frac{\partial C}{\partial \theta}, \tag{3.51}$$

which may be rewritten directly in terms of voltage,

$$\tau^{\mathrm{e}} = \frac{v^2}{2}\frac{\partial C}{\partial \theta}. \tag{3.52}$$

An alternate means of obtaining Eq. (3.52) is to use the coenergy function introduced in Section 3.2.

3.5.3 An example

As an example of a rotating capacitive transducer, consider the simple geometry of Fig. 3.22a. If the air gap d is very small compared with both the arc length $2\theta_0 R$ and height h of the electrodes, then it is justified to use a uniform electric field approximation in the gap, that is, $E_{\mathrm{r}} = v/d$. This approximation is reasonably accurate even though the radial electric field lines in the gap are not quite parallel to each other. From a simple application of Gauss's law, Eq. (A.7) of Appendix A, the capacitance takes the standard form, proportional to the quotient of overlapping area and electrode spacing:

$$C(\theta) \approx \frac{\varepsilon_0}{d}A(\theta) = \frac{2\,\varepsilon_0}{d}\,hR(\theta_0 - \theta), \quad 0 > \theta > \theta_0. \tag{3.53}$$

Note the canonical similarity of $C(\theta)$ to the variable-area capacitance, Eq. (3.35), illustrated in Fig. 3.12. From Eq. (3.52), the torque of electrical origin is

$$\tau^{\mathrm{e}} = -\frac{\varepsilon_0 h R}{d}v^2, \quad \theta > 0. \tag{3.54}$$

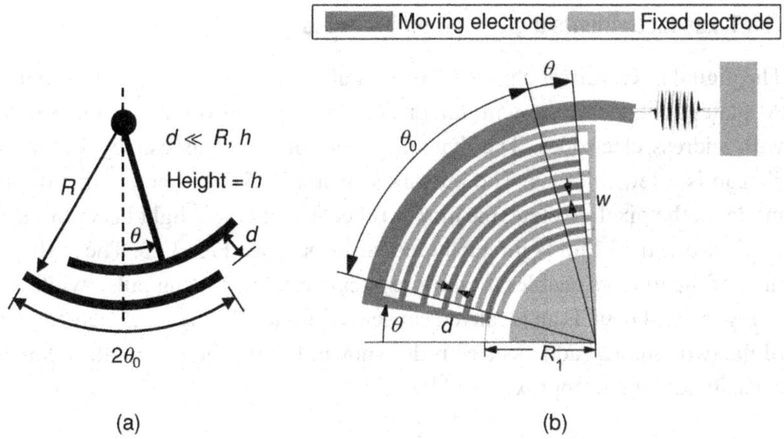

Figure 3.22 Rotational displacement capacitive transducers. (a) This geometry is similar to the variable-area capacitor. If the air gap is very small, that is, $d \ll R, w$, fringing fields may be ignored, thereby justifying a uniform field approximation to estimate capacitance. (b) Design for a torsional comb drive. A practical version of this geometry would have many more interdigitated electrode pairs than are depicted here.

Like f^{e} for the variable-area capacitor, τ^{e} is constant at fixed voltage. Furthermore, the torque of electrical origin acts to increase the capacitance.

Figure 3.22b depicts a more practical realization for a rotational, variable-area comb-drive configuration, featuring thin, coplanar interdigitated electrodes. This structure could be fabricated using surface micromachining. The rotating element, with N arc-shaped interdigitated electrodes, each of width w, is energized and the stationary electrode is at zero potential. The geometric complexity of this capacitance bank is only apparent; the capacitance depends linearly on angular displacement θ, analogous to the comb drive of Fig. 3.18:

$$C(\theta) = 2\, c_{\mathrm{e}} \underbrace{(\theta_0 + \theta)\, R_{\mathrm{eff}}}_{\text{total arc length}}, \tag{3.55}$$

where c_{e}, defined for Eq. (3.39), is the edge capacitance per unit length and

$$R_{\mathrm{eff}} = R_1 + 2(d+w)\sum_{k=1}^{N-1} k = R_1 + N(N-1)(d+w) \tag{3.56}$$

is the mean radius. The torque of electrical origin

$$\tau^{\mathrm{e}} = -c_{\mathrm{e}} R_{\mathrm{eff}} v^2, \quad \theta > 0, \tag{3.57}$$

is independent of the angular position when voltage is fixed.

By far the most important example of a rotating capacitive MEMS actuator is the digital micromirror array. Though these devices have only a relatively small range of angular displacements, typically 10° to 20°, they are legitimate examples of rotational capacitive actuators. Example 3.5 presents an electromechanical model for a micromirror array element.

Example 3.5 **The Texas Instruments digital micromirror chip**

The digital micromirror chip has a rectangular array of $>10^6$ thin, flat, highly reflective Al plates. The elements, with dimensions $\sim 15\ \mu\text{m}$, pivot above a substrate patterned with address electrodes. The dynamic range of the tilt is usually 10° to 20° and the air gap is a few microns. The mirrors, driven at $\sim 5\ V_{\text{dc}}$, operate in a digital (on–off) mode. In the pixel-on state, the mirror reflects an incident light beam to a screen, while in pixel-off, the beam is directed to an absorber. See Fig. 3.23. The $<10^{-3}$ s response time of the mirrors enables ten-bit grey-scale rendition using pulse-width modulation.

Figure 3.24 depicts an idealized capacitive model for one pixel element with only one of the two required address electrodes shown. For this geometry, the address electrode capacitance can be approximated by

$$C(\theta) = \frac{\varepsilon_0 w}{\theta} \ln \left[\frac{1 - (g/d)\theta}{1 - [(g+L)/d]\theta} \right].$$

A Taylor series expansion of $C(\theta)$ produces a useful expression for the torque τ^{e}. Assuming $\theta \ll d/(g+L)$, the expansion correct to second order in θ is

$$C(\theta) = C_0 \left[1 + \frac{2g + L}{2d}\theta + \frac{3g^2 + 3gL + L^2}{3d^2}\theta^2 + \ldots \right],$$

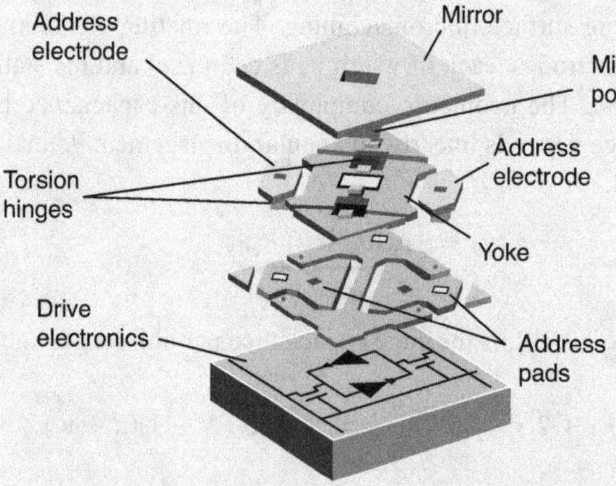

Figure 3.23 Basic construction details of digital micromirror device (DMD) developed and designed by Hornbeck and colleagues at Texas Instruments, Inc. The square Al mirror at the top, mounted to a yoke suspended by torsional hinges, tilts back and forth in response to voltages applied to the underlying address electrodes. Image provided courtesy of Texas Instruments, Inc.

The fabrication process developed to make the first micromirror array chips was revolutionary for its time. The Si substrate, containing drive electronics and address electrodes, was fabricated first. Then, a 3000 to 5000 Å thick Al layer was deposited on ~ 2 microns of sacrificial scaffolding material. Next, the Al layer was coated with the mirror finish, then patterned and etched to form the tilting pixels. Finally, the sacrificial material under the Al electrodes was removed to release the mirrors, which are supported by mechanical posts on opposite sides.

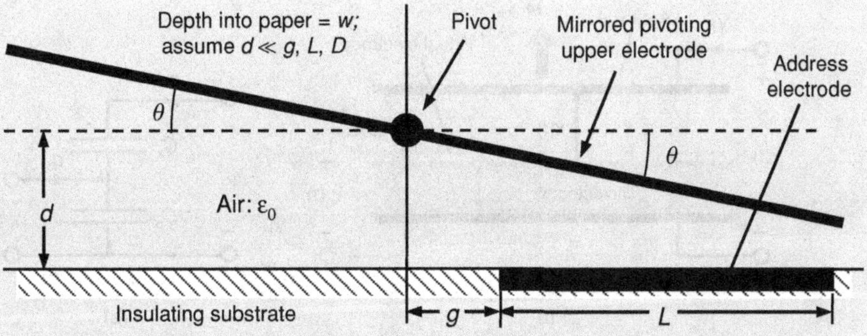

Figure 3.24 Side view of simplified capacitive model for rotating pixel element in a digital micromirror device. Not shown in the figure is the second electrode on the left side of the pivot, which facilitates the desired on–off operational behavior. Mechanical stops, also not shown, limit the tilt to prevent stictive failure.

where $C_0 \equiv C(\theta = 0) = \varepsilon_0 Lw/d$. To assess frequency-dependent behavior and stability, the second-order term is retained. Using this expression in Eq. (3.52), the torque is

$$\tau^{\mathrm{e}} = \frac{C_0 v^2}{2}\left[\frac{2g+L}{2d} + 2\frac{3g^2+3gL+L^2}{3d^2}\theta + \ldots\right].$$

The torque is positive at $\theta = 0$ and increases with tilt. To achieve on–off function, another electrode on the opposite side of the pivot, not shown, drives the mirror back the other way.

*3.6 Electrets

Electret microphones have been in use for many years. They are inexpensive and, for application in MEMS, have the advantage of being self-biased, meaning that no external voltage source is needed for their operation. Electrets are made by permanently locking electric charge in the surface or volume polarization in the bulk of thin polymer films. Their behavior is analogous in some respects to permanent magnets. Applications for electrets in MEMS include energy harvesters, electrostatic motors, and microphones.

What distinguishes electret transducers from conventional capacitive devices is their state of permanent electrification. They can be analyzed by lumped parameter analysis if proper account is taken of this permanent electrification. Thus, consider the parallel-plate structure shown in Fig. 3.25a, which has a variable air gap $x(t)$, and an electret, modeled as a layer of insulating material of dielectric constant κ and thickness d. The circuit model for this device, shown in Fig. 3.25b, consists of series-connected capacitors, $C_x = \varepsilon_0 A/x$ and $C_d = \kappa\varepsilon_0 A/d$, representing the variable air gap and the electret, respectively. The inner electrode is *virtual*, defined and utilized only to introduce the permanent electric charge onto the surface of the dielectric. We are entitled to use this virtual electrode because the surface of the dielectric happens to be an equipotential.

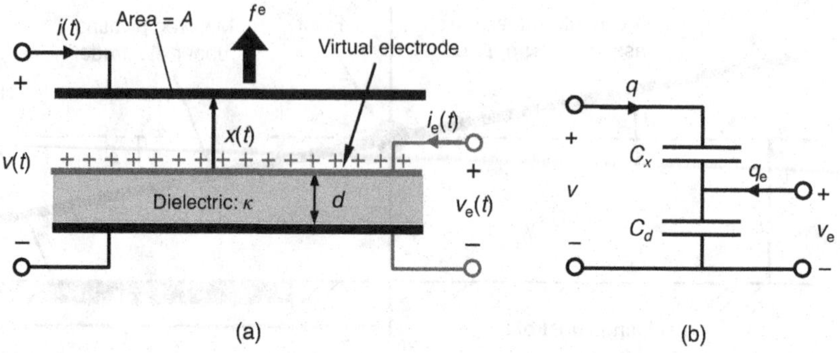

Figure 3.25 Basic electret transducer model with two electrical ports. (a) Side view showing insulating dielectric, which contains a net electrical surface charge injected via the virtual port shown on the right side. In operation, only the port on the left side is accessible. (b) Series capacitive circuit model for the transducer, consisting of the variable air gap and the dielectric layer.

Several different processes are employed to manufacture electrets. One common method is to implant free electric charge on the surface of a dielectric, usually a polymer film, such as Teflon or polycarbonate film, using an energetic electron beam or gaseous breakdown to inject charges into the immediate vicinity of the surface. For modeling purposes, knowledge of manufacturing details is unneeded, but to derive the energy function we must establish a thermodynamic process path to introduce the charge. Recall that the energy method always relies upon the assumption that the force of electrical origin is zero until the voltages or electric charges are introduced. If this condition is not met, the integration needed to obtain the energy function cannot be performed. It is for this reason that (i) the virtual electrode is introduced on the dielectric surface and (ii) the charge q_e defined in Fig. 3.25b is identified as a state variable.

Energy conservation for the device represented by Fig. 3.25b takes the form

$$\mathrm{d}W_e = -f^e\,\mathrm{d}x + v\mathrm{d}q + v_e\mathrm{d}q_e, \tag{3.58}$$

where the set (x, q, q_e) are the state variables. However, because v is a more convenient variable than q, and because q_e is constrained to be constant in the electret, one might anticipate (x, v, q_e) to be a better choice. To change variables appropriately, we use a different form for the Legendre transform to define a new "hybrid" energy function, W_e'':

$$W_e + W_e'' = qv. \tag{3.59}$$

Taking the differentials of the terms in Eq. (3.59) to eliminate $\mathrm{d}W_e$ from Eq. (3.58) yields

$$\mathrm{d}W_e'' = f^e\,\mathrm{d}x + q\mathrm{d}v - v_e\mathrm{d}q_e. \tag{3.60}$$

Note that the two electrical terms have opposite signs. Because this system is conservative, we can exploit the fact that $W_e'' = W_e''(x, v, q_e)$ to write

$$\mathrm{d}W_e'' = \left.\frac{\partial W_e''}{\partial x}\right|_{v,q_e}\mathrm{d}x + \left.\frac{\partial W_e''}{\partial v}\right|_{x,q_e}\mathrm{d}v + \left.\frac{\partial W_e''}{\partial q_e}\right|_{x,v}\mathrm{d}q_e. \tag{3.61}$$

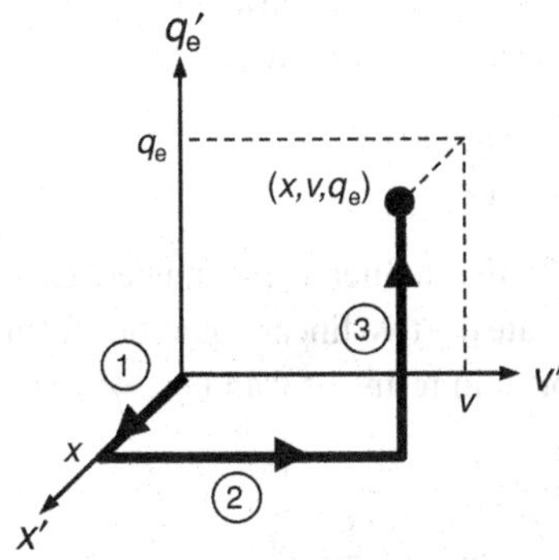

Figure 3.26 The three-dimensional state space representation of (x, v, q_e) and the integral path used to evaluate the hybrid energy function W_e''. The mechanical variable x is integrated first to take advantage of the fact that $f^e = 0$ when both the electrical variables are zero.

Comparison of Eq. (3.61) to (3.60) secures the desired relationships between the hybrid energy function and the force of electrical origin, as well as the other terminal relations:

$$f^e = \left.\frac{\partial W_e''}{\partial x}\right|_{v,q_e}, \quad q = \left.\frac{\partial W_e''}{\partial v}\right|_{x,q_e}, \quad \text{and} \quad v_e = -\left.\frac{\partial W_e''}{\partial q_e}\right|_{x,v}. \tag{3.62}$$

It remains to integrate Eq. (3.60) to obtain the energy function $W_e''(x, v, q_e)$. Figure 3.26 depicts the integral path in (x, v, q_e) state space. This path exploits the requirement that the force of electrical origin be zero when there is no electrical excitation, that is, $f^e(x, v = 0, q_e = 0) = 0$.

The order of integration of the two electric variables, v and q_e, is immaterial, but the mechanical variable must go first. Using the path defined by Fig. 3.26,

$$W_e''(x, v, q_e) = \underbrace{\int_0^x f^e(x', v' = 0, q_e' = 0)\,\mathrm{d}x'}_{=0} + \int_0^v q(x, v', q_e' = 0)\,\mathrm{d}v' - \int_0^{q_e} v_e(x, v, q_e')\,\mathrm{d}q_e'. \tag{3.63}$$

Bear in mind that the variables of integration are the primed quantities. The first integral term is zero, but the other terms require information about the electrical terminal relations. It is convenient to use the capacitance matrix because these coefficients are easily obtained by inspection of the circuit in Fig. 3.25b:

$$\begin{bmatrix} q \\ q_e \end{bmatrix} = \begin{bmatrix} C_x & -C_x \\ -C_x & C_d + C_x \end{bmatrix} \begin{bmatrix} v \\ v_e \end{bmatrix}. \tag{3.64}$$

As expected, the capacitance matrix in Eq. (3.64) is symmetric. For use in Eq. (3.63), it is necessary to recast the terminal relation into a different form:

$$\begin{bmatrix} q \\ v_e \end{bmatrix} = \begin{bmatrix} \alpha(x) & \beta(x) \\ \gamma(x) & \delta(x) \end{bmatrix} \begin{bmatrix} v \\ q_e \end{bmatrix}. \tag{3.65}$$

The coefficients α, β, γ, and δ are related to the defined capacitances, C_x and C_d. The way to proceed is to extract and then manipulate the two linear equations from Eq. (3.64). First, one can directly write the equation for v_e in terms of v and q_e,

$$v_e = \underbrace{\left(\frac{C_x}{C_d + C_x}\right)}_{=\gamma} v + \underbrace{\left(\frac{1}{C_d + C_x}\right)}_{=\delta} q_e, \tag{3.66}$$

to obtain γ and δ. Next, substitute Eq. (3.66) into the other linear relation from (3.64) to obtain

$$q = C_x v - C_x \left[\frac{C_x}{C_d + C_x} v + \frac{1}{C_d + C_x} q_e\right]. \tag{3.67}$$

After simplification, α and β may be identified:

$$q = \underbrace{\left(\frac{C_x C_d}{C_d + C_x}\right)}_{=\alpha} v + \underbrace{\left(\frac{-C_x}{C_d + C_x}\right)}_{=\beta} q_e. \tag{3.68}$$

Expressed for convenience in terms of the coefficients α, β, γ, and δ, the two non-zero integrals comprising W_e'' are

$$\int_0^v q(x, v', q_e' = 0)\,dv' = \int_0^v \alpha v' dv' = \frac{\alpha}{2} v^2 \tag{3.69}$$

and

$$-\int_0^{q_e} v_e(x, v, q_e')\,dq_e' = -\int_0^{q_e} (\gamma v + \delta q_e')\,dq_e' = -\gamma v q_e - \frac{\delta}{2} q_e^2. \tag{3.70}$$

The hybrid energy function is then

$$W_e'' = \frac{\alpha}{2} v^2 - \gamma v q_e - \frac{\delta}{2} q_e^2, \tag{3.71}$$

and

$$f^e = \frac{v^2}{2}\frac{d\alpha}{dx} - v q_e \frac{d\gamma}{dx} - \frac{q_e^2}{2}\frac{d\delta}{dx}. \tag{3.72}$$

One can now substitute the expressions for α, γ, and δ in terms of C_x and C_d from Eqs. (3.66) and (3.68) into Eq. (3.72) to obtain the desired force of electrical origin. After some manipulation,

$$f^e = -\frac{\varepsilon_0 A}{2(x + d/\kappa)^2}\left(v - \frac{q_e}{C_d}\right)^2; \tag{3.73}$$

f^e is non-zero when the external voltage $v = 0$, demonstrating why it was necessary to introduce the virtual electrical port and its associated variables, v_e and q_e, to account for the permanent surface charge. The form of this force expression is reminiscent of the result for the variable-gap capacitor, Eq. (3.33). The only differences are the replacement of x by the net (series) dielectric thickness, $x \to x + d/\kappa$, and the appearance of an effective voltage, that is, $v \to v - q_e/C_d$, which accounts for the electret's permanent charge. The force goes to zero when $v = q_e/C_d$ because, under that condition, both the electric field in the air gap and the charge on the plate (q) are zero. Typically, $q_e/C_d \sim 10^3\ V_{dc}$. The self-biasing feature of electret devices is advantageous in capacitive MEMS applications.

*3.7 Non-linear conservative electromechanical systems

The remaining chapters of the text focus mainly on a linear system theory for MEMS devices. This is justified because most microelectromechanical devices are designed to operate in the linear regime. But before launching the effort that starts with Chapter 4, it is profitable to devote some attention to non-linear electromechanics. Doing so provides a broader perspective on the utility of lumped parameter electromechanics.[3]

The key to proceeding is to consider in more depth what it means for a system – linear or non-linear – to be conservative. In any conservative electromechanical system, the statics and dynamics are characterized by an energy function that depends on a set of state variables. Knowing this energy function and the initial conditions facilitates prediction of the time-dependent behavior. Earlier in this chapter, we demonstrated that there is no inherently unique energy quantity. Depending entirely on the imposed electrical constraint, energy, coenergy, or a "hybrid" quantity may be treated as a conserved quantity.

This section introduces a method for extracting conservation laws suitable for predicting the non-linear oscillatory behavior and stability of capacitive transducers. The constraints of constant voltage and constant charge are separately considered. The resulting conservation laws are easy to interpret. Moreover, they lead to better understanding of electromechanical dynamics and can be exploited to obtain numerical solutions for the non-linear behavior. Still further, the method itself serves as a portal to the classic methods of Lagrangian and Hamiltonian mechanics, which are sometimes invoked to investigate the dynamics of MEMS devices.

3.7.1 Conservation laws for capacitive devices

Consider the dynamics of a linear system having mass $= m$ held in equilibrium at $x = 0$ by a linear mechanical spring of spring constant k, where f^e is the force of electrical origin. Without loss of generality, it is assumed that $f^e(x = 0) = 0$. Gravity and other

[3] This section is based heavily on Section 5.2 of *Electromechanical Dynamics* by H. H. Woodson and J. R. Melcher [2].

forces are not considered here. To develop the conservation law for this system, we start with Newton's first law of motion,

$$m\frac{\mathrm{d}^2x}{\mathrm{d}t^2} = -kx + f^{\mathrm{e}}. \tag{3.74}$$

Even though it may have a strong influence on the dynamics of many MEMS devices, we here ignore mechanical damping. This approximation facilitates derivation of the desired conservation law.

To achieve desirable generality and to gain insights, the energy and coenergy formulations are treated in parallel. This convenient approach allows us to keep track of and then contrast the cases of constant charge and constant voltage. Replacing f^{e} in Eq. (3.74) by the appropriate energy derivative yields

$$m\frac{\mathrm{d}^2x}{\mathrm{d}t^2} + kx + \left\{ \begin{array}{c} \partial W_{\mathrm{e}}/\partial x|_q \\ -\partial W_{\mathrm{e}}'/\partial x|_v \end{array} \right\} = 0. \tag{3.75}$$

Multiplying each term in the above by the velocity $\mathrm{d}x/\mathrm{d}t$ gives

$$m\frac{\mathrm{d}^2x}{\mathrm{d}t^2}\frac{\mathrm{d}x}{\mathrm{d}t} + kx\frac{\mathrm{d}x}{\mathrm{d}t} + \left\{ \begin{array}{c} \partial W_{\mathrm{e}}/\partial x|_q \\ -\partial W_{\mathrm{e}}'/\partial x|_v \end{array} \right\} \mathrm{d}x/\mathrm{d}t = 0. \tag{3.76}$$

Using the product and chain rules of differentiation and then performing some straightforward algebraic manipulation, Eq. (3.76) reduces to the form of an energy conservation rule:

$$\frac{\mathrm{d}}{\mathrm{d}t}\left[\frac{m}{2}\left(\frac{\mathrm{d}x}{\mathrm{d}t}\right)^2 + U(x)\right] = 0, \tag{3.77}$$

where

$$U(x) = \frac{k}{2}x^2 + \left\{ \begin{array}{c} W_{\mathrm{e}}(x, q), \\ -W_{\mathrm{e}}'(x, v), \end{array} \right\}, \quad \left\{ \begin{array}{l} \text{charge fixed} \\ \text{voltage fixed} \end{array} \right. \tag{3.78}$$

is a potential energy consisting of the sum of the mechanical potential energy of the spring and an electrical energy term. The latter component reflects, in one way or another, stored electrostatic energy. For the constant charge constraint, this quantity is the electrostatic energy, W_{e}, while for constant voltage, coenergy, W_{e}', serves the purpose.

According to Eq. (3.77), the conserved energy quantity is the sum of the kinetic and potential energies, that is,

$$\frac{m}{2}\left(\frac{\mathrm{d}x}{\mathrm{d}t}\right)^2 + U(x) = \text{constant}. \tag{3.79}$$

When q is held constant, it is reasonable to interpret the left-hand side of Eq. (3.79) as a "conventional" total system energy. In this case, the electric terminal current is zero. Because no electrical power flows into or out of the capacitor, the system is *adiabatic* in the thermodynamic sense. On the other hand, when the voltage is fixed, the electric charge is *not* constrained. Thus, the current is non-zero, electric power can flow in and out, and the system is not adiabatic. Therefore, the quantity defined by Eq. (3.78) for fixed voltage is not readily interpreted as "regular" energy. Yet, because it is a conserved

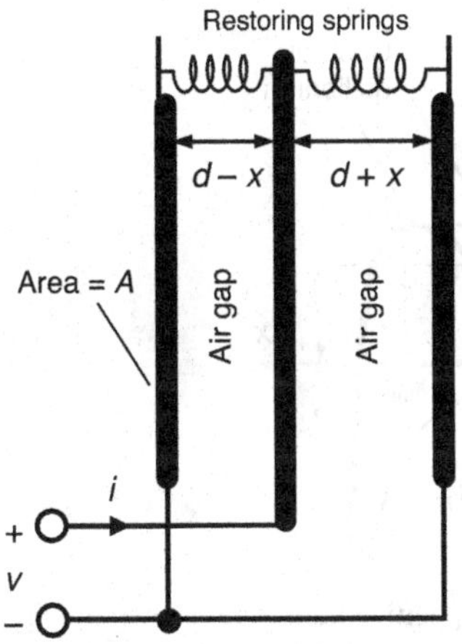

Figure 3.27 Capacitive transducer consisting of a moving central electrode of mass m and two fixed electrodes that enclose it. The fixed area overlap of the electrodes is A. In the absence of external forces, the moving electrode is constrained to equilibrium at $x = 0$ by a restoring mechanical spring with spring constant k. As it moves, the central electrode remains parallel to the fixed outer electrodes.

quantity, it is in every important respect a legitimate energy function and may be used under the condition of fixed voltage.

3.7.2 Non-linear oscillations and stability

Using the general results from Section 3.7.1, one can investigate the non-linear dynamics of the capacitive transducer consisting of three parallel electrodes of fixed overlapped area A and shown in Fig. 3.27. The moving element is the central electrode. It has mass m and is held in equilibrium at $x = 0$ by linear mechanical springs of spring constant k.

To describe the electromechanics, an expression for the capacitance is needed. From inspection, the circuit model is two parallel capacitors, corresponding to the left- and right-hand sides of the central moving electrode, that is,

$$C(x) = \frac{\varepsilon_0 A}{d - x} + \frac{\varepsilon_0 A}{d + x} = \frac{2d\varepsilon_0 A}{d^2 - x^2}. \tag{3.80}$$

Consider the constant voltage constraint first by setting $v = v_0$. The desired conservation law results from combining Eqs. (3.79) and (3.80) with Eq. (3.18) for the coenergy,

$$\frac{m}{2}\left(\frac{\mathrm{d}x}{\mathrm{d}t}\right)^2 + \frac{k}{2}x^2 - \frac{\varepsilon_0 A d v_0^2}{(d^2 - x^2)} = K_0, \tag{3.81}$$

where K_0 is a constant determined from initial conditions. For example, if the plate passes the midpoint $x = 0$ with velocity $\mathrm{d}x/\mathrm{d}t = u_0$ at time $t = 0$, then

$$K_0 = \frac{m}{2}u_0^2 - \frac{\varepsilon_0 A v_0^2}{d}, \tag{3.82}$$

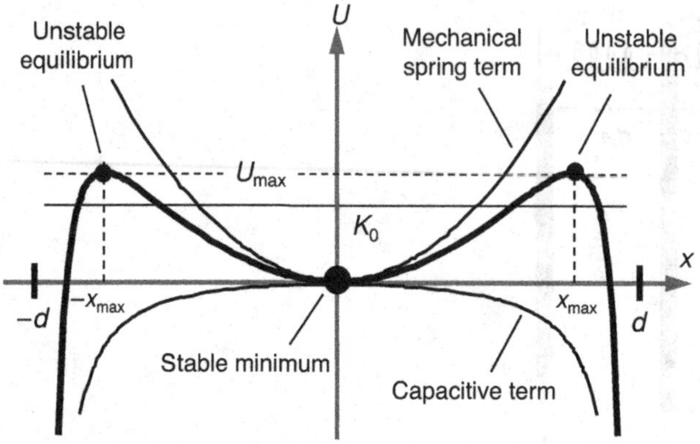

Figure 3.28 Plot of potential energy $U(x)$ for the capacitive transducer shown in Fig. 3.27 under the constraint of constant voltage. For this plot, it is assumed that $k > \varepsilon_0 A v_0^2/d^3$, so that the equilibrium at $x = 0$ is stable.

and the conservation law becomes

$$\frac{m}{2}\left(\frac{dx}{dt}\right)^2 + \left(\frac{k}{2}\right)x^2 - \frac{\varepsilon_0 A d v_0^2}{(d^2 - x^2)} = \frac{m}{2}u_0^2 - \frac{\varepsilon_0 A v_0^2}{d}, \tag{3.83}$$

or, in simpler form,

$$\underbrace{\frac{m}{2}\left(\frac{dx}{dt}\right)^2}_{\text{kinetic energy}} + \underbrace{\left(\frac{k}{2}\right)x^2 - \frac{\varepsilon_0 A v_0^2}{d}\frac{x^2}{(d^2 - x^2)}}_{U(x) = \text{potential energy}} = \frac{m}{2}u_0^2. \tag{3.84}$$

Equation (3.84) separates the kinetic energy from a conveniently defined potential energy, $U(x, v_0)$. The sum of these two quantities is constant and equal to the initial kinetic energy of the moving plate.

Plotting $U(x, v_0)$ versus x provides an interpretation of the non-linear dynamical behavior in terms of a potential well. Refer to Fig. 3.28. The two contributions to potential function are a strictly positive parabolic term, representing the linear mechanical spring, and a negative hyperbolic term, which accounts for the electromechanical coupling.

For $|v_0| < [kd^3/2\varepsilon_0 A]^{1/2}$, $U(x)$ has a local minimum at $x = 0$ and local maxima situated at x_{max} and $-x_{max}$.[4] The system is a *non-linear oscillator* with a period that depends on the initial energy. As long as the initial kinetic energy is less than $U_{max} = U(x_{max})$, the oscillation remains bounded and centered on $x = 0$. On the other hand, if $K_0 > U_{max}$, the plate will venture outside the well defined by the pair of local maxima. Outside this region, the electrical force overwhelms the linear spring and pulls the moving plate further out until it strikes one of the electrodes. Strong van der Waals forces usually cause MEMS devices to stick permanently. This type of failure may be the consequence of a sharp jolt or any unexpectedly large acceleration.

[4] The minimum at $x = 0$ is a stable equilibrium and the maxima evident in Fig. 3.28 are unstable. We will develop another, more convenient method to explore such behavior in Section 4.8.

Another failure mechanism results if the voltage is increased to the point where $|v_0| > [kd^3/2\,\varepsilon_0 A]^{1/2}$. In this case, the equilibrium at $x = 0$ becomes unstable. Any tiny force leading to an initial displacement will start the plate moving away from the equilibrium toward one of the outside plates. This is the *pull-in* phenomenon, a form of *absolute instability*. For MEMS devices, pull-in can be catastrophic, again because the strong van der Waals forces cause the electrodes to stick permanently once they come into contact. To avoid pull-in, it often suffices to incorporate mechanical stops.

3.7.3 Numerical solutions

Rearranging Eq. (3.84) by taking a square root to expose the derivative $\mathrm{d}x/\mathrm{d}t$ and paying due attention to algebraic signs yields an equation that, subject to the initial conditions, can be integrated numerically to obtain $x(t)$:

$$\frac{\mathrm{d}x}{\sqrt{u_0^2 - \left(\dfrac{k}{m}\right)x^2 + 2\dfrac{\varepsilon_0 A v_0^2}{md}\dfrac{x^2}{(d^2 - x^2)}}} = \begin{cases} +\mathrm{d}t, & u_0 > 0, \\ -\mathrm{d}t, & u_0 < 0. \end{cases} \tag{3.85}$$

Note that this equation is valid only for specific initial conditions, viz., $x = 0$ and $\mathrm{d}x/\mathrm{d}t = u_0$ at $t = 0$. The sign on the right-hand side is chosen based on the sign of u_0. Equation (3.85) can be integrated to analyze the motion of the plate. When $|v_0| < [kd^3/2\,\varepsilon_0 A]^{0.5}$, the period of the non-linear oscillator, which depends on the initial kinetic velocity u_0, can be determined. On the other hand, if $|v_0| > [kd^3/2\,\varepsilon_0 A]^{0.5}$, the time required for the moving plate to strike one of the outer electrodes can be calculated.

Equation (3.85) is inconvenient in that the solution is implicit, giving t as a function of x, rather than the other way around. Also, the numerical integral itself can be troublesome because of convergence difficulties. A more general formulation of the dynamics of this second-order system results from using *mechanical state space variables*. This approach recasts the second-order equation of motion as a pair of first-order, ordinary differential equations (ODEs). Using Eq. (3.19) for f^{e} in Eq. (3.74) and manipulating the equation to identify displacement $x(t)$ and velocity $\dot{x}(t)$ as the state variables, the result is

$$\frac{\mathrm{d}}{\mathrm{d}t}\dot{x} = -\frac{k}{m}x + \frac{v^2}{2\,m}\frac{\mathrm{d}C}{\mathrm{d}x}, \tag{3.86a}$$

$$\frac{\mathrm{d}}{\mathrm{d}t}x = \dot{x}. \tag{3.86b}$$

This formalism, familiar to those who have studied dynamic systems, is readily solved by numerical means using MATLAB or other ODE solvers. The big advantage is that these solutions naturally provide a direct, explicit solution for x as a function of t.

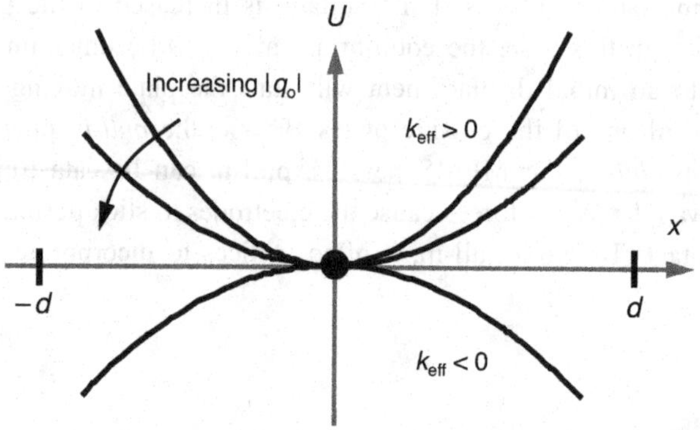

Figure 3.29 Potential energy $U(x)$ for the capacitive transducer shown in Fig. 3.27 under the constraint of constant charge. The potential energy function is parabolic. The equilibrium at $x = 0$ is stable only when $k > 2q_0^2/\varepsilon_0 A$. The arrow indicates the effect on $U(x, q)$ of increasing the magnitude of charge q_0.

3.7.4 Constant charge constraint

Now consider the constant charge constraint, that is, $q = q_0$, applied to the transducer of Fig. 3.27. Using the energy function for a linear capacitor, $W_e(x, q) = q^2/2\,C(x)$, Eq. (3.76) yields

$$\frac{m}{2}\left(\frac{\mathrm{d}x}{\mathrm{d}t}\right)^2 + \frac{k}{2}x^2 + \frac{q_0^2}{4\varepsilon_0 d A}(d^2 - x^2) = K_0. \tag{3.87}$$

Using the same initial conditions from Section 3.7.2 for the plate,

$$K_0 = \frac{m}{2}u_0^2 + \frac{q_0^2 d}{4\varepsilon_0 A}. \tag{3.88}$$

Combining Eqs. (3.87) and (3.88) and rearranging,

$$\underbrace{\frac{m}{2}\left(\frac{\mathrm{d}x}{\mathrm{d}t}\right)^2}_{\text{kinetic energy}} + \underbrace{k_{\mathrm{eff}}x^2}_{\substack{U(x)=\\ \text{potential energy}}} = \frac{m}{2}u_0^2, \tag{3.89}$$

where

$$k_{\mathrm{eff}} \equiv k - \frac{q_0^2}{4\varepsilon_0 d A} \tag{3.90}$$

is an effective spring constant. Under a constant-charge constraint, the potential energy exhibits the familiar parabolic form of a linear spring, as shown in Fig. 3.29,

$$U(x) = \frac{k_{\mathrm{eff}}}{2}x^2, \tag{3.91}$$

except that k_{eff} can change sign from positive to negative if the charge q_0 is large enough. As long as $k_{\mathrm{eff}} > 0$, the system behaves like a linear oscillator with a stable equilibrium at $x = 0$. As the charge increases, k_{eff} and the resonant frequency decrease.

When k_{eff} goes negative, the equilibrium at $x = 0$ becomes unstable. Interestingly, when this charge threshold is expressed in terms of voltage using $q_0 = C(x = 0)\, v_0$, it is identical to the stability condition obtained for the constant voltage constraint.

Example 3.6 **Pull-in of variable-gap capacitive device**

As the voltage of a variable-gap capacitive actuator is turned up, a threshold is eventually reached where the electrodes spontaneously come together. The van der Waals force then fastens the electrodes together permanently, ruining the device. This *pull-in* instability occurs when the mechanical spring can no longer compensate for the inverse x dependence of the electrostatic force. The equation of motion for the upper plate of the capacitor shown in Fig. 3.30 is

$$m\frac{\mathrm{d}^2 x}{\mathrm{d}t^2} = -k(x - x_0) + \frac{v^2}{2}\frac{\mathrm{d}C}{\mathrm{d}x},$$

where A = area, m = mass, k = spring constant of a linear spring, which is relaxed at $x = x_0$.

Using $C(x) = \varepsilon_0 A/x$ in the equation of motion above, then multiplying all terms by $\mathrm{d}x/\mathrm{d}t$ and invoking the product rule of differentiation yields

$$\frac{\mathrm{d}}{\mathrm{d}t}\left\{\frac{m}{2}\left(\frac{\mathrm{d}x}{\mathrm{d}t}\right)^2 + \frac{k}{2}(x - x_0)^2 - \frac{\varepsilon_0 A v^2}{2}\frac{1}{x}\right\} = 0.$$

The braced term is constant in time. Thus, initial conditions can be used to write an equation describing the motion. Assuming $x = x_0$ and $\mathrm{d}x/\mathrm{d}t = u_0$ at $t = 0$, then after some manipulation,

$$\frac{m}{2}\left(\frac{\mathrm{d}x}{\mathrm{d}t}\right)^2 + \underbrace{\frac{k}{2}(x - x_0)^2 - \frac{\varepsilon_0 A v^2}{2}\left(\frac{1}{x} - \frac{1}{x_0}\right)}_{U(x)} = \frac{m}{2}u_0^2.$$

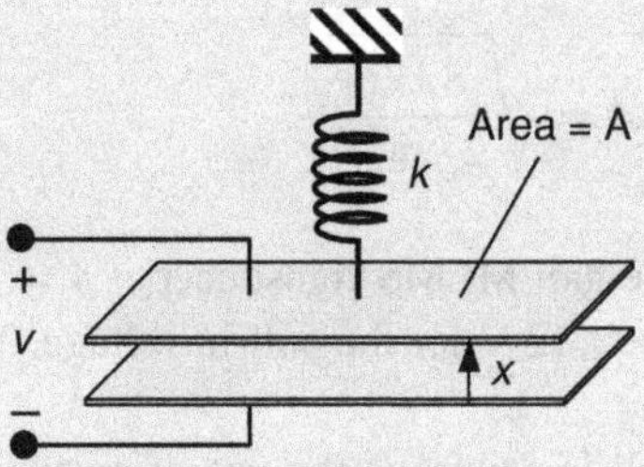

Figure 3.30 Basic variable-gap capacitive actuator with a linear mechanical spring of spring constant k attached to a moving plate, which remains parallel to the bottom electrode. The relaxed position of the spring is at $x = x_0$.

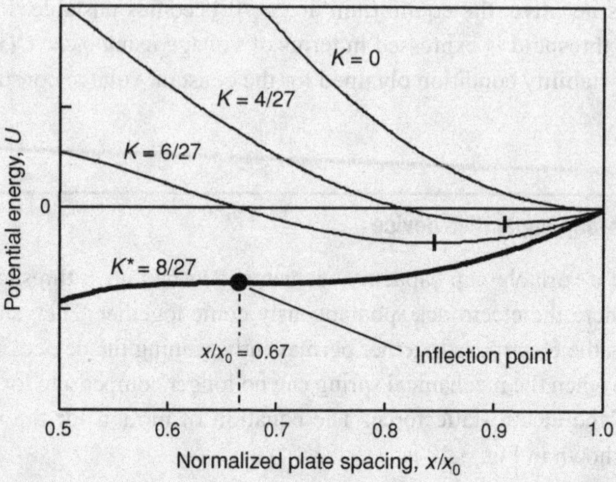

Figure 3.31 Plot of $U(\bar{x})$ in arbitrary units illustrating the influence of K on stability. At $K = 0$, there is no electrical force and equilibrium is at $\bar{x} = 1$. As voltage is raised, the equilibrium shifts to the left and the potential well becomes shallower. At $K^* = 8/27$, corresponding to $\bar{x}^* = 2/3$, the potential well changes into an inflection point and stability is lost.

The identified potential energy $U(x)$ can be exploited to investigate stability. Using the normalization $\bar{x} \equiv x/x_0$ yields

$$U(\bar{x}) = \frac{kx_0^2}{2}\left\{(\bar{x}-1)^2 - K\left(\frac{1}{\bar{x}} - 1\right)\right\},$$

where $K = \varepsilon_0 A v^2/kx_0^3$, a dimensionless modulus, compares the influences of the electrical and spring forces. Figure 3.31 plots $U(\bar{x})$ versus $\bar{x}$ for several values of K. Equilibrium, defined at $\mathrm{d}U/\mathrm{d}\bar{x} = 0$, is stable if $\mathrm{d}^2 U/\mathrm{d}\bar{x}^2 > 0$. If $v = 0$, then $K = 0$ and $\bar{x} = 1$ is a stable equilibrium. When voltage is applied, the equilibrium point shifts to the left and the potential well becomes shallower. At $K^* = 8/27 \approx 0.296$, corresponding to $\bar{x}^* = 2/3 \approx 0.67$, the minimum becomes an inflection point and *pull-in* occurs. Setting $K = K^*$, the *voltage threshold* is

$$v^* = \sqrt{\frac{kx_0^3}{\varepsilon_0 A}K^*} = \sqrt{\frac{8\,kx_0^3}{27\,\varepsilon_0 A}}.$$

Selecting parameters typical for a variable-gap MEMS transducer – $A = 62.5\ \mu\mathrm{m}^2$, $x_0 = 3\ \mu\mathrm{m}$, and $k = 10\ \mu\mathrm{N}/\mu\mathrm{m}$ – yields $v^* \approx 12$ V for the pull-in voltage. The critical spacing is $x^* = 2\ \mu\mathrm{m}$.

The most reliable protection against pull-in failure is the use of mechanical stops to limit the contact area so that the van der Waals forces cannot fasten the electrodes together permanently.

3.7.5 Discussion

The non-linear analysis given here has revealed two distinct mechanisms of capacitive device failure. One is the well-known pull-in instability. It can occur under either fixed voltage or fixed charge constraints. It is an absolute instability and it has a threshold. Keeping the voltage (or charge) magnitude below this threshold avoids the instability. The other mechanism is motion imparted to the plate due to jolts or large accelerations that can cause the plate to come into contact with one of the electrode. Any MEMS device has such vulnerability, irrespective of electrical constraint. But, from a comparison of the potential functions plotted in Figs. 3.28 and 3.29, voltage-constrained systems may be more susceptible to such failure than otherwise identical charge-constrained systems. This conclusion is based on a comparison of the potential plots, which shows a narrower range of allowed motion when the voltage is fixed, that is, $-x_{\max} < x < +x_{\max}$, as shown in Fig. 3.28. Note that this kinetic failure mode is a practical concern, and can occur even when the equilibrium itself is stable. Some MEMS devices, such as accelerometers, are very sensitive to mechanical shock, and must be potted in special gel materials for protection before being shipped, even before any voltage is applied.

The constant charge constraint has broad applicability beyond electret transducers. For example, if a DC bias is connected to a capacitive transducer through a large resistor, the effective *RC* time constant might be much longer than the time scale of the mechanical motions, in which case the electric charge is effectively fixed. There will be more to say about electrical constraints in Chapter 5.

3.8 Summary

This chapter has covered the basics of capacitive, lumped parameter electromechanics. We introduced a method, founded on the principle of virtual work, that employs an energy function to formulate the concept of the *lossless electromechanical coupling.* A thermodynamic formulation based on state variables makes it possible to derive expressions for the so-called force of electrical origin f^{e}. Essential to the derivation are the assumptions that (i) the system is conservative and (ii) $f^{\mathrm{e}} = 0$ when the electrical excitation is absent. In addition, knowledge of the electrical terminal relation, ordinarily taking the form $q = C(x)v$, is needed. Several generic types of capacitive transducer, including variable-gap and variable-area geometries, were then presented. Their behavior, as described in terms of conventional force versus displacement plots, was found to depend strongly on the external electrical constraint. Two special cases, specifically the constant voltage and constant charge limits, have been investigated. A generalization to frequency-dependent circuit constraints will come later.

The chapter concluded by presenting a general modeling methodology for analyzing the non-linear behavior of electromechanical systems, including oscillation and stability. This method nicely elucidates the influence of external electrical constraints on the dynamics. It also provides a practical way to perform numerical analyses of non-linear dynamic behavior.

Looking ahead, the next four chapters of the text focus exclusively on linear electromechanical system modeling of capacitive transducers. In particular, Chapter 4 introduces a linearization procedure for lossless couplings and then exploits it to develop the general linear electromechanical multiport system theory to be employed in the later chapters.

Problems

3.1 Over some range of the mechanical variable, $a < x < b$, a capacitor has the following x-dependence: $C(x) = \varepsilon_0 A/(x + \alpha x^2)$. Find an expression for the force of electrical origin within this range, using the energy and coenergy functions. Show that your two answers are the same.

3.2 Over some range of positive and negative displacement values, a capacitor depends on a linear mechanical variable x as follows:

$$C(x) = C_0 \mathrm{e}^{-\alpha x^2}.$$

(a) Find an expression for the force of electrical origin using the energy function.
(b) Repeat this exercise using the coenergy function.
(c) Show that your two answers are identical.
(d) Sketch the x dependencies of $C(x)$ and f^{e} (with the voltage fixed). Does the force of electrical origin tend to increase or decrease the capacitance?

3.3 The force of electrical origin exerted by a MEMS device is measured to have the following dependence on the voltage, v, and a mechanical variable, x:

$$f^{\mathrm{e}}(x) = -bv^2 \frac{x/a}{(1 - x^2/a^2)}.$$

Within a constant factor, find an expression for the system capacitance.

3.4 Assuming $L_1, L_2 \gg g, x$, and ignoring fringing fields, find an approximate expression for the capacitance $C(x)$ of the device shown below. Then, derive an expression for the x-directed force of electrical origin using coenergy.

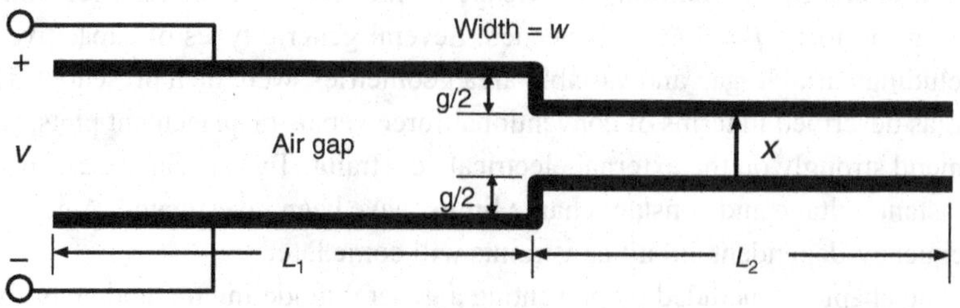

3.5 Neglect fringing fields to find an approximate, voltage-dependent expression for the force of electrical origin acting on the grounded central plate of the capacitive transducer shown.

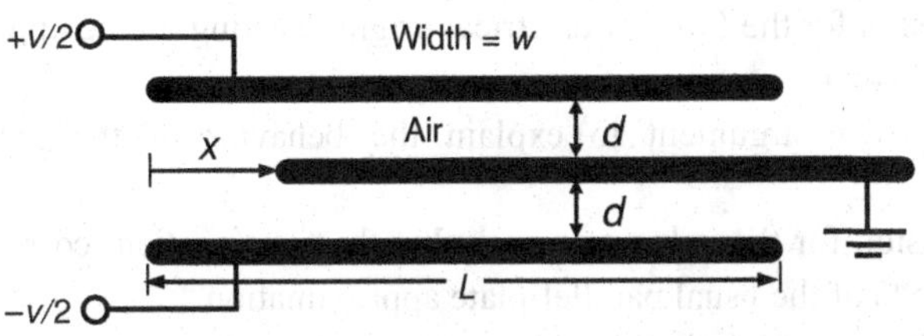

3.6 For the device below, neglect fringing and use the parallel-plate approximation to obtain the capacitance and then the force of electrical origin acting on the upper electrode plate using the energy function. Express the force in terms of electric charge.

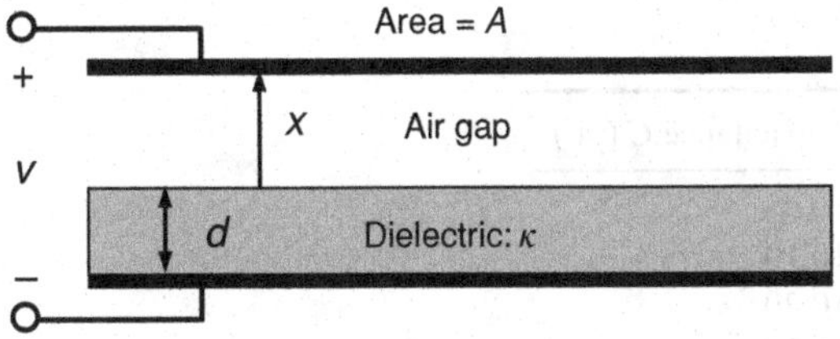

3.7 Use coenergy to find the force of electrical origin as a function of voltage acting on the upper electrode plate of the transducer shown below. The device is square and of area $A = w^2$. You should assume here that x and d are much smaller than w.

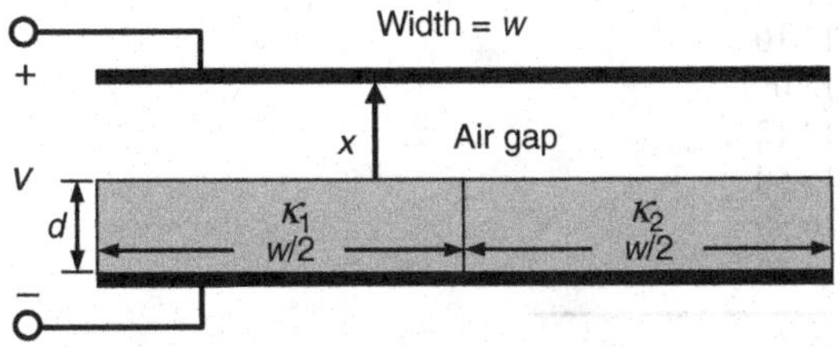

3.8 The variable air gap capacitive transducer below has square electrodes of side dimension b. The capacitance, with a correction term to account for the fringing fields, is:[5]

$$C(x) = \frac{\varepsilon_0 b^2}{x} + \frac{2\varepsilon_0 b}{\pi} \ln\left(\frac{\pi b}{x}\right), \quad x > 0.$$

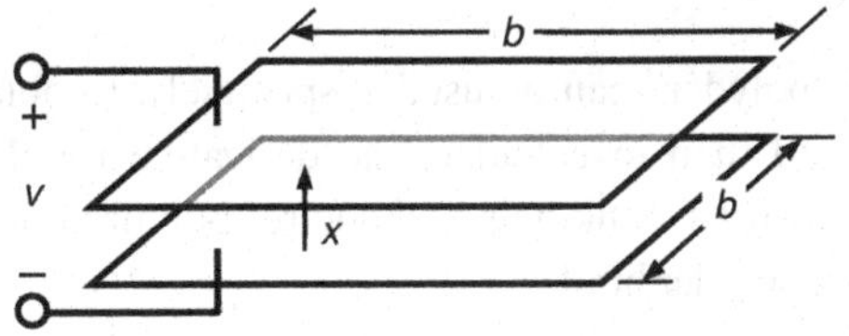

[5] R. Plonsey and R. E. Collin, *Principles and Applications of Electromagnetic Fields* (New York: McGraw-Hill, 1961) p. 162.

(a) Find an expression for the force of electrical origin f^e acting on the upper electrode as a function of voltage v and spacing x.
(b) Develop a physical argument to explain the behavior of the fringing field correction.
(c) Find an expression for the value of x such that the fringing field correction to the capacitance is 5% of the usual parallel-plate approximation.
(d) What is the value of x such that the fringing field correction amounts to 5% of the force of electrical origin as calculated by neglect of fringing?

3.9 Capacitance versus displacement data obtained from numerical analysis of a capacitive MEMS actuator are provided in the accompanying table.

Displacement x (μm)	Capacitance C (pF)
0.0	10.0
1.0	10.10
2.0	10.20
3.0	10.30
4.0	10.42
5.0	10.56
7.0	10.75
9.0	10.90
11.5	11.40
14.0	12.10
17.0	13.0
20.0	14.12
24.0	15.50
29.0	16.50
35.0	18.10

(a) Employ polynomial curve fitting to obtain an algebraic expression for the capacitance $C(x)$.
(b) Plot the capacitance data points and the fitted curve versus x.
(c) Calculate and then plot using convenient units the force f^e versus displacement x at voltage $v = 5\ V_{dc}$.
(d) Investigate how the number of polynomial terms specified for the fitted curve influence the estimate of the force.

MATLAB functions "polyfit" and "polyder" can be used, respectively, to obtain a fitted curve for the capacitance data and then to calculate the derivative for the force of electrical origin. Go to www.mathworks.co.uk/help/techdoc/ref/polyfit.html and www.mathworks.co.uk/help/techdoc/ref/polyder.html

3.10 Force versus displacement measurements for a capacitive MEMS actuator with 15 V_{dc} applied to the device are provided in the table.

Displacement x (μm)	force f^e (nN)
0.0	200
2.0	170
4.0	145
6.0	125
8.0	110
10	100
13	93
17	87
20	83
25	79
30	76
35	74
40	73
45	71
50	70.5

(a) Obtain a polynomial curve fit for the force of electrical origin. In one plot, show the data and the approximating fitted curve.
(b) Use the relationship between force and capacitance to find a polynomial expression for the capacitance. Ignore the constant of integration.
(c) Plot this capacitance versus x.
(d) Investigate how the number of polynomial terms specified for the fitted force curve influences your estimate for $C(x)$.
(e) Discuss the general nature of this capacitive transducer.

MATLAB functions "polyfit" and "polyint" can be used, respectively, to obtain a fitted curve for the capacitance data and then to calculate the integral for the capacitance. Go to www.mathworks.co.uk/help/techdoc/ref/polyfit.html and www.mathworks.co.uk/help/techdoc/ref/polyint.html.

3.11 A rotational capacitive actuator depends on the angular displacement variable θ as follows:

$$C(\theta) = C_0/[1 + (\theta/\theta_0)^2].$$

(a) Find an expression for the torque of electrical origin using the energy function.
(b) Repeat this exercise using the coenergy function.
(c) Show that your two answers are identical.
(d) Sketch the θ dependencies of capacitance $C(\theta)$ and, at fixed voltage, the torque, τ^e. Does the torque of electrical origin tend to increase or decrease the capacitance?

3.12 In the rotating actuator shown, a semicircular arc-shaped dielectric element of thickness d and dielectric constant κ moves concentrically within a pair of cylindrical electrodes.

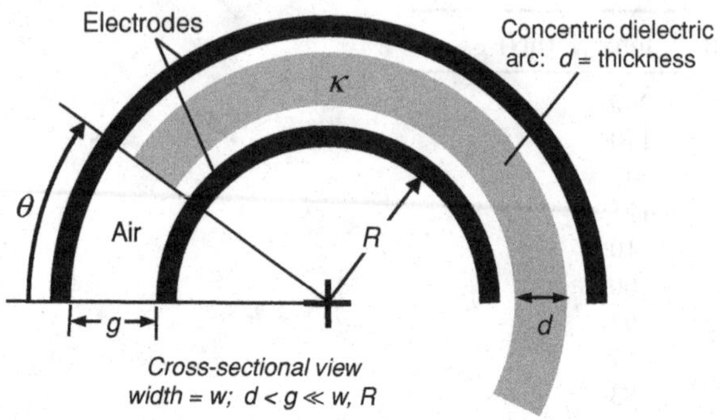

(a) Neglect fringing to propose a capacitive model for this device.
(b) From what you know about the electromechanics of capacitive transducers, predict the direction of the torque exerted on the rotating dielectric.
(c) Further simplify the lumped parameter model by assuming that $d < g \ll w, R$ to find an approximate expression for the torque of electrical origin in terms of voltage that is approximately applicable over the range $0 < \theta < \pi$.

3.13 For the capacitive network shown in Fig. 3.25b, adapt the two-port circuit analysis methods revealed in Section 2.3.1 to determine the coefficients for the capacitance matrix appearing in Eq. (3.64).

3.14 The two-port parallel-plate transducer below has a uniform, planar, dielectric layer of thickness d and a variable air gap x. Using the standard parallel-plate capacitor approximation, find an expression for the force of electrical origin acting in the x direction on the upper electrode as a function of independent voltages, v_a and v_b. Hint: use the terminal relation from Eq. (3.64).

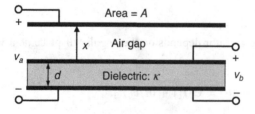

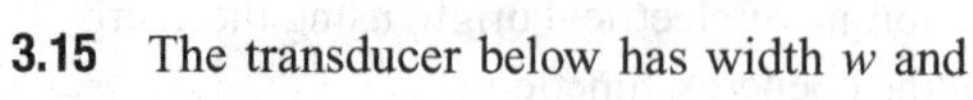

3.15 The transducer below has width w and length L. Assume that $d \ll x \ll L, w$. The inner electrode is constrained to move only in the x direction.

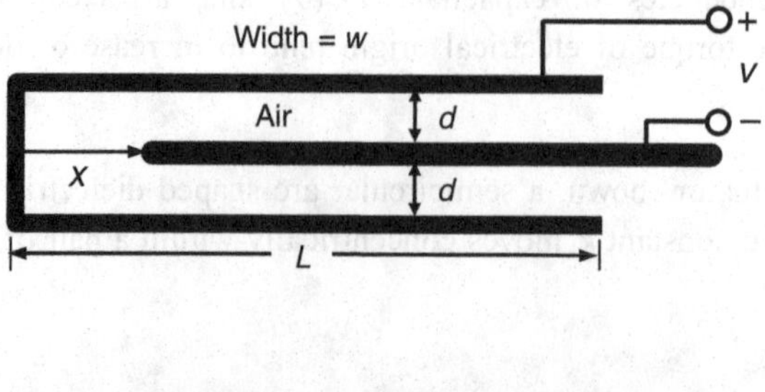

(a) Find an expression for the force of electrical origin $f^e(v, x)$ on the inner electrode.
(b) Explain the origin of the force on the inner moving electrode in terms of a Coulombic interaction with a sketch showing the interaction of electric field lines and charge.
(c) One form of Newton's third law states that if the force exerted by a body A upon another body B is F_{AB}, then the force exerted back by B on A is $-F_{AB}$. Draw a new sketch showing the Coulombic origin of the force exerted on the outer electrode by the inner electrode. This sketch should clearly reveal the opposite direction of the force.
(d) Why can the energy method be used here to determine the force of electrical origin without knowledge of the fringing field?

3.16 The lower electrode in the transducer below is free to move only in the x direction. Assume that the gap is filled with air and that $d_1 < d_2 \ll L, w$.

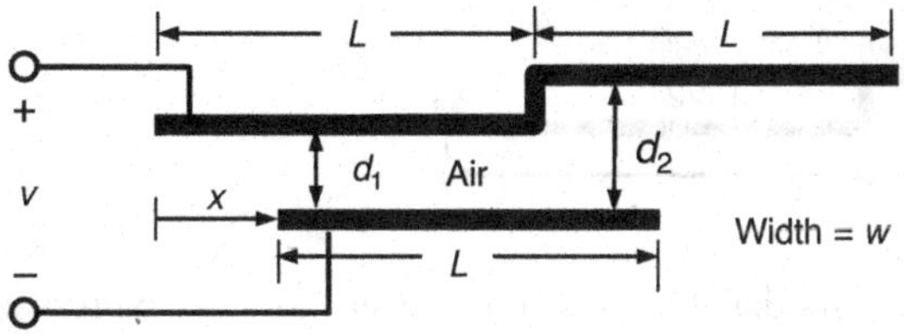

(a) Propose a capacitive model for this structure and then predict the direction of the force.
(b) Obtain an approximate expression in terms of (v, x) for the force of electrical origin acting on this electrode.
(c) Next, express this force in terms of (q, x) and explain any differences in the x dependence.
(d) Explain this force in terms of Coulombic electrostatic interactions by sketching electric field lines and charges. Your sketch should reveal the direction in which this force acts.

3.17 The parallel, outer electrodes of the device shown below are fixed and the middle electrode, also parallel, is movable. Assume that d is much smaller than the areal dimensions of the electrodes.

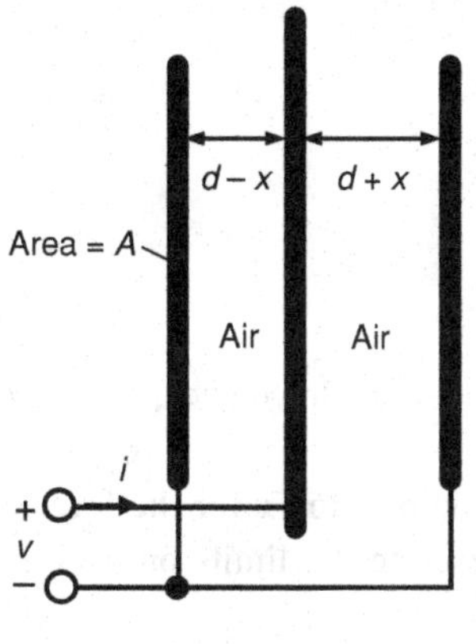

(a) Draw an equivalent circuit for this device and obtain expressions for all capacitors used in the model.
(b) Find expressions for the x-dependent force of electrical origin acting on the middle electrode for fixed voltage, $v = V_0$, and for fixed charge, $q = Q_0$.
(c) For both fixed-voltage and fixed-charge cases, draw field lines and use your sketches to explain the force expressions obtained in (b).

3.18 A dielectric slab of length L, width w, thickness d and dielectric constant κ is free to slide back and forth within the electrode structure below.

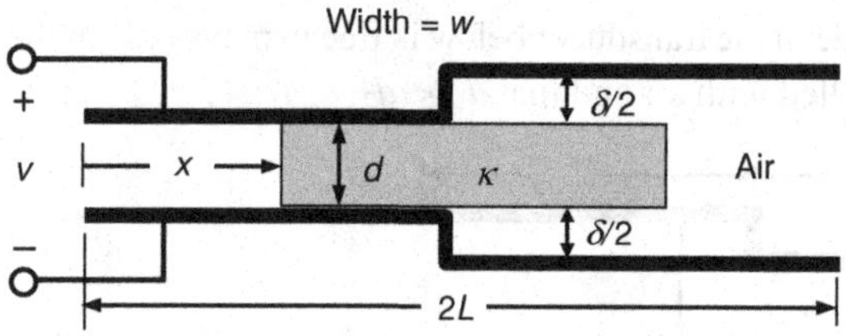

(a) Propose a capacitive model for this transducer, giving expressions for all components of the model. Use the capacitance model to predict the force direction.
(b) Use energy methods to find the force of electrical origin acting on this slab, under the assumption that d and δ are both much smaller than L and w.
(c) Explain why and under what circumstances the energy method can be used to determine the electrical force without knowing the details of the electric field non-uniformity.

3.19 A dielectric slab between two sets of parallel electrodes is free to move back and forth in the x direction under the restriction that $0 < x < L$. Assume that $d \ll g \ll L, w$.

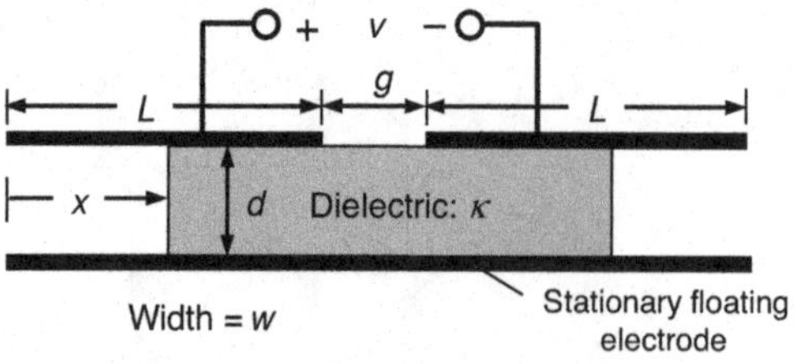

(a) Draw an approximate capacitive circuit model for this device, defining all capacitances.
(b) Find an expression in terms of voltage v and position x for the force of electrical origin acting on the dielectric slab and describe the limits on x over which it is accurate.

3.20 A capacitive transducer with capacitance $C(x)$ is excited by time-dependent voltage $v(t)$.

(a) Assuming that $x = x_0$ is a constant, find an expression for the time-dependent force of electrical origin $f^e(t)$ when $v(t) = V_0\cos(\omega t)$. What is the time-average force?
(b) Repeat part (a) for the case where $v(t)$ is a symmetric square wave of peak magnitude V_0 at the same frequency.
(c) List and discuss some relative advantages of sinusoidal and square-wave excitation of a transducer.

3.21 A dielectric slab having width w, thickness d, length L, and dielectric constant κ slides freely between parallel electrodes as shown. Assume that $d \ll g \ll L$.

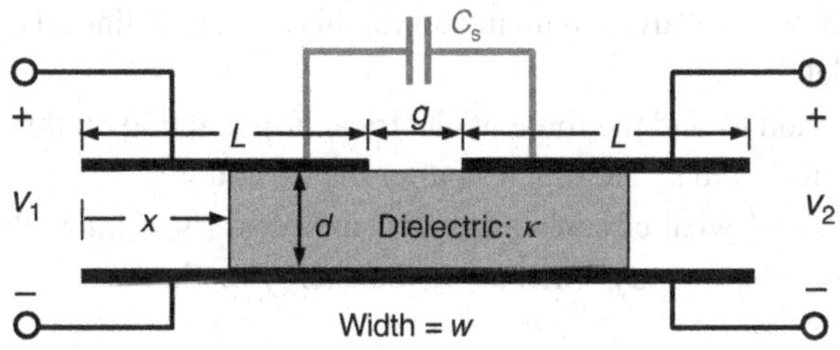

(a) Draw an equivalent capacitive circuit for this transducer, defining all capacitances in the model.
(b) Find an expression in terms of v_1 and v_2 for the net force of electrical origin acting on the dielectric.
(c) What effect on the force, if any, has C_s, the stray capacitance between the two upper electrodes, assuming that it does not depend on x?

***3.22** A position sensor consists of two coplanar, triangular electrodes and a smaller rectangular electrode mounted parallel above it at spacing d. Assume that $d \ll g \ll b, L$. An AC signal, $v_s(t) = V_0\cos(\omega t)$, is applied to the coplanar electrodes as shown and the voltage induced in the moving electrode, $v_{out}(t)$, is sensed. Neglecting fringing fields and assuming that the movable electrode always fully overlaps both triangular electrodes, what is the relation between the sensing voltage $v_{out}(t)$ and position x?

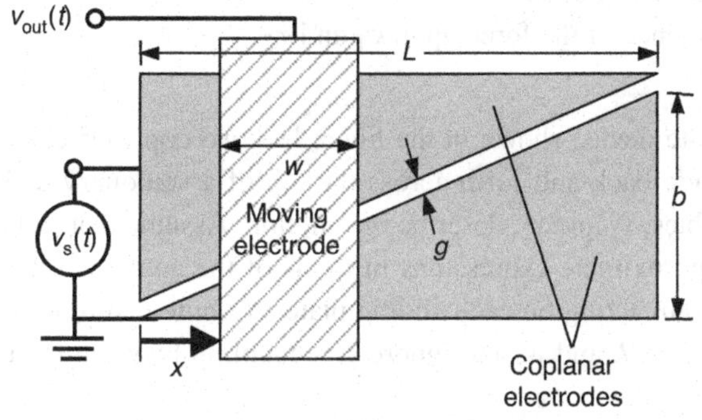

3.23 A laminated dielectric slab having width w, length L, thickness $d_1 + d_2$, and dielectric constant κ is free to slide back and forth within two parallel electrodes.

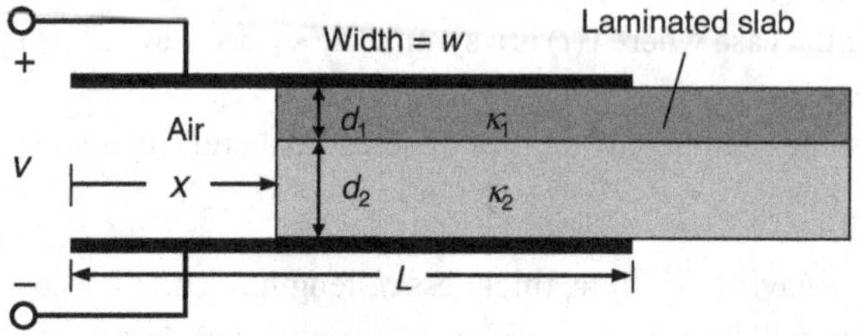

(a) Draw an approximate capacitive circuit model for this device. Define all capacitances used in your model.
(b) Use an energy method to find the force of electrical origin acting on this slab, under the assumption that d_1 and d_2 are much smaller than L and w.
(c) Explain why and under what circumstances it is unnecessary to know the details of the electric field non-uniformity when using the energy method.

3.24 A dielectric slab of width w, thickness d, and overall length $L_1 + L_2$ consists of two sections with dielectric constants κ_1 and κ_2. The slab is free to slide back and forth in between parallel electrodes under the restriction that $0 < x < L_2$.

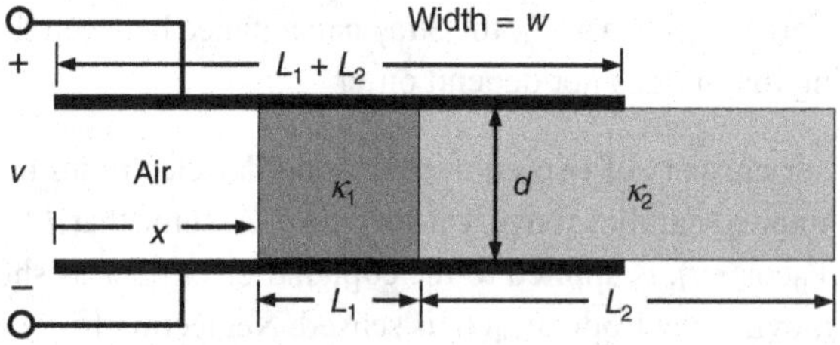

(a) Find the force of electrical origin acting on this slab, under the assumptions that $d \ll L_1, L_2$, and also that $0 < x < L_1$.
(b) Explain the dependence of the force upon κ_1 and κ_2.

3.25 The parallel-plate device shown in the figure has two coplanar electrodes, each of which is free to slide back and forth across the top of a stationary dielectric slab. Two mechanical variables, x_1 and x_2, describe this motion. Assume that $|x_1|$ and $|x_2|$ are less than $L/2$. Find approximate expressions in terms of the applied voltage v for the forces of electrical origin acting on each sliding plate. Assume that d is much smaller than the areal dimensions, L and w, and ignore any coupling between the two moving plates.

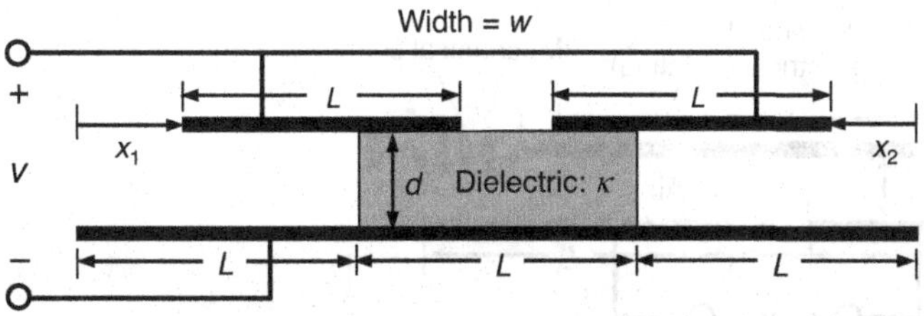

3.26 One can model microfluidic devices based on the so-called electrowetting effect using lumped parameter electromechanics. In the figure below, a moving solid conductor represents the liquid and the parallel electrodes are thinly coated with a dielectric. The perfectly conducting slab of thickness D, width w, and length L slides in the x direction between the electrodes, which are coated with a material of thickness d and dielectric constant κ. Assume that $d \ll D \ll L, w$. The $d \ll D$ condition justifies a simplified expression for capacitance. Obtain expressions for the force of electrical origin acting on the slab in terms of voltage v and in terms of electric charge q.

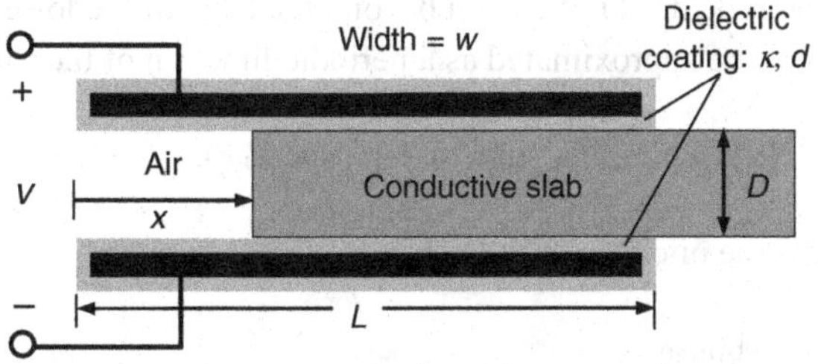

3.27 The structure below, having width w, is used to position the grounded, movable plate located between them. Assume that a linear restoring spring (not shown) exerts an x-directed force $f_x = -kx$ on the plate for both positive and negative excursions from $x = 0$. Neglect fringing to obtain an expression for x as a function of the control voltages v_{left} and v_{right}.

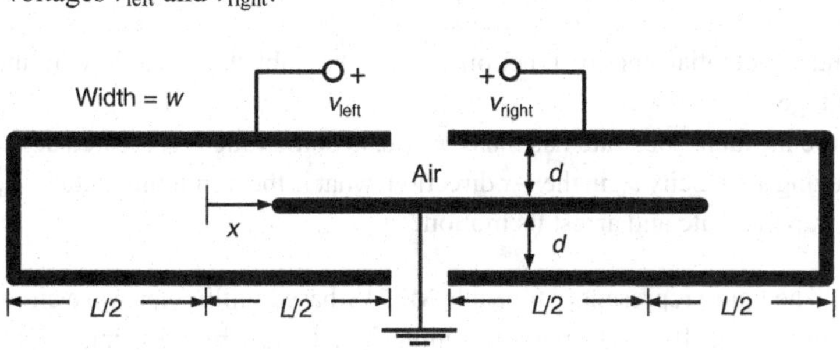

***3.28** For the simple capacitive transducer device shown, $d \ll g \ll L, w$. The upper electrode, which moves back and forth in the x direction, is electrically floating, that is, it maintains zero net electric charge. It is attached to a mechanical spring of spring constant k at rest when $x = L/2$.

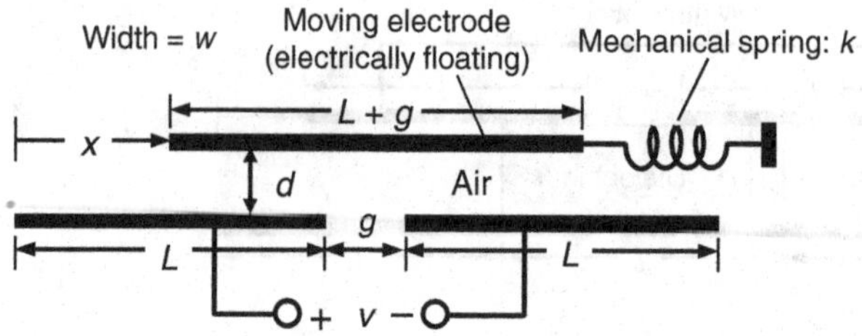

(a) Use the method of Section 3.7 to find expressions for the conserved energy $U(x)$ for fixed voltage ($v = V_0$), and for fixed charge ($q = Q_0$) conditions.
(b) Plot these two energy functions versus x and provide arguments to explain why the system is stable for both constraints.

3.29 Two planar arrays of parallel electrode strips, embedded in parallel dielectric substrates, are positioned at fixed distance. A voltage V_0 is maintained between them. The upper array is free to move in the $\pm x$ direction, that is, perpendicular to the strips. As long as the upper electrode does not extend beyond the edge of the lower electrode, the system capacitance can be approximated as a periodic function of the form:

$$C(x) \approx C_0 + C_1 \sin(kx) + C_3 \sin(3kx),$$

where $k = 2\pi/w$ and w is the pitch.

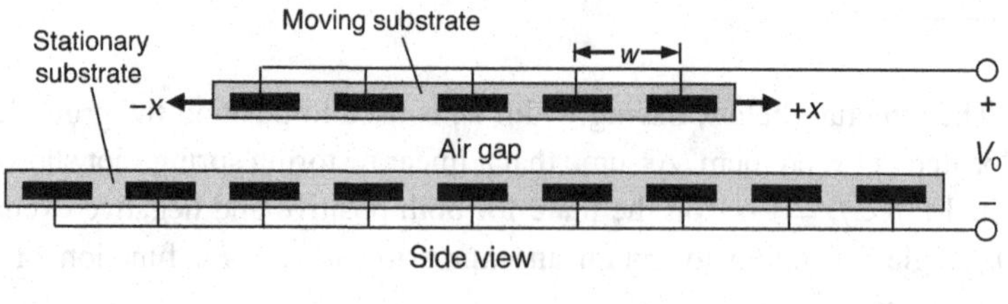

(a) Find a potential energy function useful in establishing stable equilibria at fixed voltage.
(b) If the movable substrate has mass m and, at some time t, is located at $x = w/4$ and moving at velocity u_0 in the $+x$ direction, what is the minimum voltage V_{min} required to trap the plate and arrest its motion?

*__3.30__ The figure represents a tuneable MEMS-based millimeter-wave circular antenna (operating in the 10^2 GHz range). Tuning is achieved by applying a DC voltage to change the air gap x. The electromagnetic resonant frequency of the antenna f_{res} is related to the effective dielectric constant $f_{res} \propto 1/\sqrt{\kappa_{eff}}$, where, for the simple structure shown,

$$\kappa_{eff} = (x + d)/(x + d/\kappa_d).$$

A linear spring of spring constant k maintains equilibrium at $x = x_0$ when no voltage is applied. When voltage is applied, the equilibrium changes by only a small amount, that is, $x = x_0 + \delta x(v)$, where $|\delta x| \ll x_0$.

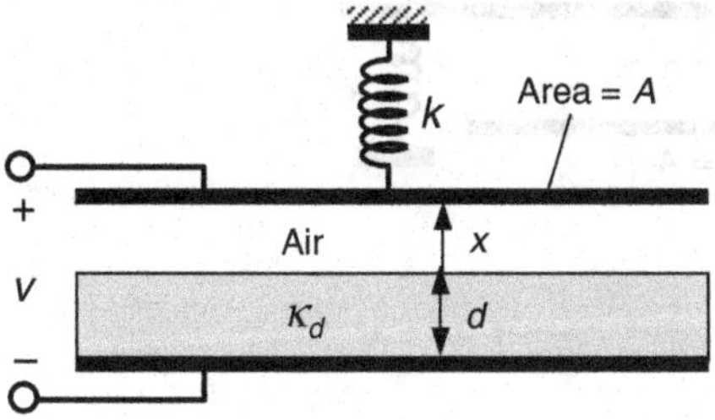

(a) Obtain an approximate expression for the antenna resonant frequency shift, $(\delta f/f)_{\text{res}}$, in terms of applied voltage v.
(b) The operation of this variable-gap MEMS actuator is limited by pull-in. Derive the pull-in voltage and use it to obtain an expression for the maximum attainable frequency shift.

***3.31** A variable air gap capacitive transducer of area A has a linear spring attached to the upper, movable plate that exerts a restoring force $f_k = -k(x - x_0)$. The electrical terminals are connected in series to a DC voltage V_0 and a controlled source v_f. In turn, v_f is controlled by a feedback system that senses plate separation x and then produces voltage $v_f = \alpha(x - x_f)$, where $x_0 > x_f$. Write the equation of static equilibrium for the movable upper plate in terms of V_0 and other system parameters. Do not attempt to solve for the value of x.

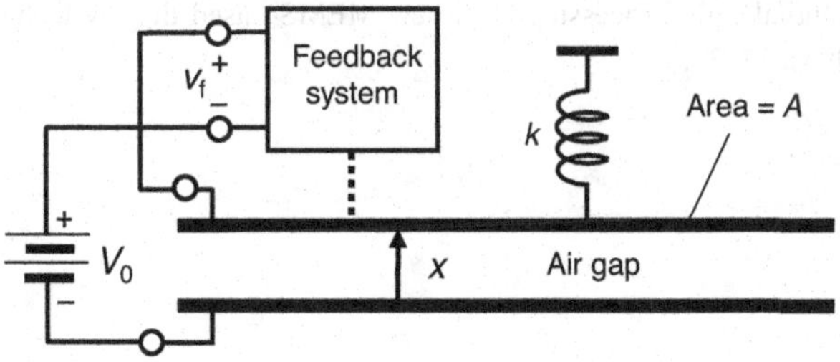

***3.32** A movable plane electrode is positioned between two parallel electrodes, both of area A. A linear restoring spring force, $f_{\text{spring}} = -kx$, establishes equilibrium at $x = 0$. The movable electrode experiences an external force f_{ext} in addition to the force of electrical origin. Assume that the gap d is much smaller than the areal dimensions of the sensor. This device employs *force feedback* to detect and measure f_{ext} while maintaining the middle plate at $x = 0$. This scheme uses a correction voltage Kf_{ext} applied to the upper electrode as shown. Using the approximation that $|Kf_{\text{ext}}| \ll |V_0|$, find an expression for the force feedback coefficient K in terms of the system parameters such that $x = 0$ for all f_{ext}.

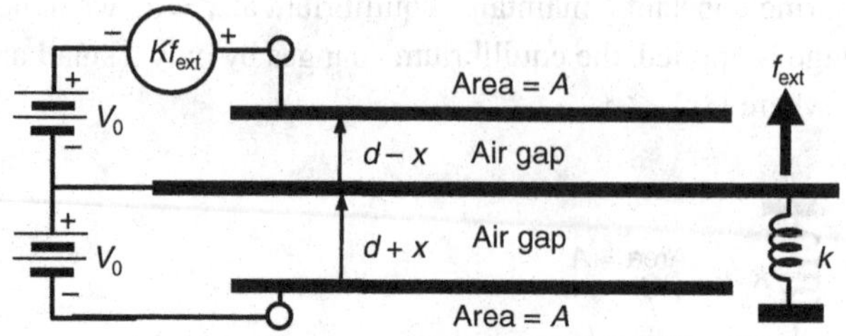

References

1. L. D. Landau and E. M. Lifshitz, *Electrodynamics of Continuous Media* (Oxford: Pergamon Press, 1960), §16.
2. H. H. Woodson and J. R. Melcher, *Electromechanical Dynamics, Part I* (Wiley: New York, 1968).
3. W. C. Tang, M. G. Lim, and R. T. Howe, Electrostatically balanced comb drive for controlled levitation, *IEEE Solid State Sensor and Actuator Workshop Technical Digest* (June, 1990), pp. 23–27.
4. K. Y. Park, C. W. Lee, Y. S. Oh, and Y. H. Cho, Laterally oscillated and force-balanced micro vibratory rate gyroscope supported by fish hook shape springs, *Proceedings of the IEEE Micro-Electro Mechanical Systems Workshop (MEMS'97)*, 1997, Japan.
5. L. S. Fan, Y. C. Tai, and R. S. Muller, IC-processed electrostatic micromotors, *Sensors and Actuators*, 20 (1989), 41–47.
6. M. Mehregany, W. H. Ko, A. S. Dewa, and C. C. Liu, *Introduction to Microelectromechanical Systems and the Multiuser MEMS Process*, Short course handbook, Case-Western Reserve University (1993).
7. L. J. Hornbeck, Digital light processing™: a new MEMS-based display technology. *IEEE Spectrum* (Nov. 1993), 27–31.

4 Small-signal capacitive electromechanical systems

4.1 Background

In this chapter, we introduce the linear *electromechanical two-port*, its multiport generalization, and equivalent circuit models for them. This MEMS representation makes it possible to invoke the so-called *cascade design* paradigm. The resulting electromechanical multiport networks and equivalent circuits broadly describe the critical behavior of microelectromechanical devices and systems. For the cascaded sensor depicted in Fig. 4.1a, the mechanical input might be a pressure perturbation acting on a diaphragm or an acceleration acting on a proof mass. In whatever form taken, this input acts on a mechanical element that possesses mass and is constrained by a restoring spring, mechanical damping, and possibly other influences. The electromechanical coupling converts this motion into an electrical signal, which is then appropriately conditioned and electronically amplified by a detector. For the actuator in Fig. 4.1b, an input electrical signal drives the electromechanical transducer, which in turn produces a force acting on the mechanical system. The system then responds with the desired motion. These schemes are typical examples of cascaded systems consisting of sequentially connected two-port networks.

The crucial assumption of the electromechanical two-port construct is that the amplitudes of the electrical and mechanical perturbations are small enough to justify linearization of the coupling mechanism. Linearization leads directly to transmission matrices similar to those introduced in Chapter 2. More importantly, the full power of *linear systems theory*, starting with AC phasors and transfer functions, becomes available. Initially, students with electrical engineering backgrounds will have the advantage because they are already adept at using these methods to design and analyze electronic amplifiers and filters. Mechanical engineering students have to be patient here and in the next chapter because, delving more deeply into MEMS, we will start to rely on the powerful analogy between electrical and mechanical systems and then, in Chapter 6, focus on the mechanics and electromechanics of elastic beams and diaphragms.

While the remainder of this chapter and the book largely restricts itself to MEMS devices operating in the linear regime, there are important examples of devices and systems that are non-linear. Because such systems are not amenable to systematic analytical methods, numerical techniques, such as the energy conservation approach

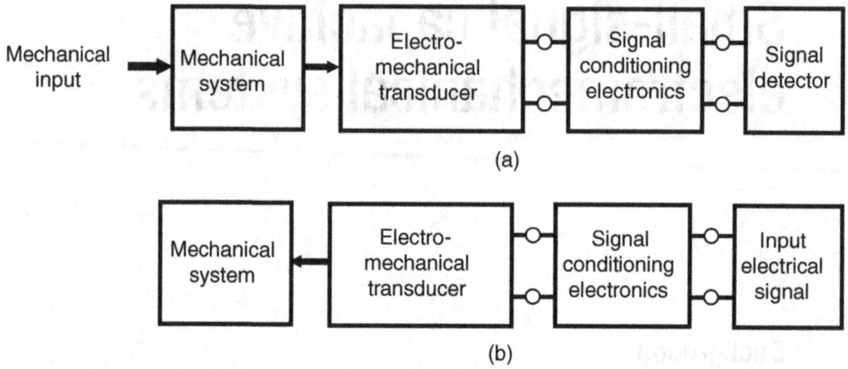

Figure 4.1 Generic cascade block diagrams of MEMS devices. (a) A sensor with mechanical force input acting on a proof mass. The resulting motion generates a signal, which is conditioned and amplified electronically. Most sensor schemes are far more complex than this simple example. (b) An actuator with amplified electrical signal serving as input. The electrical power is converted to a mechanical force to achieve the desired force or motion. Many actuator drives use feedback, which is not depicted here.

outlined in Section 3.7, must be used. Powerful software simulation tools are also available.

One limitation to the approach taken here is that far from all integrated micro-electromechanical and electronic systems strictly fit the cascaded design paradigm. In particular, some schemes employ feedback while others feature multiple mechanical or electrical ports. In a few special cases, including the three-plate sensor analyzed in Section 4.7.5, one can reduce a system to an effective cascade by appropriate grouping of electrical variables. Yet even when such tricks cannot be employed, one can always fall back on the electromechanical circuit models introduced and relied upon throughout this and the following chapters.

4.2 Linearized electromechanical transducers

This section introduces a linear formulation for electromechanical couplings. The starting point is the set of general expressions for force and either charge or voltage in terms of energy functions. Separate analyses based, respectively, on energy W_e and coenergy W_e' lead to distinct but equivalent electromechanical two-port network representations. The choice of which one to use in a specific application depends on the external electrical constraints.

4.2.1 Some preliminaries

Before introducing the linearization procedure, it is convenient make some changes in the mechanical variable notation. Doing so enforces consistency with previous texts, principally Neubert [1] and facilitates later application of the cascaded system methods

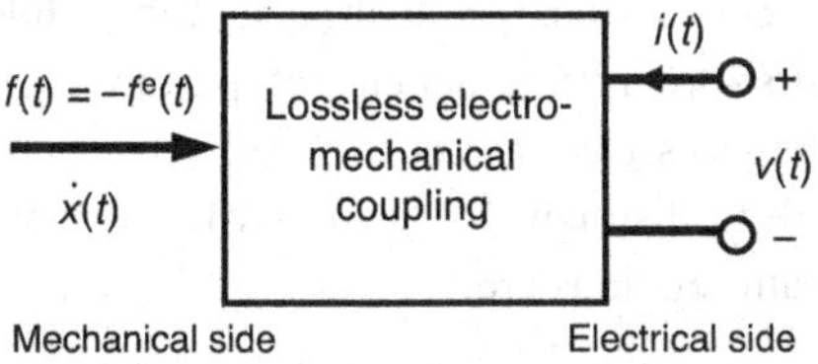

Figure 4.2 Representation for the basic lossless, electromechanical coupling with identification of current $i(t)$ and velocity $\dot{x}(t)$ as convenient terminal variables. Also note the choice of $f(t) = -f^{\mathrm{e}}(t)$ as the force variable.

introduced in Chapter 2. The changes are revealed by a comparison of Figs. 4.2 and 3.1. The mechanical and electrical terminals have switched sides and a new sign convention for the electrical force has been introduced, namely,

$$f(t) = -\,f^{\mathrm{e}}(t). \tag{4.1}$$

Further, current $i(t) = \mathrm{d}q/\mathrm{d}t$ is used in place of electrostatic charge and the velocity $\dot{x}(t) = \mathrm{d}x/\mathrm{d}t$ replaces displacement $x(t)$. These two changes are very convenient and, having no reason now to fall back on the state variables employed for force calculations, we are free to select variables directly related to instantaneous mechanical and electrical power, respectively, $p_{\mathrm{mech}}(t) = f(t)\dot{x}(t)$ and $p_{\mathrm{elec}}(t) = v(t)i(t)$.

4.2.2 Linearization in terms of energy and coenergy

The relations between the energy functions, W_{e} and W'_{e}, and their associated state variables serve as starting points for the linearization. To avoid confusion, the energy and coenergy formulations are treated separately.

4.2.2.1 Energy formulation

The energy function W_{e} uses charge q and displacement x as the independent variables, so the force of electrical origin f and the voltage v are the dependent variables. Using Eq. (4.1) in Eq. (3.4):

$$f = -f^{\mathrm{e}}(x,q) = \left.\frac{\partial W_{\mathrm{e}}}{\partial x}\right|_q \quad \text{and} \quad v(x,q) = \left.\frac{\partial W_{\mathrm{e}}}{\partial q}\right|_x. \tag{4.2}$$

As in Woodson and Melcher [2] and Tilmans [3], the linearization proceeds from the following set of assumed solution forms:

$$f(t) = f_0 + f'(t), \quad \text{where } |f'(t)/f_0| \ll 1, \tag{4.3a}$$
$$x(t) = x_0 + x'(t), \quad \text{where } |x'(t)/x_0| \ll 1, \tag{4.3b}$$
$$v(t) = v_0 + v'(t), \quad \text{where } |v'(t)/v_0| \ll 1, \tag{4.3c}$$
$$q(t) = q_0 + q'(t), \quad \text{where } |q'(t)/q_0| \ll 1. \tag{4.3d}$$

The variables subscripted with "0" are equilibrium (constant) quantities, while all primed variables, such as f' and x', denote first-order (time-dependent) perturbations. The next task is to perform first-order Taylor series expansions of f and v using the variables defined by Eqs. (4.3a–d), taking care to distinguish the first-order variables from the equilibrium quantities. The equilibrium relations are

$$f_0 = \left.\frac{\partial W_\mathrm{e}}{\partial x}\right|_{x_0, q_0} \quad \text{and} \quad v_0 = \left.\frac{\partial W_\mathrm{e}}{\partial q}\right|_{x_0, q_0}. \tag{4.4}$$

Thus, the first-order relations are

$$f' = \left.\frac{\partial^2 W_\mathrm{e}}{\partial x^2}\right|_{x_0, q_0} x' + \left.\frac{\partial^2 W_\mathrm{e}}{\partial x \partial q}\right|_{x_0, q_0} q', \tag{4.5a}$$

$$v' = \left.\frac{\partial^2 W_\mathrm{e}}{\partial q \partial x}\right|_{x_0, q_0} x' + \left.\frac{\partial^2 W_\mathrm{e}}{\partial q^2}\right|_{x_0, q_0} q'. \tag{4.5b}$$

It is important to appreciate the notation appearing in these expressions. The partial derivatives $\frac{\partial}{\partial x}$ and $\frac{\partial}{\partial q}$ are performed holding q and x constant, respectively. The subscript "x_0, q_0" is a reminder to use the equilibrium quantities for position and charge when evaluating these partial derivatives.

4.2.2.2 Coenergy formulation

The coenergy formulation starts with Eqs. (3.16), and also uses the newly defined force variable, $f(t) = -f^\mathrm{e}(t)$:

$$f = -f^\mathrm{e}(x, v) = -\left.\frac{\partial W_\mathrm{e}'}{\partial x}\right|_v \quad \text{and} \quad q(x, v) = \left.\frac{\partial W_\mathrm{e}'}{\partial v}\right|_x. \tag{4.6}$$

Here, the roles of voltage and charge are reversed:

$$f_0 = -\left.\frac{\partial W_\mathrm{e}'}{\partial x}\right|_{x_0, v_0} \quad \text{and} \quad q_0 = \left.\frac{\partial W_\mathrm{e}'}{\partial v}\right|_{x_0, v_0}. \tag{4.7}$$

Employing the linearizing variables of Eqs. (4.3a, b, c, d) and performing a Taylor series expansion yields

$$f' = -\left.\frac{\partial^2 W_\mathrm{e}'}{\partial x^2}\right|_{x_0, v_0} x' - \left.\frac{\partial^2 W_\mathrm{e}'}{\partial x \partial v}\right|_{x_0, v_0} v', \tag{4.8a}$$

$$q' = \left.\frac{\partial^2 W_\mathrm{e}'}{\partial v \partial x}\right|_{x_0, v_0} x' + \left.\frac{\partial^2 W_\mathrm{e}'}{\partial v^2}\right|_{x_0, v_0} v'. \tag{4.8b}$$

As before, the notation indicates that all partial derivatives must be taken appropriately with respect to the state variables. The subscript "x_0, v_0" refers to the equilibrium values for position and voltage used when calculating the various partial derivatives.

4.3 Electromechanical two-port networks

The linear Eqs. (4.5a,b) and (4.8a,b) use displacement x' and charge q', choices that prove inconvenient for formulation of the electromechanical two-port. Velocity, $\dot{x}' = \mathrm{d}x'/\mathrm{d}t$, and current, $i' = \mathrm{d}q'/\mathrm{d}t$, the quantities used in the mechanical and electrical power expressions, are preferred. The easiest way to accommodate these time derivatives is to assume sinusoidal steady-state time dependence for all the small-signal perturbations and then represent them using AC phasors. Employing phasors is not as restrictive as may first appear. It provides immediate access to powerful frequency analysis techniques and Fourier transforms. Furthermore, one can at any point make the substitution $\mathrm{j}\omega \to \underline{s}$, where ω = radial frequency, $\mathrm{j} = \sqrt{-1}$, and $\underline{s}$ = complex frequency, thus opening the way to Laplace transforms and transient analysis.

4.3.1 The transducer matrix

Section 2.1.5 introduced phasor representations for voltage and current. Nothing prevents us from extending their application to the small-signal mechanical variables. Thus, let

$$f'(t) = \mathrm{Re}[\underline{f}\mathrm{e}^{\mathrm{j}\omega t}], \tag{4.9a}$$

$$x'(t) = \mathrm{Re}[\underline{x}\mathrm{e}^{\mathrm{j}\omega t}], \tag{4.9b}$$

$$v'(t) = \mathrm{Re}[\underline{v}\mathrm{e}^{\mathrm{j}\omega t}], \tag{4.9c}$$

$$q'(t) = \mathrm{Re}[\underline{q}\mathrm{e}^{\mathrm{j}\omega t}], \tag{4.9d}$$

where ω is the radian frequency in rad/s and $\underline{f}$, $\underline{x}$, $\underline{v}$, and $\underline{i}$ are phasor quantities carrying magnitude and phase information. Making the substitutions $i = \mathrm{j}\omega\underline{q}$ for current and $\mathrm{j}\omega\underline{x}$ for velocity, Eqs. (4.5a,b) and (4.8a,b) can be recast as matrices in the frequency domain, one each for the energy and coenergy-based formulations.

4.3.1.1 Energy-based M-form matrix

$$\begin{bmatrix} \underline{f} \\ \underline{v} \end{bmatrix} = \underbrace{\begin{bmatrix} \underline{Z}_{\mathrm{m}} & \underline{M} \\ -\underline{M}^{*} & \underline{Z}_{\mathrm{e}} \end{bmatrix}}_{\substack{\text{M-form} \\ \text{transducer matrix}}} \begin{bmatrix} \mathrm{j}\omega\underline{x} \\ \underline{i} \end{bmatrix}, \tag{4.10}$$

where

$$\underline{Z}_{\mathrm{m}} = \frac{1}{\mathrm{j}\omega} \left. \frac{\partial^2 W_{\mathrm{e}}}{\partial x^2} \right|_{x_0, q_0}, \tag{4.11a}$$

$$\underline{M} = \frac{1}{\mathrm{j}\omega} \left. \frac{\partial^2 W_{\mathrm{e}}}{\partial x\, \partial q} \right|_{x_0, q_0}, \tag{4.11b}$$

$$\underline{Z}_{\mathrm{e}} = \frac{1}{\mathrm{j}\omega} \left. \frac{\partial^2 W_{\mathrm{e}}}{\partial q^2} \right|_{x_0, q_0}. \tag{4.11c}$$

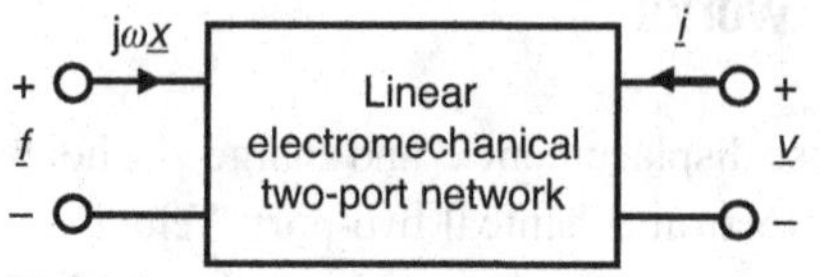

Figure 4.3 Linear electromechanical network implementation of the direct mechanical ↔ electrical analogy for the mechanical side variables. This linear two-port is to be distinguished from the electromechanical coupling of Fig. 4.2.

$\underline{Z}_\mathrm{m}$ and $\underline{Z}_\mathrm{e}$ are, respectively, purely imaginary mechanical and electrical impedances, while $\underline{M}$, also purely imaginary, is the *M-form transducer coupling coefficient.* What guarantees the reciprocal nature of this lossless, two-port network is that $\underline{M}$ is a purely imaginary quantity.

4.3.1.2 Coenergy-based N-form matrix

To use coenergy instead of energy, we start with Eqs. (4.8a, b).

$$\begin{bmatrix} \underline{f} \\ \underline{i} \end{bmatrix} = \underbrace{\begin{bmatrix} \underline{Z}'_\mathrm{m} & \underline{N} \\ -\underline{N}^* & \underline{Y}_\mathrm{e} \end{bmatrix}}_{\substack{\text{N-form} \\ \text{transducer matrix}}} \begin{bmatrix} \mathrm{j}\omega\underline{x} \\ \underline{v} \end{bmatrix}, \tag{4.12}$$

where

$$\underline{Z}'_\mathrm{m} = -\frac{1}{\mathrm{j}\omega} \frac{\partial^2 W'_\mathrm{e}}{\partial x^2}\bigg|_{x_0, v_0}, \tag{4.13a}$$

$$\underline{N} = -\frac{\partial^2 W'_\mathrm{e}}{\partial x \partial v}\bigg|_{x_0, v_0}, \tag{4.13b}$$

$$\underline{Y}_\mathrm{e} = \mathrm{j}\omega \frac{\partial^2 W'_\mathrm{e}}{\partial v^2}\bigg|_{x_0, v_0}. \tag{4.13c}$$

$\underline{Z}'_\mathrm{m}$ is another (purely imaginary) mechanical impedance, $\underline{Y}_\mathrm{e}$ is the imaginary electrical admittance, and $\underline{N}$, which is purely real, is the *N-form transducer coupling coefficient.*[1]

We can now represent the matrix expressions (4.10) and (4.12) by the electromechanical two-port shown in Fig. 4.3. This network is actually just a special case of the more general non-linear electromechanical coupling of Fig. 4.2.

The M- and N-forms, based on W_e and W'_e, respectively, are equivalent in every important respect. The choice of which one to use in a particular system is a matter of convenience usually based on the imposed electrical constraints. Straightforward manipulations of Eqs. (4.10) and (4.12) yield the following set of relationships amongst the coefficients of the two forms:

$$\underline{Z}'_\mathrm{m} = \underline{Z}_\mathrm{m} + |\underline{M}|^2/\underline{Z}_\mathrm{e}, \quad \underline{Y}_\mathrm{e} = 1/\underline{Z}_\mathrm{e}, \quad \underline{N} = \underline{M}/\underline{Z}_\mathrm{e}. \tag{4.14}$$

[1] For a more general perspective on linear coupled systems, refer to Chapter 10 of Lobontiu's text [4].

For convenience, the M- and N-form matrices and the definitions of their coefficients are summarized in Table 4.1.

The dual conditions, that $\underline{M}$ is purely imaginary and $\underline{N}$ is purely real, guarantee that the M- and N-forms are reciprocal according to Eq. (2.28).

Table 4.1 Summary of the M-form and N-form transducer matrices, circuit models, and coefficients for translational and rotational capacitive electromechanical transducers

	Translational electromechanical transducer	Rotating electromechanical transducer
M-form transducer matrix	$\begin{bmatrix} \underline{f} \\ \underline{v} \end{bmatrix} = \begin{bmatrix} \underline{Z}_{\rm m} & \underline{M} \\ -\underline{M}^* & \underline{Z}_{\rm e} \end{bmatrix} \begin{bmatrix} {\rm j}\omega \underline{x} \\ \underline{i} \end{bmatrix}$	$\begin{bmatrix} \underline{\tau} \\ \underline{v} \end{bmatrix} = \begin{bmatrix} \underline{Z}_{\rm m} & \underline{M} \\ -\underline{M}^* & \underline{Z}_{\rm e} \end{bmatrix} \begin{bmatrix} {\rm j}\omega \underline{\theta} \\ \underline{i} \end{bmatrix}$
Definitions of M-form coefficients	$\underline{Z}_{\rm m} = \frac{1}{{\rm j}\omega}\frac{\partial^2 W_{\rm e}}{\partial x^2}$, $\underline{M} = \frac{1}{{\rm j}\omega}\frac{\partial^2 W_{\rm e}}{\partial q \partial x}$ $\underline{Z}_{\rm e} = \frac{1}{{\rm j}\omega}\frac{\partial^2 W_{\rm e}}{\partial q^2} = \frac{1}{{\rm j}\omega C_0}$	$\underline{Z}_{\rm m} = \frac{1}{{\rm j}\omega}\frac{\partial^2 W_{\rm e}}{\partial \theta^2}$, $\underline{M} = \frac{1}{{\rm j}\omega}\frac{\partial^2 W_{\rm e}}{\partial q\, \partial \theta}$ $\underline{Z}_{\rm e} = \frac{1}{{\rm j}\omega}\frac{\partial^2 W_{\rm e}}{\partial q^2} = \frac{1}{{\rm j}\omega C_0}$
M-form electromechanical circuit models	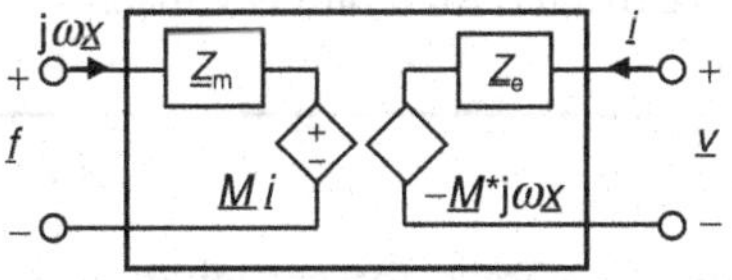	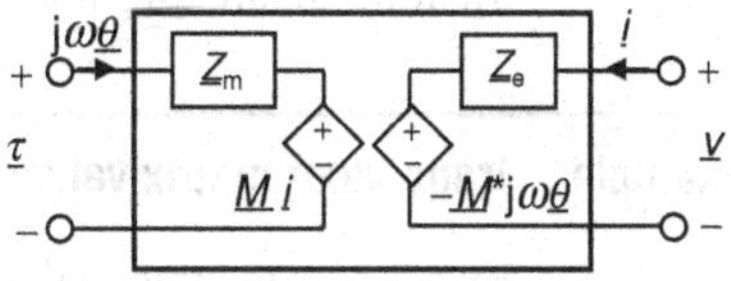
N-form transducer matrix	$\begin{bmatrix} \underline{f} \\ \underline{i} \end{bmatrix} = \begin{bmatrix} \underline{Z}'_{\rm m} & \underline{N} \\ -\underline{N}^* & \underline{Y}_{\rm e} \end{bmatrix} \begin{bmatrix} {\rm j}\omega \underline{x} \\ \underline{v} \end{bmatrix}$	$\begin{bmatrix} \underline{\tau} \\ \underline{i} \end{bmatrix} = \begin{bmatrix} \underline{Z}'_{\rm m} & \underline{N} \\ -\underline{N}^* & \underline{Y}_{\rm e} \end{bmatrix} \begin{bmatrix} {\rm j}\omega \underline{\theta} \\ \underline{v} \end{bmatrix}$
Definitions of N-form coefficients	$\underline{Z}'_{\rm m} = \frac{-1}{{\rm j}\omega}\frac{\partial^2 W'_{\rm e}}{\partial x^2}$, $\underline{N} = -\frac{\partial^2 W'_{\rm e}}{\partial v \partial x}$, $\underline{Y}_{\rm e} = {\rm j}\omega\frac{\partial^2 W'_{\rm e}}{\partial v^2} = {\rm j}\omega C_0$	$\underline{Z}'_{\rm m} = \frac{-1}{{\rm j}\omega}\frac{\partial^2 W'_{\rm e}}{\partial \theta^2}$, $\underline{N} = -\frac{\partial^2 W'_{\rm e}}{\partial v \partial \theta}$, $\underline{Y}_{\rm e} = {\rm j}\omega\frac{\partial^2 W'_{\rm e}}{\partial v^2} = {\rm j}\omega C_0$
N-form electromechanical circuit models	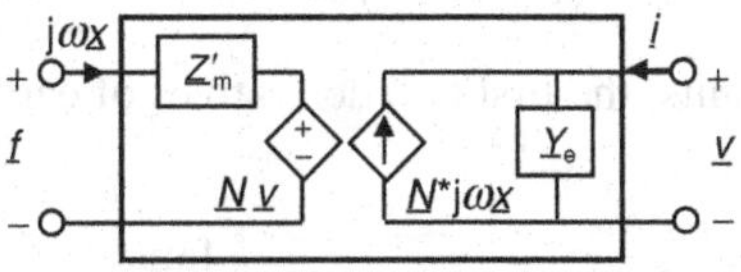	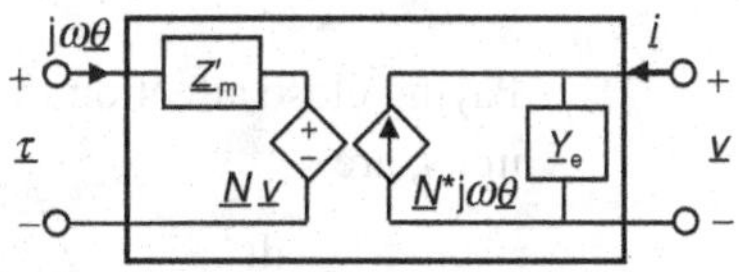
Relationships amongst M-form and N-form coefficients	$\underline{Z}'_{\rm m} = \underline{Z}_{\rm m} + \lvert\underline{M}\rvert^2/\underline{Z}_{\rm e}$, $\underline{Y}_{\rm e} = 1/\underline{Z}_{\rm e}$, $\underline{N} = \underline{M}/\underline{Z}_{\rm e}$	
Miscellany	$\underline{Z}_{\rm m}$, $\underline{Z}'_{\rm m}$, $\underline{M}$, and $\underline{Z}_{\rm e} = 1/\underline{Y}_{\rm e}$ are always purely imaginary and $\underline{N}$ is purely real because the electromechanical coupling is assumed lossless.	

4.3.2 The linear capacitive transducer

While the linear matrix relations given previously were derived under the assumption of small AC electrical and mechanical signals, there has been no assumption made about the *electrical* linearity of the capacitive transducer. Energy or coenergy expressions obtained for any capacitive device – electrically linear or otherwise – can be used in

Eqs. (4.11) or (4.13) to evaluate the matrix coefficients. Nevertheless, the most important case is the electrically linear capacitor, that is, $q = C(x)v$, and it is essential to consider it now.

4.3.2.1 M-form matrix coefficients

Using the expression for energy, $W_e = q^2/2C(x)$, in Eqs. (4.11a,b,c) yields

$$\underline{Z}_m = \frac{q_0^2}{j2\omega}\frac{d^2}{dx^2}\left[\frac{1}{C(x)}\right]_{x_0} = \frac{q_0^2}{j\omega C_0^2}\left[\frac{1}{C}\left(\frac{dC}{dx}\right)^2 - \frac{1}{2}\left(\frac{d^2C}{dx^2}\right)\right]_{x_0}, \tag{4.15a}$$

$$\underline{M} = -\frac{q_0}{j\omega}\frac{[dC/dx]_{x_0}}{C_0^2}, \tag{4.15b}$$

$$\underline{Z}_e = \frac{1}{j\omega C_0}, \tag{4.15c}$$

where $C_0 = C(x_0)$ is the equilibrium capacitance. The mechanical impedance and the transducer coefficient depend on derivatives of the capacitance function $C(x)$. The electrical impedance $\underline{Z}_e$ is just as expected for a purely capacitive element.

Example 4.1 Transducer matrix values

Expressed in practical units, picofarads for capacitance and micrometers (μm) for displacement, the capacitance versus displacement relationship for a certain comb-drive transducer is

$$C_{pF}(x) = 5[1 - 0.02\,x_{\mu m}].$$

If the equilibrium is at $x_0 = 10$ μm, then

$$C_0 = 4 \cdot 10^{-12}\ \text{F}.$$

Paying close attention to units, the first two derivatives of capacitance, expressed in SI units, are

$$\frac{dC}{dx} = 5 \cdot 10^{-12}[-0.02 \cdot 10^6] = -10^{-7}\frac{\text{farad}}{\text{meter}} \quad \text{and} \quad \frac{d^2C}{dx^2} = 0.$$

Using Eqs. (4.15a,b,c) to evaluate the M-form matrix coefficients defined by Eq. (4.10),

$$\begin{bmatrix}\underline{f}\\ \underline{v}\end{bmatrix} = \begin{bmatrix}3.91 \cdot 10^9 q_0^2/(j\omega) & 6.25 \cdot 10^{15} q_0/(j\omega)\\ 6.25 \cdot 10^{15} q_0/(j\omega) & 2.5 \cdot 10^{11}/(j\omega)\end{bmatrix}\begin{bmatrix}j\omega\underline{x}\\ \underline{i}\end{bmatrix}.$$

For the N-form matrix, one may use Eqs. (4.16a, b, c) in Eq. (4.12) to obtain

$$\begin{bmatrix}\underline{f}\\ \underline{i}\end{bmatrix} = \begin{bmatrix}0 & 10^{-7} v_0\\ -10^{-7} v_0 & j\omega 4 \cdot 10^{-12}\end{bmatrix}\begin{bmatrix}j\omega\underline{x}\\ \underline{v}\end{bmatrix}$$

Using SI units makes the numerical values in these matrices rather awkward. In practical design exercises, conversion back to more MEMS-friendly units is often advisable.

4.3.2.2 N-form matrix coefficients

Starting with Eq. (3.18) for the coenergy, that is, $W'_\mathrm{e} = C(x)v^2/2$, the coefficients for the N-form matrix are

$$\underline{Z}'_\mathrm{m} = -\frac{v_0^2}{\mathrm{j}2\omega}\left[\frac{\mathrm{d}^2C}{\mathrm{d}x^2}\right]_{x_0}, \tag{4.16a}$$

$$\underline{N} = -v_0\left[\frac{\mathrm{d}C}{\mathrm{d}x}\right]_{x_0}, \tag{4.16b}$$

$$\underline{Y}_\mathrm{e} = \mathrm{j}\omega C_0. \tag{4.16c}$$

Note once again, now in $\underline{Z}'_\mathrm{m}$ and $\underline{N}$, the critical importance of the x dependence of the capacitance.

4.3.3 Important special cases

To exemplify the M-form and N-form transducer matrices, consider now the important special cases of variable-gap and variable-area capacitive geometries. The basic variable air gap capacitor introduced in Section 3.4.1 and depicted in Fig. 3.10 has capacitance $C(x) = \varepsilon_0 A/x$. Similarly, the simple variable-area capacitor of Section 3.4.2, shown in Fig. 3.12, has a capacitance $C(x) = \varepsilon_0 w(L - x)/d$. We can use these expressions in Eqs. (4.15a,b,c) to obtain the M-form coefficients or in Eqs. (4.16a,b,c) for the N-form terms. Refer to Table 4.2.

Mechanical impedance manifests important distinctions between variable-gap and variable-area capacitive transducers. In particular, for the variable-gap capacitor,

$$\underline{Z}_\mathrm{m} = 0 \quad \text{and} \quad \underline{Z}'_\mathrm{m} \neq 0, \tag{4.17}$$

while, for the variable-area device,

$$\underline{Z}_\mathrm{m} \neq 0 \quad \text{and} \quad \underline{Z}'_\mathrm{m} = 0. \tag{4.18}$$

The electromechanical impedance is zero when the force of electrical origin is independent of x. That is why $\underline{Z}_\mathrm{m} = 0$ for the variable-gap transducer at fixed charge and $\underline{Z}'_\mathrm{m} = 0$ for the variable-area device at fixed voltage. These properties have practical performance implications, which will be examined more closely in Section 4.7 and then again in Chapter 5.

4.3.4 Transducers with angular displacement

As described in Section 3.5, some important MEMS devices, such as micromirror displays, are based on rotational rather than linear displacement. Most of these operate over a limited range of angles, meaning that linear, two-port network representations

Table 4.2 Summary of the M-form and N-form electromechanical transducer matrix coefficients for linear, capacitive devices of the variable-gap and variable-area types

	Variable-gap capacitor	Variable-area capacitor
	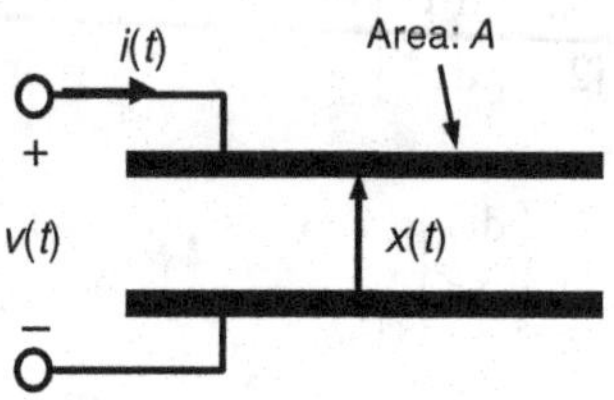	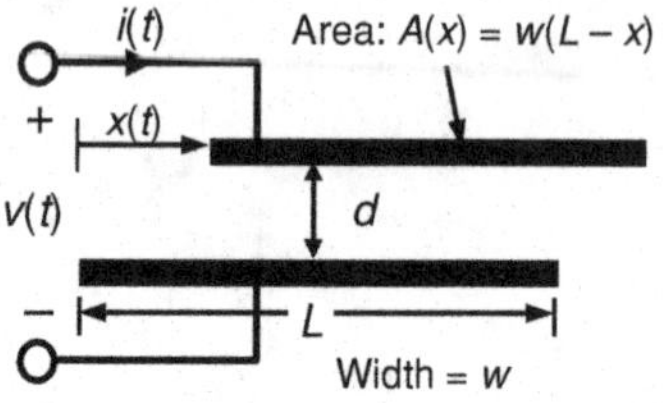
Capacitance and its derivatives	$C(x) = \varepsilon_0 A/x$ $\frac{dC}{dx} = \frac{-\varepsilon_0 A}{x^2};\ \frac{d^2C}{dx^2} = \frac{2\varepsilon_0 A}{x^3}$	$C(x) = \varepsilon_0 w(L-x)/d$ $\frac{dC}{dx} = \frac{-\varepsilon_0 w}{d};\ \frac{d^2C}{dx^2} = 0$
Definitions of equilibrium variables	$q = q_0, x = x_0, C_0 = \varepsilon_0 A/x_0,$ $E_0 = q_0/\varepsilon_0 A$	$v = v_0, x = x_0, C_0 = \varepsilon_0 w(L-x_0)/d$
M-form coefficients (based on energy: $W_e = q^2/2C(x)$)	$\underline{Z}_m = \frac{1}{j\omega}\frac{\partial^2 W_e}{\partial x^2} = \frac{1}{j\omega}\frac{\partial^2}{\partial x^2}\left[\frac{xq^2}{2\varepsilon_0 A}\right] = 0$ $\underline{M} = \frac{1}{j\omega}\frac{\partial^2 W_e}{\partial x \partial q} = \frac{1}{j\omega}\frac{\partial^2}{\partial x \partial q}\left[\frac{xq^2}{2\varepsilon_0 A}\right] = \frac{E_0}{j\omega}$ $\underline{Z}_e = \frac{1}{j\omega}\frac{\partial^2 W_e}{\partial q^2} = \frac{1}{j\omega}\frac{\partial^2}{\partial q^2}\left[\frac{xq^2}{2\varepsilon_0 A}\right] = \frac{1}{j\omega C_0}$	$\underline{Z}_m = \frac{1}{j\omega}\frac{\partial^2}{\partial x^2}\left[\frac{dq^2}{2\varepsilon_0(L-x)w}\right] = \frac{dq^2}{j\omega\varepsilon_0(L-x_0)^3 w}$ $\underline{M} = \frac{1}{j\omega}\frac{\partial^2}{\partial q \partial x}\left[\frac{dq^2}{2\varepsilon_0(L-x)w}\right] = \frac{dq}{j\omega\varepsilon_0(L-x_0)^2 w}$ $\underline{Z}_e = \frac{1}{j\omega}\frac{\partial^2}{\partial q^2}\left[\frac{dq^2}{2\varepsilon_0(L-x)w}\right] = \frac{1}{j\omega C_0}$
N-form coefficients (based on coenergy: $W'_e = C(x)v^2/2$)	$\underline{Z}'_m = \frac{-1}{j\omega}\frac{\partial^2 W'_e}{\partial x^2} = -\frac{1}{j\omega}\frac{\partial^2}{\partial x^2}\left[\frac{\varepsilon_0 A v^2}{2x}\right] = -\frac{C_0 E_0^2}{j\omega}$ $\underline{N} = -\frac{\partial^2 W'_e}{\partial v \partial x} = -C_0 E_0$ $\underline{Y}_e = j\omega\frac{\partial^2 W'_e}{\partial v^2} = j\omega C_0$	$\underline{Z}'_m = \frac{-1}{j\omega}\frac{\partial^2}{\partial x^2}\left[\frac{\varepsilon_0(L-x)wv^2}{2d}\right] = 0$ $\underline{N} = \frac{\partial^2}{\partial v \partial x}\left[\frac{\varepsilon_0(L-x)wv^2}{2d}\right] = -\frac{\varepsilon_0 w v_0}{d}$ $\underline{Y}_e = j\omega\frac{\partial^2}{\partial v^2}\left[\frac{\varepsilon_0(L-x)wv^2}{2d}\right] = j\omega C_0$

may be developed for them. The first step is to define equilibrium and perturbation variables for torque, $\tau = -\tau^e$, and displacement angle, θ:

$$\tau(t) = \tau_0 + \tau'(t), \quad \text{where } |\tau'(t)/\tau_0| \ll 1, \tag{4.19a}$$

$$\theta(t) = \theta_0 + \theta'(t), \quad \text{where } |\theta'(t)/\theta_0| \ll 1. \tag{4.19b}$$

Voltage and charge take the same form as in Eqs. (4.3c) and (4.3d). Based on the definition of the energy $W_e(\theta, q)$ from Section 3.5, the equilibrium values are

$$\tau_0 = \left.\frac{\partial W_e}{\partial \theta}\right|_{\theta_0, q_0} \quad \text{and} \quad v_0 = \left.\frac{\partial W_e}{\partial q}\right|_{\theta_0, q_0}, \tag{4.20}$$

and the linear perturbations are

$$\tau' = \left.\frac{\partial^2 W_e}{\partial \theta^2}\right|_{\theta_0, q_0} \theta' + \left.\frac{\partial^2 W_e}{\partial \theta \partial q}\right|_{\theta_0, q_0} q', \tag{4.21a}$$

$$v' = \left.\frac{\partial^2 W_e}{\partial q \partial \theta}\right|_{\theta_0, q_0} \theta' + \left.\frac{\partial^2 W_e}{\partial q^2}\right|_{\theta_0, q_0} q'. \tag{4.21b}$$

Assuming exponential time dependence, $e^{j\omega t}$, and using phasors, the linear, two-port matrix relation is

$$\begin{bmatrix} \underline{\tau} \\ \underline{v} \end{bmatrix} = \underbrace{\begin{bmatrix} \underline{Z}_m & \underline{M} \\ -\underline{M}^* & \underline{Z}_e \end{bmatrix}}_{\substack{\text{M-form} \\ \text{transducer matrix}}} \begin{bmatrix} j\omega\underline{\theta} \\ \underline{i} \end{bmatrix}, \tag{4.22}$$

where

$$\underline{Z}_m = \frac{1}{j\omega} \left.\frac{\partial^2 W_e}{\partial \theta^2}\right|_{\theta_0, q_0}, \tag{4.23a}$$

$$\underline{M} = \frac{1}{j\omega} \left.\frac{\partial^2 W_e}{\partial \theta \partial q}\right|_{\theta_0, q_0}, \tag{4.23b}$$

$$\underline{Z}_e = \frac{1}{j\omega} \left.\frac{\partial^2 W_e}{\partial q^2}\right|_{\theta_0, q_0}. \tag{4.23c}$$

By simple extension of the above derivation using coenergy $W_e'(\theta, v)$, the N-form rotational transducer matrix is

$$\begin{bmatrix} \underline{\tau} \\ \underline{i} \end{bmatrix} = \underbrace{\begin{bmatrix} \underline{Z}'_m & \underline{N} \\ -\underline{N}^* & \underline{Y}_e \end{bmatrix}}_{\substack{\text{N-form} \\ \text{transducer matrix}}} \begin{bmatrix} j\omega\underline{\theta} \\ \underline{v} \end{bmatrix}. \tag{4.24}$$

The right-hand column of Table 4.1 summarizes the definitions of $\underline{Z}'_m$, $\underline{N}$, $\underline{Y}_e$ and other important results for rotating electromechanical devices. With torque $\underline{\tau}$ and angular velocity $j\omega\underline{\theta}$ replacing force $\underline{f}$ and translational velocity $j\omega\underline{x}$ on the mechanical side, the matrix relations of Eqs. (4.22) and (4.24) are in every important respect similar to Eqs. (4.10) and (4.12).

4.3.5 Multiport electromechanical transducers

Section 3.3.1 introduced a generalization of the electromechanical coupling concept to accommodate systems with more than one electrical or mechanical port. Such systems

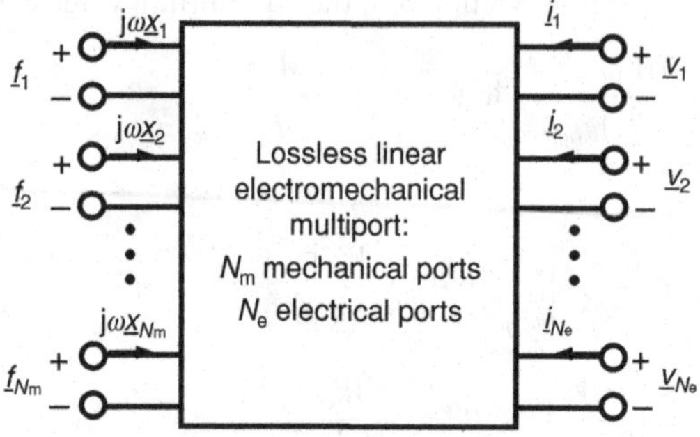

Figure 4.4 A lossless linear electromechanical coupling with N_m mechanical ports and N_e electrical ports. The general M-form and N-form matrices for this multiport port system are provided by Eqs. (4.25) and (4.27), respectively.

can be linearized to obtain multiport matrix relations. It is only necessary to keep track of the independent and dependent variables as the partial derivatives are taken to obtain the coefficients. Consider the multiport system shown in Fig. 4.4 with N_m mechanical and N_e electrical ports.

The M-form matrix is

$$\begin{bmatrix} \overline{\underline{f}} \\ \overline{\underline{v}} \end{bmatrix} = \begin{bmatrix} \overline{\overline{\underline{Z}}}_m & \overline{\overline{\underline{M}}} \\ (-\overline{\overline{\underline{M}}}^*)^t & \overline{\overline{\underline{Z}}}_e \end{bmatrix} \begin{bmatrix} j\omega\overline{\underline{x}} \\ \overline{\underline{i}} \end{bmatrix}, \tag{4.25}$$

where $\overline{\underline{f}}$, $\overline{\underline{v}}$, $\overline{\underline{i}}$, and $j\omega\overline{\underline{x}}$ are column vectors and $\overline{\overline{\underline{Z}}}_m$ and $\overline{\overline{\underline{Z}}}_e$ are square mechanical and electrical impedance matrices having dimensions N_m and N_e, respectively. $\overline{\overline{\underline{M}}}$ is an $N_e \times N_m$ rectangular matrix and $(\overline{\overline{\underline{M}}^*})^t$ is its complex conjugate transpose. The elements in these matrices are defined by various second partial derivatives of the energy function $W_e(x_1, x_2, \ldots, x_{N_m}; q_1, q_2, \ldots, q_{N_e})$:

$$(\underline{Z}_m)_{kl} = \frac{1}{j\omega}\frac{\partial^2 W_e}{\partial x_k \partial x_l}, \tag{4.26a}$$

$$\underline{M}_{kl} = \frac{1}{j\omega}\frac{\partial^2 W_e}{\partial x_k \partial q_l}, \tag{4.26b}$$

$$(\underline{Z}_e)_{kl} = \frac{1}{j\omega}\frac{\partial^2 W_e}{\partial q_k \partial q_l}. \tag{4.26c}$$

From quick reference to the above, the N-form matrix for the multiport network of Fig. 4.4 may be written down by inspection using coenergy $W_e'(x_1, x_2, \ldots, x_{N_m}; v_1, v_2, \ldots, v_{N_e})$:

$$\begin{bmatrix} \overline{\underline{f}} \\ \overline{\underline{i}} \end{bmatrix} = \begin{bmatrix} \overline{\overline{\underline{Z}}}'_m & \overline{\overline{\underline{N}}} \\ (-\overline{\overline{\underline{N}}}^*)^t & \overline{\overline{\underline{Y}}}_e \end{bmatrix} \begin{bmatrix} j\omega\overline{\underline{x}} \\ \overline{\underline{v}} \end{bmatrix}, \tag{4.27}$$

where the matrix elements are

$$(\underline{Z}'_{\mathrm{m}})_{kl} = \frac{-1}{\mathrm{j}\omega}\frac{\partial^2 W'_{\mathrm{e}}}{\partial x_k \partial x_l}, \tag{4.28a}$$

$$\underline{N}_{kl} = -\frac{\partial^2 W'_{\mathrm{e}}}{\partial x_k \partial v_l}, \tag{4.28b}$$

$$(\underline{Y}_{\mathrm{e}})_{kl} = \mathrm{j}\omega\frac{\partial^2 W'_{\mathrm{e}}}{\partial v_k \partial v_l}. \tag{4.28c}$$

Perhaps the best examples of multiport MEMS devices are the vibratory gyroscopes now available on the market. The simplest usable model for these devices is a linear electromechanical network with three electrical and two mechanical ports. Such complexity might seem daunting, yet it turns out that, with proper attention paid to the design of these gyroscopes, many of the off-diagonal terms in the matrix are negligibly small, thereby simplifying the analysis. Section 7.4 presents operational principles, design issues, and some systems considerations for vibratory gyroscopes.

4.4 Electromechanical circuit models

The electromechanical matrices in Eqs. (4.10) and (4.12) have canonical forms suggesting the linear, two-port networks reviewed in Section 2.3. The key is to think of the mechanical port on the left side of Fig. 4.3 in terms of the analogy of force and velocity, respectively, to voltage and current. For example, recall that $\underline{Z}_{\mathrm{m}} \equiv \underline{f}/(\mathrm{j}\omega\underline{x})$ is called "mechanical impedance." The analogy, covered in Appendix B, leads to a powerful analysis method for electromechanical systems.[2]

The mechanical–electrical analogy is our key to a systematic methodology for modeling MEMS sensors and actuators. Adopting it enables us to employ transmission matrices and the cascade paradigm to describe the linear behavior of many practical electromechanical systems. Because of the heavy reliance throughout this text on circuit models, the treatment of linear electromechanical two-ports adheres more closely to Tilmans [3] than to Neubert [1] from this point forward.

4.4.1 Analogous variables

In Section 4.2, we changed variables from electric charge $q(t)$ to current $i(t)$ and from mechanical displacement $x(t)$ to velocity $\dot{x}(t)$, or in phasor form, $\mathrm{j}\omega\underline{x}$. A justification offered for these changes was to facilitate direct expression of the time-average input electrical and mechanical power, respectively,

$$\langle p_{\mathrm{elec}}\rangle_{\mathrm{avg}} = \tfrac{1}{2}\mathrm{Re}[\underline{v}\,\underline{i}^*], \tag{4.29a}$$

$$\langle p_{\mathrm{mech}}\rangle_{\mathrm{avg}} = \tfrac{1}{2}\mathrm{Re}[\underline{f}(\mathrm{j}\omega\underline{x})^*]. \tag{4.29b}$$

[2] Here, the so-called *direct analogy* is invoked, such that force is analogous to voltage and velocity to current. Woodson and Melcher employ the *indirect analogy* but do not really exploit it [5].

Table 4.3 A conversion table for phasor variables and associated coefficients employed based on the direct electrical ↔ mechanical analogy represented by Fig. 4.3

Mechanical port	Units	Electrical port		Units
Force phasor: $\underline{f}$ (across variable)	N	Voltage phasor: $\underline{v}$ M-form: (across variable)	N-form: (through variable)	V
Velocity phasor: $\mathrm{j}\omega\underline{x}$ (through variable)	m/s	Current phasor: $\underline{i}$ M-form: (through variable)	N-form: (across variable)	A
Average mechanical power: $\langle p_{\mathrm{mech}}\rangle = \frac{1}{2}\mathrm{Re}[\underline{f}(\mathrm{j}\omega\underline{x})^*]$	W	Average electrical power: $\langle p_{\mathrm{elec}}\rangle = \frac{1}{2}\mathrm{Re}[\underline{v}\,\underline{i}^*]$		W
Mechanical impedance: $\underline{Z}_{\mathrm{m}}$ and $\underline{Z}'_{\mathrm{m}}$	N·s/m	Electrical impedance: $\underline{Z}_{\mathrm{e}}$		Ω
Mechanical admittance (uncommon)	m/N·s	Electrical admittance: $\underline{Y}_{\mathrm{e}}$		S
Gyrator coefficient: $-\underline{M}^*$	V·s/m	Gyrator coefficient: $\underline{M}$		N/A
Transformer coefficient: $-\underline{N}^*$	A·s/m	Transformer coefficient: $\underline{N}$		N/V
Newton's second law	—	Kirchhoff's voltage law		—
Continuity of motion	—	Kirchhoff's current law		—

Another equally compelling reason for adopting the new variables is that the analysis of general cascaded systems is simplified.

Cascaded electrical networks are readily modeled by matching voltages and currents at electrical ports. Similarly, one can employ this procedure at mechanical ports by matching forces and velocities. This method is facilitated by thinking of force and velocity in terms of their analogous electrical quantities, voltage and current, respectively. Table 4.3 identifies the analogous variables and parameters. Table B.1 of Appendix B more fully summarizes the direct analogy. The indirect analogy is less commonly used.

It is important to become comfortable with the idea of treating the mechanical force $\underline{f}$ and the velocity $\mathrm{j}\omega\underline{x}$ as analogous to voltage and current, respectively. The ultimate utility of this approach, to be exploited heavily in Section 4.6, is that it allows us to incorporate the external mechanical and electrical constraints into self-contained models for linear MEMS sensors and actuators.

4.4.2 M-form equivalent electromechanical circuit

It is easy to synthesize an equivalent circuit for an electromechanical two-port by starting with the pair of linear relations extracted from Eq. (4.10),

$$\underline{f} = \underline{Z}_{\mathrm{m}}\mathrm{j}\omega\underline{x} + \underline{M}\,\underline{i}, \tag{4.30a}$$

$$\underline{v} = -\underline{M}^*\mathrm{j}\omega\underline{x} + \underline{Z}_{\mathrm{e}}\underline{i}. \tag{4.30b}$$

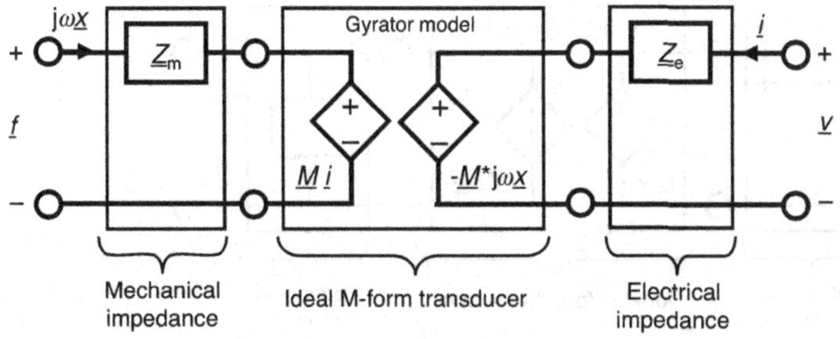

Figure 4.5 Realization of the M-form transducer equations, Eqs. (4.30a,b), as a simple electromechanical two-port using a gyrator embedded between series mechanical and electrical impedances. This circuit, while convenient, is not unique.

Treating force as a voltage-like variable and velocity as a current-like variable, the circuit shown in Fig. 4.5 results.

The left-hand and right-hand sides of this network are circuit representations of Eqs. (4.30a) and (4.30b), respectively. The dependent sources, one mechanical and one electrical, both proportional to $\underline{M}$, account for conversion of energy between mechanical and electrical forms. The electrical impedance is due to the equilibrium capacitance, $\underline{Z}_e = 1/(j\omega C_0)$, and the mechanical impedance, which we will later examine closely, derives from the force of electrical origin. Because the mechanical impedance $\underline{Z}_m$ is series-connected, one can identify velocity $j\omega\underline{x}$ as a *through variable*, analogous to the current $\underline{i}$ on the electrical side. The force $\underline{f}$ is called an *across variable*, analogous to voltage $\underline{v}$ on the electrical side.

Even in the form shown in Fig. 4.5, the equivalent circuit is invaluable for transducer modeling, but there is more that can be done with it. For example, we can partition the network into a cascade of three two-ports by simple inspection, then identify the transmission matrices for these three networks from Table 2.3, and finally use them to express $\underline{f}$ and $j\omega\underline{x}$ in terms of $\underline{v}$ and $\underline{i}$:

$$\begin{bmatrix} \underline{f} \\ j\omega\underline{x} \end{bmatrix} = \underbrace{\begin{bmatrix} 1 & \underline{Z}_m \\ 0 & 1 \end{bmatrix}}_{\substack{\text{series}\\ \text{mechanical}\\ \text{impedance}}} \underbrace{\begin{bmatrix} 0 & \underline{M} \\ -1/\underline{M}^* & 0 \end{bmatrix}}_{\substack{\text{ideal M-form}\\ \text{transducer}\\ \text{(gyrator)}}} \underbrace{\begin{bmatrix} 1 & -\underline{Z}_e \\ 0 & 1 \end{bmatrix}}_{\substack{\text{series}\\ \text{electrical}\\ \text{impedance}}} \begin{bmatrix} \underline{v} \\ \underline{i} \end{bmatrix}. \tag{4.31}$$

This partition allows us to incorporate the external mechanical and electrical constraints into an integrated transducer model. Performing the multiplication operations on the matrices of Eq. (4.31) yields a general form for the transmission matrix:

$$\begin{bmatrix} \underline{f} \\ j\omega\underline{x} \end{bmatrix} = \frac{-1}{\underline{M}^*} \begin{bmatrix} \underline{Z}_m & -(|\underline{M}|^2 + \underline{Z}_m\underline{Z}_e) \\ 1 & -\underline{Z}_e \end{bmatrix} \begin{bmatrix} \underline{v} \\ \underline{i} \end{bmatrix}. \tag{4.32}$$

Once it is recognized that this formation uses $+\underline{i}$ instead of $-\underline{i}$, it becomes obvious that this matrix form is equivalent to Eq. (2.43) for the cascaded electrical network consisting of a gyrator embedded between two series impedances.

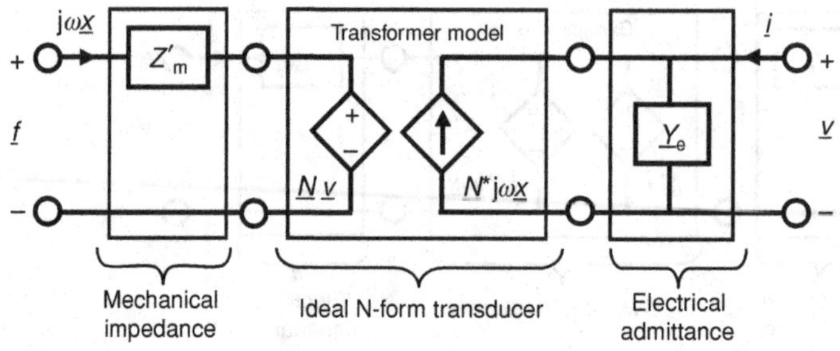

Figure 4.6 Realization of the N-form transducer equations, Eqs. (4.33a,b), as a simple electromechanical two-port employing a transformer embedded between a series mechanical impedance and a shunt-connected electrical admittance. This realization is not unique.

4.4.3 N-form equivalent electromechanical circuit

One may also derive an N-form equivalent electromechanical two-port network. From Eq. (4.12),

$$\underline{f} = \underline{Z}'_m \mathrm{j}\omega\underline{x} + \underline{N}\,\underline{v}, \tag{4.33a}$$

$$\underline{i} = -\underline{N}^* \mathrm{j}\omega\underline{x} + \underline{Y}_e \underline{v}. \tag{4.33b}$$

Figure 4.6 reveals a convenient circuit realization for these equations. The mechanical side features impedance $\underline{Z}'_m$ in series with a controlled source $\underline{N}\,\underline{v}$, while the electrical side consists of shunt-connected (capacitive) admittance $\underline{Y}_e = \mathrm{j}\omega C_0$ and a controlled current source $\underline{N}^*\mathrm{j}\omega\underline{x}$. On the mechanical side of the new circuit, velocity $\mathrm{j}\omega\underline{x}$ and force $\underline{f}$ remain through and across variables, respectively, but on the electrical side, voltage becomes the through variable and current the across variable. We can again express the relationships among the variables in terms of a multiplication of three matrices found in Table 2.3:

$$\begin{bmatrix} \underline{f} \\ \mathrm{j}\omega\underline{x} \end{bmatrix} = \underbrace{\begin{bmatrix} 1 & \underline{Z}'_m \\ 0 & 1 \end{bmatrix}}_{\substack{\text{series}\\ \text{mechanical}\\ \text{impedance}}} \underbrace{\begin{bmatrix} \underline{N} & 0 \\ 0 & -1/\underline{N}^* \end{bmatrix}}_{\substack{\text{ideal N-form}\\ \text{transducer}\\ \text{(transformer)}}} \underbrace{\begin{bmatrix} 1 & 0 \\ -\underline{Y}_e & 1 \end{bmatrix}}_{\substack{\text{shunt}\\ \text{electrical}\\ \text{admittance}}} \begin{bmatrix} \underline{v} \\ \underline{i} \end{bmatrix}. \tag{4.34}$$

The system transmission matrix for the transducer is the product of three 2×2 matrices, one each for the series mechanical impedance, the "ideal" transducer taking the form of a transformer, and a shunt-connected capacitive admittance. Performing the matrix multiplications of Eq. (4.34) yields the transmission matrix:

$$\begin{bmatrix} \underline{f} \\ \mathrm{j}\omega\underline{x} \end{bmatrix} = \frac{1}{\underline{N}^*} \begin{bmatrix} |\underline{N}|^2 + \underline{Z}'_m \underline{Y}_e & -\underline{Z}'_m \\ \underline{Y}_e & -1 \end{bmatrix} \begin{bmatrix} \underline{v} \\ \underline{i} \end{bmatrix}, \tag{4.35}$$

which is consistent with Eq. (2.45).

The ideal transducers embedded within the M- and N-form transducer networks of Figs. 4.5 and 4.6 have no separately identifiable physical significance. The inquisitive student pondering this point is challenged to figure out how *physically* to dismantle any capacitive transducer into "more fundamental" bits and pieces still capable of converting energy between electrical and mechanical forms. The exercise will fail because all realizable electromechanical transducers must have an energy-storage element. Ideal gyrators and transformers possess no energy storage mechanism.[3] In electrostatic transducers, capacitance serves the purpose, while for magnetic devices, the inductance plays this role.

Although the circuit representations presented above are not unique, the realizations depicted in Figs. 4.5 and 4.6 are probably the best choices. This is so because the M- and N-forms, along with their chosen sets of through and across variables, arise naturally from the linearizations. Throughout the remainder of this text, these equivalent circuits will be relied upon.

4.4.4 More about the cascade paradigm

The cascade paradigm, a highly systematic design approach, is utilized in the development of many complex engineered systems. MEMS are no exception. We emphasize cascading in this chapter because it allows us to build up complete electromechanical models for systems with external constraints. There are MEMS designs, however, where simple cascading cannot be used. Examples include (i) systems with multiple electrical or mechanical ports, (ii) systems using feedback, and (iii) designs in which multiple subsystems are interconnected at a common electrical or mechanical node. In all these cases, the electromechanical two-port (or multiport) models remain valid. It is the interconnections that present the challenge. To proceed, one reverts to the basics: either KCL and KVL, if the node is on the electrical side, or Newton's laws, if the node is on the mechanical side. The circuit models presented here are ideally suited for this purpose. New software tools for MEMS design and simulation have been created specifically to treat such complexity.

*4.5 Reconciliation with Neubert

The previous results are consistent with the formalism Neubert presents in his comprehensive 1967 monograph, *Instrument Transducers* [1]. Starting from a postulate that transducer action on both the electrical and mechanical sides is linear in velocity $j\omega\underline{x}$ and either voltage $\underline{v}$ or current $\underline{i}$, he introduces the concept of the *ideal transducer*:

$$\text{ideal M-form matrix:}\quad \begin{bmatrix} \underline{M} & 0 \\ 0 & -1/\underline{M}^* \end{bmatrix}, \tag{4.36a}$$

$$\text{ideal N-form matrix:}\quad \begin{bmatrix} \underline{N} & 0 \\ 0 & -1/\underline{N}^* \end{bmatrix}. \tag{4.36b}$$

[3] The fact that purely electrical transformers and gyrator circuits exist is no contradiction, as these devices do not convert energy between electrical and other forms.

While Neubert favors these forms because they are canonically identical, they are entirely equivalent to the elements identified in Eqs. (4.31) and (4.34). To maintain the reciprocal property, $\underline{M}$ is purely imaginary and $\underline{N}$ is purely real. This property is built in to the ideal transducer matrices of Eqs. (4.36a,b). Neubert invokes a straightforward argument based on continuity of time-average power flow to establish this reciprocal condition for transducers. He then invokes matrix manipulations to show that these M- or N-form networks, combined with the mechanical and electrical impedances, provide workable models for real electromechanical transducers. The more formal approach of the present text, similar to Tilmans [3], relies instead on direct derivation of the coefficients from energy or coenergy. Of course, the two methods are fundamentally the same because both are based on energy arguments.

To achieve the canonical similarity evident in Eqs. (4.36a) and (4.36b), Neubert merely reverses the order of $\underline{v}$ and $\underline{i}$ in the column matrix of his M-form matrix formulation. For consistency's sake, we adopt this convention throughout the remainder of the book. Thus,

$$\begin{bmatrix} \underline{f} \\ \mathrm{j}\omega\underline{x} \end{bmatrix} = \underbrace{\begin{bmatrix} 1 & \underline{Z}_{\mathrm{m}} \\ 0 & 1 \end{bmatrix}}_{\substack{\text{series}\\ \text{mechanical}\\ \text{impedance}}} \underbrace{\begin{bmatrix} \underline{M} & 0 \\ 0 & -1/\underline{M}^* \end{bmatrix}}_{\substack{\text{ideal M-form}\\ \text{transducer}\\ \text{(gyrator)}}} \underbrace{\begin{bmatrix} 1 & 0 \\ -\underline{Z}_{\mathrm{e}} & 1 \end{bmatrix}}_{\substack{\text{series}\\ \text{electrical}\\ \text{impedance}}} \begin{bmatrix} \underline{i} \\ \underline{v} \end{bmatrix}, \tag{4.37}$$

or

$$\begin{bmatrix} \underline{f} \\ \mathrm{j}\omega\underline{x} \end{bmatrix} = \frac{1}{\underline{M}^*} \begin{bmatrix} |\underline{M}|^2 + \underline{Z}_{\mathrm{m}}\underline{Z}_{\mathrm{e}} & -\underline{Z}_{\mathrm{m}} \\ \underline{Z}_{\mathrm{e}} & -1 \end{bmatrix} \begin{bmatrix} \underline{i} \\ \underline{v} \end{bmatrix}. \tag{4.38}$$

These minor notational changes have no effect on the form of the equivalent electromechanical circuits.

4.6 External constraints

The two-port electromechanical transducer networks depicted in Figs. 4.5 and 4.6 represent only the linear conservative (lossless) coupling. This fact, a direct consequence of the definition of the force of electrical origin f^{e}, is evident from inspection of Eqs. (4.15a,b,c) and (4.16a,b,c), which reveals that the M-form and N-form matrix coefficients depend only on capacitance $C(x)$ and its derivatives. They contain no information about circuit connections, system mass, spring constants, damping, or external mechanical forces. All such mechanical and electrical influences on the behavior of an electromechanical transducer must be treated as external constraints.

The mechanical constraints are accounted for using Newton's laws of motion. The electrical constraints, represented as circuit elements connected to the electrical port, are

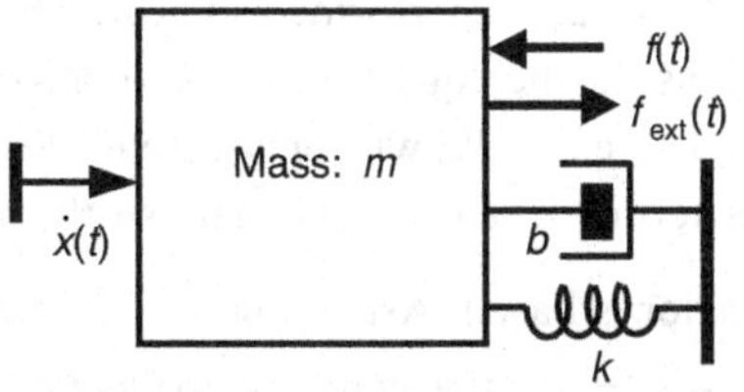

Figure 4.7 Free body diagram of a moving element of mass m in an electromechanical device with one degree of freedom x. Forces acting on the element include the electrical force f, the mechanical driving force f_{ext}, a spring force, and a damping force.

subject to Kirchhoff's laws. Recall that loss is specifically excluded from the electromechanical coupling. Thus, the constraints, mechanical and electrical, must incorporate friction, mechanical damping, and electrical resistance.

The equivalent circuit models make it possible to incorporate systematically the external mechanical and electrical constrains into a complete device model. It is at this point that the true value of the mechanical ↔ electrical analogy finally emerges. The equivalent circuits of Figs. 4.5 and 4.6 and their associated matrix partitions, as presented in Eqs. (4.31) and either (4.34) or (4.37), greatly facilitate this task.

4.6.1 Mechanical constraints

The moving element in any capacitive transducer possesses mass and responds dynamically to all forces, electrical and otherwise, acting upon it. A natural starting point for considering the mechanical constraints is Newton's second law of motion. For a typical MEMS device, there will be a linear restoring force established by a mechanical spring. These springs are integral features designed into the MEMS structure to hold the moving element in proper position and alignment and to establish an equilibrium before electrical excitation is applied. The moving electrode in a MEMS device will also experience mechanical damping due to a variety of largely unavoidable frictional mechanisms, such as the viscosity of air. For MEMS devices operated as sensors, there will often be an external force, such as acceleration or pressure. Figure 4.7 shows a free body diagram representing all the lumped parameter forces acting on the mass m.

The equation of motion for the mass m in Fig. 4.7 is

$$\begin{aligned} m\frac{\mathrm{d}^2x}{\mathrm{d}t^2} &= \sum \text{all forces} \\ &= -k(x - x_\mathrm{k}) - b\frac{\mathrm{d}x}{\mathrm{d}t} - f(t) + f_{\text{ext}}(t), \end{aligned} \tag{4.39}$$

where k is a lumped parameter spring constant, x_k is the position of the element when the spring is relaxed, b is a lumped parameter coefficient of linear damping, $f\,(= -f^\mathrm{e})$ is the force of electrical origin, and $f_{\text{ext}}(t)$ is an external force acting on the mass.

To introduce the two-port model in Section 4.2, we performed a linearization. The first step of the process was to assume solutions for the equilibrium and small-signal (first-order) variables. Refer again to Eqs. (4.3a,b,c,d). We will now extend the procedure, paying close attention to the zero-order terms because they establish the equilibrium on which the small-signal dynamic behavior is based. Assuming $\frac{\mathrm{d}^2 x}{\mathrm{d}t^2} = 0$ and $\frac{\mathrm{d}x}{\mathrm{d}t} = 0$ and defining the equilibrium position as $x = x_0$, the equation for static mechanical equilibrium is

$$0 = -k(x_0 - x_\mathrm{k}) - f_0(x_0, q_0 \text{ or } v_0) + f_{\mathrm{ext},0}, \tag{4.40}$$

where $f_{\mathrm{ext},0}$ is any time-invariant portion of the externally applied force. Equation (4.40) recognizes that the electrical force may depend on the equilibrium position x_0, and will certainly be a function of either the equilibrium electric charge q_0 or the voltage v_0. The choice between these two electrical variables depends on whether the M- or the N-form transducer matrix is to be used, respectively.[4]

Subtracting the static equilibrium described by Eq. (4.40) from the equation of motion, Eq. (4.39), exposes the small-signal, dynamic equation:

$$\frac{\mathrm{d}^2 x'}{\mathrm{d}t^2} = -kx' - b\frac{\mathrm{d}x'}{\mathrm{d}t} - f'(t) + f'_{\mathrm{ext}}(t), \tag{4.41}$$

which is linear in $x'(t)$. Assuming sinusoidal steady-state time dependence for the external force, that is, $f'_{\mathrm{ext}}(t) = \mathrm{Re}[\underline{f}_{\mathrm{ext}} \mathrm{e}^{\mathrm{j}\omega t}]$, then the displacement and the electrical signals become sinusoids represented by phasors. Accordingly, Eq. (4.41) is replaced by a frequency-dependent, linear, algebraic equation involving these phasors. Grouping the dependent variables on the left-hand side of the equation and doing some algebraic rearranging to expose the velocity $\mathrm{j}\omega\underline{x}$,

$$\underline{f}_{\mathrm{ext}} = (\mathrm{j}\omega m + b + k/\mathrm{j}\omega)\mathrm{j}\omega\underline{x} + \underline{f}. \tag{4.42}$$

Next, Eq. (4.10) or (4.12) is used to replace $\underline{f}$, with the result,

$$\text{M-form: } \underline{f}_{\mathrm{ext}} = (\mathrm{j}\omega m + b + k/\mathrm{j}\omega + \underline{Z}_\mathrm{m})\mathrm{j}\omega\underline{x} + \underline{M}\,\underline{i}, \tag{4.43a}$$

$$\text{N-form: } \underline{f}_{\mathrm{ext}} = \left(\mathrm{j}\omega m + b + k/\mathrm{j}\omega + \underline{Z}'_\mathrm{m}\right)\mathrm{j}\omega\underline{x} + \underline{N}\,\underline{v}. \tag{4.43b}$$

The form of the terms multiplying $\mathrm{j}\omega\underline{x}$ in Eqs. (4.43a,b) suggests the definition of an *external mechanical impedance*:

$$\underline{Z}_{\mathrm{m}0} = \mathrm{j}\omega m + b + k/\mathrm{j}\omega. \tag{4.44}$$

$\underline{Z}_{\mathrm{m}0}$, having the same units as $\underline{Z}_\mathrm{m}$ and $\underline{Z}'_\mathrm{m}$, accounts for the external mechanical constraints, including inertia (mass), damping, and the linear spring. Section B.1.7 of Appendix B presents some background for the analogy between $\underline{Z}_{\mathrm{m}0}$ and electrical impedance. According to the mechanical ↔ electrical analogy, we interpret $\underline{Z}_{\mathrm{m}0}\,\mathrm{j}\omega\underline{x}$ as a "voltage drop" in series with the $\underline{Z}_\mathrm{m}\mathrm{j}\omega\underline{x}$ drop of the transducer. Refer to Fig. 4.8.

$\underline{Z}_{\mathrm{m}0}$, identified without reference to the electromechanical two-port, may be used with either the M-form or the N-form transducer model. The figure suggests the recognized

[4] For a linear, capacitive element, $q_0 = C_0 v_0$, where $C_0 = C(x_0)$.

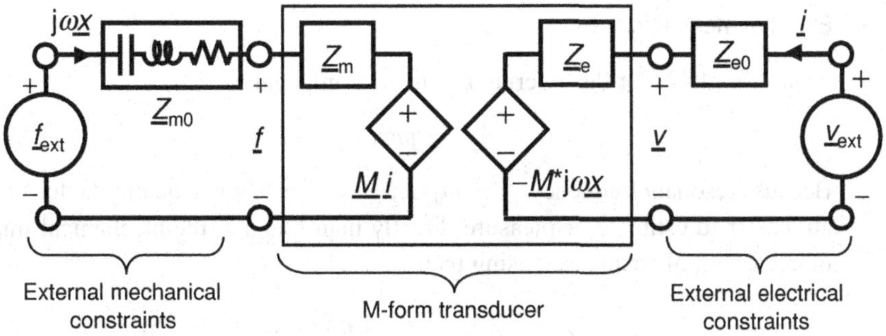

Figure 4.8 Fully constrained two-port electromechanical network: M-form. The electrical analogy is employed to represent the external mechanical constraints imposed by Newton's second law of motion on the left side of the transducer. The Thevenin equivalent circuit on the right side represents the electrical constraints. $\underline{Z}_{m0}$ and $\underline{Z}_{e0}$ are external to the electromechanical coupling.

analogies of spring force, mass, and damping to equivalent capacitance, inductance, and resistance, respectively.

A useful interpretation for $\underline{f}_{ext}$, depicted in Fig. 4.8 as a voltage source, is to consider the distinction between sensor and actuator operation. Electromechanical sensors, as conceptualized in Fig. 4.1a, provide an electrical response to an external mechanical stimulus, $\underline{f}_{ext}$. In contrast, for an actuator $\underline{f}_{ext} = 0$, so the mass m responds only to the electromechanical force.[5] In either case, the external impedance, incorporating inertia, damping, and built-in spring forces, plus possibly other mechanisms, will have a significant influence on the motion.

Example 4.2 **Typical effective parameters for a MEMS resonator**

Depending on the intended application, values for the typical effective spring constant k of the mechanical resonators in MEMS sensors range widely, at least from $\sim 10^{-3}$ to ~ 1 N/m. If the resonant frequency f_0 is known, we can quantify the effective mass. From Eq. (B.16) of Appendix B,

$$m = k/(2\pi f_0)^2.$$

Assuming $f_0 = 10$ kHz and $k = 1$ N/m, then $m = 2.53 \cdot 10^{-10}$ kg, or 0.25 micrograms. One can investigate the dimensions to see if this resonator really qualifies as a MEMS device. Assume the mass, in the shape of a square plate of side dimensions $L \times L$ and thickness $t = 2$ μm, is fabricated of polysilicon, which has mass density $\rho = 2.33$ g/cm^3. Using the mass value calculated, the dimension of the square sides is $L = 233$ μm, which is well within the normal range for surface-machined MEMS structures.

[5] With their small mass, MEMS actuators and sensors are quite susceptible to mechanical and electrical noise. Chapter 5 shows how to model noise sources as external force and voltage components superimposed on $\underline{f}_{ext}$ and $\underline{v}_{ext}$, respectively.

Example 4.3 **External mechanical impedance**

From Eq. (4.44) for the external mechanical impedance,

$$\underline{Z}_{m0} = j\omega m + b + k/(j\omega).$$

Because resonant frequency $f_0 = \omega_0/2\pi = \sqrt{k/m}/2\pi$ and quality factor $Q = \sqrt{km}/b$ are easier to estimate or measure directly than b and k, the mechanical impedance is more commonly expressed using them:

$$\underline{Z}_{m0} = j2\pi f m[1 - (f_0/f)^2 - \frac{j}{Q}(f_0/f)].$$

This convenient form highlights the commonalities of mechanical (mass–spring) and electrical (inductor–capacitor) resonators. Q is defined in Section B.2.4 of Appendix B and Section B.4 explains how it can be measured.

4.6.2 Electrical constraints

The electrical constraints imposed on an electromechanical transducer can be represented using familiar circuit elements. Recall that the assumed forms for the voltage and charge are $v(t) = v_0 + v'(t)$ and $q(t) = q_0 + q'(t)$, with $|v'(t)| \ll |v_0|$ and $|q'(t)| \ll |q_0|$. Further, remember that $i'(t) = dq'/dt$. For capacitive MEMS devices, electrical equilibrium is often established using DC voltage, so the electrical equilibrium is

$$q_0 = C_0 v_0, \tag{4.45}$$

where $C_0 = C(x_0)$ is the capacitance at the mechanical equilibrium defined by Eq. (4.40) and v_0 is the DC bias voltage applied at the electrical port.

Incorporating small-signal, external electrical constraints is straightforward. Any circuit composed of linear components and sinusoidal sources can be represented by a Thevenin equivalent, as shown on the right-hand (electrical) side of the two-port in Fig. 4.8 [6]. The Thevenin equivalent, consisting of a series impedance $\underline{Z}_{e0}$ and a voltage source $\underline{v}_{ext}$, is general enough to accommodate any linear, external circuit.

An alternate to the Thevenin circuit of Fig. 4.8 is the Norton equivalent shown on the right side of Fig. 4.9. The two are uniquely related by

$$\underline{i}_{ext} = \frac{\underline{v}_{ext}}{\underline{Z}_{e0}} \quad \text{and} \quad \underline{Y}_{e0} = \frac{1}{\underline{Z}_{e0}}. \tag{4.46}$$

It should be clear that the Thevenin equivalent on the right-hand side of Fig. 4.8 is the best match for the M-form of the electromechanical two-port circuit, while the Norton equivalent of Fig. 4.9 is more convenient when using the N-form.

4.6.3 Fully constrained electromechanical transducers

In Fig. 4.8, the mechanical impedances, $\underline{Z}_m$ and $\underline{Z}_{m0}$, and the electrical impedances, $\underline{Z}_e$ and $\underline{Z}_{e0}$, are connected in series. For the constrained N-form network of Fig. 4.9, the

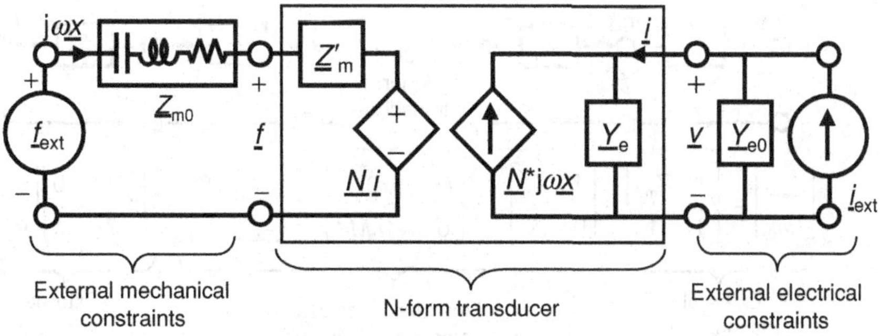

Figure 4.9 Fully constrained N-form two-port electromechanical network. The electrical analogy is employed to represent the external mechanical constraints imposed by Newton's second law of motion on the left side of the transducer. The Norton equivalent circuit on the right side represents the electrical constraints. $\underline{Z}_{m0}$ and $\underline{Y}_{e0}$ are external to the electromechanical coupling.

mechanical impedances, $\underline{Z}'_m$ and $\underline{Z}_{m0}$, are in series, while the electrical admittances, $\underline{Y}_e$ and $\underline{Y}_{e0}$, are in parallel. For convenience, Figs. 4.10a and b present fully decomposed cascade representations for these networks, along with the associated transmission matrices.

Recognizing the various series and parallel combinations of components in the electromechanical networks in Fig. 4.10 makes it easy to obtain the transmission matrices for the fully constrained transducer by simple inspection. Employing the substitutions

(i) $\underline{Z}_m \to \underline{Z}_m + \underline{Z}_{m0}$ and $\underline{Z}_e \to \underline{Z}_e + \underline{Z}_{e0}$ in Eq. (4.38) for the M-form network and
(ii) $\underline{Z}'_m \to \underline{Z}'_m + \underline{Z}_{m0}$ and $\underline{Y}_e \to \underline{Y}_e + \underline{Y}_{e0}$ in Eq. (4.35) for the N-form,

one obtains

$$\begin{bmatrix} \underline{f}_{ext} \\ j\omega\underline{x} \end{bmatrix} = \frac{1}{\underline{M}^*} \begin{bmatrix} |\underline{M}|^2 + (\underline{Z}_m + \underline{Z}_{m0})(\underline{Z}_e + \underline{Z}_{e0}) & -(\underline{Z}_m + \underline{Z}_{m0}) \\ (\underline{Z}_e + \underline{Z}_{e0}) & -1 \end{bmatrix} \begin{bmatrix} \underline{i} \\ \underline{v}_{ext} \end{bmatrix} \tag{4.47}$$

and

$$\begin{bmatrix} \underline{f}_{ext} \\ j\omega\underline{x} \end{bmatrix} = \frac{1}{\underline{N}^*} \begin{bmatrix} |\underline{N}|^2 + (\underline{Z}'_m + \underline{Z}_{m0})(\underline{Y}_e + \underline{Y}_{e0}) & -(\underline{Z}'_m + \underline{Z}_{m0}) \\ (\underline{Y}_e + \underline{Y}_{e0}) & -1 \end{bmatrix} \begin{bmatrix} \underline{v} \\ \underline{i}_{ext} \end{bmatrix}. \tag{4.48}$$

These transmission matrix relations, attributable to Neubert [1], are very general. In combination with the definitions found in Eqs. (4.11a,b,c) and (4.13a,b,c), they provide full descriptions for the linear frequency-dependent response of virtually any electromechanical sensor or actuator having one electrical and one mechanical port. Illustrative examples are presented and discussed in Section 4.7.

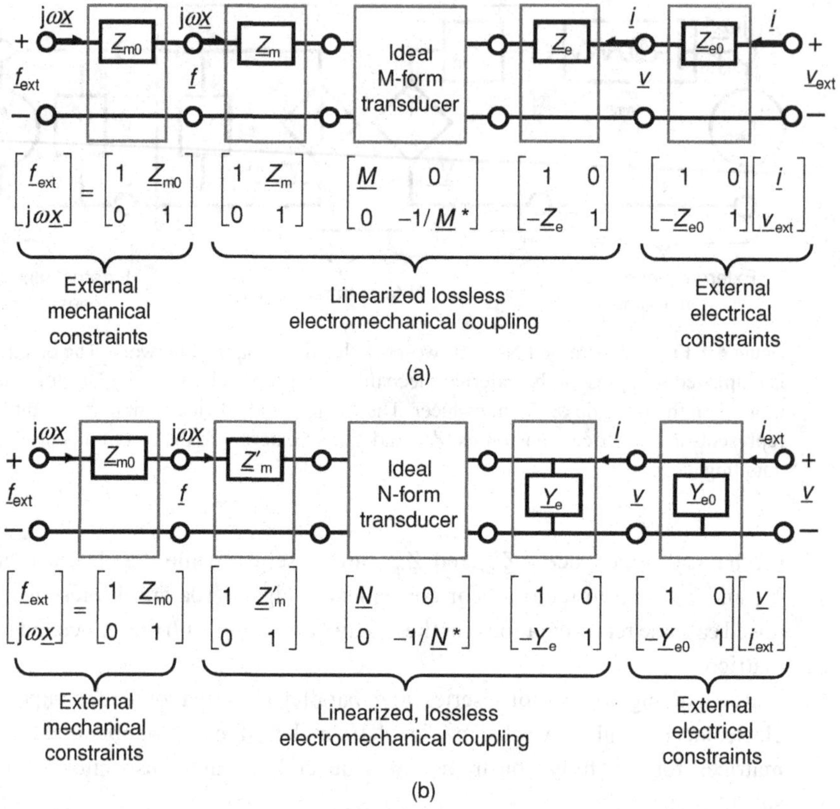

Figure 4.10 Decomposed two-port cascade models for electrically and mechanically constrained electromechanical transducers with corresponding transmission matrices. (a) M-form: ($j\omega\underline{x}$, $\underline{i}$) and ($\underline{f}_{ext}$, $\underline{v}_{ext}$) are through and across variables, respectively. (b) N-form: ($j\omega\underline{x}$, $\underline{v}$) and ($\underline{f}_{ext}$, $\underline{i}_{ext}$) are through and across variables, respectively. Adapted from Neubert [1].

4.6.4 Other useful matrix forms

Less algebraically complex matrix forms result from direct inspection of the circuits in Figs. 4.10a and b. These equations naturally express the *across* variables on the left-hand side in terms of the *through* variable on the right:

$$\text{M-form:}\quad \begin{bmatrix} \underline{f}_{ext} \\ \underline{v}_{ext} \end{bmatrix} = \begin{bmatrix} \underline{Z}_m + \underline{Z}_{m0} & \underline{M} \\ -\underline{M}^* & \underline{Z}_e + \underline{Z}_{e0} \end{bmatrix} \begin{bmatrix} j\omega\underline{x} \\ \underline{i} \end{bmatrix}, \tag{4.49}$$

$$\text{N-form:}\quad \begin{bmatrix} \underline{f}_{ext} \\ \underline{i}_{ext} \end{bmatrix} = \begin{bmatrix} \underline{Z}'_m + \underline{Z}_{m0} & \underline{N} \\ -\underline{N}^* & \underline{Y}_e + \underline{Y}_{e0} \end{bmatrix} \begin{bmatrix} j\omega\underline{x} \\ \underline{v} \end{bmatrix}. \tag{4.50}$$

In some cases, these forms are more convenient to use. For a transducer operated as a simple MEMS sensor, one typically sets $\underline{v}_{ext} = 0$, while for an actuator, $\underline{f}_{ext} = 0$ is usually appropriate. The approach outlined in Section 4.3.5 can be used to generalize to systems with more than one electrical or mechanical port.

Example 4.4 **The coupling coefficient**

Consider a sensor connected to a high impedance amplifier so that $\underline{i} = 0$. When the displacement amplitude is $|\underline{x}| = 1\ \mu\text{m}_{\text{rms}}$, the output voltage is measured to be $|\underline{v}_{\text{ext}}| = 50\ \text{mV}_{\text{rms}}$. If the operating frequency is 2 kHz, is it possible to calculate the magnitude of the coupling coefficient? From the M-form matrix of Eq. (4.10),

$$\underline{v} = -\underline{M}^{*}\text{j}\omega\underline{x} + \underline{Z}_{\text{e}}\underline{i}.$$

Using the circuit constraint $\underline{i} = 0$,

$$|\underline{M}| = \frac{|\underline{v}|}{\omega|\underline{x}|} = \frac{(0.05)}{2\pi(2000)(10^{-6})} = 3.97\,\text{V}\cdot\text{s/m}.$$

4.7 Applications of electromechanical two-port theory

This section reveals how to incorporate realistic mechanical and electrical constraints into electromechanical two-port matrices to predict the small-signal behavior of MEMS devices. First to be demonstrated is the use of impedance and source reflection to simplify electromechanical systems. Then attention is directed to deriving system transfer functions for some basic transducer types, including three-plate sensors and electret devices.

4.7.1 Application of source and impedance reflection

Impedance and source reflection make it easy to reduce electromechanical two-ports to purely mechanical or purely electrical equivalent networks. Any cascaded two-port can be manipulated in this way. Such operations often provide insights about the interplay between electrical and mechanical parameters and their parametric sensitivities. Resonance and instability can also be investigated in this way.

Consider, for example, a capacitive transducer with DC voltage bias V_0 but no AC electrical excitation, that is, $\underline{v}_{\text{ext}} = 0$. Figure 4.11a shows the small-signal, M-form cascaded model for the device. We can use the equivalent circuit of Fig. 2.22b to reduce this electromechanical two-port to the series connection of mechanical impedances depicted in Fig. 4.11b. The force–velocity relation becomes

$$\underline{f}_{\text{ext}} = \left(\text{j}\omega m + b + k/(\text{j}\omega) + \underline{Z}_{\text{m}} + \frac{|\underline{M}|^2}{\underline{Z}_{\text{e}} + \underline{Z}_{\text{e0}}}\right)\text{j}\omega\underline{x}. \tag{4.51}$$

Now assume the transducer to be a variable-gap capacitor and further that the DC bias is provided by a low-impedance supply so that $\underline{Z}_{\text{e0}} \to 0$. From Table 4.2 for the basic variable-gap capacitor geometry, $\underline{Z}_{\text{m}} = 0$, $\underline{Z}_{\text{e}} = 1/\text{j}\omega C_0$, $\underline{M} = E_0/\text{j}\omega$, and $E_0 = V_0/d$. Then, the effective mechanical impedance is

$$(\underline{Z}_{\text{m}})_v = \frac{\underline{f}_{\text{ext}}}{\text{j}\omega\underline{x}} = \text{j}\omega m + b + k_v/\text{j}\omega, \tag{4.52}$$

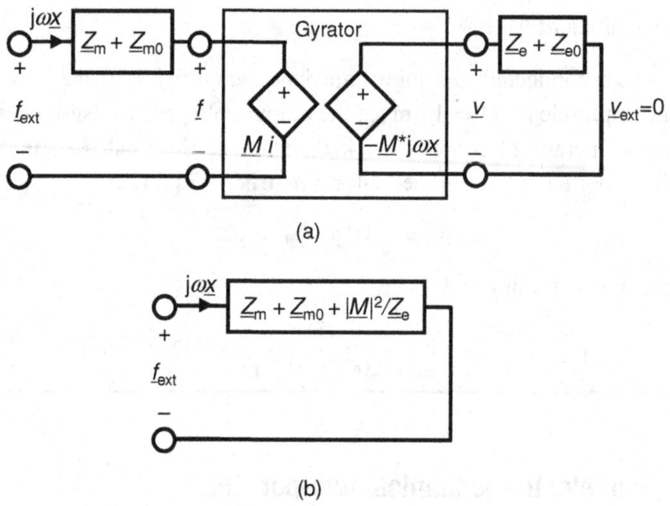

Figure 4.11 Electromechanical two-port subject to constant voltage constraint. (a) M-form cascade of a transducer with assumed DC bias V_0 (not shown) and no small-signal electrical excitation, i.e., $\underline{v}_{ext} = 0$. (b) Simplified mechanical network resulting from assuming $\underline{Z}_{e0} \sim 0$ and reflecting electrical impedance $\underline{Z}_e = 1/j\omega C_0$ to the mechanical side.

where the subscript "v" indicates the constant voltage constraint and $k_v = k - C_0 E_0^2$ is an effective spring constant. Remember that C_0 and E_0 are evaluated at the mechanical equilibrium defined by Eq. (4.40). For the voltage-constrained variable-gap transducer, the electromechanical coupling has the effect of reducing the spring constant. The mechanical resonant frequency is $(\omega_0)_v = \sqrt{k_v/m}$, so the resonant frequency goes down as voltage is increased.

Consider for comparison a variable-area device with the charge constrained. Now, it becomes more convenient to use the N-form. The small-signal model is shown in Fig. 4.12a. The N-form makes it easy to impose the fixed charge constraint by an open-circuit condition, i.e., $\underline{i}_{ext} = 0$.

The constant charge condition can be effectively realized by connecting the necessary DC bias through a large series resistor R. The constant charge condition is assured if $R \gg 1/\omega_0 C_0$. Then, because $\underline{Y}_{e0} = 1/R \to 0$, the external admittance can be neglected. Using the equivalent circuit of Fig. 2.23b to reflect the admittance across to the mechanical side yields

$$\underline{f}_{ext} = \left(j\omega m + b + k/(j\omega) + \underline{Z}'_m + N^2/\underline{Y}_e\right) j\omega \underline{x}. \tag{4.53}$$

Figure 4.12b shows the equivalent mechanical circuit. For the variable-area geometry, $\underline{Z}'_m = 0$, $N^2 = (\varepsilon_0 w E_0)^2$, $E_0 = V_0/d$, $\underline{Y}_e = j\omega C_0$, and $C_0 = \varepsilon_0 w(L - x_0)/d$. Using these expressions in Eq. (4.53), the new effective mechanical impedance is

$$(\underline{Z}_m)_q = \frac{\underline{f}_{ext}}{j\omega \underline{x}} = j\omega m + b + k_q/j\omega, \tag{4.54}$$

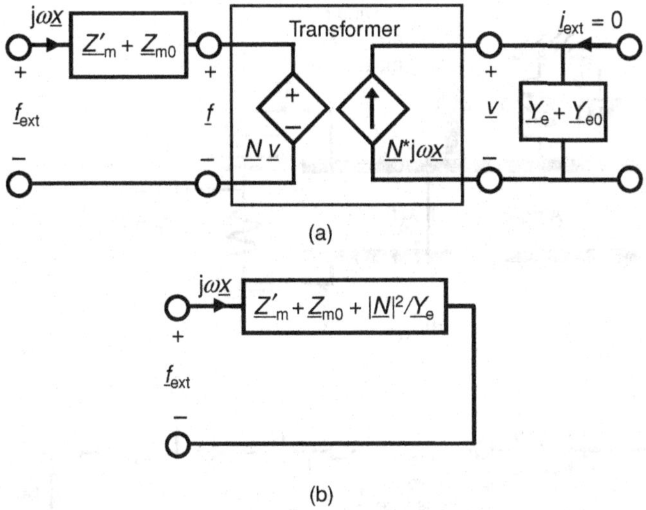

Figure 4.12 Electromechanical two-port subject to constant charge constraint. (a) N-form cascade of a transducer with assumed DC charge $q_0 = C_0 V_0$ but no small-signal current, i.e., $i_{ext} = 0$. (b) Simplified mechanical network resulting from assuming $\underline{Y}_{e0} \sim 0$ and reflecting electrical admittance $\underline{Y}_e = j\omega C_0$ to the mechanical side.

where $k_q = k + dE_0^2/(L - x_0)$ is another effective spring constant. Just as before, the equilibrium equation, Eq. (4.40), defines E_0 and x_0. The associated natural resonant frequency is $(\omega_0)_q = \sqrt{k_q/m}$. For this device, electric charge increases the stiffness of the system and the mechanical resonant frequency.

The reflection method has allowed us to group *all* the external constraints – electrical and mechanical – on the mechanical side of the network. The effective impedance expressions $(\underline{Z}_m)_v$ and $(\underline{Z}_m)_q$ incorporate the effect of the electrical bias on the resonance. For the variable-gap capacitive transducer, the resonant frequency can be adjusted downward, while for the variable-area geometry, the resonant frequency can be tuned upward. Voltage-controlled tuning is quite useful in MEMS applications where precise control of resonance is called for.

4.7.2 A capacitive microphone

Figure 4.13a depicts a simple model for a variable-gap capacitive microphone as a first example of how to use the transmission matrix to model the behavior of an electromechanical sensor. The capacitance $C(x)$ is $\varepsilon_0 A/x$, the DC voltage bias is V_0, and Eq. (4.40) defines the mechanical equilibrium. The movable plate, of mass m and area A, is constrained by linear restoring spring and damping forces. The transducer matrix coefficients appear in the left column of Table 4.2.

For operation as a microphone, the external force is $\underline{f}_{ext} = -A\underline{p}_{in}$, where the phasor $\underline{p}_{in}$ is the differential pressure acting on the moving plate. The induced output current $\underline{i}$

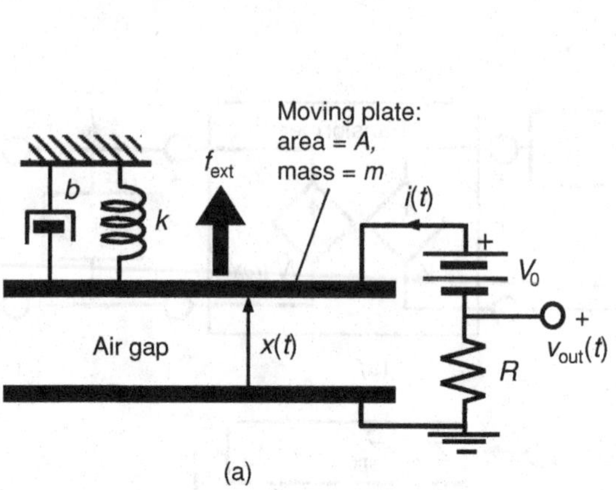

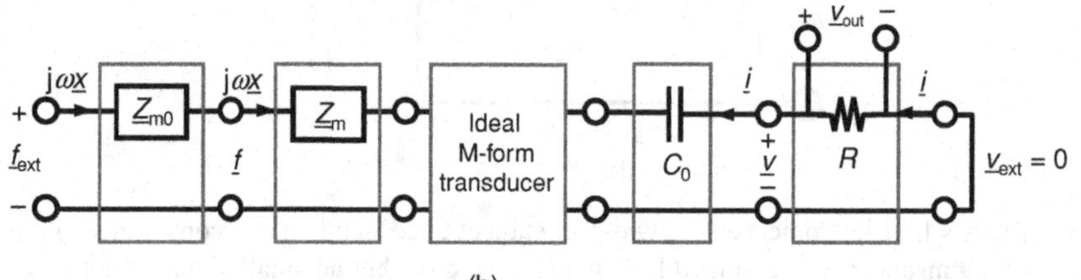

Figure 4.13 Simple model for a variable-gap capacitive MEMS microphone device. (a) Basic geometric description and electrical connections of the device. The moving plate is acted upon by the external force $\underline{f}_{\text{ext}} = -A\underline{p}_{\text{in}}$, where $\underline{p}_{\text{in}}$ is the small-signal pressure input. The output is the voltage across resistor R, that is, $\underline{v}_{\text{out}} = -R\underline{i}$. (b) Cascade model for predicting the small-signal response.

is detected as the voltage across the external load resistor R. The goal is to reduce this system to an equivalent network using the M-form electromechanical matrix formalism. The electrical constraints are $\underline{Z}_{\text{e0}} = R$ and $\underline{v}_{\text{ext}} = 0$, and the output is

$$\underline{v}_{\text{out}} = -R\underline{i}. \tag{4.55}$$

Refer to the cascade model of Fig. 4.13b. Imposing the constraints in Eq. (4.47) yields

$$\underline{v}_{\text{out}} = \frac{AR\underline{M}^*}{|\underline{M}|^2 + (\underline{Z}_{\text{m}} + \underline{Z}_{\text{m0}})\left(\dfrac{1}{\text{j}\omega C_0} + R\right)}\underline{p}_{\text{in}}. \tag{4.56}$$

The key to using the M-form has been to recognize that $\underline{v}_{\text{ext}} = 0$. An alternative method is to use the N-form transducer matrix of Eq. (4.48). Using $\underline{Y}_{\text{e0}} = 1/R$, $\underline{i}_{\text{ext}} = 0$, and $\underline{v}_{\text{out}} = \underline{v}$ in Eq. (4.48) yields

$$\underline{v}_{\text{out}} = \frac{-A\underline{N}^*}{|\underline{N}|^2 + (\underline{Z}'_{\text{m}} + \underline{Z}_{\text{m0}})\left(\text{j}\omega C_0 + \dfrac{1}{R}\right)}\underline{p}_{\text{in}}. \tag{4.57}$$

The similarity relations from Eq. (4.14) can be used to show that Eqs. (4.56) and (4.57) are identical.

4.7.3 Electromechanical transfer functions

A common way to characterize a linear system is to define its frequency-dependent transfer function:

$$\underline{H}(\mathrm{j}\omega) \equiv \frac{\underline{S}_{\text{out}}}{\underline{S}_{\text{in}}}, \tag{4.58}$$

where $\underline{S}_{\text{in}}$ and $\underline{S}_{\text{out}}$ are the input and output signal phasors. In circuit analysis these variables are usually voltages and currents but for a capacitive transducer one signal is probably electrical – voltage, current, or charge – and the other may be mechanical, that is, a force, displacement, or acceleration. We can rely on the mechanical ↔ electrical analogy, treating the mechanical side in terms of an equivalent circuit and defining $\underline{H}(\mathrm{j}\omega)$ as needed, bearing in mind that the units of the electromechanical transfer function will be mixed.

Consider the capacitive microphone treated in the previous section. Using Eq. (4.56) or (4.57), one can write the transfer function in either the M- or N-form. The M-form is

$$\underline{H}_{v/p}(\mathrm{j}\omega) \equiv \frac{\underline{v}_{\text{out}}}{\underline{p}_{\text{in}}} = \frac{-AR\underline{M}^*}{|\underline{M}|^2 + (\underline{Z}_{\text{m}} + \underline{Z}_{\text{m0}})\left(\dfrac{1}{\mathrm{j}\omega C_0} + R\right)}, \tag{4.59}$$

where $\underline{Z}_{\text{e}} = 1/(\mathrm{j}\omega C_0)$, $\underline{M} = E_0/(\mathrm{j}\omega)$, $E_0 = V_0/d$, and $\underline{Z}_{\text{m0}} = \mathrm{j}\omega m + b + k/(\mathrm{j}\omega)$. Further, note that $\underline{Z}_{\text{m}} = 0$ because the capacitor has variable-gap geometry.

Because electromechanical transfer functions may employ any of a large set of mechanical and electrical variables as inputs and outputs, we here adopt a subscripting convention to keep this information straight. Accordingly, the subscript "v/p" labeling $\underline{H}$ in Eq. (4.59) signifies pressure as the *input* signal and voltage as the *output*. The units of $|\underline{H}_{v/p}|$ are volt/pascal, or volt-meter^2 per newton.

Substituting in expressions for the matrix coefficients makes it possible to study the frequency dependence of $\underline{H}_{v/p}(\mathrm{j}\omega)$:

$$\underline{H}_{v/p}(\mathrm{j}\omega) = \frac{ARC_0E_0}{-C_0E_0^2/\mathrm{j}\omega + (\mathrm{j}\omega m + b + k/(\mathrm{j}\omega))(1 + \mathrm{j}\omega C_0 R)}. \tag{4.60}$$

The $(1 + \mathrm{j}\omega C_0 R)$ term in the denominator of Eq. (4.60) suggests identification of low- and high-frequency limits for $\underline{H}_{v/p}(\mathrm{j}\omega)$, that is, $\omega \ll 1/C_0R$ and $\omega \gg 1/C_0R$. These frequency ranges correspond, respectively, to constant voltage and constant charge constraints imposed on the transducer:

$$\underline{H}_{v/p}(\mathrm{j}\omega) \approx \begin{cases} \dfrac{\mathrm{j}\omega ARC_0E_0/k_v}{1 - \left(\dfrac{\omega}{\omega_v}\right)^2 + \mathrm{j}2\zeta_v\left(\dfrac{\omega}{\omega_v}\right)}, & \omega \ll 1/RC_0, \\[2ex] \dfrac{AE_0/k}{1 - \left(\dfrac{\omega}{\omega_0}\right)^2 + \mathrm{j}2\zeta\left(\dfrac{\omega}{\omega_0}\right) - \dfrac{E_0^2}{\mathrm{j}\omega Rk}}, & \omega \gg 1/RC_0, \end{cases} \tag{4.61}$$

where $\omega_0 = \sqrt{k/m}$, $\omega_v = \sqrt{k_v/m}$, and $k_v = k - C_0E_0^2$. $\zeta = b/(2\sqrt{mk})$, and $\zeta_v = b/(2\sqrt{mk_v})$ are damping ratios as defined in Section B.1.3 of Appendix B. For any

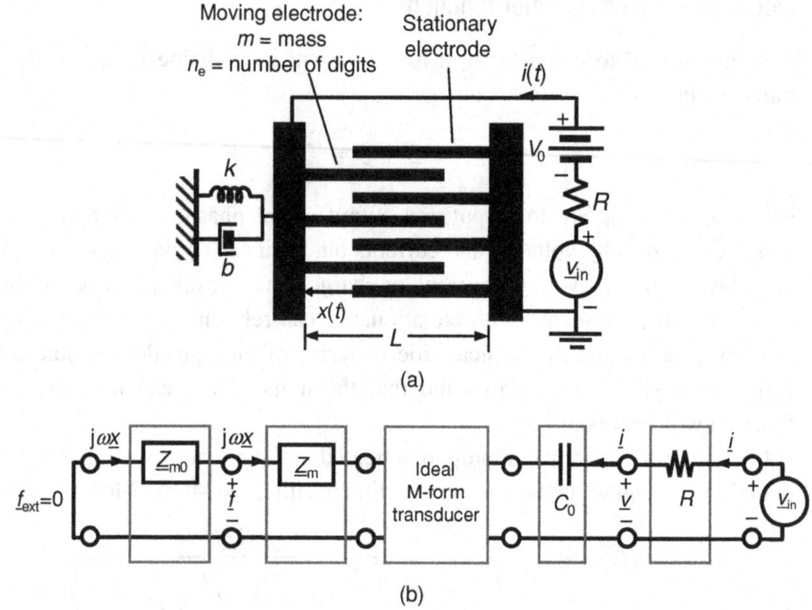

Figure 4.14 Simple model for a comb-drive capacitive MEMS actuator. (a) Top view (not to scale) with geometric details and electrical connections. The moving plate of mass m with n identical fingers is constrained by mechanical spring k and damping b. (b) Cascade model using a Thevenin equivalent for the small-signal electrical excitation. Typically, the resistor R is very small compared with the capacitive impedance.

low-frequency input pressure signal, the transducer behaves according to the constant voltage constraint; the mechanical resonance is altered (reduced) by the electromechanical coupling. High-frequency behavior corresponds to the constant charge constraint, in which limit the resonant frequency is unaffected by the electromechanics.

The important lesson here is that a capacitive transducer may operate in either the constant voltage or constant charge limit, depending on the frequency of operation. For the microphone, the critical frequency separating the two regimes is controlled by the resistance R.

4.7.4 A comb-drive actuator

Comb-drive actuators and sensors are quite common in capacitive MEMS. Like generic variable-area structures, this geometry offers large stroke and excellent response linearity, though with the penalty of weak coupling. Example 3.4 described a simple gyroscope that employs comb electrode structures both to generate motion and then to sense it.

Consider the coplanar, interdigitated structure shown in Fig. 4.14a. The capacitance function varies linearly with the displacement x:

$$C(x) = 2nc_e(L - x) + C_{\text{fringing}}, \quad \text{for} \sim w < x < \sim(L - w), \tag{4.62}$$

where n is the number of digits in the moving electrode array, c_e is capacitance per unit length, and C_{fringing} is a constant term due to fringing fields. Note that for a surface-micromachined comb drive with thin electrodes, as depicted in Fig. 3.18b, c_e would have to be calculated using finite-element software. On the other hand, for a high aspect ratio capacitor bank, c_e could be estimated using the parallel-plate approximation.

Let the equilibrium position of the moving mass be x_0 when the applied DC bias voltage is V_0. The small-signal voltage $\underline{v}_{\text{in}} = \underline{v}_{\text{ext}}$ is the input and displacement $\underline{x}_{\text{out}}$ is the output. This motion is constrained by the mass, the spring force, and damping, but there is no external force acting on the plate, that is, $\underline{f}_{\text{ext}} = 0$. Figure 4.14b shows the equivalent electromechanical network.

The N-form matrix of Eq. (4.50) turns out to be the most convenient form to use here. The electromechanical transfer function, having units of meters per volt, is

$$\underline{H}_{x/v}(\mathrm{j}\omega) \equiv \frac{\underline{x}_{\text{out}}}{\underline{v}_{\text{in}}} = \frac{-\underline{N}/\mathrm{j}\omega}{\underline{Z}'_{\text{m}} + \underline{Z}_{\text{m0}}}, \tag{4.63}$$

where $\underline{Z}'_{\text{m}} = 0$, $\underline{N} = -2nc_e V_0$, and $\underline{Z}_{\text{m0}} = \mathrm{j}\omega m + b + k/(\mathrm{j}\omega)$. The transfer function then reduces to the familiar form of a simple mechanical resonator,

$$\underline{H}_{x/v}(\mathrm{j}\omega) \approx \frac{-2nc_e V_0/k}{1 - (\omega/\omega_0)^2 + \mathrm{j}2\zeta(\omega/\omega_0)}, \tag{4.64}$$

where $\zeta = b/(2\sqrt{mk})$ is the damping ratio and $\omega_0 = \sqrt{k/m}$ is the resonant frequency. The electromechanical coupling has no influence on the mechanical resonance under the constant voltage constraint.

4.7.5 The three-plate capacitive sensor

Many capacitive sensors employ *three-plate* electrode configurations and are operated in a *half-bridge* mode. Sensors of this type are important enough to merit special attention. Chapter 5 will sustain this focus by detailing the way these devices are connected to the essential amplification and signal conditioning circuitry. The primary advantage of three-plate systems is that, when operated in a differential mode, they exhibit excellent response linearity over a large dynamic range and have some immunity to parasitic cable capacitance. Perhaps just as importantly, they virtually eliminate DC offset from the output voltage signal. Furthermore, most three-plate geometries are inherently *self-shielding*, which reduces stray coupling noise.

Both variable-gap and variable-area capacitive transducers can be configured as three-plate devices. See Figs. 4.15a and b, noting the symmetry and self-shielding properties of these geometries. A general capacitive network model for the three-plate capacitive sensor is provided in Fig. 4.15c. Inspection yields the terminal relation for this network:

$$\begin{bmatrix} q_1 \\ q_2 \end{bmatrix} = \begin{bmatrix} C_1(x) + C_{\text{m}} & -C_{\text{m}} \\ -C_{\text{m}} & C_2(x) + C_{\text{m}} \end{bmatrix} \begin{bmatrix} v_1 \\ v_2 \end{bmatrix}. \tag{4.65}$$

It is good sensor design practice to strive for linearity between input and output. Operating the three-plate sensor in a balanced mode realizes this goal if certain

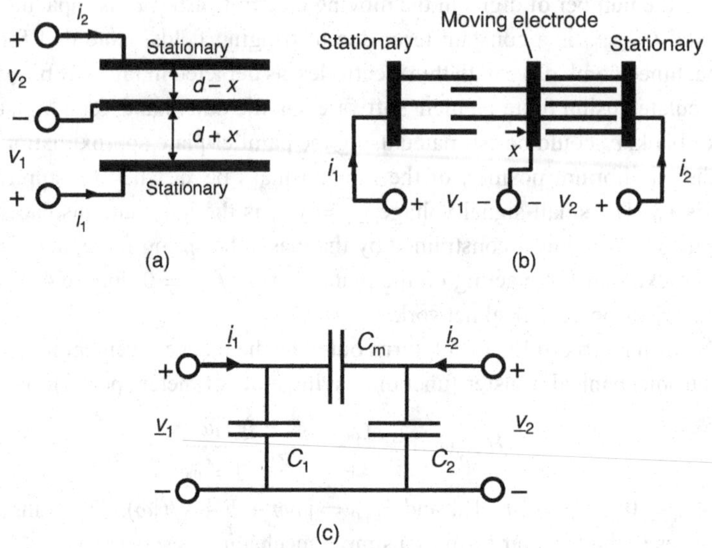

Figure 4.15 Basic geometries and an equivalent capacitive circuit model for three-plate capacitive sensors. Both structures exhibit antisymmetry as defined by Eq. (4.66) with respect to equilibrium at $x = 0$. (a) Three-plate variable-gap device. (b) Three-plate variable-area (comb-drive) device. (c) Two-port capacitive network representation for three-plate structures.

requirements are met. First, as exemplified in Figs. 4.15a and b, the self-capacitances should be antisymmetric about the equilibrium point, assumed here to be at $x = 0$. Thus,

$$C_1(x) = C_2(-x). \tag{4.66}$$

If the antisymmetry condition of Eq. (4.66) is met, then, in addition to $C_1(x = 0) = C_2(x = 0) \equiv C_0$, the following relationships exist amongst the derivatives:

$$\frac{\mathrm{d}C_1}{\mathrm{d}x} = -\frac{\mathrm{d}C_2}{\mathrm{d}x} \quad \text{and} \quad \frac{\mathrm{d}^2C_1}{\mathrm{d}x^2} = \frac{\mathrm{d}^2C_2}{\mathrm{d}x^2}. \tag{4.67}$$

Another condition, $C_\mathrm{m} \ll C_0$, can be achieved by proper design. For the present, however, we retain mutual capacitance to demonstrate the immunity of three-plate sensors to mutual (and stray) capacitance effects.

Figure 4.16 shows a biased half-bridge connection for a three-plate system. It is because the DC bias sources have opposed polarities that the output voltage signal contains no DC offset. The small-signal behavior of this device can be formulated using a linear three-port electromechanical matrix with one mechanical and two electrical ports.

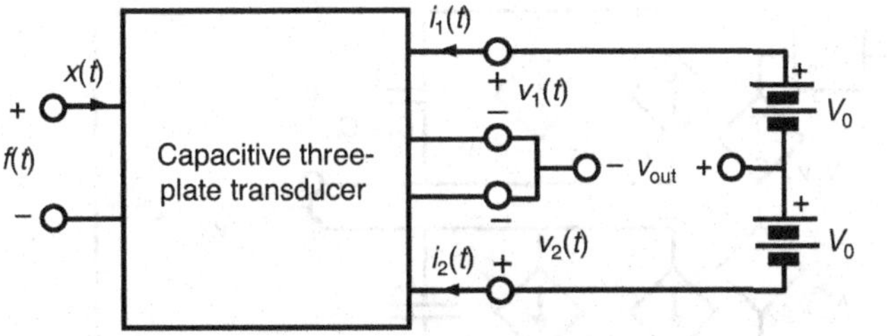

Figure 4.16 DC biasing and circuit connections for a three-plate capacitive sensor. Note that the bias voltages are opposed, so that as long as $C_1(0) = C_2(0)$, the DC offset of v_{out} is zero.

The N-form is a convenient choice here. Adapting the linearization method introduced in Section 4.2 and using phasors,

$$\begin{bmatrix} \underline{f} \\ \underline{i}_1 \\ \underline{i}_2 \end{bmatrix} = \begin{bmatrix} \underline{Z}'_m & \underline{N}_1 & \underline{N}_2 \\ -\underline{N}_1^* & \underline{Y}_1 & \underline{Y}_m \\ -\underline{N}_2^* & \underline{Y}_m & \underline{Y}_2 \end{bmatrix} \begin{bmatrix} j\omega\underline{x} \\ \underline{v}_1 \\ \underline{v}_2 \end{bmatrix}. \tag{4.68}$$

The N-form matrix uses coenergy to define the coefficients. From Eq. (3.31),

$$W'_e = \tfrac{1}{2}[C_1(x) + C_m]v_1^2 - C_m v_1 v_2 + \tfrac{1}{2}[C_2(x) + C_m]v_2^2. \tag{4.69}$$

The matrix coefficients are

$$\underline{Z}'_m = \frac{-1}{j\omega}\left[\frac{v_1^2}{2}\left(\frac{d^2C_1}{dx^2} + \frac{d^2C_m}{dx^2}\right) - v_1 v_2 \frac{d^2C_m}{dx^2} + \frac{v_2^2}{2}\left(\frac{d^2C_2}{dx^2} + \frac{d^2C_m}{dx^2}\right)\right], \tag{4.70a}$$

$$\underline{N}_1 = -v_1\left[\frac{dC_1}{dx} + \frac{dC_m}{dx}\right] + v_2\frac{dC_m}{dx} \quad \text{and} \quad \underline{N}_2 = -v_2\left[\frac{dC_2}{dx} + \frac{dC_m}{dx}\right] + v_1\frac{dC_m}{dx}, \tag{4.70b}$$

$$\underline{Y}_1 = j\omega[C_1 + C_m],\ \underline{Y}_2 = j\omega[C_2 + C_m], \quad \text{and} \quad \underline{Y}_m = -j\omega C_m. \tag{4.70c}$$

These coefficients are evaluated at the equilibrium conditions, that is, $(v_1)_0 = V_0$, $(v_2)_0 = -V_0$, and $x_0 = 0$. Introducing a new function, $C(x) \equiv C_1(x) = C_2(-x)$, simplifies these equations to

$$\underline{Z}'_m \approx -\frac{V_0^2}{j\omega}\left[\frac{d^2C}{dx^2} + 2\frac{d^2C_m}{dx^2}\right], \tag{4.71a}$$

$$\underline{N}_1 = -V_0\left[\frac{dC}{dx} + 2\frac{dC_m}{dx}\right] \quad \text{and} \quad \underline{N}_2 = -V_0\left[\frac{dC}{dx} - 2\frac{dC_m}{dx}\right], \tag{4.71b}$$

$$\underline{Y} \equiv \underline{Y}_1 = \underline{Y}_2 = j\omega[C + C_m]. \tag{4.71c}$$

Figure 4.17 shows an equivalent electromechanical network obtained using Eq. (4.68) and the circuit topology of Fig. 4.16.

This circuit considerably simplifies if the reasonable assumptions are made that (i) the mutual coupling is small, that is, $C_m \ll C_0$, and (ii) C_m is only very weakly dependent

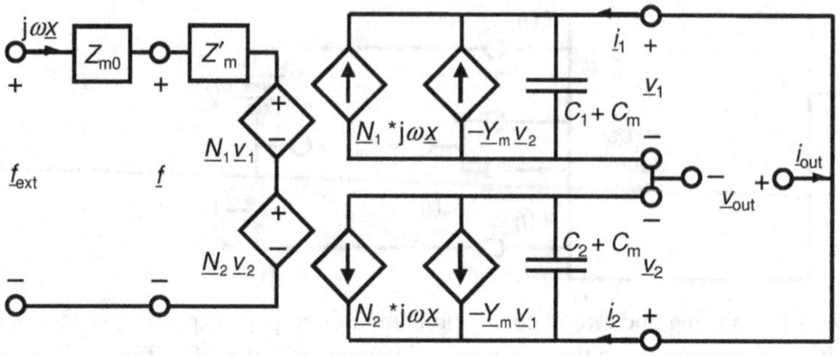

Figure 4.17 The equivalent circuit model for the three-plate sensor configuration with external mechanical impedance $\underline{Z}_{m0}$ and force $\underline{f}_{ext}$. For this circuit topography, $\underline{v}_{out} = \underline{v}_1 = \underline{v}_2$ and $\underline{i}_{out} = \underline{i}_1 = \underline{i}_2$, in which case, the network can be reduced to an electromechanical two-port network.

on x. Achieving these conditions is a matter of design. Then, $\underline{N}_1 \approx \underline{N}_2 \equiv \underline{N}$, where $\underline{N} = V_0 \mathrm{d}C/\mathrm{d}x$. By defining $\underline{v}_{out} \equiv \underline{v}_1 = \underline{v}_2$ and $\underline{i}_{out} \equiv \underline{i}_1 + \underline{i}_2$, and using Eqs. (4.71a,b,c), the network collapses to the two-port shown in Fig. 4.18.

Inspection of the circuit in Fig. 4.18 leads to a 2×2 matrix equation relating the across variables, $\underline{f}$ and $\underline{i}_{out}$, to the through variables, $\mathrm{j}\omega\underline{x}$ and $\underline{v}_{out}$:

$$\begin{bmatrix} \underline{f} \\ \underline{i}_{out} \end{bmatrix} = \begin{bmatrix} \underline{Z}'_m & 2\underline{N} \\ -2\underline{N}^* & 2\mathrm{j}\omega C \end{bmatrix} \begin{bmatrix} \mathrm{j}\omega\underline{x} \\ \underline{v}_{out} \end{bmatrix}. \tag{4.72}$$

It is a straightforward exercise to convert the transducer matrix in Eq. (4.72) to the transmission form if needed. Bear in mind that this two-port realization is correct only if the three-plate sensor is antisymmetric according to Eq. (4.66) and if the device is connected as illustrated in Fig. 4.16. Section 5.4 reveals how to connect this configuration into the *half-bridge* amplifier topology.

*4.7.6 Linear model for electret transducer

We can derive a small-signal AC circuit model for the electret transducer of Section 3.6 using the hybrid energy function $W''_e(x, v, q_e)$ defined by Eq. (3.71). The starting point

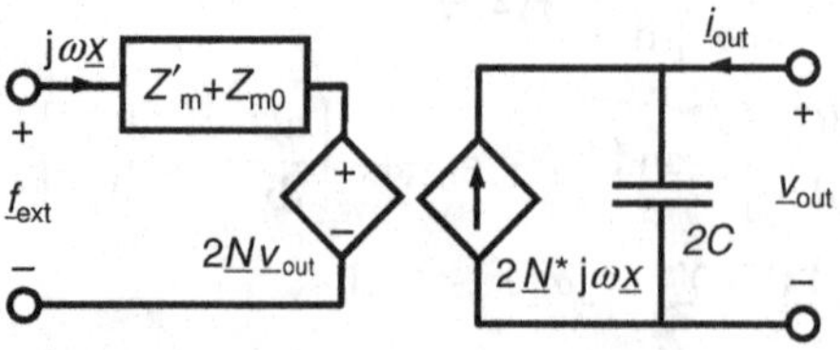

Figure 4.18 Two-port equivalent of the circuit model shown in Fig. 4.17 for a balanced three-electrode capacitive sensor. This simplification is valid under the symmetric condition $N_1 = N_2 = N$, and it exploits the circuit constraints $\underline{v}_{out} = \underline{v}_1 = \underline{v}_2$ and $\underline{i}_{out} = \underline{i}_1 = \underline{i}_2$, plus the condition that $C \gg C_m$.

is Eq. (3.62), with the substitution $f = -f^{\mathrm{e}}$:

$$f = -\left.\frac{\partial W''_{\mathrm{e}}}{\partial x}\right|_{v,q_{\mathrm{e}}}, \quad q = \left.\frac{\partial W''_{\mathrm{e}}}{\partial v}\right|_{x,q_{\mathrm{e}}}, \quad \text{and} \quad v_{\mathrm{e}} = -\left.\frac{\partial W''_{\mathrm{e}}}{\partial q_{\mathrm{e}}}\right|_{x,v}. \tag{4.73}$$

Linearizing these expressions in a Taylor series and then isolating the small-signal terms leads to the following 3×3 matrix relation:

$$\begin{bmatrix} \underline{f} \\ \underline{i} \\ \underline{v}_{\mathrm{e}} \end{bmatrix} = \begin{bmatrix} \dfrac{-1}{\mathrm{j}\omega}\dfrac{\partial^2 W''_{\mathrm{e}}}{\partial x^2} & -\dfrac{\partial^2 W''_{\mathrm{e}}}{\partial x \partial v} & \dfrac{-1}{\mathrm{j}\omega}\dfrac{\partial^2 W''_{\mathrm{e}}}{\partial x \partial q_{\mathrm{e}}} \\ \dfrac{\partial^2 W''_{\mathrm{e}}}{\partial v \partial x} & \mathrm{j}\omega\dfrac{\partial^2 W''_{\mathrm{e}}}{\partial v^2} & \dfrac{\partial^2 W''_{\mathrm{e}}}{\partial v \partial q_{\mathrm{e}}} \\ \dfrac{-1}{\mathrm{j}\omega}\dfrac{\partial^2 W''_{\mathrm{e}}}{\partial q_{\mathrm{e}} \partial x} & -\dfrac{\partial^2 W''_{\mathrm{e}}}{\partial q_{\mathrm{e}} \partial v} & \dfrac{-1}{\mathrm{j}\omega}\dfrac{\partial^2 W''_{\mathrm{e}}}{\partial q_{\mathrm{e}}^2} \end{bmatrix} \begin{bmatrix} \mathrm{j}\omega\underline{x} \\ \underline{v} \\ \underline{i}_{\mathrm{e}} \end{bmatrix}. \tag{4.74}$$

Expressing the matrix elements as derivatives of W''_{e} accentuates the origin of the coefficients from the Taylor series. See Fig. 4.19. This matrix is unnecessarily complicated. We can reduce it to an effective two-port because the port defined by the variables q_{e} and v_{e} is fictitious; the charge q_{e} is constant so that $\underline{q}_{\mathrm{e}} = 0$ and, further, $\underline{v}_{\mathrm{e}}$ is not of operational interest. Therefore, the $(\underline{q}_{\mathrm{e}}, \underline{v}_{\mathrm{e}})$ port can be ignored in the small-signal model.

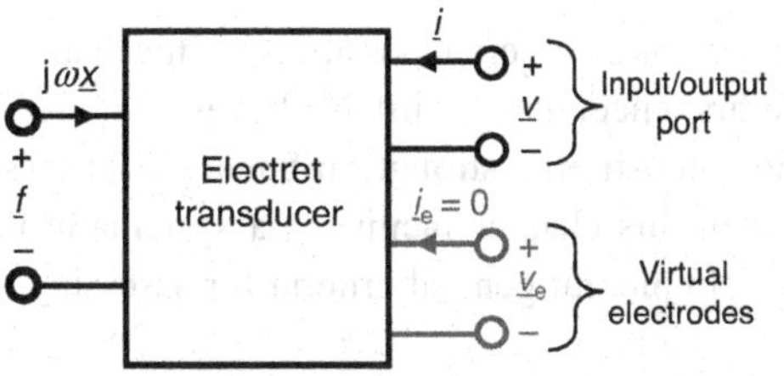

Figure. 4.19 Linear, small-signal network representation for the electret transducer of Fig. 3.25. The virtual electrodes are not accessible; in fact, $\underline{i}_{\mathrm{e}} = 0$, and $\underline{v}_{\mathrm{e}}$ is not operationally important. Thus, the small-signal model reduces to the equivalent two-port network reflected in Eq. (4.75).

The resulting 2×2 matrix relation for this electret device is

$$\begin{bmatrix} \underline{f} \\ \underline{i} \end{bmatrix} = \begin{bmatrix} \underline{Z}''_{\mathrm{m}} & \underline{N}'' \\ -\underline{N}''^{*} & \underline{Y}''_{\mathrm{e}} \end{bmatrix} \begin{bmatrix} \mathrm{j}\omega\underline{x} \\ \underline{v} \end{bmatrix}, \tag{4.75}$$

where

$$\underline{Z}''_{\mathrm{m}} = \frac{-1}{\mathrm{j}\omega}\frac{\partial^2 W''_{\mathrm{e}}}{\partial x^2}, \tag{4.76a}$$

$$\underline{N}'' = -\frac{\partial^2 W''_{\mathrm{e}}}{\partial x \partial v}, \tag{4.76b}$$

$$\underline{Y}''_{\mathrm{e}} = \mathrm{j}\omega\frac{\partial^2 W''_{\mathrm{e}}}{\partial v^2}. \tag{4.76c}$$

Referring to the geometry of Fig. 3.25a and using Eq. (3.71) for W''_e with no external bias, that is, $V_0 = 0$, yields

$$\underline{Z''}_m = \frac{1}{j\omega} \frac{\varepsilon_0 A}{(d/\kappa + x_0)^3} \left[-\frac{q_e}{C_d} \right]^2, \tag{4.77a}$$

$$\underline{N''} = \frac{\varepsilon_0 A}{(d/\kappa + x_0)^2} \left[-\frac{q_e}{C_d} \right], \tag{4.77b}$$

$$\underline{Y''}_e = j\omega \frac{C_x C_d}{C_x + C_d}, \tag{4.77c}$$

where $C_x = \varepsilon_0 A/x_0$ and $C_d = \kappa\varepsilon_0 A/d$. As expected from inspection of the schematic in Fig. 3.25b, the admittance $\underline{Y''}_e$ is the series connection of the dielectric layer and the variable air gap.

An advantage of electrets as electrostatic transducer elements is that, unlike conventional capacitive systems with air gaps, they are effectively self-biasing. The magnitude of the self-biasing voltage, $-q_e/C_d$, for common electret materials is quite high, in the range of 1000 to 2000 V_{dc}, so a blocking capacitor may be needed.

4.8 Stability considerations

Some capacitive MEMS devices possess an inherent possibility of mechanical instability and resultant stictive failure. We first encountered unstable behavior in Section 3.7 and discovered that external electrical constraints strongly influence it. The small-signal analysis methodology developed in this chapter facilitates a systematic investigation of the phenomenon and provides simple but general criteria for assessing the possible instability of a device.

4.8.1 Preliminary look at stability

A quick assessment of electromechanical instability can be gained by inspecting the effective mechanical impedances derived using impedance reflection in Section 4.7.1. $(\underline{Z}_m)_v$, the impedance expression defined by Eq. (4.52) for the variable-gap/constant voltage case, includes an effective spring constant

$$k_v = k - C_0 E_0^2, \tag{4.78}$$

where $E_0 = v_0/x_0$. Here, increasing the bias voltage always decreases k_v and reduces the resonance frequency. Eventually, k_v goes to zero and becomes negative. A negative effective spring constant means that the net spring force is no longer restoring; even the smallest initial displacement will grow exponentially until the plates make contact. For a capacitive MEMS device, this is the catastrophe known as *pull-in*. Unless mechanical stops are designed into the structure to prevent large areas of the electrodes from coming into contact, van der Waals forces tenaciously bind the surfaces together and render the device useless.

Such behavior contrasts markedly with the variable-area capacitor under the constraint of constant charge. The effective spring constant extracted from $(\underline{Z}_{\mathrm{m}})_q$ of Eq. (4.54) is

$$k_q = k + dE_0^2/(L - x_0). \tag{4.79}$$

Here, owing to the electromechanical coupling, the effective spring constant k_{q} increases, the resonant frequency rises, and there can be no instability.

These examples are instructive, but one cannot over-generalize from them. Recall the finding in Section 3.8 that the three-plate variable-gap geometry of Fig. 3.27 can be unstable under *either* constant voltage or constant charge constraints. For capacitive transducers, it turns out that fixing the voltage is no guarantee of instability and fixing the charge cannot assure its avoidance. A more systematic method is needed to predict instability in these systems.

4.8.2 General stability criteria

To uncover general criteria for predicting instability, it is convenient to introduce the complex frequency s using the substitution $\mathrm{j}\omega \rightarrow s$. The general solution form for the small-signal variables becomes e^{st}; instability occurs when $\mathrm{Re}[s] > 0$. Using s in Eq. (4.42) yields

$$(s^2 m + sb + k)\underline{x} + \underline{f} = \underline{f}_{\mathrm{ext}}. \tag{4.80}$$

We will shortly return to the important limits of constant voltage and constant charge to reduce $\underline{f}$ in Eq. (4.80), but first it is instructive to anticipate a general form of this equation, namely,

$$(s^2 m + sb + k_{\mathrm{eff}})\underline{x} = \underline{f}_{\mathrm{ext}}, \tag{4.81}$$

where k_{eff} represents either k_{v} or k_{q}, depending on whether voltage or charge is fixed. Equation (4.81) can be expressed in terms of a transfer function relating displacement $\underline{x}$ to external driving force $\underline{f}_{\mathrm{ext}}$:

$$\underline{H}_{x/f}(s) = \frac{\underline{x}}{\underline{f}_{\mathrm{ext}}} = \frac{1}{s^2 m + sb + k_{\mathrm{eff}}}. \tag{4.82}$$

The roots of the quadratic polynomial in the denominator of $\underline{H}_{x/f}(s)$ are

$$s_{1,2} = -\frac{b}{2m} \pm \mathrm{j}\sqrt{\frac{k_{\mathrm{eff}}}{m} - \left(\frac{b}{2m}\right)^2}, \tag{4.83}$$

where s_1 and s_2 are the exponential coefficients of the homogeneous solutions for the resonator; that is, $\mathrm{e}^{s_1 t}$ and $\mathrm{e}^{s_2 t}$ constitute the transient response of the system.

From Section 4.8.1, we know that k_{eff} can either increase or decrease as a function of the applied bias voltage. Instability only results when k_{eff} decreases so we

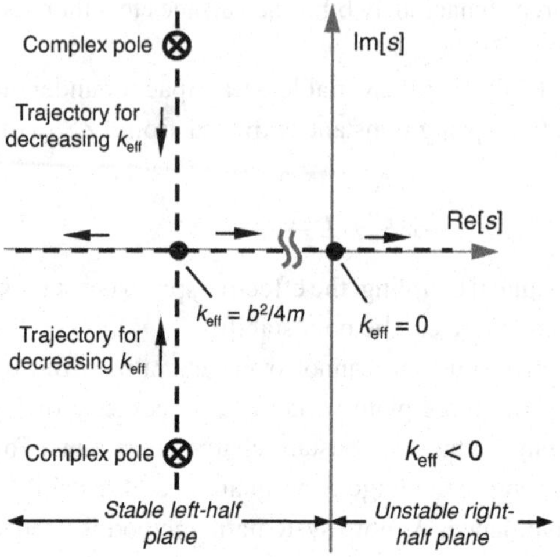

Figure 4.20 Poles of the second-order transfer function $\underline{H}_{x/f}$ defined by Eq. (4.86). The dashed lines plot the trajectories traced by these poles as k_{eff} is decreased from k. The poles are initially complex conjugates, but after they meet on the real axis, they become negative real. At $k_{eff} = 0$, the real pole on the right crosses the imaginary axis and the system becomes unstable, indicating that any small initial displacement grows exponentially.

focus on that case first. Figure 4.20 plots the trajectories of s_1 and s_2 in the complex plane as k_{eff} is decreased from $k_{eff} = k$. Initially, the poles are complex conjugates and they move straight toward each other. Eventually they merge on the real axis and then separate, one moving to the right and the other to the left. At $k_{eff} = 0$, the right-most real pole reaches the imaginary axis and for $k_{eff} < 0$ crosses into the right-half plane.

From Eq. (4.83), it is evident that, as long as $k_{eff} > 0$, the exponential transients are stable, that is, $\text{Re}[s_1] < 0$ and $\text{Re}[s_2] < 0$. As k_{eff} goes through zero and becomes negative, one of these transient solutions changes from a decaying to a growing exponential. The transition from stable resonance to purely exponential growth at a threshold condition is an example of *absolute instability*. Even the smallest displacement from equilibrium, due to a small mechanical shock or even thermal noise, grows without limit. When this happens, the capacitor plates come into contact and, without mechanical stops, stick together.

To derive general stability criteria, we separately investigate the constraints of constant voltage and constant charge imposed on capacitive transducers.

4.8.2.1 Constant voltage case

For the constant voltage constraint, the N-form transducer matrix is used with $\underline{v} = 0$. Then, the electrical force expression reduces to

$$\underline{f} = \underline{Z}'_{m} s \underline{x}, \tag{4.84}$$

where

$$\underline{Z}'_{\mathrm{m}} = -\frac{1}{s}\frac{\partial^2 W'_{\mathrm{e}}}{\partial x^2}\bigg|_v \tag{4.85}$$

is the mechanical impedance expressed in terms of s and $W'_{\mathrm{e}} = \frac{1}{2}C(x)v^2$. Now define the transfer function,

$$\underline{H}_{x/f}(s) \equiv \frac{\underline{x}}{\underline{f}_{\mathrm{ext}}} = \frac{1}{s^2 m + bs + k'_{\mathrm{eff}}}, \tag{4.86}$$

where

$$k'_{\mathrm{eff}} = k - \frac{\partial^2 W'_{\mathrm{e}}}{\partial x^2}\bigg|_v = k - \frac{v_0^2}{2}\frac{\mathrm{d}^2 C}{\mathrm{d}x^2}. \tag{4.87}$$

The threshold condition is $k'_{\mathrm{eff}} = 0$ and so the necessary and sufficient condition for instability is

$$\frac{v_0^2}{2}\left[\frac{\mathrm{d}^2 C}{\mathrm{d}x^2}\right]_{x=x_0} > k > 0. \tag{4.88}$$

Therefore, a *sufficient condition* for stability at constant voltage, dependent only on the geometry of the capacitive transducer, is

$$\frac{\mathrm{d}^2 C}{\mathrm{d}x^2} \le 0. \tag{4.89}$$

4.8.2.2 Constant charge case

Now consider the transducer subject to the constant charge constraint: $\underline{i} = 0$. Using the M-form with Eq. (4.30a),

$$\underline{f} = \underline{Z}_{\mathrm{m}} s \underline{x}, \tag{4.90}$$

where, based on Eq. (4.11a),

$$\underline{Z}_{\mathrm{m}} = \frac{1}{s}\frac{\partial^2 W_{\mathrm{e}}}{\partial x^2}\bigg|_q . \tag{4.91}$$

Using Eqs. (4.90) and (4.91) in (4.80), the transfer function becomes

$$\underline{H}_{x/f}(s) \equiv \frac{\underline{x}}{\underline{f}_{\mathrm{ext}}} = \frac{1}{s^2 m + sb + k_{\mathrm{eff}}}, \tag{4.92}$$

where

$$k_{\mathrm{eff}} = k + \frac{\partial^2 W_{\mathrm{e}}}{\partial x^2}\bigg|_q . \tag{4.93}$$

Substituting $W_{\mathrm{e}} = \frac{1}{2}q^2/C(x)$ into Eq. (4.93) yields

$$k_{\mathrm{eff}} = k + \frac{q_0^2}{2C_0^2}\left[-\frac{\mathrm{d}^2 C}{\mathrm{d}x^2} + \frac{2}{C_0}\left(\frac{\mathrm{d}C}{\mathrm{d}x}\right)^2\right]. \tag{4.94}$$

Table 4.4 The stability of basic variable-gap and variable-area capacitive transducer structures under constant charge and constant voltage constraints. Fringing fields and damping are ignored here. Care must be taken when applying these results to other capacitive geometries

Electrical constraint	Basic variable-gap capacitor: $\frac{d^2C}{dx^2} > 0$	Basic variable-area capacitor: $\frac{d^2C}{dx^2} = 0$
Constant charge: $\underline{i} = 0$	Always stable; no effect on system stiffness	Always stable; electromechanical coupling stiffens system and increases resonant frequency
Constant voltage: $\underline{v} = 0$	Electromechanical coupling decreases resonant frequency; pull-in instability if $\lvert v_0 \rvert > \sqrt{2k/d^2C/dx^2}$	Always stable; no effect on system stiffness

Note that k_{eff} will be a decreasing function of charge $|q_0|$ only if $d^2C/dx^2 > 2(dC/dx)^2/C_0$. The *necessary and sufficient condition* for instability at equilibrium position $x = x_0$ is

$$\frac{q_0^2}{2C_0^2}\left[\frac{d^2C}{dx^2} - \frac{2}{C_0}\left(\frac{dC}{dx}\right)^2\right]_{x=x_0} > k > 0. \tag{4.95}$$

A geometry-dependent *sufficient condition* for the avoidance of instability under the constant charge constraint is

$$\frac{d^2C}{dx^2} < \frac{2}{C_0}\left(\frac{dC}{dx}\right)^2. \tag{4.96}$$

When either an analytical expression for $C(x)$ is known or numerical capacitance values are available, either from measurement or finite-element analysis, Eqs. (4.89) and (4.96) provide a quick way to assess the possibility of instability for a capacitive device.

Consider these criteria for the particular cases of variable-gap and variable-area capacitors. Substituting the capacitance expressions from Table 4.2 into Eqs. (4.89) and (4.96) confirms that *only* the variable-gap device can be unstable and then only when the voltage is fixed. These results are summarized in Table 4.4.

It is safest to test the stability of other, more complex geometries according to the more general criteria of Eqs. (4.89) and (4.95), rather than relying on Table 4.4. Several end-of-chapter problems treat the stability question for other geometries; readers are urged to study them.

Example 4.5 **Electrical tuning of mechanical resonance**

Voltage controls the resonant frequency of variable-gap capacitive transducers through the electromechanical influence on the effective spring constant:

$$f_0 = \sqrt{k_{\text{eff}}/m}/2\pi,$$

where $k_{\text{eff}} = k - C_0 E_0^2$. Consider an air gap device having area $A = 10^3\ \mu\text{m}^2$, mass $m = 1.0\ \mu\text{g}$, initial gap spacing and mechanical resonant frequency, both at zero voltage, of $x_k = 3\ \mu m$ and $f_{v=0} = 3.5$ kHz, respectively. At zero voltage, $C_{v=0} = \varepsilon_0 A/x_k = 2.59$ fF. The mechanical spring constant is $k = (2\pi f_{v=0})^2 m = 0.48$ N/m. With no external force, the equilibrium condition, Eq. (4.40), may be written as

$$v_0 = \sqrt{2k\left(x_k x_0^2 - x_0^3\right)/\varepsilon_0 A}.$$

This equation, implicit in x_0, is plotted in Fig. 4.21a. As expected, the plates draw together as the voltage increases, but at $v_0 = 21.3$ V, there is a bifurcation called pull-in. This instability is the subject of Section 4.8.3.

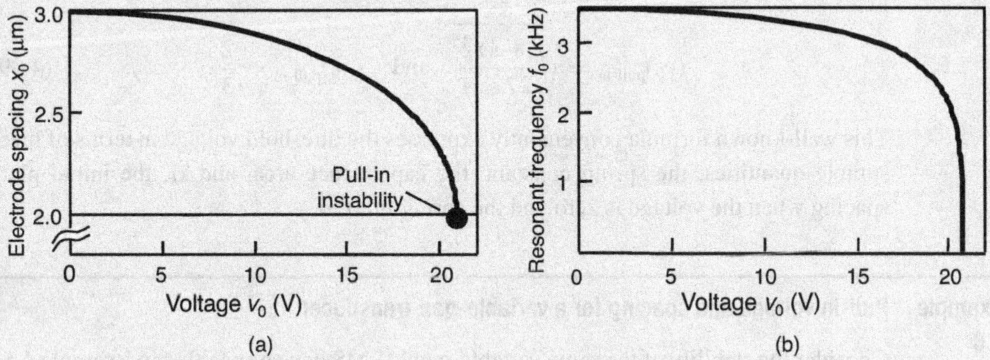

Figure 4.21 Equilibria of variable-gap capacitive MEMS device with area $A = 10^3\ \mu m^2$, mass $m = 1.0\ \mu g$, and $x_k = 3\ \mu m$. (a) Equilibrium gap spacing x_0 versus bias voltage v_0. (b) Resonant frequency f_0 versus bias voltage.

The resonant frequency can be written as

$$f_0 = f_{v=0}\sqrt{1 - \varepsilon_0 A v_0^2/\left(k m x_0^2\right)}.$$

The cubic term in the equilibrium precludes an explicit form for f_0 as a function of v_0; however, the relationship can be plotted using x_0 as a parameter. Refer to Fig. 4.21b, which reveals that the mechanical resonant frequency can be tuned (downward) by ~20% before the slope becomes so large that good resolution is lost.

4.8.3 The pull-in instability threshold

Because it can disable a MEMS device permanently, the pull-in phenomenon manifested by variable-gap capacitive transducers warrants special attention. Pull-in is an absolute instability, the important attribute of which is that it possesses a threshold, in this case voltage. Equation (4.88) defines this threshold, but we cannot use it directly to solve for this voltage because x_0 depends on the voltage. This difficulty is surmounted by considering the equilibrium balance of the spring and the force of electrical origin.

Assuming that x_k is the value of the gap spacing when the voltage is zero, static force balance requires

$$-\frac{v_0^2}{2}\frac{\mathrm{d}C}{\mathrm{d}x} - k(x_k - x_0) = 0. \tag{4.97}$$

Using $C(x) = \varepsilon_0 A/x$ for the air gap capacitor, the threshold condition is

$$\frac{v_0^2}{2}\left[\frac{\mathrm{d}^2C}{\mathrm{d}x^2}\right]_{x=x_0} = \frac{\varepsilon_0 A}{x_0^3}v_0^2 = k. \tag{4.98}$$

Combining Eqs. (4.97) and (4.98) yields

$$(v_0)_{\text{pull-in}} = \sqrt{\frac{8}{27}\frac{kx_k^3}{\varepsilon_0 A}} \quad \text{and} \quad (x_0)_{\text{pull-in}} = \frac{2}{3}x_k. \tag{4.99}$$

This well-known formula conveniently expresses the threshold voltage in terms of measurable quantities: the spring constant, the capacitance area, and x_k, the initial plate spacing when the voltage is zero and the spring relaxed.

Example 4.6 **Pull-in voltage and spacing for a variable-gap transducer**

Consider the stability of the same variable-gap MEMS device specified in Example 4.5. The threshold condition for pull-in is $k_{\text{eff}} = 0$, that is,

$$(v_0)_{\text{pull-in}} = \sqrt{\frac{8}{27}\frac{kx_k^3}{\varepsilon_0 A}} = \sqrt{\frac{8}{27}\frac{(0.48)(3\cdot 10^{-6})^3}{(8.854\cdot 10^{-12})(10^{-9})}} = 21.3\ \text{V}.$$

As this threshold is approached, the equilibrium gap spacing decreases. Pull-in occurs at

$$x_{\text{pull-in}} = 0.67x_k = 2\,\mu\text{m},$$

which is consistent with Fig. 4.21a of Example 4.5.

4.8.4 A physical interpretation of instability

To interpret the influence of the electrical constraints on stability, one can consider how the electric field and charge distribution change as the plate moves for variable-gap and variable-area structures. Figure 4.22a depicts the effect of varying the gap spacing of a simple capacitor at fixed voltage. The electrical force attracting the two parallel plates together increases above its equilibrium value when the electrodes move closer together and decreases when they move further apart. Thus, the electromechanical contribution to the net force is non-restoring; it reduces the stiffness and the resonant frequency. If the voltage is increased too much, the mechanical spring is overwhelmed and pull-in ensues. On the other hand, if charge is fixed by open-circuiting the device, the electric field does not change appreciably as plate separation is changed and the mechanical stiffness is

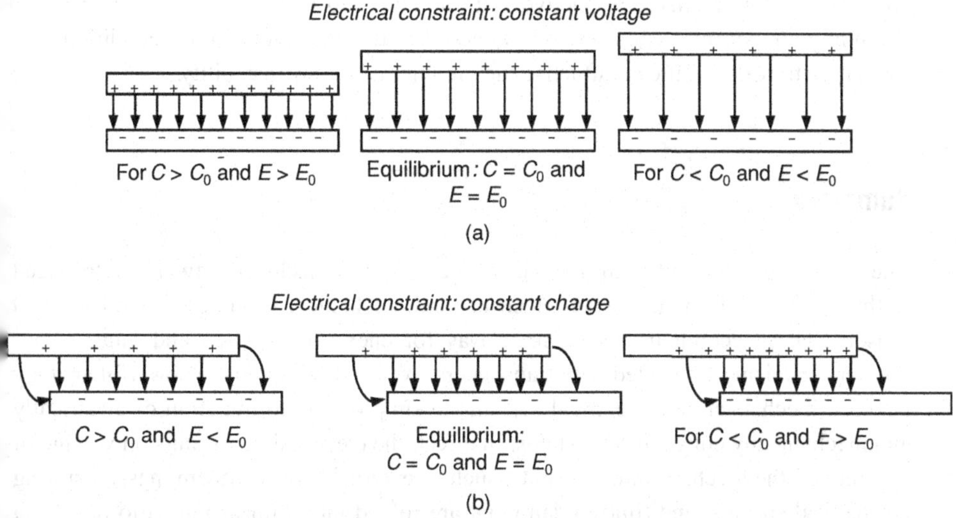

Figure 4.22 Physical interpretation for how electrical constraints influence electromechanical stability of capacitive transducers. (a) For a variable-gap capacitor at constant voltage, the electric field E_0 and the attractive electrostatic force are inversely related to the spacing. At sufficiently high voltage, this force overwhelms the mechanical spring and pull-in instability results. At constant charge, the electrostatic force is virtually constant, so there is no instability. (b) For a variable-area capacitor at constant charge, the fringing electric field responsible for the electrostatic force is inversely related to area of overlap. Because it increases the effective stiffness, the system remains stable. At constant voltage, the fringing field magnitude is unchanged as the plate moves back and forth, so stiffness is unchanged and the device remains stable.

unaltered. This case is not depicted in the figure. Though ignored here, fringing fields do not change the results.

For the variable-area device shown in Fig. 4.22b, the force is due to the fringing field. For fixed charge, the density of the surface charge, which determines the field magnitude in both the uniform and the fringing field regions, is inversely related to the area of overlap of the electrodes. Because the fringing field force always acts to increase the capacitance (and to increase the area of overlap), either an increase or a decrease in the area leads to an electrostatic restoring force contribution. Thus, the effective stiffness of the system is increased and instability can never result. When voltage is fixed instead (a case not shown in the figure), the fringing field magnitude and the electrical force remain fixed and the effective stiffness is unaffected.

Similar heuristic reasoning, with proper account taken of the electrical constraints, may be employed for geometries that are more complex, but such exercises are best confirmed with knowledge of $C(x)$ using the criteria of Eqs. (4.89) and (4.96). Finally, it is well to remember that the external electrical connections can influence the frequency dependence of the effective electrical constraint. Recall from Section 4.7.3 that, depending on the value of the series resistance R, a constant voltage or constant charge constraint can be realized, respectively, in the low- and high-frequency dynamic limits. While an external series resistor cannot alter the absolute stability condition of a capacitive transducer, it can influence the growth rate of any instability.

4.9 Summary

The linear two-port and multiport electromechanical transducer networks introduced in this chapter link the important electrical and mechanical variables in actuators and sensors. They account in a systematic way for energy conversion and mutual coupling mechanisms associated with transducers. We have exploited the powerful analogy between mechanical and electrical systems to simplify the formulation of electrically and mechanically constrained systems. In effect, the electrical constraints on voltage or current and the mechanical constraints, such as external forces, system mass, restoring mechanical springs, and (linear) damping, are folded into a linear model to provide a complete transducer system representation.

This approach provides a systematic methodology for obtaining small-signal electromechanical transfer functions, which depend on frequency and usually have one input and one output. Most often, one of these is electrical and the other is mechanical. The approach leads to a very general model for predicting the performance of MEMS devices. It also facilitates consideration of electromechanical device stability under realistic electrical and mechanical constraints.

The two-port transducer models presented in this chapter are amenable to cascading. There are clear benefits to the cascade design paradigm, though these may be largely from the viewpoint of mathematical convenience. In fact, we exploited cascade analysis effectively to develop the M- and N-form transducer models illustrated in Fig. 4.10. Many real MEMS, however, are not true cascaded systems, at least in the strict sense evoked by the diagrams in Fig. 4.1. This is why this chapter heavily emphasizes the mechanical ↔ electrical analogy and circuit models. Even when feedback or other complexities seem to preclude cascade modeling and the associated matrices, circuit models can always be relied upon to provide complete descriptions of a microelectromechanical system fully integrated with both mechanical constraints and electronics.

It is useful here to summarize a procedure that one can always rely on to obtain a linear electromechanical network model:

(i) Define and then derive the appropriate energy function.
(ii) Express the force of electrical origin and either voltage or current as derivatives of this energy function.
(iii) Use mechanical and electrical equations to establish the static equilibrium.
(iv) Linearize to obtain perturbation force and chosen electrical variables as functions of the other identified perturbation variables.
(v) Use inspection to synthesize a circuit model from the resulting set of linear coupled equations.

The network will incorporate mechanical impedances, electrical impedances or admittances, and either gyrator or transformer elements as appropriate to the chosen set of dependent variables. Often, it will be easier and safer to take this approach than to try to adapt the various, sometimes confusing, matrix forms found throughout this chapter.

Next, Chapter 5 presents some of the basic signal conditioning and amplifier circuit topologies commonly employed for MEMS sensors. We will show how these circuits can be integrated with the electromechanical two-port (or multiport) models to provide complete system-level descriptions. To avoid confusion arising from conventions for positive current flow, matrix formulations, and other distractions, we will restrict our usage to the circuit models of Figs. 4.8 and 4.9.

Problems

4.1 The electrical terminal relation for a capacitor with two mechanical degrees of freedom, x and y, is $q = C(x, y)v$.

(a) Write a general form for the 3×3 M-form electromechanical transducer matrix of this system.

(b) Use linearization to find the coefficients of this matrix in terms of $C(x, y)$ and its derivatives.

4.2 Demonstrate the validity of each of the three relationships between the electromechanical transducer coefficients for the M- and N-form matrices, as found in Eq. (4.14).

4.3 There are several alternative matrix expressions for the linear electromechanical two-port network, which are useful in different circumstances.

(a) Show that Eqs. (4.30a,b) and (4.31) for the M-form two-port network are equivalent.

(b) Repeat this exercise for Eqs. (4.33a,b) and (4.34) for the N-form.

Hint: Use the M- and N-form equivalent circuit models.

4.4 The capacitance matrix for an actuator with two electrical ports, labeled 1 and 2, and two mechanical degrees of freedom, x and y, is

$$\begin{bmatrix} C_{11}(x, y) & C_{12}(x, y) \\ C_{21}(x, y) & C_{22}(x, y) \end{bmatrix}.$$

(a) Explain on what basis we may expect $C_{12} = C_{21}$.

(b) Obtain the 4×4 N-form transducer matrix for this system, expressing all matrix elements in terms of equilibrium voltages V_1 and V_2, plus the capacitance functions and their derivatives.

4.5 Use linearization to derive the expressions found in the right column of Table 4.1 for the mechanical impedance $\underline{Z}'_{\mathrm{m}}$, coupling coefficient $\underline{N}$, and admittance $\underline{Y}_{\mathrm{e}}$ of a two-port rotating electromechanical transducer.

4.6 A rotational MEMS transducer exhibits the following dependence of capacitance on angular displacement θ:

$$C_{\mathrm{pF}}(\theta) = \frac{100}{20 + \theta_{\mathrm{deg}}} \quad \text{for} \quad -10^\circ < \theta_{\mathrm{deg}} < +20^\circ,$$

where C_{pF} is in picofarads and θ_{deg} is in degrees. Obtain frequency- and charge-dependent expressions for the M-form matrix coefficients of this transducer assuming equilibrium at $\theta_{\mathrm{deg}} = 0^\circ$ when the DC voltage is V_0. Be careful with units.

4.7 In matrix form, the electrical terminal relation for a rotating capacitive transducer with two electrical ports, a and b, and one angular mechanical degree of freedom, θ, is

$$\begin{bmatrix} q_a \\ q_b \end{bmatrix} = \begin{bmatrix} C_a(\theta) & C_{\mathrm{m}}(\theta) \\ C_{\mathrm{m}}(\theta) & C_b(\theta) \end{bmatrix} \begin{bmatrix} v_a \\ v_b \end{bmatrix},$$

where C_a and C_b are self-capacitances and C_{m} is mutual capacitance.

(a) Invoke linearization to obtain the N-form small-signal electromechanical transducer matrix.

(b) Find expressions for the coefficients of this matrix, using V_a and V_b as the equilibrium voltages.

4.8 Use basic circuit methods and admittance reflection to simplify the cascaded electromechanical network of (a) below to the form (b) consisting of an ideal transformer N, mechanical impedance $\underline{Z}_x$ and admittance $\underline{Y}_x$. Describe your derivation and explain what the equivalent network represents.

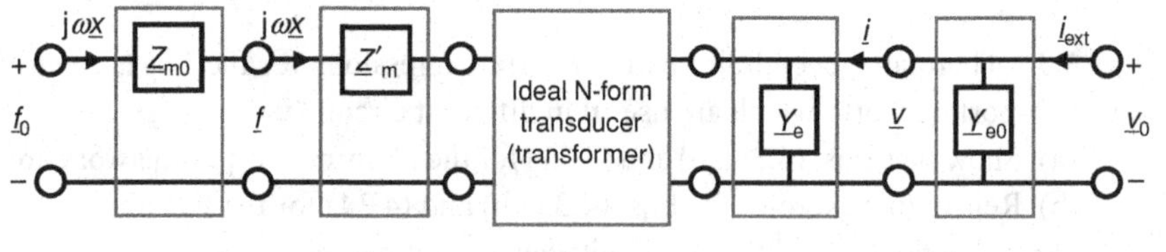

(a)

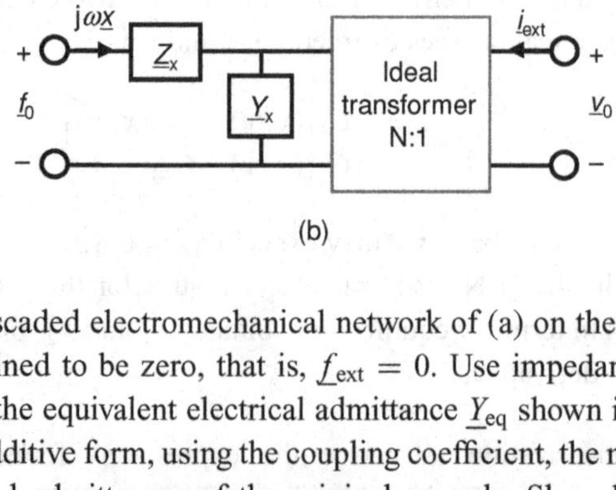

(b)

4.9 For the cascaded electromechanical network of (a) on the next page, the external force is constrained to be zero, that is, $\underline{f}_{\mathrm{ext}} = 0$. Use impedance reflection to reduce this network to the equivalent electrical admittance $\underline{Y}_{\mathrm{eq}}$ shown in (b). For convenience, express $\underline{Y}_{\mathrm{eq}}$ in additive form, using the coupling coefficient, the mechanical impedances, and the electrical admittances of the original cascade. Show how you obtained your answer and explain what the equivalent network represents.

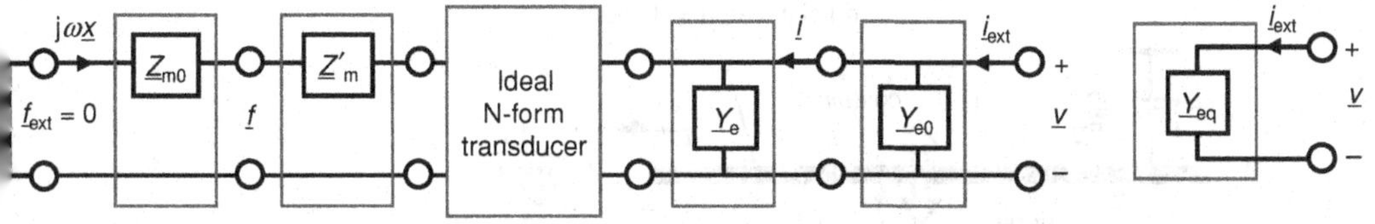

(a) (b)

4.10 A dielectric slab is free to move in the $\pm x$ direction between the electrodes of a parallel-plate structure, shown below. The slab is constrained by a linear spring of spring constant k, which is not under tension or compression when the slab is at the midpoint of the structure, $x = 0$. Assume that $d \ll g \ll L, w$, so that fringing fields and parasitic coupling between the two lower electrodes may be ignored.

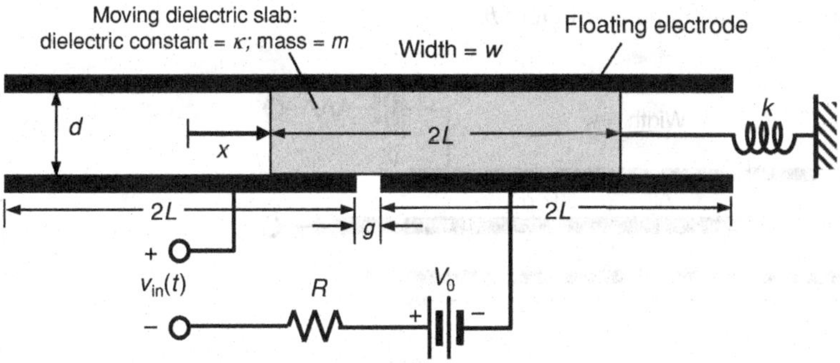

(a) Derive the force of electrical origin f^e in terms of state variables.
(b) What is the equilibrium x_0 for the applied DC voltage V_0?
(c) Can this system exhibit instability?
(d) Find an expression for the small-signal AC input electrical admittance. Note that the dielectric slab experiences no external force but is constrained by its own mass m and the spring k.
(e) What is the mechanical resonance frequency of this system?

4.11 The proof mass of the variable air gap accelerometer shown on the next page is a flat electrode plate of mass m and area A. Attached to the lower, stationary electrode is a dielectric layer of dielectric constant κ and thickness d. The upper electrode is parallel to the lower plate and is constrained by a linear spring k and linear damper b. At voltage bias V_0, the mechanical equilibrium is at $x = x_0$. The sinusoidal acceleration experienced by the proof mass is $a_{in}(t) = \ddot{x} = \alpha \cos(\omega t)$.
(a) Find an expression for the time-dependent, small-signal output voltage $v_{out}(t)$ induced by the acceleration. To avoid algebraic complexity, leave your answer in terms of equilibrium capacitance $C_0 = C(x_0)$, coupling coefficient N, and other circuit parameters.
(b) Obtain high and low frequency limits for $v_{out}(t)$ and explain how these relate to effective constant voltage and constant current constraints.

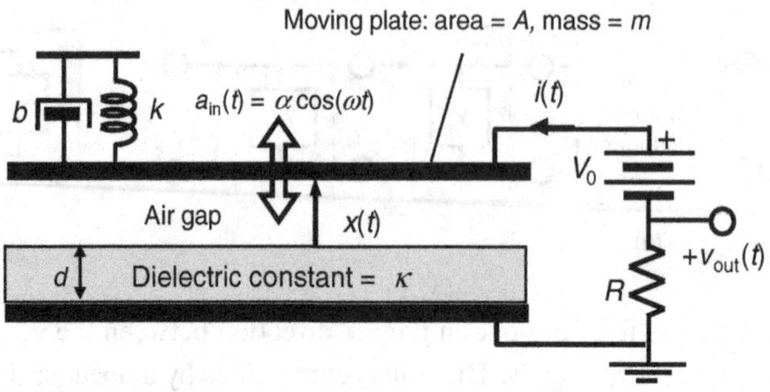

4.12 The movable center electrode of the self-shielded variable-area capacitive actuator below has mass m and is attached to a spring (not shown) of spring constant k. The spring force is zero at $x = x_k$. The moving plate is also subject to linear, velocity-dependent damping of damping coefficient b.

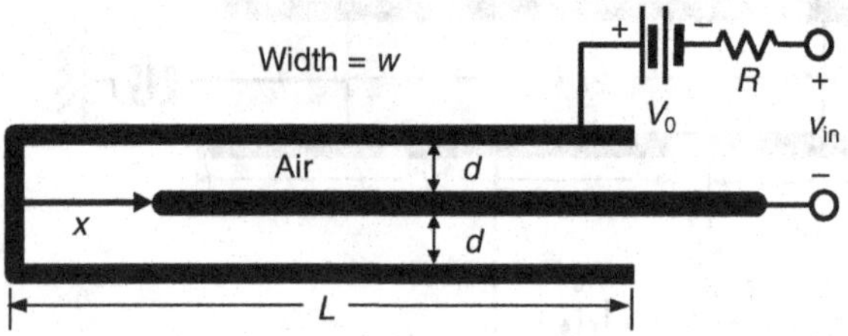

(a) Find an equation for the static equilibrium position of the plate, x_0, as a function of DC bias voltage V_0 and other system parameters.

(b) Obtain the transfer function $\underline{H}_{x/v}(j\omega) \equiv \underline{x}_{out}/\underline{v}_{in}$ for small-signal motions of this actuator.

4.13 A simple model for a capacitive sensor for monitoring the position of a movable, conductive slab of length L and width w is shown in the figure below. The parallel electrodes, also of length L and width w, are uniformly coated with an insulator of dielectric constant κ and thickness d. The spring has spring constant k and is at rest when $x = x_k$.

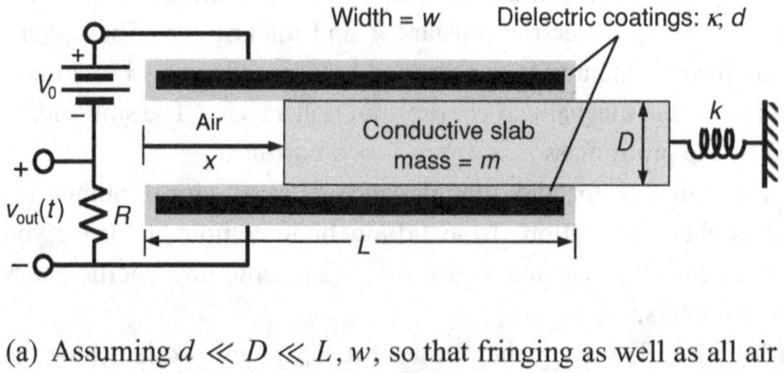

(a) Assuming $d \ll D \ll L, w$, so that fringing as well as all air gap capacitances may be ignored, write an approximate expression for the system capacitance.

(b) Write the equation defining the equilibrium position x_0.
(c) Obtain the electromechanical transfer function: $\underline{H}_{v/x}(\mathrm{j}\omega) \equiv \underline{v}_{\mathrm{out}}/\underline{x}$.
(d) Explain the effect of the resistor on this response by examining the following limits: $\omega RC_0 \ll 1$ and $\omega RC_0 \gg 1$, where $C_0 = C(x_0)$ is the capacitance at the equilibrium point.

4.14 The rotational capacitive transducer shown in the figure below is connected to a biasing voltage V_0 and a series resistor R. If θ is small, the device capacitance can be approximated by $C(\theta) = (\varepsilon_0 w/\theta)\ln(1 + b/a)$. Assume that the angular displacement is $\theta(t) = \theta_0 + \theta_1\cos(\omega t)$, where $|\theta_1| \ll \theta_0|$, to find an expression for the time-dependent voltage $v_{\mathrm{out}}(t)$ across the resistor R.

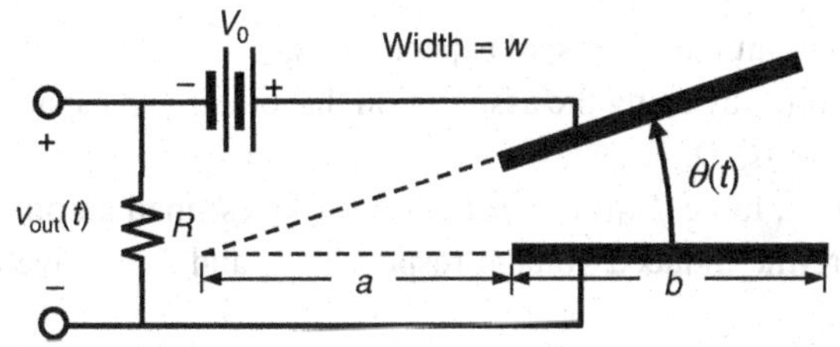

4.15 The rotational comb-drive capacitive transducer illustrated below has capacitance $C(\theta) = 2c_{\mathrm{e}}(\theta_0 + \theta)R_{\mathrm{eff}}$, where c_{e} is the capacitance per unit length of the adjacent thin electrodes and $R_{\mathrm{eff}} = R_1 + N(N-1)(d+w)$. Assuming an equilibrium at $\theta = \theta_0$ when $q = q_0$, employ the approach of Section 4.3 to find the M-form of the small-signal two-port transducer matrix:

$$\begin{bmatrix} \underline{\tau} \\ \underline{v} \end{bmatrix} = \begin{bmatrix} \underline{Z}_{\mathrm{m}} & \underline{M} \\ -\underline{M}^* & \underline{Z}_{\mathrm{e}} \end{bmatrix} \begin{bmatrix} \mathrm{j}\omega\underline{\theta} \\ \underline{i} \end{bmatrix},$$

where $\underline{\tau}$ and $\mathrm{j}\omega\underline{\theta}$ are, respectively, small-signal torque and angular velocity in radians per second.

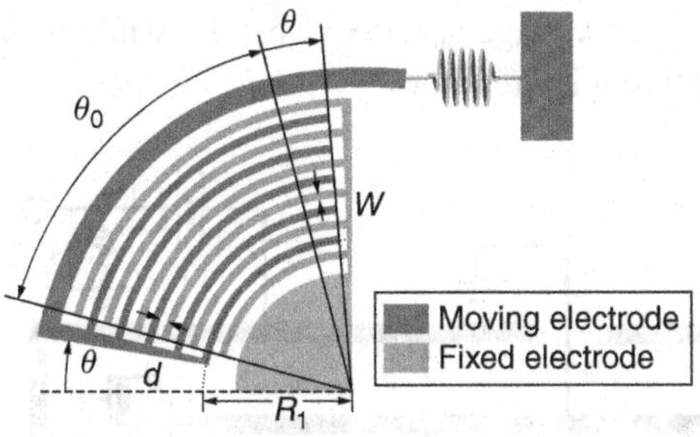

***4.16** An electrically isolated conductive slab of width w is free to move in the $\pm x$ direction between the dielectrically coated electrodes of two parallel-plate structures with electrode spacing D. The coatings all have thickness $d \ll D$ and dielectric constant κ. The slab motion is constrained by linear damping (damping constant b) and a spring of spring constant k. This spring exerts no force when the slab is at $x = 0$, the midpoint of the structure. Assume $d \ll g \ll L$, to justify ignoring fringing fields and parasitic coupling between the electrodes.

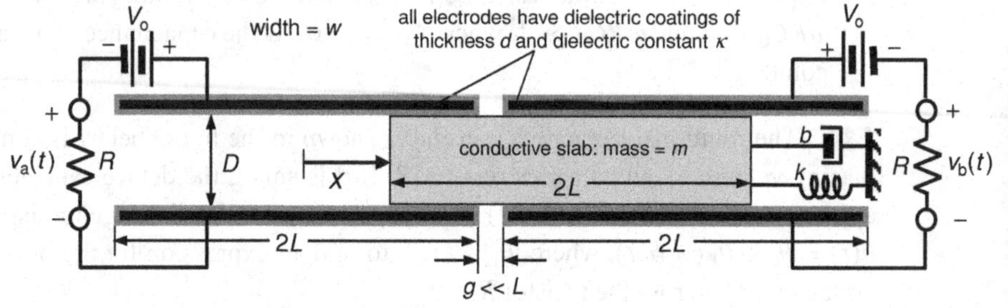

(a) Draw an equivalent circuit that shows all capacitances.
(b) Simplify the model of (a) by using the assumption that all air gap capacitances may be ignored, that is, $d/\kappa \ll D$.
(c) An external, sinusoidal, force $f_{\text{ext}}(t) = f_0 \cos(\omega t)$ induces small amplitude motion of the slab. What are the induced voltage responses $\underline{v}_a$ and $\underline{v}_b$? Leave answers in phasor form.
(d) What is the net mechanical impedance? Explain the limiting value of this quantity when $R = 0$ for both resistors.
(e) For operation of this device as an actuator, the resistors are removed and the electrical terminals are driven by two sinusoidal AC voltage sources. What is the phasing of the equal-amplitude voltages $v_a(t)$ and $v_b(t)$ required to maximize the motion?

***4.17** A flat plate of mass m, length $2L$, and width w is parallel to and free to move in the $\pm x$ direction between a pair of electrodes with spacing d. The plate, which is grounded, is constrained by a linear mechanical spring of spring constant k, which exerts no force when the plate is at $x = 0$, the midpoint of the structure. Linear velocity-dependent damping b is also present. Both electrodes are DC biased to voltage V_0. The electrode at the left is excited by the series AC voltage $v_{\text{in}}(t) = v' \cos(\omega t)$, while the electrode at the right is connected to ground through a resistor R.

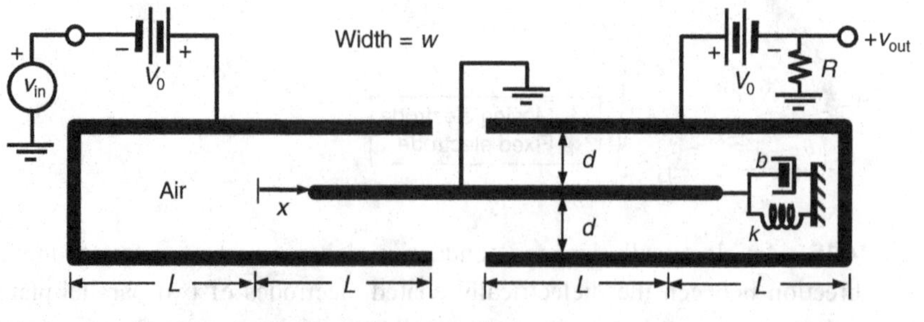

(a) Ignoring fringing fields and any parasitic capacitive coupling between the left and right electrodes, write expressions for the left-side and right-side capacitances.

(b) Draw the small-signal, electromechanical equivalent circuit for the device, which will have the form of an electrical to mechanical to electrical cascade. Provide expressions for all components of your model.

(c) With the movable plate constrained by its own mass, the attached spring, and the damper, find the voltage transfer function $H(j\omega) \equiv \underline{v}_{out}/\underline{v}_{in}$. Leave your answer in terms of the matrix coefficients already defined in (b).

4.18 The capacitance of a MEMS actuator, computed numerically using a finite-element package, is found to be well-approximated by a second-order polynomial $C(x) \approx C_0 + C'x + C''x^2$, over the range $0 < x < 10$ µm, where $C(x)$ is in femtofarads (fF), x is in microns, $C_0 = 200\,\text{fF}$, $C' = 20\,\text{fF}/\mu\text{m}$, and $C'' = -1\,\text{fF}/\mu\text{m}^2$.

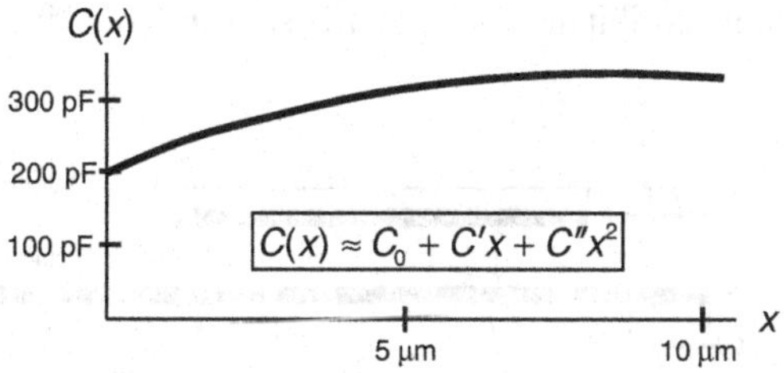

(a) Find expressions for the N-form transducer matrix coefficients of this transducer at 10 kHz, assuming DC voltage bias of 10 V. These will be polynomials in x. Hint: pay attention to units.

(b) If the device uses a restoring spring for mechanical positioning, is it possible for this transducer to exhibit instability when the voltage is fixed?

***4.19** Consider the three-plate transducer mechanism shown below. Ignore all fringing fields and assume no coupling between the two outside electrodes.

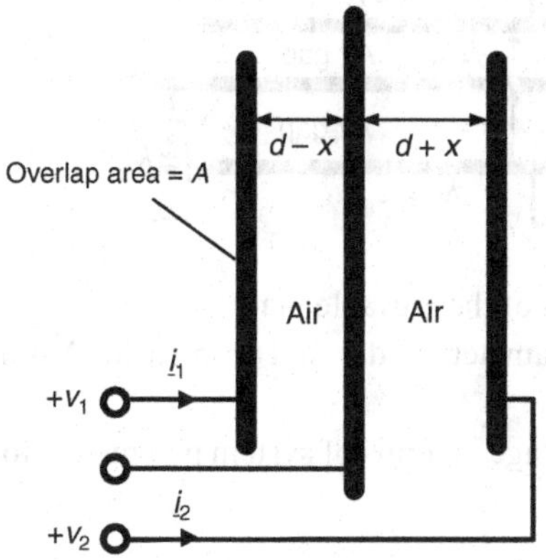

(a) Assuming equilibrium at $x = 0$ at DC bias $v_1 = v_2 = V_0$, develop a small-signal electromechanical model (M-form) for the variable-gap device shown. Note that this device has two electrical ports and one mechanical port, so your answer will be in the form of a 3 × 3 matrix. Use the transducer parameters to define the matrix coefficients.

(b) Repeat (a) for the N-form.

(c) Now assume that a mechanical spring of spring constant k holds the moving plate at $x = 0$. Consider the stability of the equilibrium at $x = 0$. What is the threshold condition on the bias voltage V_0 for the pull-in instability? Compare your result with Section 3.7.2.

4.20 Investigate the possibility of electromechanically induced instability for the two variable-area capacitive transducer geometries shown in (a) and (b). Both are held at fixed voltage V_0. The moving electrode in (b) is electrically isolated and has zero net charge. Can either of these configurations exhibit instability? Offer some physical interpretations for your answers.

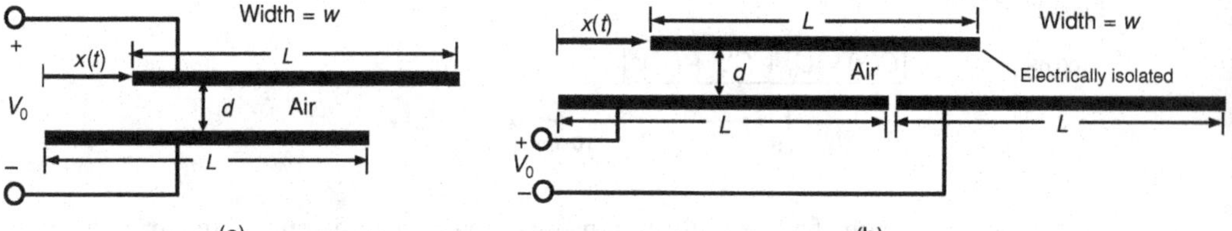

*__4.21__ A planar moving electrode is positioned between two other planar electrodes, which are parallel and of area A. Ignore fringing fields. In addition to the force of electrical origin, a restoring spring force, $f_{\text{spring}} = -kx$, always tends to push the moving plate back to $x = 0$. The electrostatic potential of this plate is controlled by a linear potentiometer, so that the moving plate voltage is $V_{\text{plate}} = \alpha V_0 x$ with respect to ground.

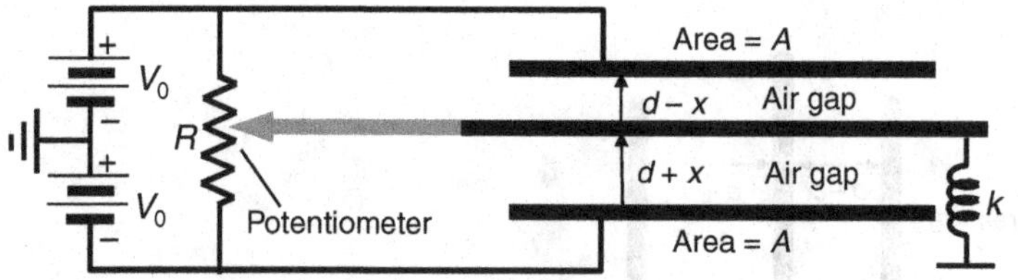

(a) Determine the equilibrium position of the movable plate x_0.

(b) Develop a capacitive lumped parameter model and from it an N-form electromechanical network.

(c) Determine the condition on the voltage in terms of system parameters for which the equilibrium is stable.

4.22 A reliable way to defeat the pull-in effect in variable-gap capacitive MEMS transducers is to use mechanical stops. Consider a variable-gap device with a restoring spring of spring constant k. With no electrical force present, that is, $V_0 = 0$, the spring establishes a mechanical equilibrium at $x = x_k$.

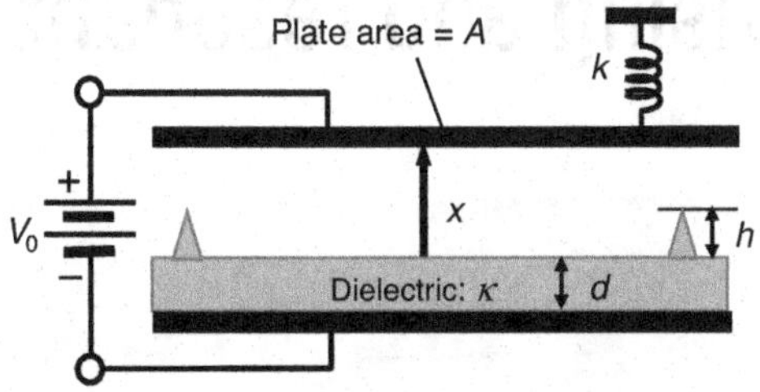

(a) Obtain an expression for the minimum height h of the mechanical stops needed to avoid the instability leading to pull-in. Ignore fringing as well as the small effect of the stops on the capacitance.
(b) Find an expression for the pull-in voltage.
(c) Given $x_k = 3$ μm, $d = 1$ μm, $A = 2500$ μm^2, $\kappa = 3$, and $k = 0.3$ N/m, calculate the pull-in voltage and the minimum value for h needed to prevent occurrence of the instability.

References

1. H. K. P. Neubert, *Instrument Transducers* (Oxford: Clarendon Press, 1975).
2. H. H. Woodson and J. R. Melcher, *Electromechanical Dynamics Part I* (New York: Wiley, 1968), Section 5.1.
3. H. A. C. Tilmans, Equivalent circuit representation of electromechanical transducers: I. Lumped parameter systems, *Journal of Micromechanics and Microengineering*, 6 (1996), 157–176.
4. N. Lobontiu, *System Dynamics for Engineering Students* (Burlington, MA: Academic Press, 2010).
5. H. H. Woodson and J. R. Melcher, *Electromechanical Dynamics Part I* (New York: Wiley, 1968), Section 2.2.
6. A. G. Bose and K. N. Stevens, *Introductory Network Theory* (New York: Harper and Row, 1965), Chapter 7.

5 Capacitive sensing and resonant drive circuits

5.1 Introduction

MEMS devices are usually exploited in practical measurement and actuation technologies by integrating them on-chip with the necessary sensing circuitry and electronic drives. While the electromechanical transducer itself is the heart of any MEMS system, its capabilities can be realized only with appropriate amplification, regulation, and signal conditioning. Electronic design has always been central to actuator and sensor technologies, but the development of microfabricated devices has presented new challenges for circuit designers. Devices with dimensions of the order of tens to hundreds of microns have very small capacitances – $C < 1$ pF – and circuits must be reliably sensitive down to $\Delta C \sim 10$ fF. Further, typical devices fabricated on chips suffer significant parasitics, requiring that serious attention be paid to electrostatic shielding and to the issues of signal strength and noise.

In this chapter, we introduce and analyze some of the basic operational-amplifier-based circuit topologies for capacitive sensing. The presentation focuses on how to integrate the two-port electromechanical models developed in Chapter 4 with simple amplifier circuits. The emphasis is on basic principles at the systems level. The starting point is a brief review of the ideal operational amplifier and its most important circuit implementation, namely, the inverting amplifier configuration. We then apply this very robust and adaptable circuit to the basic DC biased two-plate capacitive sensor, showing that such systems provide good sensitivity but entail the serious problem of large DC voltage offsets. Three-plate sensing schemes are then introduced as a way to avoid DC offsets and to take advantage of the inherent sensitivity of differential measurements. We next introduce AC sensing schemes, which eliminate the need for DC bias altogether and, when used in a signal modulation mode, provide another distinct advantage, namely, good immunity to $1/f$ noise. Next comes a brief description of switched capacitance circuits, including a simple analog circuit representation for them.

The last section of the chapter covers periodic pulsed drive circuits used for resonant actuation of MEMS devices. We emphasize the implementations commonly used for MEMS elements operated at mechanical resonance. The importance of feedback for resonant operation is illustrated in our consideration of phase-locked loop systems.

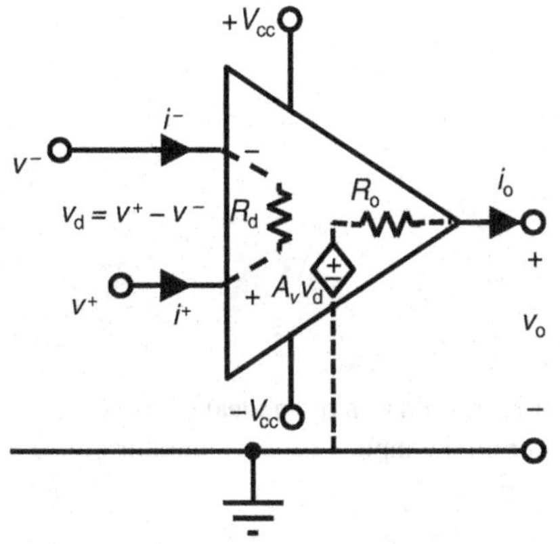

Figure 5.1 The basic operational amplifier with definitions of the inputs and the output. Also shown is the commonly used circuit model with resistors and the dependent voltage source having gain $A_v \gg 1$. Biasing voltages, typically provided by bipolar DC, as shown by $+V_{cc}$ and $-V_{cc}$, control the small-signal operation of the op-amp, but are not directly apparent in the small-signal model, which remains accurate as long as $|v_o| < V_{cc}$.

5.2 Basics of operational amplifiers

The operational amplifier (or op-amp) lends itself readily to idealization and straightforward analysis. As shown in the schematic of Fig. 5.1, op-amps have two input terminals, labeled inverting (−) and non-inverting (+), and a single output, all referenced to ground. The figure provides a basic circuit model, with input resistance R_d, output resistance R_o, and a controlled voltage source with gain A_v. Note the definition of the voltage $v_d = v^+ - v^-$. The important characteristics of an ideal amplifier are:

1. Very high input impedance ($R_d \to \infty$),
2. Very low output impedance ($R_o \to 0$),
3. Very high open-loop gain ($A_v \to \infty$),
4. Very small input current ($i^+, -i^- \to 0$),
5. Large bandwidth (depending on the op-amp).

The essential behavior of many op-amp-based circuits can be captured using approximations based on these characteristics.

While op-amps enjoy broad applications in electronics, in MEMS they are used almost exclusively for signal amplification. Accordingly, we here limit attention to robust topologies featuring negative feedback. The open-loop gain is very large; $A_v \approx 10^5$ is typical even for very inexpensive components. Thus, negative feedback is employed to constrain device operation within the bounds of linearity, that is, $|v_o| < \sim V_{cc}$. Effectively, the voltage gain of such configurations is determined by the feedback.

The usual approach to analyzing op-amp circuits uses two approximations, sometimes elevated to the status of "rules" [1]. The first rule is that the output adjusts itself so that

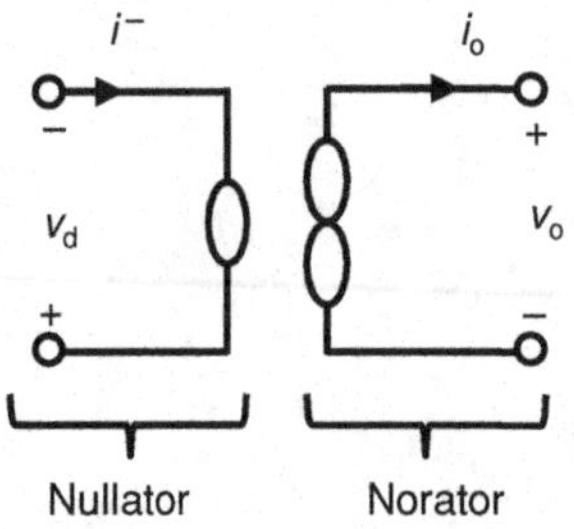

Figure 5.2 Schematic of the two-port nullor representation of an ideal operational amplifier. Refer to Eq. (5.3). The nullor consists of a nullator on the input side and a norator on the output side.

the voltage drop between the inverting and non-inverting inputs is very, very small:

$$v_d \to 0. \tag{5.1}$$

What actually drives this differential voltage to zero is the feedback coupled with the large amplification factor A_v. The second rule is that the inverting and non-inverting inputs draw virtually no current, that is,

$$\begin{aligned} i^+ &\to 0, \\ i^- &\to 0. \end{aligned} \tag{5.2}$$

We can express these rules in terms of a two-port transmission matrix:

$$\begin{bmatrix} -v_d \\ i^- \end{bmatrix} = \begin{bmatrix} 0 & 0 \\ 0 & 0 \end{bmatrix} \begin{bmatrix} v_O \\ i_O \end{bmatrix}. \tag{5.3}$$

The matrix appearing in Eq. (5.3) certainly looks strange, yet it turns out to be useful in the analysis of sensor circuit topologies based on op-amps. Equation (5.3) is the matrix terminal relation of the *nullor*,[1] the standard schematic for which appears in Fig. 5.2. The nullor consists of two elements that always appear together: the *nullator* on the left and the *norator* on the right.

5.3 Inverting amplifiers and capacitive sensing

The most important topology using an op-amp is the inverting amplifier configuration shown in Fig. 5.3(a). This workhorse of a circuit, with its negative feedback resistor R_f, is very robust. In particular, its performance can be made quite insensitive to such device parameters as the open-loop gain A_v. The simplicity and robustness of this circuit make it a popular choice in discrete electronics. For all the same reasons, it is a good choice for capacitive MEMS.

[1] The nullor was introduced by Tellegen in 1954 as *the universal active element* [2].

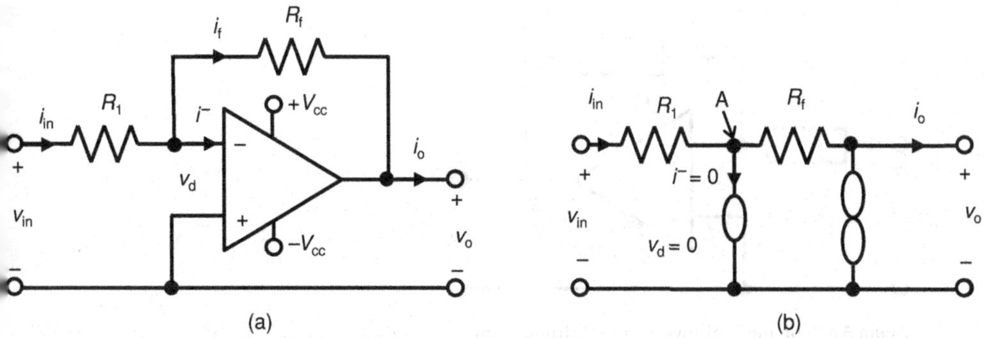

Figure 5.3 The robust inverting amplifier topology. (a) Simple inverting amplifier with resistors R_1 and R_f. (b) Two-port representation of the circuit based on the nullor.

5.3.1 Basic inverting configuration

We can put the nullor element to work right away to analyze the inverting amplifier. Refer to the equivalent circuit shown in Fig. 5.3(b). Applying Kirchhoff's current law at the node A, indicated in the figure, and then employing Eq. (5.2) tells us that the current is the same in both R_1 and R_f:

$$i_{in} = i_f. \tag{5.4}$$

Using Ohm's law to express these currents in terms of the voltages,

$$i_{in} = \frac{v_{in}}{R_1} \tag{5.5}$$

and

$$i_f = -\frac{v_0}{R_f}, \tag{5.6}$$

Eq. (5.4) becomes

$$v_0 = -\frac{R_f}{R_1} v_{in}. \tag{5.7}$$

These same expressions yield the input impedance of the inverting amplifier:

$$\underline{Z}_{in} = \frac{\underline{v}_{in}}{\underline{i}_{in}} = R_1. \tag{5.8}$$

Note that the proper definition of impedance requires use of phasor voltage and current, that is, $\underline{v}_{in}$ and $\underline{i}_{in}$, rather than time-dependent variables.

This circuit analysis, performed using time-dependent voltages and currents, led to a very simple result for gain because both external elements were resistors. One can generalize to the sinusoidal steady-state AC merely by switching over to phasor quantities and replacing R_1 and R_f with complex impedances, $\underline{Z}_1$ and $\underline{Z}_f$, as shown in Fig. 5.4. Now, Eq. (5.7) generalizes to

$$\frac{\underline{v}_{out}}{\underline{v}_{in}} = \underline{H}_{v/v}(j\omega) = -\frac{\underline{z}_f}{\underline{z}_1}, \tag{5.9}$$

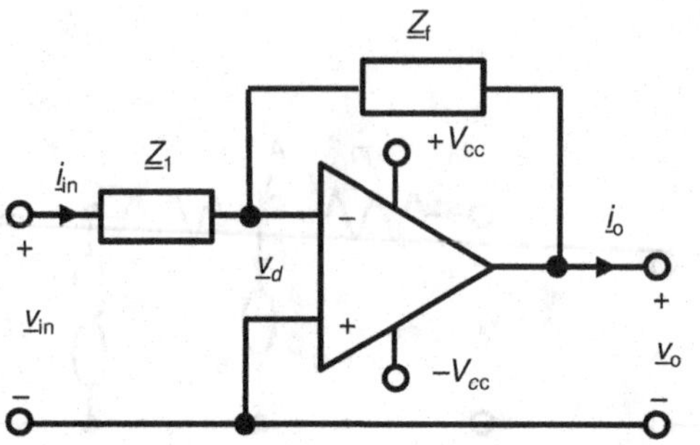

Figure 5.4 The general inverting amplifier circuit topology for sinusoidal voltages and currents (phasors). Note that complex impedances $\underline{Z}_1$ and $\underline{Z}_f$ now replace the resistances of Fig. 5.3.

where $\underline{H}_{v/v}(j\omega)$ is a voltage transfer function. The input impedance becomes

$$\underline{Z}_{in} = \frac{\underline{v}_{in}}{\underline{i}_{in}} = \underline{Z}_1. \tag{5.10}$$

Equation (5.9) suggests that sensing capacitors can be implemented directly in the configuration of Fig. 5.4, and that is what we will next consider.

5.3.2 One-sided high-impedance (charge) amplifier

The first example of a capacitive sensing scheme with an op-amp to be considered is the high-impedance (charge) amplifier configuration integrated with a two-plate sensing capacitor in the amplifier input. $\underline{Z}_1$ is the impedance of a displacement-dependent capacitance $C(x)$:

$$\underline{Z}_1 = \frac{1}{j\omega C(x)}. \tag{5.11}$$

The feedback impedance $\underline{Z}_f$ is provided by a reference capacitor C_f:

$$\underline{Z}_f = \frac{1}{j\omega C_f}. \tag{5.12}$$

The transfer function, or voltage gain, is simply the ratio of the capacitances with a minus sign:

$$\underline{H}_{v/v}(j\omega) = -\frac{\underline{Z}_f}{\underline{Z}_1} = -\frac{C(x)}{C_f}. \tag{5.13}$$

$\underline{H}_{v/v}(j\omega)$, which relates the AC output voltage phasor $\underline{v}_o$ to the AC input signal $\underline{v}_{in}$, depends on the displacement x through the capacitance $C(x)$. Modulation schemes for sensors will be considered later in the chapter. For now, attention is directed to electromechanical sensors where the input is a mechanical quantity that produces an electrical signal, which is subsequently amplified by the inverting amplifier. Before proceeding, recall that the basic types of capacitive transducer introduced in Chapter 4 use DC voltage bias and that their transducer action is based on the assumption of small signals and linearization. $\underline{M}$ and $\underline{N}$, the coupling coefficients of the M- and N-form

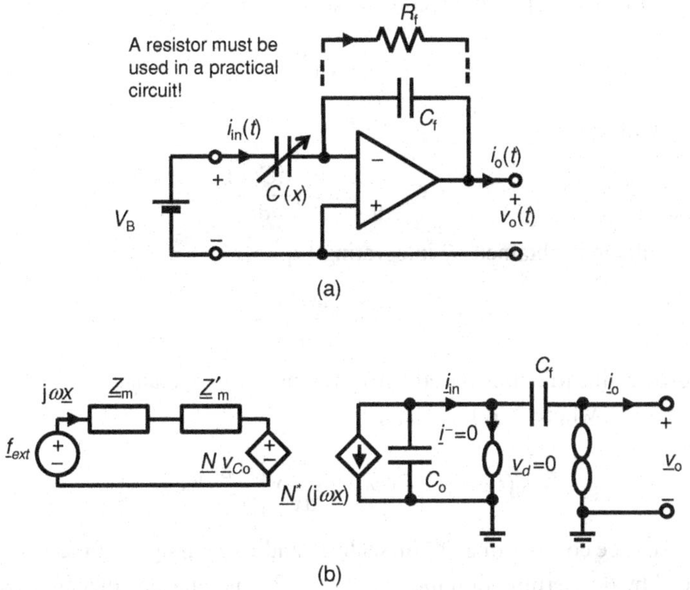

Figure 5.5 The basic, high-impedance capacitance sensing circuit topology uses a two-plate sensor in the input. (a) Schematic of the circuit showing the DC bias. For clarity, the capacitive sensing element is denoted with a diagonal arrow. (b) Equivalent small-signal AC circuit based on an N-form transducer cascaded with a nullor (op-amp) network. Note that $\underline{N}^*(j\omega\underline{x})$ points down since the reference positive electrode of $C(x)$ is shorted to ground via V_B.

transducer matrices, depend linearly on this bias voltage. Figure 5.5a shows how to connect a DC bias voltage V_B in the two-plate sensor configuration.

With only capacitive feedback, leakage currents and offsets in the op-amp always lead to op-amp saturation, which can be avoided by adding a resistor in the feedback path. A more realistic circuit will be analyzed in Section 5.3.4, but before introducing that necessary resistor and its complications, consider the simplified, idealized configuration of Fig. 5.5a.

Either the M- or N-form electromechanical two-port network introduced in Chapter 4 can be invoked to analyze the circuit shown in Fig. 5.5. The choice is based on convenience. Before starting this analysis, however, it is instructive to seek insight on the energy transduction mechanism of the system by resorting to time-domain circuit analysis. In particular, one can gain understanding of the role of the voltage bias by considering the time-dependent currents. From Eq. (5.1), the inverting input is at virtual ground, so the voltage across the sensing capacitor $C(x)$ is constant and equal to the DC bias voltage V_B. Thus, the time-dependent input current flowing through the sensing capacitor depends only on the time derivative of the capacitor $C(x)$:

$$i_{in}(t) = \frac{dC(x)}{dt} V_B. \tag{5.14}$$

The current flowing through the feedback capacitor C_f is

$$i_f = -C_f \frac{dv_0}{dt}. \tag{5.15}$$

Using KCL with $i^- = 0$,

$$-C_f \frac{dv_0}{dt} = V_B \frac{dC(x)}{dt}. \tag{5.16}$$

The output voltage is obtained by integrating Eq. (5.16):

$$v_0(t) = -V_B \frac{C[x(t)]}{C_f} + c_{intg}. \tag{5.17}$$

Next follows a linearization based on a Taylor series expansion of $C(x)$, assuming $x(t) = x_0 + x'(t)$, where $|x'(t)| \ll |x_0|$:

$$v_0(t) = -\left[C(x_0) + \frac{dC}{dx}x\right]\frac{V_B}{C_f} + c_{intg}. \tag{5.18}$$

The output voltage consists of a DC offset term and a small-signal, time-varying component induced by the perturbation motion, $x'(t)$. The integration constant c_{intg} is ignored for now, but this term is important and in practice gives rise to a saturation of the op-amp. The solution to this problem is discussed in Section 5.3.4.

This circuit analysis is straightforward, but this approach does not tell us much about the mechanical side of the device; for example, how to relate the output voltage to an applied external force, rather than the displacement. By contrast, the small-signal model of Chapter 4, that is, the M- and N-form electromechanical two-port networks and their associated matrices, provide a very general systematic methodology.

Figure 5.5b shows an equivalent circuit representation of the capacitive transducer in the N-form cascaded with the nullor representation of the inverting amplifier. It follows directly that

$$\underline{i}_{in} = -\underline{N}^*(j\omega\underline{x}), \tag{5.19}$$

$$\underline{i}_0 = -j\omega C_f \underline{v}_0. \tag{5.20}$$

Then, because $\underline{i}_{in} = \underline{i}_o$,

$$\underline{v}_0 = \frac{\underline{N}^*}{C_f}\underline{x}. \tag{5.21}$$

Using Eq. (4.16b) and letting $v_o = V_B$,

$$\underline{N} = -V_B \left.\frac{dC}{dx}\right|_{x_0}, \tag{5.22}$$

so the output voltage becomes

$$\underline{v}_0 = -\left.\frac{V_B\, dC}{c_f\, dx}\right|_{\underline{x}_0} \underline{x}, \tag{5.23}$$

which is consistent with the time-dependent small-signal component in Eq. (5.18).

Assuming a sinusoidal time-varying displacement of the form $x'(t) = x_m \cos(\omega t)$, the total output voltage response in the time domain is

$$v_0(t) = \underbrace{-\frac{C(x_0)}{C_f} V_B}_{\text{DC offset } V_0} \underbrace{- \frac{V_B}{C_f} \frac{dC}{dx}\bigg|_{x_0} x_m \cos(\omega t)}_{\text{small-signal AC}}. \tag{5.24}$$

The output consists of a sensing voltage component linear in the displacement and a DC offset that is probably much larger in magnitude.

5.3.3 Variable-gap and variable-area capacitors

As important special cases for the sensing capacitance $C(x)$, consider basic variable-gap and variable-area capacitive geometries. From Table 4.2, the variable-gap capacitor is

$$C(x) = \frac{\varepsilon_0 A}{x_0 + x'}. \tag{5.25}$$

Then the small-signal output voltage phasor expressed in terms of the displacement x phasor is

$$\underline{v}_0 = -\frac{\varepsilon_0 A}{C_f x_0^2} V_B \underline{x}. \tag{5.26}$$

For the basic variable-area capacitive transducer in Table 4.2, the capacitance is

$$C(x) = \frac{\varepsilon_0 w (L - x)}{d}, \tag{5.27}$$

so that

$$\underline{v}_0 = -\frac{\varepsilon_0 w}{C_f d} V_B \underline{x}. \tag{5.28}$$

For both variable-gap and variable-area capacitive sensors, the small-signal response depends linearly on the bias voltage and inversely upon electrode spacing. Because $L \gg x_0$, we might expect variable-gap devices to be more sensitive than variable-area geometries. Although overall sensitivity depends on factors besides the ratio $\underline{v}_0/\underline{x}$, this conclusion is still a safe one.

*5.3.4 Effect of op-amp leakage current

The previous results are predicated on the assumption of ideal operational amplifier behavior. In real capacitive MEMS systems, however, the non-ideal behavior of op-amps has a very strong influence on performance. One problem is the existence of non-zero leakage currents at the amplifier input. Though these currents are small, it is also true that MEMS capacitors are very small. Thus, in a relatively short period of time, these currents can charge up the feedback capacitor C_f and drive the op-amp into saturation, that is, the limit of linear operation, $v_o \sim V_{cc}$. Figure 5.6 shows the amplifier

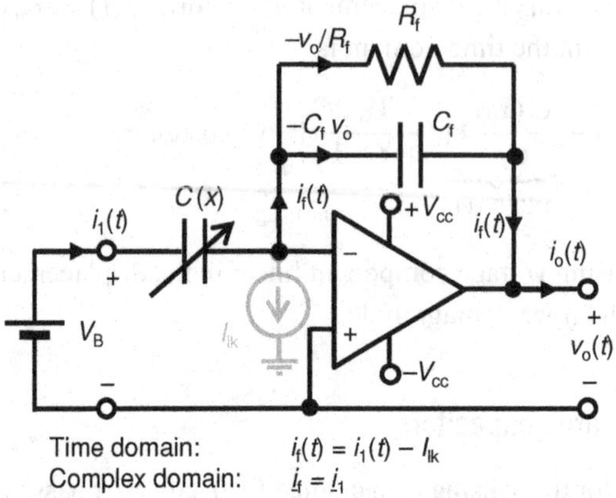

Figure 5.6 Effect of leakage current on the two-plate sensing circuit. The circuit topology employs a feedback resistor R_f to avoid saturation due to leakage current.

circuit with this leakage current represented by DC current source I_{lk}. To account for the electric charge accumulated by this small current, Eq. (5.17) requires modification:

$$v_0 = -\frac{C(x)}{C_f}V_B + \frac{1}{C_f}\int_0^t I_{lk}\,dt = -\frac{C(x)}{C_f}V_B + \frac{I_{lk}}{C_f}t. \tag{5.29}$$

The leakage current I_{lk} gives rise to a time-dependent component in the output voltage that, depending on the sign of I_{lk}, increases (or decreases) until the output voltage reaches saturation. A simple solution to this problem is to add a resistor R_f in parallel with C_f in the feedback path, as shown in Fig. 5.6. R_f must be low enough to prevent the leakage current from charging the feedback capacitor excessively but not so low as to interfere with the integrator function of the capacitor over the designed frequency range of the sensor.

One can place an upper limit on the value of R_f by calculating the steady DC voltage V_o induced by the leakage current itself. This voltage is

$$V_0 = R_f I_{lk}. \tag{5.30}$$

If $|R_f I_{bias}| \ll V_{cc}$, then saturation is avoided and linear sensor response is guaranteed.

Another way to see the necessity of the feedback resistor in a practical circuit is to note that without R_f there is no feedback path to keep the high internal amplification of the op-amp A_v under control at DC. Figure 5.7 shows the inverting amplifier with the feedback resistor, with the op-amp shown in more detail, including the leakage current. At DC, with no motion ($x =$ constant) the capacitances act as open circuits. The resistance R_f ensures the existence of the feedback at DC and low frequencies. Without R_f, the leakage current would give rise to finite v_d. Leakage current is not the only way that can lead to saturation. Without feedback, any offset voltage at the input can cause saturation due to the enormous open-loop gain A_v.

The resistor R_f solves the problem caused by the leakage current, but has an adverse effect on the frequency response. The frequency response can be determined from the

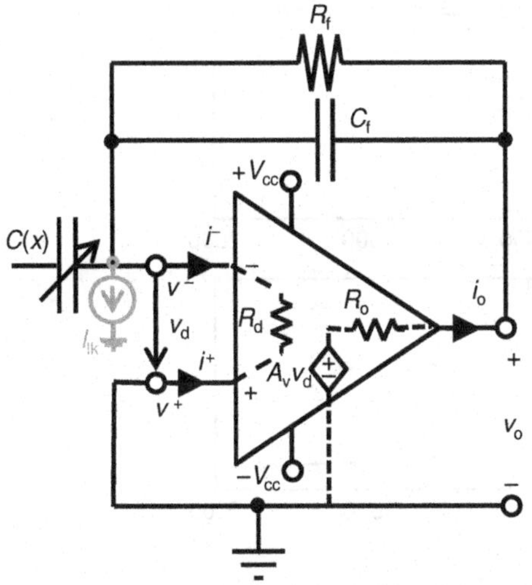

Figure 5.7 Simple amplifier model of a non-ideal behavior. As $\omega \to 0$, $1/(\mathrm{j}\omega C_f) \to \infty$, and capacitance becomes an open circuit, breaking the feedback path. The feedback resistor R_f is necessary to ensure feedback at DC. In the absence of feedback, large amplification of the op-amp A_v greatly amplifies any small input voltage, and leads to saturation.

two transfer functions of the cascaded electromechanical networks. The first of these is

$$\underline{H}_{i/x}(\mathrm{j}\omega) \equiv \frac{\underline{i}_1}{\underline{x}} = -\mathrm{j}\omega \left.\frac{\mathrm{d}C}{\mathrm{d}x}\right|_{x_0} V_B. \tag{5.31}$$

From inspection of Fig. 5.6b, the output voltage $\underline{v}_{out}$ is related to the current $\underline{i}_1$ through the transfer function $\underline{H}_{v/i}(\mathrm{j}\omega)$:

$$\underline{H}_{v/i}(\mathrm{j}\omega) \equiv \frac{\underline{v}_{out}}{\underline{i}_1} = -\frac{\underline{Z}_f \underline{i}_f}{\underline{i}_1} = \frac{-1}{\mathrm{j}\omega C_f + 1/R_f}. \tag{5.32}$$

The overall transfer function is the product of $\underline{H}_{i/x}$ and $\underline{H}_{v/i}$:

$$\begin{aligned}\underline{H}_{v/x}(\mathrm{j}\omega) &\equiv \frac{\underline{v}_0}{\underline{x}} = H_{v/i}(\mathrm{j}\omega)\, H_{i/x}(\mathrm{j}\omega)\\ &= \frac{\mathrm{j}\omega\, R_f C_f}{1 + \mathrm{j}\omega\, R_f C_f} \left(\left.\frac{\mathrm{d}C}{\mathrm{d}x}\right|_{x_0} \right) \frac{V_B}{C_f}.\end{aligned} \tag{5.33}$$

From Eq. (5.33), it is evident that the feedback resistor R_f reduces the magnitude and shifts the phase of the transfer function in the lower frequencies. In the time domain, the output consists of a superposition of the DC term from Eq. (5.30) and the AC component from Eq. (5.33), that is,

$$v_{out}(t) = \underbrace{R_f I_{lk}}_{\text{DC offset, } v_0} + \underbrace{K_{vx} x_m \cos(\omega t + \phi_{xv})}_{\text{AC component}}. \tag{5.34}$$

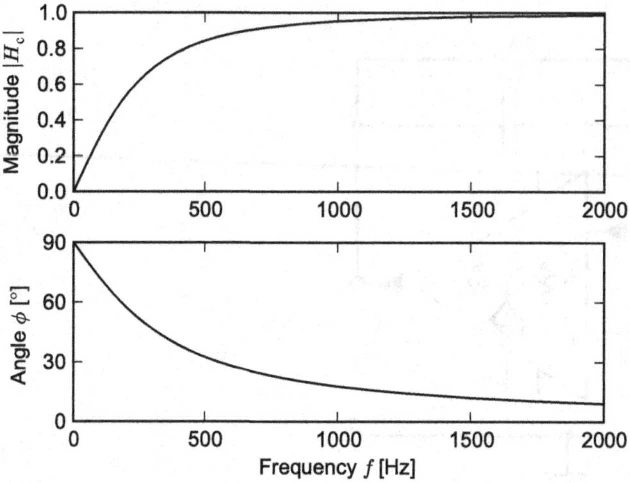

Figure 5.8 Frequency response of $H_c(\omega)$ for $R_f = 1\ \mathrm{G\Omega}$ and $C_f = 1$ pF. From the frequency response it is clear that $H_c(\omega)$ is a first-order high-pass filter. At $f = 1/\tau$, the magnitude response reaches about 95% of its maximum, and the phase drops to about 18°.

Using the definition $\tau = R_f C_f$, the gain factor K_{vx} and phase angle ϕ_{vx} are

$$K_{vx} = \frac{\omega\tau}{\sqrt{1+(\omega\tau)^2}} \frac{V_B}{C_f} \frac{\mathrm{d}C}{\mathrm{d}x}, \tag{5.35}$$

$$\phi_{vx} = \arctan\left(\frac{1}{\omega\tau}\right). \tag{5.36}$$

Frequency-independent behavior of the sensor is limited to high frequency, that is, $\omega\tau \gg 1$, in which case Eq. (5.34) reduces to

$$v_0(t) = \underbrace{R_f I_{lk}}_{\text{DC part}} - \underbrace{\left(\frac{V_B}{C_f}\frac{\mathrm{d}C}{\mathrm{d}x} x_m\right)\cos(\omega t)}_{\text{AC part}}. \tag{5.37}$$

Note that the AC terms of Eqs. (5.37) and (5.24) are identical. On the other hand, the DC values differ. The DC component appearing in Eq. (5.24) is the voltage observed immediately after the bias is applied, while the DC term in Eq. (5.37) is the final settled value observed for $t \geq \sim 3\tau$.

The effect of R_f on the frequency response is conveniently expressed as a correction factor,

$$H_c(\omega) = \frac{\mathrm{j}\omega R_f C_f}{1+\mathrm{j}\omega R_f C_f} = \frac{\mathrm{j}\omega\tau}{1+\mathrm{j}\omega\tau} = \frac{\omega\tau}{\sqrt{1+(\omega\tau)^2}} \mathrm{e}^{\mathrm{j}\arctan\left(\frac{1}{\omega\tau}\right)}, \tag{5.38}$$

which acts as a high-pass filter. For example, Fig. 5.8 shows the frequency response of $H_c(\omega)$ for $R_f = 1\ \mathrm{G\Omega}$ and $C_f = 1$ pF ($\tau = 1$ ms). The magnitude response exceeds 0.95 at 1 kHz, which corresponds to $1/\tau$. At the same frequency, the phase angle has dropped to about 18°.

Example 5.1 **Numerical example of the effect of the leakage current**

Assume the following numerical values for the parameters of the circuit shown in Fig. 5.6:

$$I_{\rm lk} = 0.1\text{pA}; R_{\rm f} = 1\ \text{G}\Omega; C_{\rm r} = C_0 = 1\ \text{pF}; V_{\rm B} = 5\ \text{V}; f = 10\ \text{kHz range}.$$

The frequency limit imposed by the integrator condition is $2\pi f R_{\rm r} C_{\rm r} \gg 1$. For the present case, $2\pi f R_{\rm r} C_{\rm r} = 2\pi(10^4)\cdot(1\times10^9)/(1\times10^{-12}) = 62.8 \gg 1$, so excellent integrator operation is assured. After the initial transient, before charge due to the leakage current has started to accumulate in the capacitor, and assuming no sensor input, the DC offset voltage from Eq. (5.24) is $V_{\rm out} = 1/1{*}5 = 5$ V. The steady-state condition accounting for the bias current from Eq. (5.30) gives a much lower offset voltage,

$$R_f I_{\rm lk} = (0.1\times10^{-12})(1\times10^9) = 0.1\,\text{mV}.$$

From Eq. (5.29), the settling time from 5 V to 0.1 μV is

$$\Delta t = \Delta V C_{\rm r}/I_{\rm lk} \approx (5)(1\times10^{-12})/(0.1\times10^{-12}) = 50\,\text{s}.$$

Thus, the integrator condition is easily satisfied. Note, however, that it may take considerable time before DC level settles to $R_{\rm f} I_{\rm lk}$. To prevent saturation, one must ensure that $V_{\rm B} C_{\rm o}/C_{\rm f}$ is within the range of the op-amp, $V_{\rm B} C_{\rm o}/C_{\rm f} < V_{\rm cc}$.

The following two examples illustrate the problems associated with the one-sided amplifiers studied so far.

Example 5.2 **Variable-gap MEMS device – the motivation for differential (three-plate) sensing**

The equilibrium capacitance is bound to be much larger than the time-varying portion due to the mechanical motion. To get some sense of the relative magnitudes, consider the variable-gap capacitive transducer shown in Fig. 5.9. Neglecting fringing fields, the perturbed capacitance is

$$C(x) = N\varepsilon_0 h/L_{\rm eff}\left(\frac{1}{g_1 - x} + \frac{1}{g_2 - x}\right) \approx \underbrace{N\varepsilon_0 h/L_{\rm eff}\frac{g_1+g_2}{g_1 g_2}}_{C_0} + \underbrace{N\varepsilon_0 h/L_{\rm eff}\frac{g_2^2+g_1^2}{g_1^2 g_2^2}}_{\mathrm{d}C/\mathrm{d}x},$$

where $x(t)$ is the displacement from equilibrium condition.

It is useful to compare the two terms on the right-hand side of the last equation:

$$\frac{\frac{\mathrm{d}C}{\mathrm{d}x}x}{C_0} = \frac{g_2 - g_1}{g_1 g_2}.$$

It is instructive to compute this ratio using some realistic values for the gaps, $g_1 = 1$ μm and $g_2 = 3$ μm, and a desired value for the sensor resolution, $x_{\rm Min} = 1$ Å. A spatial resolution of 1 Å may seem very small; however, we shall find in Chapter 7 that such values are quite realistic and attainable for inertial sensors. Using these numbers in Eq. (5.24), the result is

$$\frac{|\mathrm{d}C/\mathrm{d}x|}{C_0}|x| \sim 6.7\times10^{-5}.$$

Equation (5.24) reveals that this same very small number defines the ratio of the DC component of output voltage to the AC component of the output voltage. The conclusion is that the high-impedance (two-plate) sensor scheme is not well suited to high-sensitivity variable-gap capacitors because the resolution will be strongly affected by this ratio. In fact, the situation will be made worse by parasitic capacitances, as shown in Example 5.3.

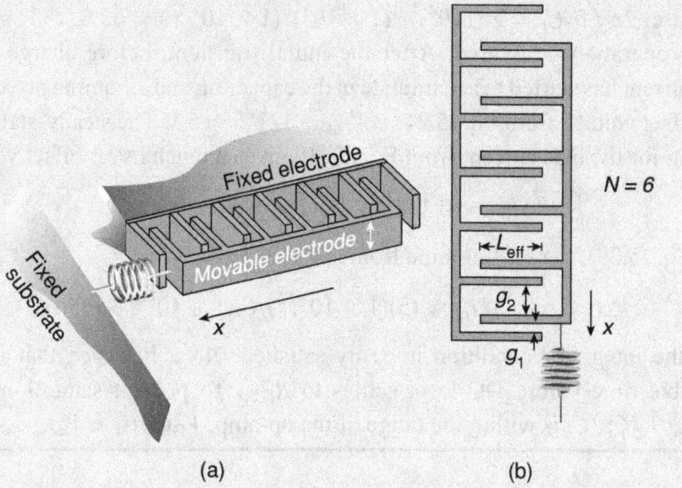

Figure 5.9 A simple single-ended variable-gap capacitive bank. (a) Perspective view showing the interdigitated electrodes. (b) Top view with definitions of the gaps.

Example 5.3 **The effect of parasitic capacitance in MEMS sensors**

Consider the capacitance bank introduced in Example 1.1 (see Fig. 1.2). The capacitance expression $C(x)$ is given by Eq. (1.9). Here, we make a comparison of the equilibrium capacitance of the sensor,

$$C_0 = 2N\varepsilon\frac{hL_{\text{eff}}}{g},$$

with the parasitic capacitance of a wire having the same width as the electrode and a length just sufficient to connect the N moving electrodes, assuming that wires are insulated from the substrate by the oxide thickness $t_{ox} = 0.1$ μm. C_0 was computed in Example 1.1 to be $C_0 = 8$ fF. The parasitic capacitance C_s is estimated by

$$C_s = \varepsilon_0\kappa\frac{w\,(2N-1)\,(w+g)}{t_{ox}} = 30.4 \text{ fF}.$$

The effective equilibrium capacitance of the sensing circuit is the sum of C_0 and C_s. Therefore, even for a variable-area capacitive sensor exhibiting highly linear behavior over a large percentage of the dynamic range of the sensor, parasitic capacitance is very significant and is typically much larger than C_0. A similar conclusion is reached for the variable-gap sensor geometries. The offset problem of the one-sided, high-impedance (charge) amplifier configuration is actually even bigger than previously demonstrated.

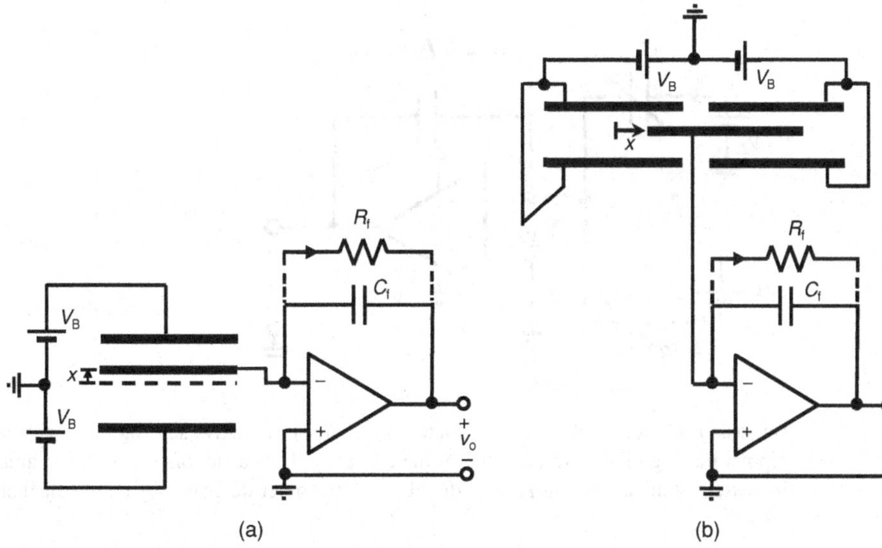

Figure 5.10 Three-plate capacitive sensing schemes based on inverting amplifier topology with capacitive feedback C_f. The moving element is connected to inverting input of the operational amplifier. For balanced half-bridge operation, the two capacitances must be equal when the moving element is in its equilibrium position at $x = 0$. (a) Variable-gap sensor. (b) Variable-area sensor. The role of the feedback resistance is to prevent saturation. It is ignored in the first analysis. Its presence is accounted for by multiplying the resulting transfer function by $H_c(\omega)$, as defined in Eq. (5.38).

5.4 Differential (three-plate) capacitance sensing

The problem of DC offset in the output of high-impedance capacitive sensor systems is serious. Often, the DC component is much greater in magnitude than the voltage signal induced by the motion that we wish to detect. This results in insurmountable noise-related difficulties. One way to avoid a DC offset in the output is to use the somewhat more complex three-plate capacitive geometries already described in Section 4.7.5. Refer to Fig. 4.15a and b, respectively, for variable-gap and variable-area implementations of three-plate sensors. Practical advantages of three-plate schemes include their inherently more sensitive, differential operation and the self-shielding feature of such geometries.

Integrated three-plate sensing schemes, for variable-gap and variable-area capacitive sensors, are shown in Fig. 5.10. In both geometries, note that the moving element is connected to the inverting input of the op-amp, while the stationary electrodes are connected to ground through the bias voltages, $+V_B$ and $-V_B$. The feedback resistance R_f, used to prevent saturation, is ignored in the first pass of our analysis, but then taken into account later.

Three-plate systems can be analyzed using either time-domain circuit analysis or the far more systematic transducer matrix approach. To gain insight, it is useful first to resort to a time-domain solution. Examination of Fig. 5.11 reveals that such systems are simply high-impedance topologies with dual inputs. Thus, superposition can be invoked

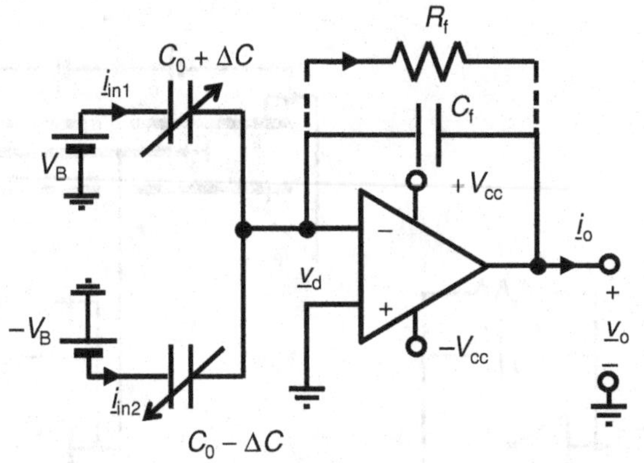

Figure 5.11 Circuit model for three-plate (differential) capacitive sensing system showing the bipolar biasing of the two branches of the capacitor. The capacitors are balanced and, when one increases by the increment $\Delta C = (\mathrm{d}C/\mathrm{d}x)_{x_0} x'$, the other decreases by an identical amount.

to write down the output voltage using Eq. (5.24):

$$\begin{aligned} v_0(t) &= -\underbrace{\left[\frac{C_0}{C_f} V_B + \frac{V_B}{C_f}\frac{\mathrm{d}C}{\mathrm{d}x} x_m \cos(\omega t)\right]}_{\text{upper branch, due to } i_{in1}} \\ &\quad - \underbrace{\left[\frac{C_0}{C_f}(-V_B) + \frac{(-V_B)}{C_f}\frac{\mathrm{d}C}{\mathrm{d}x}(-x_m \cos(\omega t))\right]}_{\text{lower branch, due to } i_{in2}} \\ &= -2\frac{V_B}{C_f}\frac{\mathrm{d}C}{\mathrm{d}x} x_m \cos(\omega t). \end{aligned} \tag{5.39}$$

Note that the DC offset terms from the two branches of the circuit cancel each other out, leaving only a term linear in the displacement and proportional to the product of the bias voltage and the derivative of the capacitance.

DC cancellation is realized if the two capacitances are matched at the mechanical equilibrium point. When not perfectly balanced, a DC component persists, though it can usually be kept very small. If the capacitance mismatch is large compared with the capacitance change, then it may affect the dynamic range, or even lead to saturation. Variable-gap configurations invariably exhibit small differences in $\mathrm{d}C/\mathrm{d}x$ between the upper and the lower branches, but such differences are not important because the two terms are additive in the output.

To eliminate the capacitor-charging problem caused by the leakage current, a shunt resistor R_f is placed in the feedback path.

Baxter [3] points out that, for balanced three-plate sensor systems, second-order corrections to the amplitude of the output voltage signal, which are proportional to $\mathrm{d}^2C/\mathrm{d}x^2$, cancel out. Three-plate systems are a good choice in applications where a highly linear response is needed over a large dynamic range, because the next term having an influence is third-order.

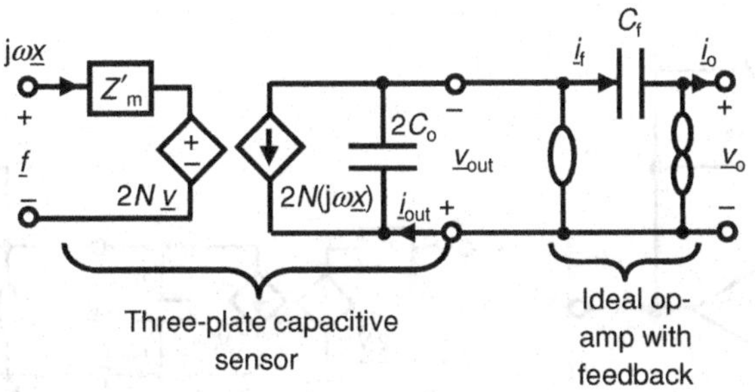

Figure 5.12 Small-signal AC equivalent electromechanical circuit model for a three-plate capacitance sensor using the nullor representation of the operational amplifier.

A systematic consideration of the AC small-signal response of the three-plate system is possible using the methods of Chapter 4. Using the formulation presented in Section 4.7.5 for these systems, in particular, the circuit model illustrated in Fig. 4.18, one can implement the cascade method. Figure 5.12 shows an equivalent circuit to be used for the small-signal AC analysis. Neglecting all mutual coupling capacitances, the two-port electromechanical transducer matrix of Eq. (4.72) is

$$\begin{bmatrix} \underline{f} \\ \underline{i}_{\mathrm{out}} \end{bmatrix} = \begin{bmatrix} \underline{Z}'_{\mathrm{m}} & 2\underline{N} \\ -2\underline{N}^* & 2\mathrm{j}\omega C_0 \end{bmatrix} \begin{bmatrix} \mathrm{j}\omega\underline{x} \\ \underline{v}_{\mathrm{out}} \end{bmatrix}. \tag{5.40}$$

From the nullor model for the operational amplifier, $\underline{v}_{\mathrm{d}} = 0$ so the voltage at the inverting input is ~0. There is also the external circuit constraint relating the total current and the output voltage:

$$\underline{i}_{\mathrm{out}} = -\frac{\underline{v}_0}{\underline{Z}_{\mathrm{f}}} = -\mathrm{j}\omega C_{\mathrm{f}}\underline{v}_0. \tag{5.41}$$

Combining Eqs. (5.40) and (5.41) with the condition on the voltage, $\underline{v}_{\mathrm{d}} = 0$, yields

$$\underline{v}_0 = 2\frac{\underline{N}^*}{C_{\mathrm{f}}}\underline{x}. \tag{5.42}$$

Mutual capacitance is negligible here, so

$$\underline{N} \approx -V_{\mathrm{B}}\frac{\mathrm{d}C}{\mathrm{d}x}. \tag{5.43}$$

Inserting Eq. (5.43) into Eq. (5.42),

$$\underline{v}_0 = -2\frac{\mathrm{d}C/\mathrm{d}x}{C_{\mathrm{f}}}V_{\mathrm{B}}\underline{x}. \tag{5.44}$$

As expected, since both approaches are based on linearization about the equilibrium, Eq. (5.44) is consistent with Eq. (5.39).

5.4.1 DC feedback for the differential configuration

It has been mentioned before that capacitance alone in the feedback path of an inverting amplifier always leads to saturation, and that a resistor added to the feedback path

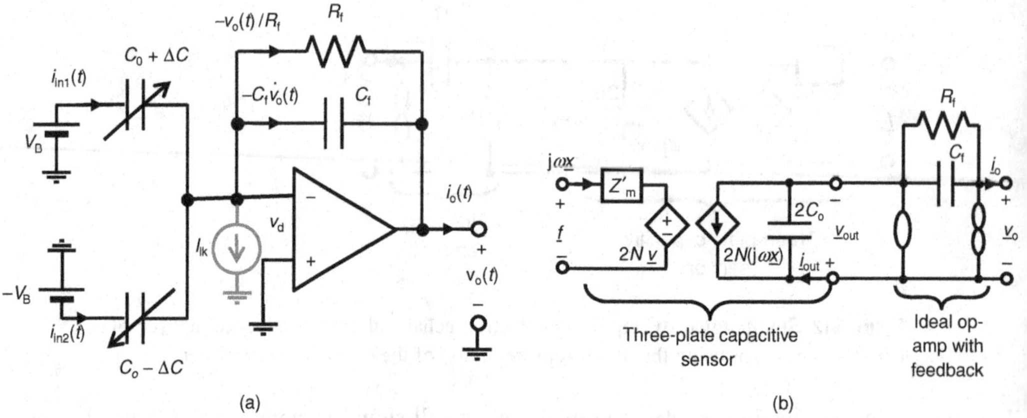

Figure 5.13 Three-plate capacitive sensor with feedback resistor introduced to eliminate saturation caused by leakage currents. (a) The standard circuit model. (b) The small-signal AC model with N-form of the three-plate transducer.

can solve the problem. Figure 5.13 shows the three-electrode capacitive sensor circuit including this necessary resistor along with the cascaded small-signal equivalent network.

The output voltage response due to mechanical motion is obtainable by inspection of the equivalent circuit:

$$\underline{v}_o = 2\underline{N}(j\omega) \underbrace{\left[R_f \parallel \frac{1}{j\omega C_1}\right]}_{\substack{\text{parallel connection}\\ \text{of resistor and capacitor}}} = -2\frac{dC/dx}{C_f} V_B \underbrace{\frac{j\omega C_f R_f}{1 + j\omega C_f R_f}}_{\underline{H}_c(\omega)} \underline{x}. \tag{5.45}$$

Not surprisingly, the term $\underline{H}_c(\omega)$ here is the same factor that appears in Eq. (5.38) for the two-plate high-impedance sensor.

5.5 AC (modulated) sensing

It turns out that DC bias is impractical as a means of energizing capacitive MEMS sensors.[2] Zero-mean time-periodic voltage excitation is often employed instead. This section introduces and analyzes such sensing schemes. An advantage of time-periodic excitation schemes is that they provide the option of using *signal modulation* and *detection*. Signal modulation makes it possible to suppress certain types of noise that can seriously degrade the performance of capacitive MEMS sensors. Both sinusoidal and square-wave voltage signals are considered, the former because it is of instructive value and relatively easy to analyze, and the latter for the practical reason that it is readily available on-chip. The AC sensors based on sinusoidal excitation were developed first and they have real operational advantages; however, square-wave systems now dominate

[2] This is because DC bias can give rise to a long-term drift due to the motion of trapped charges in oxides.

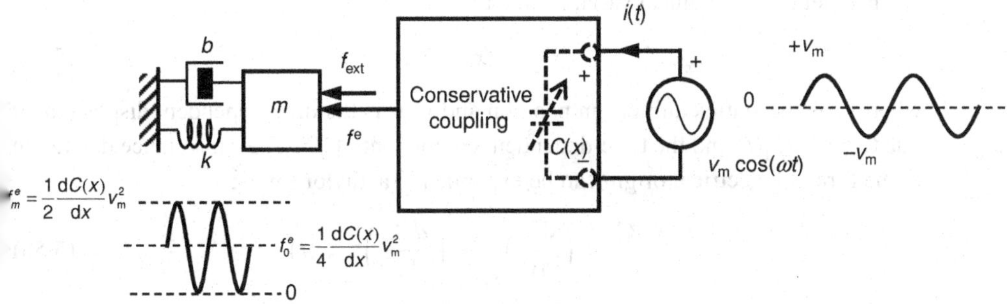

Figure 5.14 Basic capacitive sensor with sinusoidal bias voltage. In this simple configuration, the sensor input is a mechanical force $f'_{ext}(t)$, which is detected as an output current signal $i'(t)$.

for the very practical reason that modern junction transistor-based electronic systems can provide square-wave signal and pulses much more conveniently.

To treat AC capacitive sensing schemes, it is necessary to set aside the systematic formalism of the electromechanical transducer matrix of Chapter 4. This is because time-periodic voltage excitation introduces time-dependent coefficients to the lincar equations, which are not amenable to the derivation leading to the M- and N-form two-port networks. The alternative is to fall back on the basics of electromechanics first introduced in Chapter 3 and perform time-domain analysis. As long as the motion of the moving element can be accurately regarded as a perturbation, linearized expansion of the capacitance function is justified.

5.5.1 Capacitive sensor excited by zero-mean sinusoidal voltage

Before investigating circuit topologies for AC capacitive sensing, it is well to examine the effect of time-varying bias voltage on capacitive energy transduction. It might seem that the *virtual work method* introduced in Chapter 3 implies a condition of static voltage or charge, but this is not the case. The approximation that the electroquasistatic system is conservative guarantees that Eq. (3.1) will account properly for conversion and storage, irrespective of how voltage or charge vary in time. That being the case, then, once the force of electrical origin is determined, there are no restrictions on their time-dependence. We are thus entitled to assume that $v(t) = v_m \cos(\omega_e t)$, in which case,

$$f^e = \frac{1}{2}\frac{dC}{dx}(v_m \cos(\omega t))^2 = \frac{v_m^2}{4}\frac{dC}{dx}(1 + \cos(2\omega t)). \tag{5.46}$$

Consider the capacitive sensor shown in Fig. 5.14, which is designed to detect motions resulting from a force f_{ext} exerted on the proof mass m. The equation of motion is

$$m\ddot{x} + b\ddot{x} + kx = f_{ext} + \frac{1}{2}\frac{dC}{dx}v^2(t), \tag{5.47}$$

where b and k are a damping coefficient and spring constant, respectively. Let the external force f_{ext} take the form

$$f_{ext} = f_{ext,0} + f'_{ext}(t), \tag{5.48}$$

where $f_{ext,\,0}$ is constant and $f'_{ext}(t)$ is the component detected by the sensor.

Then, let the mechanical variable take the form

$$x(t) = x_0 + x'(t), \tag{5.49}$$

where x_0 is the static equilibrium position and $x'(t)$ is the time-dependent displacement due to *both* $f'_{\text{ext}}(t)$ and the time-dependent component of f^{e}. The capacitance derivative in the force of electrical origin can be expanded in a Taylor series:

$$\frac{\mathrm{d}C}{\mathrm{d}x} \approx \left(\frac{\mathrm{d}C}{\mathrm{d}x}\right)_{x_0} + \left(\frac{\mathrm{d}^2C}{\mathrm{d}x^2}\right)_{x_0} x'(t). \tag{5.50}$$

Combining these expressions, we can reduce the equation of motion, Eq. (5.47), to a static equilibrium equation for x_0:

$$kx_0 = f_{\text{ext},0} + \frac{v_{\text{m}}^2}{4}\frac{\mathrm{d}C}{\mathrm{d}x}\bigg|_{x_0}, \tag{5.51}$$

and a dynamic equation for $x'(t)$:

$$m\ddot{x}' + b\dot{x}' + k_{\text{eff}}x' - \underbrace{\left(\frac{v_{\text{m}}^2}{4}\frac{\mathrm{d}^2C}{\mathrm{d}x^2}\bigg|_{x_0}\right)\cos(2\omega_{\text{e}}t)\,x'}_{\text{high-frequency parametric term}} = \underbrace{f'_{\text{ext}}}_{\substack{\text{low-freq.}\\ \text{sensor input}}} + \underbrace{\frac{v_{\text{m}}^2}{4}\frac{\mathrm{d}C}{\mathrm{d}x}\bigg|_{x_0}\cos(2\omega_{\text{e}}t)}_{\text{high-frequency excitation}}, \tag{5.52}$$

where

$$k_{\text{eff}} = k - \frac{v_{\text{m}}^2}{4}\frac{\mathrm{d}^2C}{\mathrm{d}x^2}\bigg|_{x_0} \tag{5.53}$$

is an effective spring constant based on the RMS average of the AC excitation voltage. Now assume further that ω_{e}, the voltage excitation frequency, greatly exceeds the highest frequency component of the input force, $f'_{\text{ext}}(t)$. Then, Eq. (5.52) simplifies to an ordinary differential equation with constant coefficients:

$$m\ddot{x}' + b\dot{x}' + k_{\text{eff}}x' = f'_{\text{ext}}. \tag{5.54}$$

Equations (5.51) and (5.54) reveal that the only significant effect of the electromechanical coupling is to change the spring constant and, thus, the resonant frequency if $\mathrm{d}^2C/\mathrm{d}x^2 \neq 0$. In particular, for $\mathrm{d}^2C/\mathrm{d}x^2 > 0$, the effective spring constant k_{eff} is reduced and instability becomes possible.

There is no back action term because the voltage is fixed; but, even if a perturbation voltage term were added to account for the series source impedance, it would only add a high-frequency component to Eq. (5.52) that would be overwhelmed by the excitation term ignored to obtain Eq. (5.59).

Now consider the current flow on the electrical side of the sensor shown in Fig. 5.14:

$$\begin{aligned} i(t) &= \frac{\mathrm{d}}{\mathrm{d}t}[C(x)v(t)] \\ &= \frac{\mathrm{d}}{\mathrm{d}t}\left[\left(C_0 + \frac{\mathrm{d}C}{\mathrm{d}x}x'\right)v_{\text{m}}\cos(\omega_{\text{e}}t)\right]. \end{aligned} \tag{5.55}$$

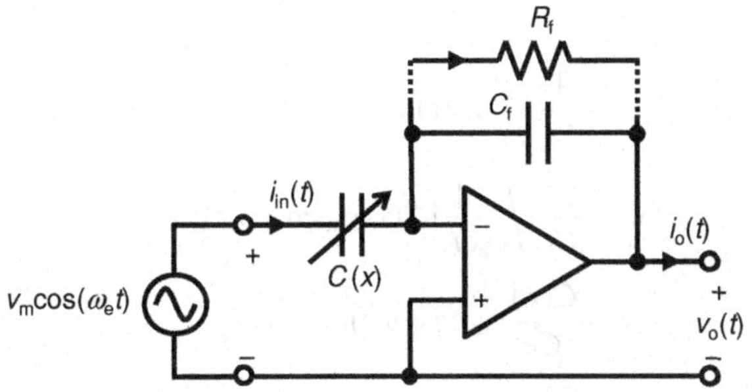

Figure 5.15 A two-plate capacitive transducer connected in the high-impedance inverting amplifier configuration using sinusoidal voltage excitation. Ignore the feedback resistance in the first pass.

Expansion of this expression yields

$$i(t) = -\omega_e C_0 v_m \sin(\omega_e t) + v_m \frac{dC}{dx}\left[-\omega_e \sin(\omega_e t)x' + \cos(\omega_e t)\frac{dx'}{dt}\right]. \tag{5.56}$$

If we again invoke the assumption that the excitation frequency is very high, the term involving dx'/dt can be ignored. Then, rearranging Eq. (5.56), we have

$$i(t) = -\omega_e v_m \sin(\omega_e t)\left[C_0 + \frac{dC}{dx}x'\right]. \tag{5.57}$$

This expression consists of the usual capacitive term plus a second, smaller, term that depends on the displacement x', which, as Eq. (5.54) teaches, is linearly proportional to the input force f'_{ext}. Therefore, $i(t)$ does contain the desired sensor information, though multiplied by the high-frequency sine function. The only way to extract the sensor information is by *demodulation*. The following subsections introduce some practical circuit configurations for AC-excited sensors and then present some fundamental principles of demodulation.

From this exercise just completed, it will be obvious that, in general, AC-excited capacitive systems are not amenable to the convenient two-port network models relied upon in the previous sections. Even though AC-excited systems can be linearized, the output is a modulated signal and conventional phasor-based AC analysis is not possible.

5.5.2 Two-plate capacitive sensing with AC excitation

Figure 5.15 shows a simple AC sensing circuit based on a two-plate capacitor. It has the same topology as the high-impedance amplifier of Fig. 5.5a, except now the source voltage is a zero-mean AC signal at frequency ω_e.

Assuming ideal behavior for the integrator, with no resistor R_f, and then invoking KCL, the output voltage is equal to the negative of the integral of the input current $i_{in}(t)$

divided by C_f:

$$\begin{aligned} v_0(t) &= -\frac{1}{C_f}\int_0^t i_{in}(t)\mathrm{d}t \\ &= -\frac{1}{C_f}\int_0^t \frac{\mathrm{d}}{\mathrm{d}t}[C(x)v_m\cos(\omega_e t)]\mathrm{d}t \\ &= -\frac{C(x)}{C_f}v_m\cos(\omega_e t). \end{aligned} \tag{5.58}$$

Next, expanding the capacitance function $C(x)$ in a Taylor series, the output voltage is

$$v_0(t) = -\underbrace{\left[1+\frac{\mathrm{d}C/\mathrm{d}x}{C_0}x'(t)\right]\frac{C_0}{C_f}v_m}_{\text{time-varying amplitude}}\cos(\omega_e t). \tag{5.59}$$

The coefficient of the cosine function includes the time-dependent factor $[1+(\mathrm{d}C/\mathrm{d}x/C_0)x'(t)]$, which has the form of an *amplitude-modulated* (AM) signal. The excitation voltage serves as the carrier signal and ω_e is the carrier frequency. The time-dependent portion of this coefficient $(\mathrm{d}C/\mathrm{d}x/C)x'(t)$, called the *modulation signal*, contains the information about the displacement that the sensor is designed to detect.

Almost invariably for capacitive two-plate sensors, $|\mathrm{d}C/\mathrm{d}x \cdot x'(t)|_{x_0} \ll C_0$, so the coefficient of the cosine function in Eq. (5.59) never goes negative. This condition, which defines conventional AM modulation, guarantees that the modulating signal can always be recovered. There is, of course, a second, practical requirement that, for reliable detection of the modulating signal, the frequency of the excitation voltage must be much greater than the maximum rate of change of the displacement $x'(t)$. Usually this condition is met comfortably, i.e., $\omega_e \gg \omega_{max}$. Representative modulation and carrier waveforms meeting these requirements are plotted in Fig. 5.16.

5.5.3 Analysis including the feedback resistance R_f

The previous analysis ignored the feedback resistor R_f; however, since any practical circuit must have the means to bleed off charge accumulating due to the leakage current, it is important to consider the effect of the feedback resistor R_f on performance. Refer now to Fig. 5.17.

The current $i_{in}(t)$ is related to the output voltage through KCL:

$$i_{in}(t) = \dot{q}(t) = -\frac{v_0(t)}{R_f} - C_f\dot{v}_0(t). \tag{5.60}$$

The Laplace transform of this equation is

$$sQ(s) - q(0) = -\frac{\underline{V_0}(s)}{R_f} - C_f[s\underline{V_0}(s) - \underline{v_0}(0)]. \tag{5.61}$$

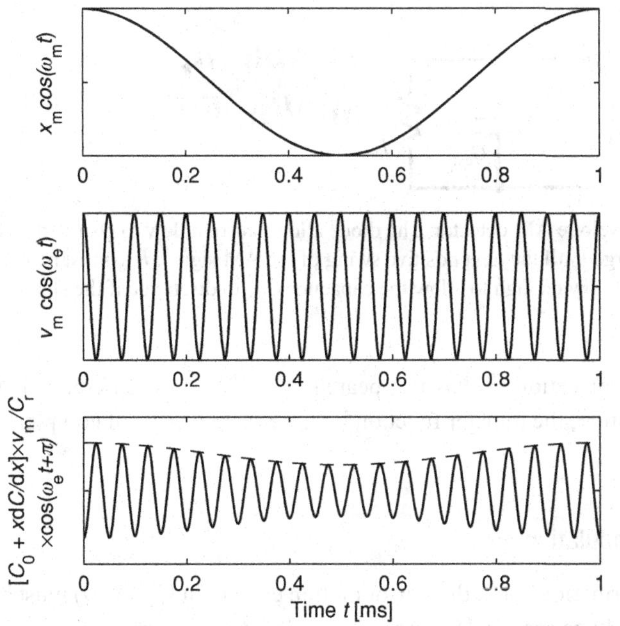

Figure 5.16 Sample waveforms for a capacitive sensor with AC excitation. (a) Modulating signal, i.e., the mechanical displacement $x(t)$ at 1 kHz. (b) The carrier signal, that is, the sinusoidal input voltage at 20 kHz. (c) The modulated (AM) output voltage signal. The maximum modulation here is $[dC/dx \cdot x_m]/C_0 = 0.3$.

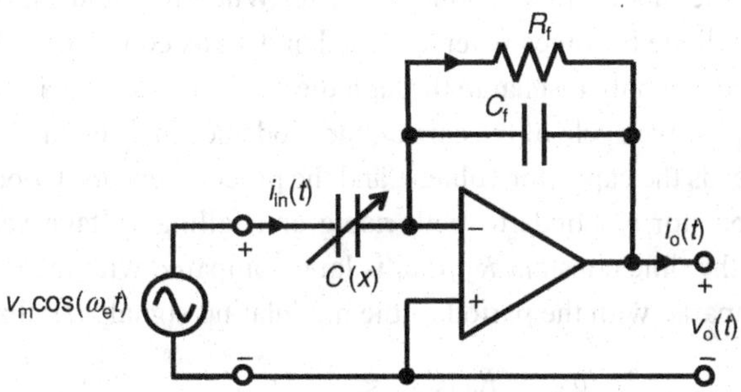

Figure 5.17 Sensor with AC modulation. The second pass with the feedback resistor R_f.

From Eq. (5.61) and the assumed initial conditions $q(0) = 0$ and $v_o(0) = 0$, the transfer function $\underline{H}_{v/q}(s)$ is

$$\underline{H}_{v/q}(s) = \frac{\underline{V_0}(s)}{\underline{Q}(s)} = -\frac{sR_fC_f}{1 + sR_fC_f}. \tag{5.62}$$

The output voltage is then

$$\underline{V_0}(s) = \frac{\underline{Q}(s)}{C_r}\underbrace{\frac{sR_fC_f}{1 + sR_fC_f}}_{\underline{H}_c(s)}. \tag{5.63}$$

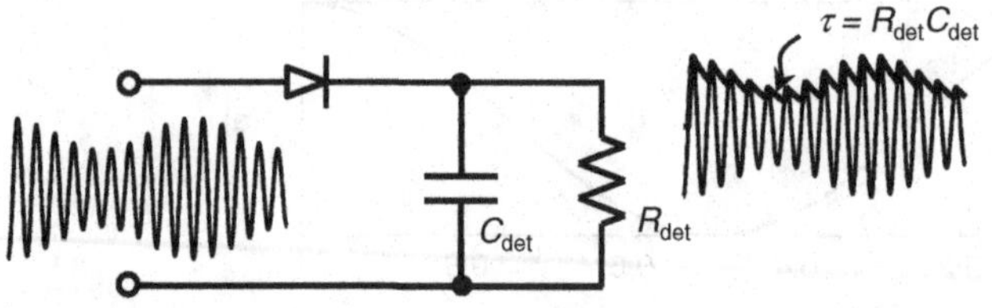

Figure 5.18 The envelope AM detector. The diode allows current flow to charge the capacitor and then hold this charge until the next positive swing of the AM signal. R_{det} must be large enough to hold the charge but low enough to allow the capacitor C_{det} to discharge if the signal amplitude decreases.

Notice that the correction $\underline{H}_c$ has reappeared here. The only difference is that, in the Laplace formulation, the transfer function is expressed in terms of complex frequency s rather than $j\omega$.

5.5.4 AM signal demodulation

The desired information about the sensor motion evident in Eq. (5.59) must be extracted from the input voltage carrier. This process is called *demodulation*. A very simple circuit for demodulating AM signals is the *envelope detection* scheme shown in Fig. 5.18. This detector consists of a non-linear diode, a capacitor C_{det}, and a resistor R_{det}. During the positive half-cycle of the signal, the diode conducts, charging up the capacitor to the peak value of the voltage signal. This charging time is extremely small, because of the low resistance of the diode in its conducting state. When the input falls below the capacitor voltage, the diode becomes reverse-biased and stops conducting. During this period, the capacitance can only discharge through the resistor R_{det} with a time constant $R_{det}C_{det}$. On the next positive cycle of the carrier, the diode becomes conducting again if the input voltage exceeds the capacitor voltage, and the process repeats. Good detection, guaranteed if the capacitor responds to both rising and falling voltage values of the signal, is achieved if the time constant $R_{det}C_{det}$ is long compared with the period of the carrier, but short compared with the period of the modulating signal $x(t)$; that is,

$$2\pi/\omega_e \ll R_{det}C_{det} \ll 2\pi/\omega_m. \tag{5.64}$$

Figure 5.19 shows a representative capacitive AC sensor system with a diode-based envelope detector cascaded with the circuit of Fig. 5.15. The envelope signal will require amplification and additional signal conditioning, both to block out the DC component and to reject any signal components at the carrier frequency. These system details are not considered here. Further, to maintain operation in the linear range of the operational amplifier, a feedback resistor R_f is needed in the feedback path to bleed off charge accumulating due to the leakage current.

This AC sensing scheme suffers the same disadvantage as the previously considered two-plate scheme with respect to DC bias. If the response of the sensor to the motion, ΔC, is small compared with C_0, then the modulation index will be small, and sensor performance will be adversely affected by noise.

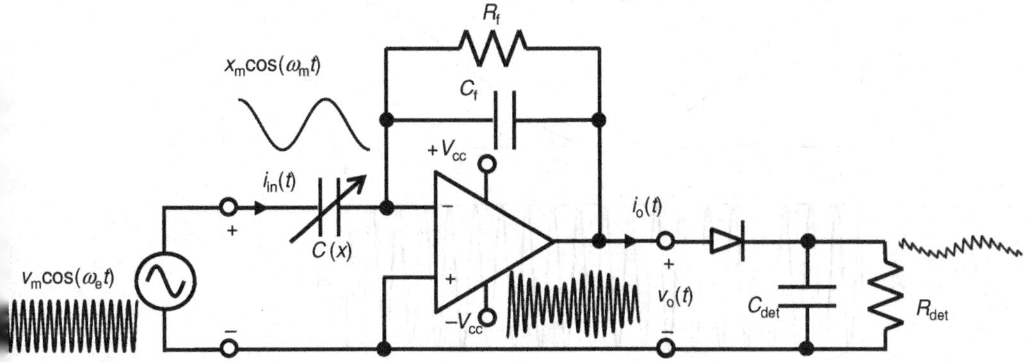

Figure 5.19 Two-plate capacitive AC sensor with simple envelope detection network. The normally required signal filtration and amplification systems are not shown in this schematic.

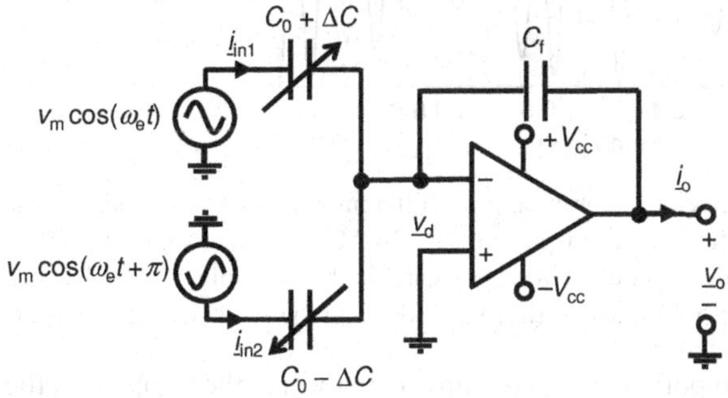

Figure 5.20 Three-plate capacitive sensor topography using zero-mean AC voltage excitation. Note that the two sinusoidal voltages are equal in magnitude and 180° out of phase. As before, $\Delta C \approx \left(\frac{dC}{dx}\right) x'$.

5.5.5 Differential AC sensing

These AC capacitive sensing systems can also be implemented using three-plate (differential) capacitive sensor geometries. And, just as they do for DC biasing, these schemes ameliorate the problem of two-plate systems by eliminating the large offsets from the output voltage. We here consider the same circuit topology studied in Section 5.4. The only difference is that the two opposed DC sources shown in Fig. 5.11 are now replaced by a pair of AC voltages that are 180° out-of-phase. Refer to Fig. 5.20.

One can exploit the superposition principle to add up the separate contributions to the output voltage from the two input currents using Eq. (5.58) to obtain

$$
\begin{aligned}
V_0(t) &= -\frac{1}{C_f}\int_0^t (i_{in1}(t) + i_{in2}(t))\,dt \\
&\approx \underbrace{-2x'(t)\frac{v_m}{C_f}\frac{dC}{dx}}_{\text{modulated amplitude}} \cos(\omega_e t). \qquad (5.65)
\end{aligned}
$$

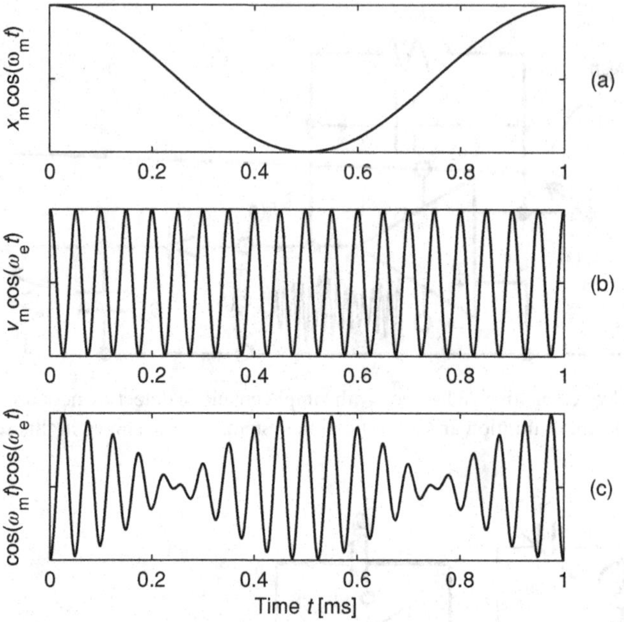

Figure 5.21 Example waveforms for three-plate differential capacitive sensing with sinusoidal AC excitation. (a) Modulating signal (displacement). (b) AC carrier signal. (c) Double-sideband suppressed carrier modulated output voltage signal. In this example, the displacement is assumed to be a cosine: $x'(t) = x_{\mathrm{m}} \cos(\omega_{\mathrm{m}} t)$ at 1 kHz and the sinusoidal carrier is at 20 kHz.

There emerges one important difference about the form of the signal from the amplitude-modulated case of Eq. (5.59), namely, that the sign of the coefficient of the sinusoidal function in Eq. (5.65) will most certainly make both positive and negative excursions from the equilibrium. Therefore, the output voltage $v_{\mathrm{o}}(t)$ will now take the new form revealed in Fig. 5.21. This type of modulation is called a *double-sideband, suppressed carrier signal* [4].

5.5.6 Synchronous demodulation

Envelope detection cannot be used to demodulate double-sideband, suppressed carrier signals. Instead, the *synchronous demodulation* scheme illustrated in Fig. 5.22 is needed.

According to this scheme, the output of the amplifier is first multiplied by a replica of the carrier signal, in this case, a cosine:

$$\begin{aligned} v_{\mathrm{O}}(t)\cos(\omega_{\mathrm{e}}t) &= -2x'(t)\frac{v_{\mathrm{m}}}{C_{\mathrm{f}}}\frac{\mathrm{d}C}{\mathrm{d}x}\cos^2(\omega_{\mathrm{e}}t) \\ &= -x'(t)\frac{v_{\mathrm{m}}}{C_{\mathrm{f}}}\frac{\mathrm{d}C}{\mathrm{d}x}\left[1+\cos\left(2\omega_{\mathrm{e}}t\right)\right]. \end{aligned} \tag{5.66}$$

Passing the resulting signal through a low-pass filter suppresses the sinusoidal component at $2\omega_{\mathrm{e}}$, leaving a signal directly proportional to $x'(t)$:

$$-\frac{v_{\mathrm{m}}}{C_{\mathrm{f}}}\frac{\mathrm{d}C}{\mathrm{d}x}x'(t). \tag{5.67}$$

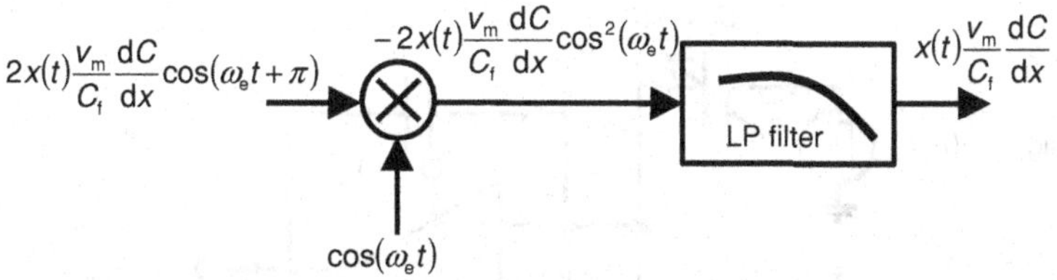

Figure 5.22 A system-level block diagram for synchronous demodulation of the double-sideband suppressed carrier signal output of a three-plate differential capacitive sensor.

This signal is dependent on the amplitude of the carrier and the derivative of the capacitance, dC/dx. In Section 5.8.7, we shall return to the subject of signal modulation and demodulation to illustrate its utility for reducing the noise problems inherent in MEMS sensors.

5.6 AC sensors using symmetric square-wave excitation

Most practical capacitive MEMS sensor systems employ square-wave excitation instead of sinusoidal AC. The motivation for doing so is entirely practical. In modern transistor-based electronic circuits, it is usually easiest to generate square waves and pulse trains. Square-wave excitation does have some disadvantages over harmonic signals, but these are outweighed by the convenience of their availability.

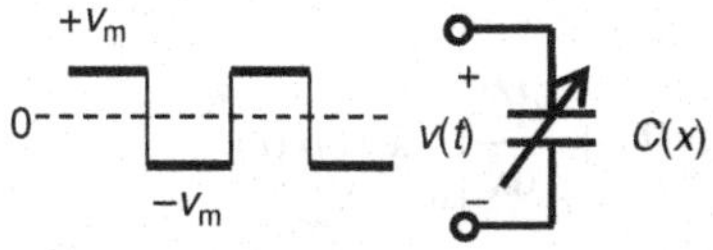

Figure 5.23 Capacitive transducer excitation by a symmetrical square-wave voltage. Square-wave signals are more readily obtained from transistor-based electronics than sinusoidal signals.

5.6.1 Transducers using square-wave excitation

Because the techniques of small-signal analysis of Chapter 4 cannot be applied, we again resort to basic lumped parameter electromechanics and energy methods. Using the coenergy formation, the force of electric origin is

$$f_x^e = -\frac{\partial W_e'}{\partial x} = -\frac{1}{2}\frac{dC}{dx}v^2(t). \qquad (5.68)$$

For the square-wave voltage excitation waveform of Fig. 5.23, the square of the voltage is constant in time, $v^2(t) = v_m^2$, and the force of electrical origin is also constant:

$$f_x^e = -\frac{1}{2}\frac{dC}{dx}v_m^2. \qquad (5.69)$$

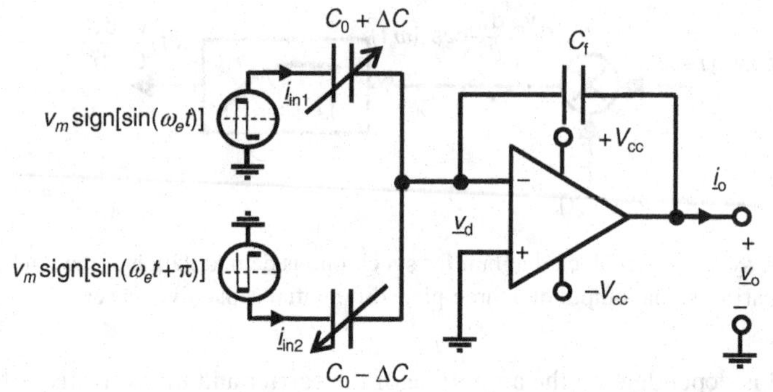

Figure 5.24 Basic three-plate differential capacitive sensing system with opposed square-wave excitation. As before, the capacitance change ΔC is approximately $x'\mathrm{d}C/\mathrm{d}x$.

Expanding the capacitance function in a Taylor series, the equation of motion becomes

$$m\ddot{x}' + b\dot{x}' + \left(k - \frac{1}{2}\frac{\mathrm{d}^2C}{\mathrm{d}x^2}\bigg|_{x_0} v_\mathrm{m}^2\right)x' = f_\mathrm{ext} + \frac{1}{2}\frac{\mathrm{d}C}{\mathrm{d}x}\bigg|_{x_0} v_\mathrm{m}^2. \tag{5.70}$$

Because it is independent of time, the zero-order component of the electrical force acting on the system influences only the equilibrium. The effective spring constant is shifted if $\mathrm{d}^2C/\mathrm{d}x^2 \neq 0$.

Using a similar linearization on the electrical side, the electric current in the sensor is

$$i = \frac{\mathrm{d}}{\mathrm{d}t}(Cv) = C(x)\dot{v}(t) + \frac{\mathrm{d}C}{\mathrm{d}x}\bigg|_{x_0}\dot{x}'(t)v(t). \tag{5.71}$$

This expression is similar to Eq. (5.55). Because the voltage of the square wave is discontinuous, there will be large positive and negative current spikes, even if the internal impedance of the square-wave source degrades the rise time of the square wave. These current pulses can give rise to additional noise, a potential disadvantage of square waves over sinusoidal excitation.

5.6.2 Three-plate sensing using square-wave excitation

For three-plate capacitance sensing, out-of-phase square-wave voltages are used. Refer to Fig. 5.24. In other respects, the circuit topology of Fig. 5.20 is unaltered.

Based on Eq. (5.58) and superposition, the output voltage is

$$\begin{aligned} v_\mathrm{O}(t) &= -\frac{1}{C_\mathrm{f}}\int_0^t (i_\mathrm{in1}(t) + i_\mathrm{in2}(t))\mathrm{d}t \\ &= 2x'(t)\,\frac{v_\mathrm{m}}{C_\mathrm{f}}\frac{\mathrm{d}C}{\mathrm{d}x}\bigg|_{x_0}\ \mathrm{sign}\left[\sin\left(\omega_\mathrm{e}t + \pi\right)\right], \end{aligned} \tag{5.72}$$

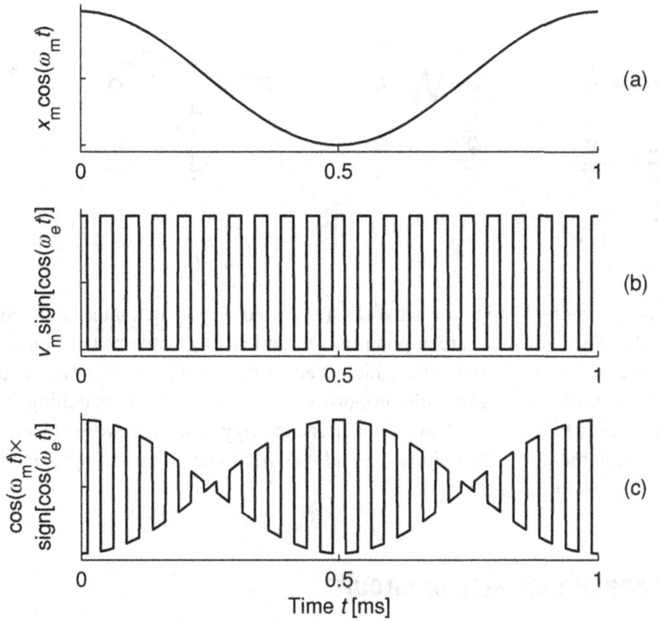

Figure 5.25 Example waveforms for a capacitive sensor using square-wave excitation. For purposes of this example, we assume that $x'(t) = x_m \cos(w_m t)$ at 1 kHz and that the carrier frequency is 20 kHz. (a) Modulating signal (displacement). (b) Square-wave carrier signal. (c) Double-sideband, suppressed carrier modulated square-wave output voltage signal.

where it has been useful to introduce the sign function:

$$\text{sign}\,(a) = \begin{cases} +1 & \text{for}\, a > 1, \\ -1 & \text{for}\, a < 1. \end{cases} \tag{5.73}$$

The output signal waveform of Eq. (5.62) is a square wave with its amplitude modulated by the time-dependent plate displacement $x'(t)$. See Fig. 5.25.

Synchronous demodulation is achieved by multiplying the output signal $v_{\text{out}}(t)$ by a square-wave signal synchronized to the carrier signal:

$$\begin{aligned} v_O(t)\,\text{sign}\,[\sin(\omega_e t + \pi)] &= 2x'(t)\frac{v_m}{C_f}\frac{dC}{dx}\underbrace{\text{sign}^2\,[\sin(\omega_e t + \pi)]}_{=1} \\ &= 2\frac{v_m}{C_f}\frac{dC}{dx}x'(t). \end{aligned} \tag{5.74}$$

According to Eq. (5.74), the desired signal is recovered directly. In practice, however, high-frequency signal components induced by the large capacitive charging current spikes due to the square-wave excitation of the capacitance must be filtered out.

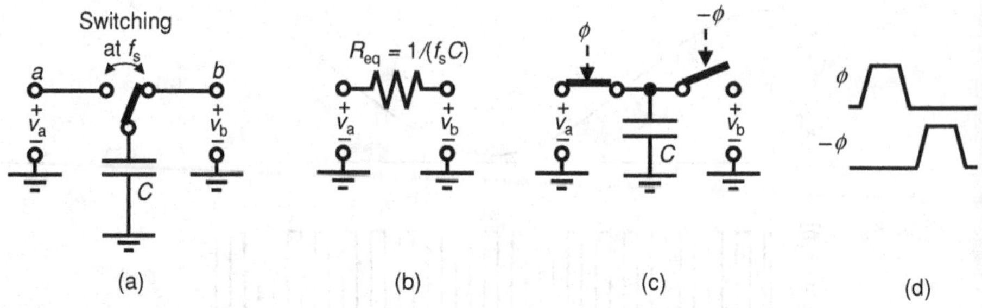

Figure 5.26 Simple switched-capacitor networks. (a) A two-way switch alternately connects the capacitor to nodes a and b, which are at potentials v_a and v_b, respectively. (b) Simple resistive model for switched capacitance network, which is accurate if the switching rate is sufficiently high. Effective resistance R_{eq} is inversely proportional to the product of switching frequency f_s and capacitance C. (c) Practical switched-capacitor topology based on two solid-state switches controlled by complementary clock signals ϕ and $-\phi$. (d) Time plot of complementary non-overlapping clock signals.

5.7 Switched capacitance sensor circuits

The feedback resistor R_f in the circuit shown in Fig. 5.6 was added to solve the problem of leakage current-driven saturation, but its introduction came at the price of eliminating the low-frequency response of capacitive sensors. This sacrifice is unfortunate because detecting low-frequency signals is a valuable attribute of any sensor, particularly for capacitive elements because their excellent long-term stability makes them otherwise ideal for operation at DC. *Switched capacitance* circuits offer an alternative that eliminates the op-amp bias current problem without sacrificing low-frequency sensitivity.

Switched-capacitor systems are examples of digital circuit schemes. They use a capacitor and electronic switches, usually FETs, controlled by a clock signal. Detailed coverage of this topic is far beyond the scope of this text. Instead, we present here a simple analysis intended to reveal the basic operational principles of switched capacitance. Proceeding in this way facilitates introduction of equivalent circuit representations similar to those so heavily relied on elsewhere in this chapter.

5.7.1 Basics of switched-capacitor circuits

The circuit shown in Fig. 5.26a consists of a capacitor and a two-way switch. If the nodes a and b are maintained at DC voltages v_a and v_b, respectively, then each time the switch is cycled from a to b and then back to a, an electric charge Δq is transferred from a to b:

$$\Delta q = C\,(v_a - v_b)\,. \tag{5.75}$$

This transferred charge will be positive or negative, depending only on the relative values of v_a and v_b.

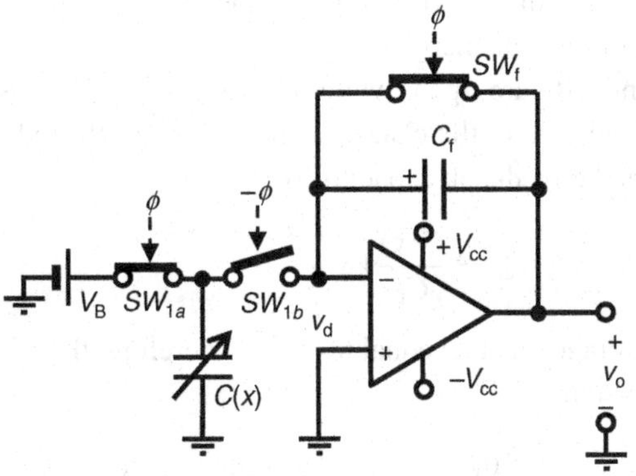

Figure 5.27 A switched-capacitor amplifier. In the first state, called *sample*, SW_{1a} and SW_{f} are closed, SW_{1b} is open, and $C(x)$ is charged to V_{B}. In the second state, called *hold*, SW_{1a} and SW_{f} are open while SW_{1b} is closed, and the charge is transferred to the output.

If f_{s} is the clock frequency of the switch, there will be an average flow of current between the two nodes:

$$I_{\mathrm{av}} = f_{\mathrm{s}} \Delta q. \tag{5.76}$$

Because $\Delta q \propto C\,(v_a - v_b)$, this two-port network behaves like a resistor connected between nodes a and b, as shown in Fig. 5.26b. The value of this resistor is

$$R_{\mathrm{eq}} = \frac{\Delta v}{I_{\mathrm{av}}} = \frac{1}{f_{\mathrm{s}} C}. \tag{5.77}$$

This simple model is adequate as long as (i) the switching frequency is much higher than the bandwidth of the signal to be detected and (ii) the voltages v_a and v_b are unaffected by the switching [5].

Realizable switched capacitance networks employ ganged solid-state switches, one on either side of the capacitor. See Fig. 5.26c. The clock signals controlling these switches are complementary in the sense that they are never on at the same time. In addition, to avoid short circuits, there is a short time interval when both switches are off. See Fig. 5.26d.

5.7.2 Simple sensor based on switched capacitance

A simple sensor using a switched-capacitor network is shown in Fig. 5.27. This topology resembles the high-impedance single-ended scheme studied earlier in this chapter. An additional switch SW_{f} in the feedback path, and driven by the same clock, serves the purpose identical to a feedback resistor R_{f} of preventing saturation of the operational amplifier.

In the *sample state*, switches SW_{1a} and SW_{f} are closed and SW_{1b} is open. The capacitor $C(x)$ is charged by voltage V_{B}:

$$q = C(x) V_{\mathrm{B}}. \tag{5.78}$$

With the output voltage shorted to the inverting input of the op-amp, which is at virtual ground, the feedback capacitor C_f is uncharged.

In the *hold state*, SW_{1a} and SW_f are open, while SW_{1b} is closed. Because the input current to the op-amp is virtually zero, the charge q on $C(x)$ is transferred to C_f. From Eq. (5.78), the voltage across the feedback capacitor is

$$v_{C_\mathrm{f}} = \frac{C(x)}{C_\mathrm{f}} V_\mathrm{B}. \tag{5.79}$$

Because $v_\mathrm{d} \approx 0$, the output voltage must be equal to $-v_{C_\mathrm{f}}$. Therefore, the output voltage depends on the state of the system:

$$v_\mathrm{O} = \begin{cases} 0, & \text{sample state,} \\ -\dfrac{C[x(t_\text{end of sample state})]}{C_\mathrm{f}} V_\mathrm{B}, & \text{hold state.} \end{cases} \tag{5.80}$$

The switching occurs at frequency f_s, so Eq. (5.80) can be written in terms of the *sign* function:

$$v_\mathrm{O}(t) \approx \frac{C(x\left(\lfloor t f_\mathrm{s} \rfloor / f_\mathrm{s}\right))}{2C_\mathrm{f}} V_\mathrm{B}\{\mathrm{sign}[\sin(\omega_\mathrm{s} t) + 1], \tag{5.81}$$

where $\omega_\mathrm{s} = 2\pi f_\mathrm{s}$ and $\lfloor . \rfloor$ is the floor operator, which we must use for selecting the instances of time at the end of the sample state. Linearizing $C(x)$ in the usual way gives the form

$$v_\mathrm{O}(t) = -\frac{1}{2C_\mathrm{f}} \left(C_\mathrm{O} + \left.\frac{\mathrm{d}C}{\mathrm{d}x}\right|_{x \to x_0} x'\left(\lfloor t f_\mathrm{s} \rfloor / f_\mathrm{s}\right) \right) V_\mathrm{B}[\mathrm{sign}(\sin(\omega_\mathrm{s} t)) + 1]. \tag{5.82}$$

This form for $v_\mathrm{O}(t)$ strongly resembles Eq. (5.72). In particular, the second term, proportional to the product of the displacement signal, $x'(.)$, and the high frequency square-wave signal, sign $[\sin(\omega_\mathrm{s} t)]$, is a conventional, AM-modulated signal. The output voltage waveform, illustrated in Fig. 5.28, shows a waveform for the case $x'(t) = x_\mathrm{m} \cos(\omega_\mathrm{m} t)$. Generally, it will be that $\omega_\mathrm{m} \ll \omega_\mathrm{s}$. To recover a replica of the mechanical signal, the envelope detection method described in Section 5.5.4 can be used.

The single-ended configuration shown in Fig. 5.27 is simple enough, but suffers the serious disadvantage of reduced resolution because the signal to be detected is superimposed on a much larger DC bias.

5.7.3 Half-wave bridge sensor using switched capacitance

We showed in Section 5.4 that the DC offset in analog sensors is eliminated using a three-plate capacitive structure. The same improvement can be made for switched-capacitor systems. Refer to Fig. 5.29.

When this system is in the sampling state, C_1 and C_2 are charged to voltages V_B and $-V_\mathrm{B}$, respectively. By virtue of their antisymmetry, the capacitors have charges of

$$q_1 = C_\mathrm{O} V_\mathrm{B} + \Delta C V_\mathrm{B} \quad \text{and} \quad q_2 = -C_\mathrm{O} V_\mathrm{B} + \Delta C V_\mathrm{B}. \tag{5.83}$$

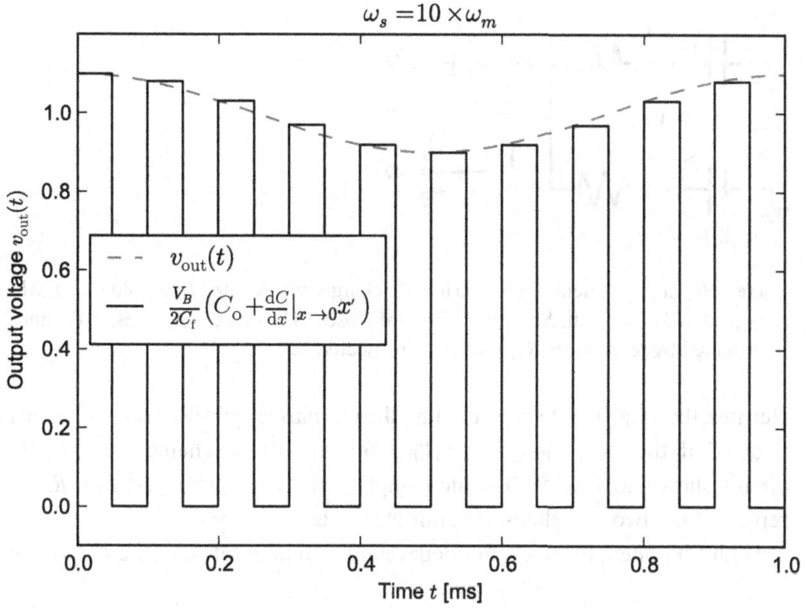

Figure 5.28 Sample output waveform, as given by Eq. (5.82), for an assumed sinusoidal displacement signal $x'(t) = x_m \cos(\omega_m t)$, with $\omega_s = 10 \times \omega_m$ and $x(dC/dx)/C_0 = 0.1$. In a practical MEMS device, $x(dC/dx)/C_0$ will be considerably smaller than 0.1.

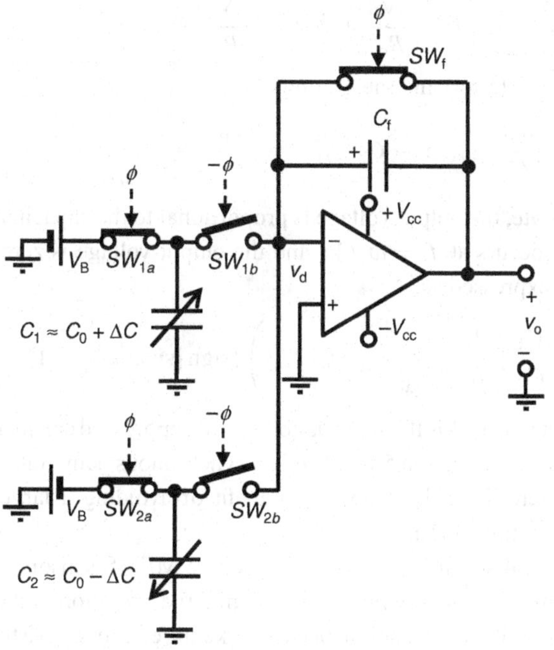

Figure 5.29 A differential switched-capacitor amplifier. In the sample state $C_1(x)$ and $C_2(x)$ are charged to V_B and $-V_B$, respectively (SW_{1a}, SW_{2a}, and SW_f are closed; SW_{1b} and SW_{2b} are open). In the hold state (SW_{1a}, SW_{2a}, and SW_f are open; SW_{1b} and SW_{2b} closed), this charge is transferred to the output. As before, $\Delta C \approx (dC/dx)_{x_0} x'$.

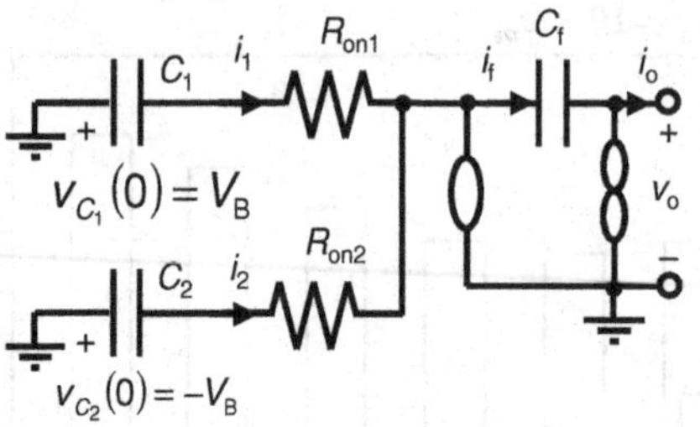

Figure 5.30 An equivalent circuit during the charge transfer step (the hold state), when SW_{1a}, SW_{2a}, and SW_f are open; SW_{1b} and SW_{2b} are closed. The closed switches, SW_{1b} and SW_{2b}, are represented by resistances R_{on1} and R_{on2}, respectively.

Because the amplifier detects the sum, the DC bias is cancelled out while the perturbation is doubled, that is, $q_1 + q_2 = 2C_0V_B$. To analyze this scheme, consider the equivalent circuit shown in Fig. 5.30, which implements equivalent resistors R_{on1} and R_{on2} to represent the two switched-capacitor elements.

With SW_f open, the amplifier behaves like an integrator, so the output voltage is

$$v_0 = -\frac{1}{C_f}\int_0^{\infty} i_f(t)\mathrm{d}t. \tag{5.84}$$

The currents $i_1(t)$ and $i_2(t)$ from the switched-capacitor branches are simple decaying exponentials. Using superposition,

$$i_f = i_1 + i_2 = \frac{V_B}{R_{on1}}e^{-\frac{t}{R_{on1}C_1}} - \frac{V_B}{R_{on2}}e^{-\frac{t}{R_{on2}C_2}}. \tag{5.85}$$

Inserting Eq. (5.85) in Eq. (5.84) and integrating,

$$v_0 = -\frac{V_B}{C_f}[C_1(x) - C_2(x)] \approx -2\frac{V_B}{C_f}\left(\frac{\mathrm{d}C}{\mathrm{d}x}\right)_{x=x_0} x'. \tag{5.86}$$

Thus, during the hold state, the output voltage is proportional to the capacitance change. Because the switching occurs at $f_s = \omega_s/2\pi$ and the output voltage is zero during the sample state, it can be expressed as

$$v_0(t) = -\frac{V_B}{C_f}\left(\left.\frac{\mathrm{d}C}{\mathrm{d}x}\right|_{x\to x_0} x'(\lfloor tf_s\rfloor / f_s)\right)[\mathrm{sign}[\sin(\omega_s t)] + 1]. \tag{5.87}$$

The waveform shown in Fig. 5.31 is the same double-sided suppressed carrier modulation signal already encountered in Section 5.6.2. While a synchronous demodulation scheme is more complex than envelope detection, the benefit of avoiding a large DC offset greatly outweighs that complication.

Virtually all commercial sensing circuits for capacitive MEMS sensors now employ switched capacitors. This section provided a glimpse into the operation of these circuits. We first showed how a switched capacitor behaves like an effective resistor when the switching frequency is much higher than the bandwidth of the signals of interest. Then a switched-capacitor single-ended amplifier was introduced, using the same basic topology employed in the previous sections, where the basic concept of MEMS sensing had been

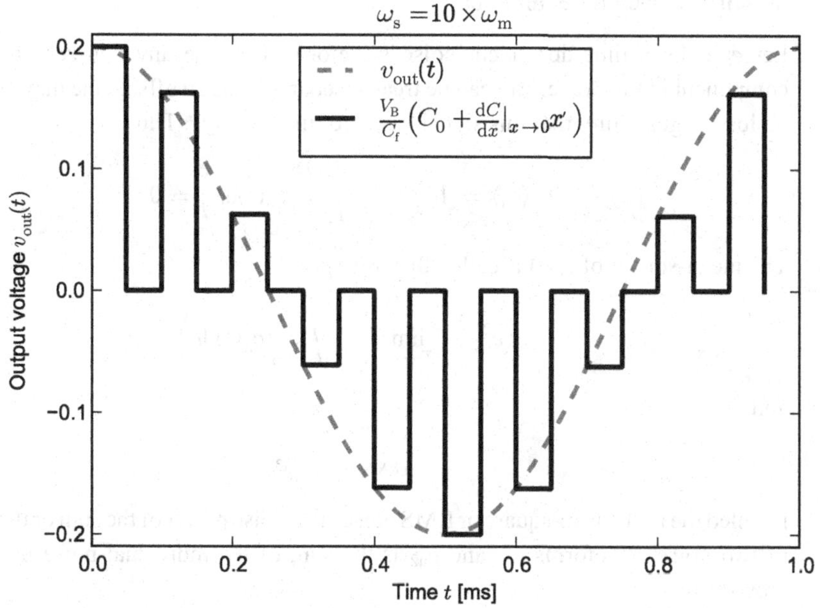

Figure 5.31 An illustration of the output waveform, given by Eq. (5.87) for a sinusoidal displacement signal of the form $x'(t) = x_m\cos(\omega_m t)$, assuming $\omega_s = 10 \times \omega_m$ and $x(dC/dx)/C_0 = 0.1$. Compare this waveform with Fig. 5.28.

introduced. As before, the single-ended amplifier was then replaced by the differential topology to remove a large DC bias. As before, differential sensing requires synchronous demodulation to extract the signal.

*5.8 Noise in capacitive MEMS

Capacitive MEMS sensors operate in an environment of noise arising from many sources. Their small size and inherently high impedance makes these devices particularly susceptible to thermal and $1/f$ noise. This section offers a very brief summary of the characteristics of the thermal and $1/f$ noise mechanisms that predominate in the input stage of the sensor electronics, and then shows how modulation schemes reduce their effects on measurement. Noise is unavoidable in MEMS and understanding how to deal with it is crucial to successful designs.

What is noise? In a broad sense, noise is any unwanted signal that corrupts a measurement we want to make. Noise has adverse effects on sensitivity, precision, and accuracy. Though noise can be broadband, here we focus upon noise signals that are in the same frequency range as the measured signal. Because it is random, there is no analytical way to describe the waveform of a noise signal. It can only be described by average statistical metrics called *noise characteristics*.

5.8.1 Common noise characteristics

Let $e_n(t)$ be a time-dependent noise waveform. Because any non-zero time-average component in this waveform can be treated separately as an offset, one may assume with no loss of generality that this signal has zero time average. Thus,

$$\langle e_n \rangle = \lim_{T\to\infty}\left(\frac{1}{T}\int_{-T/2}^{T/2} e_n(t)\mathrm{d}t\right) = 0. \tag{5.88}$$

The mean-square of $e_n(t)$ is called the *noise power*,

$$\langle e_n^2 \rangle = \lim_{T\to\infty}\left(\frac{1}{T}\int_{-T/2}^{T/2} e_n^2(t)\mathrm{d}t\right), \tag{5.89}$$

and

$$e_{n\,\mathrm{RMS}} = \sqrt{\langle e_n^2 \rangle} \tag{5.90}$$

is called the root-mean-square or RMS noise. The noise power of the sum or the difference of two noise waveforms e_{n1} and e_{n2} is the sum of the individual noise powers plus a cross-term:

$$\begin{aligned}\langle (e_{n1} \pm e_{n2})^2 \rangle &= \lim_{T\to\infty}\left[\frac{1}{T}\int_{-T/2}^{T/2} (e_{n1}(t) + e_{n2}(t))^2\,\mathrm{d}t\right]\\ &= \underbrace{\lim_{T\to\infty}\left[\frac{1}{T}\int_{-T/2}^{T/2} (e_{n1}^2(t)\mathrm{d}t)\right]}_{\langle e_{n1}^2\rangle} + \underbrace{\lim_{T\to\infty}\left[\frac{1}{T}\int_{-T/2}^{T/2} (e_{n2}^2(t)\mathrm{d}t)\right]}_{\langle e_{n2}^2\rangle}\\ &\quad \underbrace{\pm 2\lim_{T\to\infty}\left(\frac{1}{T}\int_{-T/2}^{T/2} e_{n1}(t)e_{n2}(t)\mathrm{d}t\right)}_{\langle e_{n1}e_{n2}\rangle}.\end{aligned} \tag{5.91}$$

The last term, involving the product $e_{n1}(t)e_{n2}(t)$, is the *cross-correlation*. When the cross-correlation is zero, it means that the two noise sources, e_{n1} and e_{n2}, are independent. In that important case, Eq. (5.91) simplifies to

$$\langle (e_{n1} \pm e_{n2})^2 \rangle = \langle e_{n1}^2 \rangle + \langle e_{n2}^2 \rangle. \tag{5.92}$$

The *signal-to-noise ratio*, or *SNR*, is defined as the ratio of the signal power to the noise power,

$$\mathrm{SNR} = \langle s^2 \rangle / \langle e_n^2 \rangle, \tag{5.93}$$

and is often expressed using the logarithmic decibel scale.

$$\mathrm{SNR_{dB}} = 10\log\left(\langle s^2 \rangle / \langle e_n^2 \rangle\right). \tag{5.94}$$

The two-sided (bidirectional) noise *power spectral density* (PSD) is the square of the signal per frequency band. For applications of interest in MEMS, the signal can be voltage, current, force, or velocity, so the units of PSD are $\mathrm{V^2/Hz}$, $\mathrm{A^2/Hz}$, $\mathrm{N^2/Hz}$, or $\mathrm{(m/s)^2/Hz}$, respectively. This quantity describes the spectral distribution of noise and is defined by a Fourier transform:

$$S_b(f) = \mathcal{F}\{R_{ee}(\tau)\} = \int_{-\infty}^{\infty} R_{ee}(\tau)e^{-2\pi j f\tau}d\tau, \tag{5.95}$$

where

$$R_{ee}(\tau) = \lim_{T\to\infty}\left(\frac{1}{T}\int_{-T/2}^{T/2} e_n(t)e_n^*(t+\tau)dt\right) \tag{5.96}$$

is the *autocorrelation function*. Some useful relationships between the autocorrelation function and the noise power can be obtained by inspection from their respective definitions, that is, Eqs. (5.96) and (5.89):

$$\langle e_n^2\rangle = R_{ee}(0). \tag{5.97}$$

5.8.2 Filtered noise

It is useful to examine what happens when a noise waveform $e_{n,in}$ passes through a linear filter characterized by impulse response $h(t)$. The input and output waveforms are related via the convolution integral:

$$\begin{aligned} e_{n,out}(t) &= e_{n,in}^* h \\ &= \int_{\langle e_n^2\rangle}^{\infty} e_{n,in}(\tau)h(t-\tau)d\tau. \end{aligned} \tag{5.98}$$

Inserting Eq. (5.98) into Eq. (5.96) gives the autocorrelation of the output noise:

$$\begin{aligned} &R_{out,out}(\tau) \\ &= \lim_{T\to\infty}\frac{1}{T}\int_{-T/2}^{T/2}\left[\int_{-\infty}^{\infty} h(u)e_{n,in}(t-u)du\right]\left[\int_{-\infty}^{\infty} h^*(v)e_{n,in}^*(t-\tau-v)\,dv\right]dt \\ &\times \lim_{T\to\infty}\frac{1}{T}\int_{-T/2}^{T/2}\left[\int_{-\infty}^{-\infty}\int_{-\infty}^{\infty} h(u)h^*(v)e_{n,in}(t-u)e_{n,in}^*(t-\tau-v)\right]dudvdt. \end{aligned} \tag{5.99}$$

Defining $z = t - u$,

$$\begin{aligned} R_{out,out}(\tau) &= \int_{-\infty}^{\infty}\int_{-\infty}^{\infty} h(u)\,h^*(v)\underbrace{\lim_{T\to\infty}\frac{1}{T}\int_{-T/2+u}^{T/2+u} e_{n,in}(z)\,e_{n,in}^*(z+u-\tau-v)\,dzdudv}_{=R_{e_{n,in},e_{n,in}}(\tau+v-u)} \\ &= \int_{-\infty}^{\infty} h^*(v)\underbrace{\int_{-\infty}^{\infty} h(u)R(\tau+v-u)du}_{h(\tau+v)^*R(\tau+v)}dvh^*(-\tau)^*\,h(\tau)^*\,R_{in,in}(\tau). \end{aligned} \tag{5.100}$$

Taking the Fourier transform of the last expression leads to a very useful relationship between the input and output power spectral densities of a linear system:

$$S_{be_{n\,out}}(f) = |\underline{H}(f)|^2\, S_{be_{n\,in}}(f). \tag{5.101}$$

All two-sided power spectral densities are even functions of frequency, i.e.,

$$S_b(f) = S_b(-f). \tag{5.102}$$

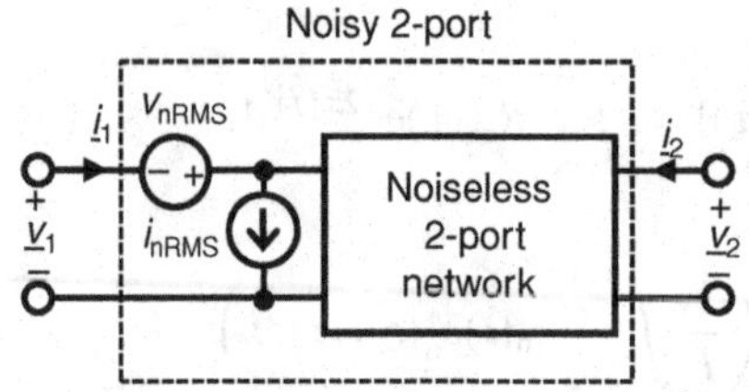

Figure 5.32 Circuit representation of a noisy linear two-port network using an ideal noiseless two-port and equivalent current and voltage noise sources.

The one-sided (unidirectional) PSD is defined as

$$S_u(f) = S_b(f) + S_b(-f) = 2S_b(f), \quad f \geq 0. \tag{5.103}$$

The noise power can also be computed from the one-sided PSD, that is,

$$\langle e_n^2 \rangle = \int_0^\infty S_u(f) df. \tag{5.104}$$

Information about other commonly used measures of noise can be found in the literature, e.g., [6].

5.8.3 Noisy two-ports

Figure 5.32 shows a representation of a two-port network with internal noise sources. Both v_{nRMS} and i_{nRMS} are needed to ensure that noise is present, irrespective of the way the two-port happens to be terminated. For example, one cannot expect the noise to disappear either when a port is open-circuited (e.g., $\underline{i}_1 = 0$) or when it is shorted ($\underline{v}_1 = 0$). The two sources must have at least a partially common origin and will therefore be correlated, that is, $\langle V_{nRMS}\, i_{nRMS} \rangle > 0$.

5.8.4 Electrical thermal noise

The thermal noise in electric circuits, often referred to as *Johnson noise*, is the consequence of the thermal agitation (Brownian motion) of the charge carriers flowing in resistors and other components. The one-sided PSD of the thermal noise in a resistor R is given by Nyquist's relation[3]

$$S_u(f) = 4k_B TR, \tag{5.105}$$

where $k_B = 1.38 \times 10^{-23}$ J/K is Boltzmann's constant, T is the absolute temperature (K). The PSD of thermal noise is usually flat over the entire bandwidth of practical interest, typically 10^{13} Hz [7]. Refer to Fig. 5.33a. Because of this frequency independence, thermal noise is also called *white noise*.

[3] The one-sided PSD S_u is most commonly used. Since this text is concerned with real signals only, the two-sided PSD S_b is obtained simply by dividing S_u by 2; $S_b = S_u/2$. To avoid possible confusion, this section consistently employs subscripts u and b to denote the one-sided (unilateral) and the two-sided (bilateral) PSD, respectively.

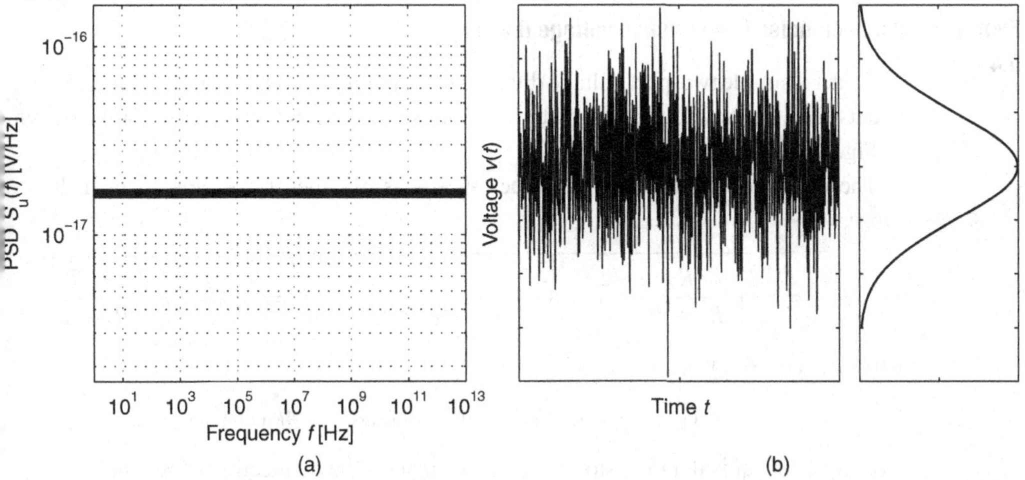

Figure 5.33 Thermal noise of an $R = 1$ kΩ resistor at $T = 300$ K. (a) Flat PSD $\approx 16 \times 10^{-18}$ V²/Hz. (b) Typical appearance of white noise in the time domain. The amplitude distribution has a Gaussian profile with a standard deviation of ~4 nV/√Hz.

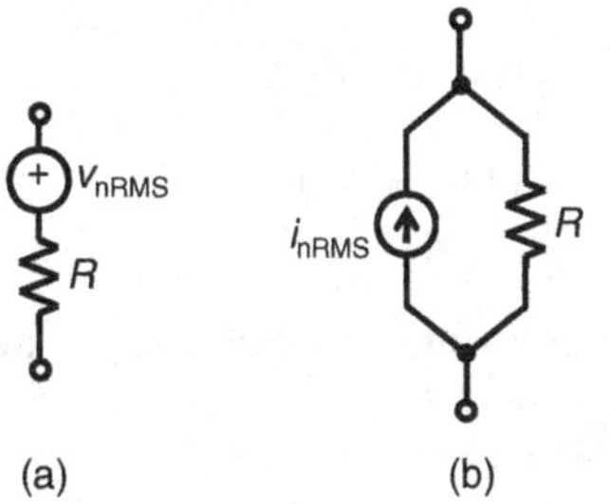

Figure 5.34 Alternative models for the noisy resistor. (a) Thevenin model with ideal noiseless resistor and series RMS noise voltage. (b) Norton model with ideal resistor and RMS noise current.

While the spectral distribution of the thermal noise is flat, its amplitude distribution is Gaussian with a mean of zero and standard deviation equal to the noise RMS given by Eq. (5.90):

$$\langle v_{\text{n}}^2 \rangle = 4k_{\text{B}} T R \Delta f, \tag{5.106}$$

where Δf is the bandwidth of interest in Hz. The noise RMS is

$$v_{\text{nRMS}} = \sqrt{4k_{\text{B}} T \Delta f R}. \tag{5.107}$$

A noisy resistor can be modeled by the series connection of an ideal resistor and a RMS noise voltage source, as shown in Fig. 5.34(a), or as the parallel connection of an ideal resistor and a RMS noise current source, as shown in Fig. 5.34(b).

Just as one might expect, these two circuits are related according to the well-known Thevenin–Norton equivalence. Thus,

$$i_{\text{nRMS}} = v_{\text{nRMS}}/R = \sqrt{4k_{\text{B}} T \Delta f / R}. \tag{5.108}$$

Example 5.4 **Johnson noise in a resistive voltage divider**

Consider a simple two-port voltage divider with two noisy resistors, shown in Fig. 5.35a, and its equivalent circuit representation with noise sources and noiseless resistors, shown in Fig. 5.35b.

The output of the two noise sources v_{noRMS} is obtained by adding them in RMS sense,

$$v_{\mathrm{noRMS}} = \sqrt{\left(\frac{R_2}{R_1 + R_2} v_{\mathrm{n1RMS}}\right)^2 + \left(\frac{R_1}{R_1 + R_2} v_{\mathrm{n2RMS}}\right)^2} = \sqrt{4 k_{\mathrm{B}} T \Delta f \frac{R_1 R_2}{R_1 + R_2}},$$

where v_{n1RMS} and v_{n2RMS} are

$$v_{\mathrm{R1nRMS}} = 4 k_{\mathrm{B}} T R_1 \quad \text{and} \quad v_{\mathrm{R2nRMS}} = 4 k_{\mathrm{B}} T R_2.$$

Note that the equivalent resistor in v_{noRMS} is the parallel connection of R_1 and R_2.

Consider the equivalent circuit in in Fig. 5.35c. For the case of a shorted input port, $\underline{v}_{\mathrm{in}} = 0$, the output noise is due to v_{nRMS} alone:

$$v_{\mathrm{noRMS}} = \frac{R_2}{R_1 + R_2} v_{\mathrm{nRMS}}.$$

To preserve the output voltage computed earlier, v_{nRMS} must be

$$v_{\mathrm{noRMS}} = \sqrt{4 k_{\mathrm{B}} T \Delta f \frac{R_1 (R_1 + R_2)}{R_2}}.$$

For the case of open input port, $\underline{i}_{\mathrm{in}} = 0$, the output voltage is due to i_{nRMS} alone:

$$v_{\mathrm{noRMS}} = R_2 i_{\mathrm{nRMS}}.$$

Thus, the noise current i_{nRMS} must be

$$i_{\mathrm{noRMS}} = \sqrt{4 k_{\mathrm{B}} T \Delta f \frac{R_1}{R_2 (R_1 + R_2)}}.$$

As stated in Section 5.8.3, while the noise powers of the two resistors are uncorrelated, $\langle V_{\mathrm{R1nRMS}}, I_{\mathrm{R2nRMS}} \rangle = 0$, the noise powers of the equivalent sources moved outside of the noiseless two-port are clearly correlated.

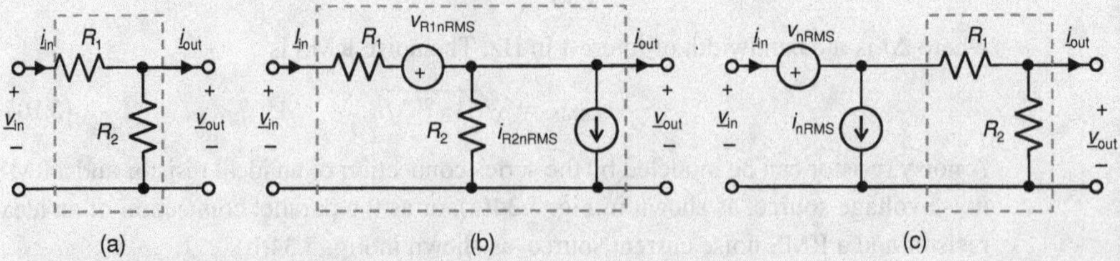

Figure 5.35 (a) A two-port representation of a voltage divider with noisy resistors. (b) Two-port with noiseless resistors. (c) Equivalent representation with noisy sources separated from the ideal noiseless two-port.

Example 5.5 **Band-limited noise of an RC filter**

Consider an RC filter consisting of a parallel connection of a noisy resistor and a capacitor. The noisy resistor is represented by an ideal component and a series noise voltage source.

The transfer function of the circuit follows from the voltage divider of the impedances:

$$H(f) = \frac{1/\mathrm{j}\omega C}{R + \mathrm{j}\omega C} = \frac{1}{1 + \mathrm{j}2\pi f RC}.$$

According to Eq. (5.101), the PSD at the output is

$$S_{\mathrm{out}}(f) = \left| \frac{1}{1 + \mathrm{j}2\pi f RC} \right|^2 S_{\mathrm{in}}(f).$$

Inserting Eq. (5.105) into the above equation, and then integrating over all frequencies:

$$\langle v_{\mathrm{Cn}}^2 \rangle = \int_0^\infty \frac{4kTR}{1 + (2\pi f RC)^2} \mathrm{d}f.$$

Introducing a new variable, $\xi = \pi f RC$, the last equation becomes

$$\langle v_{\mathrm{Cn}}^2 \rangle = \frac{4kTR}{2\pi RVC} \int_0^\infty \frac{\mathrm{d}\xi}{1 + \xi^2} = \frac{kT}{C}.$$

This is an example of a masked thermal noise [8], sometimes referred to as kT/C noise.

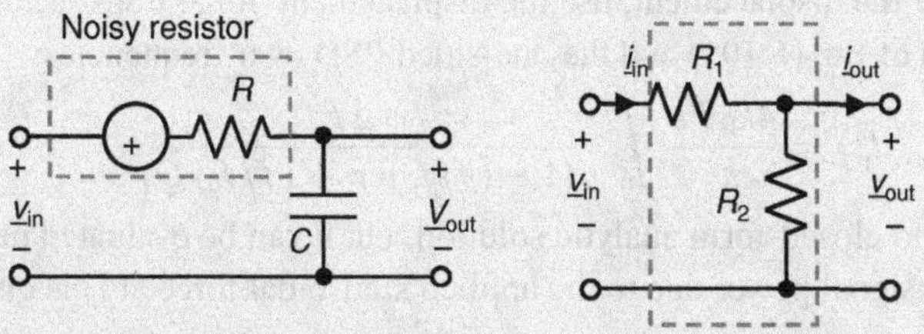

Figure 5.36 An RC filter with a noisy resistor. The equivalent circuit consists of an ideal resistor R and a series voltage source.

5.8.5 Mechanical thermal noise

Thermal agitation also influences capacitive MEMS devices, and their small size makes them quite susceptible. For example, the random impacts of gas molecules with the electrodes of a MEMS device immersed in imperfect vacuum or air give rise to small motions that are then reflected in the output signal. This noise is analogous to the thermal noise of a resistor [9, 10]. In fact, it is often convenient to invoke the mechanical ↔ electrical analogy of Section 4.4 to combine mechanical and resistor noise sources, in order to determine the net SNR of a sensing system.

For mechanical systems, damping provides the dissipative mechanism responsible for thermal noise. Relying on the mechanical ↔ electrical analogy, we can obtain the RMS noise force merely by replacing resistance R in Eq. (5.107) by the damping b, that is,

$$f_{\mathrm{nRMS}} = \sqrt{4kTb\Delta f}. \quad (5.109)$$

Alternatively, the force noise power is

$$\langle f_n^2 \rangle = 4kTb\Delta f. \qquad (5.110)$$

The force noise power is directly proportional to temperature T, damping coefficient b, and bandwidth Δf. These proportionalities all have intuitive interpretations. First, the kinetic energy of the random motion of gas particles increases with temperature, because the number of collisions and the energy they impart to an electrode both increase with T. Second, because the damping coefficient is proportional to the number of particles, it is reasonable to expect that the noise should rise proportionately with b. Experiment confirms this; mechanical noise is reduced for MEMS devices operated in high vacuum. Third, the bandwidth controls the noise power by effectively filtering out the portion of the spectrum outside the range of Δf. The power spectral density is given by

$$S_u = 4kTb. \qquad (5.111)$$

Example 5.6 **Displacement noise power of a mechanical resonator**

A mechanical resonator and its equivalent circuit are shown in Fig. 5.37. The force-to-displacement transfer function is derived in Appendix B (see Eq. (B.30)). To obtain the noise power of the displacement, use the displacement–force transfer function, the input–output PSD of Eq. (5.101) and the one-sided PSD over frequencies:

$$\langle d_n^2 \rangle = \frac{4k_B T b}{k^2} \int_0^\infty \frac{df}{(1-(f/f_0)^2)^2 + (f/f_0/Q)^2}$$

This integral has no closed-form analytic solution, but it can be evaluated numerically. The displacement signal power due to the applied sinusoidal force at f_d is given by

$$f_{ext} = F_{ext} \cos(2\pi f_d t).$$

The displacement signal power is the product of the power of the force signal and the square of the transfer function magnitude:

$$\langle d_s^2 \rangle = \underbrace{\frac{F_{ext}^2}{2}}_{\langle f_s^2 \rangle} \underbrace{\frac{1/k^2}{(1-(f_d/f_0)^2)^2 + (f_d/f_0/Q)^2}}_{|H_{d/f}(f)|^2}.$$

The signal-to-noise ratio is obtained by inserting the above expressions for d_n^2 and d_s^2 into Eq. (5.94):

$$\begin{aligned} \text{SNR} &= 10\log\left(\frac{\langle d_s^2 \rangle}{\langle d_n^2 \rangle}\right) \\ &= 10\log\left[\frac{F_{ext}^2/2}{[(1-(f_d/f_0)^2)^2 + (f_d/f_0/Q)^2]} \frac{1}{4k_B T b \int_{-\infty}^{\infty} \frac{df}{(1-(f/f_0)^2)^2 + (f/f_0/Q)^2}}\right]. \end{aligned}$$

When the range of operation is well below the resonance ($f_{\max} \ll f_0$), SNR simplifies to

$$\mathrm{SNR} \approx 10\log\left[\frac{F_{\mathrm{ext}}^2}{8k_{\mathrm{B}}Tbf_{\max}}\right].$$

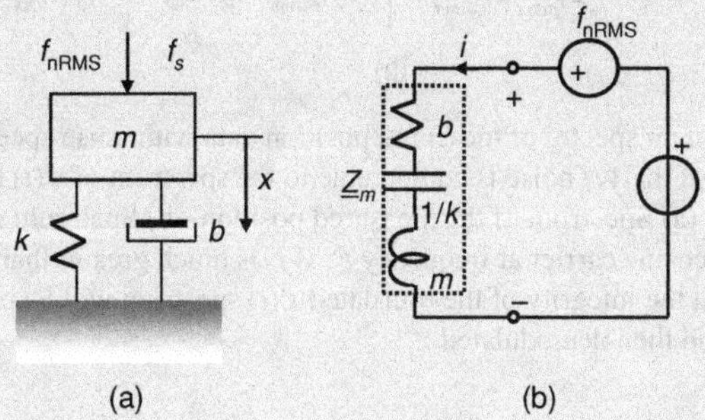

Figure 5.37 A mechanical resonator with the effective noise force. (a) Schematic. (b) Equivalent circuit.

5.8.6 1/*f* amplifier noise

Colored noise is any type of noise that does not have a flat power density spectrum. The most important type of colored noise is $1/f$ noise, also known as *flicker*. Flicker is important in MEMS sensors because of its presence in MOS-based operational amplifiers [5]. Figure 5.38 shows a typical representation of the one-sided noise PSD of a MOS semiconductor device.

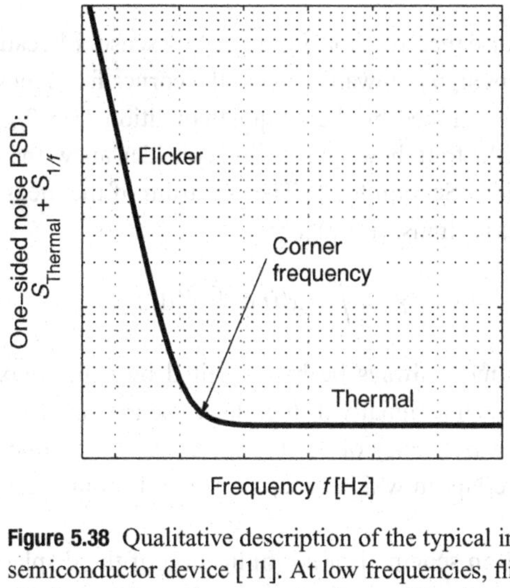

Figure 5.38 Qualitative description of the typical input noise voltage spectrum for a MOS-based semiconductor device [11]. At low frequencies, flicker noise dominates, while thermal noise is dominants above the corner frequency. The slope of the flicker noise α is defined in Eq. (5.112).

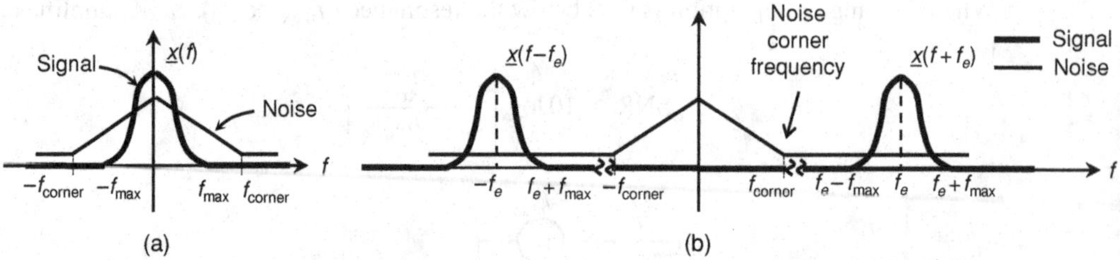

Figure 5.39 Effect of modulation on spectra of measured position data with noise spectra superimposed. It is assumed that the $1/f$ noise is comparable to the spectrum of $x'(t)$ but that the thermal noise is much smaller. (a) Spectrum of the measured position. (b) Spectrum of the AM-modulated signal using a cosine carrier at frequency f_e. If f_e is much greater than the corner frequency of the $1/f$ noise, then the integrity of the translated $x'(t)$ spectrum will be preserved. This signal can be amplified and then demodulated.

The one-sided PSD of this noise is usually represented by the following empirical relation:

$$S_{u,1/f} = K_1 I^{\gamma} / f^{\alpha}. \tag{5.112}$$

Here, K_1 is a device constant, I is the DC current flowing through the device, and the exponent γ is in the range of 0.5 to 2. The frequency exponent α is usually close to unity [7]. This type of noise is important in MEMS sensors and in fact dominates at low frequencies. The typical corner frequency f_{corner} can be as high as ~10 MHz, depending on the geometry of the transistor in the front-end pre-amplifiers. The method most commonly used to suppress employs signal modulation techniques, such as the AM or double side-band suppressed carrier schemes described in Sections 5.5.3 and 5.5.5.

5.8.7 Effect of modulation on $1/f$ noise

Modulation improves the performance of capacitive MEMS sensors because it shifts the spectrum of the displacement signal upward beyond the corner frequency of the $1/f$ noise. The minimum shift is determined by the Nyquist condition $f_e \geq 2f_{\text{corner}}$, where f_e is the carrier frequency. To illustrate how this works, consider anew the three-plate (differential) AC sensing circuit of Section 5.5.5. The spectrum of the measured signal is determined by taking the Fourier transform of $x'(t)$:

$$\underline{x}(f) = \mathcal{F}\{x'(t)\} = \int_{-\infty}^{\infty} x'(t) e^{-j2\pi f t} dt. \tag{5.113}$$

The mechanical motion will almost always be band-limited by some maximum frequency, $f_{\max}$. The spectrum of such a signal will then have the appearance depicted in Fig. 5.39a. If $f_{\max}$ and the corner frequency of the $1/f$ noise are of the same order, then the signal and noise spectra overlap, in which case, the signal becomes corrupted and sensor performance suffers.

The key to this AM modulation scheme is a multiplication of the displacement signal, proportional to $x'(t)$, by the carrier signal, that is, $x'(t)\cos(2\pi f_e t)$. The spectrum of this AM signal is revealed by using Euler's equation to replace the cosine function by exponentials, that is, $\cos(y) = (e^{jy} + e^{-jy})/2$, and then performing the Fourier transform:

$$\begin{aligned}\mathcal{F}\{x'(t)\cos(2\pi f_e t)\} &= \int_{-\infty}^{\infty} x'(t)\cos(2\pi f_e t)\,e^{-j2\pi f t}\,dt \\ &= \frac{1}{2}\int_{-\infty}^{\infty} x'(t)[e^{j2\pi f_e t} + e^{-j2\pi f_e t}]e^{-j2\pi f t} \\ &= \frac{1}{2}\underline{x}(f - f_e) + \frac{1}{2}\underline{x}(f + f_e).\end{aligned} \tag{5.114}$$

The AM signal spectrum consists of two replicas of $\underline{x}(f)$, one each centered at $+f_e$ and $-f_e$. See Fig. 5.39b. An examination of this figure shows that, by setting f_e far above the corner frequency, $1/f$ noise can be isolated from the sensor input signal in the frequency domain. Then, running the modulated signal through a high-pass amplification stage, followed by demodulation to restore a replica of $x(t)$, the $1/f$ noise is suppressed. Only the thermal noise will remain present in the output.

It is important to stress that this modulation strategy is effective only in suppressing noise sources within the electronics. Fortunately for capacitive MEMS sensors, the principle source of noise is indeed the signal conditioning electronics. Thus, modulation techniques are highly effective and very commonly used.

5.9 Electrostatic drives for MEMS resonators

In certain important capacitive MEMS devices, the system is intentionally driven at its mechanical resonance to achieve large-amplitude motion. A voltage or current signal derived from this motion is then monitored to obtain the desired information. Examples based on this scheme include gyroscopes, which are sensitive to vector velocity changes. There are also chemical and biological sensors based on resonant elements that are sensitive to very small frequency shifts. In fact, MEMS components are now starting to replace *quartz-crystal oscillators* as highly stable electronic resonators. Common to all these applications is an electronic drive that preferentially excites motion at the resonance. A critical performance requirement is for the device to operate right at the resonant peak. The only way to guarantee this condition for an electromechanical resonator is to use feedback, that is, to close the loop in the electronic drive itself.

This section explores some basic design issues for resonant MEMS drive circuitry. After reviewing mechanical resonators, we consider the drive and sense electrodes, pointing out the range-of-motion advantage of variable-area capacitive actuators over variable-gap geometries. We also emphasize the practical matter of using square-wave or pulse-train drive signals instead of sinusoids. Following these sections comes a very concise summary of the dominant feedback drive system, the *phase-locked loop*.

5.9.1 Mechanical resonators

Consider a damped linear, mechanical mass-spring system driven by a periodic driving force $f(t)$. The equation of motion is

$$m\ddot{x} + b\dot{x} + kx = f(t), \tag{5.115}$$

where $x(t)$ is displacement, m is mass, b is the damping coefficient, and k is a spring constant. For a review of mechanical resonance, refer to Appendix B. As before, we assume a solution of the form $x(t) = x_0 + x'(t)$, where the primed term is the time-dependent, small-signal variable and the subscript "0" signifies the equilibrium quantity. Although the mechanical device will resonate for any periodic driving force $f(t)$ as long as it operates at $\omega_e = \omega_0$, for now let us restrict our attention to a simple harmonic drive, that is,

$$f'(t) = \mathrm{Re}[\underline{f}e^{j\omega_e t}]. \tag{5.116}$$

Doing so allows us to use phasor notation $\underline{x}$ to represent the sinusoidal motion, $x'(t) = \mathrm{Re}[\underline{x}e^{j\omega t}]$. The transfer function relating displacement to force is

$$\underline{H}_{x/f}(\omega_e) = \frac{\underline{x}}{\underline{f}} = \frac{1/k}{1 - (\omega_e/\omega_0)^2 + \frac{j}{Q}(\omega_e/\omega_0)}, \tag{5.117}$$

where $\omega_0 = \sqrt{k/m}$ is the natural frequency and $Q = \sqrt{km}/b$ is the quality factor. $\underline{H}_{x/f}$ contains all the information needed about the phase and magnitude of $\underline{x}$ for a sinusoidal input drive force at arbitrary drive frequency ω_e.

5.9.2 Drive electrodes with sinusoidal drive

There are two critical requirements for the electrostatic drive of an electromechanical resonator, each of which imposes significant constraints on the design of electrostatic actuators. First, to accommodate the large unconstrained travel that characterizes high-Q resonant motion, variable-area electrodes are used almost exclusively. Recall that the force of electric origin is proportional to the capacitive gradient $\mathrm{d}C/\mathrm{d}x$, which is roughly constant for variable-area capacitive transducers. This attribute is a great advantage for the design of sensors with a highly linear response. The second requirement is that, to achieve the desired large-amplitude bidirectional ($\pm x$) resonant motion, two sets of drive electrodes are needed. This requirement stems from the fundamental fact, already revealed in Example 1.1, that the force of electrical origin for an individual capacitive actuator always exerts a force so as to increase the capacitance.

While it is possible to achieve effectively bidirectional motion about an equilibrium point using a single capacitive actuator by superimposing the AC drive voltage upon a DC bias, this approach is not very efficient, as we show here. Consider a simple electrostatic drive with capacitance $C(x)$ and let the drive voltage be

$$v_{d+}(t) = v_0 + v_m \sin(\omega_0 t). \tag{5.118}$$

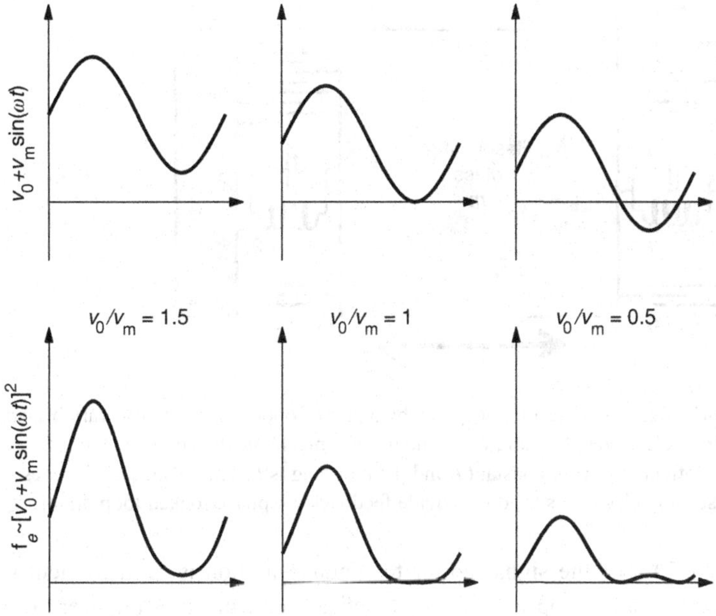

Figure 5.40 Voltage and force waveforms for various DC/AC voltage ratio values.

From Eq. (3.11), the force of electrical origin is

$$\begin{aligned} f^{\mathrm{e}}(t) &= \frac{1}{2}\frac{\mathrm{d}C}{\mathrm{d}x}v_0^2 + \frac{\mathrm{d}C}{\mathrm{d}x}v_0 v_{\mathrm{m}}\sin(\omega_{\mathrm{e}}t) + \frac{1}{2}\frac{\mathrm{d}C}{\mathrm{d}x}v_{\mathrm{m}}^2\sin^2(\omega_{\mathrm{e}}t) \\ &= \frac{1}{2}\frac{\mathrm{d}C}{\mathrm{d}x}\left(v_0^2 + \frac{1}{2}v_{\mathrm{m}}^2\right) + \frac{\mathrm{d}C}{\mathrm{d}x}v_0 v_{\mathrm{m}}\sin(\omega_{\mathrm{e}}t) - \frac{1}{4}\frac{\mathrm{d}C}{\mathrm{d}x}v_{\mathrm{m}}^2\cos^2(2\omega_{\mathrm{e}}t). \end{aligned} \tag{5.119}$$

The force includes a steady-state, DC component, which can be offset by a mechanical spring, plus AC components at both ω_{e} and $2\omega_{\mathrm{e}}$. Comparing the last two terms of Eq. (5.119), we see that as long as $|v_0| > |v_{\mathrm{m}}|$ the first harmonic of the force has a larger amplitude than the second harmonic. Figure 5.40 shows voltage and force waveforms for several values of the DC/AC voltage ratio. As the voltage ratio v_0/v_{m} decreases, higher harmonic force terms start to emerge and effective bidirectional motion is diminished.

A more quantitative measure of the inherent problem with combining DC and AC voltage excitation to achieve bidirectional motion emerges if we consider the *average signal power* content of the force:

$$\frac{1}{T}\int_0^T [f^{\mathrm{e}}(t)]^2\,\mathrm{d}t = \underbrace{\left[\frac{1}{2}\frac{\mathrm{d}C}{\mathrm{d}x}\left(v_0^2+\frac{1}{2}v_{\mathrm{m}}^2\right)\right]^2}_{\mathrm{DC}} + \underbrace{\frac{1}{2}\left[\frac{\mathrm{d}C}{\mathrm{d}x}v_0 v_{\mathrm{m}}\right]^2}_{\omega_{\mathrm{e}}} + \underbrace{\frac{1}{2}\left[\frac{1}{4}\frac{\mathrm{d}C}{\mathrm{d}x}v_{\mathrm{m}}^2\right]^2}_{2\omega_{\mathrm{e}}}. \tag{5.120}$$

For the special case of $|v_0| = |v_{\mathrm{m}}|$, the signal power of the force is ~46% of the total at ω_{e}, while ~51% is at DC and ~3% is at the second harmonic, $2\omega_{\mathrm{e}}$. Thus, less than 50% of the power is available to drive the system at resonance, as required. Utilizing

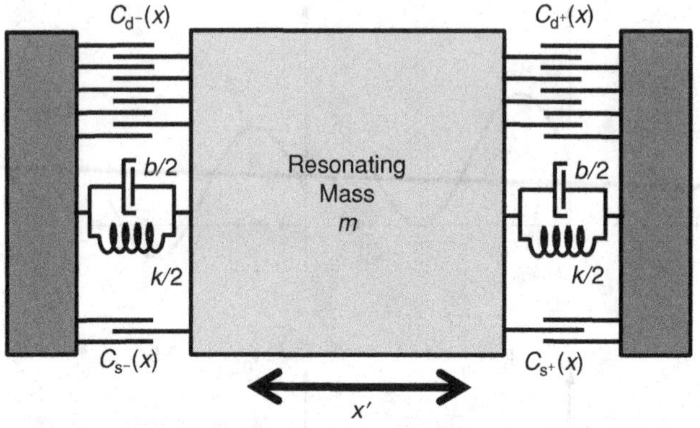

Figure 5.41 Mechanical resonator driven by a pair of opposed symmetric variable-area or comb-drive electrodes C_{d+} and C_{d-}. The mass is mechanically connected to the substrate via a spring of effective spring constant k and the damping is b. The separate electrodes C_{s+} and C_{s-} are the sensing electrodes used to provide feedback for phase-locked loop drives.

less than 50% of the signal power to produce motion is an inefficient usage of the sinusoidal voltage supply. From the practical point of view, the ever-present trend to reduce power supply voltages V_{cc} in consumer electronics makes such inefficiency less and less tolerable. V_{cc} limits the swing of the drive voltage, and the force is proportional to its square.

This power supply inefficiency can be greatly diminished by adding a symmetric bank of electrodes on the opposite side of the moving element. In fact, we will discover here an important reason for the fact that symmetry is a friend of MEMS designers. Figure 5.41 depicts a simple example of a spring-mass system with opposed sets of drive electrodes, C_{d+} and C_{d-}. These two electrode banks are driven by 180° out-of-phase sinusoidal voltages and identical DC bias. Thus, the voltage supplied to the second set of electrodes is

$$v_{d-}(t) = v_0 - v_m \sin(\omega_e t). \tag{5.121}$$

By superposition, the net force on the moving element is

$$f^e(t) = 2\frac{dC_d}{dx} v_0 v_m \sin(\omega_e t) - \frac{1}{2}\frac{dC_d}{dx} V_m^2 \sin(2\omega_e t). \tag{5.122}$$

Note that the DC component of the net force has vanished, leaving only AC components at ω_0 and $2\omega_0$. Assuming as before that $|v_0| = |v_m|$, the force term driving the resonance at the resonance $\omega_e = \omega_0$ now contains ~94% of the total signal power, while only ~6% is at $2\omega_0$. Any increased complexity of the MEMS fabrication is more than made up for by this improvement in the power supply utilization. Often, because the resonant motion at $\omega_e = \omega_0$ is so large, the second harmonic at $2\omega_e$ may be completely neglected.

It is now a straightforward exercise to adapt the electromechanical network from Section 4.7.5 to create an approximate linear model for this opposed-drive actuator system. This network, shown in Fig. 5.42, represents only the dominant, first harmonic, that is, the resonant motion at frequency ω_0. Employing the expression for the external

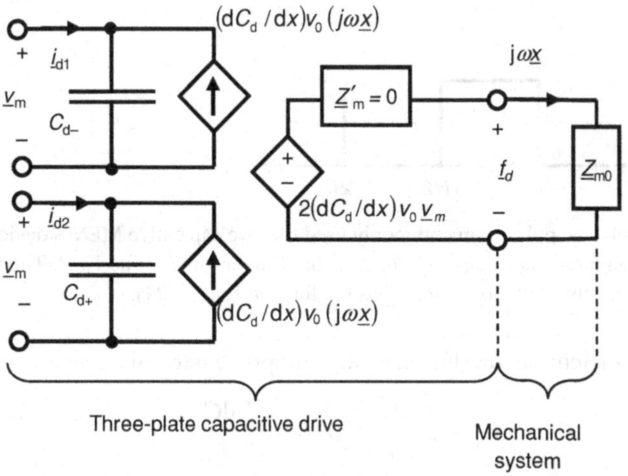

Figure 5.42 An equivalent AC circuit representation based on the electromechanical transducer matrix for the first harmonic at the (resonant) harmonic ω_0 for the case of the symmetric sinusoidal voltage drives described by Eq. (5.122) or, for square-wave drives, by Eq. (5.125).

mechanical impedance $\underline{Z}_{m0}$ from Eq. (4.44) and the impedance-reflection technique from Section 2.3.7, one can obtain a full description of the load presented to the resonant drive electronics. The biggest deficiency of this model probably stems from the assumption that the damping force is linearly dependent upon the velocity.

5.9.3 Non-harmonic drives

Harmonic (sinusoidal) voltage is not usually practical to produce, but any periodic waveform may be used as a voltage drive. The most common choice for resonant drives is the rectangular pulse train shown in Fig. 5.43. This waveform can be expressed as

$$v_{d+}(t) = \begin{cases} v_p, & kT \leq t \leq \left(k + \frac{1}{2}\right) T, \quad \text{integer } k, \\ 0, & \qquad\qquad\qquad\qquad\quad \text{elsewhere.} \end{cases} \tag{5.123}$$

The force of electric origin for a capacitive actuator driven by this waveform is then

$$f^{e}(t) = \begin{cases} \dfrac{1}{2}\dfrac{\mathrm{d}C}{\mathrm{d}x}V_p^2, & kT \leq t \leq \left(k + \frac{1}{2}\right) T, \\ 0, & \text{elsewhere.} \end{cases} \tag{5.124}$$

To compare Eq. (5.124) with the force based on sinusoidal excitation, that is, Eq. (5.119), it is convenient to expand the pulse train function in a Fourier series:

$$f^{e}(t) = A_0 + \sum_{n=1}^{\infty} \left(A_n \cos(n\omega_0 t) + B_n \sin(n\omega_0 t)\right). \tag{5.125}$$

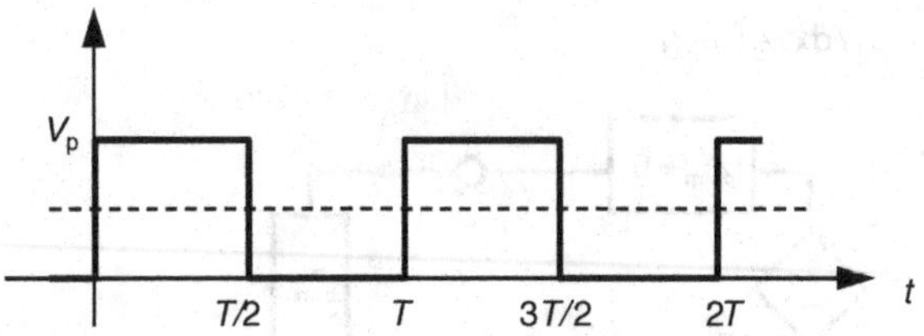

Figure 5.43 Square-wave pulse train commonly used to drive capacitive MEMS devices. The period is T, the magnitude is V_p, and the function has time-average value $V_p/2$. The force of electrical origin for this excitation voltage has the form of Eq. (5.124).

The Fourier coefficients are evaluated from appropriate integrals:

$$A_0 = \frac{1}{T}\int_0^T f^e(t)\mathrm{d}t = \frac{1}{4}\frac{\mathrm{d}C}{\mathrm{d}x}v_p^2, \tag{5.126}$$

$$A_n = \frac{2}{T}\int_0^T f^e(t)\cos(n\omega_0 t)\,\mathrm{d}t = 0, \quad \text{all } n, \tag{5.127}$$

$$B_n = \frac{2}{T}\int_0^T f^e(t)\sin(n\omega_0 t)\,\mathrm{d}t = \begin{cases} \dfrac{v_p^2 \mathrm{d}C_d}{\pi n\,\mathrm{d}x}, & n = \text{odd}, \\ 0, & n = \text{even}. \end{cases} \tag{5.128}$$

Combining Eqs. (5.126) through (5.128) with (5.125) yields the expression for the force as a series of odd sine harmonics:

$$f^e(t) = \frac{1}{4}\frac{\mathrm{d}C_d}{\mathrm{d}x}v_p^2 + \frac{1}{\pi}\frac{\mathrm{d}C_d}{\mathrm{d}x}v_p^2 \sum_{n=1,3,5,\ldots}^{\infty} \frac{\sin(n\omega_0 t)}{n}. \tag{5.129}$$

The analysis of the signal power distribution starts with evaluation of the total signal power of the force by a direct integration using Eq. (5.124):

$$\frac{1}{T}\int_0^T [f^e(t)]^2\,\mathrm{d}t = \frac{1}{8}\left(\frac{\mathrm{d}C_d}{\mathrm{d}x}\right)^2 v_p^4. \tag{5.130}$$

The sine terms are mutually orthogonal, so the net signal power of the sine series on the right-hand side of Eq. (5.129) can be expressed simply as the sum of the squares of the Fourier coefficients. Thus,

$$\frac{1}{T}\int_0^T [f^e(t)]^2\,\mathrm{d}t = A_0^2 + \frac{1}{2}\sum_{n=1,3,5}^{\infty} B_n^2. \tag{5.131}$$

Inserting Eqs. (5.126)–(5.128) into Eq. (5.131) yields

$$\frac{1}{T}\int_0^T [f^e(t)]^2\mathrm{d}t = \underbrace{\frac{1}{16}\left(\frac{\mathrm{d}C_d}{\mathrm{d}x}\right)^2 v_p^4}_{\text{DC term}} + \underbrace{\frac{1}{2}\frac{v_p^4}{\pi^2}\left(\frac{\mathrm{d}C_d}{\mathrm{d}x}\right)^2 \sum_{n=1,3,5,\ldots}^{\infty}\frac{1}{n^2}}_{\text{harmonic terms}}. \tag{5.132}$$

From Eqs. (5.130) and (5.132), it is clear that the DC term contains 50% of the signal power. The first harmonic at ω_0 contains 40.5% and the third at $3\omega_0$ has 4.5%.

As already argued in Section 5.9.2, the DC component of the force may be eliminated using a pair of symmetrically placed electrodes. The second set is driven 180° out of phase, that is,

$$v_{\mathrm{d}-}(t) = \begin{cases} V_{\mathrm{p}}, & \left(k+\dfrac{1}{2}\right)T \le t \le 2kT, \quad \text{integer } k, \\ 0, & \text{elsewhere.} \end{cases} \tag{5.133}$$

From superposition, the total driving force of electric origin now becomes

$$f^{\mathrm{e}}(t) = 2\frac{1}{\pi}\frac{\mathrm{d}C_{\mathrm{d}}}{\mathrm{d}x}v_{\mathrm{p}}^{2}\sum_{n=1,3,5,\ldots}^{\infty}\frac{\sin(n\omega_0 t)}{n}. \tag{5.134}$$

To compute the distribution of the signal power amongst the different harmonic terms, we follow the same procedure as before. The total force signal power becomes

$$\frac{1}{T}\int_0^T [f^{\mathrm{e}}(t)]^2\,\mathrm{d}t = \frac{1}{4}\left(\frac{\mathrm{d}C_{\mathrm{d}}}{\mathrm{d}x}\right)^2 v_{\mathrm{p}}^4. \tag{5.135}$$

As before, the signal power of the sine series is equal to the sum of the squares of the Fourier coefficients. Thus,

$$\frac{1}{T}\int_0^T [f^{\mathrm{e}}(t)]^2\,\mathrm{d}t = 2\frac{1}{\pi^2}\left(\frac{\mathrm{d}C_{\mathrm{d}}}{\mathrm{d}x}\right)^2 v_{\mathrm{p}}^4\sum_{n=1,3,5,\ldots}^{\infty}\frac{1}{n^2}. \tag{5.136}$$

The DC component vanishes and all power is contained in the harmonics. In particular, the first, third, and fifth harmonics contain, respectively, 81.1%, 9.0%, and 3.2% of the total signal power. So, once again, the improved efficiency of opposed drive electrodes has been demonstrated.

Because the drive operates at the resonant peak, it is usually adequate to consider only the first harmonic of the force, in which case we are entitled to identify an effective value for the electromechanical coupling coefficient N:

$$f^{\mathrm{e}}(t) \approx 2\frac{1}{\pi}\frac{\mathrm{d}C_{\mathrm{d}}}{\mathrm{d}x}v_{\mathrm{p}}^{2}\sin(\omega_0 t) = \underbrace{2\frac{\mathrm{d}C_{\mathrm{d}}}{\mathrm{d}x}\frac{v_{\mathrm{p}}}{\pi}}_{\substack{\text{effective}\\ \text{coefficient: } 2N}}\ \underbrace{v_{\mathrm{p}}\sin(\omega_0 t)}_{\substack{\text{effective}\\ \text{AC voltage}}}. \tag{5.137}$$

The associated capacitive current induced by the displacement for this system is

$$i_C(t) = \frac{\mathrm{d}}{\mathrm{d}t}\left\{\frac{\partial}{\partial v_{\mathrm{d}}}\left[\frac{1}{2}C_{\mathrm{d}}(x)v_{\mathrm{d}}^2(t)\right]\right\} = C_{\mathrm{d}}\frac{\mathrm{d}v_{\mathrm{d}}}{\mathrm{d}t} + \frac{\mathrm{d}C_{\mathrm{d}}}{\mathrm{d}x}v_{\mathrm{d}}(t)\frac{\mathrm{d}x}{\mathrm{d}t}. \tag{5.138}$$

Using the first component of the Fourier expansion of $v_{\mathrm{d}}(t)$, we identify the first harmonic of the small-signal term on the right-hand side of Eq. (5.138),

$$\frac{\mathrm{d}C_{\mathrm{d}}}{\mathrm{d}x}\frac{v_{\mathrm{p}}}{\pi}\frac{\mathrm{d}x}{\mathrm{d}t}. \tag{5.139}$$

The effective $\underline{N}$ is

$$\underline{N} = \frac{\mathrm{d}C_{\mathrm{d}}v_{\mathrm{p}}}{\mathrm{d}x\pi}. \tag{5.140}$$

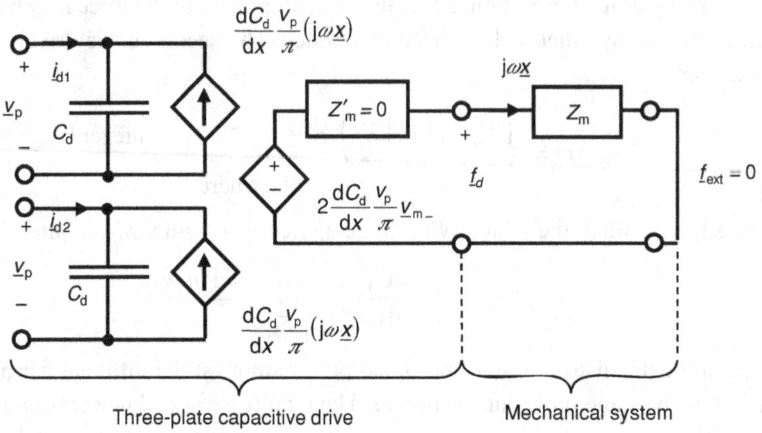

Figure 5.44 Equivalent small-signal AC electromechanical circuit representation for the first harmonic of the square-wave drive. This circuit is virtually identical to the sinusoidal AC drive shown in Fig. 5.42.

Figure 5.44 illustrates the equivalent AC circuit for a first-harmonic analysis of the square-wave drive system. The external mechanical impedance $\underline{Z}_{m0}$ is obtained from Eq. (4.44).

5.9.4 Sense electrodes

To maintain the motion of a MEMS actuator right at resonance, feedback is required. The actuator motion must be sensed and then the loop must be closed. Designs for resonant MEMS systems typically feature built-in capacitive sensing electrodes for this purpose. Note the opposed sensing capacitors shown in Fig. 5.41. To accommodate the large travel of high-Q resonant motion, the sense capacitor bank will be variable-area or comb-drive structures. This choice also assures uniform sensitivity even for large-magnitude displacements. Compared with the drive electrodes, the portion of the total device area taken up by the sense electrodes in such resonators is usually of the order of ~20%.

5.9.5 Harmonic oscillators based on MEMS resonators

Many important circuit designs rely on a fixed stable frequency reference for operation. For years, *quartz crystal resonators* have served this purpose. Quartz crystals achieve highly stable oscillatory behavior as a result of their piezoelectric property and they considerably exceed the performance of *LC* tank circuits and mechanical tuning forks [1]. They are commercially available in a broad range of frequencies from ~50 kHz to ~50 MHz at quality factors in the range of $10^4 < Q < 10^6$. But now, smaller and cheaper MEMS-based oscillator devices are starting to displace quartz crystal components. See Fig. 5.45.

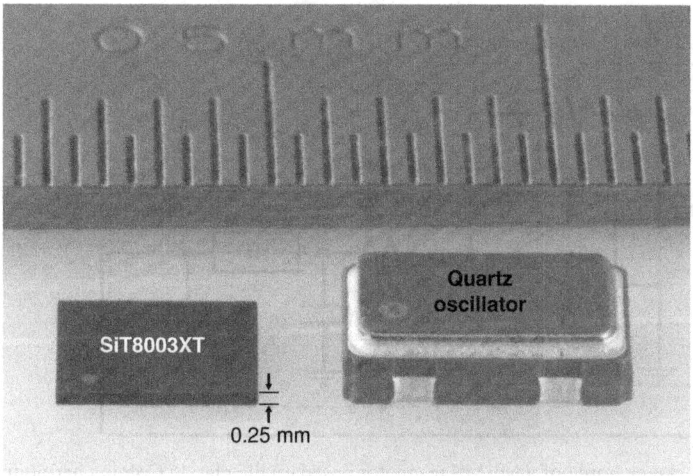

Figure 5.45 Photograph of a MEMS oscillator compared with a common quartz oscillator. Courtesy of SiTime Corporation. Used with permission.

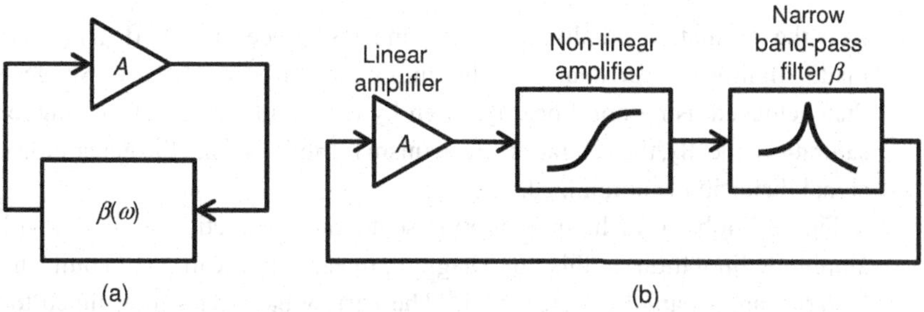

Figure 5.46 A simplified block diagram of a harmonic oscillator. (a) The highly idealized case used to identify the Barkhausen criteria for oscillation. (b) A practical implement incorporating a non-linear element to limit the oscillator amplitude.

This section presents an impedance model for a simple MEMS-based oscillator. The analysis reveals that these devices are actually quite similar in their behavior to quartz crystal devices.

Harmonic oscillators operate using positive feedback as exemplified in the flow diagram of Fig. 5.46a. Stable resonance is achieved when the product of $\underline{A}$, the closed loop amplification of the amplifier, and $\underline{\beta}(\omega)$, the frequency-selective transfer function of the feedback path, satisfy the Barkhausen criteria,

$$|A\beta| = 1, \tag{5.141}$$

$$\angle A\beta = 2k\pi, \quad k = 1, 2, 3, \ldots. \tag{5.142}$$

In a practical oscillator, $|A\beta|$ must be slightly greater than unity to ensure stable operation when circuit parameters shift as a result of temperature changes, component aging, and other factors. If, for example, $|A\beta|$ drops below one because of changes in the parameters,

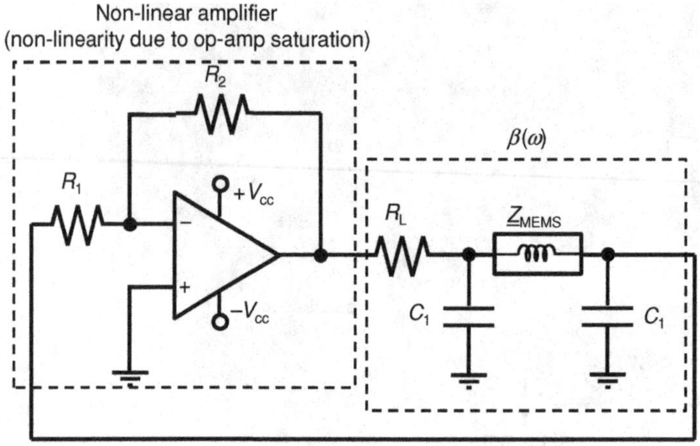

Figure 5.47 An example of a MEMS resonator based on the classic Colpitts harmonic oscillator. A MEMS device, represented by impedance $\underline{Z}_{MEMS}$, is the frequency-selective element that provides the essential function of the band-pass filter. An advantage of using a capacitive MEMS element here is that the frequency can be tuned via the bias voltage.

then the oscillation amplitude decays and resonance is lost. If $|A\beta|$ is intentionally made slightly greater than one, the amplitude starts to increase. Stable operation is then achieved using non-linearity, often by taking advantage of the inherent amplifier saturation. Ideally, the saturation mechanism is "soft," so that the generation of harmonic signal distortion is minimized.

Figure 5.46b shows the simplest representation of a practical oscillator with non-linear amplitude limitation. In this flow diagram, linear amplification and built-in circuit non-linearity are separately represented.[4] The narrow band-pass filter tuned to the desired resonant frequency suppresses the harmonics generated by the non-linear amplitude control.

A classic analog circuit, the well-known Colpitts oscillator, is shown in Fig. 5.47. A MEMS device, represented here by inductive impedance $\underline{Z}_{MEMS}$, forms part of the band-pass filter. While not commonly used in MEMS resonators, this topology serves our purposes quite well here, because we can implement the electromechanical transducer formalism directly from Chapter 4 to obtain an expression for $\underline{Z}_{MEMS}$.[5] Doing so avoids the need to introduce models for discrete transistors, another topic beyond the scope of this text.

For the Colpitts oscillator to lock onto resonance, $\underline{Z}_{MEMS}$ must be inductive, and so the original circuits employed a discrete inductor. The same circuit, using a quartz

[4] In practical oscillators, power amplification and non-linear amplitude limiting are usually achieved in the same active element.

[5] Typical oscillators operate at relatively high frequencies. Because most op-amps do not have the necessary bandwidth, transistor-based amplifier topologies are most commonly used. However, op-amps that operate in the MHz range, such as the LT1190 made by Linear Technologies, have long been available.

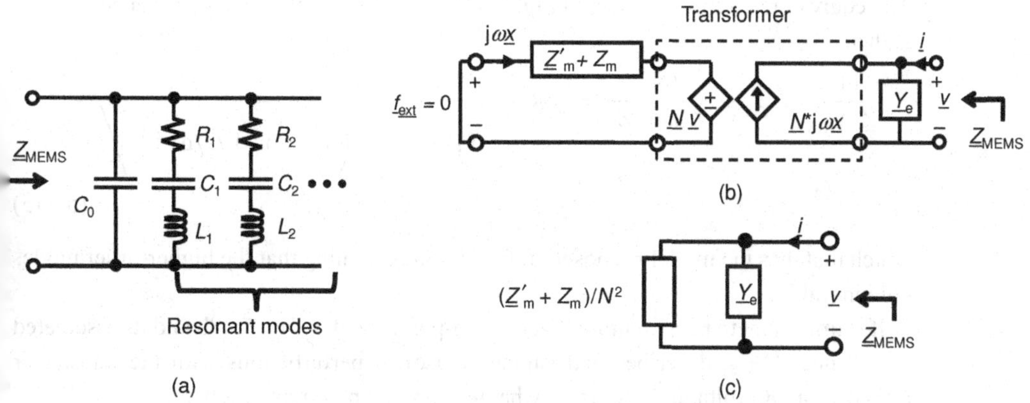

Figure 5.48 Circuit models for MEMS resonators. (a) Equivalent circuit for an electromechanical resonator with multiple modes. (b) N-form electromechanical two-port network with zero external force. (c) Equivalent circuit after reflection of mechanical impedance across to electrical side.

crystal operated in a frequency range where its impedance is inductive, provides much greater frequency control because of the high quality factor of piezoelectric elements. Capacitive MEMS devices with sufficiently strong electromechanical coupling can be used in the same way as quartz crystals. Example 5.7 investigates the requirements for successful implementation of a MEMS resonator in an oscillator circuit.

Figure 5.48a shows a phenomenological model for a network that achieves the desired frequency-selective behavior needed for $\underline{Z}_{\text{MEMS}}$ in the circuit of Fig. 5.47. C_0 is the now-familiar low-frequency capacitance, and each of the *RLC* sections connected in parallel represents a resonant mode of the device. We will return to the possible significance of multiple resonances later, but for now assume that the resonator is operated at the lowest (dominant) mode and ignore all higher resonances.

The N-form of the electromechanical two-port network shown in Fig. 5.48b is most convenient here. What we now seek is the input impedance on the electrical side, that is, $\underline{Z}_{\text{MEMS}}$. Because this system operates as a actuator, there is no external, small-signal force: $\underline{f}_{\text{ext}} = 0$. Under this constraint, one obtains $\text{j}\omega\underline{x} = -\underline{N}\underline{v}/(\underline{Z}'_{\text{m}} + \underline{Z}_{\text{m0}})$, and the dependent current source on the electrical side of Fig. 5.48b becomes

$$-\underline{N}(\text{j}\omega\underline{x}) = |\underline{N}|^2 \frac{\underline{v}}{\underline{Z}_{\text{m0}} + \underline{Z}'_{\text{m}}}. \tag{5.143}$$

Next, the mechanical impedance is reflected over to the electrical side. Finally, substituting the expression for $\underline{Z}'_{\text{m}}$ from Eq. (4.84) yields

$$\frac{\underline{Z}_{\text{m0}} + \underline{Z}'_{\text{m}}}{|N|^2} = \frac{\text{j}\omega m + b + \left(k - \frac{1}{2}v_0^2 \text{d}^2C/\text{d}x^2\right) \Big/ \text{j}\omega}{|N|^2}. \tag{5.144}$$

The equivalent circuit is shown in Fig. 5.48c. The desired, effective impedance, $\underline{Z}_{\mathrm{MEMS}}$, is then

$$\frac{1}{\underline{Z}_{\mathrm{MEMS}}} = \underline{Y}_{\mathrm{e}} + \frac{|N|^2}{Z_{\mathrm{m0}} + Z'_{\mathrm{m}}} = \mathrm{j}\omega C_0 + \frac{|N|^2}{\mathrm{j}\omega m + b + \left(k - \frac{1}{2}v_0^2 \mathrm{d}^2 C/\mathrm{d}x^2\right) \Big/ \mathrm{j}\omega}, \tag{5.145}$$

which matches the model proposed in Fig. 5.48a, assuming that the higher-order modes are ignorable.

It is important to keep in mind that the two-port model of Fig. 5.48b and its associated impedance $\underline{Z}_{\mathrm{MEMS}}$ describe small-signal sinusoidal perturbations with the transducer biased by a DC voltage. Also, from Chapter 4, $\underline{N}$ is a pure real number.

If we now assume that the MEMS device is a variable-area capacitor, so that $\mathrm{d}^2C/\mathrm{d}x^2 = 0$, simplification of Eq. (5.145) using $\omega_0 = \sqrt{k/m}$ and $Q = \sqrt{m\omega_0/b}$ is possible:

$$\frac{1}{\underline{Z}_{\mathrm{MEMS}}} = \mathrm{j}\omega C_0 + \frac{N^2/k}{\left(\mathrm{j}\frac{\omega}{\omega_0^2} + \frac{1}{Q\omega_0} + \frac{1}{\mathrm{j}\omega}\right)}. \tag{5.146}$$

Equation (5.146) can be manipulated into a form more conducive to interpretation by making the following definitions:

$$C_{\mathrm{eq}} = N^2/k, \quad R_{\mathrm{eq}} = b/N^2, \quad \text{and} \quad L_{\mathrm{eq}} = m/N^2;$$

$$\begin{aligned}\underline{Z}_{\mathrm{MEMS}} &= \frac{1}{\mathrm{j}\omega C_0} \frac{R_{\mathrm{eq}} + \mathrm{j}[\omega \underline{Z}_{\mathrm{eq}} - 1/(\omega C_{\mathrm{eq}})]}{R_{\mathrm{eq}} + \mathrm{j}\{\omega L_{\mathrm{eq}} - [1/(\omega C_0) + 1/(\omega C_{\mathrm{eq}})]\}} \\ &= \frac{1}{\mathrm{j}\omega C_0} \frac{R_{\mathrm{eq}} + \mathrm{j}\omega L_{\mathrm{eq}}\left[1 - \left(\frac{\omega_0}{\omega}\right)^2\right]}{R_{\mathrm{eq}} + \mathrm{j}\omega L_{\mathrm{eq}}\left[1 - \left(\frac{\omega_{\mathrm{A}}}{\omega}\right)^2\right]}.\end{aligned} \tag{5.147}$$

Equation (5.147) has two critical frequencies, the mechanical resonance ω_0 already identified, and another frequency,

$$\omega_{\mathrm{A}} = \omega_0\sqrt{1 + C_{\mathrm{eq}}/C_0} \approx \omega_0\left(1 + \frac{1}{2}\frac{C_{\mathrm{eq}}}{C_0}\right), \tag{5.148}$$

called the antiresonance. From examination of Eq. (5.147), it is evident that $\underline{Z}_{\mathrm{MEMS}}$ can only be inductive between ω_0 and ω_{A}, and only if the coupling coefficient is sufficiently large, or if R_{eq} is sufficiently small. The inductive nature of $\underline{Z}_{\mathrm{MEMS}}$ in the range $\omega_0 < \omega < \omega_{\mathrm{A}}$ is even more obvious for a strongly coupled, high-Q resonator, where it is permissible to let $R_{\mathrm{eq}} \to 0$. Then, Eq. (5.147) reduces to

$$\underline{Z}_{\mathrm{MEMS}} \approx \frac{1}{\mathrm{j}\omega C_0} \frac{\left[1 - \left(\frac{\omega_0}{\omega}\right)^2\right]}{\left[1 - \left(\frac{\omega_{\mathrm{A}}}{\omega}\right)^2\right]}. \tag{5.149}$$

Example 5.7 **$\underline{Z}_{\text{MEMS}}$ of a variable-gap vibrating plate**

As an example of a MEMS resonator used in an oscillator, consider the structure depicted in Fig. 5.49. The task is to obtain an expression for $\underline{Z}_{\text{MEMS}}$ using realistic parameter values to determine the frequency range and related performance measures. This structure is a silicon plate clamped around its circular periphery. The plate has radius $r_a = 200$ μm and thickness $h = 2$ μm, and the air gap between the plate and the fixed electrode is $d = 2$ μm. The resonance is assumed to have quality factor $Q = 200$ and applied DC bias $V_B = 15$ V. The plate material, silicon, has mass density $\rho_{Si} = 2330$ kg/m^3, Young modulus $Y_{Si} = 150$ GPa, and Poisson ratio $\nu_{Si} = 0.3$.

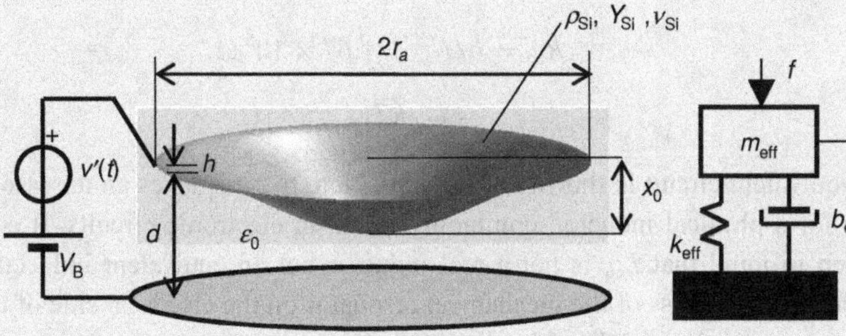

Figure 5.49 Sketch (not to scale) of a silicon resonator with plate radius r_a, thickness h, and spacing between the plate and the bottom electrode d. Consistent with normal practice, the quality factor Q is assumed to have been measured.

This distributed structure is unlike previously considered variable-gap capacitors in that the moving plate is not rigid but instead deforms like a drumhead. Nevertheless, as will be shown in Chapter 6, we can approximate this type of capacitor up to its first resonance as a lumped parameter device with effective mechanical displacement, effective mass, and effective spring constant. For simplicity, any influence of the bias voltage on the equilibrium of the plate is ignored here. From Section 6.6.1 the lumped parameter mass and spring constant for the plate are

$$k_{\text{eff}} = \underbrace{\frac{16}{3}\frac{Y_{\text{Si}}h^3}{1-\nu_{\text{Si}}^2}\frac{\pi}{r_a^2}}_{\text{for distributed force and center displacement}} - \underbrace{\varepsilon_0\frac{r_a^2\pi}{d^2}V_B}_{\text{the term due to electromech. coupling is negligible}} = 716.0\,\frac{\text{N}}{\text{m}},$$

$$m_{\text{eff}} = 0.81 m_{\text{total}} = 0.81\rho_{\text{Si}}r_a^2\pi h = 4.74\times10^{-10}\text{ kg}.$$

From Appendix B, the lumped parameter damping coefficient can be expressed in terms of Q:

$$b_{\text{eff}} = \frac{\sqrt{m_{\text{eff}}k_{\text{eff}}}}{\text{Q}} = 2.91\times10^{-6}\,\frac{\text{N s}}{\text{m}}.$$

The equilibrium capacitance C_0 is

$$C_0 = \varepsilon_0 \frac{r_a^2 \pi}{d} = 5.56 \times 10^{-13}\,\text{F}.$$

The voltage-dependent coupling coefficient $\underline{N}$ is

$$\underline{N} = \varepsilon_0 \frac{r_a^2 \pi}{d^2} V_\text{B} = 4.17 \times 10^{-6}\,\frac{\text{C}}{\text{m}}.$$

The effective spring constant, damping, and mass, once reflected over to the electrical side, take the forms of equivalent capacitance, resistance, and inductance:

$$C_\text{eq} = N^2/k_\text{eff} = 2.43 \times 10^{-14}\,\text{F},$$

$$R_\text{eq} = b/N^2 = 1.67 \times 10^5\,\Omega,$$

$$L_\text{eq} = m/N^2 = 27.2\,\text{H}.$$

The equivalent circuit is shown in Fig. 5.50. Note that L_eq takes an unreasonably high value for a physical inductor, commonly found in electronic circuits. It is important to keep in mind that L_eq is not a real inductor but an equivalent inductance due to reflection of the mass of the mechanical resonator on the electrical side of the system, within a model that employs the direct electromechanical analogy. Since the MEMS capacitor C_0 and the equivalent capacitance due to reflected spring constant mapped to the electrical side C_eq are very small, the equivalent inductance must be very large to achieve usable resonant frequencies.

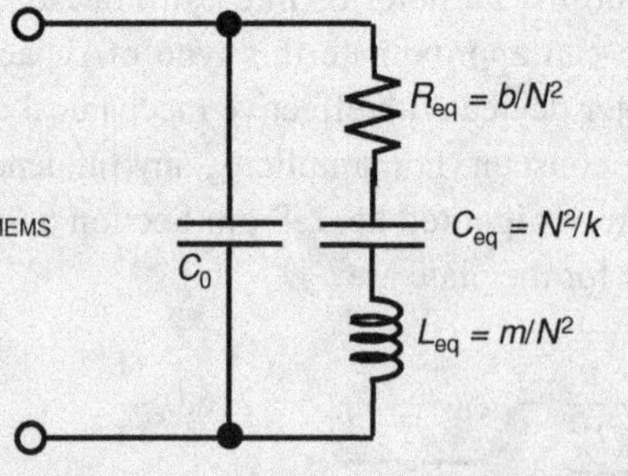

Figure 5.50 Equivalent circuit of the plate resonator, with all mechanical components reflected to the electrical side. The model applies only when DC bias voltage V_B is applied. Note that because $N \sim V_\text{B}$, R_eq and L_eq are inversely proportional to V_B^2, while C_eq is directly proportional to V_B^2. This same equivalent circuit is commonly used for piezoelectric crystals [12].

The behavior of this circuit is readily interpreted. At low frequencies, C_eq dominates the series *RLC* branch so that $\underline{Z}_\text{MEMS} \sim [\text{j}\omega(C_0 + C_\text{eq})]^{-1}$. At the mechanical resonance ω_0, the impedances of the inductor L_eq and capacitor C_eq cancel each other out, leaving the resistance R_eq in parallel with C_0. Above this resonance, the *RLC* branch is dominated by L_{eq}. If the electromechanical coupling is sufficiently strong, this inductance also dominates over C_0 so that $\underline{Z}_{MEMS}$ becomes inductive up to the antiresonance at ω_A. Above ω_A, $\underline{Z}_{MEMS}$ becomes capacitive again.

The various quantities plotted in Fig. 5.51 reveal the interplay of mechanical resonance and the conditions required for $\text{Im}[\underline{Z}_{MEMS}] > 0$. For example, Fig. 5.51a plots the mechanical transfer function defined in Eq. (5.117). The peak of this plot, at $\omega = \omega_0$, indicates the maximum amplitude of the mechanical vibration. The peak of the magnitude response corresponds to the zero of the reactive component of $\underline{Z}_{MEMS}$. The peak of the magnitude response corresponds to the zero of the reactive component of $\underline{Z}_{MEMS}$, as shown in Fig. 5.51b and Fig. 5.51c.

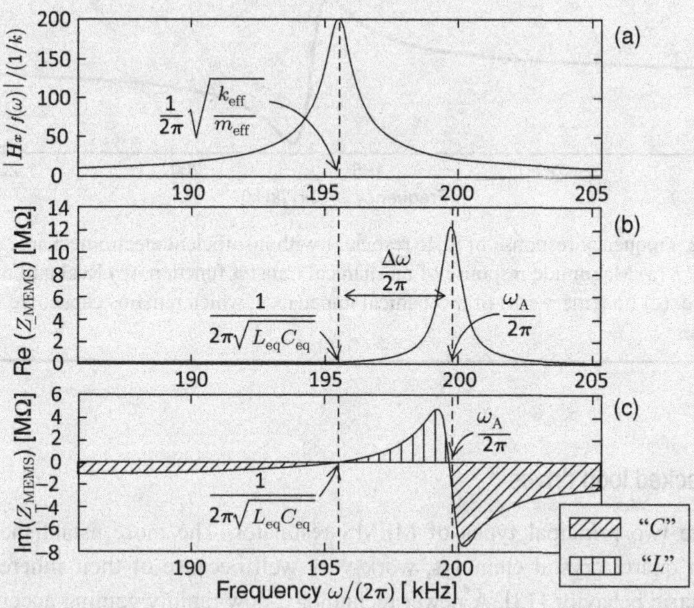

Figure 5.51 Frequency response of the plate resonator with sufficient electromechanical coupling ($V_B = 15$ V). (a) Magnitude response of the mechanical transfer function. (b) Real part of the equivalent impedance–resistance. (c) Imaginary part of the mechanical impedance–reactance. The reactance is capacitive ("C") for $\omega \le \omega_{00}$, inductive ("L") for $\omega_0 < \omega \le \omega_A$ (given sufficiently strong electromechanical coupling) and then capacitive again for $\omega_A < \omega$.

If the electromechanical coupling is not sufficiently strong, then the inductive reactance of the *RLC* branch can never cancel out the reactance of the capacitor C_0, in which case $\underline{Z}_{MEMS}$ remains capacitive for all frequencies, and the Colpitts oscillator cannot lock on to the resonance. This coupling depends on the bias voltage V_B through the coupling coefficient $\underline{N} \propto V_B$. Figure 5.52 shows what happens when the bias voltage V_B is reduced from 15 V to 7 V. The reactive component of the impedance remains capacitive and oscillation cannot be achieved.

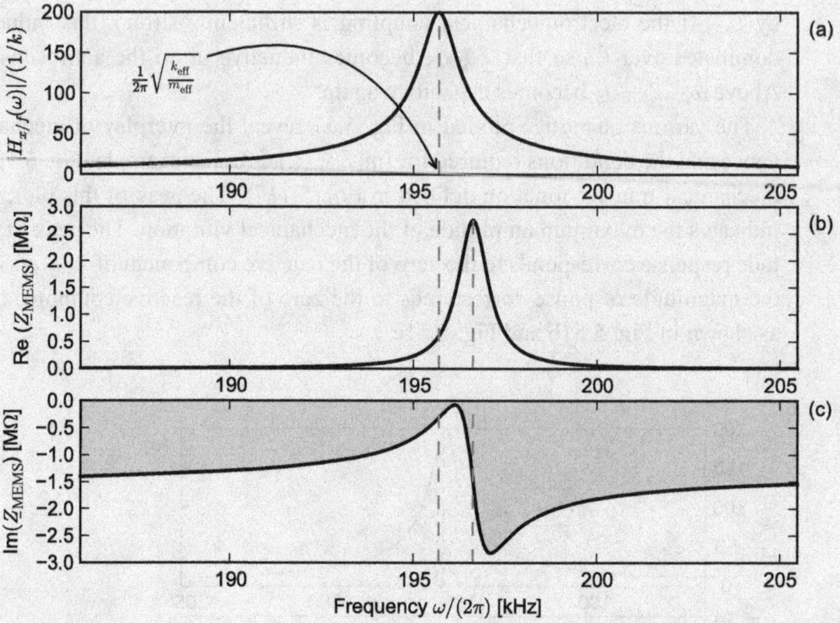

Figure 5.52 Frequency response of plate resonator with insufficient electromechanical coupling ($V_B = 7$ V). (a) Magnitude response of mechanical transfer function. (b) Real part of equivalent impedance. (c) Imaginary part of mechanical impedance, which remains capacitive for all frequencies.

*5.9.6 Phase-locked loop drives

There are two principal types of MEMS resonator. The more established of these, based on quartz crystal elements, work very well because of their inherently stable, piezoelectric behavior [13]. A newer technology, now rapidly gaining acceptance, uses electromechanical resonators driven by *phase-locked loop* (PLL) systems. Phase-locked loop schemes are more complex but also more robust. As engineering experience with them in MEMS has grown, their availability has risen while their cost has fallen. They find principal use in systems such as gyroscopes, which are considered in Chapter 7.

The non-linear lock-in mechanism of the PLL is not treated here. Instead, we limit attention to a qualitative examination of their linearized operation near the operational equilibrium point defined to exist at the mechanical resonance. This is not the same as the mechanical equilibrium introduced in Chapter 4. To facilitate a concise systems-level consideration, the presentation here is founded on the block diagram for a particular PLL realization.

The two most important elements of the system are the mechanical resonator and the *voltage-controlled oscillator* (VCO). A VCO is a device that produces a square-wave function at frequency ω_{VCO} proportional to an input voltage. In practice, the linear relationship between the input voltage and the output frequency is range-limited. Outside

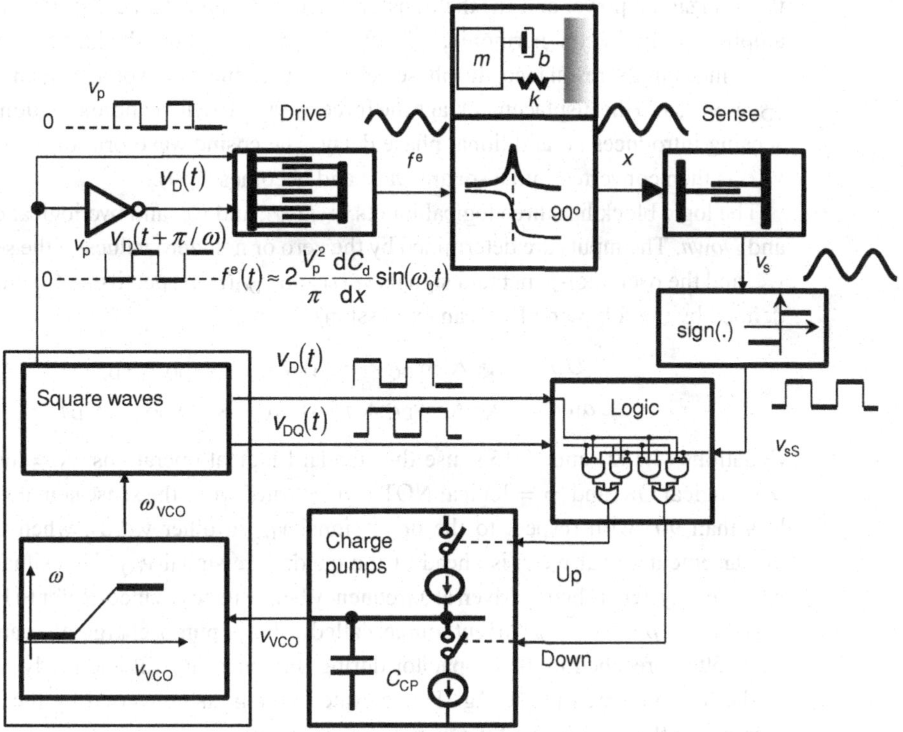

Figure 5.53 The block diagram for a phase-locked loop system.

of this range, the output frequency has hard limits. A first-order, linearized approximation for such a frequency–voltage relation is

$$\omega_{VCO} \equiv \begin{cases} \omega_{MIN}, & v_{VCO} \leq v_{MIN}, \\ \omega_{MIN} + k_{VCO}\left(v_{VCO} - v_{MIN}\right), & v_{MIN} \leq v_{VCO} \leq v_{MAX}, \\ \omega_{MAX}, & v_{MAX} \leq v_{VCO}, \end{cases} \tag{5.150}$$

where

$$k_{VCO} = \frac{\omega_{MAX} - \omega_{MIN}}{v_{MAX} - v_{MIN}}. \tag{5.151}$$

Figure 5.53 shows the block diagram of a PLL system, with waveforms at different test points of the system provided for reference.

Having examined the drive electrodes, the mechanical force–displacement relationship of the resonator, and the sense electrodes, we can now investigate the signal flow in the system. It is convenient to start at the output of the VCO and follow the signal path around and back to its input. The VCO generates the square-wave drive waveform $v_D(t)$ plus a 90°-delayed *quadrature* signal $v_{DQ}(t)$. The drive voltage is applied to the variable-area electrodes. As provided by Eq. (5.134), this drive voltage can be described as a set of odd harmonic force terms. Because (i) the first harmonic contains ~81% of the total signal power and (ii) the transfer function of a mechanical system at resonance amplifies only the first harmonic, it is often justified to ignore the higher-order terms.[6]

At mechanical resonance, the phase delay between the drive voltage v_D and quadrature v_{DQ} is 90°, and the displacement lags the force by 90°. To a first approximation, capacitive sensing introduces no additional phase delay. The cosine waveform of the sense signal $v_s(t)$ is then converted into a square wave and becomes $v_{SS}(t)$.

The logic block has three logical inputs, v_D, v_{DQ}, and v_{SS}, and two logical outputs, *Up* and *Down*. The inputs are determined by the zero or non-zero values of the sense signal, v_{SS}, and the oscillator-generated signals $v_D(t)$ and $v_{DQ}(t)$, respectively. *Up* and *Down* are defined by the following Boolean expressions:

$$Up = (v_{SS} \wedge \neg v_{DQ} \wedge \neg v_D) \vee (v_{SS} \wedge v_{DQ} \wedge v_D), \tag{5.152}$$

$$Down = (v_{SS} \wedge \neg v_{DQ} \wedge \neg v_D) \vee (\neg v_{SS} \wedge v_{DQ} \wedge \neg v_D). \tag{5.153}$$

Equations (5.152) and (5.153) use the standard logical operations: $\wedge$ = logical AND, $\vee$ = logical OR, and $\neg$ = logical NOT. *Up* = "true" when the sense signal s_S is delayed less than 90° with respect to the drive signal v_D, in other words, when the squared displacement signal $v_{SS}(t)$ is ahead of the quadrature signal v_{dQ}. This situation occurs where the system is being driven at a frequency below its resonance. Refer to Fig. 5.54(a). As long as *Up* = "true," a current source, called a charge pump, charges the capacitor C_{CP}. The voltage reached by this capacitor during this time interval is directly proportional to the net integrated phase lag. This elevated voltage is converted by the VCO to an increase in the frequency of the square-wave output.

Similarly, *Down* detects the system state when the squared sense signal s_S is delayed more than 90° with respect to the drive signal v_D, that is, when the mechanical system is being driven at a frequency above its resonance. A second charge pump circuit connected to the same capacitor ensures that, when the drive frequency is above the resonance, the voltage to the VCO is reduced and the drive frequency goes down. Figure 5.54(b) depicts this case.

*5.9.7 PLL system linearization

The essential functional behavior of a MEMS/PLL drive system operating close to the equilibrium at resonance can be modeled by linearization and incorporation the lumped parameter electromechanical formalism of Chapter 4. Consider first the subsystem comprising the electrostatic drive, the mechanical resonator, and the electrostatic sensing system with its electronics, as illustrated in Fig. 5.55. For convenience, we assume a symmetric, DC-biased, three-plate capacitive sensor.

The overall transfer function for this system, obtained by inspection, is the product of the cascaded transfer functions:

$$\begin{aligned} \underline{H}_{v/v}(\mathrm{j}\omega) &= \underline{H}_{f/v}(\mathrm{j}\omega)\underline{H}_{x/f}(\mathrm{j}\omega)\underline{H}_{v/x}(\mathrm{j}\omega) \\ &= \left[\frac{2}{\pi}\frac{\mathrm{d}C}{\mathrm{d}x}V_P\right]\left[\frac{1/k}{1-(\omega/\omega_0)^2+\mathrm{j}\omega/(Q\omega_0)}\right]\left[\frac{2}{C_r}\frac{\mathrm{d}C}{\mathrm{d}x}V_B\right]. \end{aligned} \tag{5.154}$$

[6] Because all real MEMS structures have multiple resonances, the higher-order modes must be considered in the design of practical systems.

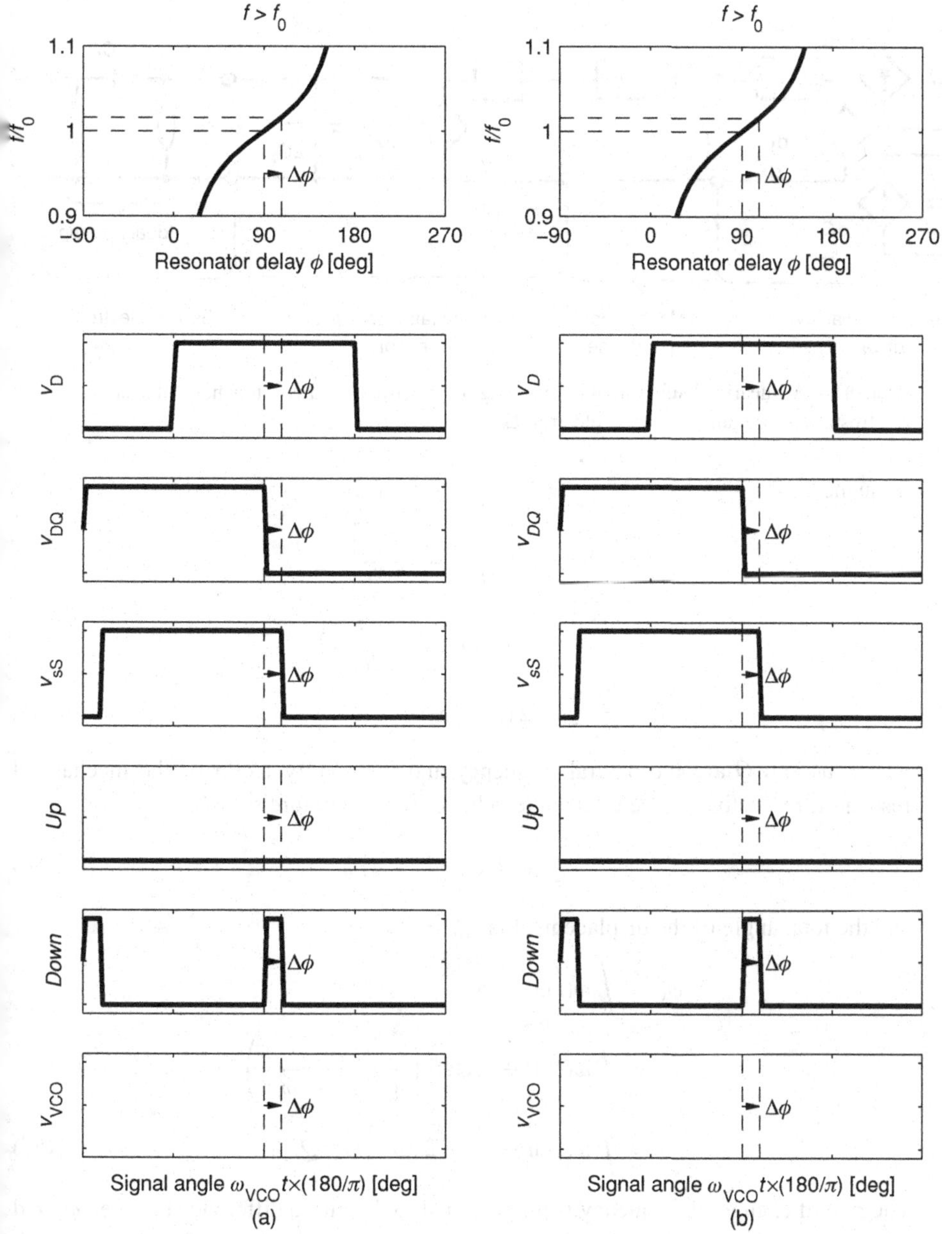

Figure 5.54 Representative control system voltage waveforms and logical variable values for a phase-locked loop resonant drive (a) when the drive frequency is below the resonance and (b) when the drive frequency is above the resonance.

A first-order analysis can be carried out by linearizing the PLL system about the equilibrium at resonance. As stated, we ignore any phase-insensitive elements in order to focus on the mechanical transfer function and the VCO. Close to the mechanical system resonance, the phase delay of the resonator is very nearly linearly related to the

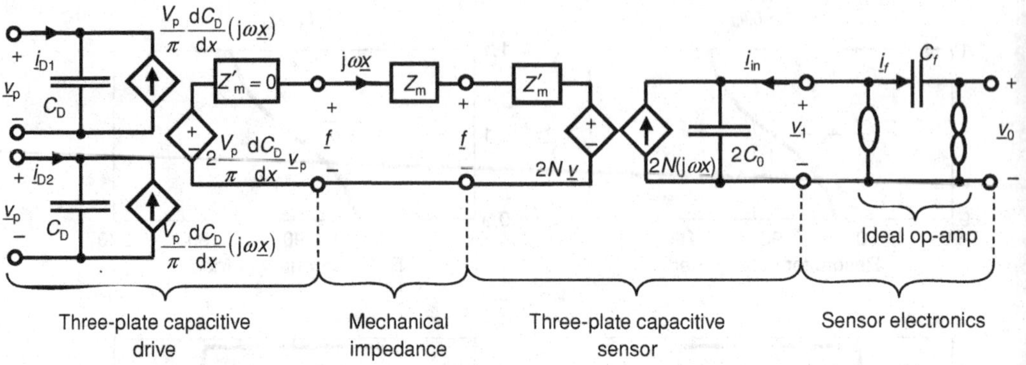

Figure 5.55 A linearized subsystem comprising the electrostatic drive, mechanical system, electrostatic sense, and the sense electronics.

frequency,

$$\begin{aligned}\phi &= \arctan\left[\frac{\omega}{Q\left(\omega^2-\omega_0^2\right)}\right]\\ &\approx \phi|_{\omega\to\omega_0} + (\omega-\omega_0)\left.\frac{\mathrm{d}\phi}{\mathrm{d}\omega}\right|_{\omega\to\omega_0}\\ &= \frac{\pi}{2} - 2(\omega-\omega_0)Q,\end{aligned} \tag{5.155}$$

where ω_0 and Q are the natural frequency and the quality factor of the mechanical resonator, respectively. The total angle value of the quadrature signal is

$$\varphi_Q = \int \omega(t)\mathrm{d}t - \pi/2 \tag{5.156}$$

and the total angle of the displacement is

$$\begin{aligned}\varphi_x &= \int \omega(t)\mathrm{d}t - \phi\\ &= \int \omega(t)\mathrm{d}t - \arctan\left(\frac{\omega}{Q\left(\omega^2-\omega_0^2\right)}\right)\\ &\approx \int \omega(t)\mathrm{d}t - \frac{\pi}{2} + 2(\omega-\omega_0)Q.\end{aligned} \tag{5.157}$$

The rate of change of frequency is proportional to the phase difference between φ_x and φ_Q:

$$\dot{\omega} = k_{\mathrm{CP}}k_{\mathrm{VCO}}(\varphi_{\mathrm{Q}} - \varphi_x) = -2k_{\mathrm{CP}}k_{\mathrm{VCO}}(\omega-\omega_0)Q, \tag{5.158}$$

where k_{CP} and k_{VCO} are the gains of the charge pump and the slope of the ω_{VCO}–v_{VCO} curve, respectively.

At the resonance, $\omega = \omega_0$, the quadrature and displacement signals are in phase and the VCO system is in equilibrium. When the resonator is driven at a frequency above its resonance, $\omega > \omega_0$, the quadrature phase integral is greater than the total phase of the displacement, $\varphi_Q > \varphi_x$, so the frequency is driven down. Similarly, when the resonator operates below its resonance, $\omega < \omega_0$, the integral is smaller than the total phase of the displacement, $\varphi_Q < \varphi_x$, so the frequency is forced back up toward ω_0.

5.10 Summary

The purpose of this chapter has been to introduce basic circuit topologies based on operational amplifiers and to show how they are integrated into a few basic working capacitive sensor systems. A linear two-port network for the op-amp, the nullor model, was presented to facilitate cascading of the electronics with the linearized, electromechanical two-ports. For the sensor applications considered here, attention has been limited to inverting amplifier configurations, an easy choice based on their robustness and wide applicability in MEMS. Other op-amp topologies, for example, the voltage follower and the non-inverting amplifier, are not covered here, but the nullor model is easily adapted to treat these circuit configurations. While this chapter is mostly devoted to standard linear amplifiers, the basics of switched-capacitor sensing are included to acknowledge that many practical MEMS devices are based on switched-capacitor circuits.

For basic DC-biased capacitive sensors connected to the inverting amplifier configuration, the M- and N-form multiport electromechanical models from Chapter 4 offer a convenient systematic analysis platform. Transfer functions displaying the frequency and parameter dependencies for both sensor and actuator configurations are readily obtained, once the proper electrical and mechanical constraints are identified and implemented. Both two-plate (high-impedance) and three-plate (differential) sensor systems submit easily to electromechanical multiport representations. One general finding of this chapter is that two-plate schemes share the major disadvantage of DC offsets that are additive to the output. On the other hand, three-plate geometries avoid these offsets and provide much more favorable signal-to-noise characteristics.

The AC sensing schemes, using periodic zero-time-average voltage excitation, represent a departure from DC-biased systems. Periodic voltage excitation avoids any need for a DC voltage. To analyze and understand AC-excited sensors, resort must be made to basic electromechanical principles and time-dependent analysis. This need arises because the matrix methods introduced in Chapter 4 are no longer applicable. These AC sensing schemes inherently fall into the classification of modulated signal sensors. In particular, the output of the basic high-impedance system using a two-plate capacitive geometry produces an amplitude-modulated output. This signal can be demodulated using simple envelope detection with a diode and capacitor. In contrast, differential (three-plate) sensors produce double-side-band (suppressed carrier) output, calling for synchronous detection. Either harmonic (sinusoidal) or square-wave carrier signals can be used in AC sensors. Because square-wave signals are available in transistor-based electronics, the latter are far more commonly employed. This is true despite certain problems created by the higher harmonic content of square-wave signals.

The biggest motivation for using sensor signal modulation is that it can be used to circumvent the $1/f$ noise arising from the electronic signal conditioning and amplification circuitry. The only critical requirement is that the frequency of the carrier signal, f_e, should be large compared with the corner frequency of the noise. Modulation does not eliminate thermal (white) noise, but, at least for capacitive MEMS devices, the noise floor established by thermal noise is usually well below the detected signal strength.

The subject of the last section of this chapter was voltage drives used to excite MEMS devices at mechanical resonance. In certain applications, resonance is exploited because of the large motion amplitudes achievable. The transducer elements used for such applications usually have the form of a pair of opposed variable-area (or comb-drive) electrode banks. These electrodes achieve strong, bidirectional, periodic forces. In fact, signal power analysis teaches that the most efficient voltage drives are based on such electrode structures. Maintaining stable large-amplitude motion of a resonant system demands tight control of the drive frequency, usually employing feedback. Probably the best example of a resonant drive system for MEMS applications is the phase-locked loop. A major topic in Chapter 7 is vibratory MEMS gyroscopes. Most such systems use PLL drives to achieve the stable large-amplitude motion needed for good sensitivity.

Problems

5.1 For a feedback amplifier with open-loop gain A and feedback attenuation β, compute the overall gain $A_f = v_{out}/v_{in}$. Then take the limit of $A \to \infty$ to show that the voltage gain is determined by feedback alone. What are the advantages of feedback systems?

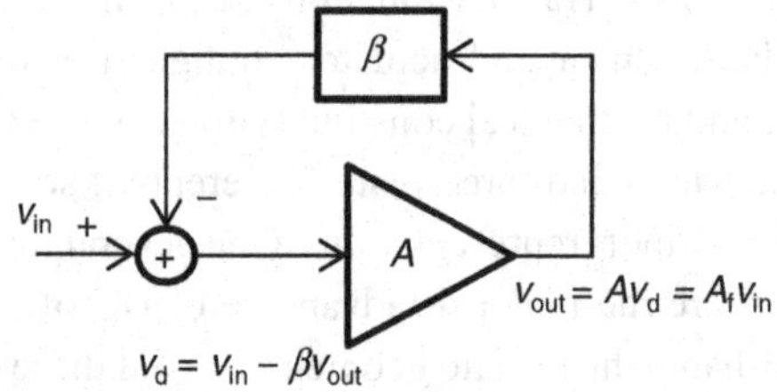

5.2 The circuit diagram shows a general operational amplifier model featuring finite differential input resistance R_d, finite open-loop gain A_v, and non-zero output resistance R_o.

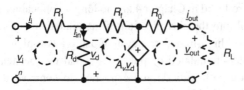

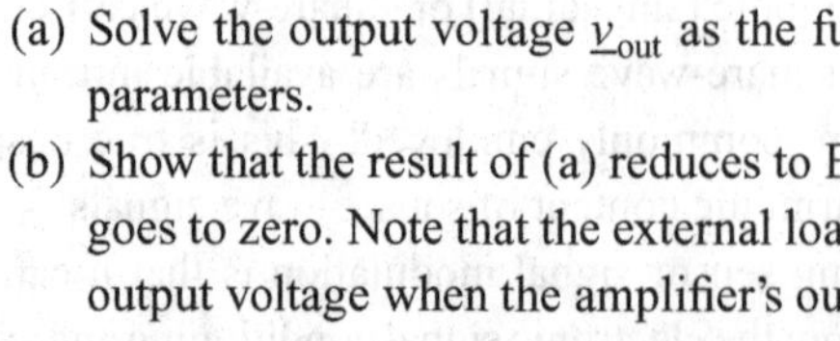

(a) Solve the output voltage v_{out} as the function of the input voltage $_{in}$ and the circuit parameters.

(b) Show that the result of (a) reduces to Eq. (5.7) when R_d and A_v go to infinity and R_o goes to zero. Note that the external load resistor R_L must be retained to compute the output voltage when the amplifier's output resistance R_o is included in the analysis.

5.3 Verify that the linear small-signal AC model shown in Fig. 5.5b is the equivalent of the circuit in Fig. 5.5a. To solve this problem, remember that a DC source is replaced by a short circuit for the small-signal circuit representation. Obtain an equivalent circuit based on the M-form of the sensor.

5.4 The circuit shown in the schematic models the effect of the leakage current I_{lk} on the time-dependent behavior of an inverting amplifier without a resistor in the feedback.

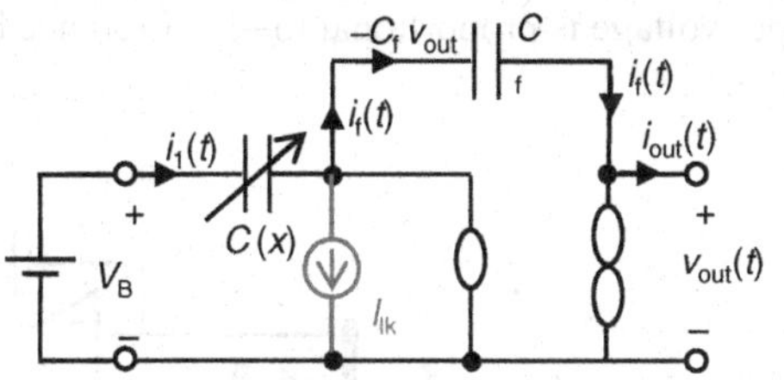

(a) Derive the first-order differential equation for $v_{out}(t)$ for the circuit in the figure. Note that there are two source terms for the voltage: one due to the leakage current I_{lk} and the other due to the time-dependence of x.

(b) Assuming that I_{lk} is turned on at $= 0$ and that the initial output voltage is zero, obtain the solution for the output voltage due to the bias current. Sketch qualitatively the behavior of the output voltage. What happens to the output voltage as time goes to infinity? Does the output reach its final value in finite time?

(c) Find the solution due to the motion x, using the Laplace transforms approach. Again, assume that the initial voltage is zero. Finally, obtain the total solution by adding the two solutions.

5.5 The circuit shown models the effect of the leakage current on the time-dependent behavior of an inverting amplifier with a resistor in the feedback path.

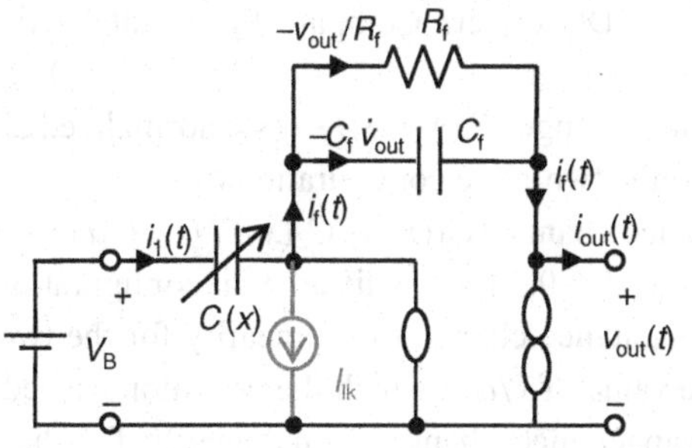

(a) Derive the first-order differential equation for $v_{out}(t)$ for the circuit in the figure. Note that there are two source terms for the voltage: one due to the leakage current I_{lk} and the other due to the time-dependence of x.

(b) Assuming that I_{lk} is turned on at $t = 0$ and that the initial output voltage is zero, obtain the solution for the output voltage due to the bias current. It will be of the form $v(t) = V_\infty + (V(0) - V_\infty)e^{-\frac{t}{\tau}}$.

(c) Find the solution due to the motion x, using the Laplace transforms approach. Again, assume that the initial voltage is zero. Finally, obtain the total solution by adding the two solutions.

5.6 Consider a two-plate, variable-gap electrode configuration (a), where the output voltage is proportional to the capacitance change and a three-plate, variable-gap electrode configuration (b), where the output voltage is proportional to the difference between C_1 and C_2. Assume equilibrium at $x = 0$.

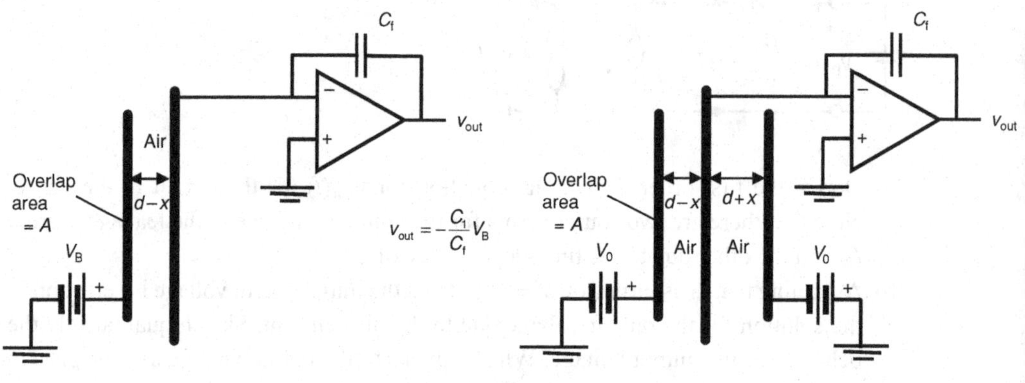

(a) Expand $C(x)$ of the single-ended configuration (a) in Taylor series. What is the sensitivity $C(x)/x$ for $|x| \ll d$? What is the overall sensitivity v_{out}/x for small signals? What is the DC offset?

(b) Express the two capacitances of the three-plate configuration (b) in Taylor series and show that the even terms of the expansion cancel out. What is the sensitivity $(C_1(x) - C_2(x))/x$ for small signals ($|x| \ll d$)? What is the overall sensitivity v_{out}/x for small signals? What is the DC offset? Compare the results with the results obtained in (a).

(c) Plot the normalized capacitance change $C(x)/(\varepsilon_0 A/d)$ vs. normalized displacement x/d for $-0.5 \leq x/d \leq 0.5$ for the two-plate configuration (a).

(d) Plot the normalized capacitance change $(C_1(x) - C_2(x))/(\varepsilon_0 A/d)$ vs. normalized displacement x/d for $-0.5 \leq x/d \leq 0.5$ for the differential configuration (b).

(e) Plot the deviation of this capacitance change from linearity for the two-plate configuration (a). Determine the value of x/d at which the deviation exceeds 5%.

(f) Plot the deviation of this capacitance change from linearity for the differential configuration (b). Determine the value of x/d at which the deviation exceeds 5%.

(g) In typical MEMS sensors, the chip area allocated for the sensor electrodes is fixed. Ignoring manufacturing details, compare sensitivity, offset, and linearity for two alternative designs: a two-plate scheme using the allotted area for one large, variable-gap structure, and a three-plate scheme where the area is divided between two variable-gap capacitors.

5.7 A DC biased, variable-area, three-plate sensor geometry is connected to an inverting op-amp integrator as shown in the figure.

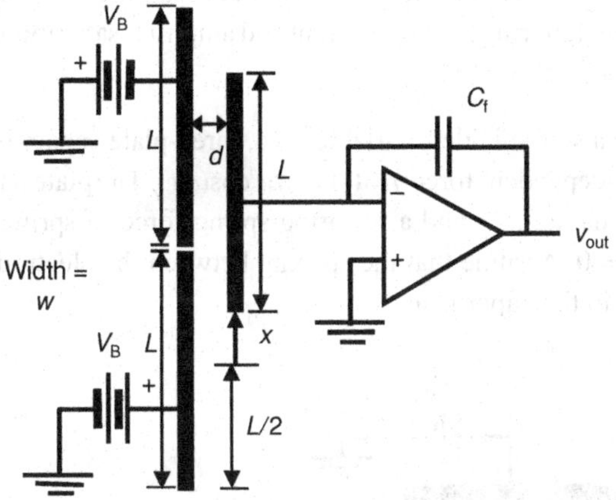

(a) Assuming $-L/2 \leq x \leq L/2$, find the static relationship between the plate displacement x and the output voltage v_{out} of the inverting amplifier circuit.
(b) Sensors that inherently feature a linear relationship between input and output are favored in some practical applications. Think of some examples and identify their common attributes.
(c) In that respect, does this device possess inherent linearity? Identify and discuss qualitatively any limits to the range.

5.8 The DC-biased, capacitive sensor system illustrated in the figure has an inverting-op-amp-based integrator connected to three independent electrodes controlled by mechanical variables: $x_1(t)$, $x_2(t)$, and $x_3(t)$. The electrodes all have identical area A and the same nominal undisplaced electrode spacing d.

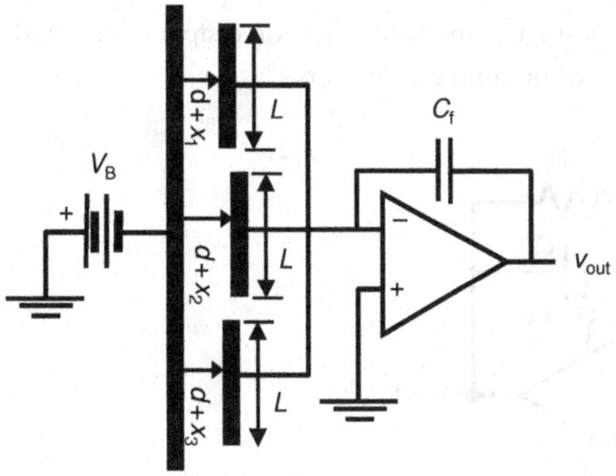

(a) Find an explicit time-domain expression for the output voltage $v_{out}(t)$ for small values of the displacements x_1, x_2, and x_3 near equilibrium by integrating the three currents. Assume here that all $\mathrm{d}C_i(x_i)/\mathrm{d}x_i$ are constant near the equilibria.

(b) Draw the equivalent circuit for small signals using the N-form.
(c) Draw the equivalent circuit for small signals using the M-form.
(d) Repeat the derivation for $v_{out}(t)$ but this time without assuming that the capacitive derivatives are constant. Ignoring leakage current and amplifier saturation, determine the DC offset of this device.

5.9 The moving plate of a self-shielded, variable-area, three-plate capacitive sensor is subject to an external time-dependent force $f_{ext}(t) = f_{in}\cos(\omega t)$. The plate has mass $= m$ and it is subject to linear damping b and a restoring spring force of spring constant k that is equal to zero at $x = 0$. Assume that the spacing between the electrodes is d and the depth of the electrode in the paper is w.

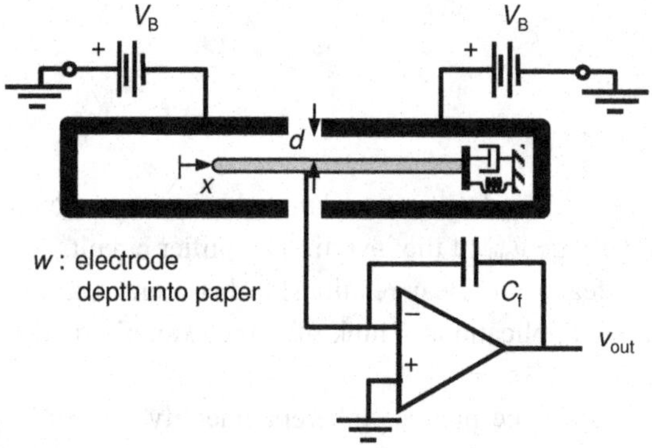

(a) Using the methods of Chapters 4 and 5, propose a cascade model for the electromechanical transducer and the inverting amplifier stage connected to it.
(b) Assuming zero damping, find an expression for the transfer function:

$$\underline{H}_{v/f}(\mathrm{j}\omega) = \underline{v}_{out}/\underline{f}_{ext}.$$

5.10 The figure shows a three-plate capacitance sensor system circuit with a feedback resistor to eliminate the effect of op-amp bias currents.

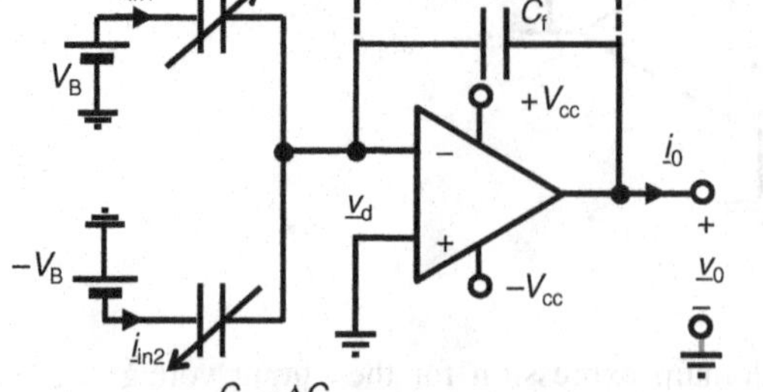

(a) Derive the expression for the output voltage and compare the result with Eq. (5.52).

(b) What is the corner frequency above which the sensor response becomes essentially frequency-independent?

5.11 Write the differential equation in terms of $v_{out}(t)$ and $x(t)$ for the output voltage for the AC differential scheme shown in the figure. Do not attempt to solve it.

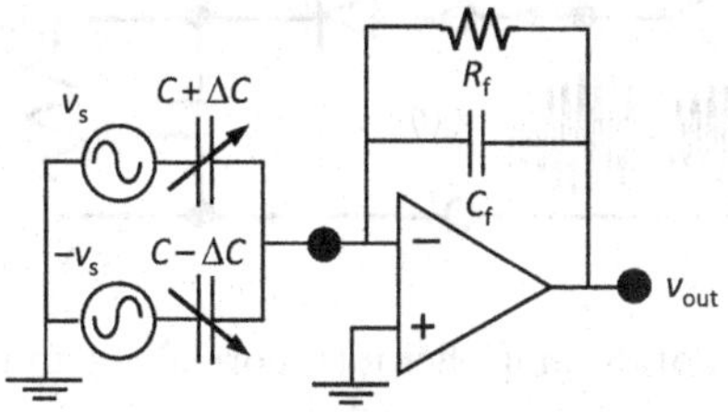

5.12 Three different AC-excited three-plate sensor systems are illustrated in the figure. For each circuit solve for the output voltage in terms of the amplitude of sinusoidal voltage v_{so}, nominal capacitance C, and capacitance perturbation ΔC. For this problem assume that ΔC is constant.

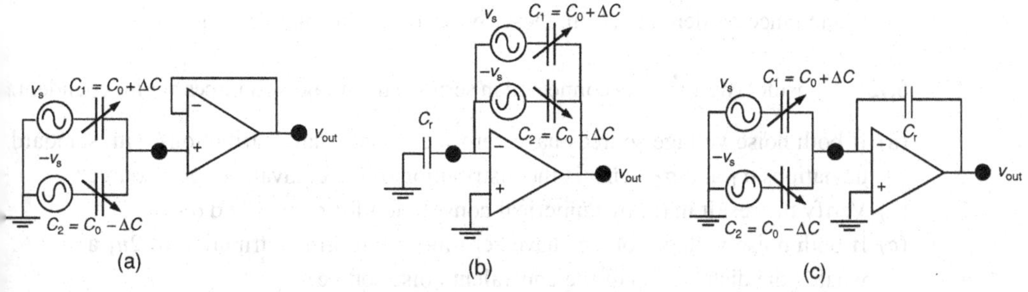

5.13 Consider the ideal switched-capacitor element of Fig. 5.26a.

(a) What is the equivalent resistance R_{eq} of Fig. 5.26b for a switching frequency of $f_s = 100$ kHz and a capacitance of 0.2 pF.

(b) Write the approximate expression for the overall DC offset and DC gain of the system for the two-plate sensor shown in Fig. 5.27, assuming ideal demodulation and an idea, unity-gain filter applied after the amplifier.

(c) Write the approximate expression for the overall DC offset and DC gain of the system based on the differential amplifier shown in Fig. 5.29, assuming ideal demodulation and an ideal unity-gain filter applied after the amplifier.

5.14 For the AC sensing system in the figure, modulated signal with $m(t) = 0.3\cos(2\pi \cdot 10^3 t)$ and a carrier frequency $\omega_e/(2\pi) = 20$ kHz.

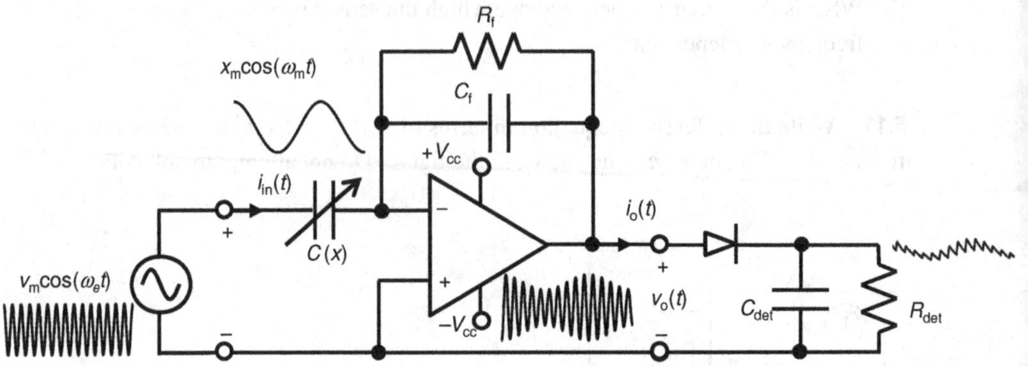

(a) Write the expression for the output of the amplifier $v_o(t)$ normalized to v_m.

(b) Show that the current flowing through capacitor C_{det} is given by

$$i_C = C_{det}\dot{v}_{det}(t) = \begin{cases} (v_0 - v_{det})/R_{diode}, & v_{det} < v_0(\text{diode conducts}), \\ -v_{det}/R_{det}, & v_0 \leq v_{det}(\text{diode is open circuit}). \end{cases}$$

(c) For $R_{det} = 2$ kΩ, $R_{diode} = 1$ Ω, and $C = 200$ nF integrate the voltage across R_{det} numerically. Plot $v_{det}(t)/v_m(t)$ and $v_o(t)/v_m(t)$ in the same plot for $0 \leq t \leq 1$ ms.

(d) Repeat the simulation for $C = 400$ nF and $C = 40$ nF to investigate the effect of capacitance on demodulation. Describe and explain what you find.

5.15 Consider two resistors connected in series. The two noise sources are independent.

(a) If both noise voltage sources have zero-mean Gaussian distributions with standard deviations σ_1 and σ_2, what is the distribution of the equivalent noise source?

(b) Verify the result in (a) by numerical convolution for $\sigma_1 = 1$ and $\sigma_2 = 3$.

(c) If both noise voltage sources have zero-mean uniform distribution of $2u_1$ and $2u_2$, what is the distribution of the equivalent noise source?

(d) Verify the result in (c) by numerical convolution for $u_1 = 1$ and $u_2 = 3$.

(e) Can you think of a MEMS device featuring additive mechanical noise?

Hint: the distribution of the sum of two independent random variables $v = v_1 + v_2$ is their convolution:

$$p_v(v) = p_{v_1} * p_{v_2} = \int_{-\infty}^{\infty} p_{v_1}(v_1)\, p_{v_2}(v - v_1)\, dv_1.$$

5.16 Consider a parallel connection of two resistors $R_1 = 1$ kΩ and $R_2 = 3$ kΩ.

(a) What are the noise voltages for each of the two resistors at room temperature ($T = 300$ K) over a bandwidth of $\Delta f = 1$ Hz?

(b) Find the values for the Thevenin equivalent of this noise source using ideal noiseless resistors. (*Hint:* simple superposition is not applicable here.)

5.17 Consider the comb-drive silicon resonator shown in the figure. All the beams are assumed to be $w = 2$ μm wide and $h = 10$ μm deep. The air gap between the adjacent electrodes is $d = 2$ μm. Assume that the range of motion is $-L/2 \leq x \leq L/2$, where $L = 50$ μm. The length of the spring straight segment $L_{\text{spring}} = 50$ μm. The resonant quality is measured to be $Q = 20$. Use the following values for the mass density and Young modulus of Si: $\rho_{\text{Si}} = 2330$ kg/m^3 and $Y_{\text{Si}} = 160$ GPa.

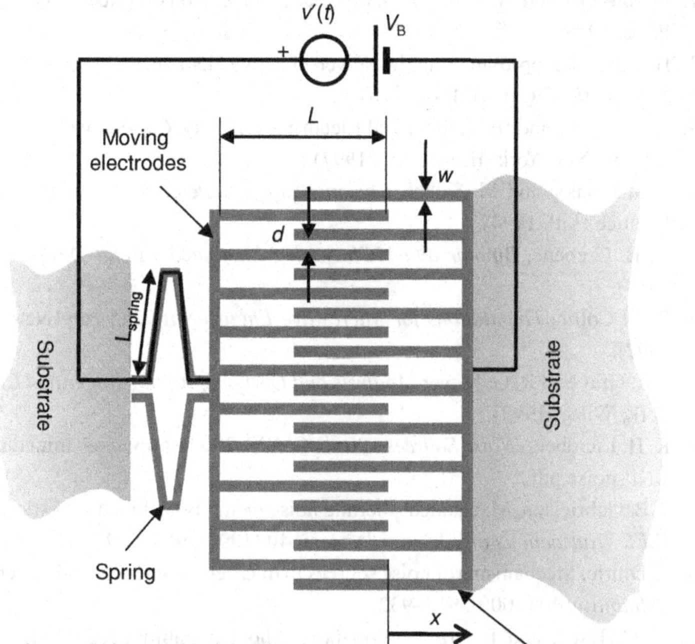

All beams have width w and height (depth into the page) h.
The spacing between the electrodes is d.

(a) Ignoring the effective mass of the spring, estimate the equivalent mass of the comb drive. Using $k_{\text{eq}} = Yh(wlL_{\text{spring}})^3$, estimate the equivalent spring constant.[7] What is the resonant frequency of this device?

(b) Write the expression for the force of electric origin.

(c) Using the state-space equations for a single-degree-of-freedom system,[8] solve numerically for the displacement $x(t)$ in the interval $0 \leq t \leq 2\pi Q/\omega_0$. Assume $V_B = 2$ kV and $v' = 300 \cos(\omega_0 t)$ V. Plot $x(t)$ and $f^e(t)$ on the same plot at different y axes. Plot $x(t)$ vs. $f^e(t)$. What is the phase between the force and displacement signals?

(d) Repeat the exercise in (c) for $V_B = 2$ kV and $v' = 1.8^* \cos(\omega_0 t)$ kV.

(e) Repeat the exercise in (c) for $V_B = 0$ V and $v' = 2 \cos(\omega_0)$ kV.

[7] More on the analysis of this type of spring is found in Section 7.3.3.

[8] Refer to Appendix B, Section B.2.5.

(f) Repeat the exercise in (c) for $V_B = 0$ V and $v' = 2\cos((\omega_0/2)t)$ kV.
(g) Discuss the results.

References

1. P. Horowitz and W. Hill, *The Art of Electronics*, 2nd edn (New York: Cambridge University Press, 1989).
2. H. Schmid, Approximating the universal active element, *IEEE Transactions on Circuits and Systems II*, 47 (2000), 1160–1169.
3. L. K. Baxter and IEEE Industrial Electronics Society, *Capacitive Sensors: Design and Applications* (New York: IEEE Press, 1997).
4. J. G. Proakis and M. Salehi, *Communication Systems Engineering* (Englewood Cliffs, NJ: Prentice Hall, 1994).
5. A. B. Grebene, *Bipolar and MOS Analog Integrated Circuit Design* (New York: J. Wiley, 1984).
6. R. E. Collin, *Foundations for Microwave Engineering*, 2nd edn (New York: McGraw-Hill, 1992).
7. P. R. Gray and R. G. Meyer, *Analysis and Design of Analog Integrated Circuits*, 3rd edn (New York: Wiley, 1993).
8. K. H. Lundberg, *Noise Sources in Bulk CMOS* (2002). http://web.mit.edu/klund/www/papers/UNP_noise.pdf.
9. T. B. Gabrielson, Mechanical-thermal noise in micromachined acoustic and vibration sensors, *IEEE Transactions on Electron Devices*, 40 (1993), 903–909.
10. Z. Djuric, Mechanisms of noise sources in microelectromechanical systems, *Microelectronics Reliability*, 40 (2000), 919–932.
11. K. C. Hsieh and P. Gray, A low-noise chopper-stabilized differential switched-capacitor filtering technique, in *IEEE International Solid-State Circuits Conference. Digest of Technical Papers. 1981* (1981), pp. 128–129.
12. K. S. Van Dyke, The piezo-electric resonator and its equivalent network, *Proceedings of the Institute of Radio Engineers*, 16 (1928), 742–764.
13. B. J. T. Boser, Electronics for micromachined inertial sensors, in *International Conference on Solid State Sensors and Actuators, TRANSDUCERS '97 Chicago* (1997), pp. 16–19.

6 Distributed 1-D and 2-D capacitive electromechanical structures

6.1 Introduction

Previous chapters of this text have relied exclusively on lumped parameter capacitance models with mechanical motion represented by one or perhaps a few discrete mechanical variables. In these models, the capacitive electrodes have been assumed to be rigid structures. In MEMS, however, many of the common designs do not fit such a description. For example, one of the most widely exploited structures is the cantilevered beam, a 1-D continuum. Also, pressure sensors and microphonic transducers are usually based on deformable 2-D continua, typically circular diaphragms. Other applications of MEMS continua exist and are growing in numbers. For example, Fig. 6.1 shows a deformable mirror, developed by Boston Micromachines Corporation, for use in laser-pulse-shaping applications where the deformable mirror modifies the phase of the spectral components of the laser pulse to achieve desired temporal pulse characteristics.

This chapter develops modeling approaches for such continua and demonstrates that it is usually possible to devise reduced-order, lumped-parameter models that capture the essential behavior and reveal important trade-offs amongst the system parameters. Section 6.2 employs this approach, using the simple example of a cantilevered beam operated as a capacitive transducer. After introducing some basics from the mechanics of continua, a familiar-looking lumped parameter model is extracted and then tested for accuracy by comparing the predicted resonant frequency with the well-known analytical solution for the cantilevered beam. In the initial exercise, certain assumptions and approximations are presented without much explanation but these are tested and justified in Section 6.3 in a detailed revisit to the problem. The effort leads naturally to the electromechanical two-port transducer representations of Chapter 4. In subsequent sections, the same treatment is extended to the important geometry of the circular diaphragm.

The strategy is to add successive layers of sophistication to achieve better accuracy and deeper understanding. Students, especially those encountering the mechanics of solids for the first time, are urged to refer often to Appendix B, which reviews the dynamics of mechanical systems, and Appendix D, which reviews the mechanics of solids.

The models derived in this chapter are relatively easy to implement in MATLAB or any similar programming environment. To illustrate the point, numerous examples of MATLAB code have been placed in examples throughout the text. These code snippets are kept short. More code details, plus executable functions and scripts, will be found on the book's website.

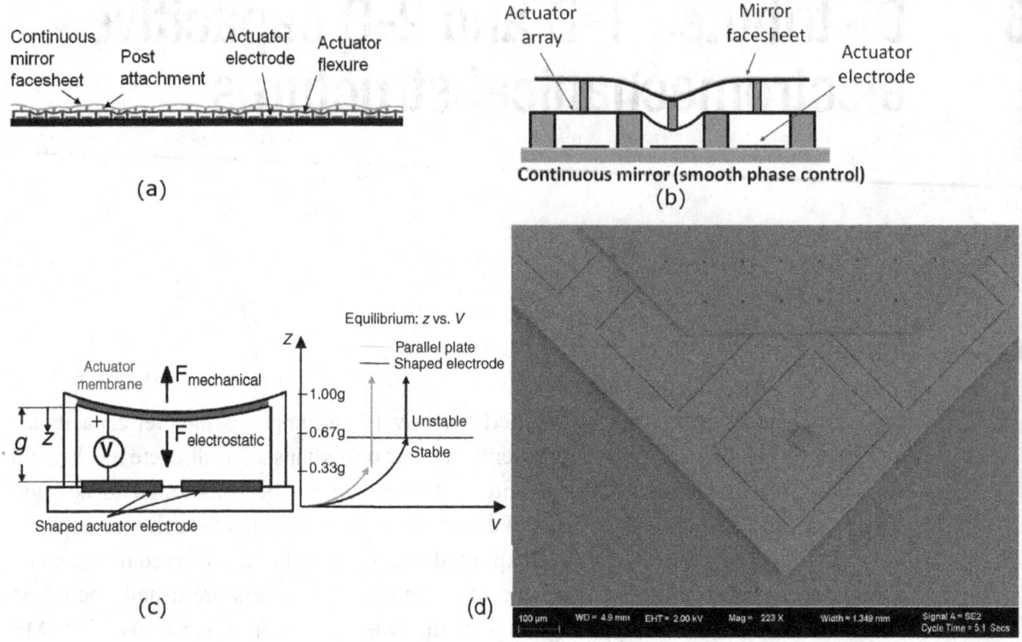

Figure 6.1 Deformable mirror for use in laser-pulse-shaping applications. The deformable mirror modifies the phase of the spectral components of the laser pulse to achieve a desired temporal pulse characteristic. (a) A schematic of the flexible mirror facesheet. The facesheet is mounted atop an array of capacitive actuators that impart the desired spatial deformation to the mirror. (b) Close-up showing three actuator elements. The middle section is turned on. (c) Schematic of electrostatic actuation of a double-clamped beam. (d) A discrete (segmented) version of the mirror. The image is credited to Steven Cornelissen of Boston Micromachines, Inc. Used with permission.

6.2 A motivating example – electrostatic actuation of a cantilevered beam

A good first choice of a structure exhibiting spatially varying displacement is the cantilevered beam. It is geometrically simple and, because of the ease with which it can be fabricated using standard processing, it enjoys wide use in MEMS technology. Under certain conditions, a representative mechanical displacement variable and an associated point force can be identified and then used to derive a lumped parameter model in a familiar form. This reduced-order model captures with fair accuracy the essential dynamics and parametric sensitivities of the structure. The derivation is based on a few simplifying assumptions and principles from the mechanics of solids, as reviewed in Section D.7 of Appendix D.

6.2.1 Problem description

The cantilevered beam depicted in Fig. 6.2 is *prismatic*, that is, uniform along its length L, and it has a rectangular cross-section of width w and thickness h. The beam and the substrate are electrically grounded. Patterned on the substrate, but insulated from it and

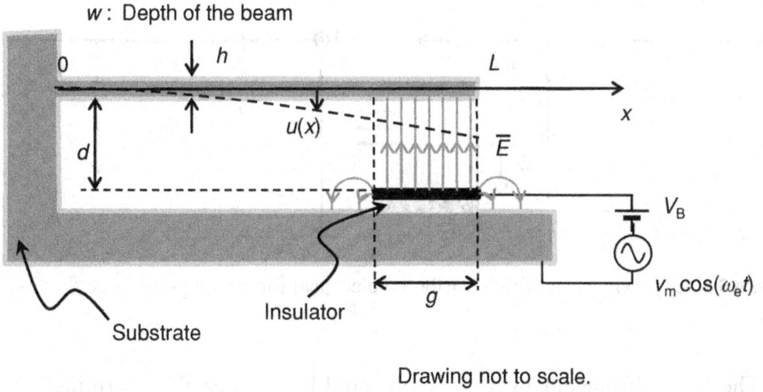

Figure 6.2 A capacitive transducer employing a bending (cantilevered) prismatic beam as the moving electrode element and an addressable electrode located below it. Drawing is not to scale: $L > g \gg w, h$.

located directly beneath the beam, is an addressable electrode of length g with its outer edge and sides aligned with the adjacent edges of the beam. The undeformed spacing between the grounded beam and this electrode is d. The time-dependent voltage applied between the electrode and the grounded beam structure is $v(t) = V_B + v_m \cos(\omega_e t)$. The DC voltage, V_B, is needed to enable energy transduction, while the AC component of magnitude v_m is a sensor drive signal.

Unlike the rigid, parallel-plate, variable-gap capacitor introduced and analyzed in Section 3.4.1, the bending beam realizes a continuum of displacement values along the x coordinate. Therefore, it becomes necessary to denote displacement along the beam as a function of x, namely, $u(x)$. This new notation reserves x, y, and z as the Cartesian spatial coordinates, and introduces u_x, u_y, and u_z to signify corresponding displacements in the three orthogonal directions. More often than not, displacement will be virtually unidirectional. In such cases, when it is clear which displacement is important, the directional subscript will be omitted.

6.2.2 Derivation of the lumped parameter model

The objective of this section is to obtain an equivalent lumped parameter electromechanical model for the cantilevered beam-based capacitive transducer of Fig. 6.2. The required values for this model are the force of electric origin f^e, an equivalent spring constant k_{eq}, an equivalent value for mass m_{eq}, and an equivalent damping coefficient, b_{eq}. Two key assumptions are required to obtain these parameters. First, the operating frequency ω_e must be comparable to or less than the frequency of the first mechanical resonance of the beam. Second, the deformations must be small. Over the range of conditions for which these assumptions are valid, it turns out that the dynamic shape of the deformed beam is virtually the same as a statically loaded beam. Though this quasistatic approximation must be accepted for now, Example 6.1 below provides a detailed examination of its validity and limitations.

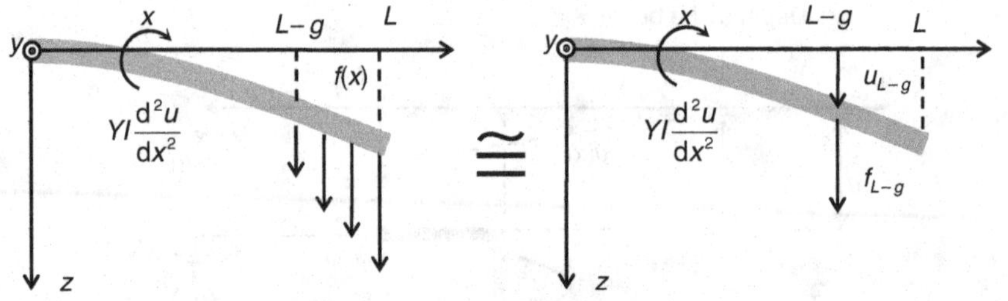

Figure 6.3 Approximation of the distributed electrical force by a point force f_{L-g} exerted at $x = L - g$.

The static displacement of a cantilevered beam is easily determined by integrating the equilibrium equation relating beam curvature to the moment $M(x)$ along the length of the beam. Refer to Eq. (D.44) of Appendix D. Approximating beam curvature by the second derivative of $u(x)$, the static beam equation is

$$YI\frac{\mathrm{d}^2u}{\mathrm{d}x^2} \approx -M(x), \tag{6.1}$$

where Y is Young modulus and I is the moment of inertia.[1] The moment of inertia of a prismatic beam, i.e., any beam with a rectangular cross-section, is

$$I = \frac{1}{12}wh^3. \tag{6.2}$$

It is obvious that this transducer is not a parallel-plate capacitor because the curvature of the bending beam distorts the electric field lines. For this reason, it is not possible simply to write down a capacitance expression as done in Chapter 3. The implication of this for the electromechanics is that the force of electrical origin is not uniformly distributed along the beam. Figure 6.3 depicts a representation (highly exaggerated) of the non-uniformly distributed electrical force. Figure 6.3 shows an approximation needed for the lumped-parameter model, with the distributed force replaced by an equivalent line force acting at $x = L - g$. The validity of this approximation will be tested later.

The next step is to define the effective displacement, u_{L-g}, and the effective force of electrical origin, f_{L-g}, acting at this point along the beam. Other than its expected dependence on voltage squared, the force is unknown because the capacitance is unknown. Clearly, a new methodology is needed. The steps to be taken are as follows:

(i) Find an expression for the beam displacement $u(x)$ due to the point force f_{L-g}, to evaluate the chosen lumped parameter displacement $u_{L-g} = u(L-g)$ and then determine the relationship between the two.

(ii) Integrate capacitance per unit length over the extent of the electrode, $L - g \leq x \leq L$, to determine the total capacitance as a function of u_{L-g}.

(iii) Invoke standard lumped parameter electromechanics, using energy or coenergy, to obtain the force of electrical origin.

[1] To avoid confusion here with the electric field $\overline{E}$, Y is used for Young modulus.

To proceed requires solution of Eq. (6.1). The moment $M(x)$ is

$$M_y = \begin{cases} -(L-g-x)f_{l-g}, & 0 \le x \le L-g, \\ 0, & L-g \le x \le L. \end{cases} \tag{6.3}$$

Plugging Eq. (6.3) into Eq. (6.1) and integrating twice gives a set of expressions with constants of integration that are evaluated using the boundary conditions:

$$u(x) = \begin{cases} \frac{f_{L-g}}{6YI}[3(L-g)x^2 - x^3] + c_0 + c_1 x, & 0 \le x \le L-g, \\ c_{0\mathrm{p}} + c_{1\mathrm{p}} x, & L-g \le x \le L. \end{cases} \tag{6.4}$$

The constants c_0 and c_1 used in the solution range $0 \le x < L - g$ must be zero to satisfy the boundary conditions at the fixed end of the beam, that is, at $x = 0$:

$$u(0) = 0; \quad \frac{\mathrm{d}u}{\mathrm{d}x}(0) = 0. \tag{6.5}$$

The constants $c_{0\mathrm{p}}$ and $c_{1\mathrm{p}}$ for the range $L-g \le x \le L$ are evaluated using the continuity conditions imposed on displacement and its first derivative at $x = L - g$, that is,

$$u(L-g^-) = u(L-g^+) \quad \text{and} \quad \frac{\mathrm{d}u}{\mathrm{d}x}(L-g^-) = \frac{\mathrm{d}u}{\mathrm{d}x}(L-g^+). \tag{6.6}$$

Substituting the evaluated constants into Eq. (6.4) gives the displacement $u(x)$ due to the point force f_{L-g}:

$$u(x) = \begin{cases} \frac{f_{L-g}}{6YI}[3(L-g)x^2 - x^3], & 0 \le x \le L-g, \\ \frac{f_{L-g}}{6YI}[3(L-g)^2 x - (L-g)^3], & L-g \le x \le L. \end{cases} \tag{6.7}$$

Thus,

$$u_{L-g} = u(x = L-g) = \frac{(L-g)^3}{3YI} f_{L-g}. \tag{6.8}$$

For small displacements, the capacitance per length C' in the region $L-g \le x \le L$ may be approximated by a Taylor series expansion of $w/(d-u)$, retaining only the first term:

$$C'(u(x)) \approx \varepsilon_0 \frac{w}{d-u(x)} \approx \varepsilon_0 \frac{w}{d}\left(1 + \frac{u(x)}{d}\right). \tag{6.9}$$

Substituting Eq. (6.7) into Eq. (6.9) and integrating over the span of the electrode yields the total capacitance in terms of the point force, f_{L-g}:

$$\begin{aligned} C(f_{L-g}) &= \int_{L-g}^{L} \varepsilon_0 \frac{w}{d}\left(1 + \frac{1}{d}\frac{f_{L-g}}{6YI}[3(L-g)^2 x - (L-g)^3]\right)\mathrm{d}x \\ &= C_0\left[1 + \frac{(L-g)^2}{12YId}(4L-g)\, f_{L-g}\right], \end{aligned} \tag{6.10}$$

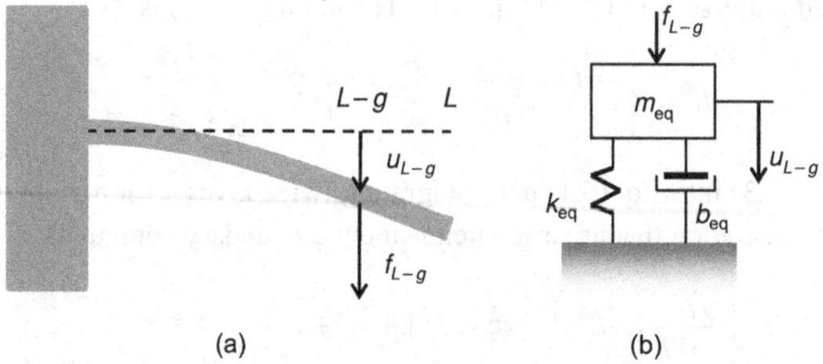

Figure 6.4 An equivalent lumped parameter model for dynamic modeling. (a) Physical model of the bending beam. (b) Dynamic model based on lumped parameter elements. The spring constant k_{eq} is usually derived analytically, while effective mass m_{eq} and damping factor b_{eq} are often obtained from experimental measurements on prototypes.

where $C_0 = \varepsilon_0 wg/d$ is the capacitance of the undeformed beam. Combining Eqs. (6.8) and (6.10) yields the desired expression of capacitance in terms of the chosen lumped parameter displacement variable u_{L-g}:

$$C(u_{L-g}) = C_0\left[1 + \frac{(4L-g)}{4(L-g)d}u_{L-g}\right]. \tag{6.11}$$

With Eq. (6.11) expressing capacitance in terms of an effective lumped parameter variable, we can now use coenergy to evaluate the force of electric origin using the energy method of Chapter 3:

$$\begin{aligned} f(u_{L-g}) &= \frac{\partial W_{\text{e}}'}{\partial u_{L-g}} \\ &= \frac{1}{2}\frac{\mathrm{d}C(u_{L-g})}{\mathrm{d}u_{L-g}}v^2 \\ &= \frac{1}{2}C_0\frac{4L-g}{4(L-g)d}v^2. \end{aligned} \tag{6.12}$$

While this derivation is based on the apparently crude approximation that the force of electric origin is concentrated in a line at $x = L - g$, it will be shown later that the results compare surprisingly well with those obtained from more sophisticated methods based on a distributed force model:

$$u_{L-g} = \frac{1}{2}C_0\frac{(4L-g)(L-g)^2}{12YId}v^2. \tag{6.13}$$

6.2.3 Evaluation of equivalent spring constant, mass, and mechanical damping

Equivalent lumped parameter values for the mechanical spring constant k_{eq} and mass m_{eq} of the cantilevered beam are not difficult to identify (Fig. 6.4). The spring constant is due to the elastic nature of the beam. Simple rearrangement of Eq. (6.8) yields

$$k_{\mathrm{eq}} = \frac{f_{L-g}}{u_{L-g}} = \frac{3YI}{(L-g)^3} = \frac{Ywh^3}{4(L-g)^3}; \tag{6.14}$$

k_{eq} depends strongly upon the beam thickness h and the reduced length $(L - g)$.

One way to determine the equivalent mass m_{eq} is to equate the kinetic energy based on the chosen displacement variable, u_{L-g}, to the total kinetic energy of the deflecting beam. The static beam deflection from Eq. (6.7) is used to calculate this energy. If ρ is the mass density of the beam, then

$$\frac{1}{2} m_{\mathrm{eq}} u_{L-g}^2 \cong \frac{1}{2} \int_0^L u^2(x) \rho w h \mathrm{d}x. \tag{6.15}$$

Note the assumption here that the displacement takes the form of the static solution, that is, Eq. (6.7). Substituting $u(x)$ and Eq. (6.8) into Eq. (6.15), integrating along the entire length of the beam, and solving for m_{eq} gives

$$m_{\mathrm{eq}} \frac{wh\rho}{u_{L-g}^2} \int_0^L u^2(x)\,\mathrm{d}x = \frac{3 + 2\hat{g} + \hat{g}^2}{8(1-\hat{g})} m_{\mathrm{beam}}, \tag{6.16}$$

where $\hat{g} = g/L$. The equivalent mass depends strongly on the location where the point force is applied. For example, if $\hat{g} = 0.2$, the equivalent mass is approximately one half of the total beam mass, that is, $m_{\mathrm{eq}} = 0.54 m_{\mathrm{beam}}$. As expected, m_{eq} is always smaller than the total beam mass. This is because the clamped end of the beam does not move at all and the displacements increase monotonically from the fixed end toward the free end. As the electrode length increases and displacement u_{L-g} moves toward the clamped end, the equivalent mass increases because it is being measured with respect to a smaller effective displacement.

6.2.4 Resonance of a cantilevered beam

Knowing the equivalent spring constant k_{eq} and mass m_{eq} makes it possible to estimate the mechanical resonant frequency ω_0, that is,

$$\omega_0 = \sqrt{\frac{k_{\mathrm{eq}}}{m_{\mathrm{eq}}}} = \left[\frac{1}{1-\hat{g}} \sqrt{\frac{2}{3 + 2\hat{g} + \hat{g}^2}}\right] \frac{h}{L^2} \sqrt{\frac{Y}{\rho}}. \tag{6.17}$$

It is worth pointing out that, in MEMS design and development, it is usually far easier to measure the resonant frequency ω_0 and then to estimate equivalent mass using

$$m_{\mathrm{eq}} = \frac{k_{\mathrm{eq}}}{\omega_0^2}. \tag{6.18}$$

The final lumped parameter quantity needed to model the cantilevered beam capacitor is the damping coefficient, b_{eq}. In MEMS devices, accurate estimation of damping is generally a difficult proposition because the principal loss mechanism is the viscosity of the fluid in which the device operates. Even using such approximations as *lubrication theory*, MEMS geometries make numerical FEM simulations a daunting prospect. In practice, measurements performed upon actual prototypes are often employed. If the resonant frequency and quality factor, k_{eq} and Q, respectively, can be measured, then

$$b_{eq} = \frac{k_{eq}}{\omega_0 Q}. \tag{6.19}$$

Refer to Section B.4 of Appendix B for practical details on implementing this method to identify the damping coefficient.

Example 6.1 **Numerical values for lumped parameter cantilevered beam model**

As an example, let us estimate k_{eq}, m_{eq}, and C_0 for a 100 × 4 × 2 highly doped Si beam separated 2 μm apart from a 20 μm long substrate electrode aligned with the tip of the beam. The structure in Section 6.2 is most likely to be fabricated via a surface micromachining process, in which case the beam is polycrystalline silicon (polysilicon). Polysilicon is isotropic, and has Young modulus of ~169 Gpa. We have

$$k_{eq} = \frac{Y_{polySi} w h^3}{4(L-g)^3} = \frac{\left(169\times10^9\right)\left(4\times10^{-6}\right)\left(2\times10^{-6}\right)^3}{4[(100-20)\times10^{-6}]^3} \approx 2.64\left[\frac{\text{N}}{\text{m}}\right].$$

If the beam is made of crystalline silicon (bulk micromachining), the Young modulus is not isotropic, i.e., it takes different values in different directions (see Appendix D). For example, in the direction 011, $Y = 169$ GPa, the same as polysilicon. On the same die, another beam, oriented 45° with respect to the 011 direction, has a Young modulus of about 130 GPa and an effective spring constant of

$$k_{eq2} = \frac{Y_{Si-010} w h^3}{4(L-g)^3} \approx 2.03\left[\frac{\text{N}}{\text{m}}\right].$$

The equivalent mass is

$$\begin{aligned} m_{eq} &= \frac{3 + 2\hat{g} + \hat{g}^2}{8(1-\hat{g})} \underbrace{whL\rho_{Si}}_{m_{beam}} \\ &= \frac{3 + 2\times0.2 + 0.2^2}{8\times0.8}(4\times10^{-6})(2\times10^{-6})(100\times10^{-6})(2330) \\ &\approx 1.00\times10^{-12}\ [\text{kg}], \end{aligned}$$

which compares with $m_{beam} = 1.86\times10^{-12}$ kg for the total beam mass.

The next example considers a practical device that employs a cantilevered beam. Special attention is given to damping.

Example 6.2 **Use of damping to control dynamic response**

A new application for MEMS-based cantilevers is in power switching. The small size of these capacitively actuated relays guarantees rapid actuation times but the current-carrying capacity is correspondingly quite limited. To circumvent this problem, large arrays of identical relays fabricated on the same chip and operated in parallel are used. Key design elements are electronic circuitry that suppresses arcing [1] and electroformed nanocrystalline Ni beams that combine good electrical conductivity with a relatively high modulus of elasticity [2]. Figure 6.5 shows the geometry of a relay element and an SEM of a beam array.

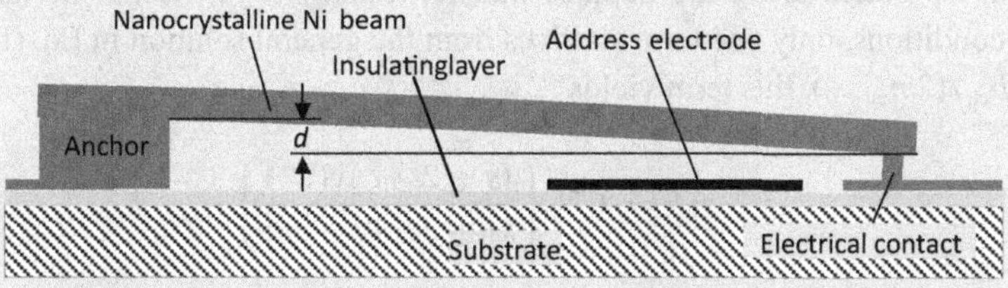

(a)

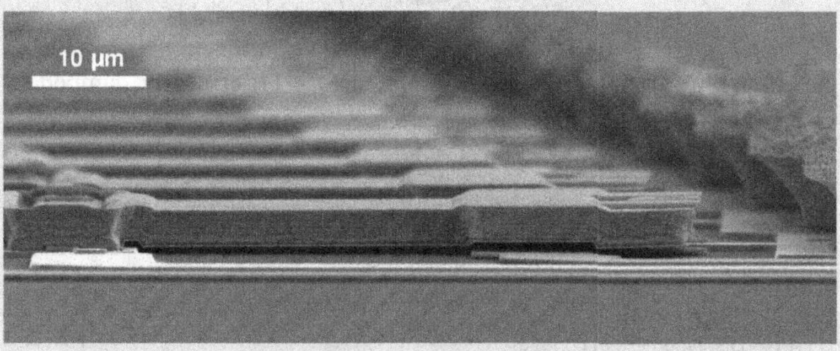

(b)

Figure 6.5 MEMS-based capacitively actuated relay for on–off control of AC power current. To achieve microsecond response time, damping is tuned by specifying the pressure inside the hermetically sealed device. (a) Side view (not to scale) showing electrical contacts and address electrode for capacitive actuation. (b) SEM of an array of relays [1]. Reproduced with permission of the IEEE©.

The current-carrying contact is at the tip of the beam, where deflection is largest. The actuation electrode patterned on the substrate exerts its force on a section of the beam well away from this contact. To control power currents effectively, mechanical actuation must be fast. For example, when actuation voltage is removed, the beam must break the contact and settle to equilibrium within a few microseconds. Furthermore, reliable operation over millions of cycles without arc-induced contact degradation, *contact bounce* or other oscillatory transients must be guaranteed. The best response is typically achieved for a damping ratio of $\zeta \approx 1$. If $\zeta < 1$, bounce can occur and if $\zeta > 1$, the response is too slow. To explore how damping affects transient behavior, consider the response of a second-order mechanical system when a constant force is suddenly removed at $t = 0$. The equation of motion, once contact is broken, is

$$m_{\text{eq}}\ddot{x} + b_{\text{eq}}\dot{x} + k_{\text{eq}}x = 0$$

and the initial conditions are

$$x = d \quad \text{and} \quad \dot{x} = 0,$$

where d is the displacement of the tip of the cantilever when in contact. We can solve this problem using the Laplace transform analysis in Section B.1.2. For these initial conditions, only one term survives from the general solution in Eq. (B.9). Using $\zeta_{\text{eq}} = b_{\text{eq}}/(2m_{\text{eq}}\omega_0)$, this term yields

$$x(t) = \mathcal{L}^{-1}\left\{\frac{(s + 2\omega_0\zeta)x(0^-)}{s^2 + 2\omega_0\zeta s + \omega_0^2}\right\}.$$

Taking the inverse Laplace transform gives the transient beam deflection $x(t)$:

$$x(t) = d\mathrm{e}^{-\omega_0\zeta t}\cos\left(\omega_0\sqrt{1-\zeta^2}t\right).$$

Figure 6.6 plots $x(t)$ for the under-damped ($\zeta < 1$), critically damped ($\zeta = 1$); and over-damped ($\zeta > 1$) regimes.

For $\zeta = 1$, the transient response reduces to

$$x(t) = d\mathrm{e}^{-\omega_0 t}.$$

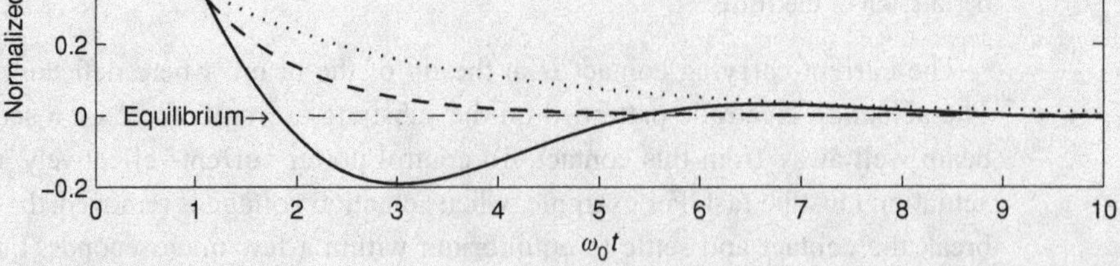

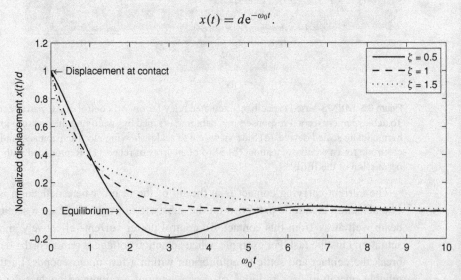

Figure 6.6 Settling transient response of electromechanically actuated cantilevered beam being turned off. Displacement is plotted versus time for three values of damping ratio. Note that the critically damped condition, $\zeta = 1$, yields the fastest response without any oscillatory component to the transient.

Defining the transient as the time required for the beam to spring back from contact at $x = x_0$ to within 1% of the undeflected state, $x = 0$, i.e., $x(t_{\text{transient}})/d = 0.01$, yields

$$t_{\text{transient}} = -\frac{\log(0.01)}{\omega_0} = \frac{4.605}{\omega_0}.$$

One can use the parameters of the cantilevered beam shown in Fig. 6.5b to estimate its mechanical response time under the assumption of critical damping. Letting $L = 50$ μm and $h = 5$ μm, and using $Y_{\text{Ni}} = 200$ GPa and $\rho_{\text{Ni}} = 10 \times 10^3$ kg/m^3, respectively, for Young modulus and mass density of the beam, gives

$$\omega_0 = 1.016\frac{h}{L^2}\sqrt{\frac{Y_{\text{Ni}}}{\rho_{\text{Ni}}}} = \sim 9.6 \times 10^6 \text{ rad/s}.$$

Combining this estimate for the resonance frequency with the above definition for transient time yields $t_{\text{transient}} \sim 0.5$ μs, which is close to the ~1 μs value needed for power relay applications.

6.2.5 Recapitulation of lumped parameter model identification procedure

The procedure for identifying lumped parameter quantities employs three key approximations. First, the shape of the dynamically bending beam is approximated by the profile of the statically deformed beam. Though in general the shape of the displacement profile depends on frequency, this quasistatic approximation is surprisingly accurate for frequencies all the way up to and somewhat above the first resonance. Second, the distributed force causing the deformation is replaced by a point force exerted at the inner edge of the electrode. See Eq. (6.12). A consequence of this assumption is that the beam curvature above the fixed electrode is zero, thereby simplifying evaluation of the capacitance integral in Eq. (6.10). In reality, however, the portion of the beam above the fixed electrode is somewhat curved. Choosing the placement of the effective line force at the inner edge of the edge becomes less accurate as the electrode length g increases. Third, the capacitance is approximated by the parallel-plate model, Eq. (6.9), despite the fact that the electric field lines are not uniform and parallel. This assumption is tenable for small displacements.

Note that retaining only the first (linear) term of the capacitance upon the displacement, as done in Eq. (6.11), means that $\partial^2 W'_e/\partial u^2_{L-g} = 0$, which, in turn, forces the effective stiffness contribution due to electromechanical coupling to zero. Thus, this model fails to capture the effect of the electromechanics on the resonant frequency, in particular the pull-in instability mechanism introduced in Section 4.8. Including the second term of the Taylor series corrects this deficiency.

Although the simplifying assumptions used here may seem overly restrictive, perhaps even questionable, they still yield confidence-building results. For example, it is possible to compare the analytical solution for the first resonance of a cantilevered beam with the estimation of Eq. (6.17). The analytical solution for the first resonance of a cantilevered beam is [3]

$$\omega_0 = 1.016 \frac{h}{L^2} \sqrt{\frac{Y}{\rho}}. \tag{6.20}$$

Assuming that the electrode is 20% of the beam length, i.e., $\hat{g} = 0.2$, Eq. (6.17) gives

$$\omega_0 = 0.953 \frac{h}{L^2} \sqrt{\frac{Y}{\rho}}, \tag{6.21}$$

which differs from Eq. (6.20) by only ~6%.

Example 6.3 **N-form formulation for the simple cantilevered beam**

With values for the equivalent stiffness k_{eq}, equivalent mass m_{eq}, and capacitance $C(u_0)$, we can formulate the N-form formulation for the simple cantilevered beam. The equivalent circuit of the N-form model is shown in Fig. 6.7.

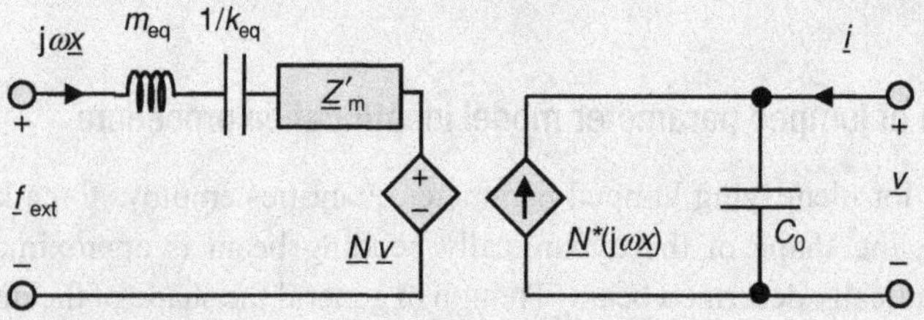

Figure 6.7 Equivalent circuit representation of beam actuated by a voltage v applied between the conductive beam and an electrode on the substrate aligned with the tip of the beam.

The coenergy function is

$$W'_e = \frac{1}{2} C(u_0) v^2 = \frac{1}{2} C_0 \left[1 + \frac{(4L - g)}{4(L - g)d} u_{L-g} \right] v^2.$$

The N-form is obtained after linearization of the force and the charge, in terms of displacement and voltage,

$$\begin{bmatrix} \underline{f} \\ \underline{i} \end{bmatrix} = \begin{bmatrix} \underline{Z}'_m & \underline{N} \\ -\underline{N}^* & \underline{Y}_e \end{bmatrix} \begin{bmatrix} j\omega\underline{x} \\ \underline{v} \end{bmatrix},$$

where

$$\underline{Y}_e = j\omega C_0 = j\omega\varepsilon_0 \frac{wg}{d},$$

$$\underline{N} = \left. \frac{\partial C}{\partial u_0} v \right|_{\substack{u_0 \to 0 \\ v \to v_0}} = \left. C_0 \frac{4L - g}{4(L - g)d} v \right|_{\substack{u_0 \to 0 \\ v \to v_0}} = \frac{C_0}{3d} v_0,$$

$$\underline{Z}'_m = \left. \frac{1}{2} \frac{\partial^2 C}{\partial u_0^2} v^2 \right|_{\substack{u_0 \to 0 \\ v \to v_0}} = 0.$$

Note again that $\underline{Z}'_{\text{m}} = 0$ is *not* a good model for a variable-gap configuration. Our Taylor series expansion of C in terms of u_{L-g} needs the second term. We will introduce a more general model in Section 6.3.

6.3 A second look at the cantilevered beam

Section 6.2 presented a quasistatic, lumped parameter, electromechanical model for a cantilevered beam with a capacitive electrode near the free end. Based on a seemingly arbitrary choice for the mechanical displacement and the effective location of the force of electric origin, expressions for the effective capacitance, spring constant, and mass were identified, and an equivalent electromechanical two-port network was derived. There was found to be a surprisingly good correspondence between the uncoupled mechanical resonant frequency predicted from the lumped parameter model and the analytical result from the classical solution.

This section revisits the lumped parameter formulation of the cantilevered beam, now using discretization of the beam to seek a more accurate representation of the dynamics. One goal is to provide a more conclusive test of the hypothesis that the dynamics of continua can be modeled accurately and effectively as lumped parameter systems. While the derivation given here is for the specific case of the cantilever, the methodology is far more general.

Consider once again a cantilevered beam. Figure 6.8 shows a structure with the entire bending beam serving as an electrode.

The capacitance of the transducer in the equilibrium state with no beam deflection is

$$C_0 = \varepsilon_0 \frac{wL}{d}. \tag{6.22}$$

Based on the assumption that $L > w \gg d$, fringing and end effects are ignored. As long as the displacement is small, the curvature of the electric field lines will be negligible. See Fig. 6.9.

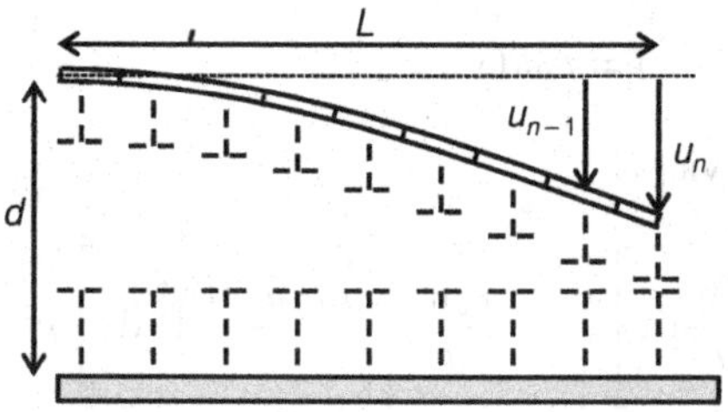

Figure 6.8 Side view of a capacitive transducer based on a bending (cantilevered) beam. The fixed electrode is at the bottom. The beam can be described either as a continuous or a discretized structure.

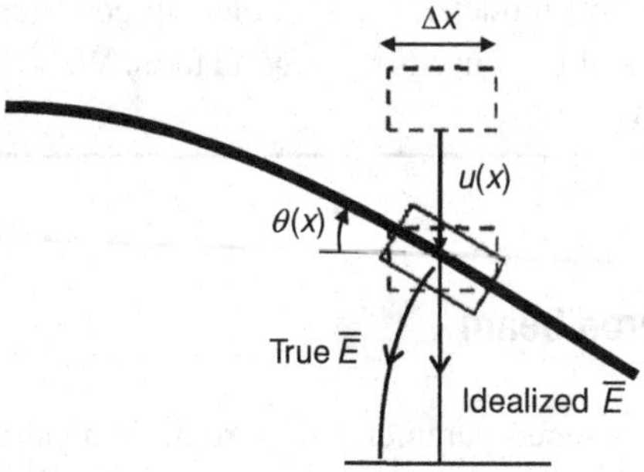

Figure 6.9 Exaggerated view of a single discrete beam element showing distortion of electric field lines due to the beam not being parallel to the electrode at the bottom. The idealized $\overline{E}$ field used for the capacitance per unit length is also shown.

6.3.1 Parameterization of distributed capacitance

Using the above assumptions, the total capacitance of the deformed beam is

$$C \approx \varepsilon_0 w \int_0^L \frac{\mathrm{d}x}{d - u(x)}. \tag{6.23}$$

Because $|u(x)| \ll d$, the integrand in Eq. (6.23) can again be approximated by a Taylor series, but this time the first two terms are retained:

$$C \approx \frac{\varepsilon_0 w}{d} \int_0^L \left(1 + \frac{u(x)}{d} + \left(\frac{u(x)}{d}\right)^2\right) \mathrm{d}x = C_0 + \frac{\varepsilon_0 w}{d} \int_0^L \left(\frac{u(x)}{d} + \left(\frac{u(x)}{d}\right)^2\right) \mathrm{d}x. \tag{6.24}$$

The second term is needed to capture the possibility of the pull-in instability.

In the useable operating range of cantilevered beams, as well as most other continua exploited in MEMS, the shape of the deformed structure is virtually independent of the magnitude of the displacement. Thus, we can express the displacement profile as the product of a scale factor, u_0 with units of length, and a dimensionless beam shape function, $\hat{u}(x)$:

$$u(x) = u_0 \hat{u}(x). \tag{6.25}$$

Inserting Eq. (6.25) into Eq. (6.24) yields

$$\begin{aligned} C &\approx C_0 + \frac{\varepsilon_0 w}{d} \underbrace{\left[\int_0^L \frac{\hat{u}(x)}{d} \mathrm{d}x\right]}_{=\xi^{(1)}} u_0 + \frac{\varepsilon_0 w}{d} \underbrace{\left[\int_0^L \left(\frac{\hat{u}(x)}{d}\right)^2 \mathrm{d}x\right]}_{=\xi^{(2)}} u_0^2 \\ &= C_0 + \frac{\varepsilon_0 w}{d} \xi^{(1)} u_0 + \frac{\varepsilon_0 w}{d} \xi^{(2)} u_0^2. \end{aligned} \tag{6.26}$$

The integrals $\xi^{(1)}$ and $\xi^{(2)}$ are weighting factors that depend on the shape of the displacement profile, but not on the displacement magnitude; $\xi^{(1)}$ is dimensionless, and $\xi^{(2)}$ has units of 1/length. There is arbitrariness in the choice of u_0, though usually it is the maximum displacement, corresponding to the most easily measured or estimated value. Once the location for u_0 is chosen, $\hat{u}(x)$ is scaled so that it takes unity value at the point of measurement. For the cantilevered beam example of Fig. 6.8, the maximum displacement occurs at the tip of the beam.

*6.3.2 Discretized capacitance model for the beam

In the design of MEMS transducers, numerical tools are essential because many configurations possess no analytical solution. In such cases, the approach is to work with a discretized model. The purpose of this section is to develop a discrete analog to the treatment presented in the previous section. The simple cantilevered beam, which certainly does not require discretization, is used to provide continuity with the previous sections. The general approach used here can be applied to more complicated structures.

A good way to introduce the discretization is to start with the capacitance integral of Eq. (6.23). Dividing the beam into n equal elements, the capacitance integral can be approximated by a finite sum

$$C \approx \varepsilon_0 w \frac{L}{n} \sum_{k=1}^{n} \frac{1}{d_0 - u_k} \approx C_0 + \frac{\varepsilon_0 w}{d_0} \frac{L}{n} \sum_{k=1}^{n} \left(\frac{u_k}{d_0} + \left(\frac{u_k}{d_0} \right)^2 \right). \tag{6.27}$$

Each element, having average displacement u_k, is regarded as a degree of freedom of the discretized system. With the displacement profile fixed, the displacements u_k are proportional to a chosen reference displacement u_0 and the shape function:

$$u_k = u_0 \hat{u}_k. \tag{6.28}$$

Inserting Eq. (6.28) into Eq. (6.27) yields

$$\begin{aligned} C &\approx C_0 + \underbrace{\frac{\varepsilon_0 w}{d_0} \frac{L}{n} \sum_{k=1}^{n} \frac{\hat{u}_k}{d_0}}_{=\xi_n^{(1)}} u_0 + \underbrace{\frac{\varepsilon_0 w}{d_0} \frac{L}{n} \sum_{k=1}^{n} \left(\frac{\hat{u}_k}{d_0} \right)^2}_{=\xi_n^{(2)}} u_0^2 \\ &= C_0 + \frac{\varepsilon_0 w}{d_0} \xi_n^{(1)} u_0 + \frac{\varepsilon_0 w}{d_0} \xi_n^{(2)} u_0^2. \end{aligned} \tag{6.29}$$

The only distinction between Eqs. (6.26) and (6.29) is that $\xi_n^{(1)}$ and $\xi_n^{(2)}$ are finite summations that approximate the integrals $\xi^{(1)}$ and $\xi^{(2)}$.

Now consider the cantilevered beam transducer operated as an actuator. We can calculate the force of electrical origin by letting u_0 take the role of a lumped parameter displacement. Thus,

$$f^{\mathrm{e}} = \frac{\partial W_{\mathrm{e}}'}{\partial u_0} = \frac{1}{2} \frac{\mathrm{d}C(u_0)}{\mathrm{d}u_0} v^2. \tag{6.30}$$

Substituting Eq. (6.26) into Eq. (6.30) yields

$$f^{\mathrm{e}} = \frac{1}{2}\frac{\varepsilon_0 w}{d}\left(\xi_n^{(1)} + 2\xi_n^{(2)} u_0\right) v^2. \tag{6.31}$$

Recalling Eq. (4.13a), the N-form mechanical impedance is

$$\underline{Z}'_{\mathrm{m}} = -\frac{1}{\mathrm{j}\omega}\frac{\partial^2 W'_{\mathrm{e}}}{\partial u_0^2} = -\frac{1}{\mathrm{j}\omega}\frac{\varepsilon_0 w}{d}\xi_n^{(2)} v^2. \tag{6.32}$$

The capacitor model can also be used to express the induced current:

$$i = \left(C_0 + \frac{\varepsilon_0 w}{d}\xi_n^{(1)} u_0 + \frac{\varepsilon_0 w}{d}\xi_n^{(2)} u_0^2\right)\dot{v} + \frac{\varepsilon_0 w}{d}\left(\xi_n^{(1)} + 2\xi_n^{(2)} u_0\right)\dot{u}_0 v. \tag{6.33}$$

For a typical beam structure, Eq. (6.33) can be further simplified because $|u_0| \ll d$:

$$i = C_0 \dot{v} + \frac{\varepsilon_0 w}{d}\xi_n^{(1)} \dot{u}_0 v. \tag{6.34}$$

The first term is the capacitive current and the second term is the induced motional current. This form differs from that of parallel-plate variable-gap capacitance only in the need for the shape factor $\xi_n^{(1)}$, which accounts for the distributed nature of the capacitance along the deformed beam.

Example 6.4 **Shape constants and capacitance application**

It is instructive now to compare the results for the discretized cantilevered beam to those for a parallel-plate variable-gap capacitor with rigid electrodes. For the parallel-plate geometry, the shape function is

$$\hat{u}(x) = 1.$$

Using Eqs. (6.25),

$$\xi^{(1)} = \int_0^L \frac{1}{d}\mathrm{d}x = \frac{L}{d}; \quad \xi^{(2)} = \int_0^L \frac{1}{d^2}\mathrm{d}x = \frac{L}{d^2}.$$

Inserting this result into Eq. (6.25) yields the approximation of the parallel-plate capacitance,

$$C(u_0) = C_0\left(1 + \frac{u_0}{d} + \frac{u_0^2}{d^2}\right).$$

Now consider the cantilevered beam, starting with Eqs. (D.50) and (D.51) of Appendix D. In the quasistatic limit, with a point force f applied at $x = L$, the deflections take the form of a third-order polynomial:

$$u(x) = \frac{f}{6YI_x}(3Lx^2 - x^3).$$

Normalizing these displacements to the maximum deflection occurring at the tip, that is, $u_0 = fL^3/(3YI_x)$, gives the shape function

$$\hat{u}(x) = \frac{u(x)}{u_0} = \frac{1}{2L^3}(3Lx^2 - x^3).$$

Then, $\xi^{(1)}$ and $\xi^{(2)}$ are

$$\xi^{(1)} = \int_0^L \frac{1}{2L^3 d}(3Lx^2 - x^3)\mathrm{d}x = \frac{3L}{8d},$$

$$\xi^{(2)} = \int_0^L \frac{1}{4L^6 d^2}(3Lx^2 - x^3)^2\mathrm{d}x = \frac{33L}{140d^2}.$$

Inserting the above expressions for $\xi^{(1)}$ and $\xi^{(2)}$ into Eq. (6.25) yields

$$C(u_0) = C_0\left(1 + \frac{3}{8}\frac{u_0}{d} + \frac{33}{140}\frac{u_0^2}{d^2}\right).$$

Comparing the two equations for C confirms the intuitive result that the rigid and cantilevered beam capacitors have qualitatively similar behavior. The differences are captured in the shape-dependent coefficients multiplying the u_0/d and u_0^2/d^2 terms. As expected, the capacitance function is more weakly dependent on u_0/d for the cantilevered beam.

When the displacement profile is known, the analysis of continua can be greatly simplified by expressing the displacements as the product of a shape function and a single measurement that scales the displacement. After the shape function is integrated over the domain of displacement, it scales the familiar rigid-body expressions. There are situations where the shape function is unknown and difficult to obtain. Furthermore, the complication arises for most continua that the shape functions depend on frequency. Section 6.4 shows how to deal with this more general case.

6.4 MDF models for beams

This section presents a systematic approach to the treatment of MEMS continua, specifically beams, as finite assemblies of coupled elements. Each element is defined as a *degree of freedom* (DOF) and the methodology is the *multiple degrees of freedom* (MDF) method. Figure 6.10 shows that each element i is assigned a displacement value u_i. The complete set of DOFs, u_1 through u_n, comprises a discrete approximate representation of the beam deformation. Arguably the simplest discrete numerical approach, the MDF method is a powerful tool for analyzing MEMS devices that are not readily

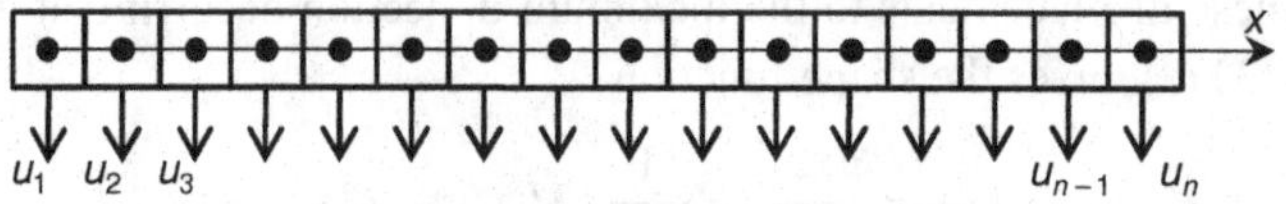

Figure 6.10 Beam divided into n elements, each of which is constrained to move up and down in the vertical direction. Each u_i value is a degree of freedom.

amenable to analytical treatment as boundary value problems. Its application is usually restricted to geometries where the displacements are unidirectional.

Because of the assumptions made in defining the degrees of freedom, MDF models cannot capture all classes of deformations of continua, such as beams. The simplicity and transparency of MDF analysis make it suitable for a first look at discretization of continuum electromechanical systems. Finite-element analysis (FEA), while more flexible and general than MDF, requires a considerable investment of time and computational resources, and it does not provide the physical interpretation inherent in MDF modeling, at least not to the newcomer to the field. Students with no prior exposure to the statics and dynamics of continua can use MDF analysis to gain insights about the behavior of MEMS structures. They can also acquire intuition, enabling them to judge whether it is possible to develop lumped parameter models appropriate for linear two-port transducer representation.

6.4.1 MDF description

The MDF method uses matrices extensively. The starting point is the individual displacements $u_1, u_2, \ldots, u_N$, which are organized in a vector as follows:[2]

$$\overline{u} = [u_1 \quad u_2 \quad \ldots \quad u_{n-1} \quad u_n]^T. \tag{6.35}$$

The MDF equations of motion are represented by a matrix-vector generalization of the scalar equation of motion:

$$\overline{\overline{m}}\,\ddot{\overline{u}} + \overline{\overline{k}}\overline{u} = \overline{f}^{\mathrm{e}}. \tag{6.36}$$

The mass becomes a diagonal matrix $\overline{\overline{m}}$, with each term equal to the mass of the corresponding individual beam element:

$$\overline{\overline{m}} = \begin{bmatrix} m_1 & 0 & \ldots & 0 & 0 \\ 0 & m_2 & \ldots & 0 & 0 \\ \ldots & \ldots & \ldots & \ldots & \ldots \\ 0 & 0 & \ldots & m_{n-1} & 0 \\ 0 & 0 & \ldots & 0 & m_n \end{bmatrix}. \tag{6.37}$$

[2] The displacement vector is defined as a column vector to preserve the same order of multiplication as in scalar equations. Scalar ku corresponds to matrix-vector $\overline{\overline{k}}\overline{u}$. To conserve space in this text, they are expressed as transposes of row vectors.

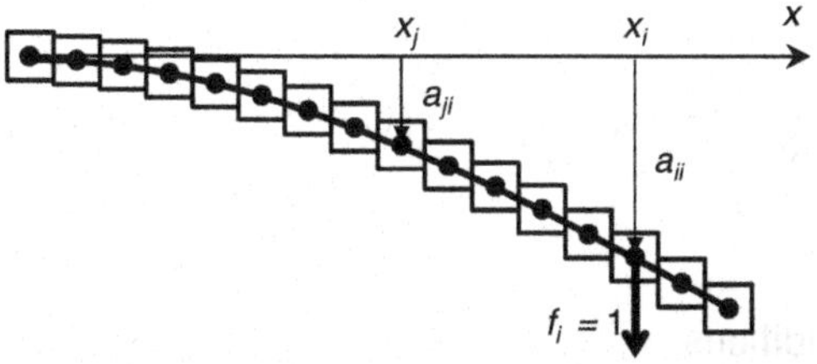

Figure 6.11 Cantilevered beam depicting the thought experiment used to define the flexibility influence coefficients. When a unit force is applied at element x_i, the resulting displacements are equal to the flexibility coefficients a_{ji}, where i and j take values from 1 to N.

The individual components sum up to the total beam mass:

$$m_{\text{beam}} = \sum_{i=1}^{n} m_i. \tag{6.38}$$

For a uniform beam, the n segments are all equal, so that

$$m_1 = m_2 = \ldots = m_n = \frac{m_{\text{beam}}}{n}. \tag{6.39}$$

The stiffness matrix $\overline{\overline{k}}$ is more complicated, because in general the terms depend both on the properties of the beam and the boundary conditions imposed upon it. A relatively easy way to obtain this matrix is to start with the flexibility matrix $\overline{\overline{a}}$, which is the inverse of $\overline{\overline{k}}$, that is, $\overline{\overline{a}} = (\overline{\overline{k}})^{-1}$. The elements in this matrix, a_{ji}, called the flexibility influence coefficients, can be defined as the displacement u_j at location x_j due to a unit force applied at location x_i. Thus,

$$a_{ji} = u_j|_{f_1=f_2=\ldots=f_{i-1}=f_{i+1}=\ldots=f_N=0 \text{ and } f_i=1}. \tag{6.40}$$

The thought experiment used to obtain element a_{ji} is depicted in Fig. 6.11.

For simple structures such as the cantilevered beam, it is convenient to obtain the a_{ij} coefficients directly from a closed-form solution. Note that the following derivation essentially repeats the first few steps followed in Section 6.2.2, but now with notational modifications appropriate for generalization to an MDF formulation. For all $x_i \geq x_j$, using the reference directions defined in Fig. 6.11, the bending beam equation due to the force applied at $x = x_i$ and with the free end boundary at $x = L$ is

$$YI\frac{\partial^2 u}{\partial x^2} = -M = f_i \underbrace{(x_i - x)}_{\text{moment arm}}, \quad x < x_i. \tag{6.41}$$

Integrating Eq. (6.41) twice yields an expression very similar in form to Eq. (6.4):

$$u = \frac{f_i}{6YI}(3x_i x^2 - x^3) + c_1 x + c_2, \quad x_i \geq x_j, \tag{6.42}$$

where c_1 and c_2 are integration constants to be determined by the boundary conditions specific to a given case. Combining Eq. (6.40) with Eq. (6.42) gives the flexibility influence coefficients:

$$a_{ij} = u\big|_{\substack{f_i=1\\ x=x_j}} = \frac{1}{6YI}\left(3x_i x_j^2 - x_j^3\right) + c_1 x_j + c_2, \quad x_i \geq x_j. \tag{6.43}$$

6.4.2 Application of boundary conditions

The cantilevered beam has one end fixed and the other end free. The boundary conditions at the fixed end, $x = 0$, are

$$u(0) = 0 \quad \text{and} \quad \left.\frac{\mathrm{d}u}{\mathrm{d}x}\right|_0 = 0. \tag{6.44}$$

For $u(x)$ from Eq. (6.42) to satisfy these two conditions, $c_1 = 0$ and $c_2 = 0$, so the influence flexibility coefficients are

$$a_{ij} = \frac{1}{6YI}\left(3x_i x_j^2 - x_j^3\right), \quad x_i \geq x_j. \tag{6.45}$$

Assuming the beam is divided into n elements of equal length,

$$x_i = \frac{i}{n}L. \tag{6.46}$$

Combining Eqs. (6.46) and (6.45) yields

$$a_{ij} = \frac{L^3}{6YIn^3}\left(3ij^2 - j^3\right), \quad i \geq j. \tag{6.47}$$

The coefficients expressed by Eq. (6.47) populate the main diagonal of the flexibility matrix and the elements below it. Proof is provided in Section 6.4.3 that the flexibility matrix is symmetric, that is, $a_{ij} = a_{ji}$. Therefore, the coefficients for the upper half of the matrix are also known, and with the flexibility matrix fully determined, the stiffness matrix is also determined:

$$\overline{\overline{k}} = \overline{\overline{a}}^{-1}. \tag{6.48}$$

Example 6.5 MDF – illustration with a few degrees of freedom

In this example, we compute the flexibility and stiffness matrices and the mass matrix for a simple 2 μm × 4 μm × 100 μm silicon cantilever using $n = 4$ degrees of freedom. The MATLAB code in Fig. 6.12b starts by defining the geometry and the mechanical material properties of the Si beam. Then it proceeds to calculate the flexibility coefficients "$a(i, j)$" in two "for" loops. Finally, the mass matrix "m" and the stiffness matrix "k" are computed in single statements.

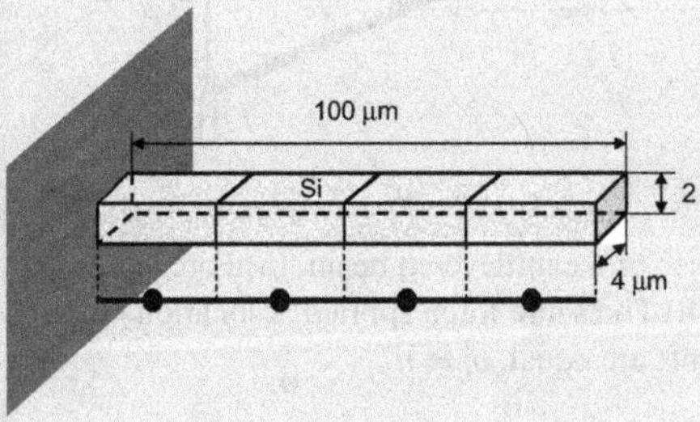

(a)

```
n = 4;
% - beam geometry ----------------------
L = 100e-6;     % m, length
w = 4e-6;       % m, width
h = 2e-6;       % m, height
% - Si material properties -------------
Y = 169e9;      % Pa
rho = 2330;     % kg
% - calculations -----------------------
I = h^3*w/12;       % area moment
a = zeros(n,n)
B = L^3/(6*n^3*Y*I)   % a constant
for i  = 1:n
    for j = 1:i
        a(i,j) = B*(3.*(i)*(j)^2-(j)^3);
        a(j,i) = a(i,j)
    end
end
m = L*w*h*rho/n*eye(n);% mass matrix
k = inv(a);              % stiffness matrix
```

(b)

Figure 6.12 (a) A silicon cantilevered beam approximated by four degrees of freedom. (b) MATLAB code.

The 4 × 4 flexibility matrix is given by

$$\bar{\bar{a}} = \frac{L^3}{6n^3 Y_{\text{polySi}} I}\begin{bmatrix} 2 & 5 & 8 & 11 \\ 5 & 16 & 28 & 40 \\ 8 & 28 & 54 & 81 \\ 11 & 40 & 81 & 128 \end{bmatrix} \approx (5.78 \times 10^{-3}) \times \begin{bmatrix} 2 & 5 & 8 & 11 \\ 5 & 16 & 28 & 40 \\ 8 & 28 & 54 & 81 \\ 11 & 40 & 81 & 128 \end{bmatrix} \left[\frac{\text{m}}{\text{N}}\right],$$

where $Y_{\text{polySi}} = 169$ GPa and the area moment of inertia is given by

$$I = \frac{1}{12}h^3 w = \frac{8}{3} \times 10^{-24} m.$$

The stiffness matrix is the inverse of the flexibility matrix (note the symmetry):

$$\bar{\bar{k}} = \bar{\bar{a}}^{-1} = \begin{bmatrix} 542.4 & -340.8 & 128.4 & -21.4 \\ -340.8 & 413.9 & -276.5 & 74.9 \\ 128.4 & -276.5 & 285.4 & -105.3 \\ -21.4 & 74.9 & -105.3 & 46.4 \end{bmatrix} \left[\frac{\text{N}}{\text{m}}\right].$$

The mass matrix is

$$\bar{\bar{m}} = wh\frac{L}{N}\rho_{\text{Si}} \begin{bmatrix} 1 & 0 & 0 & 0 \\ 0 & 1 & 0 & 0 \\ 0 & 0 & 1 & 0 \\ 0 & 0 & 0 & 1 \end{bmatrix} = (4.66 \times 10^{-13}) \begin{bmatrix} 1 & 0 & 0 & 0 \\ 0 & 1 & 0 & 0 \\ 0 & 0 & 1 & 0 \\ 0 & 0 & 0 & 1 \end{bmatrix} [\text{kg}].$$

The sum of the elements of the mass matrix equals the total mass of the beam.

*6.4.3 Maxwell's reciprocity theorem

The MDF model exploits the symmetry of the flexibility influence coefficients. This symmetry is a consequence of a reciprocity property attributed to Maxwell. Consider the two cases of a cantilevered beam shown in Fig. 6.13a and Fig. 6.13b. In case 1, a normal force f applied at location i gives rise to displacement u_j' at location j. In case 2,

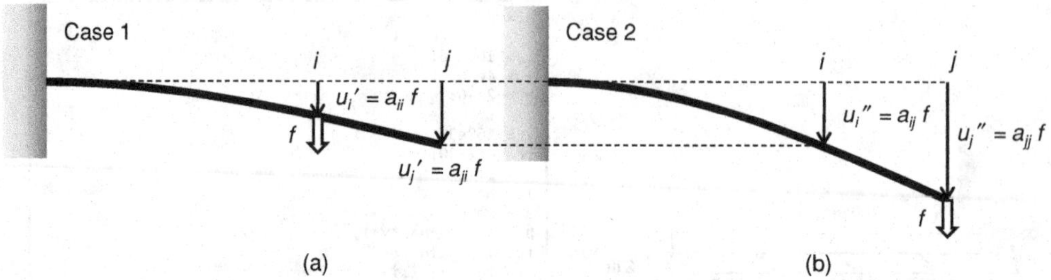

Figure 6.13 Demonstration of the reciprocal response of a cantilevered beam. (a) Force f applied at location i causes displacement u_j' at location j. (b) The same force applied at location j causes displacement u_i'' at location i. The two displacements are equal, $u_j' = u_i''$.

the same force f is applied at location j giving rise to displacement u_i'' at location i. Using the flexibility matrix,

$$u_j' = a_{ji} f \quad \text{and} \quad u_i'' = a_{ij} f. \tag{6.49}$$

If the flexibility matrix is indeed symmetric, then $u_j' = u_i''$. Analogous reciprocity properties are common to all linear systems, including circuits, antennas, and microwave systems.

This reciprocity property can be proved using a simple energy argument. Consider a beam with force f applied at position i. The elastic energy stored in the beam is

$$W_1 = \int_0^{u_{ii}} k_{ii} \upsilon \, d\upsilon = \frac{1}{2} k_{ii} u_{ii}^2 = \frac{1}{2} a_{ii} f^2, \tag{6.50}$$

where the convention is adopted here that u_{ii} is the displacement at i due to force f applied at i. Now, with f still applied at i, an *additional* force of the same magnitude f is applied at location j. The additional stored elastic energy becomes

$$\begin{aligned} W_2 &= \int_{u_{ji}}^{u_{ji}+u_{jj}} k_{jj}(\upsilon - u_{ji}) \, d\upsilon + \int_{u_{ii}}^{u_{ii}+u_{ij}} f \, d\upsilon \\ &= f^2 (a_{jj}/2 + a_{ij}). \end{aligned} \tag{6.51}$$

Thus, the net energy stored in the beam is

$$W_1 + W_2 = f^2 [a_{ii}/2 + a_{ji} + a_{jj}/2]. \tag{6.52}$$

Now we repeat the process, this time applying the force first at j and then at i. It should be obvious that the new result for the total elastic energy stored is

$$W_1' + W_2' = f^2 [a_{ii}/2 + a_{ij} + a_{jj}/2]. \tag{6.53}$$

From superposition, the final state of deflection of the beam must be the same for both experiments, so the energies must be identical:

$$W_1 + W_2 = W_1' + W_2'. \tag{6.54}$$

Therefore, $a_{ij} = a_{ji}$, showing that the flexibility matrix is indeed symmetrical. While proof is provided only for the case of a cantilevered beam, the theorem applies to any linear elastic continuum and any type of elastic deformation.

6.4.4 Applications of static MDF model

The problems worked in this section exemplify use of the MDF model. In the first example, a force is applied to one element on the beam and the full displacement vector is then computed. In the next example, the displacement vector is determined for a force distributed along the beam, similar to the problem worked approximately in Section 6.2. The last example shows how a moment can be represented as a *couple*, that is equal and opposite forces applied to adjacent elements on the beam. This last example is relevant to the important case of a piezoelectric element attached to the surface of a cantilevered beam, discussed in Section 8.4.

6.4.4.1 Static displacements due to normal force

For the case of a force f_0 applied normally to the beam at $x = x_0$, at the ith element, the force vector is given by

$$\overline{f} = f_0[0 \quad \ldots \quad 0 \quad 1 \quad 0 \quad \ldots \quad 0]^{\mathrm{T}}. \tag{6.55}$$

To handle the discrete nature of the MDF model, one defines the index i as the nearest integer to $(x_0/L)n$:

$$i = \mathrm{round}\left(\frac{x_0}{L}n\right). \tag{6.56}$$

The static displacement vector is equal to the product of the flexibility matrix and the force vector:

$$\overline{u} = \overline{\overline{a}}\,\overline{f}. \tag{6.57}$$

A MATLAB implementation of the model is given in Fig. 6.14.

The numerical solution of any structural analysis problem includes:

(i) Geometry definition;
(ii) Material property specification;
(iii) Meshing;
(iv) Solution; and
(v) Post-processing.

It is important to remember that, in an MDF formulation, the boundary conditions are built into the model. Thus, if the boundary conditions are altered, a new flexibility matrix must be found. This requirement is a real disadvantage of MDF modeling.

```
function cantBeamMDF
      g = setGeometry;
     mp = setMatProp;
      n = 40;                     % discretization (mesh)
  mBeam = mp.rho*g.L*g.h*g.w;     % beam mass
      a = zeros(n,n);             % flexibility matrix a
      B = g.L^3/(6*mp.Y*g.Ix*n^3);% convenient constant
      for j = 1:n
          for i = 1:n
              if i>= j
                  a(i,j) = B*(3*i*j^2-j^3);
              else
                  a(i,j) = a(j,i);
              end;
          end;
      end;
         f = zeros(n,1);% a zero vector is init. First
      f(n) = 10e-6;     % non-zero comp. specified
         u = a*f;       % solution is simply matrix. mult.
           vizDisp(u);% visualize displacements
return

function g = setGeometry
     g.L = 40e-6;          % length in meters
     g.h =  2e-6;          % height in meters
     g.w =  4e-6;          % width in meters
    g.Ix = g.h^3*g.w/12; % Ix for prismatic beam
return

function mp = setMatProp
     mp.rho = 2330;          % kg/m^3
     mp.Y = 150e9;           % GPa
return

function vizDisp( u, N )
```

Figure 6.14 MATLAB implementation of MDF model for a cantilevered beam subjected to concentrated static force. Some common elements of numerical descriptions of mechanical continua are evident in this code: *geometry definition* specifies beam length, width, and thickness; *material properties* specifies Young modulus and mass density; *meshing* is the discretization level of the beam (number of elements); *solution* for this example is a matrix multiplication $\overline{\overline{a}}\,\overline{f}$; *post-processing* displays beam deformation.

Figure 6.15 shows solutions for the force applied at three different points along the beam: $i = n/2$, $i = 3n/4$, and $i = n$ with $n = 40$. As expected, the displacements are larger when the force is applied closer to the free end. Also, the beam exhibits zero curvature to the right of the point where the force is applied.

6.4.4.2 Distributed electrostatic force

Now reconsider the problem of Section 6.2, where the force is distributed along the beam between $x = L - g$ and the end, $x = L$. Using an iterative approach, the MDF model can accurately predict the beam response. The electrostatic force is non-zero along the

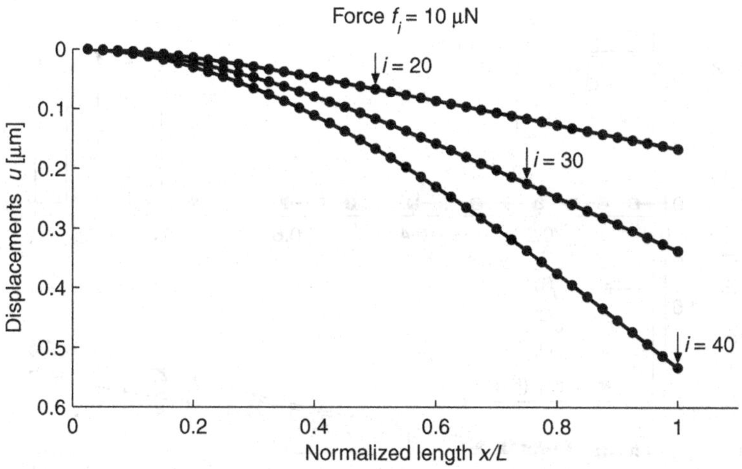

Figure 6.15 MDF solution for static displacement due to a concentrated force applied at three locations $i = n/2$, $i = 3n/4$, and $i = n$, with $n = 40$. The parameters used are those for a crystalline silicon beam of length $L = 40$ μm, width $w = 4$ μm, and thickness $h = 2$ μm.

portion of the beam directly above the electrode, that is,

$$i \geq n\left(\frac{L-g}{L}\right). \tag{6.58}$$

In the initial trial solution, the force is assumed to be uniform:

$$f_i = \begin{cases} 0, & i < n(L-g)/L, \\ \frac{1}{2}\varepsilon_0 \frac{wL/n}{d^2} V^2, & i \geq n(L-g)/L, \end{cases} \tag{6.59}$$

and the displacements are then calculated according to Eq. (6.57). If the displacements are small compared with the gap d, this approximation is sufficient. If not, then an iterative method is invoked using the displacement vector solution obtained from the trial solution to revise the force vector as follows:

$$f_i = \begin{cases} 0, & i < n(L-g)/L, \\ \frac{1}{2}\varepsilon_0 \frac{wL/n}{(d-u_i)^2} V^2, & i \geq n(L-g)/L. \end{cases} \tag{6.60}$$

The second iteration is obtained using the modified force terms from Eq. (6.60) in Eq. (6.57). The procedure is repeated until convergence is achieved, which is typically very rapid. Figure 6.16 shows three iterations for a crystalline silicon beam having dimensions $L = 300$ μm, $w = 10$ μm, $h = 2$ μm, with electrode/beam separation $d = 2$ μm and applied voltage $V = 5$ V. The superscripts in the legend identify the number of the iteration.

Note how the magnitude of the distributed force increases the most at the free end of the beam. As voltage is increased, this effect becomes steadily more pronounced. In

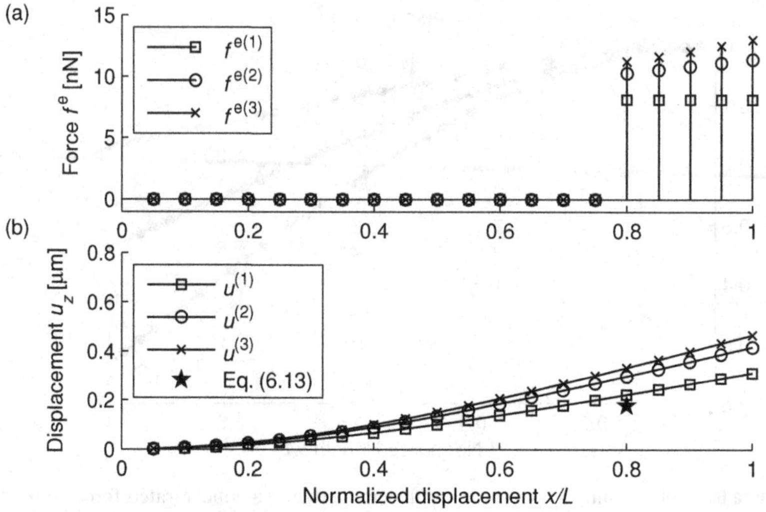

Figure 6.16 Convergence of static MDF model computation using iteration of Eqs. (6.60) and (6.57). Superscripts signify the number of the iteration. The starting profile is the displacement vector calculated using the method of Section 6.2.

fact, convergence failure and displacements growing larger than d are indicative of the electrostatic pull-in instability first presented in Section 4.8.

For comparison with the analytical model presented in Section 6.2, the displacement at $x = 0.8L$ as calculated using Eq. (6.13) is indicated in Fig. 6.16b. As expected, the approximation of putting the entire force at a single point underestimates the beam displacement.

It is important to keep in mind that the MDF model is applicable only for small displacements. In particular, the beam theory used to derive the influence coefficients and the parallel-plate approximation used for the capacitance calculations apply only in the small deformation limit.

6.4.4.3 Beam response to a force moment

The previous examples all considered the response of a cantilevered beam to simple arrays of normal forces applied to the beam. There are, however, MEMS transducers using beams where a moment rather than a normal force is exerted to get a response. One important example is when an extensible transducer element, such as a piezoelectric chip, is affixed to one face of a cantilevered beam. See Section 8.4. The MDF method is readily adapted to such a case.

Figure 6.17a depicts the case of a moment M_0 applied along a beam at $x = x_0$. The MDF method can be adapted to this problem if the ideal moment is replaced by a *force couple*, that is, a set of two equal and opposite, normal forces of magnitude,

$$f_0 = \frac{n}{L} M_0, \tag{6.61}$$

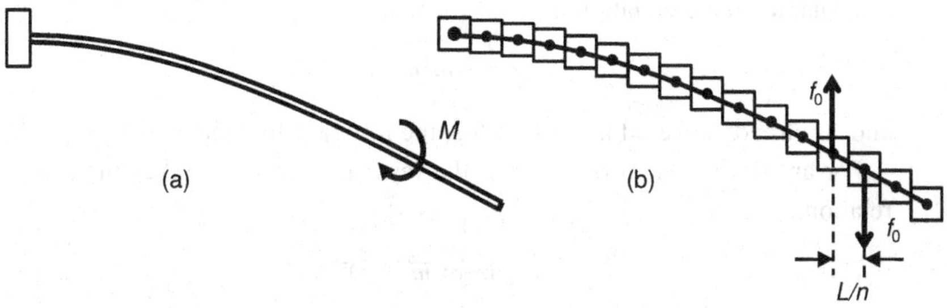

Figure 6.17 (a) Moment M_0 acting on a cantilevered beam and (b) the MDF model, which approximates M_0 as a pair of equal, normal forces, f_0, acting in opposite directions upon adjacent elements.

acting at two adjacent elements, as shown in Fig. 6.17b. The *moment arm*, that is, the distance from the midpoint to the point where the forces are applied, is $L/2n$. In the case of finite discretization of a beam, moment M_0 acting at location x_0 is approximated by the following force vector:

$$\overline{f} = [0 \quad \ldots \quad 0 \quad -1 \quad 1 \quad 0 \quad \ldots \quad 0] f_0. \tag{6.62}$$

The forces are applied at the elements i and $i+1$, where i is given by

$$i = \text{round}\left(\frac{1}{2} + \frac{x_0}{L} n\right). \tag{6.63}$$

6.4.5 Modal analysis

As pointed out in Example 6.3, the shape of the beam deformation depends on the frequency at which the beam is excited. For continua, such as beams and plates, the frequency dependence of the deformation is related directly to the set of resonant modes of the geometry. The only requirement for using the MDF formulation is to know the shapes of the important modes. These can be obtained using *modal analysis*, also referred to as *structural eigenanalysis*.[3] Beams and other mechanical continua have an infinite number of resonances, and, for many MEMS devices, it is important to know how closely spaced they are.

For lossless beams (and other mechanical continua), resonance is the condition that the elastic energy stored when the beam is at maximum deflection (with zero motion) is exactly equal to the kinetic energy at maximum motion (at zero deflection). At any resonant frequency, the inertial component of the mechanical impedance exactly cancels the elastic component. Thus, for the single-degree-of-freedom systems considered in Chapter 4 with no loss, the resonance condition is obtained from the equation of motion by assuming periodic response and setting the force to zero:

$$(-\omega^2 m + k)\underline{x} = 0. \tag{6.64}$$

[3] Readers encountering eigenanalysis for the first time should read Appendix B in parallel to this section.

The characteristic equation is

$$-\omega^2 m + k = 0, \tag{6.65}$$

and, as already covered in Section 6.2, the resonant frequency is $\omega = \sqrt{k/m}$.

For an MDF system of order n, the resonant condition takes the form of a matrix relation:

$$(-\omega^2 \overline{\overline{m}} + \overline{\overline{k}})\underline{x} = 0. \tag{6.66}$$

Recall that $\overline{\overline{k}}$ depends on both the beam properties and the boundary conditions. Equation (6.66) predicts a set of n resonant frequencies, one associated with each of n mode shapes. To investigate these modes, the first step is to find their resonant frequencies using the characteristic equation, which is obtained by setting the determinant of $-\omega\overline{\overline{m}} + \overline{\overline{k}}$ to zero and solving for ω:

$$|-\omega^2 \overline{\overline{m}} + \overline{\overline{k}}| = 0. \tag{6.67}$$

As shown in Sections B.5.2 and B.5.3 of Appendix B, eigenanalysis can be done by hand for matrices of low order. For a system of a high order, there are efficient computational algorithms built into certain widely available commercial and open-source software packages.[4] For example, MATLAB™ has the convenient function **eig** for the purpose, which computes the eigenvectors $\overline{\overline{\Phi}}$ and the eigenvalues $\overline{\overline{w}}$ of a square input matrix $\overline{\overline{R}}$ as

$$[\overline{\overline{\Phi}}, \overline{\overline{W}}] = \text{eig}(\overline{\overline{\text{R}}}), \tag{6.68}$$

where $\overline{\overline{\Phi}}$ is called the modal matrix. It is a full matrix whose columns are the eigenvectors. In structural mechanics, the eigenvectors are referred to as the mode shapes because they describe the relative amplitude of the deformations at the given resonance. $\overline{\overline{W}}$ is a diagonal matrix, the diagonal elements of which are the corresponding eigenvalues, that is, the squares of natural frequencies. Therefore, the following matrix equation holds true:

$$\overline{\overline{R}}\,\overline{\overline{\Phi}} = \overline{\overline{R}}\,\overline{\overline{w}}. \tag{6.69}$$

To use the **eig** function efficiently, it is convenient to pre-multiply Eq. (6.66) by the flexibility matrix $\overline{\overline{a}}$. Thus,

$$(-\omega^2 \overline{\overline{a}}\,\overline{\overline{m}} + \overline{\overline{I}}_{n\times n})\overline{x} = \overline{0}, \tag{6.70}$$

where $\overline{\overline{a}}\,\overline{\overline{k}} = \overline{\overline{I}}_{n\times n}$ is the identity matrix of order n. Using the fact that $\overline{\overline{m}}$ is diagonal and assuming the elements are identical, according to Eq. (6.39), Eq. (6.70) further reduces to

$$\left(\frac{1}{\omega^2}\overline{\overline{I}}_{n\times n} - \overline{\overline{a}}\frac{m_{\text{beam}}}{n}\right)\overline{x} = \overline{0}. \tag{6.71}$$

[4] Refer to [4] for general information on efficient eigensolvers, and to [5] for the efficient eigensolvers for vibration applications.

```
function cantBeamMDF
      g = setGeometry;
     mp = setMatProp;
      n = 20;                          % discretization (mesh)
  mBeam = mp.rho*g.L*g.h*g.w;          % beam mass
      a = zeros(n,n);                  % flexibility matrix A (A = inv(K))
      B = g.L^3/(6*mp.Y*g.Ix*n^3);     % convenient constant
      for j = 1:n
          for i = 1:n
              if i>= j
                  a(i,j) = B*(3*i*j^2-j^3);
              else
                   a(i,j) = a(j,i);
              end;
          end;
      end;
      [v,lam] = eig(a*mBeam/n);     % solution: Modal analysis
            w = 1./sqrt(lam);
return

function g = setGeometry
    g.L = 40e-6;             % length in meters
    g.h =  2e-6;             % height in meters
    g.w =  4e-6;             % width in meters
   g.Ix = g.h^3*g.w/12;           % area moment o inertia of a prismatic
beam
return

function mp = setMatProp
    mp.rho = 2330;           % kg/m^3
    mp.Y = 150e9;            % GPa
return
```

Figure 6.18 MATLAB implementation of an MDF model for a cantilevered beam.

The eigenvalues of $\bar{\bar{a}} m_{\text{beam}}/n$ are the desired $1/\omega^2$ values and the associated eigenvectors are the mode shapes. Figure 6.18 provides a very simple MATLAB implementation of this eigenmode extraction procedure.

As a test of the MDF analysis of beams, one can compare predictions for the resonant frequencies with those obtained from analytical solutions. For thin (Euler–Bernoulli) beams, the resonant frequencies are given by the following expression [3]:

$$f_k = \frac{w_k}{2\pi} = \beta_k^2 \sqrt{\frac{YI_x}{\rho A}}. \tag{6.72}$$

Here, L, I_x, A, Y, and ρ are the beam length, area moment of inertia, cross-sectional area, Young modulus, and mass density, respectively, and β_k is a mode parameter arising from the general solution that depends on the boundary conditions. For the cantilevered beam, all β_k values satisfy the following transcendental equation:

$$1 + \cosh(\beta_k L)\cos(\beta_k L) = 0. \tag{6.73}$$

Figure 6.19 plots this expression and reveals the first three zeros, 1.875, 4.694, and 7.854, which correspond to the first three modes.

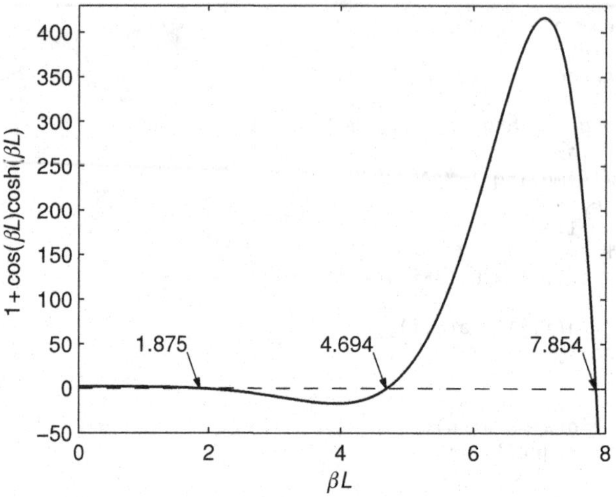

Figure 6.19 Plot of the transcendental function of Eq. (6.73) for a uniform cantilevered beam, $[1 + \cos(\beta L)\cosh(\beta L)]$, revealing its first three zeros. Each β value corresponds to a modal resonance.

The first three resonant frequencies, calculated using these zeros in Eq. (6.72), are found in Table 6.1. For comparison, values obtained from the MDF model for $n = 10$ and $n = 0$ elements are also provided. The error for the MDF resonance frequency predictions is ~6% for $n = 10$ and ~2% for $n = 40$. For reference, Fig. 6.20 plots the mode shapes of these three lowest resonances.

Table 6.1 The first three resonant frequencies in MHz of a homogenous cantilevered beam with a rectangular cross-section calculated using the analytical solution, Eq. (6.72), and separately obtained using the MDF model. The MDF predictions are more accurate for higher values of n, though even $n = 10$ gives a reasonably close answer. The parameters used are for a crystalline Si beam of length $L = 40$ μm, width $w = 4$ μm, and thickness $h = 2$ μm

		MDF model calculation	
Mode	Analytical solution	$n = 10$ elements	$n = 40$ elements
f_1[MHz]	1.6	1.5	1.6
f_2[MHz]	10.1	9.3	9.9
f_3[MHz]	28.4	26.1	27.8

6.4.6 Decoupling of the equation of motion

Once the mode shapes and resonance frequencies are known, the matrix equation of motion, describing the system of n coupled differential equations,

$$\overline{\overline{m}}\ddot{\overline{x}} + \overline{\overline{k}}\overline{x} = \overline{f}, \tag{6.74}$$

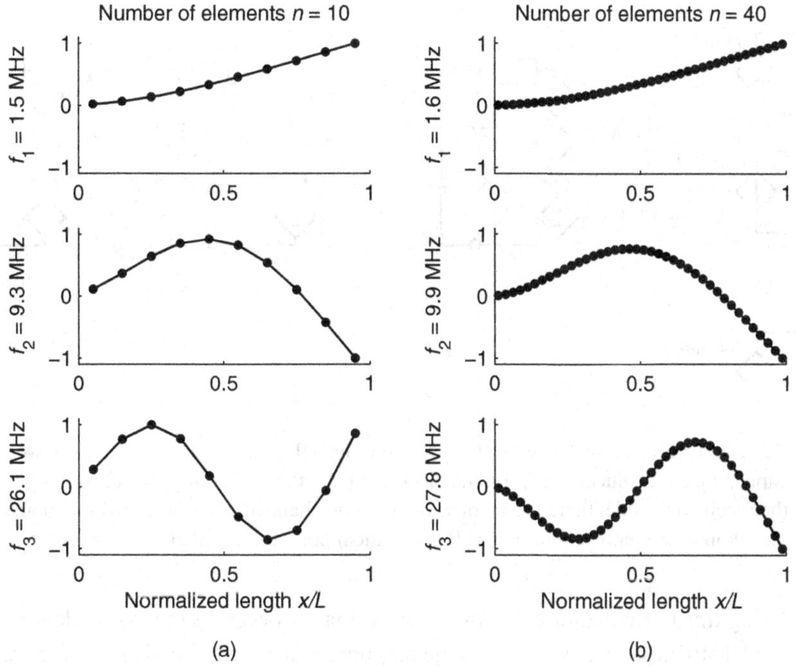

Figure 6.20 First three resonant modes of a cantilevered beam obtained from MDF analysis. (a) $n = 10$ elements, (b) $n = 40$ elements. Figure D.10, Appendix D, compares the mode shapes and resonances with those obtained using finite-element analysis (FEA). The parameters used are for a crystalline silicon beam of length $L = 40$ μm, width $w = 4$ μm, and thickness $h = 2$ μm.

can be replaced by a set of n decoupled equations. Refer to Appendix B for more details. The decoupling is achieved in two steps. First, the coupled matrix equation of motion is pre-multiplied by the transpose of the modal matrix $\overline{\overline{\Phi}}^{\mathrm{T}}$. Then, the generalized displacements are introduced implicitly:

$$\overline{x} = \overline{\overline{\Phi}}\,\overline{\eta}. \tag{6.75}$$

After performing these operations, the matrix equation of motion takes the form

$$\overline{\overline{M}}\,\ddot{\overline{\eta}} + \overline{\overline{K}}\,\overline{\eta} = \overline{\overline{\Phi}}^{\mathrm{T}}\overline{f}, \tag{6.76}$$

where $\overline{\overline{M}} = \overline{\overline{\Phi}}^{\mathrm{T}}\overline{\overline{m}}\,\overline{\overline{\Phi}}$ and $\overline{\overline{K}} = \overline{\overline{\Phi}}^{\mathrm{T}}\overline{\overline{k}}\,\overline{\overline{\Phi}}$ are the mass and stiffness modal matrices, respectively. Note that while the components of the displacement vector $\overline{x}$ inform us how much each of the degrees of freedom participate in the collective motion of the continua, the components of the vector of modal displacement $\overline{\eta}$ quantify the relative magnitudes of all the modes contributing to the collective motion of the continua.

Both $\overline{\overline{M}}$ and $\overline{\overline{K}}$ are diagonal matrices, as shown in Appendix B. Within a multiplicative constant, eigenvectors are uniquely determined. Any eigenvector scale will satisfy Eq. (6.76), but perhaps the most common normalization is with respect to the mass matrix

$$\overline{\phi}_k^{\mathrm{T}}\overline{\overline{m}}\,\overline{\phi}_k = 1. \tag{6.77}$$

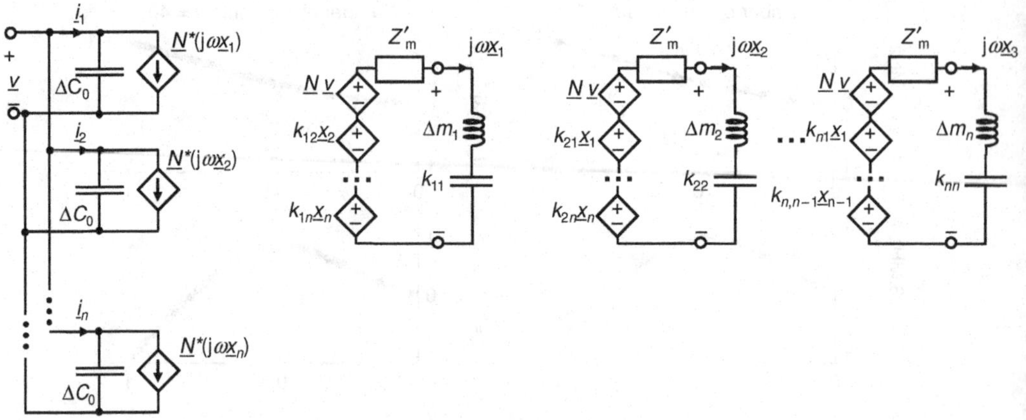

Figure 6.21 Equivalent circuit representation of an MDF model. Each DOF is viewed as a small capacitance transducer. Thus, on the electrical side, there are n transducer primary circuits. On the mechanical side, there are n corresponding mechanical ports. The stiffness matrix is a full matrix in MDF representation, so all mechanical ports are coupled to one another.

Using this normalization, the modal mass matrix becomes the $n \times n$ identity matrix, the modal stiffness matrix becomes the diagonal matrix with its diagonal elements equal to the respective squares of the natural frequencies, and the equation of motion becomes

$$\overline{\overline{I}}_{n\times n}\,\ddot{\overline{\eta}} + \begin{bmatrix} \omega_1^2 & 0 & \dots & 0 \\ 0 & \omega_2^2 & \dots & 0 \\ \dots & \dots & \ddots & \dots \\ 0 & 0 & \dots & \omega_n^2 \end{bmatrix} \overline{\eta} = \overline{\overline{\Phi}}^{\mathrm{T}}\overline{f}. \tag{6.78}$$

When a mechanical continuum is excited close to any isolated resonance, the displacement profile of the motion assumes the form of the mode shape. The MDF analysis allowed us to compute these resonant frequencies and mode shapes.

6.4.7 Equivalent circuit using modal analysis

The MDF model can be cast in the form of an attractive and useful equivalent electromechanical network model by paying proper attention to the results of the modal analysis. The starting point is the equivalent circuit representation for the discretized beam model. This circuit, shown in Fig. 6.21, is not simple at all. Assuming that the entire beam is an electrode, each DOF has capacitance

$$(\Delta C_0)_i = \varepsilon_0 \frac{wL/n}{d - u_i}. \tag{6.79}$$

These elements are all connected in parallel. There are also n mechanical ports, one for each degree of freedom. Because the stiffness matrix $\overline{\overline{k}}$ is full, having no zero elements, the mechanical ports are all elastically coupled. Even for small values of n,

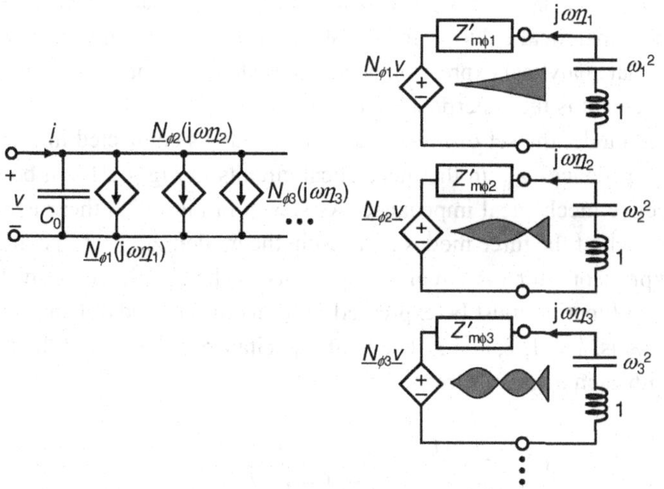

Figure 6.22 Simplified n-DOF system approximated by the first three modes. For practical reasons, only those modes likely to be encountered in operation are retained. The mechanical ports are no longer mutually coupled. The normalization is with respect to the mass matrix, which yields ones for modal masses, and ω^2 for modal spring constants.

e.g., $n = 10$, the system seems dauntingly complex. The electromechanical coefficients for the N-form transducer matrix are obtained from the coenergy:

$$\underline{N} = \left.\frac{\partial(\Delta C_0)_i}{\partial u_i} v\right|_{u_i=0, v=c_0} = \varepsilon_0 \frac{wL/n}{d^2} v_0, \tag{6.80}$$

$$\underline{Z}'_m = \frac{1}{j\omega}\left(\frac{v^2}{2}\frac{\partial^2 \Delta C_0(u_k)}{\partial u_i^2}\right)_{u_i=0, v=v_0} = \frac{\varepsilon_0 v_0^2}{j\omega}\frac{wL/n}{d^3}. \tag{6.81}$$

Note that, because the beam has been divided into n identical elements, the electromechanical coefficient values are the same.

Fortunately, simplification of the network is achieved by taking account of the modal behavior of the cantilevered beam. The first few beam modes are well separated, so by restricting attention to the lower frequencies – usually a good approximation in MEMS – it is justified to take advantage of the fact that, near any one resonant frequency, the beam behaves like a simple, second-order resonator. Thus, effective mass, spring constant, and damping coefficients can be defined for each individual mode. The modal approach thus opens the door to an easy to interpret network consisting of the superposition of the modes deemed important. In the present example, only the first three modes are retained and the simplified, equivalent circuit of Fig. 6.22 is the result.

The key to this simplification is recognizing that all n degrees of freedom for any one mode are interrelated through the mode shape. Thus, the set of n displacements for arbitrary mode k can be represented by the modal displacement, η_k:

$$\overline{u}_k = \overline{\phi}_k \eta_k. \tag{6.82}$$

This procedure is, in fact, the MDF equivalent of the approach taken in Section 6.2, which used the analytical solution for the static profile of the cantilevered beam. The equivalent of that analytical expression is the mode shape vector $\overline{\phi}_k$. Of course, the total vector displacement is the superposition of all these modes.

For any one mode, the set of $n-1$ dependent sources connected in series with the mechanical impedances $\underline{Z}'_{\mathrm{m}}$ in the mechanical circuits in Fig. 6.21 can be summed to form an effective mechanical impedance. As shown in Fig. 6.22, there is one of these networks for each of the three modes retained in the model: $\underline{Z}'_{\mathrm{m}\phi 1}$, $\underline{Z}'_{\mathrm{m}\phi 2}$, and $\underline{Z}'_{\mathrm{m}\phi 3}$.

To find expressions for the circuit components of the simplified equivalent circuit, the capacitance function must be expressed in terms of the modal displacement η_k for each mode, that is, $k = 1$, 2, and 3. The total capacitance is the sum of the capacitances associated with each degree of freedom:

$$\begin{aligned} C &= \varepsilon_0 \frac{w(L/n)}{d} \sum_{i=j}^{n} \frac{1}{1-(u_i/d)} \\ &= C_0 + \frac{C_0}{nd} \sum_{i=1}^{n} u_i + \frac{C_0}{nd^2} \sum_{i=1}^{n} u_i^2 + \ldots \end{aligned} \tag{6.83}$$

Combining Eq. (6.82) and Eq. (6.83) gives the capacitance in terms of the modal displacement η_k associated with the kth mode:

$$C(\eta_k) = C_0 + \frac{C_0}{nd}[1 \quad 1 \quad \ldots \quad 1]^{\mathrm{T}} \overline{\phi}_k \eta_k + \frac{C_0}{nd^2} \overline{\phi}_k^{\mathrm{T}} \overline{\phi}_k \eta_k^2 + O\left(\eta_k^3\right). \tag{6.84}$$

This equation expresses the dependence of beam capacitance on η_k when only the kth mode is present. The associated coenergy is

$$W'_{\mathrm{e}} = \frac{1}{2} C(\eta_k) v^2. \tag{6.85}$$

This coenergy may be used to determine the electromechanical transducer network parameters for this mode in the usual way, taking partial derivatives with respect to η_k:

$$\underline{N}_k = \left.\frac{\partial^2 W'_{\mathrm{e}}(\eta_k)}{\partial v \partial \eta_k}\right|_{\eta_k=0, v=v_0} = v_0 \left.\frac{\mathrm{d}C(\eta_k)}{\mathrm{d}\eta_k}\right|_{\eta_k=0} = \frac{C_0 v_0}{nd}[1 \quad 1 \quad \ldots \quad 1] \overline{\phi}_k, \tag{6.86}$$

$$\underline{Y}_{\mathrm{e}} = \mathrm{j}\omega \left.\frac{\partial^2 W'_{\mathrm{e}}}{\partial v^2}\right|_{\eta_k=0, v=v_0} = \mathrm{j}\omega C_0, \tag{6.87}$$

$$\underline{Z}'_{m\phi_k} = \left.\frac{1}{\mathrm{j}\omega} \frac{d^2 W'_{\mathrm{e}}}{\partial \eta_k^2}\right|_{\eta_k=0, v=v_0} = \left.\frac{1}{\mathrm{j}\omega} \frac{1}{2} \frac{\partial^2 C(\eta_k)}{\partial \eta_k^2} v_0^2\right|_{\eta_k=0} = \frac{1}{\mathrm{j}\omega} \frac{C_0}{nd^2} \overline{\phi}_k^{\mathrm{T}} \overline{\phi}_k v_0^2. \tag{6.88}$$

Terms to second order in η_k must be retained in the capacitance expression, to obtain an expression for $\underline{Z}'_{m\phi_k}$.

6.4.8 Damping

Mass and stiffness matrices $\overline{\overline{m}}$ and $\overline{\overline{k}}$ were introduced for the MDF model in Sections 6.4.1 and 6.4.2, respectively, as generalizations of the familiar scalars: m and k. Obtaining a damping matrix $\overline{\overline{b}}$ is far more difficult. The practical approach for MEMS devices with a single degree of freedom is to obtain the resonant quality factor Q from the measured frequency response and then to extract b from it using the method described in Section B.4 of Appendix B. Implementing this procedure for MEMS continua is less straightforward. One commonly utilized empirical method for determining $\overline{\overline{b}}$ relies on measurements of only the resonances falling within the operating range of the device. Approximation is possible because for small damping, say $Q > 10$, the resonant frequency values themselves are virtually unaffected. Furthermore, the system response away from the resonance is hardly affected at all.

There are several important damping mechanisms in MEMS devices, including (i) viscous damping of the ambient gas, (ii) structural or hysteretic energy dissipation inherent to the material (structural damping), and (iii) vibrational (parasitic) energy transfer to the substrate, sometimes called *clamping damping*. Viscous damping, which dominates in most MEMS devices, is controlled by the air or gas pressure. For high-Q applications, commercial devices are sealed hermitically in a package at reduced pressure. To further increase Q in demanding applications, clamping damping can be alleviated by mechanical isolation of moving element from the substrate. Section 6.8.2 treats the case of supporting a beam to isolate it from the substrate.

The *Rayleigh damping* model is an empirical approach that expresses the damping matrix as a linear combination of the mass and the stiffness matrices, that is,

$$\overline{\overline{b}} = \alpha_1 \overline{\overline{m}} + \alpha_2 \overline{\overline{k}}. \tag{6.89}$$

There is no physical motivation for this form. It is used for modal analyses of beams because now the damping matrix can be diagonalized just like the mass and stiffness matrices. With $\overline{\overline{b}}$ diagonalized damping can, at least close to any resonance, be related in a simple way to the measured quality factor Q.[5] For the cantilevered beam model, if the frequency range of interest includes the first three resonances, the quality factor associated with these resonances, Q_1, Q_2, and Q_3, can be measured and the damping matrix takes a diagonal form:

$$\overline{\overline{b}} = \overline{\overline{\Phi}} \begin{vmatrix} \omega_1/Q_1 & 0 & 0 & \dots & 0 \\ 0 & \omega_2/Q_2 & 0 & \dots & 0 \\ 0 & 0 & \omega_3/Q_3 & \dots & 0 \\ \dots & \dots & \dots & \dots & \dots \\ 0 & 0 & 0 & \dots & \omega_n/Q_n \end{vmatrix} \overline{\overline{\Phi}}^{\mathrm{T}}, \tag{6.90}$$

where the columns of matrix $\overline{\overline{\Phi}}$ are the eigenvectors. The quality factors for $n \geq 3$ do not affect the frequency below the third resonant frequency and can be assigned arbitrary values, usually non-zero to avoid singularities.

[5] Refer to Section B.4 of Appendix B.

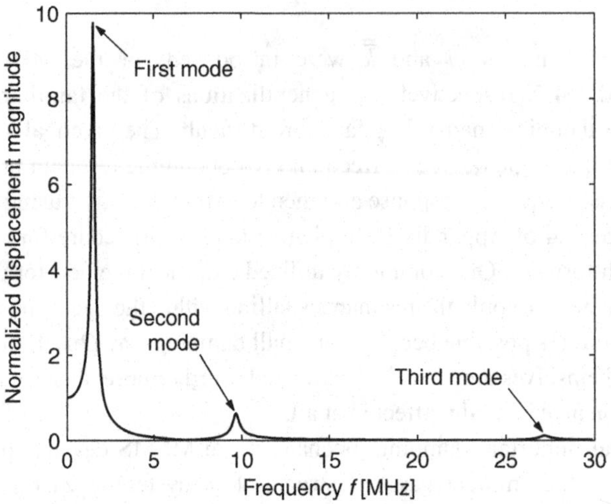

Figure 6.23 MDF model simulation of the first three modes of the frequency response of a cantilevered beam assuming $Q_1 = 10$, $Q_2 = 30$, and $Q_3 = 50$. The observed displacement is at the tip of the beam. The parameters used are for a crystalline silicon beam of length $L = 40$ μm, width $w = 4$ μm, and thickness $h = 2$ μm.

To illustrate this method, consider a second-order system. For a single degree of freedom, the frequency response is

$$\underline{u}(\omega) = \frac{\underline{f}(\omega)}{-\omega^2 m + j\omega b + k}. \tag{6.91}$$

Then, for an MDF with n degrees of freedom, Eq. (6.91) generalizes to

$$\overline{\underline{u}}(\omega) = [-\omega^2 \overline{\overline{m}} + j\omega \overline{\overline{b}} + \overline{\overline{k}}]^{-1} \overline{\underline{f}}(\omega). \tag{6.92}$$

In the case where a force is applied at only one element and the displacement then monitored at another element, Eq. (6.92) reduces to a scalar equation. As an example, let the force be applied at the tip of the beam and the displacement observed at the same point:

$$\underline{u}_n(\omega) = [0 \quad 0 \quad \ldots \quad 0 \quad 1][-\omega^2 \overline{\overline{m}} + j\omega \overline{\overline{b}} + \overline{\overline{k}}]^{-1} [0 \quad 0 \quad \ldots \quad 0 \quad 1]^{\mathrm{T}} \underline{f}_N(\omega). \tag{6.93}$$

Figure 6.23 shows the magnitude response of a cantilevered beam with length $L =$ 40 μm, width $w = 4$ μm, and thickness $h = 2$ μm, modeled by the MDF method with $n = 40$ degrees of freedom. The quality factors of the first three modes are assumed to be $Q_1 = 10$, $Q_2 = 30$, and $Q_3 = 50$. The displacement magnitude is normalized with respect to the first mode, which has modal stiffness ω_1^2. The higher resonances have lower peaks inspite of higher Q values simply because $b_n = \omega_n^2 / Q_n$.

In summary, modal analysis makes it possible to diagonalize the Rayleigh damping matrix, Eq. (6.89), thereby decoupling the modes so that they can be treated individually. This diagonalization is achieved because the Rayleigh model assumes the damping matrix to be a linear combination of the mass and stiffness matrices, both of which become diagonal after modal analysis. Then, for measurements of the quality factors of individual modes the damping matrix is obtained using Eq. (6.90). This method assumes the quality factors of the higher-order modes to be large.

6.5 Using the MDF model for dynamics

The second-order MDF matrix equation of motion in the time domain is

$$\bar{\bar{m}}\ddot{\bar{u}} + \bar{\bar{b}}\dot{\bar{u}} + \bar{\bar{k}}u = \bar{f}. \tag{6.94}$$

To solve this *ordinary differential equation* (or ODE) in the time domain, the most common approach is to recast it in state-space form. Section B.2.5 of Appendix B shows the state-space formulation of an SDF system. For an MDF system, the state vector is formed by concatenating the displacement and velocity vectors:

$$\bar{q} = \begin{bmatrix} \bar{u} \\ \dot{\bar{u}} \end{bmatrix}. \tag{6.95}$$

The next step is to express the derivative $\dot{\bar{q}}$ using the definition of $\bar{q}$, Eq. (6.95), and the equation of motion, Eq. (6.94). Thus,

$$\dot{\bar{q}} = \begin{bmatrix} \bar{\bar{0}}_{N\times N} & \bar{\bar{I}}_{N\times N} \\ -\bar{\bar{m}}^{-1}\bar{\bar{k}} & -\bar{\bar{m}}^{-1}\bar{\bar{b}} \end{bmatrix} \bar{q} + \begin{bmatrix} \bar{0}_{N\times 1} \\ \bar{\bar{m}}^{-1}\bar{f} \end{bmatrix}. \tag{6.96}$$

This set of coupled linear equations can be solved using a numerical ODE solver. For the electrostatically coupled beam, the force terms $\bar{f}$ depend on the displacement $\bar{q}$, but if the displacements are small, one can ignore this dependence. In that limit, Eq.(6.96) is linear. On the other hand, if this approximation is not justified, then the system non-linearity must be confronted:

$$\dot{\bar{q}} = \begin{bmatrix} \bar{\bar{0}}_{N\times N} & \bar{\bar{I}}_{N\times N} \\ -\bar{\bar{m}}^{-1}\bar{\bar{k}} & -\bar{\bar{m}}^{-1}\bar{\bar{b}} \end{bmatrix} \bar{q} + \begin{bmatrix} \bar{0}_{N\times 1} \\ \bar{\bar{m}}^{-1}\bar{f}(q) \end{bmatrix}. \tag{6.97}$$

Equation (6.97) enables us to calculate the evolution in time of the individual degrees of freedom of the MDF model in time. Some applications are presented in Examples 6.5 and 6.6.

Example 6.6 Solving for the dynamic response from initial conditions

One can use the MDF formulation and an ordinary differential equation (ODE) solver to solve for the transient response due to initial conditions. Consider two cases, shown in Fig. 6.24: (i) when the beam is initially displaced in the first mode (this situation occurs if somehow the beam gets plucked); (ii) when the beam is initially displaced in the second mode (it would be difficult to displace the beam statically in this fashion, but the response occurs when the device is operated at the second resonance and the excitation force is abruptly turned off).

```
function [K,invM,B,n,w,v] = getMDF
    g = setGeometry;
   mp = setMatProp;
    n = 20; % discretization level (mesh)
    M = eye(n)*mp.rho*g.L*g.h*g.w/n;
    a = zeros(n,n);
    B = g.L^3/(6*mp.E*g.Ix*n^3)
    for j = 1:n
        for i = 1:n
            if i>= j
               a(i,j) = B*(3*i*j^2-j^3);
            else
               a(i,j) = a(j,i);
            end;
        end;
    end;
    [v,lam] = eig(a*M);
    w = flipud(sqrt(1./diag(lam)));
    v = fliplr(v);
    K = inv(a);
    zeta = .002;
    B = M*diag(w)*zeta;
    invM = M^(-1);
return
function showResults(n, uo, t, y)…
function g = setGeometry …
function mp = setMatProp …
```

(a)

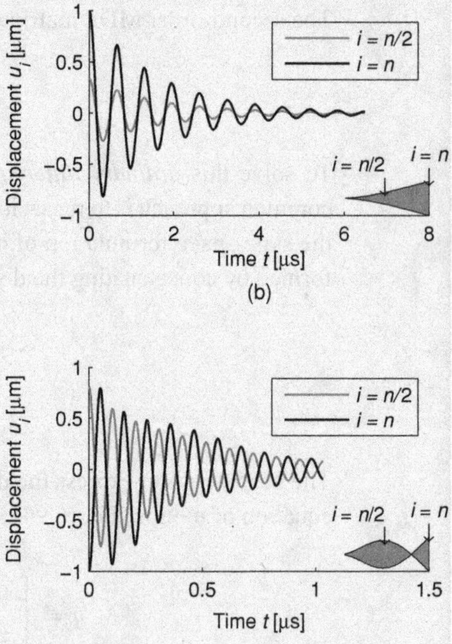

Figure 6.24 (a) Simplified MATLAB code. (b) Response for first mode. (c) Response for second mode. Previously defined functions described elsewhere in the text are suppressed in the code for clarity. Note that the responses in the middle and at the tip of the beam are in phase for the first case, which shows the damped vibrations due to initial displacement of the first mode. These two displacements are out of phase when the beam is initially displaced as the second vibration mode. Note also the difference in frequencies for the two responses.

Example 6.7 MDF ODE – initial transient due to sine force at tip of beam

This example (see Fig. 6.25) explores the transient response to a sine driving force, starting at zero initial conditions. The frequency of the drive force is near the first natural frequency of the beam. The objective is to obtain the response when the force is applied first at the midpoint of the beam and then at the tip of the beam.

```
function MDFodeForce

    global K invM B n w
    [K,invM,...
    B,n,w,v] = getMDF;
           k = 1;
          uo =  zeros(2*n,1);
       [t,y] = ode45(@MDF,...
                  [0, 10*2*pi/w(k)],uo);
              showResults(n,t,y)

return

function qDot = MDF(t,q)

    global K invM B n w

              D = [zeros(n,n),eye(n);...
                   -invM*K, -invM*B];
         fvec = zeros(2*n,1);
    fvec(end) = 1e-6;
         qDot = D*q...
                +[zeros(n,1);invM*fvec]...
                *cos(1*w(1)*t);

return

function [K,invM,B,n,w,v] = getMDF...

function showResults(n, t, y)...

function g = setGeometry ...

function mp = setMatProp ...
```

(a)

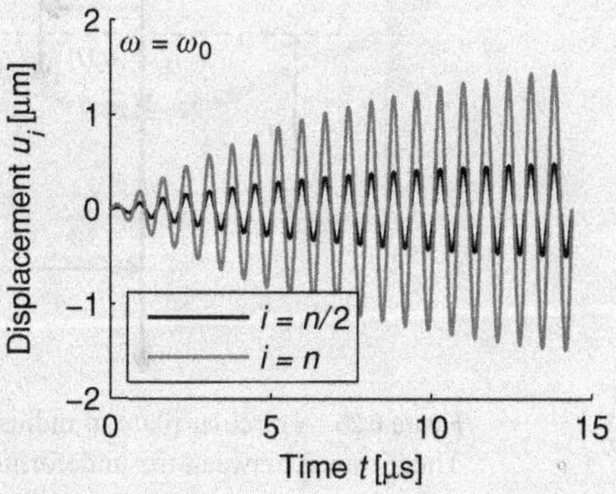

(b)

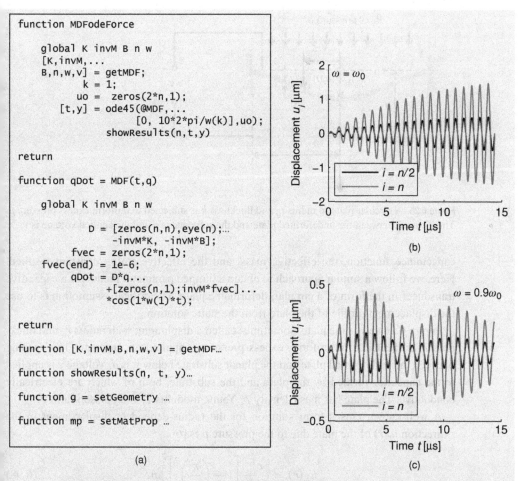

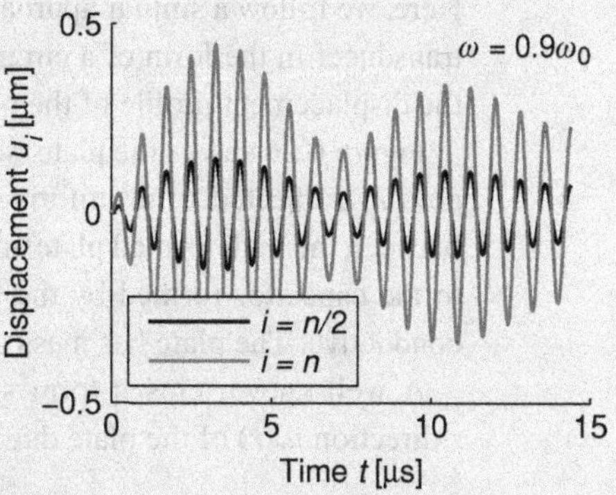

(c)

Figure 6.25 The response due to the force at the tip of the beam at and near the first resonance. The displacements at $n/2$ and n are in phase. (a) Simplified MATLAB code. (b) The force is exactly at resonance $\omega = \omega_0$ – the response increases monotonically. (c) $\omega = 0.9\omega_0$ – beating phenomenon.

Because the beam is operated either at or near the first resonance, the responses in the middle and at the tip of the beam are in phase. When $\omega = \omega_0$, the response monotonically increases. When driven slightly off resonance, the beating phenomenon occurs. See Appendix B, Section B.2 for analytical solution of the SDF resonator.

6.6 A first look at plates

Section 6.2 introduced the cantilevered beam as a first example of a mechanical continuum relevant to MEMS. The analysis used a quasistatic model for the perturbation dynamics based on the static beam deflection and made a somewhat arbitrary choice for the operative mechanical displacement variable. Approximate expressions for the

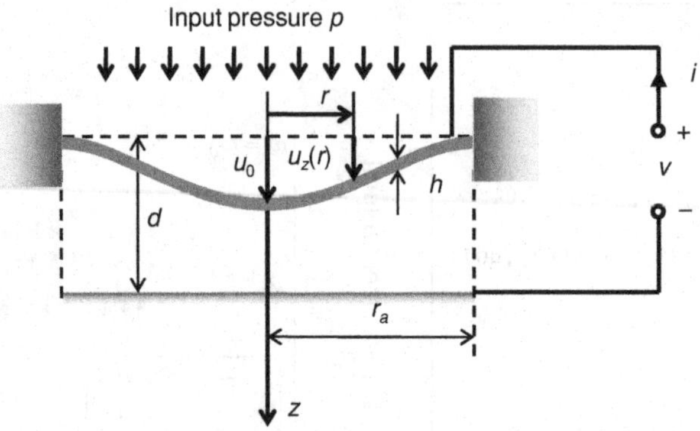

Figure 6.26 A circular plate of radius r_a and thickness h is subjected to uniform excess pressure p. The distance between the undeformed plate and the substrate is d and the applied voltage is v.

capacitance function, the effective mass, and the effective spring constant resulted. Here, we follow a similar approach to obtain a lumped parameter model for a capacitive transducer in the form of a circular, deformable plate. The critical assumption is to use the displacement profile of the plate from the static solution.

Figure 6.26 shows the plate, sometimes called a diaphragm, with radius r_a and thickness h, subjected to a uniform, excess pressure p applied on one side. The spacing between the undeformed plate and the planar substrate below it is d. Voltage v is applied to the capacitor formed by the plate and the substrate, both of which are electrically conductive. The plate has mass density ρ, Young modulus Y, and Poisson ratio ν.

A well-known closed-form solution for the radius-dependent displacement in the z direction $u_z(r)$ of the plate due to the pressure p is [6]

$$u_2(r) = \frac{r_a^4}{64D}\left[1 - \left(\frac{r}{r_a}\right)^2\right]^2 p, \tag{6.98}$$

where D is the plate modulus or flexural rigidity discussed in Appendix D, Eq. (D.83),

$$D = \frac{Yh^3}{12(1-\nu^2)}. \tag{6.99}$$

See Section D.11. The displacement at the center of the plate, clearly the most convenient and natural choice for the required lumped parameter displacement variable, is

$$u_0 = \frac{r_a^4}{64D}p. \tag{6.100}$$

Inserting Eq. (6.100) into Eq. (6.98) yields the displacement function $u_z(r)$ in terms of the displacement in the center of the plate,

$$u_z(r) = u_0(1 - (r/r_a)^2)^2. \tag{6.101}$$

This selection determines the equivalent mass, the equivalent spring constant, and the capacitance function.

6.6.1 Equivalent spring constant and mass

The equivalent spring constant is the ratio of the total force to the lumped parameter displacement:

$$k_{\mathrm{eq}} = \frac{f}{u_0} = \frac{16}{3}\frac{Yh^3}{1-\nu^2}\frac{\pi}{r_a^2}. \tag{6.102}$$

The equivalent mass is estimated by equating the kinetic energy of the lumped parameter system to the same quantity distributed over the plate:

$$\frac{1}{2}m_{\mathrm{eq}}\dot{u}_0^2 = \frac{1}{2}\int_0^{r_a} \dot{u}_z^2(r)\underbrace{\rho(2\pi r h)\mathrm{d}r}_{\mathrm{d}m} = \dot{u}_0^2 m_{\mathrm{total}}\int_0^1 (1-(r_n)^2)^4 2r_n \mathrm{d}r_n, \tag{6.103}$$

where $r_n = r/r_a$. An integration yields

$$m_{\mathrm{eq}} = \frac{256}{315}m_{\mathrm{total}} = 0.81 m_{\mathrm{total}}, \tag{6.104}$$

where $m_{\mathrm{total}} = \rho h \pi r_a^2$. As expected, $m_{\mathrm{eq}} < m_{\mathrm{total}}$.

6.6.2 Capacitance

Using the parallel-plate approximation and ignoring fringing, the total capacitance of the displaced plate is

$$C(u_0) = \int_0^{r_a} \varepsilon_0 \frac{2\pi r}{d}\frac{1}{1-u_z(r)/d}\mathrm{d}r. \tag{6.105}$$

This expression is only accurate when $|u_0| \ll d$. By expanding the integrand of Eq. (6.105) in a Taylor series and retaining the first three terms, the result is

$$C(u_0) \approx 2C_0 \int_0^{r_a}\left(1+\frac{u_0}{d}\left(1-\frac{r^2}{r_a^2}\right)^2+\frac{u_0^2}{d^2}\left(1-\frac{r^2}{r_a^2}\right)^4\right)\frac{rdr}{r_a^2}, \tag{6.106}$$

where $C_0 = \varepsilon_0 \pi r_a^2/d$. Upon integration,

$$C(u_0) \approx C_0\left(1+\frac{u_0}{3d}+\frac{u_0^2}{5d^2}\right). \tag{6.107}$$

Retaining terms up to u_0^2 is essential if the stability of this capacitive transducer is to be considered.

6.6.3 Resonance

Once k_{eq} and m_{eq} are known, the mechanical resonant frequency ω_0 can be estimated:

$$\omega_0 = \sqrt{k_{\mathrm{eq}}/m_{\mathrm{eq}}} = \frac{8.87}{r_a^2}\sqrt{\frac{D}{\rho h}}. \tag{6.108}$$

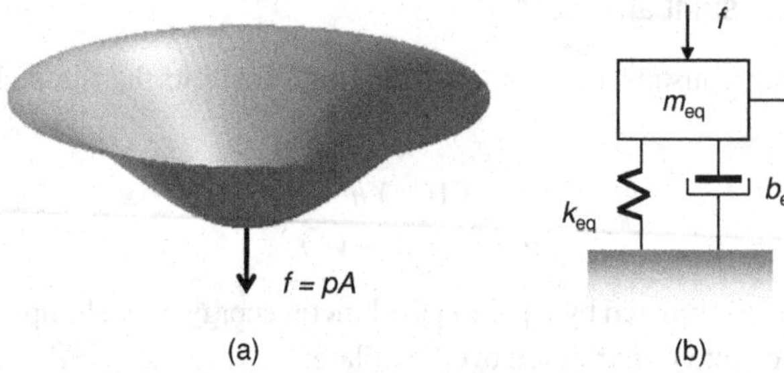

Figure 6.27 Equivalent lumped parameter model for dynamic modeling. (a) Bending-plate physical model. (b) Dynamic model based on lumped parameter elements. The spring constant k_{eq} is usually derived analytically, while the equivalent mass m_{eq} and damping factor b_{eq} are often obtained from experimental measurements on prototypes.

A more accurate expression for the resonant frequency obtained from analytical theory is [7]

$$\omega_0 = \frac{10.22}{r_a^2}\sqrt{\frac{D}{\rho h}}. \tag{6.109}$$

The lumped parameter SDF model underestimates the resonant frequency by approximately 13%.

Figure 6.27 shows the deformed circular plate and its associated SDF model, which is good for frequencies up to the first resonance.

Example 6.8 **Small-signal lumped parameter model for a circular plate**

This example derives the N-form for the circular plate subjected to uniform pressure and draws the equivalent circuit.

Once the equivalent spring constant k_{eq}, the equivalent mass m_{eq}, and the capacitance $C(u_0)$ are known, a derivation of the lumped parameter model is easily obtained using the methods of Chapter 4. The coenergy function is

$$W'_e = \frac{1}{2}C(u_0)v^2 = \frac{1}{2}C_0\left(1 + \frac{u_0}{3d} + \frac{u_0^2}{5d^2}\right)v^2.$$

The charge and force are obtained by differentiating coenergy:

$$f = -f^e = -\frac{\partial W'_e}{\partial u_0},$$

$$q = \frac{\partial W'_e}{\partial v}.$$

The N-form is obtained after linearization of the force and the charge, then using $\underline{i} = j\omega \underline{q}$ for the small-signal current:

$$\begin{bmatrix} \underline{f} \\ \underline{i} \end{bmatrix} = \begin{bmatrix} \underline{Z}'_m & \underline{N} \\ -\underline{N}^* & \underline{Y}_e \end{bmatrix} \begin{bmatrix} j\omega \underline{x} \\ \underline{i} \end{bmatrix},$$

where

$$\underline{Y}_{\rm e} = {\rm j}\omega \left.\frac{\partial^2 W'_{\rm e}}{\partial v^2}\right|_{u_{L-g}\to 0} = {\rm j}\omega C_0 = {\rm j}\omega\varepsilon_0 \frac{\pi r_a^2}{d},$$

$$\underline{N} = \left.\frac{\partial^2 W'_{\rm e}}{\partial v \partial u_0}\right|_{\substack{u_0\to 0\\ v\to v_0}} = \left.\frac{\partial C}{\partial u_0}v\right|_{\substack{u_0\to 0\\ v\to v_0}} = C_0\left(\frac{1}{3d}+\frac{2u_0}{5d^2}\right)v\Bigg|_{\substack{u_0\to 0\\ v\to v_0}} = \frac{C_0}{3d}v_0,$$

$$\underline{Z}'_{\rm m} = \left.\frac{\partial^2 W'_{\rm e}}{\partial u_0^2}\right|_{\substack{u_0\to 0\\ v\to v_0}} = \left.\frac{1}{2}\frac{\partial^2 C}{\partial u_0^2}v^2\right|_{\substack{u_0\to 0\\ v\to v_0}} = \frac{1}{2}C_0\left(\frac{2}{5d^2}\right)v^2\Bigg|_{\substack{u_0\to 0\\ v\to v_0}} = \frac{C_0}{5d^2}v_0^2.$$

Figure 6.28 shows an equivalent circuit representation of the small-signal lumped parameter model.

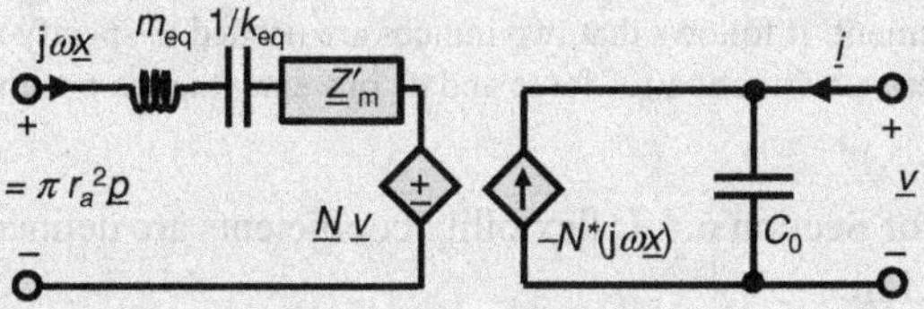

Figure 6.28 An equivalent electromechanical circuit representation of a circular plate subjected to uniform pressure $\underline{p}$. External circuit constraints are not shown.

6.7 MDF modeling of plates

This section describes a way to extend the MDF modeling approach to two-dimensional (2-D) continua, viz., plates. Though in principle any complex geometry can be simplified and reduced to an SDF system, the case of 2-D geometries is not trivial. First, two simple ways to parameterize and discretize a plate are presented. Then, the MDF approach is applied to the important case of a circular plate, where device symmetry permits 1-D discretization.

6.7.1 Uniform discretization of rectangular 2-D plates

For a plate, the forces and displacements depend on two coordinates, that is, $u = u(x, y)$ and $f = f(x, y)$. Thus, specification of the discretized plate for an MDF model requires two indices to identify each element. The conventional way to do this, illustrated in Fig. 6.29, uses separate indices to specify the force and the element where displacement is observed, respectively, (i_1, i_2) and (j_1, j_2). The discretization is assumed to be uniform for a plate having $n_1 \times n_2$ elements. So the indices i_1 and j_1 run from 1 to n_1, and i_2 and j_2 run from 1 to n_2. The indices can be expressed using a *ceiling function*:

$$i_1 = \lceil x_i/n \rceil, i_2 = \lceil y_i/n_2 \rceil, j_1 = \lceil x_j/n_1 \rceil, i_2 = \lceil y_j/n_2 \rceil, \tag{6.110}$$

where the operator $\lceil . \rceil$ is defined as the smallest integer not less than the argument.

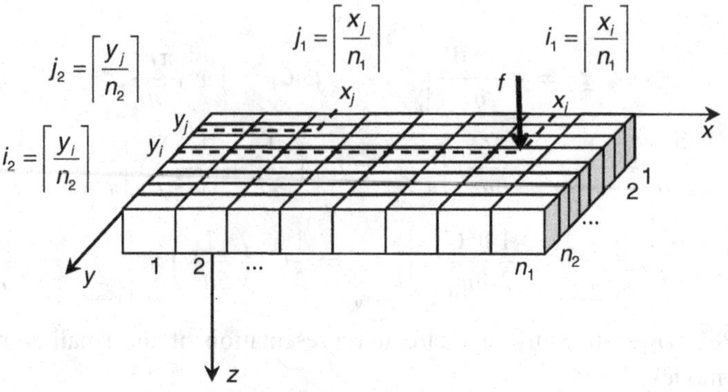

Figure 6.29 Two-dimensional elastic discretization of a plate using spatial coordinates, x and y, to localize force f and displacement. It follows that two indices are needed to specify an element in the discrete model. The reference direction for force and displacement is the positive z direction.

Following the method of Section 6.4.1, flexibility coefficients are defined as the ratio of displacement and unit force:

$$a_{i_1 i_2 j_1 j_2} = \left. \frac{u_{j_1 j_2}}{f_{i_1 i_2}} \right|_{f_{i_1 j_2}=1}. \tag{6.111}$$

The flexibility coefficients for beams have two indices, but for 2-D geometries they require four indices, thereby creating bookkeeping difficulties because there is no simple matrix representation for them.

To secure the advantage of matrix representations of the flexibility coefficients, it is necessary to rearrange the indices and to introduce a scheme that maps 2-D index arrays into 1-D arrays. One way to do this is to concatenate the rows of a 2-D index arrays as follows:

$$i = i_1 + i_2 n_2, \tag{6.112}$$

$$j = j_1 + j_2 n_2. \tag{6.113}$$

This one-to-one mapping converts the 2-D array into a 1-D array. The **mod** function[6] is then employed for the inverse mappings (from 1-D array to 2 array);

$$i_1 = \mathbf{mod}\,(i, n_1) + 1, \tag{6.114}$$

$$i_2 = \left\lceil \frac{i}{n_2} \right\rceil, \tag{6.115}$$

$$j_1 = \mathbf{mod}\,(j, n_1) + 1, \tag{6.116}$$

$$j_2 = \left\lceil \frac{j}{n_2} \right\rceil. \tag{6.117}$$

[6] The **mod**() function gives the remainder of division. For example **mod**(13,7) = 6, **mod**(14,7) = 0, etc.

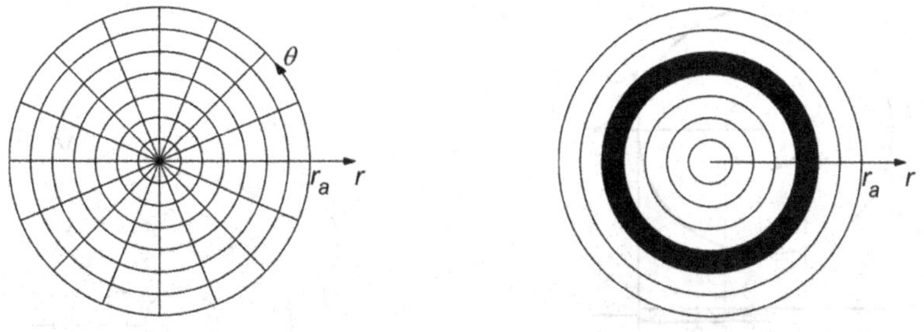

Figure 6.30 Circular geometries are most easily parameterized and discretized using polar coordinates. (a) In the general case, one uses two coordinates: radial distance r and azimuthal angle θ. (b) As long as the plate is uniform, so that the forces acting on it are radially symmetric, that is, independent of θ, the plate can be discretized into a set of concentric rings.

Using the 1-D indices i and j defined by Eqs. (6.112) and (6.113), Eq. (6.111) becomes

$$a_{ij} = \left.\frac{u_j}{f_i}\right|_{f_i=1}, \tag{6.118}$$

so the desired rectangular matrix representation of the flexibility coefficients is achieved.

The same index mapping is used for mass. Each element in Fig. 6.29 has its own mass, and can be addressed either by two indices, according to Eq. (6.110), or by a single index, according to Eq. (6.112). With this 1-D array indexing scheme, the equation of motion takes the familiar form of Eq. (6.66).

6.7.2 2-D discretization of circular plates

Polar coordinates using radius r and azimuthal angle θ are the most convenient choice for describing circular plates. See Fig. 6.30a. The relationship between polar and Cartesian coordinates is

$$r = \sqrt{x^2 + y^2}, \tag{6.119}$$

$$\theta = \arctan\left(\frac{y}{x}\right), \tag{6.120}$$

or

$$x = r\cos(\theta), \tag{6.121}$$

$$y = r\cos(\theta). \tag{6.122}$$

The discretization for an MDF model of a circular plate is usually based on polar coordinates. An important consequence of this choice is that the elements are not uniform in size.

Pressure sensors are probably the most common of all MEMS devices that use the circular-plate structure. Because pressure is uniformly distributed in this application, it is often justified to assume radial symmetry for the elements, and thus to discretize the plate into concentric ring-shaped elements, as shown in Fig. 6.30b.

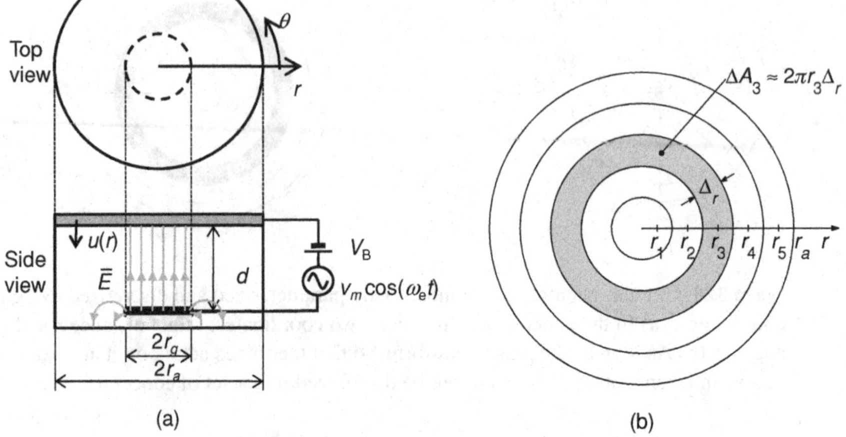

Figure 6.31 A practical circular geometry. (a) Deformable circular plate of radius r_a electrostatically coupled to a smaller, concentric, electrode of radius r_g. (b) Example of plate discretization into $n = 5$ ring-shaped elements.

6.7.3 2-D example: electrostatic actuation of a circular plate

Consider a uniform circular conductive plate mounted a distance d above a circular electrode. As shown in Fig. 6.31a, the radii of the plate and fixed electrode are r_a and r_g, respectively. Assuming that all distributed forces acting on the plate are radially symmetric, then the associated displacements will be independent of θ. Thus, the plate can be discretized as a set of concentric rings, as shown in Fig. 6.31b. The plate is divided into n ring elements having equal radial widths:

$$\Delta_r = r_a/n. \tag{6.123}$$

6.7.3.1 Discrete capacitance model

The device capacitance is the sum of the capacitances of all the displaced, ring-shaped elements:

$$C \approx \varepsilon_0 \sum_{i=1}^{l} \frac{2\pi r_i \Delta_r}{d - u(r_i)}, \tag{6.124}$$

where r_i are discrete (averaged) radii defined in Fig. 6.24b,

$$r_i = \frac{i - 0.5}{n} r_a, \tag{6.125}$$

and l is the number of elements inside radius b,

$$l = \text{round}\left(\frac{r_g}{L} n\right). \tag{6.126}$$

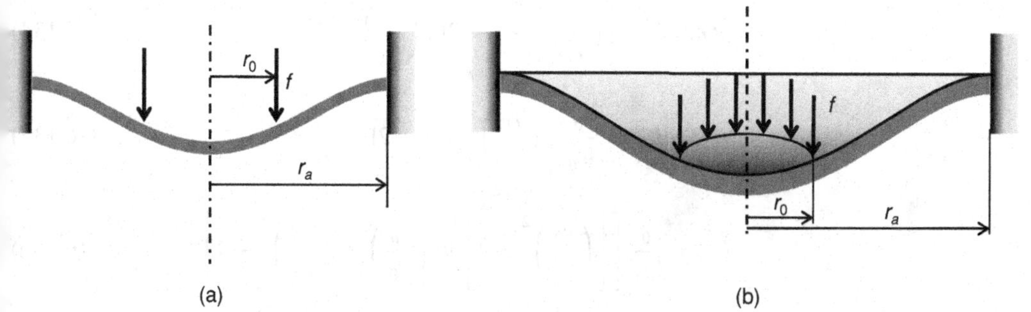

Figure 6.32 Plate deformation due to ring distributed symmetrically around a ring at radius r_0. (a) Thin slice through the middle of a plate, adapted from [8]. (b) Cross-section view.

Inserting Eqs. (6.125) and (6.126) into Eq. (6.124) gives

$$C = \frac{2\varepsilon_0 \pi r_a^2}{n^2} \sum_{i=1}^{\text{round}(nr_g/r_a)} \frac{i - 0.5}{d - u\left((i - 0.5) r_a/n\right)}. \tag{6.127}$$

In the limit of no plate displacements, the equilibrium capacitance is

$$C_0 = 2\varepsilon_0 \frac{r_a^2}{n^2} \frac{1}{d} \frac{[\text{round}(nr_g/r_a)]^2}{2}. \tag{6.128}$$

Unless there is some truncation error due to rounding, i.e., round$(nr_g/r_a) \neq nr_g/r_a$,

$$C_0 = \varepsilon_0 \frac{\pi r_g^2}{d}. \tag{6.129}$$

6.7.3.2 Generalized flexibility matrix

Consider the displacement profile of the plate due to force f on one element as shown in Fig. 6.32. Bear in mind that this force is uniformly distributed around the ring.

A derivation of displacement profiles for circular plates is beyond the scope of this book. Instead, a solution from the literature is used by adaptation of the chosen notation. From Table 11.2 (Case 9b) of the text by Roark and Young [8], the circular plate displacement due to the force f is

$$u(r) = u(0) + \frac{M_c r^2}{2D(1+v)} + LT_y \tag{6.130}$$

where[7]

$$u(0) = \frac{f' r_a^3}{2D} (L_6 - 2L_3), \tag{6.131}$$

$$M_c = -f' r_a (1+v) L_6, \tag{6.132}$$

$$LT_y = \frac{f' r^2}{D} G_6, \tag{6.133}$$

[7] The notation from Roark and Young is modified only where the original notation is in conflict with ours.

$$G_6 = \frac{r_0}{4r}\left[\left(\frac{r_0}{r}\right)^2 - 1 + 2\ln\left(\frac{r}{r_0}\right)\right]\delta^{-1}(r - r_0), \tag{6.134}$$

$$L_6 = \frac{r_0}{4r_a}\left[\left(\frac{r_0}{r_a}\right)^2 - 1 + 2\ln\left(\frac{r_a}{r_0}\right)\right], \tag{6.135}$$

$$L_3 = \frac{r_0}{4r_a}\left[\left[\left(\frac{r_0}{r_a}\right)^2 + 1\right]\ln\left(\frac{r_a}{r_0}\right) + \left(\frac{r_0}{r_a}\right)^2 - 1\right]. \tag{6.136}$$

In Eq. (6.134),

$$\delta^{-1}(z) = \begin{cases} 1, & z \geq 0 \\ 0, & z < 0 \end{cases} \tag{6.137}$$

is the unit step function. Each term on the right-hand side of Eq. (6.130) has f', a force per unit length, as a multiplier. The flexibility coefficients are obtained by setting $f' = 1$ and replacing r and r_0 by their discrete MDF counterparts:

$$r \to \frac{j - 0.5}{n} r_a \quad \text{and} \quad r_0 \to \frac{i - 0.5}{n} r_a. \tag{6.138}$$

Thus,

$$a_{ij} = u(r \to r_a(j - 0.5)/n \quad \text{and} \quad r_0 \to r_a(i - 0.5)/n,\ f' = 1). \tag{6.139}$$

Note that the units of the a_{ij} coefficients are force per unit length.

The mass matrix $\overline{\overline{m}}$ is diagonal, with element values equal to the masses of the corresponding rings:

$$m_i = \underbrace{2\pi r_i \Delta_r}_{\Delta A} h\rho. \tag{6.140}$$

A MATLAB implementation of the stiffness and mass matrices is provided in Fig. 6.33. Note that a particular implementation does not require explicit substitutions $r \to r_a(j - 0.5)/n$ and $r_0 \to r_a(i - 0.5)/n$. In fact, keeping the substitutions implicit often leads to a simpler, more readable, code.

For the case of $n = 5$ rings,

$$\overline{\overline{a}} = 10^{-4} \times \begin{bmatrix} 0.0661 & 0.0494 & 0.0289 & 0.0115 & 0.0014 \\ 0.1481 & 0.1220 & 0.0756 & 0.0310 & 0.0038 \\ 0.1445 & 0.1260 & 0.0889 & 0.0399 & 0.0051 \\ 0.0807 & 0.0724 & 0.0559 & 0.0399 & 0.0051 \\ 0.0125 & 0.0114 & 0.0093 & 0.0060 & 0.0017 \end{bmatrix}. \tag{6.141}$$

The flexibility matrix is not symmetric because the generalized flexibility coefficients are the ratios of displacements to force *per unit length*. Symmetry is restored by scaling the flexibility coefficients to the midline circumferences of the ring elements. Refer to

```
n =5;                                     % number of elements
r_a = 200e-6;       h = 1e-6;              % geometry (radius L and thickness h)
  Y = 150e9;       nu = .3;               % mat. properties
rho = 2330;         D = Y*h^3/12/(1-nu^2); % more mat. properties
  a = zeros(n,n); m = zeros(n,n);          % initiate flexibility and mass matrix
for i = 1:n
     ro = (i-0.5)*r_a/n;
     for j = 1:n
          r = (j-0.5)*r_a/n;
          if r > ro
              G3 = ro/(4*r)*(((ro/r)^2+1)*log(r/ro)+(ro/r)^2-1);
                                          % Roark's formulas, 7th edition p. 457
          else
              G3 = 0;
          end;
          L3 = ro/(4*r_a)*(((ro/r_a)^2+1)*log(r_a/ro)+(ro/r_a)^2-1);
                                          % Roark's formulas, 7th edition p. 457
          L6 = ro/(4*r_a)*((ro/r_a)^2-1+2*log(r_a/ro))  ;
                                          % Roark's formulas, 7th edition p. 457
          uo = r_a^3*(L6-2*L3)/2/D;    ?  % Roark's formulas, 7th edition p. 488
          Mc = -r_a*(1+nu)*L6;            % Roark's formulas, 7th edition p. 488
      a(i,j) = (uo+Mc*r^2/2/D/(1+nu)+r^3/D*G3);
                                          % Roark's formulas, 7th edition p. 487
      %a(i,j) = (uo+Mc*r^2/2/D/(1+nu)+r^3/D*G3)/(2*ro*pi);   % use this expression for
                                                             % symmetric a-matrix
     end;
     m(i,i) = (2*pi*ro)*(r_a/N)*h*rho;
end;
```

Figure 6.33 MATLAB implementation of the MDF model for a circular plate. This code yields an asymmetric flexibility matrix $\overline{a}$. To recover symmetry, use the alternative expression for $a(i, j)$, commented out in the code.

the MATLAB code of Fig. 6.33 for the code needed to obtain the symmetric coefficients. The flexibility matrix rescaled for symmetry is

$$(\overline{a})_{\text{scaled}} = 10^{-2} \times \begin{bmatrix} 5.2600 & 3.9300 & 2.3000 & 0.9200 & 0.1100 \\ 3.9300 & 3.2400 & 2.0100 & 0.8200 & 0.1000 \\ 2.3000 & 2.0100 & 1.4200 & 0.6400 & 0.0800 \\ 0.9200 & 0.8200 & 0.6400 & 0.3500 & 0.0500 \\ 0.1100 & 0.1000 & 0.0800 & 0.0500 & 0.0100 \end{bmatrix}. \quad (6.142)$$

The symmetric form of the flexibility matrix is much more convenient and will be used from this point on.

6.7.3.3 Static displacements

Pressure sensors and microphones are by far the most common MEMS applications for plate structures. For these cases, with pressure difference as input or output, respectively, the external force is uniformly applied to the plate.

The flexibility matrix and the force vector are needed to determine the static displacement of the circular plate. Thus, consider a circular plate transducer operating as a microphone, converting an electrical signal into sound. The distributed electromechanical force depends on radius r because of the r dependence of displacement: $u(r)$. An

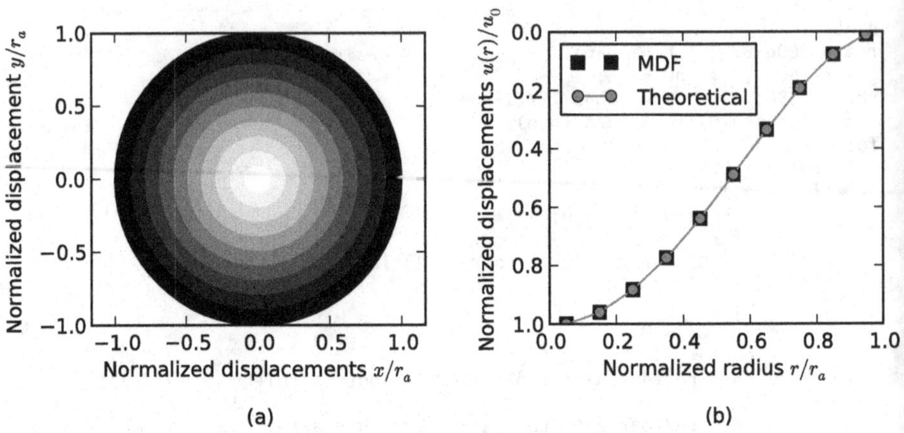

Figure 6.34 Deformation of circular Si plate, with radius $r_a = 200$ μm and thickness $h = 1.0$ μm discretized into $n = 10$ rings of equal Δr. The electrode separation is $d = 2.0$ μm and the applied voltage is $V = 5$ V. The resulting displacement at the center of the plate is $u_0 = 48$ nm. (a) Contour plot of circular plate displacements. (b) Comparison of normalized displacement profile, $u(r)$: data obtained from MDF model with $n = 10$ elements and analytical prediction from Eq. (6.98).

iterative approach can be employed here. The first step is to formulate the electrical force terms for zero plate displacement:

$$f_f^{(0)} = \begin{cases} \varepsilon_0 \dfrac{v^2 r_a}{d^2 n} (2\pi r_0), & i \le n_0, \\ 0, & i > n_0, \end{cases} \tag{6.143}$$

where n_0 is the degree of freedom corresponding to the edge of the electrode. Refer to Fig. 6.31 for the geometry. The displacements are

$$\overline{u}^{(1)} = \overline{\overline{a}}\, \overline{f}^{(1)}. \tag{6.144}$$

Values obtained from Eq. (6.144) are then used to revise the force vector to account for the reduced gap to recalculate the displacement vector. For the qth iteration, the force and displacement vectors are

$$f_i^{(q)} = \begin{cases} \varepsilon_0 \dfrac{2\pi r_a^2}{n^2} \dfrac{(i - 0.5)}{\left(d - u_i^{(q-1)}\right)^2} v^2, & i \le n_0, \\ 0, & i > n_0, \end{cases} \tag{6.145}$$

$$\overline{u}^{(q)} = \overline{\overline{a}}\, \overline{f}^{(q)}. \tag{6.146}$$

This iterative approach is identical to that described in Section 6.4.4.2 for cantilevered beams.

When this or any other numerical model is first implemented, testing and verification are crucial. The best way to achieve such verification is to rely on existing closed-form solutions. Even if they are applicable only in certain limits, they lend confidence to the exercise. For the circular plate with a clamped boundary subjected to uniform pressure

P, Eq. (6.98) serves the purpose. In the first iteration, assuming zero displacements, the effective pressure P due to electrostatic actuation is

$$P = \varepsilon_0 v^2 / 2d^2. \tag{6.147}$$

The MDF model and the electromechanical force distribution are illustrated in Fig. 6.34a using $n = 10$ ring-shaped elements. The plate displacements are depicted as contours. Figure 6.34b compares the analytical model with the computed shape. The error at the center, $r = 0$, is ~0.3%.

The parameters used for this example lead to predicted displacements sufficiently small that iteration is not needed. Were this not the case, the iterative method described previously would be used.

6.7.3.4 Modal analysis

Modal analysis of 2-D systems is performed in the same way as for 1-D systems. Using such programming environments as MATLAB, it is only necessary to call the **eig** function,

$$\lfloor \overline{\overline{\Phi}}, \overline{\overline{W}} \rfloor = \text{eig}(\overline{\overline{a}}\, \overline{\overline{m}}), \tag{6.148}$$

and then to compute the resonant frequencies as before, using

$$\omega_i = \frac{1}{\sqrt{W_{ii}}}. \tag{6.149}$$

The previous analysis assumed azimuthal symmetry. It is this θ symmetry that facilitates 1-D discretization (in the r-direction). Figure 6.35 plots normalized radial displacement versus r for the three lowest modes obtained from a MATLAB routine.

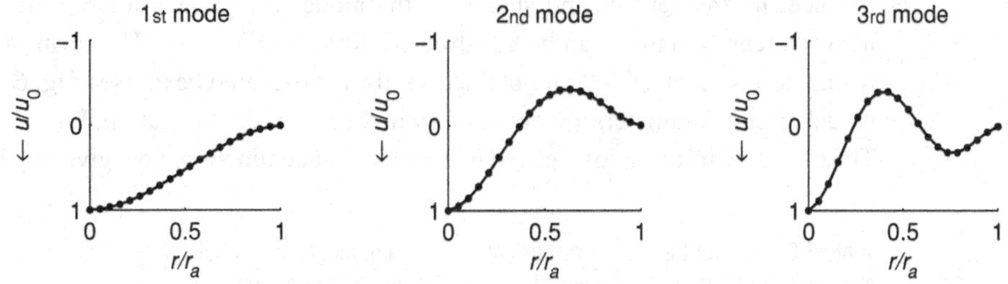

Figure 6.35 The first three mode shapes: normalized displacement u/u_0 versus the normalized radius r/r_a. The discretization level is $n = 20$ DOF.

Just as for the cantilevered beam, we can compare the MDF model results with the existing analytical solution. Derivation of the plate resonances and mode shapes is beyond the scope of this text. Instead, we call upon published solutions. The characteristic equation for the azimuthally symmetric modes is [7, 9]

$$I_1(\beta_k r_a) J_0(\beta_k r_a) + J_1(\beta_k r_a) I_0(\beta_k r_a) = 0. \tag{6.150}$$

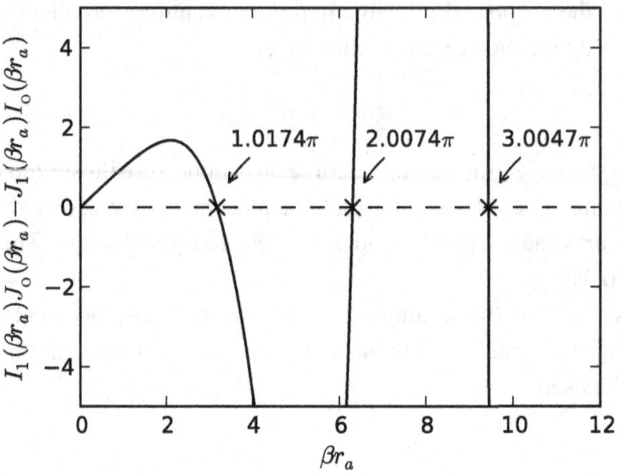

Figure 6.36 Zeros of the characteristic equation, Eq. (6.150), for a circular plate of radius r_a attached all around its periphery.

As with Eq. (6.73) for the resonant modes of the cantilevered beam, the zeros of this characteristic equation are found numerically. The first three are $\beta r_a = 1.0174\pi$, 2.0074π, and 3.0047π. Refer to Fig. 6.36.

The resonant frequencies are proportional to the squares of the zeros of the characteristic equation:

$$\omega_n = \beta_n^2 \sqrt{D/\rho h}. \tag{6.151}$$

Table 6.2 compares the MDF model predictions (with $n = 40$ degrees of freedom) with the analytical solutions given by Eq. (6.151).

Once the frequencies and shapes of the modes are known, a simplified equivalent network representation can be constructed. Refer to Fig. 6.37. The form of this circuit realization is identical to that obtained for the cantilevered beam (see Fig. 6.21), and the modal circuit parameters are obtained from the same set of equation, Eqs. (6.86)–(6.88). The only new information needed is the capacitance function, now given by Eq. (6.127).

Table 6.2 The first three resonant frequencies of a homogenous circular plate calculated using the analytical solution, Eq. (6.151), and then independently using the MDF model with 40 degrees of freedom. The plate parameters are radius $r_a = 200$ μm and height $h = 1$ μm and the material properties of Si are $\rho = 2330$ kg/m^3, $Y = 160$ GPa, and $\nu = 0.3$

Mode	Analytical solution	MDF model
f_1 [kHz]	98.70	98.68
f_2 [kHz]	384.22	384.13
f_3 [kHz]	860.83	860.49

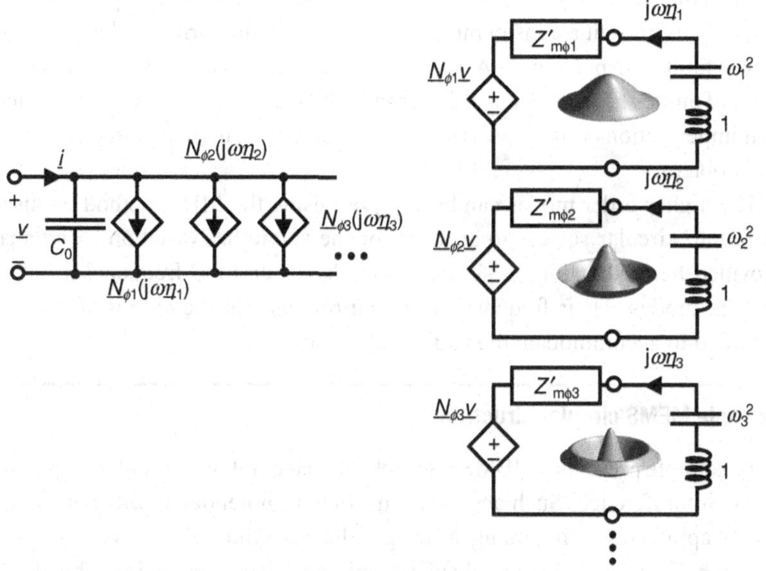

Figure 6.37 Equivalent electromechanical network representation of a circular plate, accounting for the first three mechanical resonances. The circuit parameters are obtained using Eqs. (6.86), (6.87), and (6.88), plus the capacitance function, Eq. (6.127).

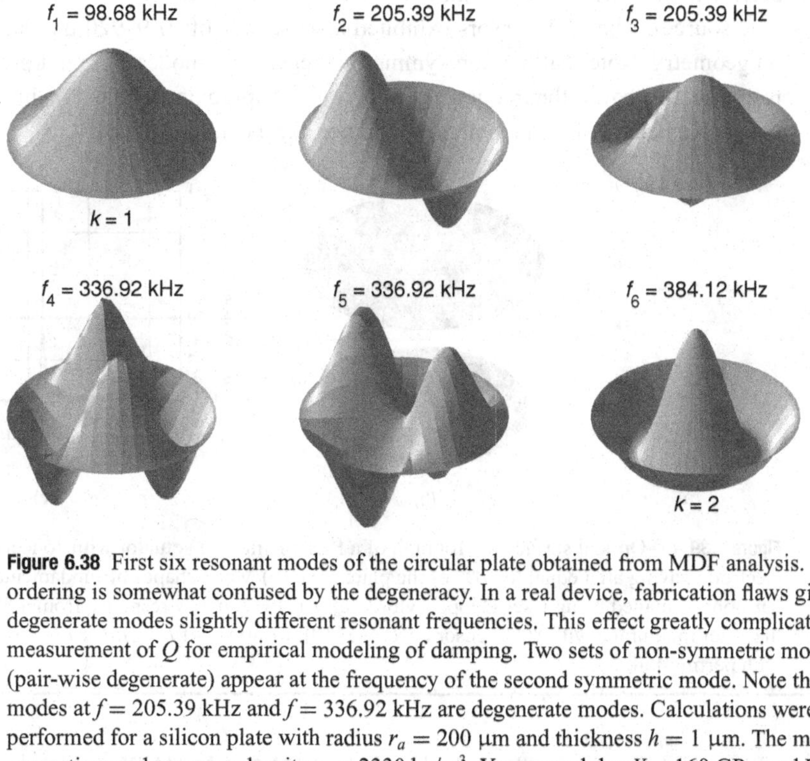

Figure 6.38 First six resonant modes of the circular plate obtained from MDF analysis. The ordering is somewhat confused by the degeneracy. In a real device, fabrication flaws give the degenerate modes slightly different resonant frequencies. This effect greatly complicates the measurement of Q for empirical modeling of damping. Two sets of non-symmetric modes (pair-wise degenerate) appear at the frequency of the second symmetric mode. Note that the modes at $f = 205.39$ kHz and $f = 336.92$ kHz are degenerate modes. Calculations were performed for a silicon plate with radius $r_a = 200$ μm and thickness $h = 1$ μm. The material properties used are mass density $\rho = 2330$ kg/m^3, Young modulus $Y = 160$ GPa, and Poisson ratio $\nu = 0.3$. The plate is discretized with $n = 40$ degrees of freedom in radial direction.

Up to this point, only radially symmetric modes of the circular plate have been considered. Plates present a considerable complication in the form of addition resonant modes that exhibit θ-dependence. A problem arises here because some of these previously ignored modes have resonances lower than the second ($k = 2$) symmetric mode. Fabrication imperfections often give rise to unintended structural asymmetries that exacerbate the problem.

The higher-order modes can be analyzed using the MDF method by discretizing the plate into circular sectors to account for the azimuthal variations. The analysis is not provided here,[8] but Fig. 6.38 shows the shapes and the frequencies obtained for the first six modes. Their frequency ordering reveals that the circuit of Fig. 6.37 must be modified to accommodate the additional modes.

Example 6.9 **Example MEMS circular structure**

This example presents a MEMS microphone that employs a circular diaphragm clamped at its circumference. Such acoustic sensors have potential applications in a wide variety of applications, including hearing aids, surveillance, and heart monitoring [10]. Figure 6.39a shows the optical surface micrograph of the device. This device is based on piezoelectric transduction (refer to Chapter 8) using $PbZr_{0.52}Ti_{0.48}O_3$ (PZT). The device was fabricated on a silicon wafer by silicon deep reactive ion etching (DRIE). Sensors ranging from 500 to 2000 microns in diameter have been fabricated and characterized with the use of scanning laser Doppler vibrometry and calibrated acoustic tone sources. The PZT sensors exhibited a sensitivity of 97.9–920 nV/Pa, depending on geometry. Note that the non-symmetric, degenerate modes were not present in the response. Moreover, the frequencies were ordered approximately as $1/r$, indicating that the device behaves like a membrane, and not like a clamped plate.

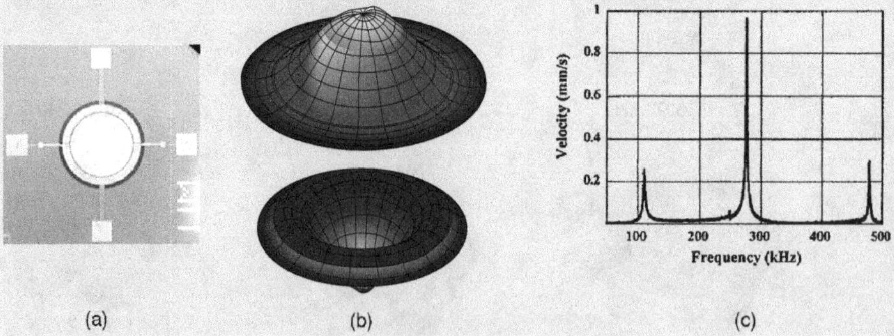

Figure 6.39 (a) Optical surface micrograph of a PZT membrane actuator with concentric electrode having area equal to 80% of the plate area. (b) Mode shapes of fundamental and third harmonics imaged using laser Doppler vibrometry. (c) Frequency response from a 500-μm diameter membrane with 80% sensor coverage [10]. Courtesy of *Integrated Electronics*, used with permission.

[8] Interested readers are referred to the book's website for a MATLAB-based MDF model.

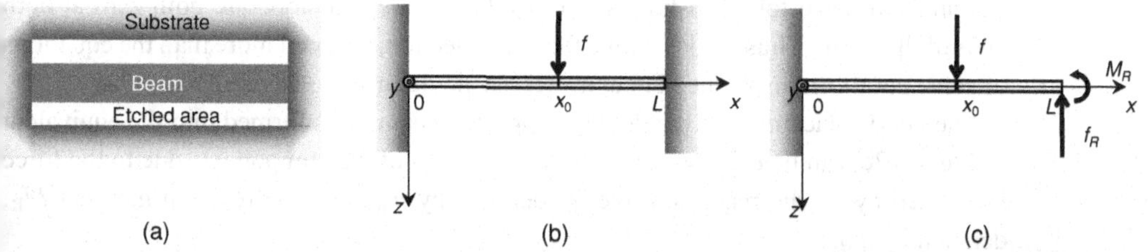

Figure 6.40 Double-clamped beam. (a) Top view of a MEMS structure. (b) Side view of the beam of length L acted upon by a concentrated force at x_0. (b) Side view showing the equivalent cantilever with right-hand-side clamp replaced by equivalent reaction force and reaction moment.

*6.8 Additional beam configurations

While the cantilevered beam is undoubtedly the most prevalent, other beam configurations are also found in MEMS devices. Two of these are the double-clamped beam and the beam with simple supports. Double-supported beams are commonly used when there is a possibility of stiction. Simply supported beams, which feature special placement of the supports, are used to achieve good vibration isolation for high-Q resonators and to suppress noise.

The MDF models for these structures require us to introduce some new concepts from classical beam mechanics. For the *doubly clamped beam*, the notion of statically indeterminate systems is introduced. For the *simply supported beam*, the MDF model requires use of a piecewise linear and continuous bending moment.

6.8.1 Doubly clamped beam

Figure 6.40a shows the top view of a beam clamped at both ends. The flexibility coefficients for this configuration are derived using essentially the same method used for the cantilevered beam, that is, imposing a normal force at arbitrary position x_0 and then solving for the displacement. Refer to Fig. 6.40b. The boundary condition at $x = L$ is represented by the superposition of an equivalent reaction force f_R (that maintains zero displacement at $x = L$), and an equivalent moment M_R (that fixes the slope of the beam to zero). See Fig. 6.40c. Using this method, the doubly clamped beam problem effectively reduces to a cantilevered beam with f_R and M_R acting at the free end in addition to f acting at x_0. Note that f_R and M_R are functions of geometry and linearly proportional to the externally applied force f.

The doubly clamped beam is an example of a *statically indeterminate* structure. It gains this characterization because it has more supports, two more in this case, than are needed to maintain the beam in static equilibrium. The simplest way to appreciate this fact is to recognize that, even if the clamp at the right end is entirely removed, so that f_R and M_R go to zero, the beam retains an equilibrium. Alternatively, simple supports, one at each end, can be used. In this case, the beam shape has non-zero slopes at both ends, but it still displays equilibrium. By definition, a statically indeterminate beam has more than the minimum number of supports needed to achieve equilibrium. In the present example, we have four constraints: the displacements and slope are both zero at both ends of the beam. Thus, to determine the net deflection, we need more than the equations of static equilibrium. We need to consider elastic deformation as well.

The net displacement of the doubly supported beam, transformed into the equivalent of Fig. 6.32c, can be expressed as the superposition of the components due to the force f, denoted by u_f, the reaction force f_R, denoted by u_{f_R}, and the reaction moment M_R, denoted by u_{M_R}:

$$u(x) = u_f(x) + u_{f_R}(x) + u_{M_R}(x). \tag{6.152}$$

Superposition is used to evaluate the terms of Eq. (6.152). The first term, $u_f(x)$, due to the force f, has been encountered already in this chapter. Using the definitions from Fig. 6.32 in Eq. (6.7) yields

$$u_f(x) = \frac{f}{6YI}\begin{cases} x^2(3x_0 - x), & 0 \le x \le x_0, \\ x_0^2(3x - x_0), & x_0 < x \le L. \end{cases} \tag{6.153}$$

The response due to f_R is a special case of Eq. (6.153) with $x_0 \to L$:

$$u_{f_R}(x) = \frac{f_R}{6YI}x^2(3L - x). \tag{6.154}$$

The final term, due to the moment M_R acting at the free end of the beam, is obtained by twice integrating the beam equation, Eq. (6.1), subject to the boundary conditions:

$$u_{M_R}(x) = \frac{M_R}{2YI}x^2. \tag{6.155}$$

Combining Eqs. (6.152)–(6.155) and then imposing the boundary conditions at $x = \mathrm{L}$,

$$u(x = L) = 0 \quad \text{and} \quad \left.\frac{\mathrm{d}u}{\mathrm{d}x}\right|_{x=L} = 0, \tag{6.156}$$

provides the necessary relations to solve for f_R and M_R in terms of f:

$$f_R = \frac{f}{L^3}\left(3x_0^2 L - 2x_0^3\right), \tag{6.157}$$

$$M_R = \frac{f}{L^2}\left(x_0^3 - Lx_0^2\right). \tag{6.158}$$

The displacement function can now be expressed in terms of the applied force f. It is only necessary to obtain the solution form in the region $0 \le x \le x_0$, because the flexibility coefficients for the remainder of the beam, $x_0 \le x \le L$, can be obtained from symmetry. Thus,

$$u(x) = \frac{fx^2}{6YI}\left[(3x_0 - x) - \frac{(3L - x)\left(3x_0^2 L - 2x_0^3\right)}{L^3} - 3\frac{x_0^3 - Lx_0^2}{L^2}\right], x \le x_0. \tag{6.159}$$

```
function doublyClampedBeamMDF

    g = setGeometry;
   mp = setMatProp;
    n = 10; % discretization level (mesh)
    b = 1e-6;
    meff = (33/140+1/70*b/g.L+3*g.L*1/4/(g.L-b)^2)*mp.rho*g.L*g.w*g.h;
    keff = mp.E*g.w*g.h^3/4/(g.L-b);
    wAp = sqrt(keff/meff)/2/pi;
    % compute mass matrix
    M = eye(n)*mp.rho*g.L*g.h*g.w/(n-1);
    %compute the flexibility matrix A (A = inv(K))
    A = zeros(n,n);
    for j = 0:n-1
        for i = 0:n-1
           if i >= j
               A(i+1,j+1) = g.L^3/(6*mp.E*g.Ix*(n-1)^3)*j^2*...
                       ((3*i-j)-...
                        (3*(n-1)-j)*(3*i^2*(n-1)-2*i^3)/(n-1)^3-...
                        3*(i^3-(n-1)*i^2)/(n-1)^2);
            else
                A(i+1,j+1) = A(j+1,i+1);
            end;
        end;
    end;
    % solution: Modal analysis
    [v,w] = eig(A*M);
    visModal(diag(w), v,n);
return
```

Figure 6.41 An MDF implementation for a doubly clamped beam in MATLAB. For clarity, previously defined functions are suppressed in this listing.

The solution for $x_0 < x < L$ is obtained from Eq. (6.159) by substituting $x_0 \to L - x_0$ and $x \to L - x$. For the MDF model, the beam is discretized into n points, such that $x \to jL/n$, $x_0 \to iL/n$. The result for the flexibility coefficient is then

$$a_{ij} \equiv \frac{u}{f} = \frac{fL^3}{6YI}\frac{j^2}{n^3}\left[(3i - j) - \frac{(3n - j)(3i^2 n - 2i^3)}{n^3} - 3\frac{i^3 - ni^2}{n^2}\right], i \geq j. \quad (6.160)$$

The MATLAB routine for this is given in Fig. 6.41.

Results of modal analyses of a doubly clamped beam discretized into $n = 10$ and $n = 40$ elements are shown in Fig. 6.42. The geometry and material properties of the beam are the same as those used for the cantilevered beam previously considered. Refer to Fig. 6.20.

Resonant frequency values obtained from MDF analyses of the cantilevered beam and the circular diaphragm compare reasonably well to analytical models. The same is found for the doubly clamped beam. Recall the very general expression for the modal resonances of uniform beams, Eq. (6.72). In this equation, only values for β_k are

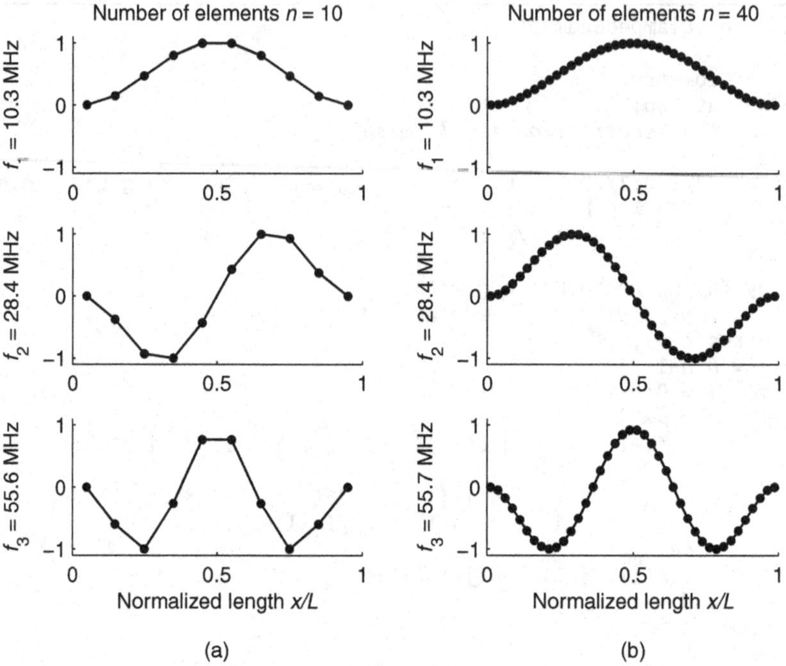

Figure 6.42 First three resonant modes of a doubly clamped beam. (a) $n = 10$ DOF (b) $n = 40$ DOF While the mode shapes become smoother with more DOF, resonant frequencies are not significantly affected. Notice also that the number of nodes is one less than the mode order (first mode has zero nodes, second mode has one, etc.). The dimensions of the Si beam are $L = 40\ \mu\text{m}$, $w = 4\ \mu\text{m}$, $h = 2\ \mu\text{m}$.

needed to calculate the frequencies. The characteristic equation for the doubly clamped configuration is

$$1 - \cosh(\beta_k L)\cos(\beta_k L) = 0. \tag{6.161}$$

Figure 6.43 reveals the first three zeros of this equation. Using these in Eq. (6.72) gives analytic predictions for the resonance frequencies. These values are tabulated in Table 6.3, along with results obtained with the MDF model for $n = 10$ and $n = 40$ elements. The MDF values agree closely with the analytical values.

Table 6.3 Analytical predictions and MDF modeling calculations for first three resonances of a $40 \times 4 \times 2\ \mu\text{m}$ homogeneous, doubly clamped Si beam. The three modes are bending in the direction of the smallest beam dimension

	Analytical solution	MDF $n = 10$	MDF $n = 40$
f_1 [MHz]	10.3	10.3	10.3
f_2 [MHz]	28.4	28.4	28.4
f_3 [MHz]	55.7	55.6	55.7

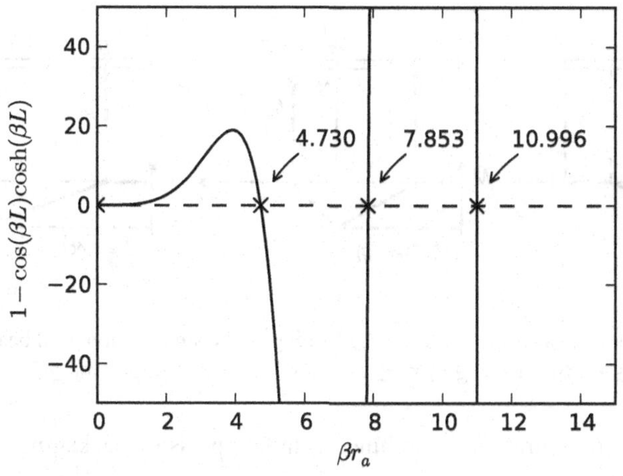

Figure 6.43 Plot of transcendental characteristic equation for the doubly clamped beam of Eq. (6.161), showing its first three zeros.

6.8.2 Simply supported beam

Another beam structure important in MEMS is the simply supported configuration depicted in Fig. 6.44a. The side view in Fig. 6.44b shows the supports and an applied normal force f at $x = x_0$. The advantage of this structure is that, if the locations of the supports are properly chosen, the beam becomes mechanically isolated from the substrate, at least when vibrating at its lowest resonance. Mechanical isolation is desirable in a resonator because it greatly reduces susceptibility to environmental mechanical

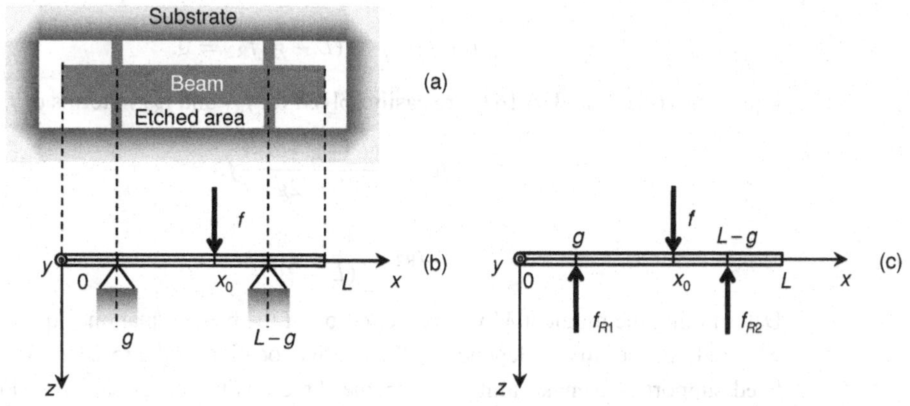

Figure 6.44 Simply supported beam. (a) Top view of bulk micromachined beam of length L. (b) Standard schematic of the simply supported beam acted on by a concentrated force at x_0. (c) Schematic with supports replaced by equivalent reaction forces.

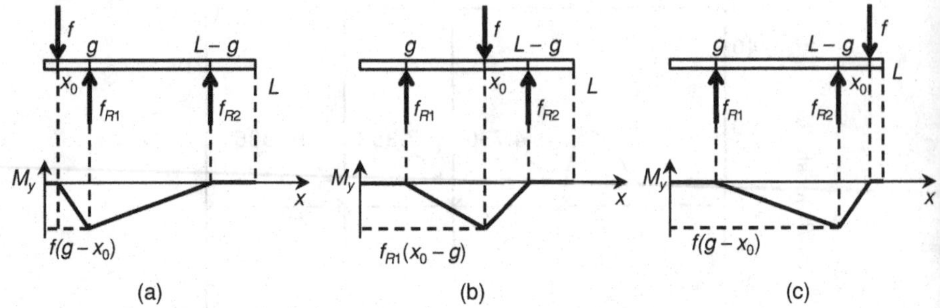

Figure 6.45 Three distinct cases for normal force loading of the simply supported beam. (a) Case 1: $0 < x_0 \leq g$. (b) Case 2: $g \leq x_0 \leq L - g$. (c) Case 3: $L - g \leq x_0 \leq L$.

noise.[9] When this configuration is combined with low-pressure packaging, the result is a high-Q resonator.

Modal analysis of a free beam reveals that the third mode is symmetric and possesses two nodes. Thus, if the beam is supported at these nodes, it is as if the beam is vibrating in free space without any supports at all. This approximation provides a convenient way to analyze the simply supported beam. Such behavior is similar in all important respects to a tuning fork, which is held at the node while the two tines vibrate out of phase. When struck, the tuning fork vibrates for tens of seconds, showing that clamping damping has been virtually eliminated. If that were not so, one would feel the vibrations.

The MDF modeling of the simply supported beam starts in the same way as the double-clamped beam, namely, by replacing the supports with equivalent reactions. Here, the system is not statically indeterminate, so the reaction forces f_{R1} and f_{R2} can be obtained from the static equilibrium. One relation comes from static force equilibrium in the z direction,

$$f - f_{\mathrm{R1}} - f_{\mathrm{R2}} = 0, \tag{6.162}$$

and the second is obtained from the moment about $x = 0$ in the positive y direction,

$$g f_{\mathrm{R1}} - x_0 f + (L - g) f_{\mathrm{R2}} = 0. \tag{6.163}$$

Equations (6.162) and (6.163) are easily solved for f_{R1} and f_{R2} in terms of f:

$$f_{\mathrm{R1}} = \frac{(L - g - x_0)}{(L - 2g)} f, \tag{6.164}$$

$$f_{\mathrm{R2}} = \frac{(x_0 - g)}{(L - 2g)} f. \tag{6.165}$$

Determining the moment $M(x)$ for integration of the beam equation, Eq. (6.1), is somewhat tedious, because it depends on the relative location of the force with respect to the fixed supports. To make things easier, the three distinct cases shown in Fig. 6.45 are defined.

9 Vibration isolation is crucial in the vibrational MEMS gyroscope. Refer to Section 7.4.

The moment equations for the three cases follow directly from inspection of the three figures. For case 1, $0 < x_0 \le g$, shown in Fig. 6.45a:

$$M = \begin{cases} 0, & 0 \le x \le x_0, \\ -f(x - x_0), & x_0 \le x \le g, \\ -f_{R2}(L - g - x), & g \le x \le L - g, \\ 0, & L - g \le x \le L. \end{cases} \tag{6.166}$$

For case 2, $g \le x_0 \le L - g$, shown in Fig. 6.45b:

$$M = \begin{cases} 0, & 0 \le x \le g, \\ -f_{R1}(x - g), & g \le x \le x_0, \\ -f_{R2}(L - g - x), & x_0 \le x \le L - g, \\ 0, & L - g \le x \le L. \end{cases} \tag{6.167}$$

For case 3, $L - g \le x_0 \le L$, shown in Fig. 6.45c:

$$M = \begin{cases} 0, & 0 \le x \le g, \\ -f_{R1}(x - g), & g \le x \le L - g, \\ -f_{R2}(x - L + g), & L - g \le x \le x_0, \\ 0, & x_0 \le x \le L. \end{cases} \tag{6.168}$$

As in the previous examples, the flexibility coefficients are obtained by the following steps:

- Integrate the moment equations twice.
- Use the boundary and continuity equations to determine the unknown integration constants.
- Divide the displacements by the applied force f.

For case 1, where $0 < x_0 \le g$, two integrations of Eq. (6.166) yield

$$u = \begin{cases} c_1 + c_2 x, & 0 \le x \le x_0, \\ \dfrac{fx^2}{6YI}(x - 3x_0) + c_3 + c_4 x, & x_0 \le x \le g, \\ -\dfrac{f_{R2}x^2}{6YI}[3(L - g) - x] + c_5 + c_6 x, & g \le x \le L - g, \\ c_7 + c_8 x, & L - g \le x \le L. \end{cases} \tag{6.169}$$

The procedure of using the boundary conditions to obtain the eight coefficients, while straightforward, is quite messy. Further, no attractive simplified expression awaits us at the other end of the manipulation. Thus, it is best to leave some of the expressions in implicit form, keeping in mind that analytic results in implicit form are easy to code.

There are three boundaries within the domain: $x = x_0, x = g$, and $x = L - g$. At each of these, there are two continuity equations, one for the displacement and one for the slope. Two additional equations arise from the constraints imposed at $x = g$ and $x = L - g$. The system of eight equations in terms of the eight constants of integration is conveniently written in matrix form:

$$\begin{bmatrix} 1 & x_0 & -1 & -x_0 & 0 & 0 & 0 & 0 \\ 0 & 1 & 0 & -1 & 0 & 0 & 0 & 0 \\ 0 & 0 & 1 & g & -1 & -g & 0 & 0 \\ 0 & 0 & 0 & 1 & 0 & -1 & 0 & 0 \\ 0 & 0 & 0 & 0 & 1 & L-g & -1 & -L+g \\ 0 & 0 & 0 & 0 & 0 & 1 & 0 & -1 \\ 0 & 0 & 1 & g & 0 & 0 & 0 & 0 \\ 0 & 0 & 0 & 0 & 0 & 0 & 1 & L-g \end{bmatrix} \begin{bmatrix} c_1 \\ c_2 \\ c_3 \\ c_4 \\ c_5 \\ c_6 \\ c_7 \\ c_8 \end{bmatrix}$$
$$= \begin{bmatrix} -fx_0^3/(3YI) \\ -fx_0^2/(2YI) \\ -g^2\left[f(g-3x_0) + f_{R2}(3L-4g)\right]/(6YI) \\ -g[f(g-2x_0) - f_{R2}(3g-2L)]/(2YI) \\ f_{R2}(L-g)^3/(3YI) \\ f_{R2}(L-g)^2/(2YI) \\ -fg^2(g-3x_0)/(6YI) \\ 0 \end{bmatrix}. \quad (6.170)$$

The left-hand side of Eq. (6.170) is written by inspection; the right-hand side requires a bit of algebra. Note that the first two rows are the displacement and slope continuity equations at $x = x_0$, respectively; the third and fourth rows are displacement and slope continuity equations at $x = g$, and so forth. The last two rows are the constraints imposed at $x = g$ and $x = L - g$.

Implementation of an MDF solution in a numerical computational environment, such as MATLAB, makes it unnecessary to solve Eq. (6.170) for the constants or even to substitute them into Eq. (6.169). Instead, we leave the matrix inversions and the substitutions to the computer.

For case 2, where the force is applied between the two supports, that is, $g \le x_0 \le L - g$, double integration of Eq. (6.167) yields

$$u = \begin{cases} c_1' + c_2' x, & 0 \le x \le g, \\ -\dfrac{f_{R1}x^2}{6YI}(x - 3g) + c_3' + c_4' x, & g \le x \le x_0, \\ -\dfrac{f_{R2}x^2}{6YI}[3(L-g) - x] + c_5' + c_6' x, & x_0 \le x \le L - g, \\ c_7' + c_8' x, & L - g \le x \le L. \end{cases} \quad (6.171)$$

The same procedure already outlined for case 1 works for Eq. (6.171) and its eight unknown integration constants. The resulting matrix is

$$\begin{bmatrix} 1 & g & -1 & -g & 0 & 0 & 0 & 0 \\ 0 & 1 & 0 & -1 & 0 & 0 & 0 & 0 \\ 0 & 0 & 1 & x_0 & -1 & -x_0 & 0 & 0 \\ 0 & 0 & 0 & 1 & 0 & -1 & 0 & 0 \\ 0 & 0 & 0 & 0 & 1 & L-g & -1 & -L+g \\ 0 & 0 & 0 & 0 & 0 & 1 & 0 & -1 \\ 1 & g & 0 & 0 & 0 & 0 & 0 & 0 \\ 0 & 0 & 0 & 0 & 0 & 0 & 1 & L-g \end{bmatrix} \begin{bmatrix} c_1' \\ c_2' \\ c_3' \\ c_4' \\ c_5' \\ c_6' \\ c_7' \\ c_8' \end{bmatrix} = \begin{bmatrix} f_{R1} g^3/(3YI) \\ f_{R1} g^2/(2YI) \\ -x_0^2[f_{R1}(3g - x_0) - f_{R2}(3g - 3L + x_0)]/(6YI) \\ -x_0[f_{R1}(2g - x_0) + f_{R2}(2L - x_0 - 2g)]/(2YI) \\ f_{R2}(L-g)^3/(3YI) \\ f_{R2}(L-g)^2/(2YI) \\ 0 \\ 0 \end{bmatrix}. \quad (6.172)$$

From symmetry considerations, case 3, with the force applied at $L - g \leq x_0 \leq L$, need not be considered separately. Taking advantage of the left–right symmetry of the boundary conditions and the symmetry of the flexibility matrix about the secondary diagonal[10] yields

$$a_{i,j} = a_{n+1-i,n+1-j}. \quad (6.173)$$

An algorithm for computing the flexibility coefficients of the simply supported beam is provided in Fig. 6.46 in the form of a pseudo code.[11] Fig. 6.47 shows the three lowest mode shapes and natural frequencies for a $40 \times 4 \times 2$ μm Si beam supported at $0.1L$ and $0.9L$. Results are given for two levels of discretization: $n = 10$ and $n = 40$. Again, it is found that the $n = 10$ model describes the first three modes fairly well. Because the computational complexity of these MATLAB codes is actually low, there is no real cost incurred in using $n \geq 40$ elements to assure accuracy.

6.8.3 Vibration isolation of the simply supported beam

The motivation for considering the simply supported beam is the vibration isolation achieved when a beam is supported at the resonance nodes. The model presented in Section 6.8.2 can be readily modified to investigate the effectiveness of this type of

[10] The secondary diagonal consists of the matrix elements located along the line that runs from the upper-right corner to the lower-left corner of the matrix, i.e., $a_{i,N+1-i}$ for $i = 1, \ldots, N$.

[11] The complete MATLAB code is available for download from the book's website.

```
for i = 0:n-1
    if i < round(g/L*(n-1))
        Use Eq. (6.170) to compute coefficients c with xo = L*i/(n-1)
    else
        Use Eq. (6.172) to compute coeffients c' with xo = L*i/(n-1)
    end;
```

$$f_{R2m} = \frac{\frac{i}{n-1}L - g}{L - 2g} \quad \text{\% fR2m is } f_{R2} \text{ multiplier, i.e. } f_{R2} = \text{fR2m*f}$$

```
    for j = 0:(N-1)
        if i ≤ j
            if i ≤ round(g/L*(n-1))
```

$$a_{i+1,j+1} = \begin{cases} \left(\frac{L}{n-1}\right)^3 \frac{j^2(j-3i)}{6YI} + c_3 + c_4\left(\frac{j}{n-1}L\right) & , \quad j < \text{round}\left(\frac{g}{L}(n-1)\right) \\ -f_{R2M}\left(\frac{j}{n-1}L\right)^2 \frac{3(L-g) - \frac{j}{n-1}L}{6YI} + c_5 + c_6\left(\frac{j}{N-1}L\right) & , \quad \text{round}\left(\frac{g}{L}(n-1)\right) \le j < \text{round}\left(\frac{L-g}{L}(n-1)\right) \\ c_7 + c_8\left(\frac{j}{n-1}L\right) & , \quad \text{round}\left(\frac{L-g}{L}(n-1)\right) \le j \end{cases}$$

```
            elseif round(g/L*(n-1))≤ i ≤ round((L-g)/L*(n-1))
```

$$a_{i+1,j+1} = \begin{cases} -f_{R2m}\left(\frac{j}{n-1}L\right)^2 \frac{3(L-g) - \frac{j}{n-1}L}{6YI} + c'_5 + c'_6\left(\frac{j}{n-1}L\right) & , \quad j < \text{round}\left(\frac{L-g}{L}(n-1)\right) \\ c'_7 + c'_8\left(\frac{j}{n-1}L\right) & , \quad \text{round}\left(\frac{L-g}{L}(n-1)\right) \le j \end{cases}$$

```
            else
```

$$a_{i+1,j+1} = a_{n-i,n-j} \quad \text{\% symmetry of the boundary conditions, about sec. diagonal}$$

```
            end;
        else
```

$$a_{i+1,j+1} = a_{j+1,i+1} \quad \text{\% symmetry about the main diagonal}$$

```
        end;
    end;
end;
```

Figure 6.46 A MATLAB-based pseudo code of MDF implementation of a 40 × 4 × 2 μm simply supported Si beam. The book's website contains the listing of the complete MATLAB code.

isolation. The only change that has to be made is to replace the fixed boundary conditions at $x = g$ and $x = L - g$ by a set of springs, as shown in Fig. 6.48. This modification means that instead of forcing the displacement to zero at these points, the beam is attached to two identical springs having spring constant k:

$$f_{R1} = ku(x = g), \tag{6.174}$$

$$f_{R2} = ku(x = L - g), \tag{6.175}$$

where f_{R1} and f_{R2} are the reaction forces now due to the springs.

These conditions alter only the last two rows of the matrix of Eq. (6.172), so as to reflect the influence of the springs. Thus, this rather simple change leads to a much more general model. By varying the spring constant k from far below the stiffness of the beam to far above it, the full range of behavior from the free beam to the simply supported

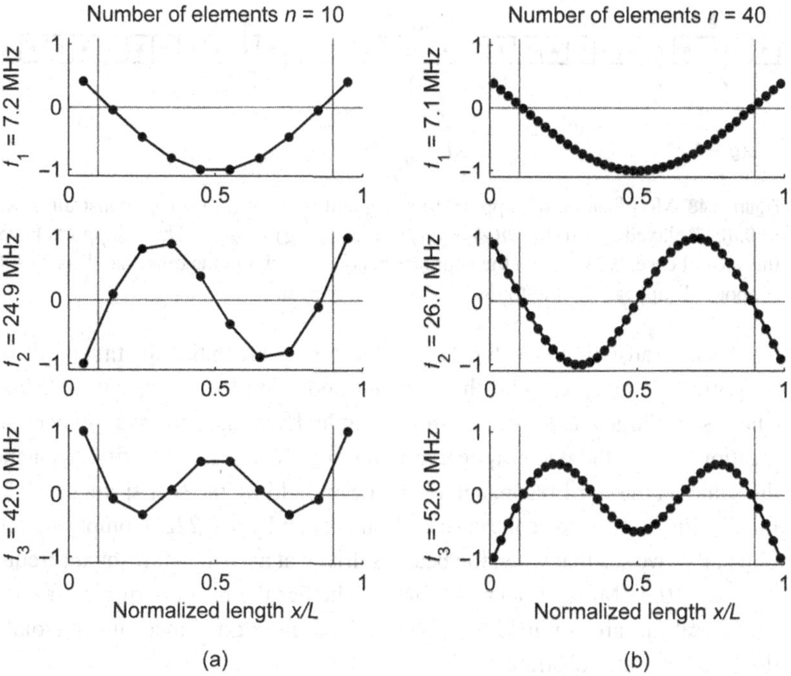

Figure 6.47 First three resonant modes of a simply supported beam held at points $0.1L$ from each of the free ends. Calculations are for a Si beam with $L = 40$ μm, $w = 4$ μm, $h = 2$ μm. (a) $n = 10$ DOF. (b) $n = 40$ DOF.

beam can be accessed:

$$\begin{bmatrix} 1 & g & -1 & -g & 0 & 0 & 0 & 0 \\ 0 & 1 & 0 & -1 & 0 & 0 & 0 & 0 \\ 0 & 0 & 1 & x_0 & -1 & -x_0 & 0 & 0 \\ 0 & 0 & 0 & 1 & 0 & -1 & 0 & 0 \\ 0 & 0 & 0 & 0 & 1 & L-g & -1 & -L+g \\ 0 & 0 & 0 & 0 & 0 & 1 & 0 & -1 \\ 1 & g & 0 & 0 & 0 & 0 & 0 & 0 \\ 0 & 0 & 0 & 0 & 0 & 0 & 1 & L-g \end{bmatrix} \begin{bmatrix} c_1' \\ c_2' \\ c_3' \\ c_4' \\ c_5' \\ c_6' \\ c_7' \\ c_8' \end{bmatrix}$$
$$= \begin{bmatrix} f_{R1} g^3/(3YI) \\ f_{R1} g^2/(2YI) \\ -x_0^2[f_{R1}(3g - x_0) - f_{R2}(3g - 3L + x_0)]/(6YI) \\ -x_0[f_{R1}(2g - x_0) + f_{R2}(2L - x_0 - 2g)]/(2YI) \\ f_{R2}(L-g)^3/(3YI) \\ f_{R2}(L-g)^2/(2YI) \\ f_{R1}/k \\ f_{R2}/k \end{bmatrix}. \tag{6.176}$$

The small k limit is illustrated in Fig. 6.49a. As before, the beam is attached to the springs at $0.1L$ and $0.9L$. The first two resonant modes are a rigid body rotation and a

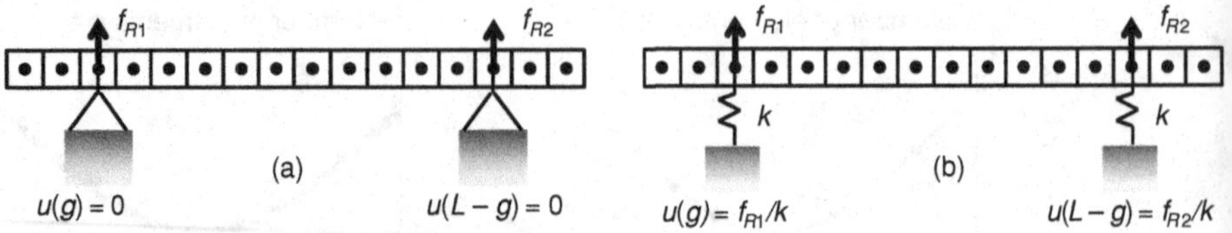

Figure 6.48 Modification of supports at $x = g$ and $x = L - g$. (a) Fixed constraints $u(g) = u(L - g) = 0$. (b) Relaxed constraints $u(g) = f_{R1}/k$ and $u(L - g) = f_{R2}/k$. These supports have case (a) as the special case. As $k \to \infty$, the supports become fixed. On the other hand, as $k \to 0$, the supports disappear leading to the model of a free beam.

rigid body translation. The third mode looks very much like the first mode of the simply supported beam except that the resonant nodes are located at $x \approx 0.22L$ and $\sim 0.78L$. The case of large k is illustrated in Fig. 6.49b. Here, the springs have been moved to the locations where the nodes appeared in the free beam case. The first resonance is almost the same as the third resonance of the beam held by the soft springs at $0.1L$ and $0.9L$. Finally, Fig. 6.49c shows the case with small k and $g = 0.22L$. Comparing Fig. 6.49b and Fig. 6.49c, we see that when the beam is driven at a particular resonant frequency, in this case $f = 10.23$ MHz, it does not matter whether the beam is rigidly or softly attached. The vibrations are confined to the beam, indicating good mechanical isolation between the beam and the substrate.

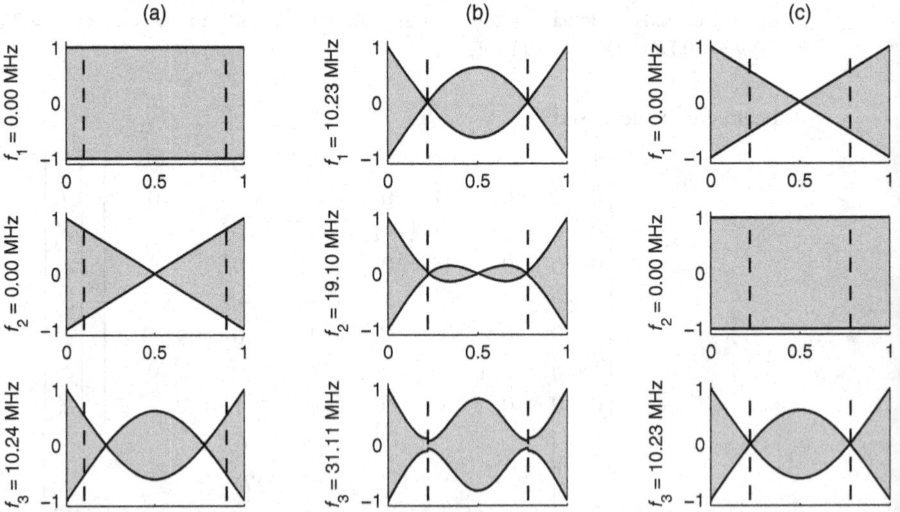

Figure 6.49 First three resonant modes of a beam supported by springs at $x = 0.1L$ and $0.9L$. Parameters are those for a Si beam with $L = 40$ μm, $w = 4$ μm, $h = 2$ μm. (a) Simply supported beam held by two soft springs at $0.1L$ and $0.9L$. The first two modes are rigid-body rotation and rigid-body translation. The third mode looks like the first mode of a simply supported beam held at $\approx 0.22L$ and $\approx 0.78L$. (b) Simply supported beam held at $0.22L$ and $0.78L$ by a hard spring. All resonances are bending modes. (c) Simply supported beam held at $0.22L$ and $0.78L$ by a soft spring.

6.8.4 Closure

As revealed by their accurate prediction of mode shapes and resonant frequencies for cantilevers and doubly clamped beams, MDF models offer a convenient and effective modeling approach for bending modes. Its chief limitation is that at least some information about the expected form of the deformation must be known. In the design of new MEMS structures, this state of knowledge often does not exist. Unwelcome surprises, particularly in the form of torsional modes, are sometimes encountered. The safest way to check for the existence of other vibrational modes is to use finite-element analysis (FEA). The powerful eigenmode tools built into FEA software are able to discover virtually all resonant modes, including their shapes and frequencies, that exist over a prescribed frequency range. For an example, refer to the resonance beam modes shown in Fig. D.11 of Appendix D. Note in particular that twisting and orthogonal bending modes are found to be interspersed with the set of modes for which the MFD analysis has been programmed. Finite-element analysis plays a critical role when investigating continua in the search for all the resonant modes. On the other hand, MDF analysis, with its virtues of simplicity and adequate accuracy, is of value in the design and system-level modeling of MEMS devices.

6.9 Summary

This chapter has introduced the important topic of electromechanical continua. In such structures, the mechanical displacement is a continuous function of one or two spatial coordinates. In many circumstances, this evident complexity can be reduced to a single lumped parameter system. The chapter has shown how that can be done for the very important case of MEMS devices based on cantilevered beams and circular plates. Extraction of lumped parameter models depends not only on identification of effective mass and spring constant values, but also on finding a useable capacitance function. For success, the system capacitance must be formulated in terms of a value for beam or plate displacement, selected at some location chosen for convenience and observability. There is some arbitrariness about the choice of this location, which affects the values of the equivalent parameters m_{eq}, k_{eq}, b_{eq}, $\underline{N}$, $\underline{Z}'_m$, and $\underline{Z}_e$. In any case, we have found that, subject to some restrictions, the essential dynamics of these continua up to the first resonance can be captured in terms of a lumped parameter model.

Proper representation of the electromechanical behavior of these continua means that the 1-D or 2-D aspects of the deformation must be treated. One vital assumption used in the implementation of the method is that deformation profiles obtained from static analyses adequately represent the dynamic shapes, even near resonance. Models predicated on existing analytical expressions provide rather good accuracy and encourage the use of MDF models to generalize the technique.

The MDF models provide the means to generalize the solution method described previously, and make it possible to consider higher-order modes if the eigensolution functions of MATLAB or equivalent programming tools are used. We have found that MDF analyses generally yield predictions for the shapes and resonant frequencies that are very close to theory. The method is effective for both cantilevered beams and plates.

A cantilevered beam served as a vehicle to show how a quasistatic model is formulated to derive a 1-DOF model. We then revisited the problem in more generality and emphasized that a mechanical continua can be parameterized with a single displacement when the shape is known.

Next, an MDF model was derived. This simple modeling approach, central for this chapter, provides more insight into the dynamics of mechanical continua. The equivalent circuit representation for the MDF beam model and simplification achieved by virtue of modal analysis were introduced. In addition, MDF formulations for a double-clamped beam and a simply supported beam were derived. These structures are of considerable importance for MEMS. Analysis of these two structures introduced important concepts from the mechanics of solids, viz. statically indeterminate systems.

The MDF method was then applied to modeling of plates, first assuming zero variation in the θ-direction, and then allowing displacement to vary along θ.

Problems

6.1 Section 6.2 compares an estimate of the fundamental resonance of a cantilevered beam (ω_0) obtained from the lumped parameter model based on effective mass and spring constant expressions to the result obtained from beam mechanics. For $g/L = 0.2$, the relative error is ~6%. Calculate and plot this relative error versus g/L over the range from 0 to 0.5. Discuss these results.

6.2 For the capacitive transducer depicted in Fig. 6.2, assume that the silicon cantilevered beam has length $L = 200$ μm, width $w = 5$ μm, and thickness $h = 2$ μm. Let the electrode length be $g = 40$ μm and the equilibrium spacing be $d = 3$ μm. Assume that the measured quality factor of the mechanical resonator is $Q = 20$ and that the bias voltage is $v_B = 5$ V. Further, the beam is oriented in the [110] direction, which means that the Young modulus and Poisson ratio are $Y_{Si} = 168$ GPa and $\nu = 0.06$ (Section D.12). Finally, the mass density for Si is $\rho_{Si} = 2330$ kg/m^3.

(a) Using the results of Section 6.2.2, obtain numerical values for the components in the circuit model of Fig. 6.5.
(b) Estimate the fundamental mechanical resonant frequency ω_0 using the component values for the lumped parameter model obtained in (a).
(c) Assume that the transducer operates as a sensor and the electrical port is terminated with a 30 GΩ resistor. Plot the magnitude of the transfer function $|\underline{H}_{v/x}(j\omega)| = |\underline{v}/\underline{x}|$ versus frequency over the range $0.05 \le \omega/\omega_0 \le 0.8$. Does the response depend on Q?

(d) Repeat the exercise in (c) for a different transfer function $|\underline{H}_{v/f}(j\omega)| = |\underline{v}/\underline{f}|$. Ignoring mechanical damping, plot the magnitude response versus frequency over the same range $0.05 \le \omega/\omega_0 \le 0.8$. On the same plot, show the magnitude response for the case of $Q = 20$.

6.3 Employ the discretized MDF model and in particular Eq. (6.34) to determine the sinusoidal current amplitude for a variable-gap capacitive transducer in the case where the moving electrode is a vibrating cantilevered beam as shown in Fig. 6.8. Assume that the substrate electrode extends along the entire length of the beam. The beam length and width are $L = 300$ μm and $w = 10$ μm, respectively, the electrode spacing $d = 2$ μm, the bias voltage is $V_B = 10$ V and the peak amplitude of sinusoidal motion at the tip is $u_0 = 0.3$μm vibrating at $\omega/(2\pi) = 30$ kHz.

(a) Assume that the displacement is $\hat{u}(x) = (3Lx^2 - x^3)/(2L^3)$ for $n = 10$ elements to compute $\zeta_n^{(2)}$.
(b) Repeat for $n = 40$ elements and compare your results with (a).
(c) Plot the current for n in the range $1 \le n \le 99$.

6.4 What changes must be in Eq. (6.34) to handle the case where the substrate electrode does not extend over the entire length of the beam, but instead has length $g < L$, as shown in Fig. 6.2? Modify the necessary parameters of this equation.

6.5 The objective of this problem is to verify symmetry for the flexibility matrix of the cantilevered beam. Proceed by deriving a_{ij} based on Eq. (6.7) in the region $L - g \le x \le L$ and then comparing these terms with the a_{ij} coefficients obtained from Eq. (6.47). Hint: let $L - g \to i/n$ and $x \to j/n$, then normalize the force variable by f_{L-g}.

6.6 This problem considers the influence of adjustably compliant boundary conditions on the flexibility coefficients. As shown in the figure, the clamped condition is replaced by translational and torsional springs. The approach is to implement a numerical MDF model to investigate the influences of k_{lin} and k_θ on natural frequencies and mode shapes by considering in turn three special cases:

(i) both k_{lin} and k_θ large;
(ii) k_θ small and k_{lin} still large;
(iii) both k_{lin} and k_θ small.

It will be necessary to experiment with the values of these spring constants to find the desired limiting cases. Assume a Si beam, with Young modulus $Y_{\text{Si}} = 168$ GPa, mass density $\rho_{\text{Si}} = 2330$ kg/m^3, length $L = 300$ μm, width $w = 10$ μm, and height $h = 2$ μm. (Hint: start with Eq. (6.43) noting that c_1 and c_2 are not constants any more. After defining the two springs, only one line of the code in Fig. 6.18 has to be modified.)

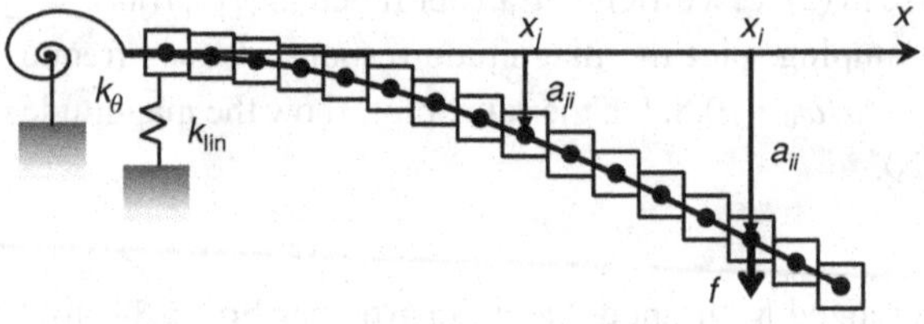

(a) To consider the clamped beam, assign large values to both k_{lin} and k_θ. Plot the first four modes.
(b) To emulate the hinged beam, with zero displacement and non-zero slope at both ends, reduce k_θ significantly. Plot the first four mode shapes. The lowest mode will be a rigid-body rotation. Though the resonant frequency of this first mode should be zero for a true hinged beam, the MDF model will return a small non-zero value.
(c) To emulate the free beam, with non-zero displacement and non-zero slope at both ends, use small values for both k_{lin} and k_θ. Plot the first four mode shapes. The first two modes for a free beam should resemble rigid-body translation and rigid-body rotation, respectively.

6.7 Consider the electrostatically actuated cantilevered beam of Section 6.4.4.2. Assume beam length $L = 300$ μm, width $w = 10$ μm, thickness $h = 2$ μm, and electrode–beam separation $d = 2$ μm. The electrode length is $g = 60$ μm. Compute the first 20 iterations of the static displacement due to the electrostatic force by alternately using Eqs. (6.57) and (7.60). Remember that each iteration uses a more accurate set of values for the electrostatic force acting on individual elements of the beam and thus convergence should be expected.

(a) For bias voltage $V = 5$ V, plot the maximum difference between modally adjacent displacements, that is, $\max(u^{(2)} - u^{(1)})$, $\max(u^{(3)} - u^{(4)})$, etc., on a linear and log y-scale.
(b) Repeat (a) using higher voltages to find the instability threshold. For what threshold voltage value does the system become unstable? (Hint: instability will manifest itself as a divergence of the values of $\max(u^{(2)} - u^{(1)})$, $\max(u^{(3)} - u^{(2)})$, etc., in successive iterations.)
(c) Discuss problems that arise with this model for the case of large displacements. Does this approach over- or under-estimate pull-in voltage? Explain.

6.8 Estimate the natural frequency and damping for a second-order system using the magnitude plot of the transfer function, $|\underline{H}_{x/f}(\mathrm{j}\omega)|$, using the scales on the axes and a ruler. There are two ways to estimate damping from a plot of this type; try both of them and compare the values thus obtained. Which of these approaches is more practical in the case of a MEMS actuator? Why? (Hint: refer to Section B.4 of Appendix B.)

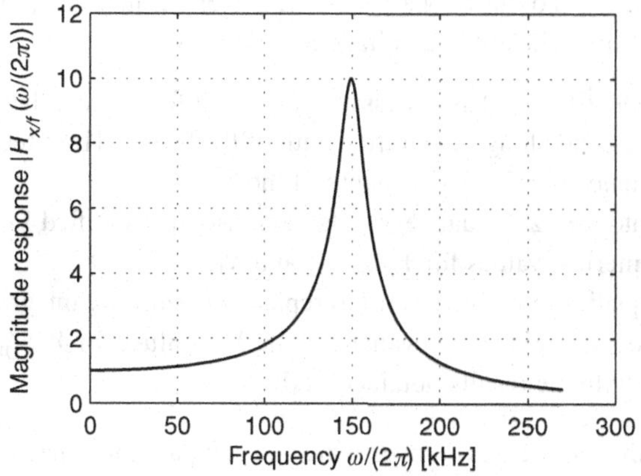

6.9 Use the closed-form expressions for k_{eq} and m_{eq} of Eq. (D.52) and Eq. (D.61), respectively, to estimate the lowest resonant frequency of a Si beam if the direction of the beam (the axis along its length) is aligned with the [110] direction of the crystal. The beam parameters are given as follows: length $L = 130$ μm, width $w = 7$ μm, and thickness $h = 3$ μm. What is the reference displacement u_0 implied for k_{eq} and m_{eq}? (Hint: refer to Section D.12 of Appendix D.)

6.10 Estimate the spring constant at the free end of a cantilevered beam by equating the elastic energy of a uniform beam given by $W_{\text{elast}} = \frac{YI}{2}\int_0^L \left(\frac{\partial^2 u}{\partial x^2}\right)^2 dx$ with the energy stored in the equivalent spring. Assume that the displacements are given by $u(x) = (3Lx^2 - x^3)/(2L^3)u_0$ where u_0 is the displacement of the free end of the beam. Compare the result with the one given by Eq. (D.52).

6.11 The transient response of a second-order mechanical SDF system is

$$x(t) = \underbrace{\sqrt{x_0^2 + \left(\frac{\dot{x}_0 + \zeta\omega_0 x_0}{\omega_d}\right)}}_{X_0} e^{-\zeta\omega_0 t}\cos\left(\omega_d t - \underbrace{\arctan\left(\frac{\dot{x}_0 + \zeta\omega_0 x_0}{\omega_d x_0}\right)}_{\phi}\right),$$

given generalized initial conditions $x(t=0) = x_0$ and $dx/dt\,|_{t=0} = \dot{x}_0$. Show that the damping factor ζ can be expressed in terms of the logarithmic decrement δ_n, that is, $\zeta = \delta_n/[\delta_n^2 + (2\pi n)^2]^{1/2}$, where $\delta_n \equiv \ln[x(t_0)/x(t_0 + nT_d)]$ and n is a time index for the discrete sampled values of $x(t)$.

6.12 Find numerical values for the circuit components $\underline{Y}_e$, $\underline{Z}_{m\phi 1}$, and $\underline{N}_1$ appearing in Fig. 6.22 for the variable-gap capacitive transducer of Fig. 6.2. Assume that the mode shape for the first mode ($k = 1$) can be approximated by the static displacement

$\phi_1(x) = (3Lx^2 - x^3)/(2L^3)$. For the Si beam, assume $L = 300$ μm, $w = 5$ μm, $h = 2$ μm, $d = 3$ μm, and $g = 100$ μm. The applied voltage is $v_0 = 10$ V.

(a) First, modify Eqs. (6.86)–(6.88) as necessary to account for the fact that the bottom electrode does not extend along the entire length of the beam. (Hint: recognize that the sums do not extend over the entire length of the beam.)
(b) Divide the beam into $n = 20$ equal segments, then use the modified Eqs. (6.86) to (6.88) to obtain numerical values for $\underline{Y}_e$, $\underline{Z}_{m\varphi 1}$, and $\underline{N}_1$.
(c) Now modify the equations obtained in (a) to replace the sums by integrals.
(d) Use the equations obtained in (c) to evaluate numerical values for $\underline{Y}_e$, $\underline{Z}_{m\phi 1}$, and $\underline{N}_1$ and compare these with the results obtained in (b).

6.13 The purpose of this problem is to investigate and compare the behavior of the two indeterminate beam examples shown in the figure.

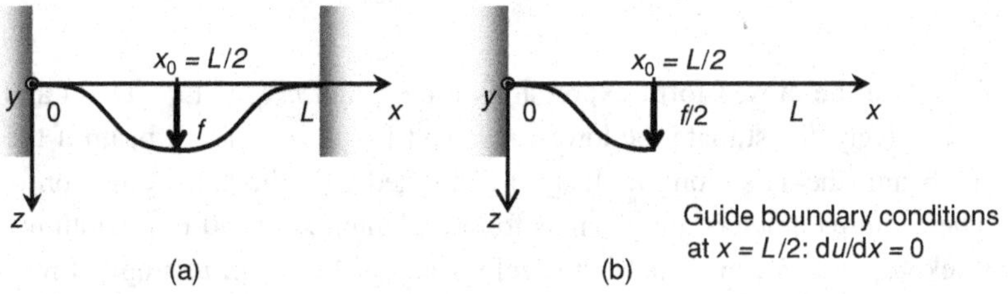

(a) Doubly clamped beam of length L. (b) Guided beam of length $L/2$.

(a) Use Eq. (6.159) to evaluate the reaction forces f_R and moments M_R when a normal force f acts at $x = L/2$, the midpoint of a doubly clamped beam. Use symmetry arguments to confirm your results.
(b) Repeat (a) for a cantilever of length $L/2$ clamped at $x = 0$ and with the guided boundary condition $du/dx = 0$ at $x = L/2$, with a normal force $f/2$ acting at the end.
(c) Employ a physical argument to explain the relationship of (a) and (b).

6.14 This problem guides you through an alternate derivation of the doubly clamped beam problem using the MDF model and certain properties of the flexibility matrix. For this problem, assume a Si beam having length, width, and thickness $L = 40$ μm, $w = 4$ μm, and $h = 2$ μm, respectively. Use $N = 40$ degrees of freedom. For this problem use the mass density $\rho_{Si} = 2330$ kg/m³ and the Young modulus $Y_{Si} = 168$ GPa.

(a) Starting with the flexibility matrix $\overline{\overline{a}}$ for a cantilevered beam clamped on the left-hand side, obtain $\overline{\overline{a}}$ for a beam clamped at the right-hand side by reflecting the coefficients about the secondary diagonal, viz., $a_{i,j} \to a_{N+1-i,N+1-j}$, where N is the number of elements in the MDF model. Verify your result by plotting the first three mode shapes.
(b) Examine the intuitive notion that the stiffness of the doubly clamped beam is the sum of the stiffnesses of the left-hand side and right-hand side clamped beams by

inverting the two flexibility matrices to obtain the stiffness matrices $\overline{\overline{k}}_{\text{left}} = \overline{\overline{a}}_{\text{left}}^{-1}$ and $\overline{\overline{k}}_{\text{right}} = \overline{\overline{a}}_{\text{right}}^{-1}$, then summing: $\overline{\overline{k}}_{\text{doubly clamped}} = \overline{\overline{k}}_{\text{left}} + \overline{\overline{k}}_{\text{right}}$. Use this stiffness matrix to find the lowest three resonant frequencies and compare the values obtained with the frequencies using the model of Fig. 6.41.

(c) As defined in (b), $\overline{\overline{k}}_{\text{doubly clamped}}$ properly accounts for the boundary conditions, but double-counts for the stiffness of the beam itself. This error can be compensated for very simply by dividing all the elements of the stiffness matrix by 2, that is, $\overline{\overline{k}}_{\text{doubly clamped}} = (\overline{\overline{k}}_{\text{left}} + \overline{\overline{k}}_{\text{right}})/2$. Use this new definition to compute the lowest three resonant frequencies and compare them with the values obtained from the model of Fig. 6.41.

6.15 Determine the flexibility influence coefficients for the doubly supported beam shown in the figure, where the clamp constraints are replaced by pairs of linear and torsional springs. These constraints have the effect of "softening" the boundary conditions. This model is more general than that presented in Problem 6.6 because here the boundary conditions at the ends can be specified independently.

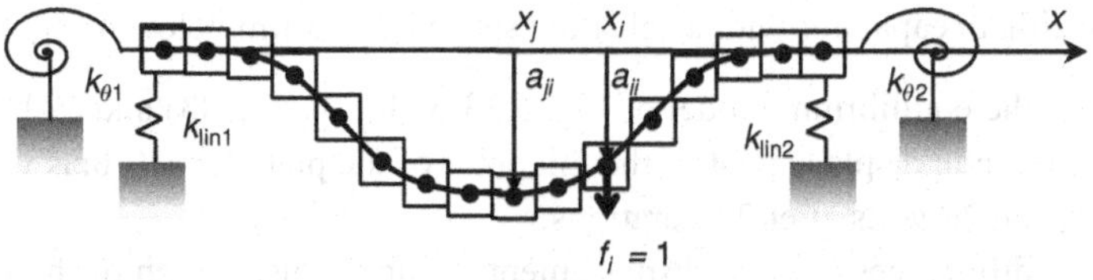

(a) Assuming that $k_{\text{lin1}} \neq k_{\text{lin2}}$ and $k_{\theta 1} \neq k_{\theta 2}$, derive the flexibility coefficients for this constrained beam.

(b) Modify the MATLAB code listed in Fig. 6.41 according to the results obtained for (a) and then use it to identify a set of numerical values for $\{k_{\text{lin1}}, k_{\theta 1}, k_{\text{lin2}}, k_{\theta 2}\}$ that emulates a cantilevered beam clamped on the left-hand side. Plot the first three mode shapes, indicating their associated resonant frequencies.

(c) Repeat (c) for cantilevered beam clamped on the right-hand side.

(d) Adjust the values for $\{k_{\text{lin1}}, k_{\theta 1}, k_{\text{lin2}}, k_{\theta 2}\}$ to achieve the limiting case of a simply supported beam displaying zero displacement and non-zero slope at both ends. Plot the first three modes and provide their resonant frequencies.

(e) Adjust the values for $\{k_{\text{lin1}}, k_{\theta 1}, k_{\text{lin2}}, k_{\theta 2}\}$ to achieve the limit of a doubly clamped beam. Plot the first three modes and indicate their resonant frequencies

(f) Find a set of numerical values for $\{k_{\text{lin1}}, k_{\theta 1}, k_{\text{lin2}}, k_{\theta 2}\}$ that emulates a *free* beam. Plot the first three modes and indicate their resonant frequencies.

6.16 The diameter and thickness of a circular Si diaphragm are 400 μm and 2 μm, respectively. The plate forms a variable-gap capacitance with the substrate, which is spaced $d = 3$ μm below it. The measured damping ratio is $\zeta = 1.0$.

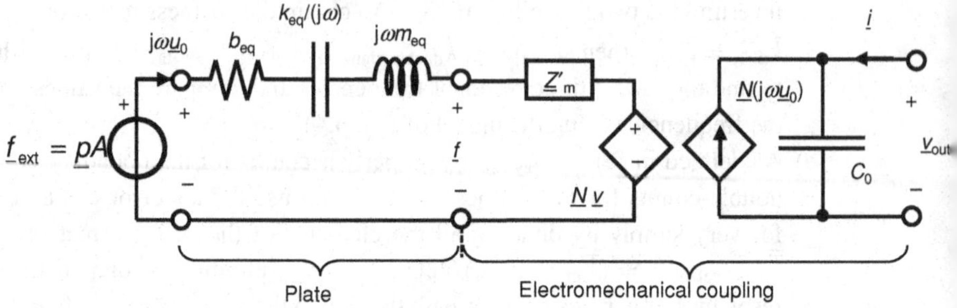

(a) Assuming that the applied bias voltage is $V_B = 10$ V, obtain numerical values for all components of the linear equivalent circuit model provided in the figure. Use mass density $\rho = 2330$ kg/m^3, Young modulus $Y = 150$ GPa, and Poisson ratio $\nu = 0.3$.
(b) Compute the resonant frequency ω_0.
(c) Plot $|\underline{H}_{v/p}(\omega)|$ vs. ω for $0 \leq \omega \leq 1.2\omega_0$.

6.17 A circular silicon plate with radius $r_a = 150$ μm and thickness $h = 1.5$ μm forms a variable air gap capacitor with a substrate spaced $d = 3$ μm below it.

(a) Employ the equilibrium model of Fig. 6.33 with Eqs. (6.143) and (6.144) to determine the static displacement of the midpoint of the plate for DC bias voltages of 5, 10, 15, and 20 volts after 20 iterations.
(b) Plot the differences in max. displacement vs. iterations. Which of these equilibria, if any, are unstable?

6.18 Compute the equivalent spring constant of a circular plate by equating the elastic energy of the effective spring constant associated with the midpoint of the plate with the total strain energy. To evaluate the total strain energy in the axially symmetric plate deformation, use

$$W_{\text{elast}} = \frac{D\pi}{2} \int_0^{r_a} \left[\left(\frac{\partial^2 u_z(r)}{\partial r^2} + \frac{1}{r} \frac{\partial u_z(r)}{\partial r} \right)^2 - 2(1-\nu) \frac{\partial^2 u_z(r)}{\partial r^2} \left(\frac{1}{r} \frac{\partial u_z(r)}{\partial r} \right) \right] r dr,$$

where $u_z(r) = u_0(1 - (r/r_a)^2)^2$ and u_0 is the displacement of the midpoint of the plate. Compare the result with Eq. (6.102). (Hint: consider using a symbolic solver.)

References

1. C. Keimel, G. Claydon, B. Li, J. Park, M. E. Valdes, Micro-electromechanical systems (MEMS) based switches for power application. *IEEE Transactions on Industry Applications*, 48 (2012), 1163–1169.

2. C. Keimel, M. Aimi, A. Detor, and A. Corwin, *New Class of MEMS Based Devices Enabled by Nanocrystalline Material*. Presented at the Technologies for Future Micro-Nano Manufacturing, Napa CA, USA (2011).
3. A. D. Dimarogonas, *Vibration for Engineers*, 2nd edn (Upper Saddle River, NJ: Prentice Hall, 1996).
4. G. H. Golub and C. F. Van Loan, *Matrix Computations*, 3rd edn (Baltimore: Johns Hopkins University Press, 1996).
5. A. A. Shabana, *Vibration of Discrete and Continuous Systems*, 2nd edn (New York: Springer, 1997).
6. S. Timoshenko and S. Woinowsky-Krieger, *Theory of Plates and Shells*, 2nd edn (New York: McGraw-Hill, 1959).
7. L. Meirovitch, *Principles and Techniques of Vibrations* (Upper Saddle River, NJ: Prentice Hall, 1997).
8. R. J. Roark and W. C. Young, *Roark's Formulas for Stress and Strain*, 6th edn (New York: McGraw-Hill, 1989).
9. A. W. Leissa, *Vibration of Plates*, reprinted edn [SI] (Acoustical Society of America through the American Institute of Physics, 1993).
10. R. G. Polcawich, M. Scanlon, J. Pulscamp, *et al.*, Design and fabrication of a lead zirconate titanate (PZT) thin film acoustic sensor, *Integrated Ferroelectrics*, 54 (2003), 595–606.

7 Practical MEMS devices

7.1 Introduction

In the previous chapters, we have investigated lumped parameter electromechanical conversion, linear multiport representations for actuators and sensors, and the effects of external constraints on transducer response. We used a powerful electromechanical analog to synthesize circuit models for transducers, examined the signal conditioning and amplification stages needed to turn MEMS devices into practical systems, and finally developed a methodology to represent mechanical continua, viz., beams and plates, as lumped parameter systems. We are now ready to put these modeling tools to use in the analysis of some practical MEMS devices.

This chapter provides concise system-level technical presentations of four important MEMS applications: pressure sensors, accelerometers, gyroscopes, and energy harvesters. These applications have been selected according to diverse criteria. For example, pressure sensors, relatively uncomplicated as MEMS go, were among the first types of sensors and actuators to be miniaturized. MEMS accelerometers, more complicated than pressure sensors, are very widely used in automotive airbag systems. Micromechanical gyroscopes, now available commercially, are considerably more complex than accelerometers, with respect to both the dynamics and the system-level electromechanical drive–sense scheme needed to make them work. Finally, the MEMS energy harvester is included because it is a concept that shows great promise but needs further development before real market penetration can occur.

These four applications are each treated in a similar fashion, with reliance placed on basic theoretical foundations and, where possible, on the multiport system methodologies and equivalent electromechanical circuit models of Chapters 5 and 6. Each section first briefly describes the device function, introduces the principles of operation, describes a realistic MEMS example, introduces realistic specifications, and discusses higher-order effects, all the while emphasizing system-level modeling. Domain-specific knowledge, primarily associated with the mechanical system side, is provided as needed within each section. As a convenience, the sections are largely self-contained. They can be studied in any order, based on the interests and needs of the reader.

We do not delve into layout considerations here, because layout is strongly affected by the fabrication process. Ideally, the design process would begin with selection of the fabrication process, but in reality it is far more common that the designer is locked into a given process, and therefore must make design decisions within the constraints of the process.

7.2 Capacitive MEMS pressure sensors

Pressure sensors became the first mass-produced MEMS devices. As of 1995, their market had reached 1 billion US$ [1]. Their success guided the research and development of other commercial MEMS devices that took much more time to develop.

The three dominant types of pressure sensor are: (1) displacement-based, where pressure changes the deflection of a deformable structure leading to a capacitive change (or an optical signal); (2) strain-based, where pressure alters strain, which is detected by a piezoresistive element; and (3) frequency-based, where pressure changes the frequency of a mechanical resonator.

This section is concerned exclusively with the first of these, viz., displacement-based pressure sensors employing capacitive transduction to measure the pressure-induced displacements. These devices convert pressure into deflection of a mechanical element, convert this deflection into capacitive change, and then convert the capacitance change into an output voltage signal.

It is important to note that strain-based piezoresistive pressure sensors are actually the dominant type. To learn about these and other pressure sensor types, the reader is referred to other texts [2–4]. Compared with their piezoresistive counterparts, capacitive sensors have distinct advantages: higher sensitivity, greater long-term stability, reduced temperature effects, and lower power consumption [5]. The disadvantage of capacitive sensors is that they exhibit higher response non-linearity.

Microphones, also not considered here, are a special case of displacement-based pressure sensor that respond to time-varying pressure, that is, sound waves. The challenge in the design of microphones is associated with the frequency and dynamic signal ranges.[1]

7.2.1 Basic displacement-based capacitive pressure sensor

Probably the most commonly employed geometry for pressure sensors is the circular diaphragm. Figure 7.1 shows such a structure subjected to the pressure difference between a reference pressure p_0 and measured pressure p and using a simple inverting amplifier to detect capacitance change. The figure provides analytical expressions that can be used to develop a simple performance model for the transducer. We will now examine these expressions in turn.

As presented in Section 6.6, a pressure difference causes the diaphragm to deflect toward the region of lower pressure. The displacement is a fourth-order polynomial function of the radius:

$$u_z(r) = \frac{r_a^4}{64D}(p - p_o)\left[1 - \left(\frac{r}{r_a}\right)^2\right]^2. \tag{7.1}$$

[1] The accepted audible frequency range is 20 Hz to 20 kHz, and the dynamic range spans six orders of magnitude (120 dB), starting from the standard lower limit of audible pressure, measured at 1 kHz.

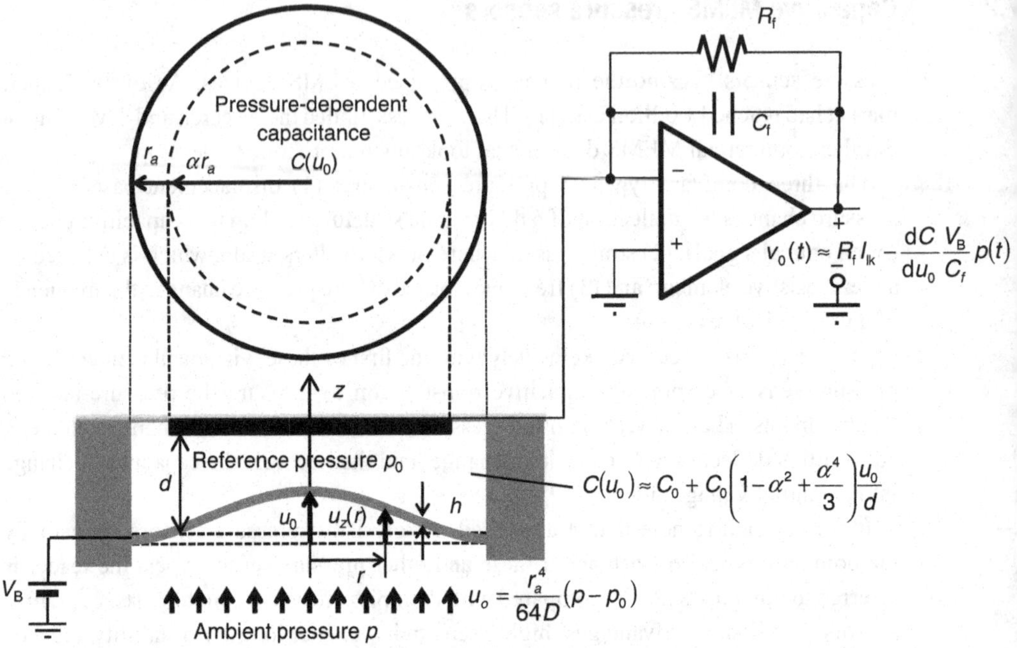

Figure 7.1 A simple circular-plate pressure sensor. The displacement at the middle of the pressure sensor diaphragm u_0 is directly proportional to pressure: $u_0 = (p - p_0)r_a^4/(64D)$. Capacitance $C(u_0)$ is a non-linear function of u_0. Using the op-amp with capacitance C_f in the feedback path, the output voltage is proportional to the pressure difference $(p - p_0)$ through the capacitance change.

Here, $u_z(r)$ is the r-dependent displacement of the plate, r_a is the radius of the diaphragm, D is the flexural rigidity of the plate, p_0 is the reference pressure, and p is the measured pressure. Using the approach outlined in Chapter 6, we select a particular displacement to characterize the magnitude of deformation and then express the capacitance as a function of this quantity to develop a lumped parameter mechanical model. The center of the diaphragm is the obvious choice for this displacement:

$$u_0 \equiv u_z(r=0) = \frac{r_a^4}{64D}(p - p_0). \tag{7.2}$$

This equation relates pressure difference to mechanical displacement u_0. The capacitance change due to this mechanical displacement is [6]

$$\begin{aligned} C(u_0) &\approx \int_0^{\alpha r_a} \varepsilon_0 \frac{2\pi r}{d - u_0(1-(r/r_a)^2)^2} \\ &\approx \frac{C_0}{2\alpha^2\sqrt{u_0/d}} \ln\left(\frac{1+\alpha^2\sqrt{u_0/d}+\alpha^2 u_0/d - u_0/d}{1+\alpha^2\sqrt{u_0/d}+\alpha^2 u_0/d - u_0/d}\right), \end{aligned} \tag{7.3}$$

where $C_0 = \varepsilon_0(\alpha r_a)^2\pi/d$ is the capacitance of the undeformed structure. The area of the capacitance here is smaller than the total area of the plate by the factor α^2. The closed-form solution of Eq. (7.3) applies for small displacements only and is based on the elastic deformation model given by Eq. (7.1).

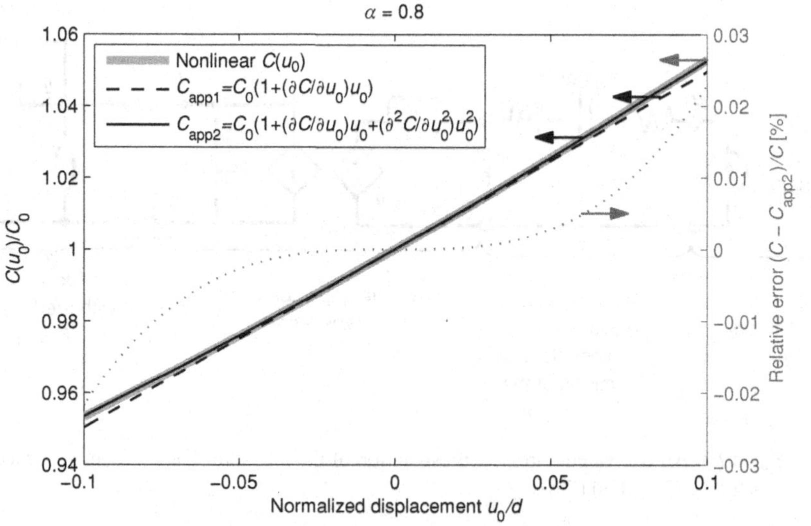

Figure 7.2 Plot of normalized capacitance of the circular-plate sensor as a function of u_0. The left axis is for capacitance $C(u_0)$ and its first-order and second-order Taylor approximations. The expanded scale on the right side shows the percentage error of the second-order Taylor approximation.

We rely on linearized models, which are more readily interpreted and parameterized. Linearization promotes systemic thinking and facilitates inclusion of the first-order effects of electromechanical coupling in a tractable manner. Thus, Taylor expansion of Eq. (7.3) is convenient for our purposes. The expansion is easy to obtain, using the approach of Section 6.6, by first expanding the integrand into a Taylor series and integrating each term:

$$C(u_0) \approx C_{\text{app2}} = C_0\left[1 + \left(1 - \alpha^2 + \frac{\alpha^4}{d}\right)\frac{u_0}{d} + \left(1 - \alpha^2 + 2\alpha^4 - \alpha^6 + \frac{\alpha^8}{5}\right)\frac{u_0^2}{d^2}\right]. \tag{7.4}$$

As it must, Eq. (7.4) reduces to Eq. (6.107) for $\alpha = 1$. Figure 7.2 plots $C(u_0)$ for $-0.1 \le u_0/d \le 0.1$ plus the linear approximation, and the second-order Taylor expansion, as given by Eq. (7.4). The second-order Taylor expansion is indistinguishable from $C(u_0)$ in this plot. The relative error is also plotted using the secondary y-scale on the right.

The output voltage for the op-amp circuit in Fig. 7.1 was derived in Section 5.3.4 Starting with Eq. (5.37), and substituting $\mathrm{d}C/\mathrm{d}u_0$ for $\mathrm{d}C/\mathrm{d}x$ and u_0 for x expressed in terms of pressure from Eq. (7.1), yields

$$v_0(t) = R_{\text{f/lk}} - \frac{\mathrm{d}C}{\mathrm{d}u_0}\frac{V_B}{C_\text{f}}\frac{\omega R_\text{f} C_\text{f}}{\sqrt{1+(\omega R_\text{f} C_\text{f})^2}}\frac{r_a^4}{64D}p_\text{m}\cos\left(\omega t + \arctan\left(\frac{1}{\omega R_\text{f} C_\text{f}}\right)\right), \tag{7.5}$$

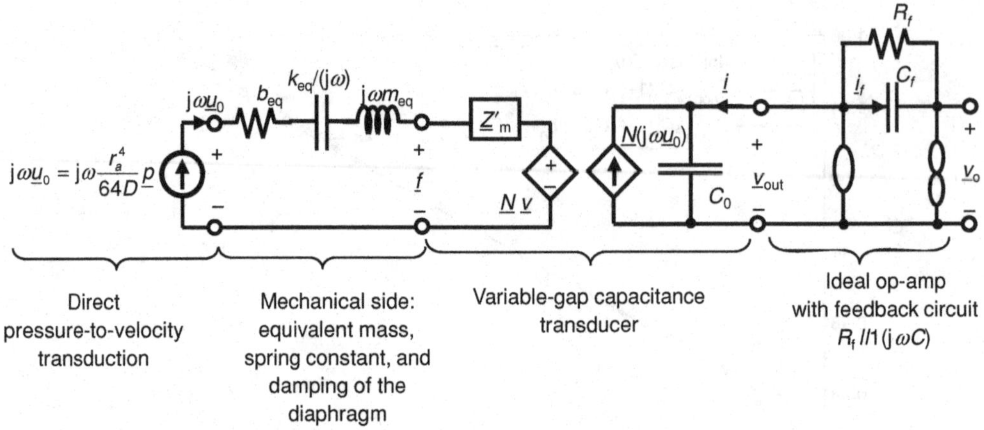

Figure 7.3 An equivalent circuit representation of the pressure sensor model governed by Eqs. (7.2), (7.4), and (7.6).

where the small-signal input pressure is of sinusoidal form:

$$p'(t) = p(t) - p_0 = p_m \cos(\omega t).$$

At high frequency, $\omega R_f C_f \gg 1$, this expression simplifies to

$$v_O t(t) = R_{f/lk} - \frac{dC}{du_0}\frac{V_B}{C_f}\frac{r_a^4}{64D} p_m \cos(\omega t). \tag{7.6}$$

At low frequency, that is, $\omega R_f C_f \leq 1$, the resistor R_f needed to avoid op-amp saturation attenuates the signal. Thus, for slowly varying pressure changes, a different sensing circuit is required. A switched-capacitance sensing circuit is capable of amplifying the DC component. An example of switched-capacitance sensing is presented in Section 7.2.3.

7.2.2 System-level model

For the analytical model of the pressure sensor given by Eqs. (7.2), (7.4), and (7.6), the displacement u_0 and, consequently, the velocity $j\omega u_0$ are directly controlled by pressure. To relate this model to an electromechanical two-port representation, the dynamics of the plate must be taken into account. To do so, we can take advantage of the general equivalent circuit approach developed in Chapters 4 and 5 to propose the model shown in Fig. 7.3.

Examination of this circuit reveals that the coupling is entirely controlled by the perturbation pressure through the controlled current source $\underline{N}(j\omega\underline{u}_0)$. The plate mass m_{eq}, equivalent spring constant k_{eq}, and damping b_{eq} have no influence, making possible considerable simplification of the electromechanical network. Refer to Fig. 7.4.

The pressure–displacement relationship given by Eq. (7.2) is derived using elasticity alone without considering the effect of the bias voltage on the spring constant via the electromechanical coupling. In fact, the equivalent spring constant k_{eq} is built into

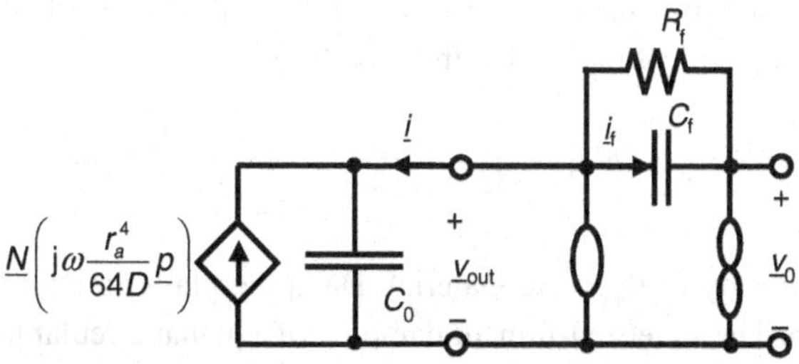

Figure 7.4 A simplified equivalent circuit representation of the pressure sensor model governed by Eqs. (7.2), (7.4), and (7.6).

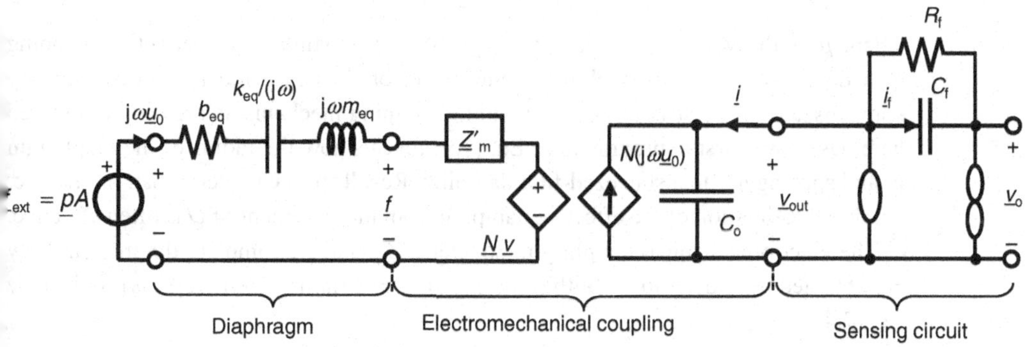

Figure 7.5 More general equivalent circuit representation: the pressure is converted to force which acts on an idealized diaphragm represented by a lumped parameter model and electromechanically bilaterally coupled with the sensing circuit. This model is valid up to the first resonance frequency of the plate. It also takes into account the effect of electromechanical coupling on the mechanical side of the transducer.

Eq. (7.2). The equivalent circuit representation showed this explicitly. In general, whenever velocity is imposed on the mechanical side of an electromechanical system, the coupling effects, including the pull-in instability, are suppressed because the effect of the bias voltage on the effective stiffness of the transducer is not taken into account.

The simple model of Fig. 7.4 is quite useful because the time variation of the pressure is often very slow compared with the first resonance of the diaphragm. In this case, the effects of electromechanical coupling can be regarded as second-order.

A more general model, valid up to the first resonance frequency of the circular plate and accounting for the effect of electromechanical coupling on the mechanical side of the transducer, is shown in Fig. 7.5. Here, the pressure difference is converted to a force acting on the idealized diaphragm, which is modeled as a lumped parameter spring-mass system, having equivalent mass m_{eq}, spring constant k_{eq}, and damping b_{eq}. The mechanical and electrical sides are mutually coupled in the usual way. The back-action of the sensor is accounted for by $\underline{N}v$, and the effect of electromechanical coupling on the dynamics of the mechanical subsystem is represented by $\underline{Z}'_m$.

From Section 6.6.1, the equivalent stiffness of the circular plate is

$$k_{eq} = \frac{16}{3}\frac{Yh^3}{1-\nu^2}\frac{\pi}{r_a^2}, \tag{7.7}$$

where Y is the Young modulus for the plate material, h is the plate thickness, and ν is the Poisson ratio. The equivalent mass, also from Section 6.6.1, is

$$m_{\text{eq}} = \frac{256}{315} \underbrace{\rho \pi r_a^2 h}_{m_{\text{tot}}}, \tag{7.8}$$

where ρ is the mass density of the plate material. Damping plays an important role in pressure sensor design. The squeezed-film air damping of a planar circular plate moving normal to a parallel wall is [4]

$$b_{\text{eq}} = \frac{3\pi}{2d^3} \mu r_a^4, \tag{7.9}$$

where μ is the viscosity of air. This expression overestimates the effective damping of a deforming circular diaphragm, but any error thus incurred is at least partially compensated for by structural and clamping damping mechanisms present but ignored here. Used as a first approximation, Eq. (7.9) reveals how the radius of the diaphragm r_a and spacing d affect squeezed-film damping. Recall from Chapter 6 that, in practice, empirical estimation of the effective damping from measurement of Q is often effective.

The electromechanical coupling parameters, $\underline{Z}'_{\text{m}}$ and $\underline{N}$, complete the model. They are obtained from derivatives of the coenergy $\underline{W}'_{\text{m}} = C(u_0)v^2/2$, where $C(u_0)$ is given by Eq. (7.4):

$$\underline{Z}'_{\text{m}} = \left. \frac{\partial^2 W'_e}{\partial u_0^2} \right|_{\substack{v \to V_{\text{B}} \\ u_0 \to 0}} = \frac{1}{j\omega} \left(1 - 2\alpha^2 + 2\alpha^4 - \alpha^6 + \frac{\alpha^8}{5} \right) \frac{C_0 V_{\text{B}}^2}{d^2}, \tag{7.10}$$

$$\underline{N} = \left. \frac{\partial^2 W'_e}{\partial u_0 \partial v} \right|_{\substack{v \to V_{\text{B}} \\ u_0 \to 0}} = C_0 \left(1 - \alpha^2 + \frac{\alpha^3}{3} \right) \frac{V_{\text{B}}}{d}. \tag{7.11}$$

In this section we examined two models: one is based on the quasistatic elastic deformation of the diaphragm and the other takes into account the dynamics of the diaphragm and electromechanical coupling. Both models are linear and applicable only in the case of small signals. The second model is useable over a wider frequency span, making it suitable for microphones.

7.2.3 A differential configuration

The modified pressure sensor design shown in Fig. 7.6, and based on Bao *et al.* [4, 6] uses a second reference capacitor fabricated on the chip adjacent to the pressure-dependent sensor unit. This second capacitor serves as a fixed reference, making it possible to achieve the desirable capability of differential operation. The fixed dummy capacitor enables three-plate sensing, first considered in Section 5.4. Because these two capacitances are fabricated side-by-side, their capacitive ratio is well controlled. They can be designed to be equal ($\alpha_1 = \alpha_2$), though usually the ratio between α_1 and α_2 is used to tune the output to be zero for a specific pressure value, i.e., to zero out any sensor

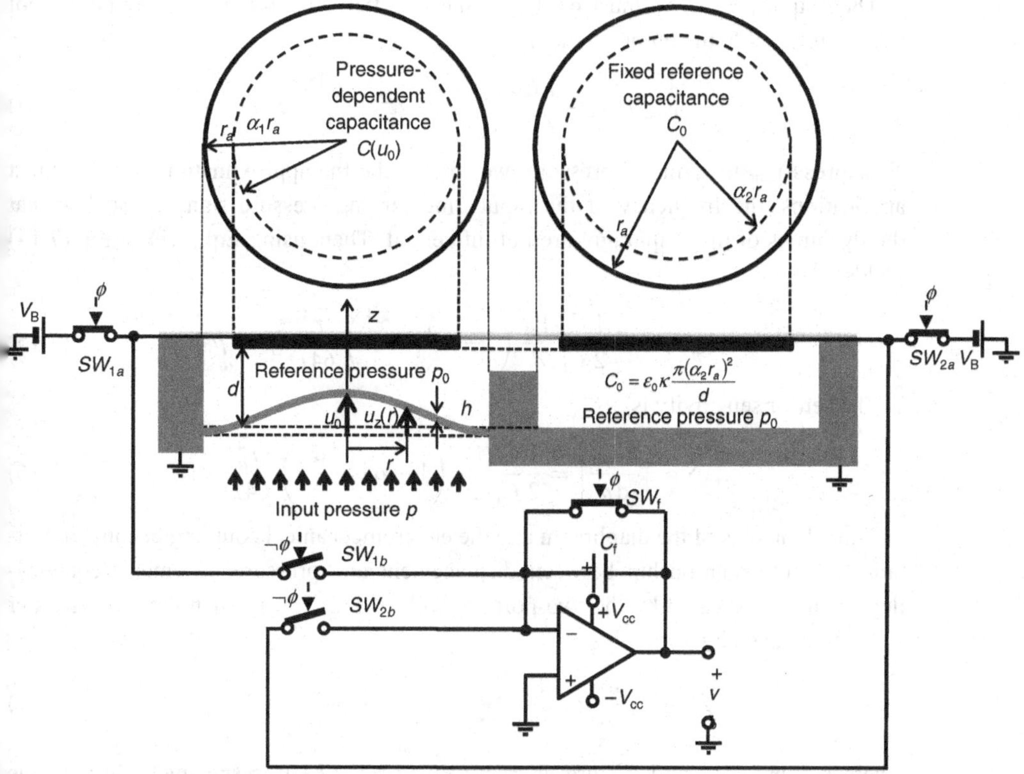

Figure 7.6 A differential pressure sensor with switched-capacitance sensing circuit. The fixed capacitance on the right serves as the reference capacitance and the capacitance on the left serves as the sensing element, dependent on pressure p.

offset. For convenience in the following analysis, we assume with no loss of generality that $\alpha_1 = \alpha_2$.

Switched-capacitance sensing circuits were presented in Section 5.7.3.[2] From Eq. (5.87), under the assumption that $\alpha_1 = \alpha_2$, the output voltage is

$$v_0(t) = -\frac{1}{2}\frac{V_B}{C_f}\left(\left.\frac{dC}{du_0}\right|_{u_0\to 0} u_0(t)\right)[\text{sign}[\sin(\omega_s t)] + 1], \tag{7.12}$$

where, from Eq. (7.4),

$$\left.\frac{dC}{du_0}\right|_{u_0\to 0} = \frac{C_0}{d}\left(1 - \alpha^2 + \frac{\alpha^4}{3}\right). \tag{7.13}$$

Equation (7.12) contains a factor 1/2 not present in Eq. (5.87), resulting from the fact that the device of Fig. 7.6 is not a true three-plate differential system. Only one of the capacitors depends on u_0.

[2] Bao's pressure sensor [4] and [6] employed a different sensing circuit. We replaced it here with the familiar topology already introduced in Chapter 5.

The output voltage v_{O} must be demodulated. After synchronous demodulation, not shown in Fig. 7.6, the output signal is

$$v_{\mathrm{out}}(t) = -\frac{1}{2}\frac{C_0}{C_{\mathrm{f}}}\frac{V_{\mathrm{B}}}{d}\left(1-\alpha^2+\frac{\alpha^4}{3}\right)u_{\mathrm{O}}(t). \tag{7.14}$$

To express v_{out} in terms of pressure, we can invoke the approximation that, for most applications, the frequency of the input time-varying pressure signal is so low that the dynamics of the diaphragm are not influential. Then, using Eq. (7.1) in Eq. (7.14) yields

$$v_{\mathrm{out}}(t) = -\frac{1}{2}\frac{C_0}{C_{\mathrm{f}}}\frac{V_{\mathrm{B}}}{d}\left(1-\alpha^2+\frac{\alpha^4}{3}\right)\frac{r_a^4}{64D}(p-p_0). \tag{7.15}$$

The sensor sensitivity is

$$S = \left|\frac{\mathrm{d}v_{\mathrm{out}}}{\mathrm{d}p}\right| = \frac{1}{2}\frac{C_0}{C_{\mathrm{f}}}\frac{V_{\mathrm{B}}}{d}\left(1-\alpha^2+\frac{\alpha^4}{3}\right)\frac{r_a^4}{64D}. \tag{7.16}$$

If the dynamics of the diaphragm and the electromechanical coupling become important, then the relationship between displacement and pressure becomes frequency-dependent, as revealed by the two-port network of Fig. 7.5. In such case the transfer function is given by

$$\underline{H}_{u/p} = \frac{\underline{u_{\mathrm{O}}}}{\underline{p}} = \frac{1}{\mathrm{j}\omega}\frac{A}{\mathrm{j}\omega m_{\mathrm{eq}}+b_{\mathrm{eq}}+k_{\mathrm{eq}}/(\mathrm{j}\omega)+\underline{Z}'_{\mathrm{m}}}, \tag{7.17}$$

where k_{eq}, m_{eq}, b_{eq}, and $\underline{Z}'_{\mathrm{m}}$ are given by Eqs. (7.7)–(7.10), respectively. In the time domain, for a sinusoidal pressure signal the displacement is

$$u_{\mathrm{O}}(t) = |H_{u/p}(\omega)|\,p_{\mathrm{m}}\cos(\omega t-\phi_{u/p}), \tag{7.18}$$

where $|H_{u/p}(\omega)|$ and $\phi_{u/p}(\omega)$ are the magnitude and phase responses of the transfer function $\underline{H}_{u/p}$:

$$|H_{u/p}(\omega)| = \frac{A}{\sqrt{(\omega b_{\mathrm{eq}})^2+\left(k_{\mathrm{eq}}+\underbrace{(1-2\alpha^2+2\alpha^4-\alpha^8/5)C_0C_{\mathrm{B}}^2/d^2}_{\underline{Z}'_{\mathrm{m}}/\mathrm{j}\omega}-\omega^2 m_{\mathrm{eq}}\right)^2}} \tag{7.19}$$

and

$$\phi_{u/p}(\omega) = -\arctan\left(\frac{k_{\mathrm{eq}}+(1-2\alpha^2+2\alpha^4-\alpha^8/5)C_0V_{\mathrm{B}}^2/d^2-\omega^2 m_{\mathrm{eq}}}{\omega b_{\mathrm{eq}}^2}\right). \tag{7.20}$$

Equation (7.18) reduces to Eq. (7.2) when $\omega \to 0$, and $k_{\mathrm{eff}} \gg \underline{Z}'_{\mathrm{m}}/(j\omega)$. Then, the sensitivity reduces to

$$S = \frac{1}{2}\frac{C_0}{C_{\mathrm{f}}}\frac{V_{\mathrm{B}}}{d}\left(1-\alpha^2+\frac{\alpha^4}{3}\right)|H_{u/p}(\omega)|. \tag{7.21}$$

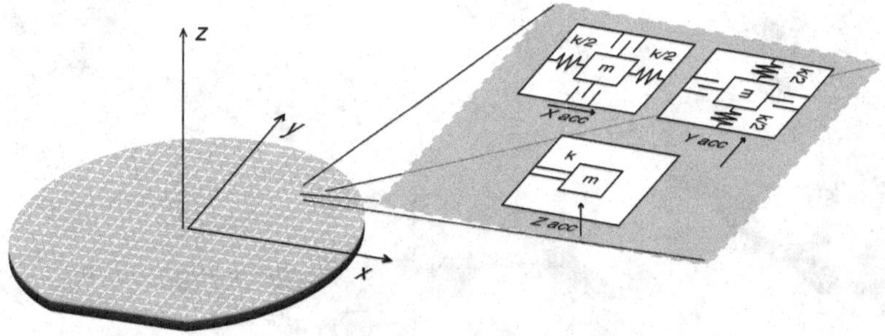

Figure 7.7 Wafer for a 3-D accelerometer with three defined orthogonal axes: two in the plane of the wafer and one out of the plane. Device symmetry is easily achieved in the two perpendicular directions in the plane of the wafer (x and y), merely by orienting two devices at right angles to each other. Design and fabrication of the third unit intended to monitor out-of-plane motion is much more difficult.

7.2.4 Closure

Two alternative models for pressure sensors have been presented. The first, quasistatic, model is simpler and captures the first-order effects well. The second is more complex, but it is valid over a wider frequency range and accounts for the effect of electromechanical coupling on the mechanical side of the sensor.

While piezoresistive technology at present dominates the pressure sensor market, capacitive MEMS pressure sensors have several distinct advantages: lower power consumption, better long-term stability, good temperature performance, and higher sensitivity. Issues such as cost will determine whether or not these devices someday overtake their piezoresistive counterparts.

7.3 MEMS accelerometers

The two types of linear motion sensor are accelerometers and displacement sensors. Accelerometers, sensitive to the rate of change of velocity, are typically employed to sense motion at medium and high frequencies, while displacement sensors, which detect positional change, are low-frequency devices. Accelerometers are in fact a type of *inertial sensor* that detects linear acceleration or deceleration. *Gyroscopes*, the subject of Section 7.4, detect rotational motion.

An accelerometer senses acceleration along a particular axis. Sometimes two (or three) accelerometers are oriented along perpendicular axes and packaged as two- (or three-) axis devices, respectively. In the macroscopic world, sets of two (or three) identical accelerometers can be mounted orthogonally on the same platform, but in the microscopic world this is not so simple. It is easy to achieve device symmetry in two directions in the plane of the wafer, but, owing to fabrication limits, extending this symmetry to motions perpendicular to the plane of the wafer is very challenging. Refer to Fig. 7.7.

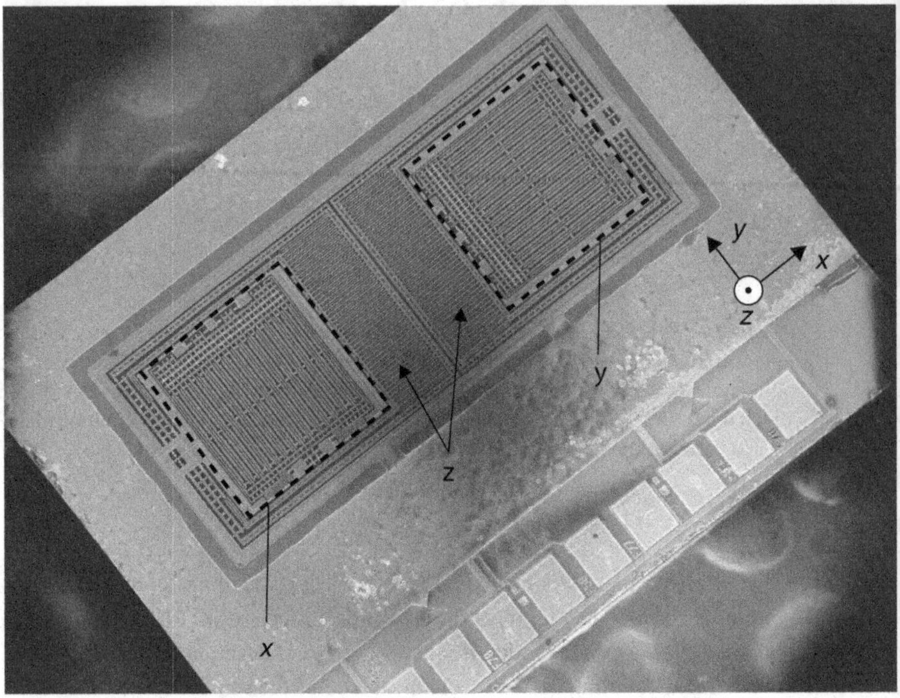

Figure 7.8 Commercial tri-axial accelerometers from Kionix. In-plane accelerometers (x and y) are the same structure rotated 90 degrees, employing variable-gap electrodes. The out-of-plane (z) accelerometer employs a different sensing scheme. The image is credited to Mollie Devoe of Kionix Inc. Used with permission.

A commercial tri-axial accelerometer is shown in Fig. 7.8. The two variable-gap devices for sensing acceleration in the x and y directions are readily identified. The z accelerometer employs variable-area electrodes. Because variable-area capacitors are less sensitive than variable-gap geometries, a larger electrode area is needed to achieve comparable sensitivity.

This section considers only capacitive MEMS devices. For accelerometers on the macroscopic scale, above ~0.5 cm, piezoelectric elements are used almost exclusively. Their frequency range extends from a few Hz up to ~70 kHz. In MEMS, on the other hand, capacitive electromechanical devices dominate. Using switched-capacitor electronics, capacitive accelerometers can respond down to DC, but present-day technology limits the highest frequencies to 50–200 Hz for high-resolution systems and a few kHz for threshold devices, such as airbag sensors.

7.3.1 Principles of operation

An accelerometer can be thought of as a spring-mass system acted upon by an inertial force. The starting point for its analysis is the equation of motion; however, because the frame of reference for this equation is not inertial, a brief review of the fundamental of

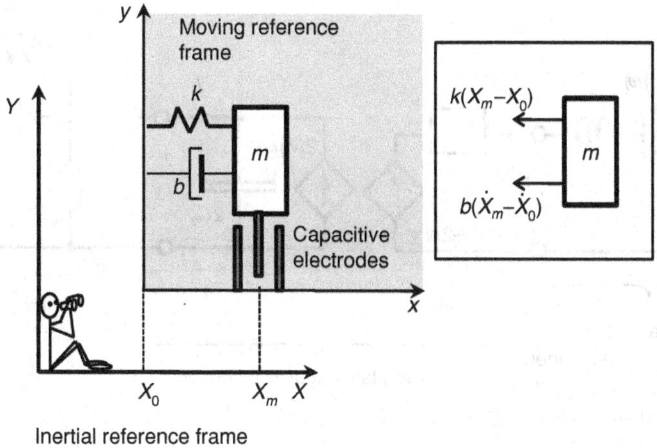

Figure 7.9 An accelerometer is a spring-mass system attached to a moving reference frame xy. XY is the inertial reference frame. The inset shows a rigid-body diagram for the mass m.

Newtonian mechanics is warranted. For an *inertial frame*, Newton's second law states that the vector sum of forces $\overline{F}$ acting on a rigid body is proportional to acceleration and the constant of this proportionality is the mass m:

$$\overline{F} = m\overline{a} = m(\ddot{x}\overline{i}_x + \ddot{y}\overline{i}_y + \ddot{z}\overline{i}_z). \tag{7.22}$$

In Eq. (7.22), $\overline{i}_x$, $\overline{i}_y$, and $\overline{i}_z$ are the unit vectors along the three perpendicular axes. For $\overline{F}$ pointing in the x direction, this vector equation reduces to a scalar expression:

$$F_x = ma_x = m\ddot{x}. \tag{7.23}$$

In practice, virtually no useful frame of reference is strictly inertial, but often they can be approximated as such. An example is the Earth moving through the Solar System. From the point of view of a bicycle rider, the roadway behaves like an inertial frame.

To understand accelerometers, one must carefully distinguish between moving and (approximately) inertial frames of reference. For example, a room can be regarded as the inertial reference, and a cell phone equipped with an inertial accelerometer as the moving reference. A simple model for an accelerometer includes a proof mass held in position by a linear mechanical spring, some form of damping, and pick-off electrodes mounted on the mass. The system is attached to a moving reference frame designated xy. Refer to Fig. 7.9. Coordinates XY signify a global frame of reference, which is assumed to be inertial.

Without loss of generality, assume XY to be both inertial and stationary, and then consider motions of xy with respect to it. X_0 and X_m are, respectively, the locations of the moving (xy) frame and the mass (m) measured in the frame XY. We express the locations of the mass and the moving frame in the global coordinate system because Newton's second law holds only in that inertial reference frame. The free-body diagram using the

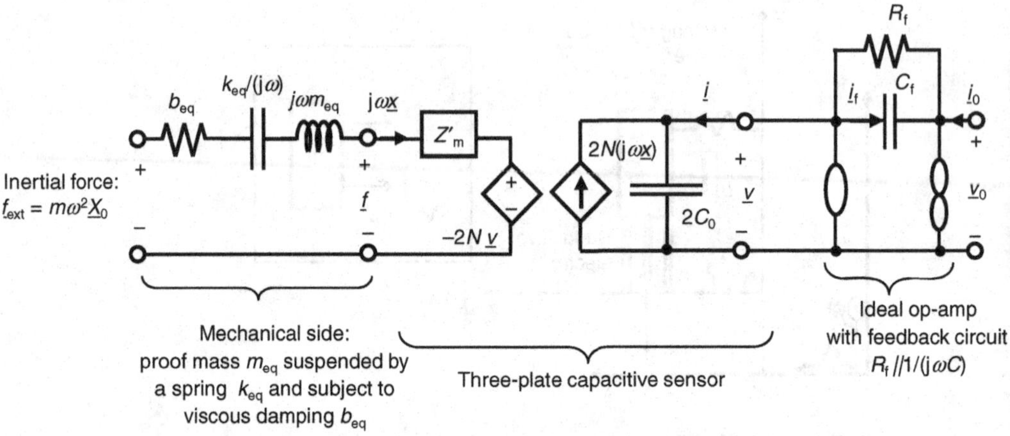

Figure 7.10 An equivalent circuit of the accelerometer system comprising a suspended proof mass, capacitive electrodes in a differential configuration and simple op-amp-based pick-off electronics.

global coordinates is shown in the inset in Fig. 7.9. The equation of motion for the mass m is

$$-k(X_m - X_0) - b(\dot{X}_m - \dot{X}_0) = m\ddot{X}_m. \tag{7.24}$$

Adding $-m\ddot{X}_0$ to both sides of Eq. (7.24), then expressing the location of the mass in terms of the moving frame coordinates,

$$x = X_m - X_0, \tag{7.25}$$

and doing some rearranging yields

$$m\ddot{x} + b\dot{x} + kx = -m\ddot{X}_0. \tag{7.26}$$

On the left-hand side of Eq. (7.26) appear the familiar inertial, damping, and linear spring force terms of the equation of motion for the proof mass, all expressed in terms of the coordinates of the moving reference frame. On the right-hand side appears a force-like term, $-m\ddot{X}_0$, that accounts for the acceleration of the moving reference frame xy with respect to XY. This is the inertial force, often misleadingly referred to as the "apparent" or "fictitious" force.

7.3.2 System transfer function and sensitivity

The simple capacitive accelerometer shown in Fig. 7.9 consists of the proof mass, restoring spring, damping element, and sensing electrodes. If the middle plate of the capacitor is connected to the mass and the outer plates are attached to the moving frame xy and individually addressable, then the three-plate model introduced in Section 4.7.5 can be used. Furthermore, the half-bridge differential amplification network presented in Section 5.4 is applicable. Figure 7.10 shows a linear network for the complete accelerometer system based on the half-bridge scheme and the N-form transducer formulation. The external mechanical input, $\underline{f}_{\text{ext}}$, is the inertial term appearing in Eq. (7.26):

$$\underline{f}_{\text{ext}} = m\omega^2 \underline{X}_0. \tag{7.27}$$

This is the acceleration term the sensor is designed to measure.

The electronic circuitry for commercial accelerometers is quite sophisticated. Typically, an application-specific integrated circuit (ASIC) is designed to handle not only sensing, but also ratiometric and temperature compensation (to be discussed in Section 7.3.7). It often converts the output to various digital formats. Furthermore, it will almost certainly use switched capacitance to mimic resistors and will employ a modulation scheme to suppress $1/f$ noise, as described in Section 5.5. Because these details of the electronics tend to obscure the basic workings of capacitive accelerometers, we here use the simple op-amp circuit with feedback capacitance and resistance described in Section 5.4.

The system can be broken down into three cascaded subsystems [7]. The overall transfer function is the product of the mechanical, transducer, and amplifier transfer functions:

$$\underline{H}_{v_0/a}(\omega) = \underline{H}_{\mathrm{j}\omega x/a}(\omega)\underline{H}_{i/\mathrm{j}\omega x}(\omega)\underline{H}_{v_0/i_0}(\omega). \tag{7.28}$$

These three terms are readily obtained from inspection of the equivalent circuit. The mechanical transfer function relating velocity of the moving electrode to the acceleration force is

$$\underline{H}_{\mathrm{j}\omega x/a}(\omega) = \frac{\mathrm{j}\omega\underline{x}(\omega)}{m\underline{X}_0\omega^2} = \frac{\mathrm{j}\omega}{-\omega^2 + \mathrm{j}\omega\dfrac{b_{\text{eq}}}{m_{\text{eq}}} + \dfrac{k_{\text{eq}} + \mathrm{j}\omega\underline{Z}'_{\text{m}}}{m}} = \frac{j\omega/\omega_0^2}{1-(\omega/\omega_0)^2 + \mathrm{j}2(\omega/\omega_0)\zeta}, \tag{7.29}$$

where $\omega_0 = (k_{\text{eq}}/m)^{1/2}$ is the mechanical resonant frequency of the suspended proof mass and $\zeta = b_{\text{eq}}/(4km)^{1/2}$ is the damping ratio. Note that the voltage-controlled source in the mechanical loop makes no contribution because the input stage to the operational amplifier is at virtual ground, that is, $\underline{v} = 0$.

The transfer function relating velocity $\mathrm{j}\omega\underline{x}$ and current $\underline{i}$ is

$$\underline{H}_{i/\mathrm{j}\omega x}(\omega) = \frac{\underline{i}(\omega)}{\mathrm{j}\omega\underline{x}(\omega)} = -2N = -2V_{\text{B}}\frac{\mathrm{d}C}{\mathrm{d}x}. \tag{7.30}$$

Finally, the transfer function relating the current $\underline{i}$ to the output voltage $\underline{v}_0$ is

$$\underline{H}_{v_0/i_0}(\omega) = \frac{\underline{v}_0(\omega)}{\underline{i}(\omega)} = \frac{R_{\text{f}}}{1+\mathrm{j}\omega R_{\text{f}}C_{\text{f}}}. \tag{7.31}$$

Thus,

$$\underline{H}_{v_0/a}(\omega) = -\frac{\mathrm{j}\omega/\omega_0^2}{1-(\omega/\omega_0)^2 + 2\mathrm{j}\zeta(\omega/\omega_0)}\left(-2V_{\text{B}}\frac{\mathrm{d}C}{\mathrm{d}x}\right)\frac{R_{\text{f}}}{1+\mathrm{j}\omega R_{\text{f}}C_{\text{f}}}. \tag{7.32}$$

Recall that R_{f} is a large resistor, the purpose of which is to prevent op-amp saturation. Thus, over the entire usable frequency range of practical interest, $\omega R_{\text{f}}C_{\text{f}} > 1$. As long as this condition is met, $\underline{H}_{v_0/a}(\omega)$ can be expressed as the product of the mechanical transfer $\underline{H}_{x/a}$ function relating acceleration to displacement of the moving plate and $\underline{H}_{v_0/x}(\omega)$, the transfer function relating the output voltage to the electrode displacement:

$$\underline{H}_{v_0/a}(\omega) \approx -\underbrace{\frac{1/\omega_0^2}{1-(\omega/\omega_0)^2+2\mathrm{j}\zeta(\omega/\omega_0)}}_{\underline{H}_{x/a}}\underbrace{\left(-2V_\mathrm{B}\frac{\mathrm{d}C}{\mathrm{d}x}\right)\frac{1}{C_\mathrm{f}}}_{\underline{H}_{v_0/x}}. \tag{7.33}$$

Typical accelerometers operate well below any inherent mechanical resonance, i.e., $\omega \ll \omega_0$, in which case,

$$\underline{H}_{x/a}(\omega) \rightarrow \frac{1}{\omega_0^2} = \frac{m}{k}, \tag{7.34}$$

so the overall transfer function reduces to

$$\underline{H}_{v_0/a}(\omega) \approx -2\frac{1}{\omega_0^2}V_0\frac{\mathrm{d}C}{\mathrm{d}x}\frac{1}{C_\mathrm{f}}. \tag{7.35}$$

The SI units of $\underline{H}_{v_0/a}$ are V/(m/s^2). In the accelerometer industry, sensitivity is usually specified in units of mV/g, where $g = 9.81$ m/s^2, the acceleration force due to gravity. To change units from standard SI units to this convention, multiply $H_{v_\mathrm{out}/a}(\omega)$ by the factor $10^3/9.81 \approx 101.94$.

According to Eq. (7.35), accelerometer sensitivity is directly proportional to $\mathrm{d}C/\mathrm{d}x$ and the bias voltage V_B, and inversely proportional to resonant frequency squared and feedback capacitance C_f. A practical constraint imposed on the MEMS industry is that, for any given fabrication process and its associated design rules, the nominal capacitive air gap is fixed. Thus, the only way to increase the derivative $\mathrm{d}C/\mathrm{d}x$ is to increase the net area of the capacitive element, which increases the proof mass. Increasing the mass decreases the resonant frequency, which further increases sensitivity, but also limits the bandwidth.

7.3.3 Basic construction of an accelerometer

Having achieved a general form for the transfer function of a MEMS accelerometer, we can consider some structural details and then examine the mechanical and electrical subsystems. Figure 7.11 illustrates the main components of a representative device.

The capacitive pick-off is a three-plate variable-gap structure connected in the standard half-bridge configuration. The middle plate, attached to the proof mass, comprises many individual electrodes connected together. Directly opposite these are two sets of stationary electrodes: $C_1(x)$ and $C_2(x)$. Assuming equilibrium at $x = 0$, the condition essential for balanced half-bridge operation is $\mathrm{d}C_2(x)/\mathrm{d}x = -\mathrm{d}C_1(x)/\mathrm{d}x$. Other three-plate configurations for capacitive MEMS accelerometers have been designed and fabricated, but this antisymmetric feature of C_1 and C_2 is common to almost all configurations.

The variable-gap configuration considered here is prone to pull-in failure. This serious problem is avoided by designing mechanical *bump-stops* (see the inset of

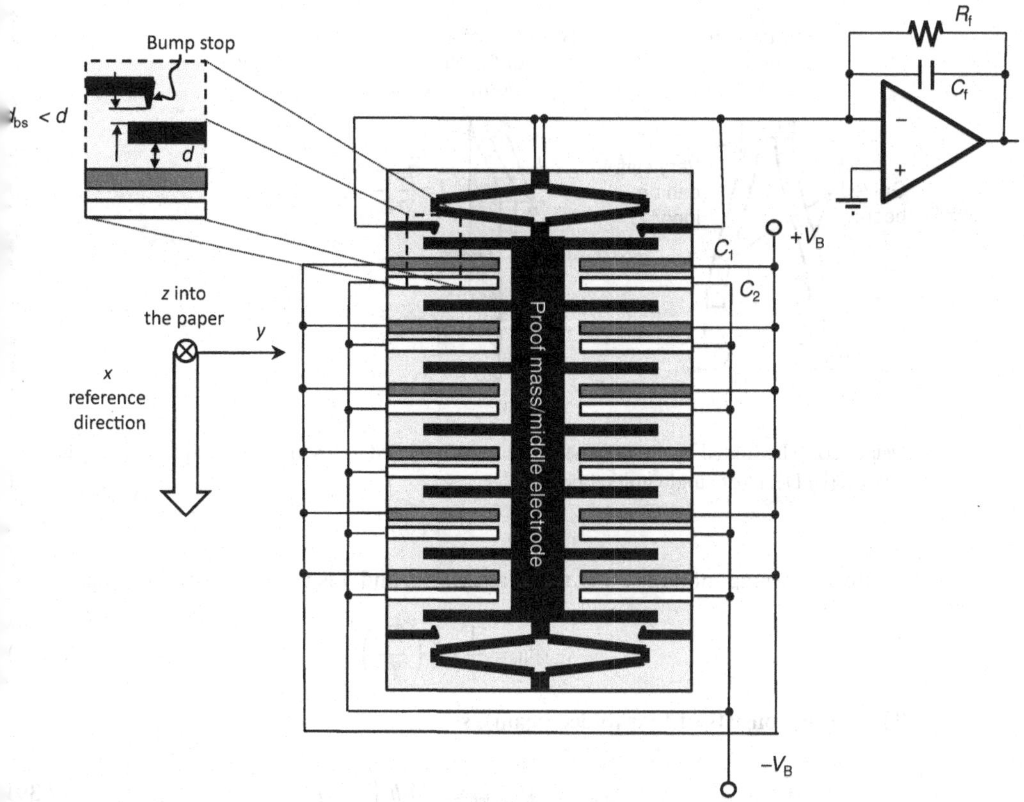

Figure 7.11 Basic variable-gap accelerometer. The principal components are the proof mass (which includes the middle electrode), the stationary electrodes, spring, and the mechanical stops. A simplified sensing circuit is also shown. The inset shows a bump stop.

Fig. 7.11) that prevent the gap from getting smaller than the pull-in spacing, i.e. $d > d_{bs} < d^*$, where d^* is the pull-in spacing. Because the bump stops have a small contact area, van der Waals forces cannot prevent the spring from restoring the moving electrode to its equilibrium.

In Fig. 7.11, a mechanical spring structure called a *folded beam* holds the movable proof mass in equilibrium. Though in practice the effective spring constant of such a structure is usually computed using finite-element analysis, there are approximate formulas for some folded-beam configurations. One simple approach is to represent the folded beam as the serial connection of two guided beams, as shown in Fig. 7.12. Then,

$$\frac{1}{k_{\text{folded}}} = \frac{1}{k_{\text{guided}}} + \frac{1}{k_{\text{guided}}}. \tag{7.36}$$

From Eq. D.55 of Appendix D, the guided beam stiffness is

$$k_{\text{guided}} = Yh\left(\frac{W}{L}\right)^3, \tag{7.37}$$

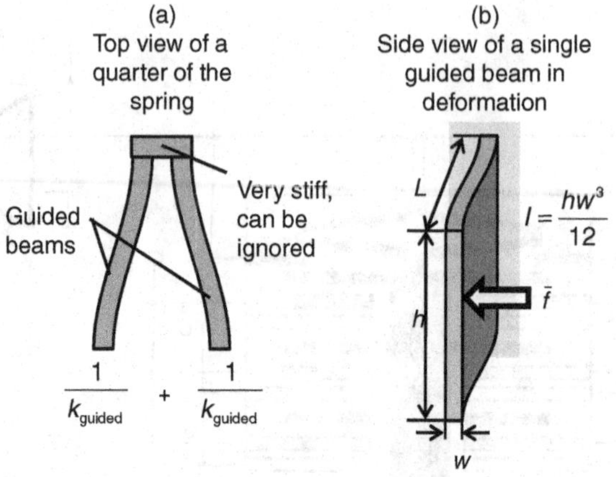

Figure 7.12 The folded beam. (a) The structure is modeled as a series connection of two guided beams. (b) Detail of single guided beam.

where Y is Young modulus, w is the beam width, and h is the beam height. Thus,

$$k_{\text{folded}} = \frac{1}{2} Yh \left(\frac{W}{L} \right)^3. \tag{7.38}$$

The spring consists of four folded beams, so

$$k_{\text{total}} = 4k_{\text{folded}} = 2Yh \left(\frac{W}{L} \right)^3. \tag{7.39}$$

An important observation about the relative sensitivities of bulk- and surface-micromachined accelerometers can be made here. From Eq. (7.38), it is apparent that the stiffness to deflection in the plane of the wafer is proportional to h. Because the mass is also proportional to h, $|\underline{H}_{x/a}| \propto 1/\omega_0{}^2$ is independent of h. However, because the capacitance $C \propto h$, $\mathrm{d}C/\mathrm{d}x \propto h$, so that $\underline{H}_{v_0/a} \propto h$. Therefore, high-aspect ratio, bulk-micromachined, MEMS accelerometers, as depicted in Fig. 7.11, inherently provide more sensitivity than surface-micromachined devices.

Figure 7.13 reveals details of the x-axis unit of the tri-axial accelerometer of Fig. 7.8. In this design, the middle electrode is fixed and the C_1 and C_2 electrodes are moving. The spring design is far more complex than in the sketch of Fig. 7.12.

7.3.4 Mechanical transfer function and mechanical thermal noise

The mechanical subsystem represented by the transfer function $\underline{H}_{x/a}(\omega)$ is a resonant second-order system. Because accelerometer operation is normally restricted to frequencies well below the resonance, that is, $\omega \ll \omega_0 \approx \sqrt{k/m}$, then $\underline{H}_{x/a}(\omega) \propto 1/\omega_0^2$. Therefore, there exists a design trade-off between the requirements of good sensitivity and large bandwidth. Figure 7.14 plots the magnitude and phase response of $\underline{H}_{v_0/a}$

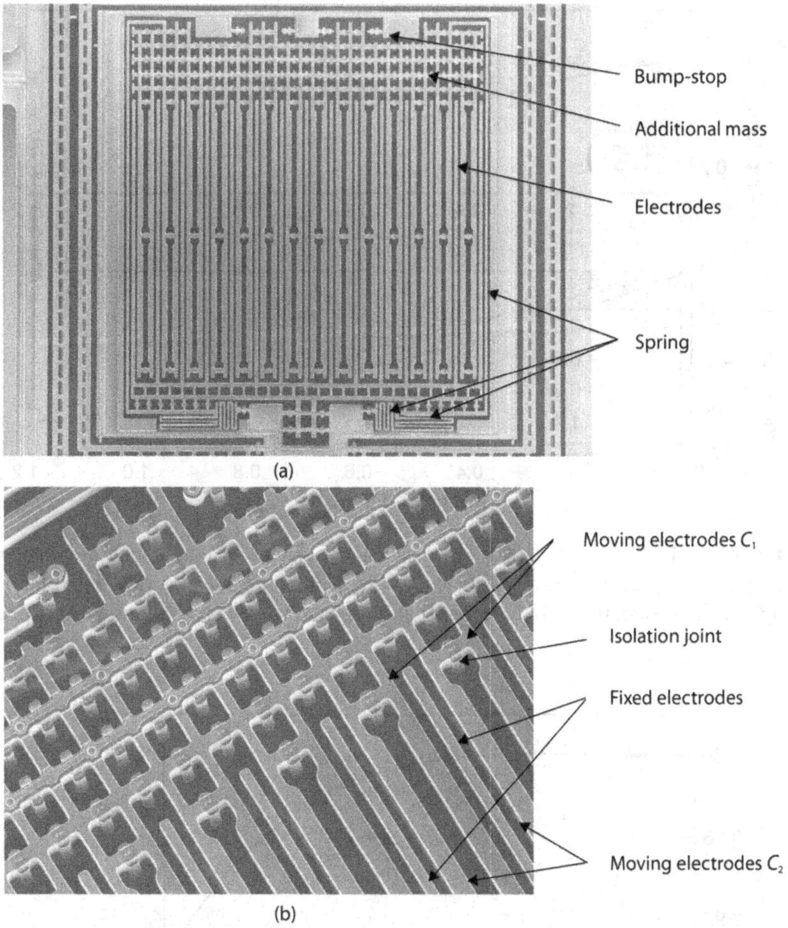

Figure 7.13 Zoom view of the x-axis device of the tri-axial accelerometer of Fig. 7.8. (a) Entire device including springs, electrodes, additional mass, and wire traces. (b) Close-up of the electrode bank, showing mechanical and electrical connections. In particular, it should be noted that isolation joints are electrically isolate electrodes. The isolation singled out in the plot isolates C_1 and C_2 electrodes. The images are credited to Mollie Devoe of Kionix Inc. Used with permission.

for several values of damping ratio ζ, ranging from 0.5 to 1.1.[3] Accelerometers are not high-Q devices. Typically, damping is adjusted to maximize the flatness of the amplitude response, which, as shown in Fig. 7.15, is obtained at $\zeta = 0.7$.[4] Refer to Appendix B. In practice, damping can be controlled by specifying the air density (pressure) within

[3] Recall that $\zeta = 1/(2Q)$. See Appendix B for more details.

[4] In some applications, the critical damping condition, $\zeta = 1$, is a better choice because it gives the best transient response.

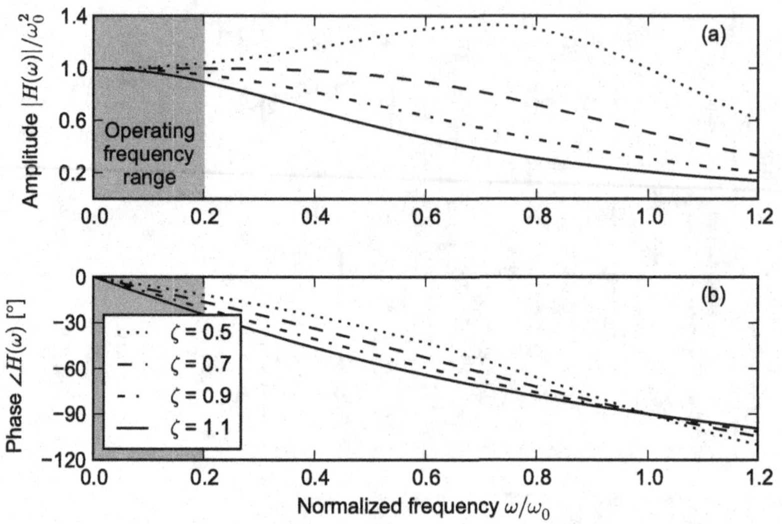

Figure 7.14 Frequency response of mechanical subsystem for four values of damping ratio ζ. The gray shade signifies the usable frequency range. (a) Amplitude plot. (b) Phase plot. For frequency approaching zero, the amplitude response, proportional to $1/\omega_0^2$, exhibits no ζ dependence.

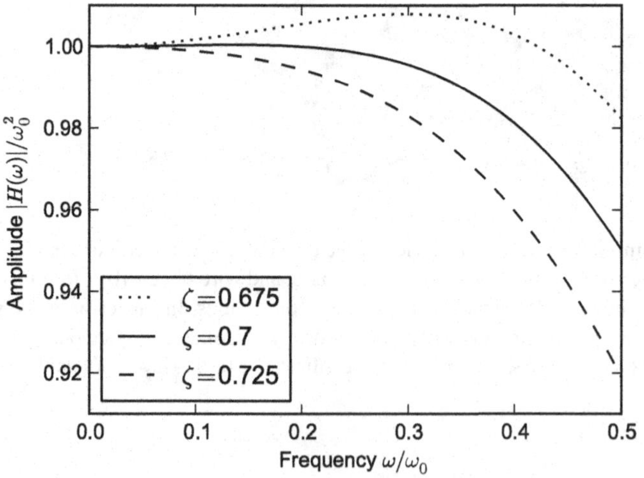

Figure 7.15 Amplitude response of the mechanical side of the system is maximally flat for $\zeta = 0.7$. Typically damping is controlled through hermetic sealing to fix the pressure. Packaging is an integral part of design and development of MEMS devices.

the hermitically sealed device package. Packaging is indeed an integral part of MEMS design and development, not an afterthought.

Mechanical (white) noise of thermal origin influences the resolution of accelerometers. Section 5.8.5 revealed that this noise can be represented as an external force

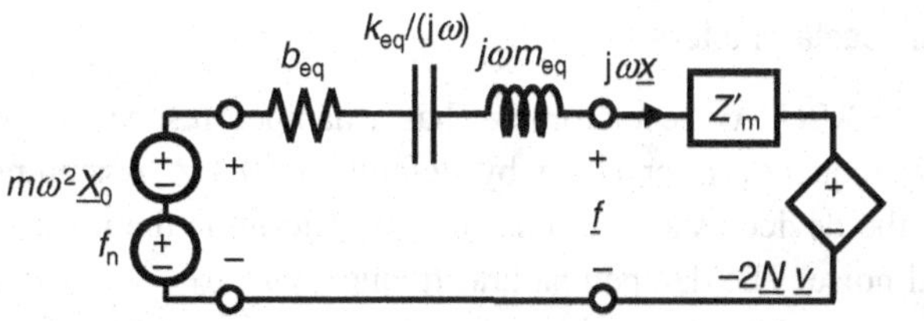

Figure 7.16 Equivalent circuit of the mechanical side including mechanical thermal noise.

contribution related to an integral of the noise density over the bandwidth of the accelerometer:

$$f_n = \sqrt{4k_B T b_{eq} \frac{\Delta\omega}{2\pi}}, \tag{7.40}$$

where $b_{eq} = 2\zeta\omega_0 m$ is the equivalent damping, $k_B = 1.38 \times 10^{-23}$ J/K is Boltzmann's constant, T is the absolute temperature in kelvin and $\Delta\omega/2\pi$ is the device bandwidth in Hz. As shown by the transfer function of Eq. (7.29), the mechanical subsystem behaves like a filter. However, an accelerometer system often employs an additional low-pass filter on the output side of the amplifier. Typically, this filter, not shown in Fig. 7.10, has a corner frequency ω_{LP} considerably lower than the natural frequency of the spring-mass system ω_0, that is, $\omega_{LP} \ll \omega_0$. It is appropriate to base the RMS value of the mechanical thermal noise given by Eq. (7.40) on this corner frequency, i.e. $\Delta\omega \to \omega_{LP}$ because any noise above ω_{LP} is rejected by the filter. Mechanical thermal noise is an additive force, as shown in Fig. 7.16.

For accelerometers, noise is specified as an acceleration quantity. Thus,

$$a_n \equiv \frac{f_n}{m} = \sqrt{\frac{4k_B T(2\zeta\omega_0)\Delta\omega/(2\pi)}{m}}. \tag{7.41}$$

Expressing the system bandwidth as a fraction of the natural frequency, that is, $\Delta\omega = \omega_{LP} = \alpha\omega_0$, provides a more convenient form for this quantity:

$$a_n = \omega_0 \sqrt{\frac{4k_B T\zeta\alpha}{\pi m}}, \tag{7.42}$$

where α is typically chosen by the requirements for the flatness of the frequency response. Equation (7.42) shows that acceleration noise is proportional to the resonant frequency of the spring-mass system. High-frequency accelerometers not only have intrinsically lower sensitivity than their low-frequency counterparts, but also lower signal-to-noise ratio. Unlike sensitivity, which depends only on ω_0, the noise depends on the mass, specifically, $a_n \propto 1/\sqrt{m}$. Thus, an accelerometer with a smaller proof mass is more susceptible to Brownian motion.

Example 7.1 **Thermal noise of an accelerometers**

The Kionix model #KXSC7 *xy* accelerometer has a damped resonance frequency of 4 kHz and its low frequency filter is set by default to 50 Hz. Assuming $\zeta = 0.7$ and approximating the device as a Si beam, one can determine the noise floor of the mechanical thermal noise. The damped natural frequency is $\omega_d = (1 - \zeta^2)^{1/2}\omega_0$ (see Appendix B). Thus

$$\omega_0 = \frac{2\pi f_d}{\sqrt{1-\zeta^2}} = \frac{2\pi \times 4 \times 10^3}{\sqrt{1-0.7^2}} = 3.519 \times 10^3 \quad \text{rad/s},$$

$$\alpha = \frac{\Delta f}{f_d} = \frac{50}{4 \times 10^3}\sqrt{1-.7^2} = 8.927 \times 10^{-3},$$

$$\begin{aligned} a_n &= \omega_0 \sqrt{\frac{4k_B T \zeta \alpha}{\pi m}} \\ &= (3.519 \times 10^3)\sqrt{\frac{4(1.38 \times 10^{-31})(300)(0.7)(8.927 \times 10^{-3})}{\pi}}\frac{1}{\sqrt{m}} \\ &\approx \frac{2 \times 10^{-12}}{\sqrt{m[\text{kg}]}}\left[\frac{\text{m}}{\text{s}^2}\right] \\ &\approx \frac{2 \times 10^{-13}}{\sqrt{m[\text{kg}]}}[g]. \end{aligned}$$

For an estimate of the mass, consider a small Si beam having length, width, and height of 100 μm, 2 μm, and 10 μm, respectively. The proof mass of a real accelerometer is considerably bigger, as can be readily seen from the SEM in Fig. 7.13, but we can use this figure nonetheless to show that even a small mass gives rise to a virtually negligible level of acceleration noise:

$$\begin{aligned} \sqrt{m_{\text{beam}}} &= \sqrt{\rho L w h} \\ &= \sqrt{(2.329 \times 10^3)(100 \times 10^{-6})(2 \times 10^{-6})(10 \times 10^{-6})} \\ &\approx 2 \times 10^{-6}\sqrt{\text{kg}}, \end{aligned}$$

$$a_n \ll 10^{-7} g.$$

Acceleration less than 10^{-7} g is far below the resolution of present MEMS accelerometers. Thus, thermal noise is not a limiting factor for accelerometers of this type.

7.3.5 Selection of the electrode types

Equation (7.35) teaches that accelerometer sensitivity is directly proportional to dC/dx. For a parallel-plate variable-gap structure having electrode length L, electrode height h, and air gap d,

$$\left.\frac{dC}{dx}\right|_{\substack{\text{variable}\\ \text{gap}}} = \varepsilon_0 \frac{hL}{d^2}, \tag{7.43}$$

while, for a variable-area configuration,

$$\left.\frac{\mathrm{d}C}{\mathrm{d}x}\right|_{\substack{\text{variable}\\ \text{area}}} = \varepsilon_0 \frac{hL}{d^2}. \tag{7.44}$$

A MEMS design engineer might need to choose between these two capacitor types for an accelerometer. The relative advantages of the two configurations can be assessed by recognizing that, once the MEMS fabrication process has been selected, the design rules will prescribe a certain minimum value for the gap spacing d and a maximum value for h. If one therefore assumes the same values for d and h in the two design alternatives, a comparison of Eqs. (7.43) and (7.44) reveals that the variable-gap configuration is L/d times more sensitive than the variable-area configuration. Typical accelerometer structures, e.g., the device shown in Fig. 7.6, feature a large number of electrodes and a correspondingly large effective length. In fact, $L/d \sim 10^3$ is typical. Thus, for high-sensitivity (low-g) applications, the variable-gap configuration may be the best choice. For low-sensitivity (high-g) applications, for example, detecting mechanical jolts and shocks, the variable-area structure may be preferred because sensitivity can be traded in favor of a larger stroke. Indeed, the accelerometers in automotive airbag systems are variable-area (comb-drive) structures.

7.3.6 Force-feedback configuration

Just like amplifiers, the performance of sensors can benefit considerably from negative feedback. The main benefits of negative feedback for amplifiers include less sensitivity to gain variation, larger bandwidth, immunity to noise, and reduction of non-linear distortion [8]. The cost for these benefits, reduced gain, is hardly a trade-off in electronic circuit design because op-amps have enormous inherent open-loop gain. Negative feedback affords similar benefits to sensors. Variable-gap electrodes offer excellent sensitivity: $S \sim 1/d^2$. However, this sensitivity comes at the price of inherent non-linearity, which becomes more and more important as d is reduced.

A particular form of negative feedback, called *force feedback*, is a common way to reduce non-linear distortion in accelerometers.[5] Force feedback uses the electrostatic force to maintain the moving element at the equilibrium position. When motion is sensed, a feedback signal drives the moving element back to this equilibrium. The output signal is derived from this feedback. Refer to the general block diagram in Fig. 7.17. The analog signal output v_O is amplified to produce the feedback voltage v_{ff}. This voltage, applied via a separate set of *feedback electrodes*, exerts an additional force f_f on the proof mass to counteract the inertial force. The difference between the inertial force and the feedback force $f_m - f_{ff}$ is referred to as the *force residual*.

Instead of performing a general analysis on the block diagram of Fig. 7.17, we can retain focus on capacitive MEMS accelerometers by using the small-signal electromechanical formalism. Consider the system of Fig. 7.18, where both the sense and

[5] In MEMS design, avoiding non-linearity is not the only factor influencing choice of the gap d. Often the fabrication process imposes a lower limit on d because of the difficulty of achieving small gaps reliably.

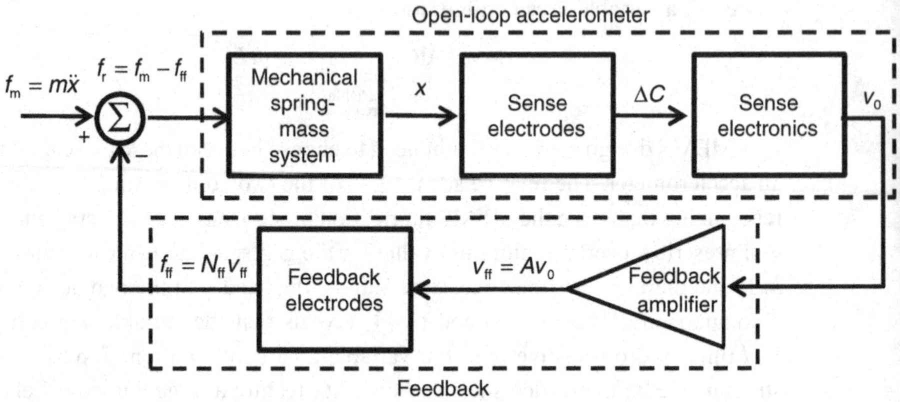

Figure 7.17 A block diagram of a force-feedback accelerometer.

feedback capacitor banks are three-plate variable-gap electrodes. The sketches of the electrode configuration found in the insets should remind the reader that precious wafer area must be reserved for these electrodes and their wiring traces. The area available on the chip must be proportioned correctly between feedback and sense electrodes.

Kirchhoff's voltage law for the equivalent circuit of the mechanical system gives

$$\underline{f}_m = \left[\mathrm{j}\omega m + b + \frac{1}{\mathrm{j}\omega}(k + \underline{Z}'_{\mathrm{m}} + \underline{Z}'_{\mathrm{mff}})\right](\mathrm{j}\omega\underline{x}) + 2\underline{N}\,\underline{v} + 2\underline{N}_{\mathrm{ff}}\underline{v}_{\mathrm{ff}}. \tag{7.45}$$

From Eq. (5.45), the feedback voltage is

$$\underline{v}_{\mathrm{ff}} = 2A\underbrace{(-1)\frac{(\mathrm{d}C/\mathrm{d}x)_{X\to 0}}{C_{\mathrm{f}}}V_{\mathrm{B}}}_{\underline{N}/C_{\mathrm{f}}}\underbrace{\frac{\mathrm{j}\omega C_{\mathrm{f}}R_{\mathrm{f}}}{1+\mathrm{j}\omega C_{\mathrm{f}}R_{\mathrm{f}}}}_{\underline{H}_{\mathrm{c}}(\omega)}. \tag{7.46}$$

In a first-order analysis, we can approximate the high pass transfer function as $\underline{H}_{\mathrm{c}}(\omega) \approx 1$.[6] Then, the feedback voltage becomes

$$\underline{v}_{\mathrm{ff}} \approx 2A\frac{\underline{N}}{C_{\mathrm{f}}}\underline{x}. \tag{7.47}$$

The electromechanical coupling of the feedback electrodes is

$$\underline{N}_{\mathrm{ff}} = -\left(\frac{\mathrm{d}C_{\mathrm{ff}}}{\mathrm{d}x}\right)_{X\to 0} V_{\mathrm{Bff}}. \tag{7.48}$$

where V_{Bff} is the bias voltage of the feedback electrodes and C_{ff} is the capacitance. In general, V_{Bff} is independent of the sense-electrode bias V_{B}.

Because $|\underline{N}_{\mathrm{ff}}\underline{v}_{\mathrm{ff}}|$ is always much greater than $|\underline{N}\,\underline{v}|$, we ignore $\underline{N}\,\underline{v}$. With this simplification, and after inserting Eq. (7.47) and Eq. (7.48), Eq. (7.45) becomes

$$\underline{f}_m = \left[\mathrm{j}\omega m + b + \frac{1}{\mathrm{j}\omega}(k + \underline{Z}'_{\mathrm{m}} + \underline{Z}'_{\mathrm{mff}} + 4A\underline{N}\,\underline{N}_{\mathrm{ff}})\right](\mathrm{j}\omega\underline{x}). \tag{7.49}$$

[6] The simplified sensing electronics here employs an op-amp with feedback resistance, but a practical solution would rely on a switched-capacitance circuit to achieve a good DC response.

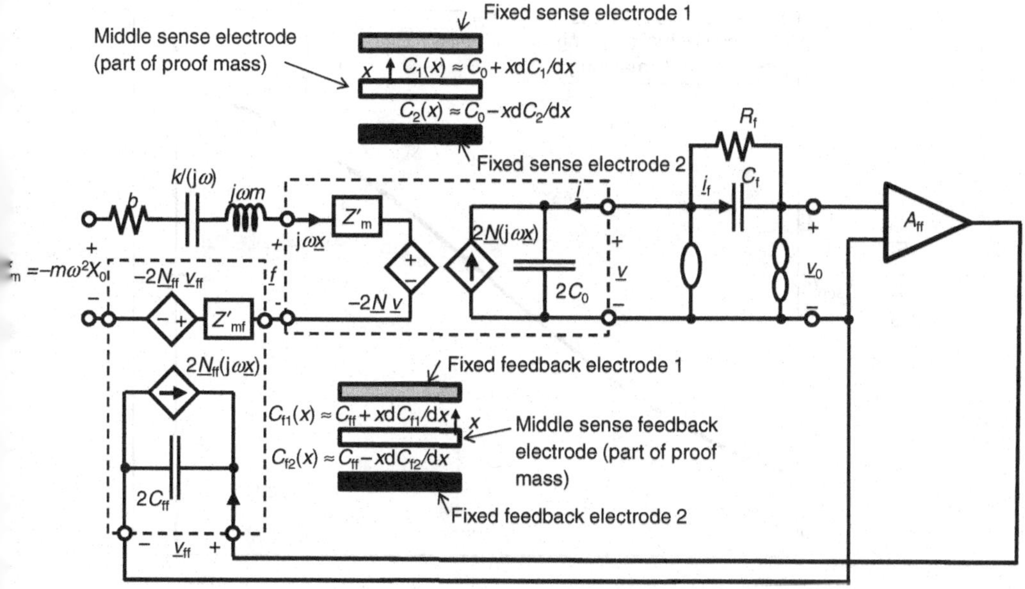

Figure 7.18 Equivalent circuit of a force-feedback accelerometer. The sense and force-feedback electrodes are both differential variable-gap structures.

In practical force-feedback designs, the term $4A\underline{N}\,\underline{N}_{\rm ff}$ dominates the total stiffness, i.e., $|4A\underline{N}\,\underline{N}_{\rm ff}| \gg k \gg |\underline{Z}'_{\rm m}|, |\underline{Z}'_{\rm mff}|$. Thus,

$$\underline{f}_{\rm m} \cong \left[{\rm j}\omega m + b + \frac{k}{{\rm j}\omega}(1+\beta) \right] ({\rm j}\omega \underline{x}), \tag{7.50}$$

where

$$\beta = \frac{\underline{Z}'_{\rm m} + \underline{Z}'_{\rm mff} + 4A\underline{N}\,\underline{N}_{\rm ff}}{k} \approx \frac{4A\underline{N}\,\underline{N}_{\rm ff}}{k}. \tag{7.51}$$

From the above inequalities, $\beta \gg 1$. For simple systems, β is a constant. In more sophisticated force-feedback systems, β is a variable proportional to the input inertial force. Such systems require an additional feedback loop. Being able to adjust A makes it possible to drive the residual force $f_{\rm r}$ virtually to zero so that the displacement of the proof mass remains very small.

The chief benefit of force feedback is a reduction of the displacement of the proof mass by a factor of $1/(1+\beta)$. As a consequence, non-linear distortion is greatly reduced. Also, the bandwidth increases, because the effective force-feedback resonance $\omega_{\rm off}$ is

$$\omega_{\rm off} \approx \sqrt{k(1+\beta)/m} = \omega_0\sqrt{1+\beta}, \tag{7.52}$$

though there are limits to this improvement [4].

An important concern in electromechanical accelerometer design is susceptibility to mechanical shock, which is accentuated for narrow electrodes. Force feedback attenuates displacement imparted by shock, but to be effective, the feedback response must be

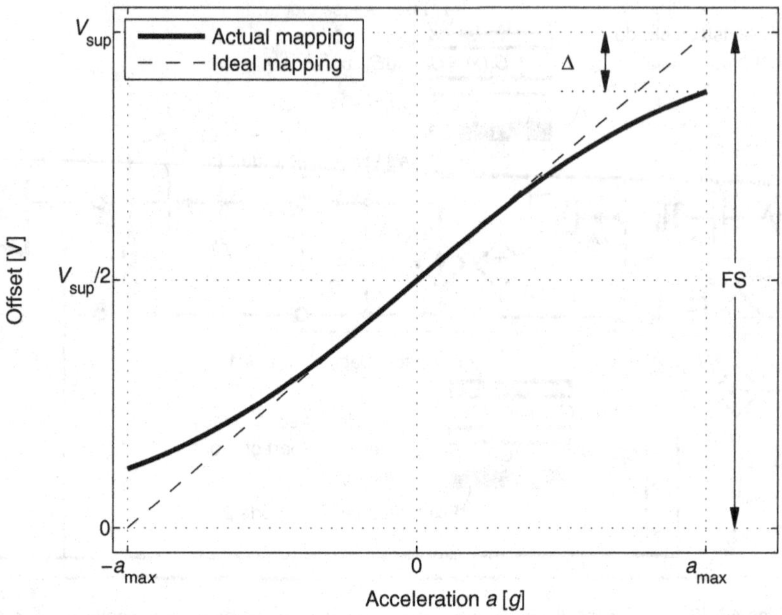

Figure 7.19 Illustration of the typical non-linearity measure of an accelerometer.

sufficiently fast. In addition, the feedback voltage must have sufficient voltage reserve (above the voltage needed in normal operation) to prevent saturation of the variable-gain amplifier. This is because shock magnitudes are apt to be large compared with the normally expected signal. Early work on the subject provides more detail on the force feedback behavior [9, 10].

7.3.7 Higher-order effects

Practical accelerometers are much more complex than hinted at before. Many higher-order effects, detailed coverage of which is beyond the scope of this book, become important in a real device. This subsection briefly introduces some higher-order effects and the way they are specified. Some new terminology is needed to present these specifications.

The zero-g accelerometer analog output, referred to as *offset*, is quantified in volts. Typically, the offset is about one-half of the supply voltage. *Sensitivity* quantifies the conversion of the input acceleration to the final output voltage and is measured in mV/g. *Resolution* is the smallest acceleration that the sensor can detect.

Non-linearity measures the maximum deviation from the ideal, linear relationship between input acceleration and output voltage, scaled by the voltage range. It is expressed as a percentage of *full range*, FS:

$$\text{nonlinearity} = \frac{\Delta}{\text{FS}} \times 100\%. \tag{7.53}$$

Refer to Fig. 7.19.

Cross-axis sensitivity: the mechanical springs of an x-axis accelerometer must be flexible in the x direction, but very rigid in the y and z directions. In addition, the electrodes must be insensitive to motion in the y and z directions. Consider the diagram in Fig. 7.11: any motion in the y or z directions will give rise to equal increments in C_1 and C_2. However, owing to unavoidable asymmetries (e.g., slight bending of the beams due to fabrication), off-axis accelerations may cause deflection of the spring in the perpendicular direction and imperfect cancellation of the capacitance increments, leading to what is called cross-axis sensitivity. In the design phase, finite-element modeling is often used to analyze the response of the structure due to off-axis acceleration. Once known, this response can be used to compute the capacitance change, which in turn is converted to equivalent acceleration. Cross-axis sensitivity is measured in %.

Temperature sensitivity: a particularly important issue in accelerometer design is temperature sensitivity. In bulk micromachining, silicon has excellent temperature characteristics [11], but other materials, such as the metals and oxides, have different coefficients of thermal expansion. Composite beams tend to bend as the temperature changes, resulting in a spurious response that mimics an acceleration. A good design employs symmetry to cancel such temperature effects.

Example 7.2 **Bending of silicon-SiO_2 composite beam**

This example illustrates the common thermo-elastic phenomenon of bending of a structure composed of different materials due to different coefficients of thermal expansion. Such composite structures are initially processed at elevated temperature. As they cool, the materials contract at different rates, giving rise to bending.

Consider a 200 × 2 × 10 μm silicon beam with 0.25 μm thick oxide on the top, shown in Fig. 7.20. Assume that this oxide is deposited using *low-pressure chemical vapor deposition* (LPCVD) at a temperature of ~200 °C [3]. The beam is stress-free at the processing temperature of 200 °C. Because silicon dioxide has a lower linear coefficient of thermal expansion than silicon, $\alpha_{SiO_2} = 0.5$ ppm/K versus $\alpha_{Si} = 2.6$ ppm/K, it shrinks less than silicon during cooling, causing the beam to bend downward.

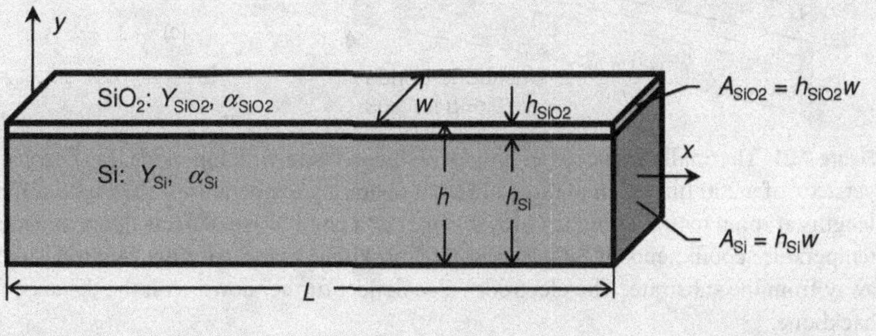

Figure 7.20 A simple 200 × 2 × 10 μm beam with 0.25 μm oxide layer on the top surface.

From Eq. (D.71) of Appendix D the curvature of the beam κ of two layers with dissimilar material is

$$\kappa = \frac{1}{\rho} = \frac{(\alpha_{Si} - \alpha_{SiO_2})\Delta T}{\frac{h}{2} + \frac{2(Y_{Si}I_{Si} + Y_{SiO_2}I_{SiO_2})}{h}\left(\frac{1}{Y_{Si}h_{Si}} + \frac{1}{Y_{SiO_2}h_{SiO_2}}\right)},$$

where ρ is the radius of curvature, Y_{Si} and Y_{SiO_2} are the moduli of elasticity of silicon and silicon dioxide, respectively, and I_{Si} and I_{SiO_2} are the area moments of inertia of the silicon and the silicon dioxide parts of the beam.

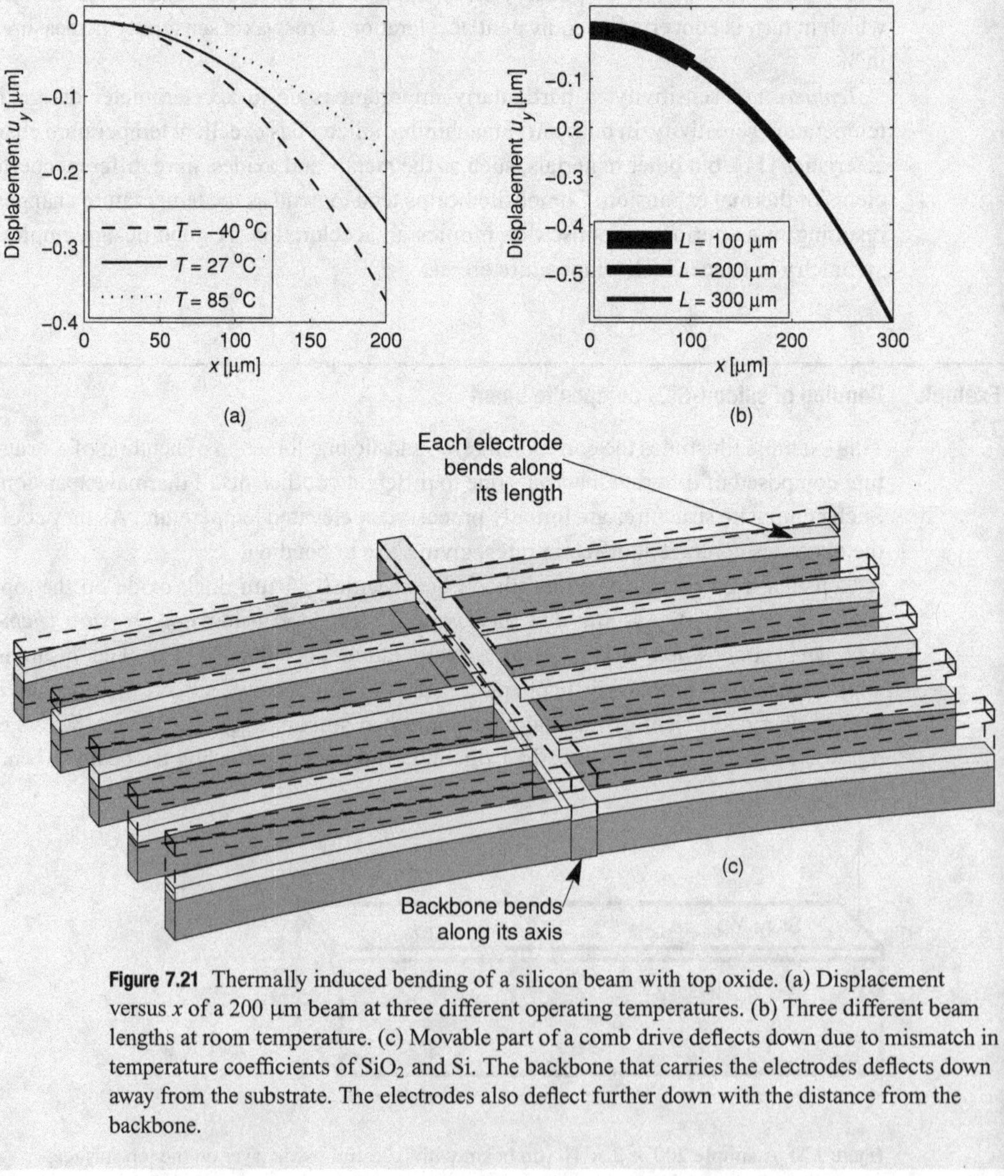

Figure 7.21 Thermally induced bending of a silicon beam with top oxide. (a) Displacement versus x of a 200 µm beam at three different operating temperatures. (b) Three different beam lengths at room temperature. (c) Movable part of a comb drive deflects down due to mismatch in temperature coefficients of SiO_2 and Si. The backbone that carries the electrodes deflects down away from the substrate. The electrodes also deflect further down with the distance from the backbone.

The x-dependent displacement of the cantilever due to differential contraction is

$$u_y(x) = \frac{\kappa}{2}x^2.$$

Refer to Fig. 7.21a. The downward bending due to temperature is ~0.27 μm at room temperature. Over an operating temperature range of −40 °C to 85 °C, the bending of the tip varies between ~0.37 μm and ~0.18 μm. The dependence on temperature is linear. The temperature effect is more important in larger structures, because the deflections are proportional to x^2. Refer to Fig. 7.21b. Figure 7.21c illustrates the deflections of the movable electrode of the comb drive in an exaggerated manner. These temperature-induced deflections give rise to capacitance change.

To predict temperature behavior, the entire structure must be considered in the analysis. Design iterations based on FEA analysis are usually needed to achieve good temperature performance.

Any remaining dependence on temperature can be compensated for by a sophisticated ASIC, which enables temperature correction through a calibration. For some critical applications, each accelerometer is individually calibrated. For less demanding applications where cost precludes such a procedure, the compensation is based on expected average temperature behavior. More specifically, two orthogonal calibrations are often used: one for the offset, and another for sensitivity. Zero-offset variation from room temperature is a specification expressed in mg/°C. Sensitivity variation is expressed in %/°C.

Ratiometricity is the ability of a sensor to track changes in supply voltage. This property is of central importance when the sensor interfaces to a microcontroller or an analog-to-digital converter. Ideally, sensitivity scales linearly with the change of the supply voltage. For example, if the voltage increases by 10%, the sensitivity is expected to increase by the same amount, as revealed in Eq. (7.35) from first-order analysis. Ratiometricity offset and sensitivity are measured in %.

Self-test capability: some applications require a device to respond to a voltage stimulus to facilitate potential sensor fault detection. A small chip area is reserved for the self-test drive electrodes. In the sensor fault function, a pulse voltage is applied to the self-test drive electrodes. The accelerometer responds to this stimulus with a transient response that is processed to verify that the structure is functional, specifically that the sensor is not *stuck*, without necessarily inducing motion.

7.4 MEMS gyroscopes

A rigid body has six degrees of freedom: three linear translations and three rotations. Accelerometers measure linear motion. Gyroscopes, on the other hand, measure angular motion. The first MEMS-scale gyroscope designs emerged in the 1990s. In the first

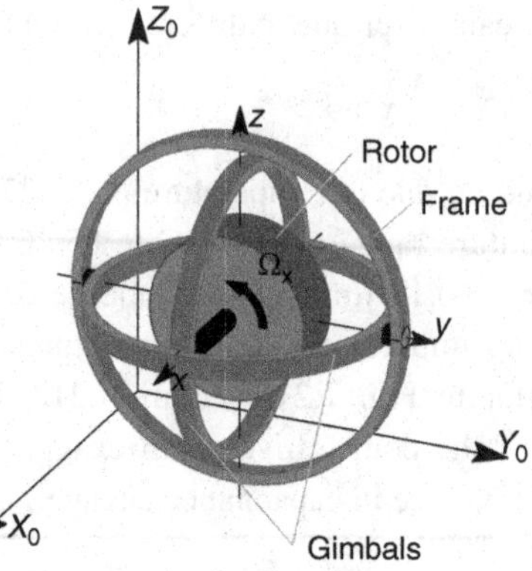

Figure 7.22 A simplified gyrocompass. The rotor spins about the x axis, parallel to the global X_0 axis. When the frame slowly changes its orientation, the gimbals adjust so that the rotor remains oriented along the same global axis. For example, when the frame is slowly rotated so that its plane is no longer perpendicular to the global X_0 axis, the gimbals adjust so that the rotor's local axis x stays parallel to X_0.

decade of the twenty-first century, several distinct design types have been developed, and commercial devices are now available on the market.

7.4.1 A qualitative description of mechanical gyroscopes with some historical notes

Though mechanical vibratory gyroscopes are the most important family for MEMS, historically speaking, rotational (*spinning rate*) gyroscopes came first, in the late nineteenth century. Most of us are familiar with the spinning top gyroscopic toy. In fact, these simple toys provide valuable qualitative understanding of gyroscopic action. Consider the simplified version of a gyrocompass shown in Fig. 7.22. A wheel (rotor) rotates at a high angular speed about its axis. When the spatial orientation of the frame changes, the *gyroscopic effect* causes the gimbals to rotate so that the direction of the spinning axis remains fixed with respect to the inertial frame. This effect is a consequence of the *conservation of angular momentum*, which gives rise to various apparent forces. One of these, the *Coriolis force*, comprises the principle of operation of all mechanical gyroscopes. Angular momentum will be considered in Section 7.4.5.1, but for now it suffices to think in terms of the inertia of a rigid-body mass spinning about its axis.

A common demonstration of angular momentum conservation is the ability to ride a bicycle along a straight path without holding the handlebars as long as the speed is sufficient [12]. When the bicycle starts to tip, the spin axis of the wheel tilts upward and the gyroscopic effect causes the handlebars to turn slightly in the same direction, averting a fall.

In the vibratory rate gyroscope, forced steady vibration in one direction induces vibration in a perpendicular direction when the device is subjected to rotation about

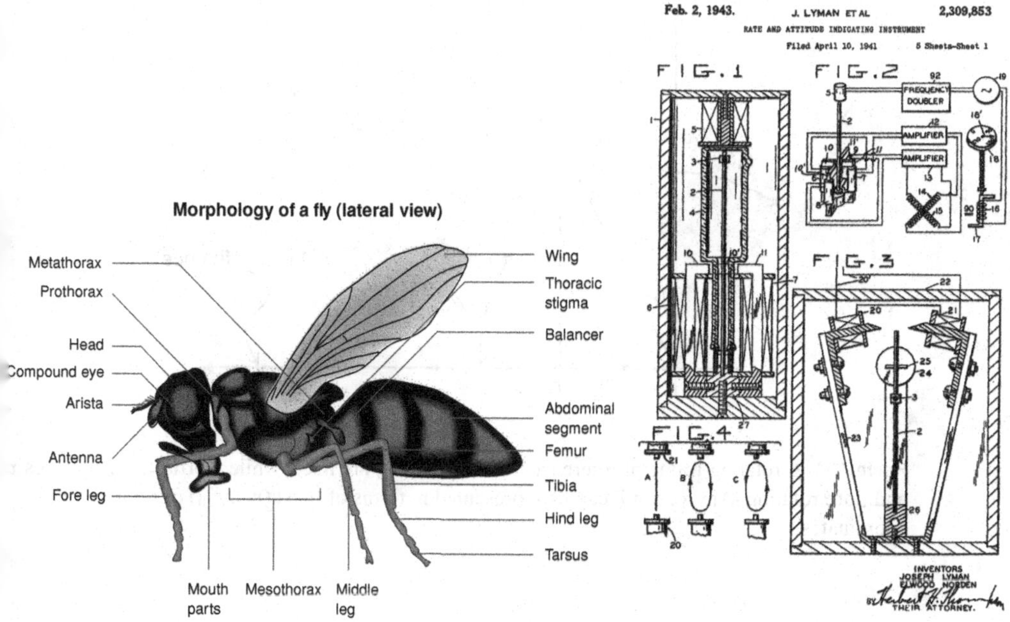

(a) (b)

Figure 7.23 Diverse examples of vibratory gyroscopic action. (a) The housefly has a pair of halteres (balancers) attached to each side of its thorax, which vibrate in a plane perpendicular to the fly's axis. As the insect's flight changes direction, the tips of the halteres experience a Coriolis force. The fly is incapable of controlled flight without them. The image is from Bernard Dery, InfoVisual©. Used with permission. (b) A 1943 vibratory gyroscope (US patent 2,309,853) [13].

the third, perpendicular axis. Vibratory gyroscopes are biologically inspired. They are essential to the flight control system in insects of order *Diptera*, to which the housefly belongs. See Fig. 7.23a. Virtually all MEMS gyroscopes are vibratory. While MEMS vibratory gyroscopes are relatively new, the early macro counterpart shown in Fig. 7.23b dates back to 1943 [13].

Man-made gyros require high symmetry and manufacturing precision. While the scale of these devices and their means of fabrication have changed considerably over time, the required mechanical precision has remained a constant challenge. Early macroscopic designs employed adjustment screws to compensate for inevitable machine imbalances. Twenty years after the first patents were granted for macroscopic scale vibratory rate gyroscopes, these devices were still being perfected for commercial application [14]. Nowadays, some MEMS designs employ laser trimming to achieve balanced operation. Tuning out imperfections in this way is no mere nuisance, but a serious obstacle to economical manufacturability.

The following subsections present a first-order model for the operation of a vibratory MEMS gyro. Higher-order effects, critically important in the design and development of practical devices, are presented later.

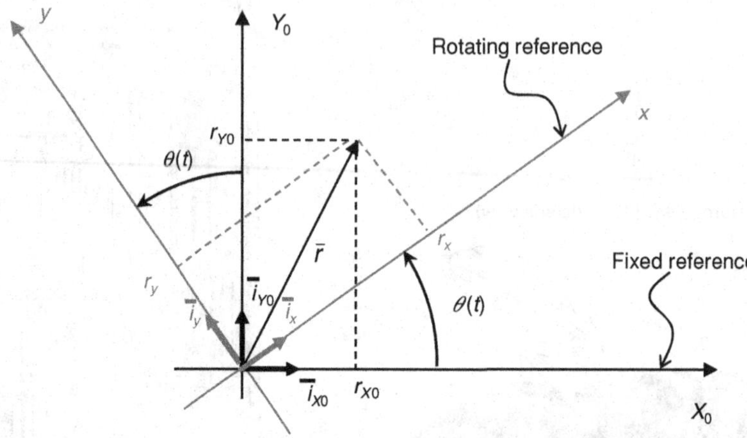

Figure 7.24 A rotating frame of reference. Axes X_0 and Y_0 are fixed, while inertial axes and axes x and y are rotating. Any vector $\bar{r}$ can be represented in terms of fixed (r_{X0}, r_{Y0}) or rotating (r_x, r_y) coordinates.

7.4.2 Rotating reference frames

Rotating frames of reference are not as common to our daily experience as translating reference frames.[7] Their kinematics and dynamics are less intuitive than the corresponding results from translating frames.

Consider the frame of reference xy rotating with respect to the fixed reference XY as shown in Fig. 7.24. Any vector $\bar{r}$ can be expressed in terms of either fixed (r_{X_0}, r_{Y_0}) or rotating coordinates (r_x, r_y):

$$\bar{r} = \overbrace{[r_x \quad r_y] \underbrace{\begin{bmatrix} \bar{i}_x \\ \bar{i}_y \end{bmatrix}}_{\text{time-dependent}}}^{\text{rotating frame of reference}} = \overbrace{[r_{X_0} \quad r_{Y_0}] \underbrace{\begin{bmatrix} \bar{i}_{X_0} \\ \bar{i}_{Y_0} \end{bmatrix}}_{\text{constant}}}^{\text{inertial frame of reference}}. \tag{7.54}$$

To take a time derivative in the rotating frame, it is important to realize that the unit vectors of the rotating frame, $\bar{i}_x$ and $\bar{i}_y$, are time-dependent. Thus, the time derivative of a vector expressed in rotating coordinates has two sets of terms: (i) the derivative of the coordinates multiplied by the unit vectors, and (ii) the coordinates multiplied by the derivative of the unit vectors, that is,

$$\frac{\mathrm{d}\bar{r}}{\mathrm{d}t} = \frac{\mathrm{d}}{\mathrm{d}t}([r_x \quad r_y]) \begin{bmatrix} \bar{i}_x \\ \bar{i}_y \end{bmatrix} + [r_x \quad r_y] \frac{\mathrm{d}}{\mathrm{d}t}\left(\begin{bmatrix} \bar{i}_x \\ \bar{i}_y \end{bmatrix}\right). \tag{7.55}$$

[7] Remember, for instance, your last attempt to walk on a merry-go-round.

An easy way to find the derivatives of the unit vectors in the rotating frame is to express them in terms of the fixed unit vectors:

$$\begin{bmatrix} \bar{i}_x \\ \bar{i}_y \end{bmatrix} = \begin{bmatrix} \cos(\theta) & \sin(\theta) \\ -\sin(\theta) & \cos(\theta) \end{bmatrix} \begin{bmatrix} \bar{i}_{X_0} \\ \bar{i}_{Y_0} \end{bmatrix}. \tag{7.56}$$

Refer to Fig. 7.24 for definition of θ. Then,

$$\frac{\mathrm{d}}{\mathrm{d}t}\left(\begin{bmatrix} \bar{i}_x \\ \bar{i}_y \end{bmatrix}\right) = \begin{bmatrix} -\dot{\theta}\sin(\theta) & -\dot{\theta}\sin(\theta) \\ -\dot{\theta}\sin(\theta) & \dot{\theta}\sin(\theta) \end{bmatrix} \begin{bmatrix} \bar{i}_{X_0} \\ \bar{i}_{Y_0} \end{bmatrix} = \dot{\theta} \begin{bmatrix} \bar{i}_y \\ \bar{i}_x \end{bmatrix}. \tag{7.57}$$

Recognizing that angular velocity is a vector in the z direction,

$$\dot{\bar{\theta}} = \dot{\theta}\bar{i}_z = \Omega_z \bar{i}_z \tag{7.58}$$

yields an expression in terms of the unit vectors of the rotating coordinate system:

$$\frac{\mathrm{d}}{\mathrm{d}t}\left(\begin{bmatrix} \bar{i}_x \\ \bar{i}_y \end{bmatrix}\right) = (\Omega_z \bar{i}_z) \times \begin{bmatrix} \bar{i}_x \\ \bar{i}_y \end{bmatrix}. \tag{7.59}$$

Thus, the velocity vector expressed in the rotating coordinates becomes

$$\dot{\bar{r}} = [\dot{r}_x \quad \dot{r}_y] \begin{bmatrix} \bar{i}_x \\ \bar{i}_y \end{bmatrix} + [r_x \quad r_y] \left((\Omega_z \bar{i}_z) \times \begin{bmatrix} \bar{i}_x \\ \bar{i}_y \end{bmatrix} \right). \tag{7.60}$$

More generally, the velocity vector of a point in any reference frame xy rotating at angular velocity $\overline{\Omega}$ about an arbitrary axis is

$$\dot{\bar{r}}_{\text{inert}} = \dot{\bar{r}}_{\text{rot}} + \overline{\Omega} \times \bar{r}_{\text{rot}}. \tag{7.61}$$

So the rate operator $\mathrm{d}/\mathrm{d}t()_{\text{inert}}$ in the inertial reference frame has two terms when based on observations in the rotating frame: a local rate operator $\mathrm{d}/\mathrm{d}t()_{\text{rot}}$ term and a cross-product $\overline{\Omega} \times ()_{\text{rot}}$,

$$\frac{\mathrm{d}}{\mathrm{d}t}()_{\text{inert}} = \frac{\mathrm{d}}{\mathrm{d}t}()_{\text{rot}} + \overline{\Omega} \times ()_{\text{rot}}. \tag{7.62}$$

Now recall that Newton's second law holds only in inertial reference frames:

$$m\ddot{\bar{r}}_{\text{inert}} = \sum \bar{f}. \tag{7.63}$$

Therefore, to express momentum conservation in the inertial reference frame $\ddot{\bar{r}}_{\text{inert}}$ based on observations in the rotating frame ($\bar{r}_{\text{rot}}$, $\dot{\bar{r}}_{\text{rot}}$, and $\ddot{\bar{r}}_{\text{rot}}$), one applies the operator of Eq. (7.62) to the velocity expression of Eq. (7.61):

$$\ddot{\bar{r}}_{\text{inert}} = \ddot{\bar{r}}_{\text{rot}} + 2\overline{\Omega} \times \dot{\bar{r}}_{\text{rot}} + \dot{\overline{\Omega}} \times \bar{r}_{\text{rot}} + \overline{\Omega} \times (\overline{\Omega} \times \bar{r}_{\text{rot}}). \tag{7.64}$$

Inserting Eq. (7.64) into Eq. (7.63) yields the equation of motion in terms of the rotating frame observables:

$$m\ddot{\bar{r}}_{\text{rot}} + 2m\overline{\Omega} \times \dot{\bar{r}}_{\text{rot}} + m\dot{\overline{\Omega}} \times \bar{r}_{\text{rot}} + m\overline{\Omega} \times (\overline{\Omega} \times \bar{r}_{\text{rot}}) = \sum \bar{f}. \tag{7.65}$$

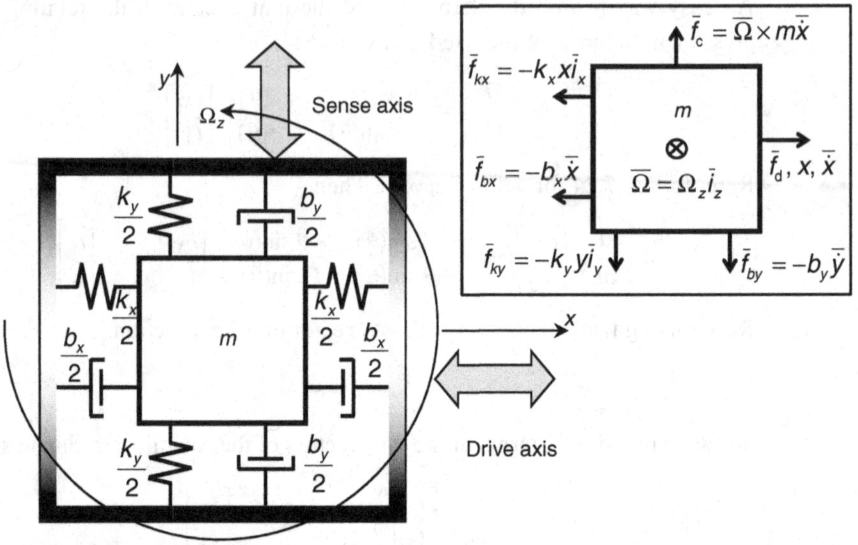

Figure 7.25 Rigid body driven at resonance in the x direction. When the reference frame experiences rotation about the z axis, the Coriolis force induces motion in the y direction. The inset is a rigid body diagram showing the mass subject to drive force f_d, Coriolis force f_c and reaction forces f_{kx}, f_{bx}, f_{ky}, and f_{by}.

Moving the acceleration terms containing $\bar{r}_{\text{rot}}$ and $\dot{\bar{r}}_{\text{rot}}$ to the right-hand side, the equation of motion takes the form

$$m\ddot{\bar{r}}_{\text{rot}} = \sum \bar{f} + \underbrace{(-2m\bar{\Omega} \times \dot{\bar{r}}_{\text{rot}})}_{\text{Coriolis force}} + \underbrace{(-m\dot{\bar{\Omega}} \times \bar{r}_{\text{rot}})}_{\text{Euler force}} + \underbrace{(-m\bar{\Omega} \times (\bar{\Omega} \times \bar{r}_{\text{rot}}))}_{\text{centrifugal force}}. \quad (7.66)$$

The $\bar{r}_{\text{rot}}$-dependent terms on the right-hand side of Eq. (7.66) are referred to as the *inertial* or *apparent* forces: the Coriolis force, the Euler force, and the centrifugal force.

7.4.3 A simple z axis rate vibratory gyroscope

The above basics of rotational dynamics for a point mass allow us to analyze a very simple yet instructive example of a z axis gyroscope. The mechanical system is examined first. The necessary electrodes and electronic control are then incorporated to build up a system-level model.

7.4.3.1 Mechanical subsystem

The equation of motion in the rotating frame serves as the starting point for an investigation of the response of a vibrating mass subjected to the Coriolis force. Consider a mass attached to a set of x and y axis restoring springs and dampers, as shown in Fig. 7.25. To maximize the amplitude of its motion, the mass is driven at its resonance in the

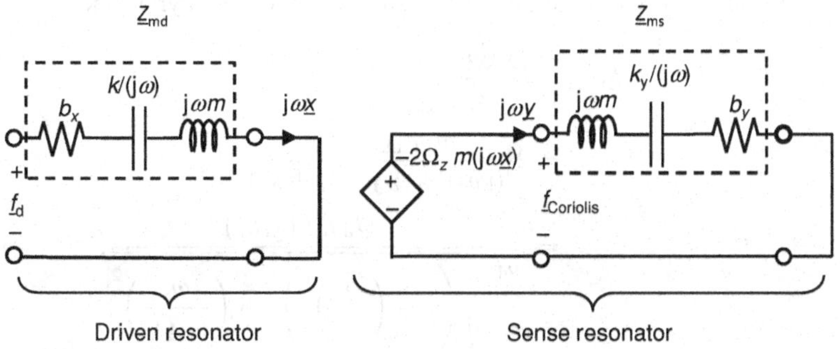

Figure 7.26 Equivalent circuit of inertially coupled resonators. The feedback from the sense resonator is ignored in this model. Strictly speaking, the input rate Ω_z is time-dependent, but in this first-order analysis, it is treated as a constant.

x direction. In the absence of any rotation, the position, velocity, and acceleration of the mass expressed in the rotating coordinates are

$$\overline{r}_{\text{rot}} = x\overline{i}_x, \ \dot{\overline{r}}_{\text{rot}} = x\overline{i}_x, \ \ddot{\overline{r}}_{\text{rot}} = \ddot{x}\overline{i}_x. \tag{7.67}$$

The equation of the driven motion is

$$m\ddot{x} + b_x\dot{x} + k_x x = f_x \cos(\omega_d t). \tag{7.68}$$

Now, under the assumption that the vibration has reached steady-state, subject the reference frame xy to rotation about the z axis at constant angular velocity $\overline{\Omega} = \Omega_z\overline{i}_z$. The angular velocity is assumed to be much smaller than the rate of vibration, i.e., $\Omega_z \ll \omega_d$. Application of Eq. (7.66) yields the following set of equations:

$$m\ddot{x} + b_x\dot{x} + k_x x = f_x \cos(\omega_d t) - 2m\Omega_z\dot{y} + m\Omega_z^2 x + m\dot{\overline{\Omega}}_z y, \tag{7.69}$$

$$m\ddot{y} + b_y\dot{y} + k_y y = -2m\Omega_z\dot{x} + m\Omega_z^2 y + m\dot{\Omega}_z x. \tag{7.70}$$

Comparing the three inertial terms on the right-hand side of Eq. (7.69) with $m\ddot{x} \sim m\omega_d^2 x$ on the left-hand side tells us that the Coriolis back action $-2m\Omega_z^2\dot{y}$, centrifugal force $m\Omega_z^2 x$, and angular acceleration $m\dot{\Omega}_z y$ may all be ignored because $\Omega_z \ll \omega_d$ and $|y| \ll |x|$. In addition, the centrifugal and angular acceleration terms are much smaller than the Coriolis terms in Eq. (7.70) and may be ignored. Subject to these approximations, the system simplifies to the following set of coupled equations:

$$m\ddot{x} + b_x\dot{x} + k_x x = f_x \cos(\omega_d t), \tag{7.71}$$

$$m\ddot{y} + b_y\dot{y} + k_y u = -2m\Omega_z\dot{x}. \tag{7.72}$$

Equation (7.72) reveals that rotation about the z axis Ω_z inertially couples the driven vibration in the x direction to motion in the y direction. Figure 7.26 provides a useful electric circuit representation for the coupled system described by Eqs. (7.71) and (7.72).

In general, the Coriolis force is a product of two time-dependent terms, but, as long as the bandwidth of the angular speed Ω_z is very small compared with the resonant frequency of the vibrating mass, one may consider Ω_z to be constant in a first-order analysis.

A useful sensitivity function for this simple point-mass gyroscope[8] is

$$\begin{aligned} S_{y/\Omega} &= \left|\frac{\underline{y}(\mathrm{j}\omega)}{\Omega_z}\right|_{\omega=\omega_\mathrm{d}} \\ &= 2\frac{\omega_\mathrm{d} m |\underline{x}_\mathrm{d}(\mathrm{j}\omega_\mathrm{d})|}{|(\mathrm{j}\omega_\mathrm{d})^2 m + \mathrm{j}\omega_\mathrm{d} b_y + k_y|} \\ &= \frac{2}{\omega_\mathrm{d}} \frac{Q_\mathrm{d} f_\mathrm{d}/(k_x k_y)}{\sqrt{\left(1-\left(\frac{\omega_\mathrm{s}}{\omega_\mathrm{d}}\right)^2\right)^2 + \left(\frac{\omega_\mathrm{s}}{\omega_\mathrm{d} Q_\mathrm{s}}\right)^2}}, \end{aligned} \tag{7.73}$$

where ω_d, ω_s, Q_d, and Q_s are the drive resonant frequency, sense resonant frequency, drive quality factor, and sense quality factor, respectively. The mass m does not appear in the sensitivity expression because the rigid body vibrates in the driven x direction and responds in the y direction. In many MEMS gyroscope designs, however, the drive and sense resonator masses are not the same. Such a case is considered in Section 7.4.6.

$S_{y/\Omega}$ is proportional to the magnitude of the drive velocity, which, in turn, is directly proportional to the quality factor Q_d and drive force f_d. The response also depends on the ratio of the two resonance frequencies. Clearly, when $\omega_\mathrm{s} = \omega_\mathrm{d}$, the sensitivity is maximized:

$$S_{y/\Omega}|_{\omega_\mathrm{s}=\omega_\mathrm{d}} = 2\frac{Q_\mathrm{s} Q_\mathrm{d} f_\mathrm{d}}{k_x k_y}. \tag{7.74}$$

This condition is referred to as *resonant gyro operation*. Because of the difficulty in achieving this condition and, moreover, maintaining it over the entire operating temperature range of a gyroscope, resonant gyro operation is seldom used in practical designs. Certain designs with inherent symmetry, such as wine-glass resonators, are good candidates for resonant gyro operation.[9]

A second problem with resonant gyro operation stems from the finite bandwidth of Ω_z. Equation (7.74) is derived assuming constant rotation Ω_z. In practice, however, though small compared with ω_d and ω_s, the bandwidth of the Ω_z input signal is finite. Therefore, the resonant peaks must be sufficiently flat so that the magnitude response does not vary appreciably over the bandwidth of the input signal. This limit imposes restrictions upon the quality factors Q_s and Q_d.

It is important to distinguish resonant gyro operation from driving the mass at resonance. All vibratory gyros employ resonance to achieve large-amplitude driven motion.

Because the challenges associated with designing a gyro for $\omega_\mathrm{s} = \omega_\mathrm{d}$ are so difficult, a question arises of how to order the two resonances: $\omega_\mathrm{d} > \omega_\mathrm{s}$ or $\omega_\mathrm{d} < \omega_\mathrm{s}$?

[8] A rigid body subjected to translation alone acts like a point mass. While this gyro is a rigid-body device, this first-order analysis only considers translation of the body due to the rotation of the frame, i.e., it employs the point-mass model.

[9] Wine-glass resonators, also known as hemispherical resonant gyros (HRG), take advantage of the mode degeneracy for the drive and sense resonator. Refer to Loper and Lynch [15] for more details on a macro HRG. A MEMS example of a ring gyroscope is described by Ayazi and Najafi [16].

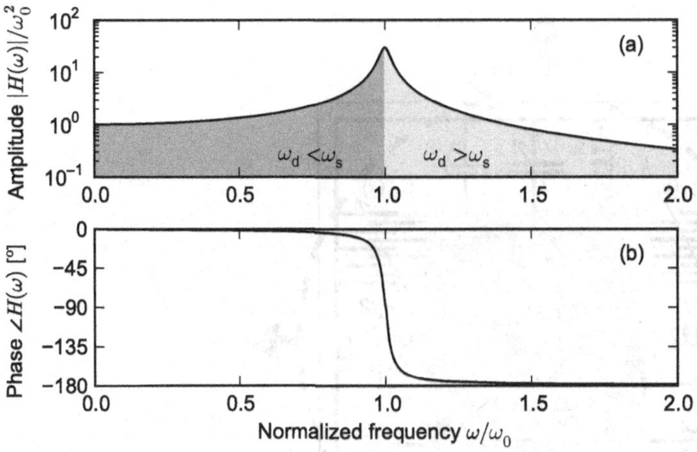

Figure 7.27 Frequency response of the sense resonator. The sense resonator is driven very nearly at ω_d. The resonator has a flatter magnitude response for $\omega_d < \omega_s$, than for $\omega_d > \omega_s$.

Figure 7.27 shows that, while the phase response of a resonator is symmetric about the resonance, the magnitude response is much flatter at frequencies below the resonance than above. This observation suggests that $\omega_d < \omega_s$ is the better choice. A practical point for MEMS gyros is that, owing to unavoidable processing imperfections on the wafer, the frequency separation $\omega_s - \omega_d$ varies from die to die. In addition, for each gyro, this separation is affected by temperature and other operating conditions. It should be clear that such variations in frequency separation will have a smaller effect on sensitivity when $\omega_d < \omega_s$.

Having discussed the important design issues influencing the relative values of ω_d and ω_s, some comments on practical frequency ranges are in order. Though greatly affected by the fabrication process, typical values are fairly well established. Larger devices operating in the kHz range feature large electrodes and a rigid frame that confines the flexure via carefully designed springs. Such structures exhibit large-amplitude motion and high sensitivity, but they are vulnerable to shock and vibration and also to the Euler force term $(m\dot{\Omega}_z x)$ ignored in the first-order analysis. For smaller devices operating at frequencies ≥ 10 kHz, it is easier to isolate the gyro mechanically to reduce shock and vibration sensitivity. Protection from shock and vibration comes at the price of reduced overall sensitivity, because a smaller device operating at higher frequency also has smaller electrodes and consequently reduced electromechanical coupling.

7.4.3.2 Electromechanical coupling and the system-level model

Many of the concepts and ideas needed to develop a system-level model for a gyroscope have already been presented. It is only necessary now to assemble all the elements. In particular, Example 3.4 introduced an apparently simple idealization for a gyroscope consisting of two orthogonal sets of variable-area capacitive electrodes attached to the driven resonant structure and a separate set of variable-gap electrodes fixed to the sense resonator. The variable-area drive and sense electrodes on the driven resonator

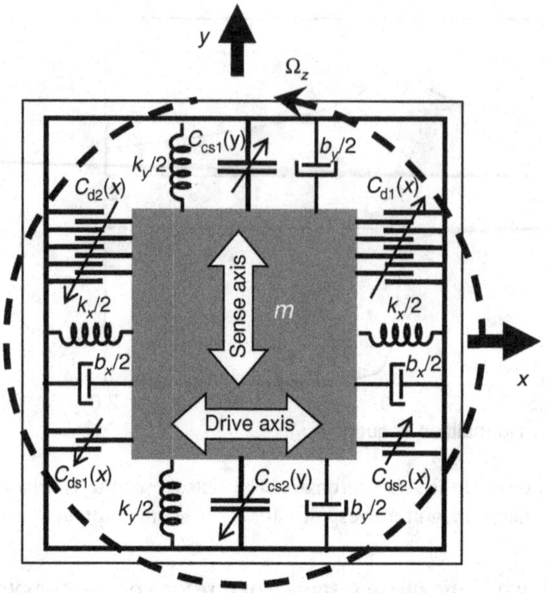

Figure 7.28 Conceptual design of a z-axis gyroscope driven at resonance in the x direction with the Coriolis sense in the y direction. C_{d1} and C_{d2} are located on opposite sides, to provide pull in the $+x$ and $-x$ directions. C_{ds1} and C_{ds2} form a three-plate differential sensing structure. C_{cs1} and C_{cs2} are also arranged in the three-plate, differential scheme. The drive electrodes, C_{d1} and C_{d2}, and drive sense electrodes, C_{ds1} and C_{ds2}, are variable-area structures, while the Coriolis sense electrodes, C_{cs1} and C_{cs2}, are variable-gap structures.

accommodate large-amplitude motion. The variable-gap sense electrodes can detect the small-amplitude induced motion and provide feedback to the control system that supplies voltage for the drive resonator.

Another place in the text that the reader might profitably review before proceeding further is Section 5.9, which introduced and analyzed resonant drives in some detail. The different pieces to be assembled here enable closed-loop resonant vibration of the driven resonator.

Figure 7.28 shows a conceptual schematic for a z-axis gyroscope. The mass m is driven at resonance in the x direction. The input Ω_z couples the driven motion in the x direction to the response motion in the y direction. The drive electrodes, C_{d1} and C_{d2}, and drive sense electrodes, C_{ds1} and C_{ds2}, are variable-area structures, while the Coriolis sense electrodes, C_{cs1} and C_{cs2}, are variable-gap structures. C_{d1} and C_{d2} are located on the opposite side to provide pull in $+x$ and $-x$ directions. C_{ds1} and C_{ds2} form a three-plate differential sensing structure. C_{cs1} and C_{cs2} are also arranged in the three-plate differential scheme. The damping and spring constants are generally different in the x and y directions. In most designs, the stiffness in a particular direction is achieved with pairs of springs placed symmetrically with respect to the mass, just as shown here.

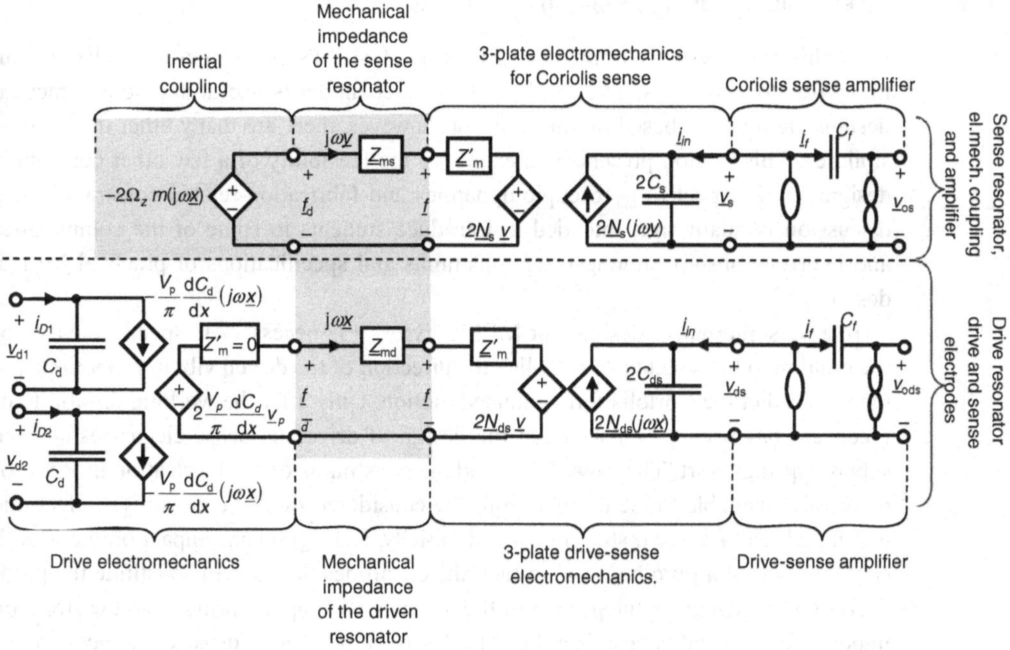

Figure 7.29 Equivalent electromechanical network for a capacitive gyroscope. The gyro consists of two inertially coupled mechanical resonators (identified by the shaded area) and three sets of electrodes. Two electrode sets (representing the drive and drive-sense electromechanics) are coupled to the driven resonator to enable closed-loop vibration of the driven resonator. The third electrode set senses the y-directed Coriolis force-induced motion and converts it to an electrical signal.

Figure 7.29 shows the equivalent electromechanical system network for a single-axis gyroscope. A system like this would use closed-loop resonant vibration, but the feedback loop between the drive sense and the drive electrodes is not shown here. Refer to Section 5.9.6 for the discussion of feedback. The shaded area marks the equivalent circuit of the two inertially coupled resonators from Fig. 7.26, and the driven resonator of Fig. 5.55. The drive and drive-sense electrode sets are coupled with the driven resonator to enable closed-loop operation. The Coriolis motion-sensing capacitors, C_{cs1} and C_{cs2}, convert the Coriolis-induced y-directed motion into an electrical signal $\underline{v}_{os}$.

The variable-area drive electrodes, C_{d1} and C_{d2}, accommodate large travel, which provides the best sensitivity to the Coriolis force. Since the gyro is driven at resonance, the motion is relatively large anyway, so the drive sense electrodes, C_{ds1} and C_{ds2}, need not be especially sensitive. Moreover, they also have to accommodate large travel. Consequently, the drive sense electrodes are also of the variable-area type. Ordinarily, more chip area is allocated to the drive electrodes than to the drive sense electrodes.

The variable-gap Coriolis sense electrodes provide the sensitivity necessary to detect the much smaller y motion. Refer to Section 7.3.5 for a discussion of the relative sensitivities of variable-gap and variable-area capacitive electrodes.

7.4.4 Other examples of MEMS-based gyroscopes

Up to this point, we have limited consideration of MEMS gyroscopes to idealized point mass vibratory systems. The resulting first-order model is useful and some practical designs are actually based on this concept; however, there are many other design possibilities. This section presents and examines the feasibility of a few other conceptual designs with respect to gyroscopic dynamics and fabrication-related constraints. The discussion is qualitative, intended to introduce students to some of the complexities and interrelationships amongst the constraints and specifications of practical MEMS designs.

To assess alternative designs for MEMS gyros, it is necessary to specify the axis of the rotation to be detected, to establish the direction of the driven vibratory motion, and then to predict the Coriolis force-induced motion. Only after the working feasibility of a concept has been established can the design of drive and sense electrodes and the support springs start. The capabilities and the constraints of the fabrication process (or processes) available to the designer must be considered from the outset. The electrode design, which may seem straightforward initially, has significant impact on the overall configuration of a gyro because, in fact, the electrodes themselves constitute the proof mass of a MEMS gyro. Integration of the mechanical support springs into the structure imposes further challenges; significant FEA modeling effort is usually expended in this task.

The first alternative designs to be examined again use a translating rigid mass. The z-axis gyro considered in Section 7.4.3 experiences Coriolis force-induced motion in the y-axis perpendicular to driven vibratory motion in the x-axis. If the device is simply rotated in the plane of the wafer, the drive and sense axes are merely exchanged, and the device still senses rotation about the normal to the plane of the wafer. But now consider instead the design of a gyro sensitive to rotation in the x- or y-axis. A solution to this problem is essential to make a working, three-axis MEMS gyro. In fact, such a device requires comparable sensitivities along all three orthogonal axes: x, y, and z.

7.4.4.1 Rigid-body mass x-/y- MEMS gyroscope

Consider a vibrating mass intended to respond to input rotation about the x-axis. Figure 7.30a shows such a scheme where mass m is driven at resonance in the y direction and rotation of the entire die about the x-axis produces a Coriolis force along the z-axis. For this configuration, it is not difficult to achieve large-amplitude driven motion in the plane of the wafer because variable-area drive electrodes can be created with either bulk or surface micromachining processes. To achieve the required high-sensitivity capacitive sensing of the small-amplitude out-of-plane motion, surface micromachining, with its ability to fabricate closely spaced, variable-gap structures, is a better choice than any bulk processes.[10] With the decision now made to use a surface micromachining process, two new challenges arise: (i) to design mechanical springs that provide good flexibility

[10] Refer to Appendix C for a review of micromachining.

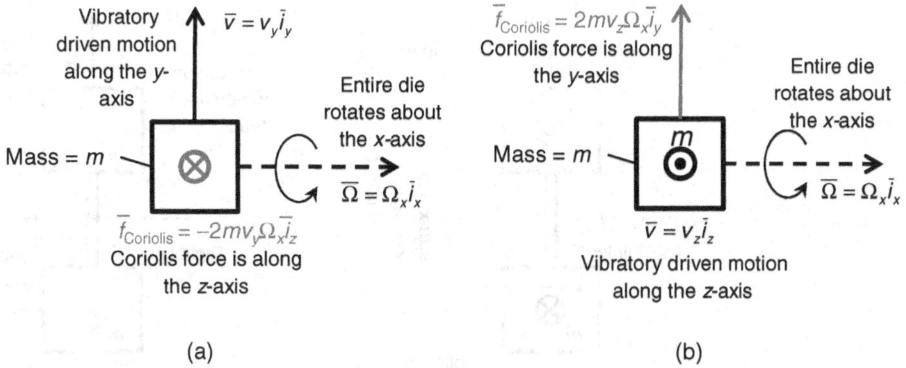

Figure 7.30 Two x-axis gyro designs based on a vibrating point mass m. The driven motion and Coriolis motion are both vibratory, but only their reference directions are indicated. (a) The driven motion is in the y direction and the Coriolis sense is in the z direction (perpendicular to wafer plane). (b) The driven motion is in the z direction (perpendicular to wafer plane), and the Coriolis sense is in the y direction.

along one dimension and (ii) to create electrodes that are rigid so that the structure behaves like a rigid body.

Figure 7.30b shows an alternative solution for detecting rotation about the x-axis. Here, point mass m is driven at resonance in the z direction, that is, perpendicular to the wafer plane, and rotation about the x-axis induces a Coriolis force in the y direction. The requirement of large-amplitude resonant motion in the z direction rules out surface micromachined structures. In contrast, high aspect ratio structures fabricated by bulk micromachining processes, such as deep reactive ion etching, accommodate large travel in the z direction and can also afford high sensitivity in the plane of the wafer. Designing the springs and attaining sufficient mass rigidity remain difficult challenges, depending on the size of the structure and the operating frequency.

We have learned here that the two MEMS gyro schemes illustrated in Figs. 7.30a and b are better suited for surface and bulk micromachining processes, respectively. The important lesson is that, from the outset, the constraints imposed by fabrication processes factor into design decisions.

7.4.4.2 Beyond single rigid-body mass design

Designing the mechanical components of MEMS gyros is just as important as proper specification of the electrodes and the fabrication process. For example, the positioning springs must confine the motion to the specified direction, providing sufficient rigidity to minimize orthogonal displacements that can be mistaken for a Coriolis-induced response. But another less obvious mechanical requirement is to make sure that the electrodes themselves are very rigid. The individual electrodes in a capacitor bank are invariably beams with high but finite stiffness. Normally the stiffness-to-mass ratio of individual electrodes is sufficiently high and individual electrodes can be considered rigid. On the other hand, the stiffness-to-mass ratio that results when large numbers of electrodes are assembled in an electrode bank is considerably lower. If these banks flex too much,

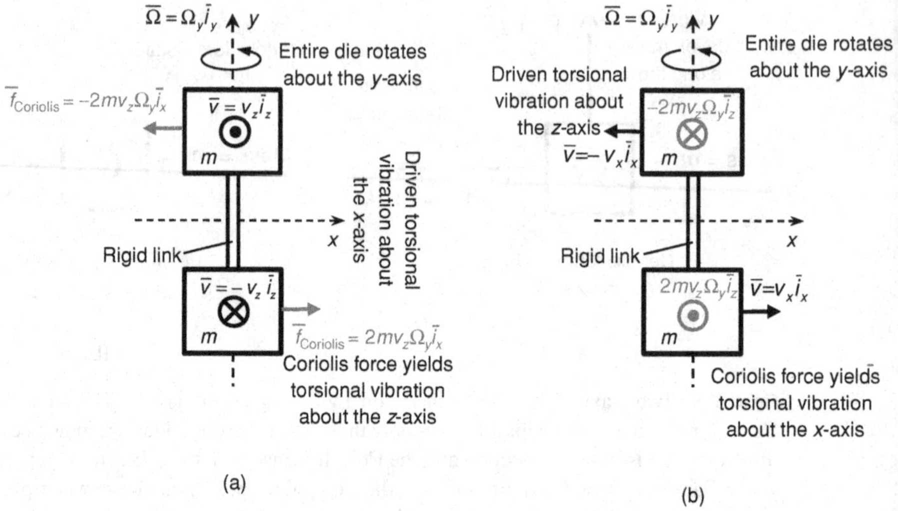

Figure 7.31 Examples of driven teeter-totter vibration and resulting torsional Coriolis-induced vibrational motions. (a) Two identical masses driven in teeter-totter resonant vibration about the *x*-axis. Input rotation about *y*-axis (Ω_y) yields teeter-totter (Coriolis) motion about the *z*-axis. (b) Identical masses driven in teeter-totter resonant vibration about the *z*-axis. Input rotation about *y*-axis now yields teeter-totter Coriolis motion about the *x*-axis.

device performance suffers. In general, larger structures are more susceptible to such unwanted flexure. On the other hand, larger electrode banks afford better sensitivity. This trade-off is further complicated by the effect of size on susceptibility to mechanical shock.

A common way to alleviate the difficulties posed by electrodes of large size is to divide their mass into smaller sections, which are inherently more rigid. Consider dividing the mass into two equal parts and then driving them out of phase in what is called a *teeter-totter* mode.

Figure 7.31a shows a scheme with two masses vibrating out-of-phase about the *x*-axis. The response to a rotation about the *y*-axis is a Coriolis force-induced torsional motion about the *z*-axis. This configuration requires large-amplitude driven motion out of the plane of the substrate. As previously argued in Section 7.4.4.1, such motion is only possible with devices fabricated using bulk micromachining methods. Sensing the small amplitude motions induced by the Coriolis force in the plane of the wafer is not a problem with such structures.

Using very similar reasoning, one discovers that the scheme shown in Fig. 7.31b is better suited for surface micromachining. The two masses are driven in torsional vibration about the *z*-axis and the Coriolis forces experienced by them due to input rotation about the *y*-axis cause small-amplitude motions normal to the wafer plane. The dual requirements of large-amplitude driven motions in the plane of the wafer and high sensitivity to small-amplitude motions perpendicular to the wafer plane indicate that indeed this scheme is best realized using a surface-micromachined design.

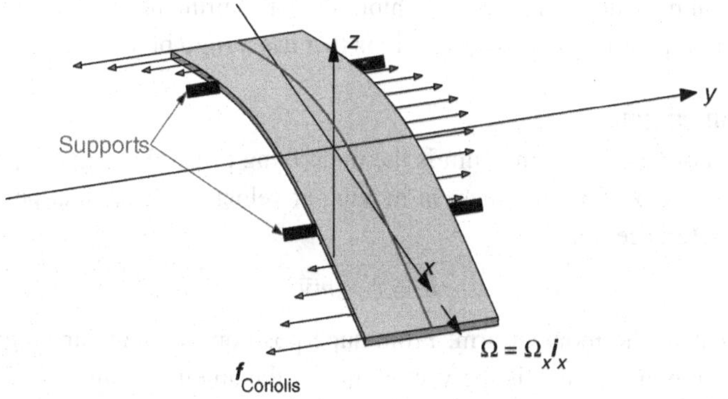

Figure 7.32 Gyro scheme based on free vibrating beam held at nodes and driven at resonance in the z direction vibrates in the y direction when subject to rotation about the x-axis.

Teeter-totter structures have the additional, very desirable, feature that the proof mass is attached to the substrate at a resonant node where the reaction forces of the two masses cancel each other out. The consequence is good vibration isolation. The design of the mechanical springs for these structures is still challenging, but generally less so than that for a single-mass gyro.

As a final example of a teeter-totter design, consider the free beam geometry shown in Fig. 7.32. Refer back to Section 6.8.3 for more information about free beams. This beam, supported at the resonance nodes, is driven at resonance in the z direction and when subjected to an input rotation Ω_x about the x-axis, it responds with a deflection in the y direction due to the Coriolis force. It is important to recognize that the proof mass is a continuum rather than a set of discrete elements. In practice, understanding the desirable properties of the free beam would be just the starting point. The actual structure will consist of a free beam with the required drive, drive-sense, and Coriolis-sense capacitive banks, the design of which is not a trivial exercise. Once all these electrodes have been attached, the basic resonance free beam will be significantly altered. In particular, the mass loading will have shifted the locations of the nodes, which will have to be located using FEM modal analysis. The supports are attached at these nodes.

These few examples illustrate that there are many options for the geometry and type of vibrational drive that can be employed in a design of vibratory MEMS gyroscopes. There are other designs based on continua, such as tuning-fork [17] and ring gyros [18]. The spring and electrode design, as well as the fabrication process, must be considered from the outset.

7.4.5 Background material

The qualitative analysis presented in the last section allowed us to test the feasibility of several gyro design concepts. To investigate two teeter-totter configurations in more depth and to compare them quantitatively, it is necessary to consider angular momentum and the dynamics of rigid bodies.

This section presents, in heuristic fashion, the basic principles needed to understand the operation of practical gyroscope designs that use a rigid body as the proof mass.

7.4.5.1 Angular momentum

Conservation of angular momentum is the underlying principle of gyro operation. The angular momentum of a point mass m moving at velocity $\overline{v}$ located at position $\overline{r}$ with respect to a reference point is

$$\overline{L} = \overline{r} \times m\overline{v}, \tag{7.75}$$

where $\overline{r}$ is called the moment arm. From superposition, the angular momentum of a collection of n point masses is the vector sum of the angular momenta of the discrete elements in the collection:

$$\overline{L} = \sum_{i=1}^{n} \overline{r}_i \times m\overline{v}_i. \tag{7.76}$$

For a solid body, the summation is replaced by an integral:

$$\overline{L} = \int_M \overline{r} \times \overline{v}\mathrm{d}m\,(\overline{r}) = \int_V \rho(\overline{r})\overline{r} \times \overline{v}\mathrm{d}V(\overline{r}), \tag{7.77}$$

where ρ, M, V, and $\mathrm{d}V$ are the mass density, total mass, total volume, and differential volume of the body, respectively. For a rigid body rotating about the reference point used to define $\overline{r}$, the velocity vector $\overline{v}$ is

$$\overline{v} = \overline{\Omega} \times \overline{r}. \tag{7.78}$$

Inserting Eq. (7.78) into Eq. (7.77) yields

$$\begin{aligned}\overline{L} &= \int_V \rho(\overline{r})\overline{r} \times \left(\overline{\Omega} \times \overline{r}\right) \mathrm{d}V(\overline{r}) = \int_V \rho(\overline{r})[r^2\overline{\Omega} - (\overline{r}\overline{\Omega})\overline{r}]\mathrm{d}V \\ &= \underbrace{\left[\int \rho(\overline{r})[r^2\overline{\overline{E}}_{3\times 3} - \overline{r}\,\overline{r}^T]\mathrm{d}V(\overline{r})\right]}_{\equiv \overline{\overline{I}}_{\mathrm{inert}}} \Omega = \overline{\overline{I}}_{\mathrm{inert}}\overline{\Omega},\end{aligned} \tag{7.79}$$

where $\overline{\overline{E}}_{3\times 3}$ is the 3×3 identity matrix and $\overline{\overline{I}}_{\mathrm{inert}}$ is the inertial moment tensor.[11] The elements are

$$\overline{\overline{I}}_{\mathrm{inert}} = \begin{bmatrix} \int_V \rho(y^2+z^2)\mathrm{d}V & -\int_V \rho xy\mathrm{d}V & -\int_V \rho xz\mathrm{d}V \\ -\int_V \rho xy\mathrm{d}V & -\int_V \rho(x^2+z^2)\mathrm{d}V & -\int_V \rho yz\mathrm{d}V \\ -\int_V \rho xz\mathrm{d}V & -\int_V \rho yz\mathrm{d}V & \int_V \rho(x^2+y^2)\mathrm{d}V \end{bmatrix}. \tag{7.80}$$

The moment of inertia is a geometrical tensor, meaning that it can be diagonalized by an appropriate choice of the coordinate system. For any rigid body, the set of axes that diagonalizes this tensor depends only on the body's geometry and mass distribution. These axes are called the *principal axes* and the diagonal elements corresponding to

[11] In Chapter 6, identity matrices are signified by $\overline{\overline{I}}$. Here, we use $\overline{\overline{E}}$ because $\overline{\overline{I}}$ is so commonly used for moment of inertia.

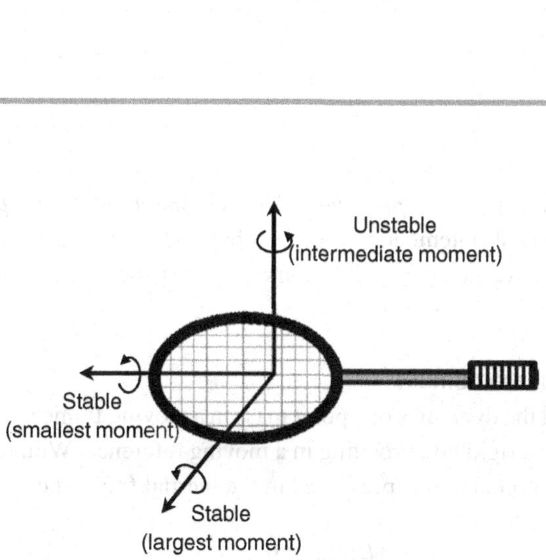

Figure 7.33 An illustration of stability and instability about principal axes of a rigid body. Adapted from [19].

these axes are called the *principal components*. Mathematically, the principal axes are the eigenvectors of the inertial moment tensor and the principal components are the associated eigenvalues.

A diagonalized inertial moment tensor takes the form

$$\overline{\overline{I}}_{\text{inert}} = \begin{bmatrix} I_1 & 0 & 0 \\ 0 & I_2 & 0 \\ 0 & 0 & I_3 \end{bmatrix}. \tag{7.81}$$

Torque-free rotation about the principal axes associated with the smallest and the largest eigenvalues are stable, while rotation about the axis of the middle eigenvalue is inherently unstable. A classic example identified by Crandall [19] is the tennis racquet illustrated in Fig. 7.33. If you throw a tennis racquet up in the air and impart rotation about one of its stable axes, it spins stably during the flight. If, on the other hand, rotation is imparted about the unstable axis, the motion during the flight is quite irregular.

In the design of any gyroscope, forced resonance vibration about the unstable axis should be avoided because any external noise tends to be dynamically amplified. This means that the drive torque has to work harder to maintain the desired driven motion. Moreover, shocks can give rise to large undesired motions in the Coriolis sense direction.

7.4.5.2 Perpendicular axis theorem

MEMS structures virtually always feature dimensions considerably larger in the plane of the wafer (x and y directions) than normal to the wafer (z direction). Thus, I_x and I_y can be approximated as

$$\begin{aligned} I_x &= \int_V (y^2 + z^2)\rho \mathrm{d}V \approx \int_V y^2 \rho \mathrm{d}V, \\ I_y &= \int_V (x^2 + z^2)\rho \mathrm{d}V \approx \int_V x^2 \rho \mathrm{d}V. \end{aligned} \tag{7.82}$$

Comparing I_z from Eq. (7.80) with I_x and I_y from Eq. (7.82) shows that the moment of inertia about the z axis is approximately equal to the sum of moments of inertia about the x- and the y axis.

$$I_z \approx I_x + I_y. \tag{7.83}$$

Equation (7.83) is known as the *perpendicular axes theorem*, or the *plane figure theorem*. It is a mathematical statement of a readily observable fact that MEMS devices, being flat and shallow, have larger inertial moments about the z axis than about the x- and y-axes.

7.4.5.3 Conservation of angular momentum

Section 7.4.3 introduced the dynamics of a point mass in a moving frame of reference. In this section, we consider a rigid body rotating in a moving reference. With no externally applied torque, angular momentum is preserved in the inertial frame, i.e.,

$$\frac{\mathrm{d}}{\mathrm{d}t}(\overline{L})_{\mathrm{inert}} = \overline{0}. \tag{7.84}$$

Using Eq. (7.79) and the rate of change operator from Eq. (7.62) yields

$$\frac{\mathrm{d}}{\mathrm{d}t}(\overline{L})_{\mathrm{inert}} = \frac{\mathrm{d}}{\mathrm{d}t}(\overline{\overline{I}}\,\overline{\omega})_{\mathrm{rot}} + \overline{\Omega} \times (\overline{\overline{I}}\,\overline{\omega}), \tag{7.85}$$

where $\overline{\omega}$ is the angular velocity in the moving reference frame and $\overline{\Omega}$ is the angular velocity vector of that frame defined with respect to the fixed reference. Assuming without loss of generality that $\overline{\omega}$ is expressed in the coordinates of the principal axes of the inertial tensor, so that $\overline{\overline{I}}$ is given by Eq. (7.81), Eq. (7.85) becomes

$$\frac{\mathrm{d}}{\mathrm{d}t}(\overline{L})_{\mathrm{inert}} = \overline{\overline{I}}\dot{\overline{\omega}} + \begin{bmatrix} 0 & -\Omega_3 & \Omega_2 \\ \Omega_3 & 0 & -\Omega_1 \\ -\Omega_2 & \Omega_1 & 0 \end{bmatrix} \begin{bmatrix} I_1\omega_1 \\ I_2\omega_2 \\ I_3\omega_3 \end{bmatrix}. \tag{7.86}$$

Finally, combining Eq. (7.86) and Eq. (7.84) yields

$$\begin{bmatrix} I_1\dot{\omega}_1 \\ I_2\dot{\omega}_2 \\ I_3\dot{\omega}_3 \end{bmatrix} = \begin{bmatrix} I_2\omega_2\Omega_3 - I_3\omega_3\Omega_2 \\ I_3\omega_3\Omega_1 - I_1\omega_1\Omega_3 \\ I_1\omega_1\Omega_2 - I_2\omega_2\Omega_1 \end{bmatrix}. \tag{7.87}$$

This matrix equation completely accounts for the coupling among rotations observed in the fixed reference due to rotation of the $\overline{\Omega}$ reference.

7.4.6 Torsional-vibration gyroscope

In Section 7.4.4, we described a gyro with resonant torsional vibration about the x-axis. See Fig. 7.31a. By determining the direction of the Coriolis force, we established the nature of the Coriolis response, which was found to be torsional vibration about the z-axis. With the background provided by Section 7.4.5, we can now return to this example and analyze it in more detail.

Consider a rigid body vibrating in a torsional (teeter-totter) mode about the x-axis in the moving reference frame. In the absence of any input rotation, the equation of motion takes the form of a torsional spring-mass system

$$I_x\ddot{\theta}_x + b_{\theta x}\dot{\theta}_x + k_{\theta x}\theta_x = \tau_x, \tag{7.88}$$

where I_x is the moment of inertia about the x-axis, which, by design, coincides with one of the principal axes of the rigid body, θ_x is the angular deflection of the proof mass about the x-axis, $b_{\theta x}$ is the damping, $k_{\theta x}$ is the rotational stiffness, and τ_x is the electrostatically driven drive torque. The amplitude of the driven motion is maximized at the resonance:

$$\omega_{\mathrm{d}} = \sqrt{k_{\theta x}/I_x}. \tag{7.89}$$

At resonance, the magnitude of the response is

$$|\underline{\theta}_x(\mathrm{j}\omega_{\mathrm{d}})| = Q_{\theta x}\frac{\tau_x}{I_x\omega_{\mathrm{d}}^2}, \tag{7.90}$$

where $Q_{\theta x}$ is the quality factor,

$$Q_{\theta x} = \frac{\sqrt{I_x k_{\theta x}}}{b_{\theta x}}. \tag{7.91}$$

When the moving frame of reference rotates about its y axis, inertial coupling transfers some the vibratory motion about the x axis into vibratory motion about the local z axis. Refer to Fig. 7.31a. While Section 7.4.4 employed the Coriolis force to establish the nature of the response, here we use the conservation of angular momentum. Substituting $\overline{\omega} = \left[\dot{\theta}_x \quad 0 \quad 0\right]^{\mathrm{T}}$ and $\overline{\Omega} = [0 \quad \Omega \quad 0]^{\mathrm{T}}$ into the right-hand side of Eq. (7.87) yields the torque due to inertial coupling:

$$\tau_{\mathrm{Coriolis}} = \lfloor 0 \quad 0 \quad I_x\dot{\theta}_x\Omega_y \rfloor. \tag{7.92}$$

It is the z component of this torque that induces small-amplitude vibration about the z-axis:

$$I_z\ddot{\theta}_z + b_{\theta z}\dot{\theta}_z + k_{\theta z}\theta_z = I_x\dot{\theta}_x\Omega_y. \tag{7.93}$$

In practice Ω_y is time-dependent, but with a frequency range well below the driven resonance. Thus, in effect it amplitude-modulates $\dot{\theta}_x$, just as discussed in Section 7.4.3.1. Under this quasistatic approximation, the sensitivity is

$$S_{\theta_z/\Omega_y} = \left|\frac{\underline{\theta}_z(\mathrm{j}\omega)}{\Omega_y}\right|_{\omega=\omega_{\mathrm{d}}} = \frac{\omega_{\mathrm{d}} I_x |\underline{\theta}_x(\mathrm{j}\omega_{\mathrm{d}})|}{|(\mathrm{j}\omega_{\mathrm{d}})^2 I_z + \mathrm{j}\omega_{\mathrm{d}} b_{\theta z} + k_{\theta z}|}, \tag{7.94}$$

which may be rewritten as

$$S_{\theta_z/\Omega_y} = \frac{1}{\omega_{\mathrm{d}}}\frac{I_x}{I_z}\frac{|\underline{\theta}_x(\mathrm{j}\omega_{\mathrm{d}})|}{\sqrt{\left(1-\left(\frac{\omega_{\mathrm{s}}}{\omega_{\mathrm{d}}}\right)^2\right)^2 + \left(\frac{\omega_{\mathrm{s}}}{\omega_{\mathrm{d}} Q_{\mathrm{s}}}\right)^2}}. \tag{7.95}$$

Note that the sensitivity is proportional to the moment ratio I_x/I_z. Since the lateral dimensions of a MEMS gyro are considerably larger thanits height, the perpendicular theorem, given by Eq. (7.83) applies, so $I_x/I_z < 1$. Based on this inequality, and comparing the coefficient $(I_x/I_z)/\omega_{\mathrm{d}}$ of Eq. (7.95) to the coefficient $2/\omega_{\mathrm{d}}$ of Eq. (7.73), the important conclusion can be made that x- or y-axis gyros of this type are inherently less sensitive than the z-axis device subjected to linear vibration. Notice, however, that if the driven motion is in the plane of the wafer and the Coriolis sense is out of the plane of the wafer, as shown in Fig. 7.31b, then $I_z/I_x > 1$. Thus, the gyro of Fig. 7.31a suffers from an effective mechanical attenuation, while the gyro of Fig. 7.31b enjoys mechanical amplification. To illustrate this point further, for a device with $I_x \approx I_y$, the moment ratio of the design in Fig. 7.31b will be approximately four times larger than the design in Fig. 7.31a.

Possibly overriding any considerations stemming from an asymmetric (flattened) MEMS structure, the device performance will be strongly influenced by the fabrication process. Bulk-micromachined devices are more suited for sensing very small Coriolis motion in the plane of the wafer than out of the plane, simply because variable-gap electrodes can be readily manufactured in the plane. The increased sensitivity of variable-gap electrode designs often outweighs the four-times mechanical amplification. Thus, for bulk micromachining, the design of Fig. 7.31a is still a good design option. In contrast, surface-micromachined devices can utilize sensitive out-of-plane variable-gap electrodes. Moreover, because surface-micromachined devices only achieve large travel in the plane of the wafer, the design in Fig. 7.31b is not only much better, but probably the only practical choice.

Example 7.3 **Practical teeter-totter gyroscope**

The conceptual design of Fig. 7.31 is useful for understanding the dynamics of operation. A capacitive MEMS gyro is considerably more complex because it must be equipped with springs for both drive and sensed motions and three sets of electrodes: drive, drive sense, and Coriolis sense. An SEM of a rather complex MEMS capacitive gyroscope is shown in Fig. 7.34. The springs and the electrode banks are annotated in the figure. In the

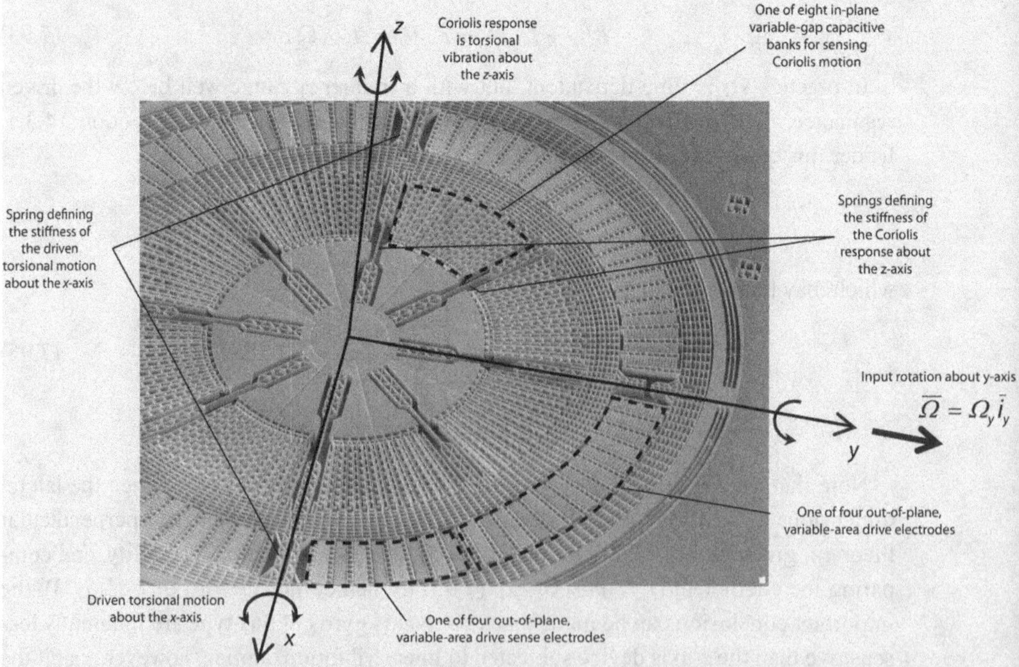

Figure 7.34 A teeter-totter vibratory gyroscope. The driven motion is torsional vibration about the x-axis. The Coriolis motion is torsional vibration about the z-axis. The image is credited to Scott Adams of Kionix Inc. Used with permission.

driven motion, the outer ring vibrates as a teeter-totter about the x-axis. The inner ring remains stationary in the absence of input rotation. When the die rotates about its y-axis, the Coriolis force couples the torsional vibration of the drive to this rotation and induces torsional vibration about the z-axis. Even without knowing I_x and I_z, the symmetry of the device enables us to make a good estimate of the moment ratio: $I_x/I_z \sim 0.5$. The drive and drive-sense electrodes are of the same type, but the drive electrodes occupy a larger area and they are placed further from the x-axis to maximize the drive torque. The sense electrodes need not be too sensitive to detect the large driven motion.

7.4.7 Higher-order effects

The analog output of a gyro with no rotation is referred to as *bias* and it is measured in volts.[12] Typically, bias is about one-half of the supply voltage. The input rate is typically measured in degrees per second, or °/s. The *sensitivity* quantifies the conversion of the input rotation to the final output voltage and is specified in mV/(°/s). Sensitivity of a digital input is count/(°/s). *Resolution*, the smallest input rate that the sensor can detect, is expressed in °/s.

For the definition of non-linearity, cross-axis sensitivity, temperature sensitivity, and self-test, the reader is referred to Section 7.3.7, because these terms are essentially the same as for accelerometers.

Shock sensitivity: without proper attention paid in the design phase, MEMS gyroscopes tend to be very sensitive to shock and vibration motion. The simple suspended mass scheme of Fig. 7.25, with driven vibration along one axis and sense axis along the other, is particularly vulnerable. In Section 7.4.3.1, it was stated that elements vibrating at higher frequencies are generally less susceptible to shock. Designs that have very little or no coupling to the substrate also have good shock immunity. Schemes for eliminating coupling between the substrate and a beam are discussed in Sections 6.8.2, 6.8.3, and the end of 7.4.4.2.

Quadrature: it is important to emphasize that the Coriolis force is proportional to the time derivative of the sinusoidal driven resonant motion. Thus, the phase difference between the two forces is

$$\angle(\underline{f}_{\mathrm{d}}, \underline{f}_{\mathrm{Coriolis}}) = 3\pi/2. \tag{7.96}$$

On the other hand, alignment imperfections between the drive and the x axis can cause vibration in the y direction. The response of the Coriolis sense electrodes to direct drive motion is called *quadrature*. This name indicates the phase difference with respect to Coriolis response. This phase difference makes it possible to eliminate the quadrature signal by synchronous demodulation.

Quadrature arises from subtle imperfections that usually vary across the wafer. One approach to reduce the quadrature effect for the z-axis gyro is to introduce a second mass

[12] Traditionally, accelerometers use the term *offset* while gyros use the term *bias* for the zero-input response.

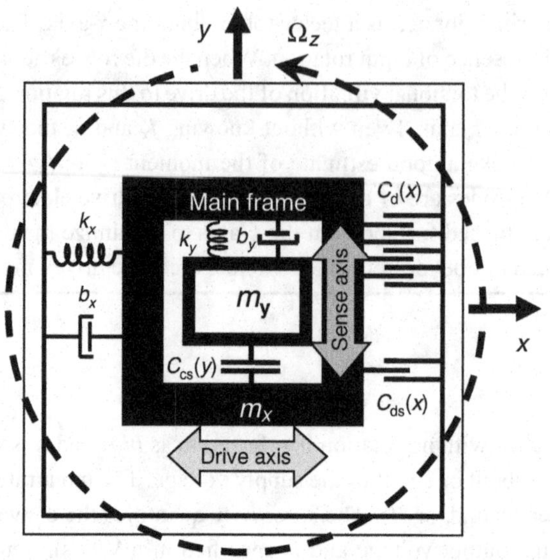

Figure 7.35 A *z*-axis gyroscope featuring different masses associated with the driven and sense motion. The main frame is constrained to move in the *x* direction to periodic electrostatic force. The mass inside the main frame is rigidly attached to the main frame in the *x* direction, but can move in the *y* direction. The driven vibration mass is $m_x + m_y$ and the sense mass is m_y.

inside the main frame, as shown in Fig. 7.35. This mass is rigidly attached to the main frame in the driven *x* direction, but can move in the *y* direction.[13] The main frame is constrained to move only in the *x* direction, while the vibration induced by the Coriolis force is constrained to the *y* direction. Alternatively, or in addition, a second pair of electrodes can be added to the system to nullify quadrature. Nullifying or aligning the driven vibration precisely with the *x*-axis can be achieved by applying static voltage on these additional electrodes. Alternatively, sinusoidal voltages can also be applied at the quadrature nullifying electrodes to reduce quadrature. Sinusoidal voltages at the drive frequency ω_d used for nullifying require lower voltage levels than DC nullifying, because of the resonance dynamic amplification. Refer to Clark *et al.* [20] for more details on quadrature nullifying.

7.4.8 Closure

Collectively accelerometers and gyroscopes are referred to as inertial sensors. By combining three gyroscopes and three accelerometers aligned along three mutually perpendicular axes and integrating them with the essential electronics and computational processing, an *inertial measurement unit* (IMU) can be created. An IMU measures and tracks the six degrees of freedom of a rigid body. The emergence of MEMS microsensors

[13] The springs are designed to be flexible in the *y* direction and much stiffer in the *x* direction.

has enabled a proliferation of inertial sensing devices in consumer devices, such as cell phones and portable game consoles.

Only capacitive accelerometers and gyroscopes have been considered here, though some successful gyroscopes based on piezoelectric sensors have been built and commercialized [2]. At the time of writing this text, the most sensitive gyros are optical systems, such as *ring laser* and *fiber optic* devices, but these are not MEMS scale devices. Refer to Barbour and Schmidt [21] for a comprehensive review of the state of inertial sensing at the beginning of the twenty-first century.

7.5 MEMS energy harvesters

Unlike pressure sensors and inertial sensors, energy harvesters have not enjoyed significant commercial success to date. Even so, their long-term promise has attracted research attention for some time. Some of the most attractive applications are in enabling distributed wireless sensor networks and battery-free portable electronics. Sensors equipped with wireless communication links that require neither wiring to deliver power nor inconvenient battery replacements have great potential in *condition-based maintenance* (CBM) of machinery and *structural-health monitoring* (SHM) of buildings and bridges [22]. In particular, cost considerations preclude remote placement of sensors that require battery replacement. Furthermore, batteries are expensive, have limited shelf life, can be bulky, and their ultimate disposal poses environmental hazards.

Scavenging of other forms of energy, such as thermal or solar, while interesting, is not considered here. This section considers only some basic principles of MEMS-based mechanical energy harvesting from ambient mechanical vibrations. Such devices can be classified into three groups based on the main transduction mechanism: piezoelectric, magnetic, and electrostatic. Only electrostatic devices are examined here.

7.5.1 Basic principle of capacitive energy harvesting

The majority of capacitive devices considered in this text require a bias voltage to enable electromechanical transduction. While some energy harvesters use DC biasing [22], an attractive alternative is to employ electrets, which were first introduced in Section 3.6. The reader is encouraged to review this section and also Section 4.7.6, which presented the linearized electret transducer model invoked in the following analysis.

Like accelerometers, capacitive energy scavengers rely on motion induced in a proof mass. Refer to Fig. 7.36, which contains two insets: one showing the rigid-body diagram of the proof mass and the other the circuit model of the electret.

Just as for the accelerometer modeled in Section 7.3.1, the coordinate x is defined as the difference between the coordinate of the moving frame X_m and the fixed frame X_0,

$$x = X_m - X_0. \tag{7.97}$$

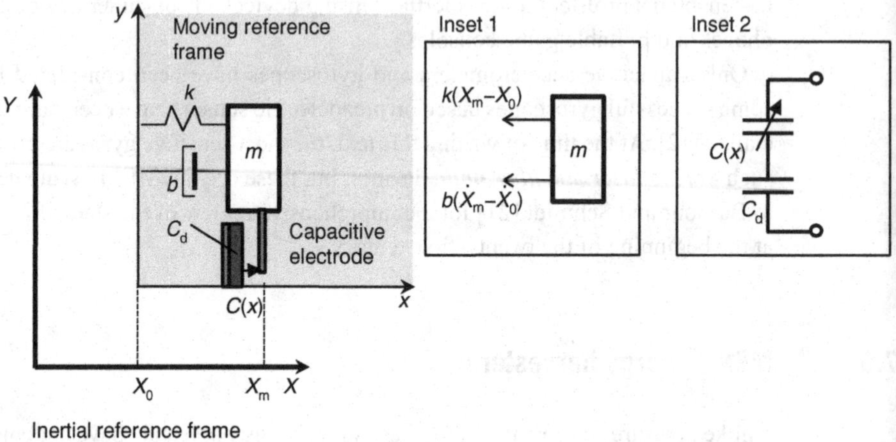

Figure 7.36 Schematic of electrostatic energy harvester showing proof mass m with spring, damping, and attached electrodes. Inset 1 shows a rigid-body diagram of the mass. Inset 2 shows the capacitive model the electret microgenerator.

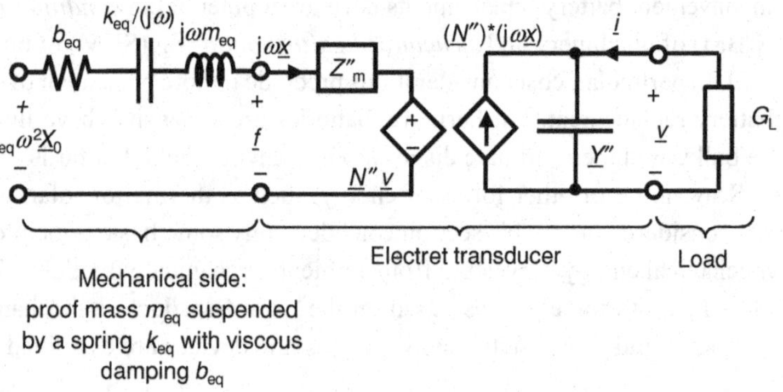

Figure 7.37 Linear model of electret-based resonant-energy harvester in the form of an equivalent electromechanical network; $\underline{f}_{ext}$ is the inertial force imparted by the ambient vibration.

The equation of motion, for convenience duplicated from that section, is

$$m\ddot{x} + b\dot{x} + kx = -m\ddot{X}_0. \tag{7.26}$$

From Section 4.7.6, the linear two-port relation for the electromechanical circuit is

$$\begin{bmatrix} \underline{f} \\ \underline{i} \end{bmatrix} = \begin{bmatrix} Z''_m & \underline{N}'' \\ -\underline{N}''^* & Y''_e \end{bmatrix} \begin{bmatrix} j\omega\underline{x} \\ \underline{v} \end{bmatrix}. \tag{4.75}$$

The overall cascaded system can be represented by the network of Fig. 7.37.

Also duplicated from Section 4.7.6 for convenience, the components of the electret network are

$$\underline{Z}''_m = \frac{1}{j\omega}\frac{\varepsilon_0 A}{(d/\kappa + x_0)^3}\left[V_0 - \frac{q_e}{C_d}\right]^2, \tag{4.77a}$$

$$\underline{N}'' = \frac{\varepsilon_0 A}{(d/\kappa + x_0)^2}\left[V_0 - \frac{q_e}{C_d}\right], \tag{4.77b}$$

$$\underline{Y}''_e = j\omega\frac{C_x C_d}{C_x + C_d}. \tag{4.77c}$$

If no battery is used in the system, $V_0 = 0$. Note that the mechanical variable x signifies the gap between the virtual and moving electrodes.

7.5.2 Power considerations and efficiency

Because MEMS energy harvesters are smaller than the structures to which they are attached, we may assume the vibrational power source to be infinite and thus unaffected by the transducer. Two key performance measures of these transducers are output power and efficiency. Both must be considered in the context of the nature of the ambient vibrations to be exploited.

The harvested mechanical power is the product of the force and the complex conjugate of the velocity:[14]

$$P_m = \frac{1}{2}\mathrm{Re}(\underline{f}_{ext}(j\omega\underline{x})^*). \tag{7.98}$$

The power delivered to the resistive load, represented by conductance G_L, is

$$P_e = \underbrace{\frac{j\omega\underline{x}\,\underline{N}''}{\underline{Y}'' + G_L}}_{\underline{v}}\underbrace{\left(\frac{j\omega\underline{x}\,\underline{N}''}{\underline{Y}'' + G_L}G_L\right)^*}_{\underline{i}^*} = \frac{\omega^2|\underline{x}|^2|\underline{N}''|^2}{G_L^2 + \left(\dfrac{1}{\omega C_t}\right)^2}G_L, \tag{7.99}$$

where C_t is

$$C_t = \frac{C_x C_d}{C_x + C_d}. \tag{7.100}$$

The transducer efficiency is the ratio of the electrical and the mechanical power:

$$\eta = P_e/P_m. \tag{7.101}$$

To proceed, we need an expression for the velocity $j\omega\underline{x}$ in terms of the system parameters. An easy way to derive it is to reflect the electrical components from Fig. 7.37 over to the mechanical side of the network, as shown in Fig. 7.38.

From this equivalent circuit,

$$j\omega\underline{x} = \frac{m_{eq}\omega^2\underline{X}_0}{b_{eq} + k_{eq}/(j\omega) + j\omega m_{eq} + \underline{Z}''_m + |\underline{N}''|^2/(\underline{Y}'' + G_L)}. \tag{7.102}$$

In their analysis, Williams and Yates ignored $\underline{Z}''$ and $\underline{Y}''$ [23]. These simplifications, justified as long as $k_{eq} \gg j\omega\underline{Z}''_m$ and $|Y''| \gg G_L$, are reflected in the circuit of Fig. 7.39.

[14] This expression is analogous to the circuit relationship $P = \frac{1}{2}\mathrm{Re}[\underline{v}\,\underline{i}^*]$.

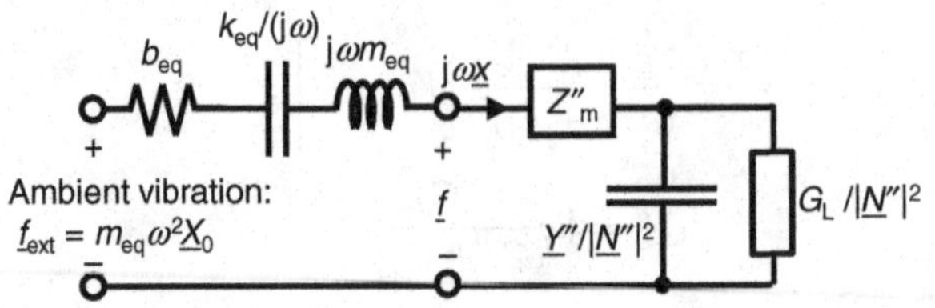

Figure 7.38 Linear mechanical model of electret-based energy harvester in equivalent circuit form with electrical components of Fig. 7.37 reflected over to the mechanical side.

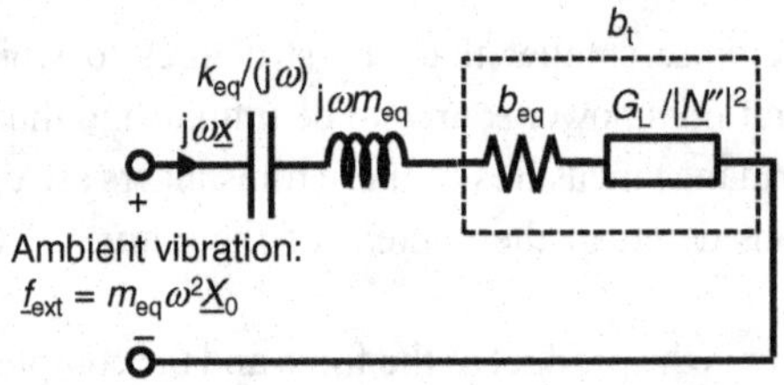

Figure 7.39 Simplified equivalent circuit for $k_{eq} \gg j\omega \underline{Z}''_m$ and $G_L \ll |Y''|$.

According to the simplified model, the input mechanical power is

$$P_m = \frac{1}{2}\mathrm{Re}(\underline{f}_{ext}(j\omega\underline{x})^*) = \frac{m_{eq}\zeta_t|\underline{a}_{in}|^2\left(\dfrac{\omega}{\omega_0}\right)^4}{\omega_0\left[1-\left(\dfrac{\omega}{\omega_0}\right)^2\right]^2+\left[2\zeta_t\dfrac{\omega}{\omega_0}\right]^2}, \tag{7.103}$$

where $\omega_0 = \sqrt{k_{eq}/m_{eq}}$ is the undamped natural frequency, $\underline{a}_{in} = -\omega_0^2\underline{X}_0$ is the input acceleration, and

$$\zeta_t = \frac{b_{eq} + N^2/G_L}{2m_{eq}\omega_0}, \tag{7.104}$$

the total damping ratio, incorporates both mechanical losses and the power converted to electrical form

Figure 7.40 plots normalized mechanical power versus frequency for four different values of damping ratio. The strong influence of damping on energy capture is apparent in this plot. In particular, when ambient vibrations are strongly band limited, a higher Q (low damping) captures the most power. On the other hand, when the ambient vibrational power is distributed over a wide frequency range, a lower Q (higher damping) may be better.

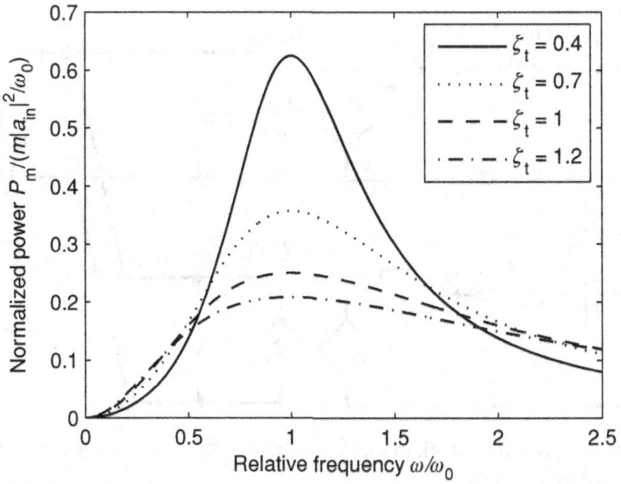

Figure 7.40 Normalized input mechanical power as a function of frequency, for several values of damping ratio ζ.

P_m is the mechanical power captured from the environment. The electrical output power P_e delivered to the load G_L is

$$P_e = \frac{1}{2}|j\omega\underline{x}|^2\frac{|\underline{N}''|^2}{G_L} = \frac{\dfrac{|\underline{N}''|^2|\underline{a}_{in}|^2}{2G_L\omega_0^2}\left(\dfrac{\omega}{\omega_0}\right)^4}{\left[1-\left(\dfrac{\omega}{\omega_0}\right)^2\right]^2+\left[2\zeta_t\dfrac{\omega}{\omega_0}\right]^2}. \tag{7.105}$$

Then, an efficiency measure may be defined:

$$\eta = \frac{P_e}{P_m} = \frac{|\underline{N}''|^2/G_L}{|\underline{N}''|^2/G_L + b_{eq}}. \tag{7.106}$$

Getting the maximum power to the load G_L depends on capturing as much mechanical power in the resonator as possible and then converting a large portion of it to electrical form. Note that the basic principle of maximum power transfer from the circuit theory demands that $|\underline{N}''|^2/G_L = b_{eq}$.

7.5.3 Multiple resonators

Usually, vibrational energy is distributed over a band of frequencies. In MEMS energy harvesters, the net recovery of this distributed energy can be enhanced using an array of cantilevered beams with slightly different resonant frequencies. It is a straightforward exercise to use the electromechanical network model of Fig. 7.37 to investigate such a scheme. Thus, consider the equivalent circuit of Fig. 7.41, which represents a model for two resonators connected to the load G_L.

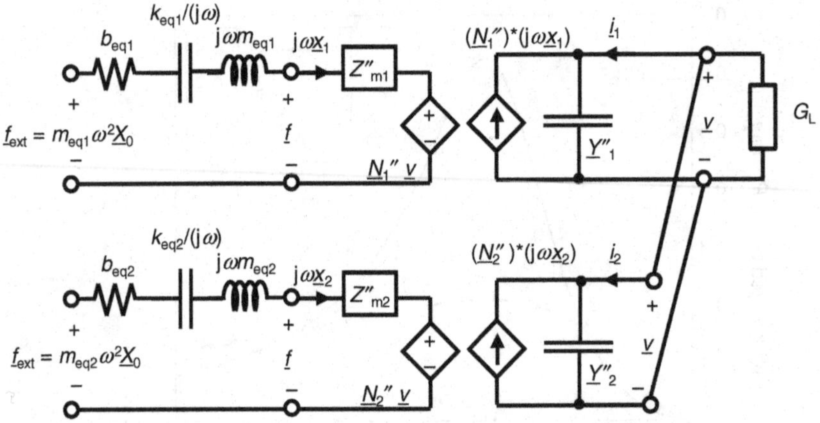

Figure 7.41 Electromechanical network model for an energy harvester consisting of two resonators connected to a load G_L.

Making the following set of simplifying assumptions, $k_{eq1} \gg j\omega \underline{Z}''_{m1}$, $k_{eq2} \gg j\omega \underline{Z}''_{m2,}$ $G_L \ll Y_1''$, $G_L \ll Y_2''$, and $\underline{N}_1'' = \underline{N}_2'' = \underline{N}''$, reduces the equivalent circuit to the form of Fig. 7.42.

The total electrical power harvested by this system is

$$P_e = \frac{1}{2}|j\omega\underline{x}_1 + j\omega\underline{x}_2|^2 \frac{|\underline{N}''|^2}{G_L}. \tag{7.107}$$

We can solve the following set of coupled equations to obtain the velocities $j\omega\underline{x}_1$ and $j\omega\underline{x}_2$:

$$\begin{bmatrix} \underline{Z}_{me1} + |\underline{N}''|^2/G_L & |\underline{N}''|^2/G_L \\ |\underline{N}''|^2/G_L & \underline{Z}_{me2} + |\underline{N}''|^2/G_L \end{bmatrix} \begin{bmatrix} j\omega\underline{x}_1 \\ j\omega\underline{x}_2 \end{bmatrix} = \begin{bmatrix} \omega^2\underline{X}_0 m_{eq1} \\ \omega^2\underline{X}_0 m_{eq2} \end{bmatrix}. \tag{7.108}$$

The resulting sum is

$$j\omega\underline{x}_1 + j\omega\underline{x}_2 = \frac{(m_1\underline{Z}_{me2} + m_2\underline{Z}_{me1})\,\omega\underline{X}_0}{\underline{Z}_{me1}\underline{Z}_{me2} + (\underline{Z}_{me1} + \underline{Z}_{me2})\dfrac{|\underline{N}''|^2}{G_L}}. \tag{7.109}$$

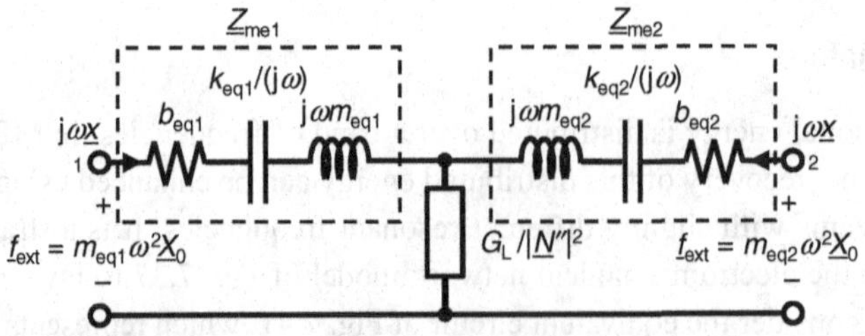

Figure 7.42 Simplified form for two mechanical resonators connected to a single load based on the assumptions: $k_{eq1} \gg j\omega\underline{Z}''_{m1}$, $k_{eq2} \gg j\omega\underline{Z}''_{m2}$, $G_L \ll \underline{Y}_1''$, $G_L \ll \underline{Y}_2''$, and $\underline{N}_1'' = \underline{N}_2''$.

```
m1      = 1;                k1 = 1;              % thinner beam is normalized
m2      = 1.3;              k2 = (1.3)^3;        % 30 % thicker beam
w1      = sqrt(k1/m1);      w2 = sqrt(k2/m2);    % natural frequencies
wr      = linspace(0,2*w2,200);                  % freq. Range; this is normalized frequency
zeta    = [.1,.2,.5,1];                          % normalized damping
b1      = 2*zeta*w1*m1;     b2 = b1;             % two beams are assumed to have the same loss
Np2_GL = b1/2;                                   % due to maximum power transfer
f1      = m1;               f2 = m2;             % normalized wrt acceleration
for i = 1:length(zeta)
    Z1 = 1i*wr*m1+b1(i)+k1./(1i*wr);             % mechanical impedances
    Z2 = 1i*wr*m2+b2(i)+k2./(1i*wr);
   jwx = (Z1.*f2+Z2.*f1)...
      ./ (Z1.*Z2+(Z1+Z2)*Np2_GL(i));             % sum of velocities
     P = 0.5*abs(jwx).^2*Np2_GL(i);              % power on the load
        plot(wr,P);hold on;                      % plot decorations are
                                                 % suppressed
end
```

Figure 7.43 MATLAB code for computing and plotting normalized power vs. frequency.

To find the power P_e, it is only necessary now to insert Eq. (7.109) into Eq. (7.107); however, presenting the results in a useful graphic form from which some value can be extracted requires further assumptions and a choice to be made for normalization.

The first assumption is that the two beams are located side by side on the chip. The resonances of these beams must be different, but instead of varying the lengths as is more common, we here let one beam be 30% thicker than the other. Doing so sets the resonances about 15% apart but keeps the damping constants equal, an advantage for our analysis and for presenting normalized results. The thinner beam serves as the reference with normalized mass $m_{eq1} = 1$ and stiffness $k_{eq1} = 1$. Then, for the thicker beam, $m_{eq2} = 1.3$ and $k_{eq2} = (1.3)^3$.

Assuming that the mechanical loss is dominated by the viscous damping due to the ambient gas, $b_{eq1} = b_{eq2}$ because they have the same area. Finally, keeping consistent with the basic principle of maximum power transfer, we assume that the power delivered to the load is the same as the mechanical dissipation in the resonators, that is,

$$|(\underline{N})^2|/G_L = b_{eq1}b_{eq2}/(b_{eq1} + b_{eq2}) = b_{eq1}/2.$$

Introducing these equalities into Eq. (7.109) and then using the result in Eq. (7.107) provides the output power P_e. The MATLAB code provided in Fig. 7.43 performs these calculations.

Figure 7.44 plots output power P_e normalized to the captured power input to the first resonator. Because the input acceleration is assumed constant over the frequency range, the inertial force on the larger mass is higher than the force on the lower mass because $f_{inertial} = m\omega^2\underline{X}_0$. Moreover, because the total damping b_{eq} is the same for the two resonators, the resonator with the higher mass and natural frequency has lower damping ratio $\zeta = b_{eq}/(m\omega_0)$. The reduced normalized damping and stronger force are partially cancelled by the higher stiffness of the second resonator.

As measured by its frequency response, the performance of the two-beam harvester is an improvement over a device with a single resonant beam. Adding more beams further improves energy capture over the bandwidth of the ambient vibration.

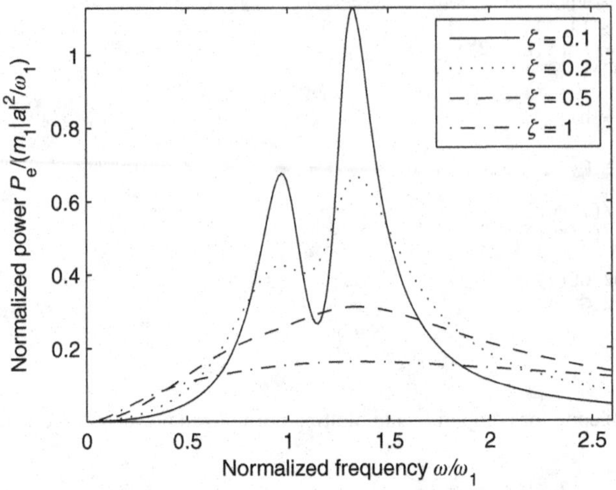

Figure 7.44 Normalized power delivered to the resistive load G_L from two side-by-side resonant beams having the same length and width but 30% different thicknesses. Two beams capture vibratory power over a wider frequency range than a single beam but coverage is still far from optimal.

Consider then an energy harvester with n resonators under the assumption that the magnitude of the equivalent, parallel capacitive admittance is much larger than the load admittance, $|\underline{Y}''|/n \gg |\underline{N}''|^2/G_L$. Further, all electromechanical impedance contributions to resonance are assumed negligible compared with the elastic beam stiffness, that is, $j\omega\underline{Z}_i'' \ll k_{eqi}$. Then, Eq. (7.108) becomes

$$\underbrace{\begin{bmatrix} \underline{Z}_{me1} + |\underline{N}''|^2/G_L & |\underline{N}''|^2/G_L & \cdots & |\underline{N}''|^2/G_L \\ |\underline{N}''|^2/G_L & \underline{Z}_{me2} + |\underline{N}''|^2/G_L & \cdots & |\underline{N}''|^2/G_L \\ \cdots & \cdots & \cdots & \cdots \\ |\underline{N}''|^2/G_L & |\underline{N}''|^2/G_L & \cdots & \underline{Z}_{men} + |\underline{N}''|^2/G_L \end{bmatrix}}_{\overline{\overline{\underline{Z}}}_m}$$

$$\times \underbrace{\begin{bmatrix} j\omega\underline{x}_1 \\ j\omega\underline{x}_2 \\ \cdots \\ j\omega\underline{x}_n \end{bmatrix}}_{j\omega\overline{\underline{x}}} = \underbrace{\begin{bmatrix} \omega^2 \underline{X}_0 m_{eq1} \\ \omega^2 \underline{X}_0 m_{eq2} \\ \cdots \\ \omega^2 \underline{X}_0 m_{eqn} \end{bmatrix}}_{\overline{\underline{f}}}. \tag{7.110}$$

We can obtain $j\omega\overline{\underline{x}}$, an n-element column vector, by inverting the impedance matrix $\overline{\overline{\underline{Z}}}_m$ and then multiplying by the force vector $\overline{\underline{f}}$:

$$j\omega\overline{\underline{x}} = \overline{\overline{\underline{Z}}}_m^{-1}\overline{\underline{f}}. \tag{7.111}$$

The electrical power delivered to the load G_{L} is proportional to the square of the vector sum of the velocities of all the elements:

$$P_{\mathrm{e}} = \frac{1}{2}\left|\left(\sum_{i=1}^{n} \mathrm{j}\omega \underline{x}_i\right)\right|^2 \frac{|\underline{N}''|^2}{G_{\mathrm{L}}}. \tag{7.112}$$

Figure 7.46 shows an example of an array of resonant cantilevers for a MEMS energy harvester. In this device, each beam has a slightly different length and corresponding resonant frequency. The patterned Au masses attached to each beam serve to reduce the resonant frequency, in order to capture ambient vibration energy at frequencies below ~100 Hz.

Example 7.4 **An array of beam resonators with different thicknesses**

Expanding on the previous example, consider now six beam resonators manufactured side by side. Assume that the beam thickness increases in 5% increments. Again, the thinnest beam is used for reference by setting $m_{\mathrm{eq1}} = 1$, $k_{\mathrm{eq1}} = 1$. A MATLAB implementation of the model given by Eqs. (7.110)–(7.112) is given as follows:

```
---------------------------------------------------------------------
n = 6;                               % number of resonators
p = .15;                             % thickness increment
zeta = [.1,.2,.5,1];                 % zeta of the first resonator
m = ones(n,1); k = ones(n,1);        % initialization
f = ones(n,1); w = ones(n,1);        %
b = repmat(2*zeta,n,1);              % assignment of b based on b1,
                                        since w1 = 1 and m1 = 1
Np2_GL = b/n;
for i = 2:n                          % assign the parameters of higher
                                       resonator in a loop
    m(i) = 1+p*(i-1);
    k(i) = (1+p*(i-1))^3;
    w(i) = sqrt(k(i)/m(i));
    f(i) = m(i);                     % normalized with respect to
                                       acceleration w^2Xo
end
wr = linspace(.01,2*w(end),200);
for i = 1:length(zeta)
    P = zeros(1,length(wr));
    for j = 1:length(wr)                 % loop over frequency range
        Zmat = Np2_GL(i)*ones(n,n);      % initial assignment of the
                                           impedance matrix
```

```
        for ifor = 1:n                        % continue assignment of
                                                the diagonal elements
            Zmat(ifor,ifor) = Zmat(ifor,ifor)+...
                1i*wr(j)*m(ifor)+b(ifor,i)+k(ifor)./(1i*wr(j));
        end
        jwx = sum(Zmat\f);                          % sum of the currents
        P(j) = .5*abs(jwx)^2*Np2_GL(i);           % power across the load
    end
    plot(wr,P); hold on;                        % plot decorations
end
-------------------------------------------------------------------------
```

Figure 7.45 plots the normalized power for different values of the damping ratio.

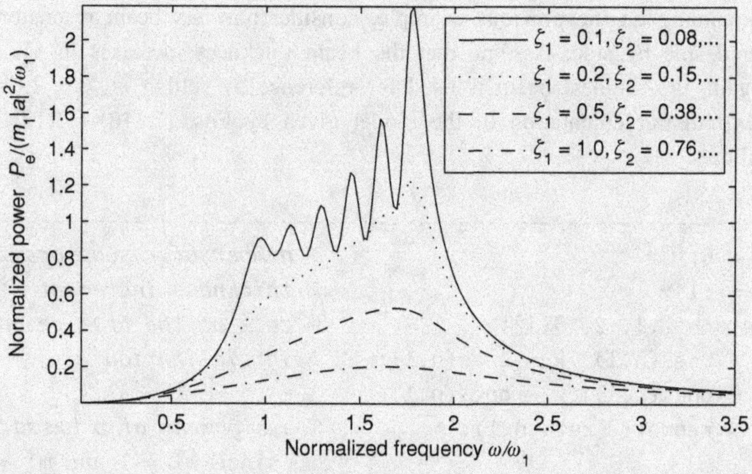

Figure 7.45 Normalized power on the admittance load G_L generated by an array of six beams of the same lengths and widths and different thicknesses, manufactured side by side.

7.5.4 Capacitive energy harvesters with bias voltage

Although capacitive MEMS energy harvesters based on electrets avoid the need for a DC bias, devices that use a fixed voltage source are still of practical interest. An example might be a harvester that collects energy and stores it in a rechargeable battery that then supplies the bias. The energy delivered by a transducer does not come from the battery, and, furthermore, any losses associated with bias can be kept very small. What is needed to make a system consisting of an energy harvester and a rechargeable battery is a smart control system that monitors the charging and discharging of the battery and also protects it. There are now off-the-shelf IC chips that accomplish this function. After making a few remarks on the construction of a capacitive MEMS energy harvester, a system consisting of a harvester and a rechargeable battery will be illustrated.

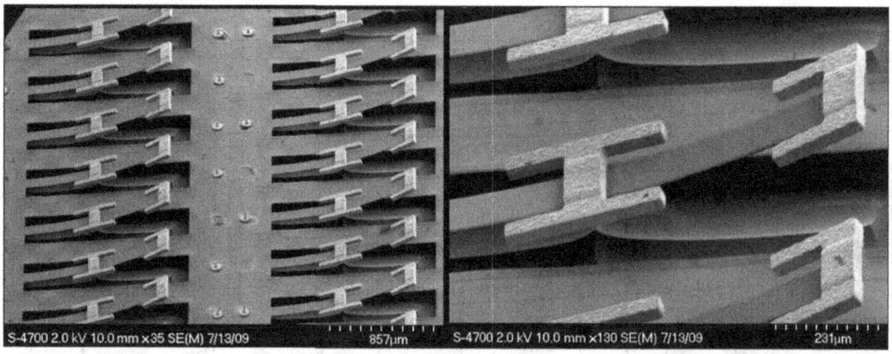

Figure 7.46 Array of cantilevered Si beams of varied length for a vibrational energy harvester using piezoelectric transduction. The clearly visible pairs of small, patterned Au mass elements on each beam serve to reduce their resonant frequencies, in order to capture ambient vibrational energy at frequencies below ~100 Hz. The beams are effectively pre-stressed with the beneficial result that the upward bending increases clearance and allows larger amplitude motion. Used with permission of the Smart System Technology & Commercialization Center.

Consider the block diagram of Fig. 7.47. This integrated circuit is designed to manage battery charging from poorly regulated power sources with the outputs in 1 μW to 100 mW range [24]. It also controls battery discharging and provides protection from over-charge and deep discharge events. This circuit can work with high-voltage sources, such as a mechanical energy harvester, and low-voltage sources, including solar cells and thermo-electric generators. A battery provides the power to the regulated load (pin REG) when the power is demanded via an event detector (pin AE). In a practical application, a self-biased device, such as an electret or piezoelectric element, monitors some parameter. When an event of interest is detected (such as vibration at a particular frequency), a high pulse signal is generated at the AE pin. This signal instructs the chip to produce regulated power, "waking up" the transmitter and sending information via a wireless channel to the nearest hub. After the transmission, the transmitter sends a low-pulse signal to the AE pin, which returns the device to the *harvest standby mode*.

7.5.5 Practical electrostatic energy harvesters

Beeby *et al.* [22] distinguish three different types of MEMS energy harvester:[15] (i) in-plane (of the wafer) variable-area capacitors; (ii) in-plane variable-gap capacitors; and (iii) out-of-plane variable-gap capacitors. Variable-gap and variable-area configurations have been examined in great detail in Chapters 3 and 4 and do not need an additional review here. Appendix C teaches that surface micromachining is generally better suited for defining efficient variable-gap out-of-plane capacitors, while the advantage of bulk micromachining is its capability to produce much larger proof masses. Silicon-on-insulator manufacturing that combines bulk and surface micromachining processing steps and is capable of producing large proof masses has excellent potential for manufacturing energy harvesters.

[15] Beeby's terminology for the three capacitor types has been modified for consistency with this text.

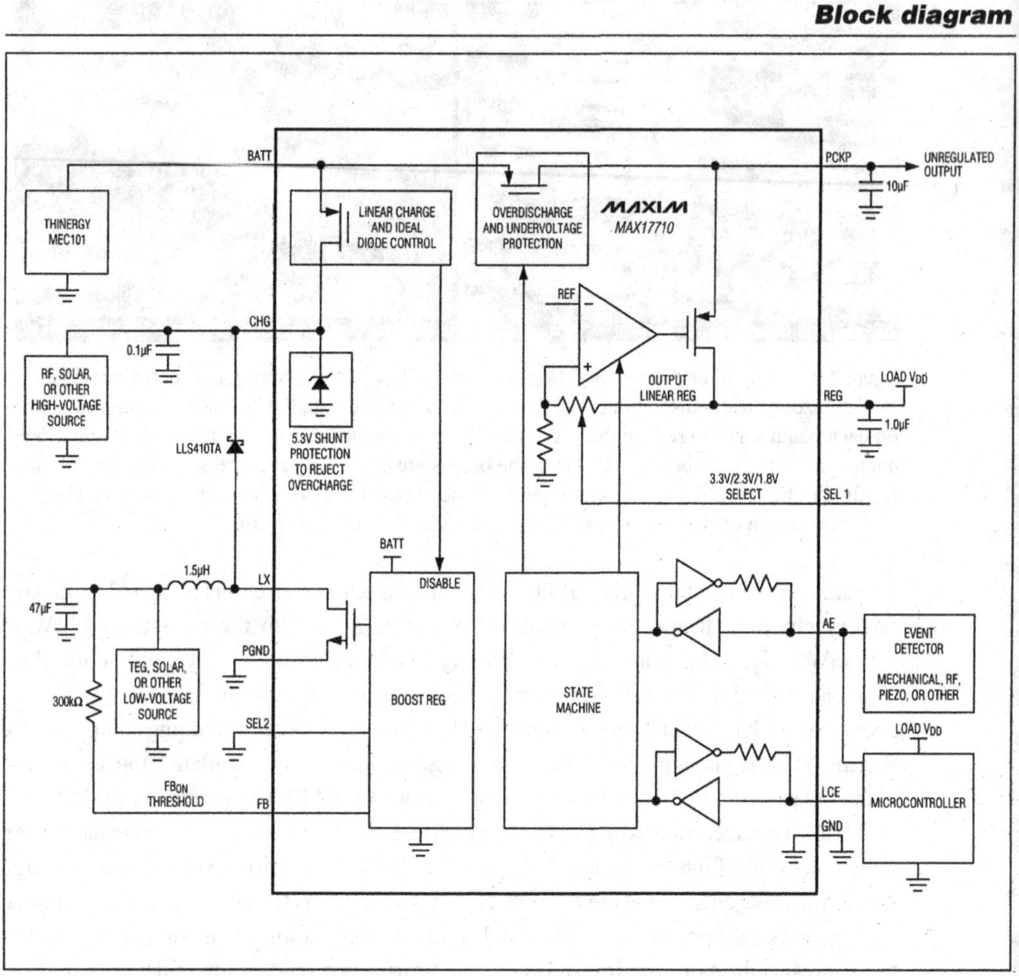

Figure 7.47 Block diagram of the MAX17710 chip, an integrated circuit system for managing, charging, and discharging a storage battery [24]. Ambient vibration harvesters are considered high-voltage energy harvesters. An additional energy source for low-voltage harvesters, such as a solar cell or thermo-electric generator (TEG), is also available. Copyright Maxim Integrated Products (www.maxim-ic.com). Used by permission.

Table 7.1 contains a summary based largely on the work of Beeby *et al.* [22], who collected and tabulated the important features of practical, representative, electrostatic generator devices. The chief parameters and attributes are power, frequency, mass, volume, and materials. In the realm of larger systems, piezoelectric-based harvesters dominate the market.

It is important to note that interest in devices with power outputs of the order of ~1 μW is now rapidly growing. Just a few years ago, such devices were dismissed as being of no real practical use, but now the power requirements of newer, more energy-efficient IC designs are steadily dropping, while the power capabilities of MEMS energy harvesters are rising. One may presume that acceptance of MEMS energy harvesters depends as

Table 7.1 Summary of electrostatic energy harvesters. Adapted from Beeby *et al.* [22]

Reference	P (μW)	f (Hz)	a (m s^{-2})	M (g)	Vol. (mm^3)	Details
Tashiro, Terumo Corp (JP) [25]	58	4.76	0.5	1200	–	Al/polyester
Meninger, MIT (US) [26]	64	2520	–	–	75	Si
Tashiro, Terumo Corp (JP) [27]	36	6	1	780	–	Al/polyester
Mitcheson, Imperial College (UK) [28]	3.7	30	50	0.1	750	Si/Pyrex
Sterken, IMEC (B) [29, 30]	100[a]	1200	1137	–	–	Si/Pyrex
Roundy, Berkeley University (US) [31]	110[a]	120	2.25	–	1000	Si
Chetwynd, Warwick University (UK) [32]	97 mV	28000	1395	–	–	Al/Au/glass
Miyazaki, Hitachi (JP) [33]	0.12	45	0.08	–	–	Metal
Ma, Hong-Kong University (China) [34]	0.065	4200	766	2×10^{-4}	–	Au/Si
Arakawa, Tokyo University (JP) [35]	6	10	3.9	–	800	Polymer/glass
Basrour, LETI (F) [36]	1052	50	8.8	104	1800	Tungsten
Basrour, LETI (F) [36]	70	50	9.2	2	32.4	Si
Peano, Turin Polytecnicco (I) [37]	50[a]	911	164	–	–	Si

[a] Simulated results.

much on the success of new, lower-power ICs as it does on further development of the devices themselves.

One of the principal challenges of implementing capacitive energy harvesters is their high capacitive output impedance. Associated problems are that the output voltage is rather high, typically ~100 V, and the output current is limited.

7.6 Summary

This chapter introduced four different capacitive MEMS devices: pressure sensors, accelerometers, gyroscopes, and energy harvesters. The presentations stressed systemic views of these devices and utilized linearized models whenever possible.

Although piezoresistive pressure sensors are dominant and piezoelectric energy harvesters are now more well-developed, this chapter has restricted itself to capacitive MEMS devices. Pressure sensors with the familiar circular diaphragm structure were examined using both traditional quasistatic and dynamic models. Capacitive sensing with an op-amp feedback resistor was employed first to obtain a simple, linear, system-level model. Then, to answer the criticism that such elements are not operative at DC, a solution based on a switched-capacitance circuit was also considered.

Capacitive MEMS accelerometers were examined both from the systems perspective and from the design point of view. After a brief introduction of the moving and reference frames, we derived a systems-level linear model for a generic accelerometer. The simple op-amp circuit with feedback resistor was implemented first. The transfer function and sensitivity were examined in detail. Electrode design, control of damping to meet various performance specifications, and certain higher-order effects were also studied.

Because gyroscopes require significant domain-specific knowledge, we started this section with a review of rotating frames of reference and the Coriolis force. The simple point-mass vibratory gyroscope analyzed first led to a useful linear system-level model. More background from kinematics and dynamics of rigid bodies was then introduced to facilitate quantitative analysis of more practical gyro designs. Design considerations, such as electrode structures, resonant frequency, damping, and higher-order effects were explored.

Finally, we examined power generated by capacitive energy harvesters. Because they do not require bias, electret-based energy harvesters are best suited for this application. A linearized system-level model was introduced first. Then, after adopting a few simplifying assumptions motivated by practical considerations, the frequency dependence of energy harvesting was studied. The use of resonator arrays to improve energy capture was also analyzed. A brief discussion of capacitive energy harvesters based on rechargeable batteries was offered and the basic power management function of a commercially available chip was described. The section concluded with comprehensive table summarizing devices that have been demonstrated or simulated.

Problems

7.1 Consider a circular silicon plate of thickness $h = 2$ μm and radius $r_a = 200$ μm employed in a pressure sensor, as shown in Fig. 7.1. The Young modulus is $Y_{\mathrm{Si}} =$ 160 GPa and the Poisson ratio is $\nu_{\mathrm{Si}} = 0.12$. The plate forms a capacitor with the circular electrode of the same size ($\alpha = 1$) patterned on the substrate. The spacing between the electrodes is $d = 2$ μm. The capacitive sensor is connected to an operational amplifier with feedback resistor $R_{\mathrm{f}} = 10$ GΩ and capacitor $C_{\mathrm{f}} = 10$ pF. The bias voltage is $V_{\mathrm{B}} = 5$V.

(a) Evaluate numerically the sensitivity of the displacement at the midpoint of the plate to the changes in pressure $\partial u_0/\partial p$ in units of μm/kPa.
(b) Use the linearized capacitance model to compute its sensitivity to the displacement of the midpoint of the plate $\partial C/\partial u_0$ in units of fF/μm. Compare $\partial C/\partial u_0$ of the flexible plate with $\partial C/\partial u_0$ of a rigid plate.
(c) Find the transfer function $\underline{v}_0/\underline{u}_0$ and determine its magnitude (the gain) for $\omega \gg 1/R_{\mathrm{f}}C_{\mathrm{f}}$.
(d) Compute the overall sensitivity of the sensor for $\omega \gg 1/R_{\mathrm{f}}C_{\mathrm{f}}$: $\partial v_0/\partial p$.
(e) What is the sensitivity in the limit of $\omega \to 0$?

7.2 For the pressure sensor of Problem 7.1, explore the parameter space r_a, h, and d by the following plots:
(a) $\partial u_0/\partial p$ vs. plate radius r_a for the range 100 μm $\leq r_a \leq$ 00 μm with $h = 2$ μm.
(b) $\partial u_0/\partial p$ vs. plate thickness h for the range 0.5 μm $\leq h \leq 3$ μm with $r_a = 300$ μm.
(c) $\partial C/\partial u_0$ vs. electrode spacing d for the range 0.5 μm $\leq d \leq 3$ μm with $r_a = 300$ μm.

(d) $\partial C/\partial u_0$ vs. plate radius r_a for the range 100 μm $\leq r_a \leq$ 500 μm with $d = 2$ μm.

Use the plots to determine whether an optimum parameter set can be found. How does the overall sensitivity depend on the geometric parameters r_a, h, and d? Are there any constraints that impose limits on this parameter space? What are the origins of these constraints?

7.3 For the pressure sensor of Problem 7.1, examine how the electrical parameters, specifically C_f and V_B, affect device sensitivity. Note that because the resistor R_f is used to compensate for op-amp leakage currents, it cannot be thought of as an adjustable component. Does an optimum set of values exist for C_f and V_B? Are there any constraints affecting this choice?

7.4 Using the parameters for the Si circular plate specified in Problem 7.1, calculate numerical values for the elements of the model of Fig. 7.5. Assume mass density $\rho_{Si} = 2330$ kg/m^3 and dynamical viscosity of air $\mu_{air} = 1.983 \times 10^{-5}$ N s/m to answer the following questions.

(a) Express the effective spring constant k_{eff} in terms of the equivalent elastic spring constant k_{eq} of the plate. Discuss qualitatively the trade-offs between sensitivity and stability, considering that k_{eff} must remain positive for stable operation of the electromechanical sensor. At what bias voltage V_B^* does k_{eff} become zero? Plot k_{eff} vs. V_B for $0 \leq V_B \leq V_B^*$.

(b) Derive the transfer function $\underline{H}_{v/p}(\omega)$ for this model. What is the operating frequency range for this sensor? What is the overall sensitivity of the pressure sensor according to this model? Compare this sensitivity to the result of Problem 7.1d, if applicable.

7.5 Derive Eq. (7.4) for the capacitance of a circular plate with radius r_a and electrode gap d as a function of the displacement of the plate midpoint u_0, up to the quadratic term.

7.6 Calculate estimates for the lowest resonant frequency, expressed in kHz, for a circular silicon plate of radius $r_a = 200$ μm and thickness $h = 2$ μm. Using Table 6.2 and Fig. 6.38, estimate the resonant frequency of the second and third non-degenerate modes and the resonant frequencies of the first two degenerate modes. The material properties for Si are: Young modulus $Y_{Si} = 160$ GPa, Poisson ratio $\nu_{Si} = 0.12$, and mass density $\rho_{Si} = 2330$ kg/m.

7.7 In this problem you are to find a small-signal circuit representation of the form shown in Fig. 7.3 for a circular-plate pressure sensor operating near the kth resonance, where $k > 1$. You are to assume that the shape of this mode is known and has radial symmetry such that $\phi_k(r)$.

(a) Obtain an integral expression for capacitance as a function of the displacement $u_0 = u(r = 0)$. Do not attempt to solve the integral.

(b) Expand $C(u_0)$ in a Taylor series retaining terms up to the quadratic, leaving the coefficients as integrals of ϕ_k over the plate area.

(c) Find expressions for $\underline{N}$ and $\underline{Z}'_m$ using the result from (b).

(d) Find an approximate expression for m_{eq}, the equivalent lumped mass associated with plate velocity at $r = 0$, using the kinetic energy argument found in Section 6.6.1.

(e) Find an approximate expression for k_{eq}, the equivalent spring constant associated with plate displacement at $r = 0$, by equating the total elastic energy and the elastic energy of the equivalent spring. Use

$$W_{\text{elast}} = \pi D \int_0^{r_a} \left[\left(\frac{\partial^2 w}{\partial r^2} + \frac{1}{r}\frac{\partial w}{\partial r} \right)^2 - 2(1-\nu)\frac{\partial^2 w}{\partial r^2}\frac{1}{r}\frac{\partial w}{\partial r} \right] r\,dr,$$

where D is the plate modulus, ν is the Poisson ratio, and w is the radially symmetry displacement, i.e., $w(r) = u_0\phi_k(r)$.

(f) Express the effective spring constant in terms of k_{eq} and $\underline{Z}'_m$.

7.8 Consider a square plate with side $a = 200$ μm and thickness $h = 2$ μm, clamped all around its perimeter. Use the following approximate expression for the displacements profile:

$$u_z(x, y) = u_0 \left(\frac{2}{a}\right)^8 \left(x - \frac{a}{2}\right)^2 \left(x + \frac{a}{2}\right)^2 \left(y - \frac{a}{2}\right)^2 \left(y + \frac{a}{2}\right)^2.$$

This very simple expression, while not exact, is consistent with the boundary conditions and easy to use. For the lumped parameter displacement, use the value at the middle of the plate, $u_0 = u_z(x = 0, y = 0)$.

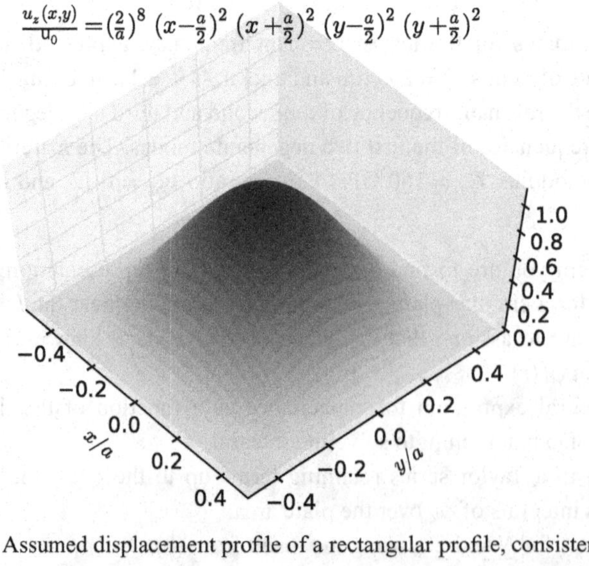

Assumed displacement profile of a rectangular profile, consistent with the boundary conditions.

(a) Find expressions for the equivalent lumped parameter mass and equivalent stiffness, using the elastic energy of a clamped rectangular plate:[16]

$$W_{\text{elast}} = \frac{D}{2}\int_{-a/2}^{a/2}\int_{-a/2}^{a/2}\left(\frac{\partial^2 u(x,y)}{\partial x^2} + \frac{\partial^2 u(x,y)}{\partial y^2}\right)^2 \mathrm{d}x\mathrm{d}y.$$

(b) Find an expression for the lowest resonant frequency and compare it with the analytical solution: $f_{00} = 1.654 c_{\mathrm{L}} h/a^2$, where $c_{\mathrm{L}} = \sqrt{[Y/(1-\nu^2)]/\rho}$ is the speed of sound in the plate material.[17]

(c) Find the capacitance $C(u_0)$ and expand it in a second-order Taylor series expansion to obtain expressions for $\underline{Z}'_{\mathrm{m}}$ and $\underline{N}$. Using the bias voltage $V_{\mathrm{B}} = 10$ V, determine their numerical values.

(d) Determine the effective stiffness. What bias voltage yields instability?

Accelerometer problems

7.9 This problem explores the condition of the so-called "maximally flat" amplitude response. Figure 7.15 reveals that the condition is achieved at $\zeta = 0.7$. For simplicity, the amplitude response is normalized to unity at $\omega_{\mathrm{r}} = 0$.

(a) Find the value for the damping ratio ζ that minimizes absolute error as defined by $\mathrm{err}_{\mathrm{abs}}(\zeta, \omega_{\mathrm{r}}) = |1 - 1/\sqrt{(1-\omega_{\mathrm{r}}^2)^2 + (2\zeta\omega_{\mathrm{r}})^2}|$ at $\omega_{\mathrm{r}} = 0.2$, where $\omega_{\mathrm{r}} = \omega/\omega_0$.

(b) Compute values of ζ that minimize the absolute error in the range $0 \le \omega_{\mathrm{r}} \le 0.3$ and plot them vs. ω_{r}.

(c) Plot the cumulative error, i.e., $\mathrm{err}_{\text{abs-cum}} = \int_0^{.25} \mathrm{err}_{\mathrm{abs}}(\zeta, \omega_{\mathrm{r}})\mathrm{d}\omega_{\mathrm{r}}$, for $0 \le \zeta \le 0.4$.

7.10 The objective of this problem is to gain familiarity with some numerical values typical of the components on the mechanical side of small-signal accelerometer models. Consider the high-aspect ratio Si device in Fig. 7.11 manufactured using a DRIE process. The Si electrodes are 2 μm wide, 12 μm tall, and 450 μm long. Use the mass density $\rho_{\mathrm{Si}} = 2330$ kg/m^3 and the Young modulus $Y_{\mathrm{Si}} = 160$ GPa.

(a) Assuming that the electrode mass constitutes 50% of the total mass, estimate m_{eq}, the equivalent proof mass.

(b) If each guided beam in the restoring spring structure is 400 μm long, 2 μm wide, and 12 μm tall, estimate the total equivalent spring constant k_{eq} and the fundamental resonant frequency.[18]

(c) What is the displacement-to-acceleration sensitivity in the DC limit? Express your result in μm/g, where g is terrestrial gravitational acceleration.

16 A. S. Saada, *Elasticity Theory and Applications*, 2nd edn. (Malabar, FL: Krieger, 1993).

17 A. W. Leissa, *Vibration of Plates*, Reprinted edn. [SI]. (Acoustical Society of America through the American Institute of Physics, 1993).

18 The spring beams are taller than the electrode beams because the DRIE process yields taller structures for larger features and the etched areas surrounding the springs are larger than the etched areas near electrodes (see Appendix C).

7.11 For the accelerometer of Problem 7.10:

(a) Plot $\Delta C(x) = C_1(x) - C_2(x)$ over the range $-d/4 \leq x \leq d/4$, assuming $d = 2$ μm is the gap between the electrodes.

(b) Determine the acceleration range corresponding to this displacement range specified in (a).

(c) Ignoring all other non-linearities except that due to the variable-gap capacitance dependence, determine the range of acceleration that limits the non-linearity error to less than 5%.

(d) Compute the capacitive sensitivity $(C_1[x(a)] - C_1[x(a)])/a$ for low frequencies, assuming small accelerations. Express your numerical result in fF/g, where $g = 9.81$ m/s^2 is the terrestrial gravitational acceleration.

(e) Using the number of electrodes and the width of the spring beams as design parameters, attempt to double the sensitivity.

7.12 Consider the accelerometer shown in the figure, which is a variable-area version of the variable-gap accelerometer presented in Fig. 7.11. Assume the electrodes are 2 μm wide, 12 μm tall, and 450 μm long Si structures separated by a 2 μm gap.

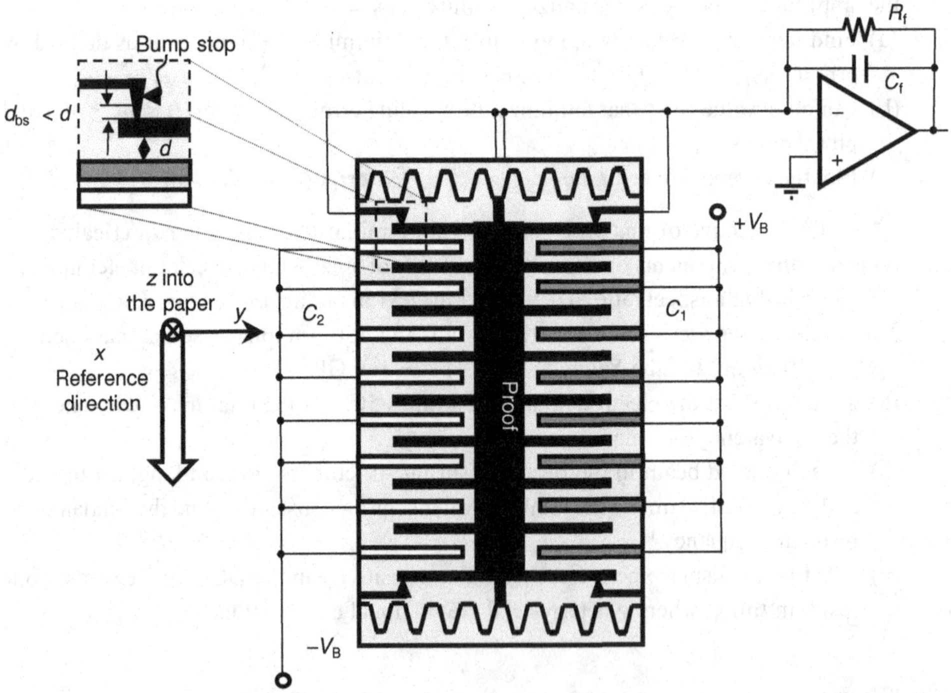

Variable-gap accelerometer.

(a) If the electrode mass is 50% of the total mass, calculate the total equivalent proof mass m_{eq}. Use the mass density $\rho_{Si} = 2330$ kg/m^3.

(b) Assuming that the length of the cantilevered beam in the restoring spring is 150 μm and that the height and the width are the same as the electrodes, estimate the equivalent stiffness k_{eq}, using $Y_{Si} = 160$ GPa.

(c) Compute dC/dx in pF/μm and compare it with the same quantity for the variable-gap structure of Fig. 7.11. Does the electrode length affect sensitivity?

7.13 Because the terrestrial gravitational field is indistinguishable from acceleration, capacitive accelerometers can be used to determine spatial orientation. Such capability is often incorporated in electronic cameras and cell phones. When an accelerometer is oriented as in the sketch of Fig. 7.11, it responds to an input force equal to the weight of the proof mass.

(a) For the device specified in Problem 7.10, compute a numerical value for the displacement x in microns for input acceleration $a = 1g$.

(b) Use this displacement to calculate the capacitance change $\Delta C(x) = C_1(x) - C_2(x)$ for a gap of $d = 2$ μm.

(c) What are ΔC_1 and ΔC_2 for $a = -1g$? How is the die oriented for $a = -1g$?

(d) How is the die oriented with respect to the gravity vector for $a = 0$? Explain why we have to use a switched-capacitor sensing circuit here.

7.14 Consider the sensing circuit of Fig. 5.29 connected to the accelerometer device of Problem 7.10. Assuming $C_f = 2C_1(0)$, $V_B = 3.3$ V, $f_s = 500$ kHz, and perfect demodulation, calculate the low-frequency sensitivity in mV/g of this sensor.

7.15 Determine the effective input force noise of an accelerometer with equivalent mass $m_{eq} = 10^{-10}$ kg, resonant frequency $f_0 = 1$ kHz, and damping ratio $\zeta = 0.7$ at room temperature, $T = 300$ K, connected to an ideal low-pass filter with corner frequency of 100 Hz.

7.16 Using the simple bimorph model of Section 7.3.7, determine the maximum temperature-induced bending deflection at room temperature ($T = 293$ K) of a 2 μm wide, 10 μm tall, and 100 μm long Si cantilever with a uniform 0.5 μm aluminum layer deposited on top at $T = 400$ °C. The Young modulus and the coefficient of thermal expansion for Si are $Y_{Si} = 160$ GPa and $\alpha_{Si} = 2.6$ ppm/K. The Young modulus and the coefficient of thermal expansion for Al are $Y_{Al} = 70$ GPa and $\alpha_{Al} = 23.1$ ppm/K.

Gyroscope problems

7.17 Rigid-body rotations have some interesting properties important for gyro applications. This problem first introduces the formula for rotation about an arbitrary axes and how these rotations combine. You will find that while finite rotations do not commute, infinitesimal rotations do. For a torsional teeter-totter gyro, drive and sense motions can be considered infinitesimal rotations, while the input rate is always a finite rotation.

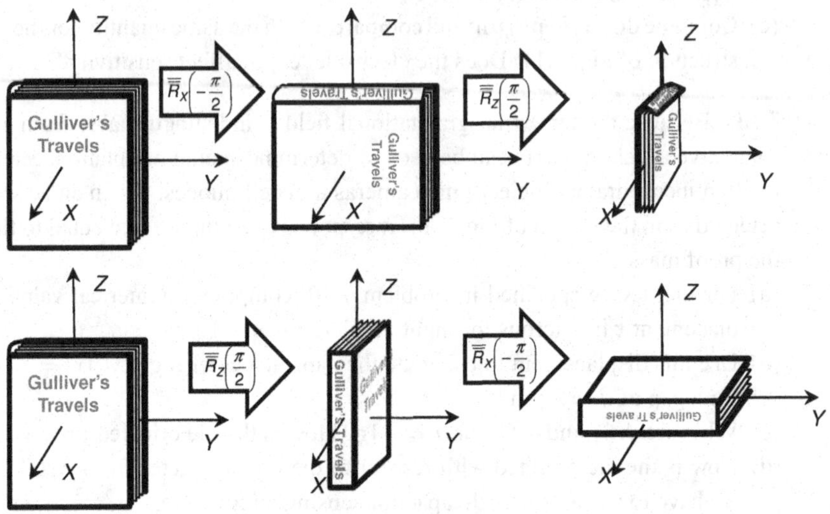

Example of the non-commutative nature of finite rigid-body rotations. A book first rotated by $\pi/2$ radians about the X-axis and then rotated by $\pi/2$ radians about the Z-axis ends up in a different final orientation than when the order of the rotations is reversed.

(a) Verify Rodrigues's formula for rotation of a vector $\overline{v}$ about an arbitrary unit vector $\overline{u} = [u_1 \quad u_2 \quad u_3]^T$ given by

$$\overline{v}_{\mathrm{R}} = \left[\cos(\phi)\overline{\overline{I}}_{3\times 3} + \sin(\phi)\begin{bmatrix} 0 & -u_3 & u_2 \\ u_3 & 0 & -u_1 \\ -u_2 & u_1 & 0 \end{bmatrix} + (1-\cos(\phi))\begin{bmatrix} u_1u_1 & u_1u_2 & u_1u_3 \\ u_2u_1 & u_2u_2 & u_2u_3 \\ u_3u_1 & u_3u_2 & u_3u_3 \end{bmatrix}\right]\overline{v} = \overline{\overline{R}}_{\overline{u}}(\phi)\overline{v},$$

where $\overline{v}_{\mathrm{R}}$ is the resulting vector and ϕ is the angle of rotation. Then evaluate the rotation matrices for the three perpendicular axes, that is, obtain

$$\overline{\overline{R}}_X(\phi),\ \overline{\overline{R}}_Y(\theta),\quad \text{and} \quad \overline{\overline{R}}_Z(\psi).$$

(a) Finite rotations generally do not commute. Perform the matrix transformations to show that

$$\overline{\overline{R}}_Z(\pi/2)\overline{\overline{R}}_X(-\pi/2) \neq \overline{\overline{R}}_X(-\pi/2)\overline{\overline{R}}_Z(\pi/2).$$

(b) In MEMS gyroscopes, the magnitudes of torsional vibration angels are very small. Show that 3-D rotation becomes virtually commutative in the limit of infinitesimal rotational displacements by comparing $\overline{\overline{R}}_Z(\psi)\overline{\overline{R}}_X(\phi)$ and $\overline{\overline{R}}_X(\phi)\overline{\overline{R}}_Z(\psi)$ for small ϕ and ψ.

7.18 Derive expressions for the velocity and acceleration vectors in polar coordinates, starting with the position vector $\overline{r} = r\overline{i}_r$, and recalling that the unit vectors in polar coordinates are

$$\begin{bmatrix} \overline{i}_r \\ \overline{i}_\theta \end{bmatrix} = \begin{bmatrix} \cos(\theta) & \sin(\theta) \\ -\sin(\theta) & \cos(\theta) \end{bmatrix} \begin{bmatrix} \overline{i}_X \\ \overline{i}_Y \end{bmatrix},$$

where $\overline{i}_x$ and $\overline{i}_y$ are unit vectors of the inertial frame of reference. Identify all acceleration terms.

7.19 Consider the rigid-body mass gyroscope shown in Fig. 7.28. Let the resonant frequency of the drive be $f_d = \omega_d/2\pi = 4$ kHz. Further, assume the relationship between the two spring constants is $k_x = 0.9k_y$ and the quality factor of the sense resonator is $Q_s = 30$.

(a) For a driven motion amplitude of $(x_d)_{max} = 12$ μm, compute the Coriolis response in μm for an input rate of 40 degrees per second. Note that the input rate of rotation is commonly specified in degrees per second for gyroscope technology.
(b) What is the sensitivity of the Coriolis sense motion expressed in nm-s/degree?
(c) Repeat (a) for several values of the ratio k_x/k_y, in the range $0.1 \le k_x/k_y \le 1.6$. Plot the calculated Coriolis response versus k_x/k_y, to determine the influence of this ratio upon the gyroscope output.

7.20 Assume that the drive frequency of a rigid-body gyroscope is $f_d = \omega_d/2\pi = 4$ kHz, and the total equivalent mass $m_{eq} = 9 \times 10^{-9}$ kg. For each of the two drive banks, C_{d1} and C_{d2}, the number of electrodes is $n = 50$, the electrode height is $h = 10$ μm, and the gap spacing is $d = 2$ μm. The two electrode banks are driven at resonance with two 10 V-amplitude voltages: $v_{d1} = 10\cos(\omega_R t)$ and $v_{d2} = 10\sin(\omega_R t)$.

(a) Determine the net force acting on the mass m_{eq}. What value of ω_R gives rise to the resonance response?
(b) Compute the amplitude of the driven motion in microns if the quality factor of the driven motion is $Q_d = 100$.
(c) Assuming that the mass of the drive electrodes for $n = 50$ contributes 20% of the total gyro mass, determine the number of electrodes required to achieve 50% larger driven motion. What is the new resonance frequency of the modified structure?

7.21 The driven (teeter-totter) motion of the gyroscope in the figure is about the y-axis. If the device is intended to detect rotations about the y-axis, determine the direction of the Coriolis force on each mass element. Present your answer in the form of a force diagram.

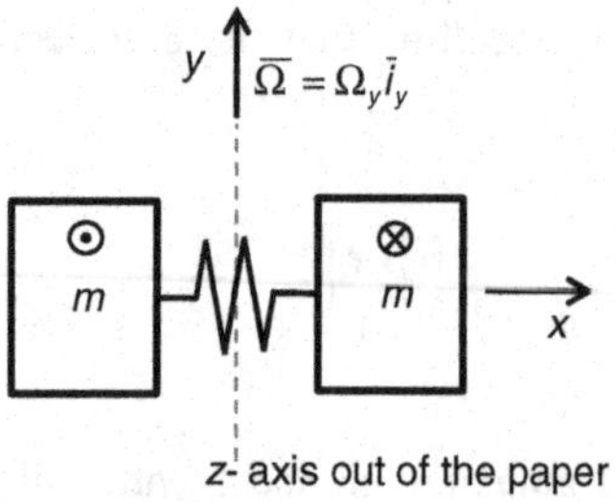

Gyroscope with driven teeter-totter motion about the y-axis. The device is intended to detect rotation about the y-axis.

7.22 Consider a silicon resonator with $\omega_0 = 10$ kHz, $Q = 50$, and $m_{eq} = 10^{-10}$ kg. The electrodes are of variable-area type with spacing of $d = 2$ μm and depth of the electrode $h = 20$ μm.

(a) Using an ODE solver, compute the displacements in the $0 \leq t \leq 4\pi Q/\omega_0$ range if the applied voltage is a square wave with 50% duty cycle $v(t) = 100\,[\text{sign}\,(\cos\,(\omega_0 t)) + 1]\,/2$. What is the maximum response $\max[x(t)]$ in that interval?

(b) Repeat (a) for a 75% duty cycle

$$v(t) = 100[\text{sign}\,(\cos(\omega_0 t) + 1/\sqrt{2}) + 1]/2.$$

(c) Repeat (a) for a 25% duty cycle

$$v(t) = 100[\text{sign}\,(\cos(\omega_0 t) - 1/\sqrt{2}) + 1]/2.$$

(d) Run ~20 simulations for duty cycle in the range [0, 100]% and plot maximum displacement vs. duty cycle. What duty cycle corresponds to the maximum response? Explain.

Energy harvester problems

7.23 Plot the input P_m of an energy harvester normalized with respect to $m_{eq}\omega_0^3|X_0|^2$ as a function of normalized frequency, $\omega_r = \omega/\omega_0$,

$$\frac{P_m}{m_{eq}\omega_0^3|\underline{X}_0|^2} = \frac{\zeta_t \omega_r^6}{\left(1-\omega_r^2\right)^2 + (2\zeta_t\omega_r)^2}$$

for several values of the damping ratio ζ_t near unity. Explain the behavior at high frequency.

7.24 For Fig. 7.39, show formally that

$$P_m = \frac{1}{2}\text{Re}\{\underline{f}_{ext}\,(j\omega\underline{x})^*\} = \frac{1}{2}\,|j\omega\underline{x}|^2\,b_{total}.$$

Can you obtain this relationship by inspection? (Hint: use electromechanical analogies and recall that storage elements, such as inductors and capacitors, do not consume power.)

7.25 This problem investigates maximum power conversion conditions, first for a single resonator and then for a coupled resonator.

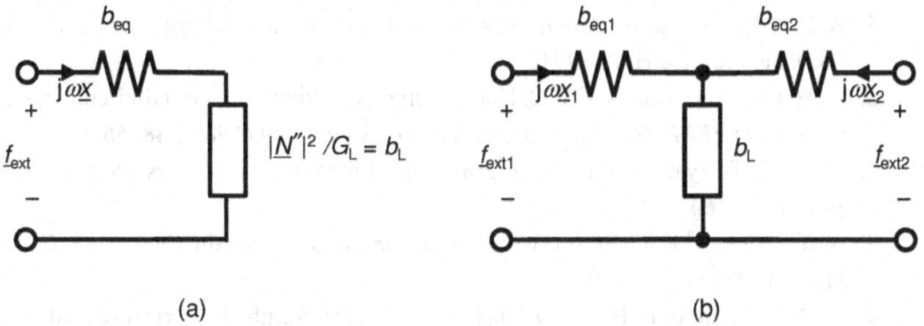

(a) Simplified equivalent circuit representation for simple single-resonator energy harvester.
(b) The equivalent circuit representation for derivation of the maximum power transfer of two resonators.

(a) Using the simplified circuit shown in part (a) of the figure, derive the principle of maximum power transfer for the resonant energy harvester with a single resonator. One way to proceed is to derive the power delivered to b_L in terms of b_{eq}, $\underline{f}_{ext}$, and b_L and then find the condition on b_L that maximizes this quantity.
(b) By inspection of the equivalent circuit of part (b) of the figure, and using the result in (a), find the expression for b_L that maximizes energy transfer for the special case $b_{eq1} = b_{eq2} = b_{eq}$ and $\underline{f}_{ext1} = \underline{f}_{ext2} = \underline{f}_{ext}$.
(c) What is the optimum value for b_L for an energy harvester with n identical resonators, each with associated individual loss represented by b_{eq}?

7.26 Repeat the analysis of the two-beam resonant energy harvester of Section 7.5.3, but now making the more practical assumption that the lengths rather than the thickness are different for the adjacent beams. Let two beams differ in length by 10%. This scheme is actually more realistic for MEMS because, while the lengths of cantilevered beams are determined by the mask design, the thicknesses of deposited layers are almost always fixed by a microfabrication process across the entire area of a die. For this problem, assume that the quality factor of the cantilever is related to length according to $Q \sim L^{-1.65}$.[19]

[19] J. H. Lee, S. T. Lee, C. M. Yao, and W. Fang, Comments on the size effect on the microcantilever quality factor in free air space, *Journal of Micromechanics and Microengineering*, 17 (2007), 139.

References

1. W. P. Eaton and J. H. Smith, Micromachined pressure sensors: review and recent developments. *Smart Materials and Structures*, 6 (1997), 530–539.
2. S. D. Senturia, *Microsystem Design*. (Boston: Kluwer Academic Publishers, 2001).
3. G. T. A. Kovacs, *Micromachined Transducers Sourcebook*. (Boston, MA: WCB, 1998).
4. M.-H. Bao, *Micro Mechanical Transducers: Pressure Sensors, Accelerometers, and Gyroscopes*. (Amsterdam: Elsevier, 2000).
5. A. D. Khazan, *Transducers and Their Elements: Design and Application*. (Englewood Cliffs, NJ: PTR Prentice Hall, 1994).
6. W. H. Ko, M. H. Bao, and Y. D. Hong, A high-sensitivity integrated-circuit capacitive pressure transducer. *IEEE Transactions on Electron Devices*, 29 (1982), 48–56.
7. N. Yazdi, F. Ayazi, and K. Najafi, Micromachined inertial sensors. *Proceedings of the IEEE*, 86 (1998), 1640–1659.
8. A. S. Sedra and K. C. Smith, *Microelectronic Circuits*, 4th edn. (New York: Oxford University Press, 1998).
9. M. A. Lemkin, B. E. Boser, D. Auslander, and J. H. Smith, A 3-axis force balanced accelerometer using a single proof-mass. *International Conference on Solid State Sensors and Actuators, TRANSDUCERS '97 Chicago*, 2 (1997), 1185–1188.
10. K. Jono, K. Minami, and M. Esashi, An electrostatic servo-type three-axis silicon accelerometer. *Measurement Science and Technology*, 6 (1995), 11.
11. K. E. Petersen, Silicon as a mechanical material. *Proceedings of the IEEE*, 70 (1982), 420–457.
12. J. P. Den Hartog, *Mechanics*. (New York: Dover Publications, 1961).
13. J. Lyman and E. Norden, *Rate and Attitude Indicating Instrument*. United States Patent US 2,309,853. 1943
14. R. N. Arnold and L. Maunder, *Gyrodynamics and its Engineering Applications*. (New York: Academic Press, 1961).
15. E. J. Loper and D. D. Lynch, *Hemispherical Resonator Gyro- Status Report and Test Results*, Institute of Navigation, National Technical Meeting, San Diego, CA. (17–19 Jan. 1984), pp. 105–107.
16. F. Ayazi and K. Najafi, A HARPSS polysilicon vibrating ring gyroscope. *Journal of Microelectromechanical Systems*, 10 (2001), 169–179.
17. J. Bernstein, S. Cho, A. T. King, *et al.*, A micromachined comb-drive tuning fork rate gyroscope. *Micro Electro Mechanical Systems, 1993, MEMS '93, Proceedings and Investigation of Micro Structures, Sensors, Actuators, Machines and Systems. IEEE*. (1993), 143–148.
18. H. Johari and F. Ayazi, High-frequency capacitive disk gyroscopes in (100) and (111) silicon. *MEMS, IEEE 20th International Conference on Micro Electro Mechanical Systems* (2007), 47–50.
19. S. H. Crandall, *Dynamics of Mechanical and Electromechanical Systems*. (New York: McGraw-Hill, 1968).
20. W. A. Clark, T. Juneau, and R. T. Howe, Micromachined vibratory rate gyroscope. United States Patent US 6,067,858, 2000.
21. N. Barbour and G. Schmidt, Inertial sensor technology trends, *IEEE Sensors Journal*, 1 (2001), 332–339.
22. S. P. Beeby, M. J. Tudor, and N. M. White, Energy harvesting vibration sources for microsystems applications. *Measurement Science and Technology*, 17 (2006), R175.

23. C. B. Williams and R. B. Yates, Analysis of a micro-electric generator for microsystems. *Sensors and Actuators A*, 52 (1996), 8–11.
24. I. Maxim, *Energy-Harvesting Charger and Protector*. MAX17710 (2011).
25. R. Tashiro, N. Kabei, K. Katayama, *et al.*, Development of an electrostatic generator that harnesses the motion of a living body: use of a resonant phenomenon. *JSME International Journal. Series C, Mechanical Systems, Machine Elements and Manufacturing*, 43 (2000), 916–922.
26. S. Meninger, J. O. Mur-Miranda, R. Amirtharajah, A. Chandrakasan, and J. H. Lang, Vibration-to-electric energy conversion. *IEEE Transactions on Very Large Scale Integration (VLSI) Systems*, 9 (2001), 64–76.
27. R. Tashiro, N. Kabei, K. Katayama, E. Tsuboi, and K. Tsuchiya, Development of an electrostatic generator for a cardiac pacemaker that harnesses the ventricular wall motion. *Journal of Artificial Organs*, 5 (2002), 239–245.
28. P. D. Mitcheson, B. H. Stark, P. Miao, *et al.*, Analysis and optimisation of MEMS electrostatic on-chip power supply for self-powering of slow-moving sensors. *Proc. Eurosensors XVII, Guimaraes, Portugal.* (21–24 Sept 2003), pp. 492–495.
29. T. Sterken, K. Baert, R. Puers, and S. Borghs, Power extraction from ambient vibration. *Proc. SeSens (Workshop on Semiconductor Sensors)* (2002), pp. 680–683.
30. T. Sterken, P. Fiorini, K. Baert, G. Borghs, and R. Puers, Novel design and fabrication of a MEMS electrostatic vibration scavenger. *Proceedings PowerMEMS* (2004), pp. 18–21.
31. S. J. Roundy, *Energy Scavenging for Wireless Sensor Nodes with a Focus on Vibration to Electricity Conversion*, Ph.D. thesis, University of California at Berkeley (2003).
32. M. Mizuno and D. G. Chetwynd, Investigation of a resonance microgenerator. *Journal of Micromechanics and Microengineering*, 13 (2003), 209.
33. M. Miyazaki, H. Tanaka, G. Ono, *et al.*, Electric-energy generation using variable-capacitive resonator for power-free LSI: Efficiency analysis and fundamental experiment. *Proceedings of the 2003 International Symposium on Low Power Electronics and Design* (2003), pp. 193–198.
34. W. Ma, M. Wong, and L. Rufer, Dynamic simulation of an implemented electrostatic power micro-generator, *Proceedings of Design, Test, Integration and Packaging of MEMS/ΣMEOM* (2005), pp. 380–385.
35. Y. Arakawa, Y. Suzuki, and N. Kasagi, Micro seismic power generator using electret polymer film, *Proceedings PowerMEMS*, (2004), pp. 187–190.
36. S. Basrour, J. J. Chaillout, B. Charlot, *et al.*, Fabrication and characterization of high damping electrostatic micro devices for vibration energy scavenging, *Symposium on Design, Test, Integration and Packaging of MEMS/MOEMS (DTIP 2005), June 1–3, Montreux: Switzerland* (2005), pp. 386–90.
37. F. Peano and T. Tambosso, Design and optimization of a MEMS electret-based capacitive energy scavenger, *Journal of Microelectromechanical Systems*, 14 (2005), 429–435.

8 Electromechanics of piezoelectric elements

8.1 Introduction

The piezoelectric effect is widely exploited in actuators and sensors larger than about a millimeter. A familiar example is the crystal oscillator, which is heavily relied on as a stable frequency standard in electronics. Migrating piezomaterials into smaller scale devices has been stymied until fairly recently by serious fabrication challenges. The main problem is that the common piezoelectric solids are either ceramics or crystals, neither of which is amenable to the surface and bulk microfabrication processes used for MEMS. This situation is now starting to change. New materials and the associated microfabrication processes needed to incorporate them into submillimeter structures are being developed. Examples include thin aluminum nitride films sputtered on such substrates as Pt and crystalline Si [1]. Figure 8.1 shows some interesting and novel structures that have now been fabricated. Piezoelectric-based MEMS product lines are on the market and further entries may be anticipated.

For use as a MEMS material, piezoelectrics have compelling advantages. First, their response is linear over a large dynamic range. This attribute simplifies the requirements placed on the signal-conditioning electronics. Perhaps more importantly, piezoelectrics possess high energy densities. Because of their favorable scaling for thin film-based structures, MEMS-scale actuators and sensors using the piezoelectric effect deliver, respectively, large forces and strong signals. Furthermore, they have the practical advantage of being self-biasing, obviating the need for a DC voltage source. Finally, piezoelectric materials are generally inexpensive.

The broad subject of piezoelectricity and the technology based on it is well-covered by some excellent volumes to which the reader may wish to refer [2, 3]. This chapter provides a basic introduction to the intrinsic behavior of piezoelectric solids and introduces lumped parameter electromechanical network models for each of the three principal piezoelectric actuation mechanisms. This treatment is limited to low-frequency behavior, where a quasistatic model is sufficient. Within the confines of the approximation, certain similarities arise between these circuit models and those derived in Chapter 4 for capacitive transducers. A detailed analysis is presented for a piezoelectric element used to drive a cantilevered beam. The chapter concludes by introducing a model for a piezoelectric sensor connected to an op-amp based charge amplifier.

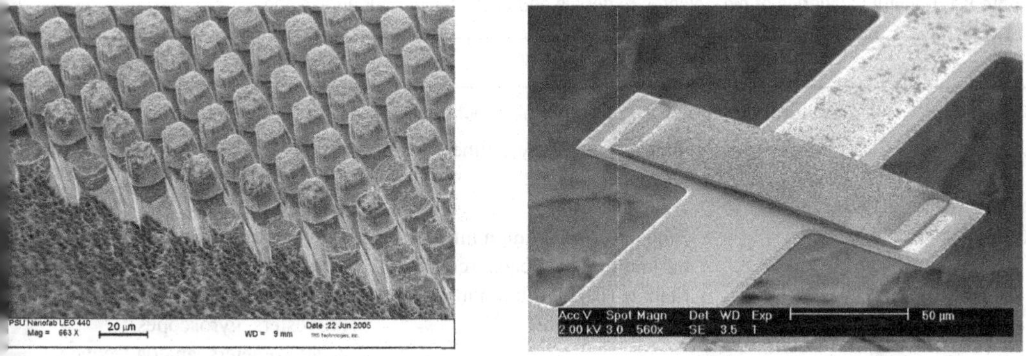

(a) (b)

Figure 8.1 Examples of novel piezoelectric structures for MEMS transducers. (a) High-aspect-ratio square pillars consisting of a Ni layer capping PZT (lead zirconium titanate) columns [1] © Institute of Physics. (b) Buckling Al beam supported at both ends and actuated from beneath by a PZT film deposited on a silicon nitride bar [4] © IEEE.

8.2 Electromechanics of piezoelectric materials

The piezoelectric effect is an intrinsic electromechanical mechanism. It arises in certain classes of crystalline and semicrystalline materials from asymmetries in the spatial distribution of the ions forming the lattice structure. Both natural and man-made crystals exhibit the effect. *Semicrystalline polymers* also exhibit piezoelectricity believed to arise from interactions between free charge groups in adjacent long-chain molecules within the polymer matrix.

When a piezoelectric material undergoes strain, individual charges within the structure are displaced from their normal lattice sites, inducing bound electric surface charge and volume-distributed dielectric polarization. This electrical response to tensile, compressive, or shear mechanical strain is called the *direct effect*. The mechanism is harnessed in MEMS sensors, energy harvesters, and other devices. Conversely Inversely, mechanical stress and resultant motion arise when piezoelectric materials are subjected to an externally applied electrical excitation. This, the so-called *indirect effect*, is exploited in actuators, such as linear positioners, mechanical resonators, gyros, and ultrasonic transducers. Piezoelectric actuation is stronger than capacitive transducer actuation and, furthermore, requires no external voltage bias.

8.2.1 Piezoelectric phenomenology

Figure 8.2 depicts the three most commonly recognized manifestations of piezoelectricity – longitudinal "L," transverse "T," and shear deformations. Crystalline anisotropy determines which of these effects will dominate in a given material.

Table 8.1 collects some basic information about the four main classes of piezoelectrics – crystalline, ceramic, polymer, and natural materials. For the obvious

Table 8.1 General categories of piezoelectric materials and examples of their application

Material type		Examples	MEMS applications
Crystalline[a]	Natural	Quartz, tourmaline, Rochelle salt	Highly stable, high frequency oscillators, gyroscopes, accelerometers, analog frequency dividers
	Man-made	Quartz, GaAs, crystalline langasite	
Ceramics[b]		Zinc oxide, aluminum nitride, barium titanate, lead zirconium titanate (PZT), lithium niobate, lithium tantalate	Can be fabricated in thin films for ultrasonic transducers, high-Q mechanical resonators, energy harvesters, gyroscopes, accelerometers, analog frequency dividers
Polymer films		Polyvinylidene fluoride (PVDF)[c]	Can be fabricated in many shapes and configurations: biomedical applications, low-frequency operation (near DC)
Natural materials		Bone[d], wood, silk	Future application in bio-implants?

[a] Integration of crystals into MEMS structures is more expensive so their application is limited to precision devices where the cost can be justified.
[b] Most piezoceramics require high-temperature processing, thus complicating their integration into MEMS fabrication.
[c] PVDF is a semicrystalline material.
[d] Collagen affords bone with its piezoelectric response.

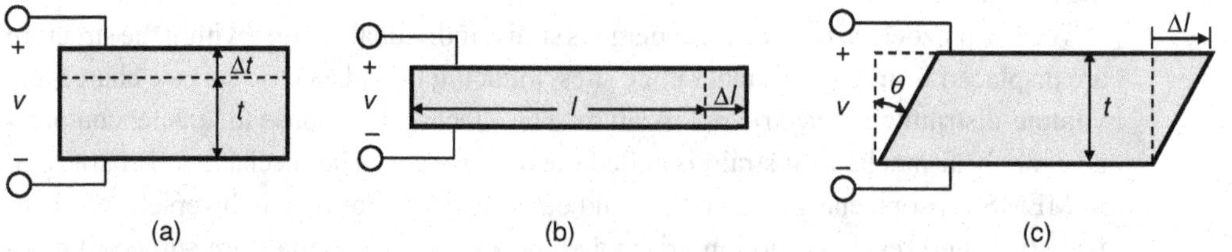

Figure 8.2 Each of the three basic piezoelectric effects has an associated d_{mn} coefficient. The relative importance of the different effects depends on crystalline anisotropy. (a) The longitudinal "L" effect (d_{33}), where voltage modulates the thickness of a thick sample. (b) The transverse "T" effect, where voltage modulates the length of a thin sample (d_{31} or d_{32}). (c) Shear coupling, where the voltage induces shear strain (d_{24} or d_{15}).

reason that most microelectromechanical devices have planar geometries and are fabricated using surface-machining methods, thin-film piezoelectric actuators are favored in MEMS. Ceramic and polymer materials are commonly employed because standard surface-microfabrication techniques can be adapted to deposit films and then pattern them into desired structures. Nevertheless, some piezoelectric devices are fabricated using bulk processes and bonding-based assembly.

8.2.2 Piezoelectric properties

Piezoelectric materials are elastic dielectric solids with zero free-volume electrical charge. This net charge neutrality masks a complex distribution of positive and negative ions, typically centered at crystalline vertices (or, in the case of the ceramics, at the boundaries of domain-like structures). The requirement for a piezoelectric response is some form of charge distribution asymmetry. For certain man-made piezoelectrics, this asymmetry is often enhanced by *poling*, that is, pre-treatment of the material with a combination of a DC electric field and heating to orient and then lock in the polarization. Poling of piezoelectric materials, while somewhat similar to forming electrets from dielectrics, is more complex and not treated here.

When piezoelectric materials are mechanically deformed, *bound surface charge* appears, accompanied by an electric field in the bulk. Despite the fact that they are highly linear, the constitutive laws for piezoelectrics are far more complicated than simple dielectrics or elastic media. This is because, being anisotropic, these relations involve the displacement and electric field vectors, $\overline{D}$ and $\overline{E}$, and the stress and strain tensors, $\overline{\overline{\sigma}}$ and $\overline{\overline{e}}$, all respectively:

$$\overline{D} = \overline{\overline{\overline{d}}} \cdot \overline{\overline{\sigma}} + \varepsilon_0 \overline{\overline{\kappa}} \cdot \overline{E}, \tag{8.1a}$$

$$\overline{\overline{e}} = \overline{\overline{\overline{\overline{s}}}} \cdot \overline{\overline{\sigma}} + \overline{\overline{\overline{d}}} \cdot \overline{E}, \tag{8.1b}$$

where $\overline{\overline{\kappa}}$, $\overline{\overline{\overline{d}}}$, and $\overline{\overline{\overline{\overline{s}}}}$ are, respectively, the second-, third-, and fourth-rank tensors for the dielectric constant, the piezoelectric constant, and the compliance matrices, and $\varepsilon_0 = 8.854 \cdot 10^{-12}$ F/m is the permittivity of free space. Equations (8.1a) and (8.1b) are *point-form* descriptions of the electromechanical behavior of piezoelectric materials using *field variables*. In particular, Eq. (8.1a) generalizes the constitutive relation for dielectrics introduced in Appendix A to account for polarization anisotropy and the direct piezoelectric effect of certain crystals. Likewise, Eq. (8.1b) generalizes the elastic constitutive law (Hooke's law as presented in Appendix D) to account for the indirect effect.

From purely geometric considerations, both $\overline{\overline{\sigma}}$ and $\overline{\overline{e}}$ must be symmetric. This symmetry reduces the number of independent coefficients in the material matrices to six, thereby simplifying the constitutive laws. It is common to transform $\overline{\overline{\sigma}}$ and $\overline{\overline{e}}$ into six-dimensional vectors as follows:

$$\overline{\overline{\sigma}} \rightarrow \overline{\sigma} = \begin{bmatrix} \sigma_x \\ \sigma_y \\ \sigma_z \\ \tau_{yz} \\ \tau_{zx} \\ \tau_{xy} \end{bmatrix} \quad \text{and} \quad \overline{\overline{e}} \rightarrow \overline{e} = \begin{bmatrix} \varepsilon_x \\ \varepsilon_y \\ \varepsilon_z \\ \gamma_{yz} \\ \gamma_{zx} \\ \gamma_{xy} \end{bmatrix}. \tag{8.2}$$

For consistency with Section D.2 of Appendix D, we employ the notational convention that $\sigma_{xx} = \sigma_x$, $\sigma_{yy} = \sigma_y$, and $\sigma_{zz} = \sigma_z$. Also from the appendix, the normal strains are $\varepsilon_x = \varepsilon_{xx} \equiv \partial u_x/\partial x$, $\varepsilon_y = \varepsilon_{yy} \equiv \partial u_y/\partial y$, and $\varepsilon_z = \varepsilon_{zz} \equiv \partial u_z/\partial z$, where u_x, u_y, and u_z are material displacements along the axes. Equation (8.2) uses the standard *engineering shear strains*: $\gamma_{xy} \equiv (\partial u_x/\partial y + \partial u_y/\partial x)$, $\gamma_{yz} \equiv (\partial u_y/\partial z + \partial u_z/\partial y)$, and $\gamma_{zx} \equiv (\partial u_z/\partial x + \partial u_x/\partial z)$.

Employing this nomenclature, the constitutive relation for the direct piezoelectric effect manifested by many materials reduces from Eq. (8.1a) to

$$\begin{bmatrix} D_x \\ D_y \\ D_z \end{bmatrix} = \underbrace{\underbrace{\begin{bmatrix} 0 & 0 \\ 0 & 0 \\ d_{31} & d_{32} \end{bmatrix}}_{\text{length expansion}} \underbrace{\begin{bmatrix} 0 \\ 0 \\ d_{33} \end{bmatrix}}_{\text{thickness expansion}} \underbrace{\begin{bmatrix} 0 & d_{15} & 0 \\ d_{24} & 0 & 0 \\ 0 & 0 & 0 \end{bmatrix}}_{\text{shear modes}}}_{\bar{\bar{d}}} \begin{bmatrix} \sigma_x \\ \sigma_y \\ \sigma_z \\ \tau_{yz} \\ \tau_{zx} \\ \tau_{xy} \end{bmatrix} + \varepsilon_o \underbrace{\begin{bmatrix} \kappa_x & 0 & 0 \\ 0 & \kappa_y & 0 \\ 0 & 0 & \kappa_z \end{bmatrix}}_{\bar{\bar{\kappa}}} \begin{bmatrix} E_x \\ E_y \\ E_z \end{bmatrix}. \tag{8.3}$$

The relatively sparse 3×6 piezoelectric matrix $\bar{\bar{d}}$ defined in Eq. (8.3) adequately describes most ceramics, such as PZT, as well as the polymer PVDF.[1] The individual coefficients d_{mn} that quantify the three piezoelectric effects depicted in Fig. 8.2 are denoted within the matrix.

The indirect effect, Eq. (8.1b), relating the mechanical strains to the stresses and the electric field, becomes

$$\begin{bmatrix} \varepsilon_x \\ \varepsilon_y \\ \varepsilon_z \\ \gamma_{yz} \\ \gamma_{zx} \\ \gamma_{xy} \end{bmatrix} = \underbrace{\begin{bmatrix} s_{11} & s_{12} & s_{13} & 0 & 0 & 0 \\ s_{12} & s_{22} & s_{23} & 0 & 0 & 0 \\ s_{13} & s_{23} & s_{33} & 0 & 0 & 0 \\ 0 & 0 & 0 & s_{44} & 0 & 0 \\ 0 & 0 & 0 & 0 & s_{44} & 0 \\ 0 & 0 & 0 & 0 & 0 & \frac{1}{2}(s_{11} - s_{12}) \end{bmatrix}}_{\text{elastic compliance matrix}} \begin{bmatrix} \sigma_x \\ \sigma_y \\ \sigma_z \\ \tau_{yz} \\ \tau_{zx} \\ \tau_{xy} \end{bmatrix}$$

$$+ \underbrace{\begin{bmatrix} 0 & 0 & d_{31} \\ 0 & 0 & d_{32} \\ 0 & 0 & d_{33} \\ 0 & d_{24} & 0 \\ d_{15} & 0 & 0 \\ 0 & 0 & 0 \end{bmatrix}}_{\text{piezoelectric matrix}} \begin{bmatrix} E_x \\ E_y \\ E_z \end{bmatrix}. \tag{8.4}$$

Equations (8.3) and (8.4) employ an economical matrix nomenclature scheme called the Voigt notation to describe the electromechanical coupling contributions to the constitutive relations for piezoelectric materials. This notation uses numerical indices for the coefficients d_{mn} and s_{mn}, and the convention for their definitions is evident from the equations. Though not immediately apparent from the forms of these matrices, Eqs. (8.3) and (8.4) indeed reflect material reciprocity, just as they must because of the

[1] The piezoelectric matrix $\bar{\bar{d}}$ takes different forms for quartz and other crystalline materials.

Table 8.2 Physical properties of some piezoelectric materials for MEMS [1, 3, 4]

Physical property	MKS units	α-quartz	PZT	PVDF
d coefficients	C/N or m/V	$d_{11} = -d_{12} = -d_{26}/2 = -2.31 \cdot 10^{-12}$	$d_{33} = 300 \cdot 10^{-12}$	$d_{33} = -25 \cdot 10^{-12}$
		$d_{14} = -d_{25} = -0.67 \cdot 10^{-12}$	$d_{32} = d_{31} = -150 \cdot 10^{-12}$	Uniaxial form: $d_{31} = 15 \cdot 10^{-12}$ $d_{32} = 3 \cdot 10^{-12}$ Bi-axial form: $d_{31} = d_{32} = 3 \cdot 10^{-12}$
		$d_{15} = 0$	$d_{15} = 500 \cdot 10^{-12}$	$d_{15} = 0$
$e_{mn} = d_{mn}/s^E$	C/m^2	$e_{11} = 0.17 \cdot 10^{-12}$ $e_{14} = 0.04 \cdot 10^{-12}$	$e_{31} = -7.5 \cdot 10^{-12}$	$e_{31} = 0.025 \cdot 10^{-12}$
κ^σ	–	Slight anisotropy: $\kappa_{11} = 4.5$, $\kappa_{33} = 4.6$	$\sim$1800	$\sim$10
ρ	kg/m^3	2650	7600	1780
$1/s^E$	N/m^2	Strongly anisotropic	$50 \cdot 10^9$	$2.5 \cdot 10^9$

assumption that the material is lossless. Material reciprocity does not always mean that $d_{mn} = d_{nm}$. For example, while $d_{31} \neq 0$ for ceramic materials, there exists no d_{13} term in the piezoelectric matrix.

Table 8.2 contains the important piezoelectric properties for the materials commonly used in MEMS applications, and it reveals that a given material is indeed often far better suited to one mode of electromechanical transduction than to another. For example, PZT appears a better choice for L-type (longitudinal-coupled) transducers than either quartz or PVDF. Also, PZT seems to be more effective than PVDF for L-type (thickness-coupled) devices; however, this apparent advantage must be weighed against the lower cost and greater microfabrication flexibility possessed by PVDF.

In Table 8.2 and the development following, $Y_E = 1/s^E = 1/s\,|_{E=0}$ is the Young modulus under the constraint that the applied electric field is zero, while $\kappa^\sigma = \kappa|_{\sigma=0}$ is the dielectric constant under the condition of zero applied stress.

8.2.3 The L-type piezoelectric transducer

Figure 8.3a shows the side view of a basic, longitudinal (L-type) piezoelectric transducer. The lower surface is specified to be rigidly fastened to the electrode substrate. Further, we assume that the device is one-dimensional: (i) the electric field is uniform and z-directed, $\overline{E} = E_z\,\overline{z}$, and (ii) there are no mechanical constraints imposed on either the x or y faces of the material, that is, $\sigma_x = \sigma_y = 0$. These constraints allow us to simplify the constitutive laws of Eqs. (8.3) and (8.4) to a 2×2 matrix,

$$\begin{bmatrix} D_z \\ \varepsilon_z \end{bmatrix} = \begin{bmatrix} \kappa_z^\sigma \varepsilon_0 & d_{33} \\ d_{33} & s_{33}^E \end{bmatrix} \begin{bmatrix} E_z \\ \sigma_z \end{bmatrix}, \tag{8.5}$$

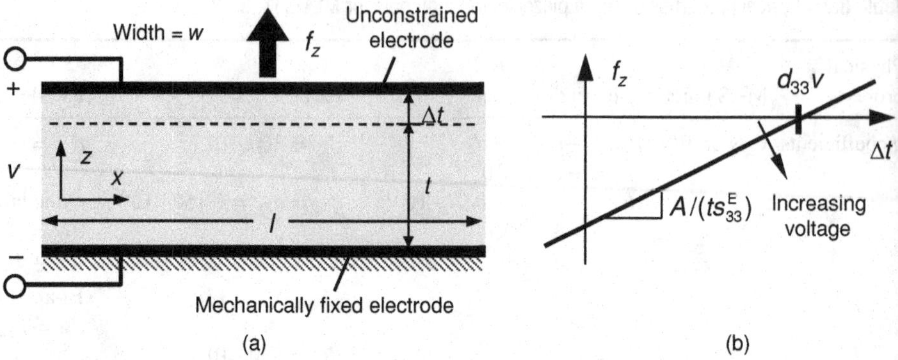

Figure 8.3 One-dimensional piezoelectric transducer operating in the longitudinal L-mode at a frequency well below any acoustic resonance. (a) The geometry of the structure. (b) Force versus displacement curve, Eq. (8.7), at fixed voltage $v > 0$.

where $\varepsilon_z = \partial u_z/\partial z \approx \Delta t/t$ is the uniform, longitudinal, z-directed strain, and it is understood that $\kappa_z^\sigma = \kappa_z|_{\sigma_z=0}$. A convenient alternative form exchanges stress and strain:

$$\begin{bmatrix} D_z \\ \sigma_z \end{bmatrix} = \begin{bmatrix} \kappa_z^\sigma \varepsilon_0 \left[1 - d_{33}^2/\left(s_{33}^E \kappa_z^\sigma \varepsilon_0\right)\right] & d_{33}/s_{33}^E \\ -d_{33}/s_{33}^E & 1/s_{33}^E \end{bmatrix} \begin{bmatrix} E_z \\ \varepsilon_z \end{bmatrix}. \tag{8.6}$$

By relating D_z and σ_z to their lumped parameter equivalents, respectively, electric charge and force, Eq. (8.6) can be converted to a more convenient lumped parameter transducer formulation [5]. From Gauss's law, Eq. (A.7), the charge on the upper electrode is $q = -AD_z$, where $A = wl$ is the area. In the low-frequency quasistatic limit, where acoustic resonances may be ignored, the force exerted on the piezoelectric element by the external system is $f_z = A\sigma_z$. Thus,

$$\begin{bmatrix} q \\ f_z \end{bmatrix} = \begin{bmatrix} C_L & Ad_{33}/\left(s_{33}^E t\right) \\ -Ad_{33}/\left(s_{33}^E t\right) & k_L^E \end{bmatrix} \begin{bmatrix} v \\ \Delta t \end{bmatrix}. \tag{8.7}$$

Here, $C_L = \kappa_z^\sigma \varepsilon_0[1 - d_{33}^2/(s_{33}^E \kappa_z^\sigma \varepsilon_0)]A/t$ is the capacitance, $k_L^E = A/s_{33}^E t$ is an effective spring constant arising from the elastic nature of the solid material, and s_{33}^E is the reciprocal of Young's modulus. Remember that ordinarily piezoelectric transducers require no DC bias; v is the applied voltage. The voltage-dependent component of f_z arises from the interaction between the applied electric field and the bound charge within the piezoelectric material.

The force f_z, plotted versus displacement Δt at fixed voltage v in Fig. 8.3b, exhibits the combined influences of the piezoelectric effect and the elastic nature of the solid. The piezoelectric contribution displaces the linear elastic response curve, downward or upward depending on the polarity of v.

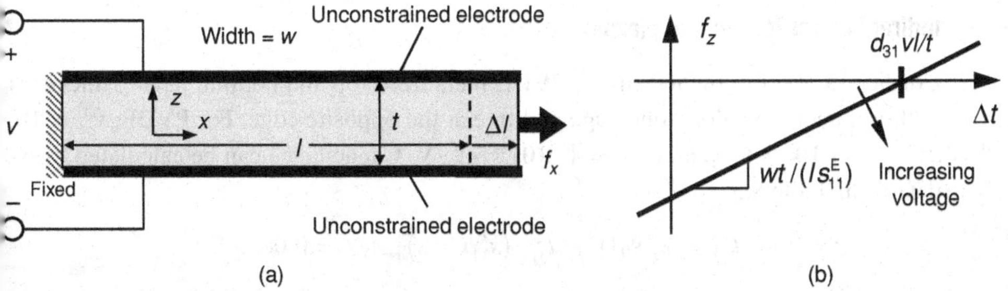

Figure 8.4 Thin piezoelectric transducer element designed to utilize the transverse T-mode at frequencies well below any acoustic resonance. (a) Side view of geometry of the structure. (b) Force versus displacement curve, Eq. (8.10), at fixed voltage $v > 0$.

8.2.4 The T-type piezoelectric transducer

Figure 8.4a shows a side view of a transverse (T-type) piezoelectric transducer. The piezo element in such transducers is typically thin. Assume here that the left edge is immobilized while the right is free. The configuration is one-dimensional; strains in the x direction predominate. The electric field is uniform and z-directed, $\overline{E} = E_z \overline{z}$, and no mechanical constraints are imposed on either the top or the bottom face, so that $\sigma_z = 0$. Likewise, there is no mechanical constraint on the front or back sides, so that $\sigma_y = 0$. Once again, at low frequency the constitutive laws, Eqs. (8.3) and (8.4), reduce to a simple pair of linear relations,

$$\begin{bmatrix} D_z \\ \varepsilon_x \end{bmatrix} = \begin{bmatrix} \kappa_z^\sigma \varepsilon_0 & d_{31} \\ d_{31} & s_{11}^{\mathrm{E}} \end{bmatrix} \begin{bmatrix} E_z \\ \sigma_x \end{bmatrix}, \tag{8.8}$$

where now $\kappa_z^\sigma = \kappa_z|_{\sigma_x=0}$. This expression can be rewritten as

$$\begin{bmatrix} D_z \\ \sigma_x \end{bmatrix} = \begin{bmatrix} \kappa_z^\sigma \varepsilon_0 \left[1 - d_{31}^2/\left(s_{11}^{\mathrm{E}} \kappa_z^\sigma \varepsilon_0\right)\right] & d_{31}/s_{11}^{\mathrm{E}} \\ -d_{31}/s_{11}^{\mathrm{E}} & 1/s_{11}^{\mathrm{E}} \end{bmatrix} \begin{bmatrix} E_z \\ \varepsilon_x \end{bmatrix}, \tag{8.9}$$

where the uniform strain is $\varepsilon_x = \partial u_x/\partial x \approx \Delta l/l$.

As before, a lumped parameter model can be formulated by relating the electric displacement vector to the charge, that is, $q = -wtD_z$, and the normal stress to an x-directed force, $f_x = wt\sigma_x$:

$$\begin{bmatrix} q \\ f_x \end{bmatrix} = \begin{bmatrix} C_{\mathrm{T}} & wd_{31}/s_{11}^{\mathrm{E}} \\ -wd_{31}/s_{11}^{\mathrm{E}} & k_{\mathrm{T}}^{\mathrm{E}} \end{bmatrix} \begin{bmatrix} v \\ \Delta l \end{bmatrix}. \tag{8.10}$$

Here, $C_{\mathrm{T}} = \kappa_z^\sigma \varepsilon_0[1 - d_{31}^2/(s_{11}^{\mathrm{E}} \kappa_z^\sigma \varepsilon_0)]\, A/t$ is the capacitance under zero mechanical strain and $k_{\mathrm{T}}^{\mathrm{E}} = wt/(ls_{11}^{\mathrm{E}})$ is an effective elastic spring constant describing elongation in the x direction. The low-frequency, quasistatic relation between force f_x and displacement Δl at fixed voltage is plotted in Fig. 8.4b. Like the L-type device, piezoelectric coupling here displaces the linear, elastic response curve down or up, depending on whether v is positive or negative, respectively.

Example 8.1 **Indirect effect for T-mode piezoelectric**

Consider a 500 μm by 500 μm of PVDF, metallized top and bottom, having thickness 100 μm, anchored along one edge and free at the opposite edge. For PVDF: $\kappa_z^{\mathrm{T}} = 10$, $d_{31} = 15 \cdot 10^{-12}$ m/V, and $s_{11}^{\mathrm{E}} = 4 \cdot 10^{-10}$ m^2/V. Capacitance can be calculated using data from Table 8.3:

$$C_{\mathrm{T}} = \kappa_z^{\sigma} \varepsilon_0 \left[1 - d_{31}^2/(s_{11}^{\mathrm{E}} \kappa_z^{\sigma} \varepsilon_0)\right] A/t = 0.0025 \text{ F},$$

where area $A = wl$. From Eq. (8.10), the governing equation for the indirect T-mode piezoelectric effect is

$$f_x = -\left(w d_{31}/s_{11}^{\mathrm{E}}\right) v + k_{\mathrm{T}}^{\mathrm{E}} \Delta l,$$

where $k_{\mathrm{T}}^{\mathrm{E}} = wt/ls_{11}^{\mathrm{E}}$. If the device is mechanically unconstrained, that is, $f_x = 0$, while $v = 10$ V$_{\mathrm{dc}}$ is applied to the electrodes, the change in the length is

$$\Delta l|_{f_x=0} = \left[w d_{31}/(s_{11}^{\mathrm{E}} k_{\mathrm{T}}^{\mathrm{E}})\right] v = (l d_{31}/t) v = 7.5 \cdot 10^{-10} \text{ m}.$$

While a 7.5 Å displacement might seem inconsequentially small, it is sufficient in many MEMS applications. If the piezoelectric element were somehow constrained so that length could not change, that is, $\Delta l = 0$, then for the same applied voltage of 10 V$_{\mathrm{dc}}$, the rather large force required to constrain it would be

$$f_x|_{\Delta l=0} = \left(w d_{31}/s_{11}^{\mathrm{E}}\right) v = 1.9 \cdot 10^{-4} \text{ N}.$$

The important lesson is that piezoelectric displacements are small but the realizable forces are large. These calculated values are ideal in the sense that the DC response of real piezoelectric transducers is virtually zero, owing to finite internal resistivity. Thus, the amplitudes calculated here are relevant to the quasistatic situation when the applied voltage is varied slowly in time.

8.2.5 Shear mode piezoelectric transducer

Figure. 8.5 shows a configuration harnessing the direct piezoelectric effect to create shear deformation. So as to avoid the need to alter the constitutive laws of Eqs. (8.3) and (8.4), the x and z coordinates are reversed from Figs. 8.3 and 8.4. The sides of the sample are assumed to be unconstrained and, as before, the electric field is x-directed and uniform.

The reduced, one-dimensional constitutive relationship for this mode of piezoelectric excitation becomes

$$\begin{bmatrix} D_x \\ \gamma_{zx} \end{bmatrix} = \begin{bmatrix} \kappa_{\mathrm{x}}^{\sigma} \varepsilon_0 & d_{15} \\ d_{15} & s_{44}^{\mathrm{E}} \end{bmatrix} \begin{bmatrix} E_x \\ \tau_{zx} \end{bmatrix}, \qquad (8.11)$$

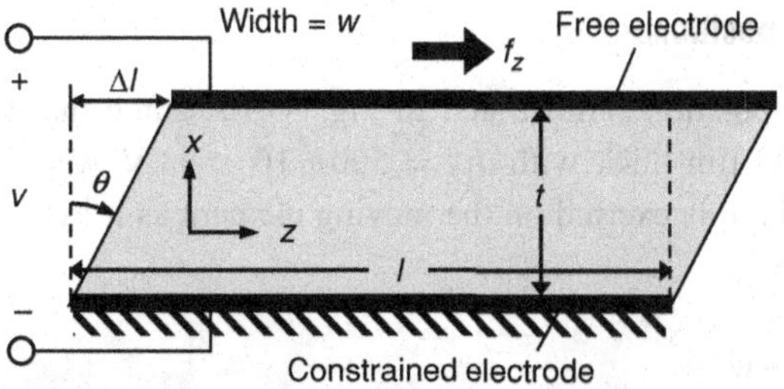

Figure 8.5 Side view of piezoelectric transducer that exploits the shear strain mode, which can be characterized by either the angle θ or the shear displacement Δl, where $\theta \approx \tan^{-1}(\Delta l/t)$.

where now $\kappa_x^\sigma = \kappa_x|_{\tau_{zx}=0}$. In modified form, Eq. (8.11) becomes

$$\begin{bmatrix} D_x \\ \tau_{zx} \end{bmatrix} = \begin{bmatrix} \kappa_x^\sigma \varepsilon_0 \left[1 - d_{15}^2/\left(\kappa_x^\sigma \varepsilon_0 s_{44}^{\mathrm{E}}\right)\right] & d_{15}/s_{44}^{\mathrm{E}} \\ -d_{15}/s_{44}^{\mathrm{E}} & 1/s_{44}^{\mathrm{E}} \end{bmatrix} \begin{bmatrix} E_x \\ \gamma_{zx} \end{bmatrix}. \tag{8.12}$$

For small deformations, the shear strain can be expressed in terms of angle $\theta = \tan^{-1}(\partial u_x/\partial z) \approx \tan^{-1}(\Delta l/t)$. In the quasistatic limit, Eq. (8.12) may be cast in terms of the lumped parameter variables $q = -AD_x$ and $f_z = A\tau_{zx}$:

$$\begin{bmatrix} q \\ f_x \end{bmatrix} = \begin{bmatrix} C_{\mathrm{shear}} & Ad_{15}/\left(s_{44}^{\mathrm{E}} t\right) \\ -Ad_{15}/\left(s_{44}^{\mathrm{E}} t\right) & k_{\mathrm{shear}}^{\mathrm{E}} \end{bmatrix} \begin{bmatrix} v \\ \Delta l \end{bmatrix}, \tag{8.13}$$

where $C_{\mathrm{shear}} = \kappa_x^\sigma \varepsilon_0[1 - d_{15}^2/(s_{44}^{\mathrm{E}} \kappa_x^\sigma \varepsilon_0)]\, A/t$ is the capacitance and $k_{\mathrm{shear}}^{\mathrm{E}} = A/(t s_{44}^{\mathrm{E}})$ is another spring constant, this one describing elastic shear displacement.

8.2.6 Summary

Equations (8.7), (8.10), and (8.13), respectively, establish linear terminal relations for the low-frequency electromechanics of longitudinal-, transverse-, and shear-type piezotransducers. These lumped parameter models, valid at frequencies well below any acoustical resonance of the solid, have been derived directly from the constitutive relations, subject to appropriate approximations and definitions. It has been unnecessary to invoke energy methods to obtain them because the material constitutive laws inherently incorporate the conservative electromechanical coupling mechanism and are consistent with energy conservation. The only requirement to develop these models has been to identify the lumped parameter quantities q and f in terms of the appropriate components of the displacement vector $\overline{D}$ and the stress $\overline{\overline{\sigma}}$.

Example 8.2 **Shear-mode-based linear positioner**

Consider the PZT linear positioner illustrated in Fig. 8.6. Assume that the element is 300 μm square and 100 μm thick with $d_{15} = 500 \cdot 10^{-12}$ m/V. Also assume that no external mechanical force is exerted on the moving element as it moves in the $\pm z$ direction.

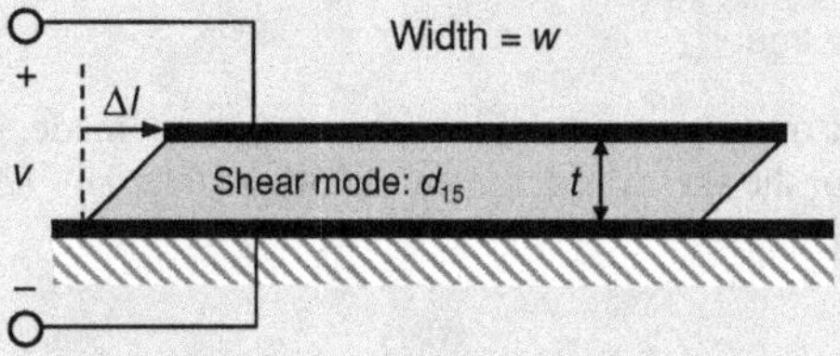

Figure 8.6 Mechanism of a simple piezoelectric positioner based on the shear strain interaction mode.

From Eq. (8.13), the general expression for the indirect shear-mode effect exhibited by a piezoelectric element is

$$f_z = -\left[Ad_{15}/\left(s_{44}^{\mathrm{E}}t\right)\right] v + k_{\mathrm{shear}}^{\mathrm{E}}\,\Delta l,$$

where $k_{\mathrm{shear}}^{\mathrm{E}} = A/(ts_{44}^{\mathrm{E}})$. The unconstrained displacement Δl versus voltage is

$$\Delta l|_{f_z=0} = \left[Ad_{15}/\left(s_{44}^{\mathrm{E}}k_{\mathrm{shear}}^{\mathrm{E}}t\right)\right] v = d_{15}\,v,$$

a result directly attainable from Eq. (8.11) when $\sigma_x = 0$. The piezoelectric effect usually goes to zero for DC excitation, but if the applied voltage v is slowly varying in time, this equation can be used. For example, consider a periodic sawtooth voltage as the input. The displacement will closely replicate this waveform, though the return part of the wave may not match the voltage for steepness. Figure 8.7 plots the ideal displacement Δl versus time for a sawtooth of peak magnitude 20 volts.

The ± 10 μm displacements predicted here are quite adequate for such applications as *atomic force microscopy*, where the typical resolution scale is 0.1 μm.

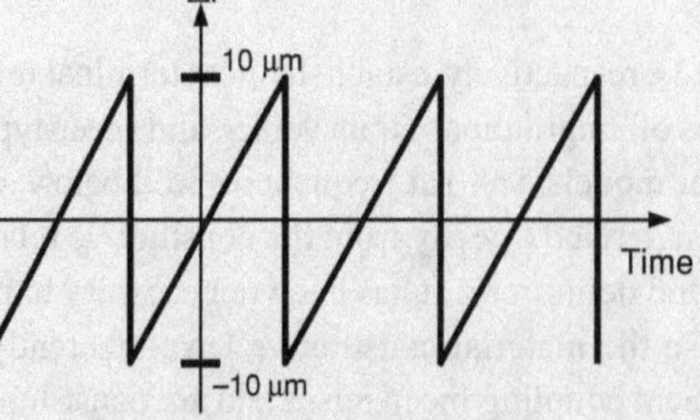

Figure 8.7 Ideal displacement response of a 300 μm square, 100 μm thick, piezoelectric positioner made of PZT and operated in the shear mode to a sawtooth voltage waveform of peak magnitude 20 V.

Table 8.3 The variables and coefficients of the piezoelectric electromechanical matrix, Eq. (8.14), for the different modes illustrated in Figs. 8.3, 8.4, and 8.5

Matrix terms	Longitudinal L-mode (Fig. 8.3)	Transverse T-mode (Fig. 8.4)	Shear mode (Fig. 8.5)
Force: f_χ	f_z	f_x	f_z
Displacement: Δ_χ	Δt	Δl	Δl
Capacitance: C_χ	$C_L = \kappa_z^\sigma \varepsilon_0 \left(1 - \frac{d_{33}^2}{s_{33}^E \kappa_z^\sigma \varepsilon_0}\right) A/t$	$C_T = \kappa_z^\sigma \varepsilon_0 \left(1 - \frac{d_{31}^2}{s_{11}^E \kappa_z^\sigma \varepsilon_0}\right) A/t$	$C_{\text{shear}} = \kappa_x^\sigma \varepsilon_0 \left(1 - \frac{d_{15}^2}{s_{44}^E \kappa_x^\sigma \varepsilon_0}\right) A/t$
Coupling: Φ_χ	$\Phi_L = -Ad_{33}/\left(s_{33}^E t\right)$	$\Phi_T = -wd_{31}/s_{11}^E$	$\Phi_{\text{shear}} = -Ad_{15}/\left(s_{44}^E t\right)$
Elastic spring constant: k_χ	$k_L^E = A/\left(ts_{33}^E\right)$	$k_T^E = wt/\left(ls_{11}^E\right)$	$k_{\text{shear}}^E = A/\left(ts_{44}^E\right)$

8.3 Two-port models for piezoelectric systems

The inherent linearity of Eqs. (8.7), (8.10), and (8.13) makes it easy to synthesize electromechanical two-port networks directly. Unlike capacitive MEMS transducers, no linearization is needed. Because the matrices for the three piezoelectric modes are canonically identical, for economy's sake we introduce here a canonical form to represent all of them. Assuming $e^{j\omega t}$ dependence for all time-dependent variables, the charge and displacement phasors, $\underline{q}$ and $\underline{\Delta}_\chi$ respectively, are replaced by current $\underline{i} = j\omega\underline{q}$ and velocity $j\omega\underline{\Delta}_\chi$. After manipulation and rearrangement, the general matrix form is

$$\begin{bmatrix} \underline{f}_\chi \\ \underline{i} \end{bmatrix} = \begin{bmatrix} k_\chi/j\omega & \Phi_\chi \\ -\Phi_\chi & \underline{Y}_\chi \end{bmatrix} \begin{bmatrix} j\omega\underline{\Delta}_\chi \\ \underline{v} \end{bmatrix}. \tag{8.14}$$

Here, the subscript χ denotes the piezoelectric mode: L, T, or shear. The capacitive admittance is $Y_\chi = j\omega C_\chi$ and Table 8.3 contains expressions for C_χ, k_χ, and Φ_χ for the three modes. This matrix has the same form as the N-form capacitive transducer matrix introduced in Section 4.3.1.

8.3.1 General transformer-based two-port network model

Figure 8.8 shows a linear, two-port realization of the 2×2 transducer matrix, Eq. (8.14). This realization, which is not unique, uses a transformer as the "ideal" transducer element.[2] The coefficient Φ_χ depends on the relevant piezoelectric coefficient and device dimensions. The transformer element is embedded between the capacitive admittance $\underline{Y}_\chi$ on the electrical side and the mechanical impedance $k_\chi^E/(j\omega)$, an effective linear spring,

[2] A gyrator-based, two-port network could be proposed instead. See Section 2.3.5.

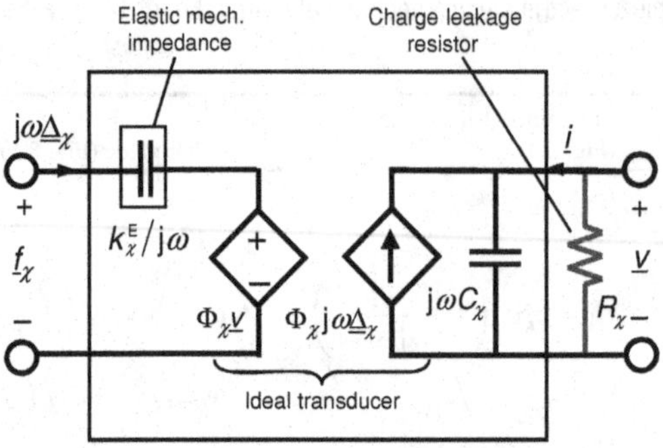

Figure 8.8 Electromechanical two-port model for a piezoelectric transducer. This non-unique realization uses a transformer as the ideal transducer element. In accord with the direct analogy, the elastic impedance is analogous to effective mechanical capacitance. Table 8.3 summarizes the coefficients for the three piezoelectric effects modes. External resistor R_χ accounts for charge leakage, which becomes important at very low frequencies.

on the mechanical side. This latter term accounts for the elastic (Hooke's law) behavior of the solid material.

The shunt resistor R_χ included in the circuit represents dissipative mechanisms that leak charge at very low frequencies and effectively short out any DC piezoelectric response. In a circuit realization, the proper place for this resistor is outside the ideal electromechanical coupling, where it can be treated as an external constraint. Practically speaking, R_χ can be ignored, except at very low frequencies.

8.3.2 External constraints

The cascade-modeling paradigm, extensively utilized for capacitive electromechanical systems, can also be employed for piezoelectric transducers. Starting with the two-port network of Fig. 8.8, Thevenin or Norton equivalent circuits are used to represent the electrical constraints. Accounting for the constraints on the mechanical side, however, is a far more challenging proposition. The difficulty is that the mechanical constraints do not readily reduce to a set of lumped parameter relationships. Part of the problem is simply that the constitutive laws of Eq. (8.1) incorporate the dielectric, elastic, and piezoelectric natures of the solid. This is why the elastic mechanical impedance, $k_\chi^E/(j\omega)$, which depends only on Young modulus and device geometry, appears in Eq. (8.14). In contrast, the M- and N-form electromechanical matrices introduced in Chapter 4 exclude all external constraints, mechanical and electrical.

Piezoelectric transducers fit into two principal classes of application. The class most relevant to MEMS includes drives and motion detectors in accelerometers, gyros, and mechanical resonators and is restricted to frequencies well below ~100 kHz. The other class of devices, primarily actuators, exploits longitudinal, acoustic, *standing-wave* resonances excited within the solid by the piezo effect. The acoustical resonance frequencies are inversely related to the thickness of the piezo element. For most MEMS devices,

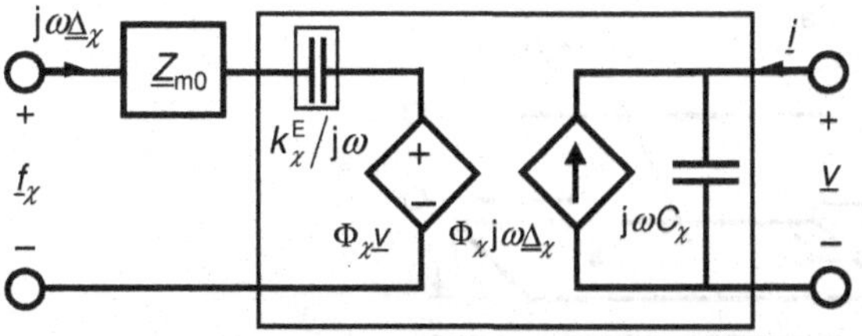

Figure 8.9 Low-frequency electromechanical network model for a piezoelectric element driving a lumped parameter mechanical system having effective mass m, spring constant k, and damping b. The mechanical impedance, $\underline{Z}_{m0} = j\omega m + b + k/(j\omega)$, is external to the piezo coupling, while the internal element $k_\chi^E/(j\omega)$ reflects the elastic property of the piezoelectric solid. Note that the charge leakage resistor R_χ is ignored in this model.

thickness dimensions are usually so small that all acoustical resonances lie far above any practical frequency range. Refer to Preumont [4] for coverage of coupled resonant-wave operation of piezoelectric devices.

It is because virtually all piezoelectric-based MEMS devices operate in the low-frequency regime that we are entitled to use the circuit of Fig. 8.8. In this limit, conventional external-lumped mass, linear springs, and damping can be identified and represented by a mechanical impedance term: $\underline{Z}_{m0} = j\omega m + b + k/(j\omega)$, and cascaded with the transducer two-port, as shown in Fig. 8.9.

One especially interesting low-frequency application, the subject of Section 8.4, exploits T-mode piezo chips mounted on the surface of deformable continua, such as vibrating beams and diaphragms. For such resonant devices, effective lumped parameter models can readily be created. Depending on the placement of the piezo element, the frequency can be adjusted to excite any of a set of discrete resonances of the structure. For this class of device, the equivalent circuit takes the phenomenological form shown in Fig. 8.10, where each shunt-connected element on the left side represents an individual mechanical resonance.

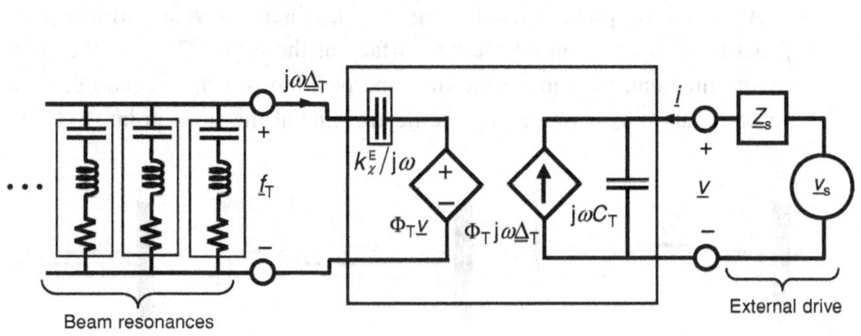

Figure 8.10 Phenomenological network model for a piezoelectric actuator coupled to a mechanical continuum having multiple resonances. Each series RLC section represents one mechanical resonant mode of the structure. The mechanical impedance term $k_\chi^E/(j\omega)$ reflects the elastic property of the piezoelectric solid.

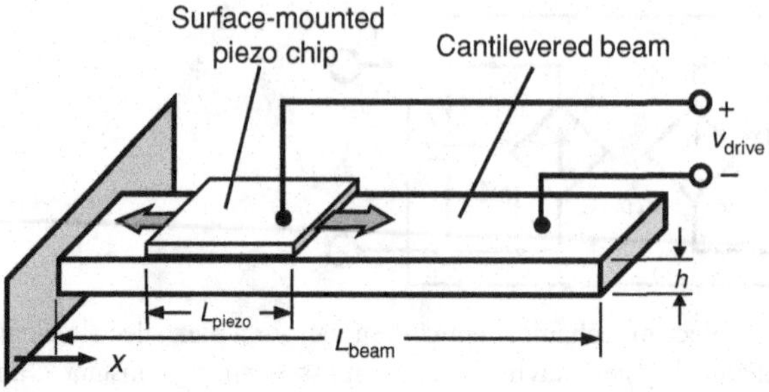

Figure 8.11 Actuation of a cantilevered beam with a thin surface-mounted piezoelectric chip operated in the T-mode of Fig. 8.2b. The piezo element responds to AC excitation by lengthening and contracting along the beam axis. The beam deflects up and down because of the asymmetric surface mounting of the chip.

*8.4 Piezoelectric excitation of a cantilevered beam

One way to exploit piezoelectric actuation in MEMS is to mount the piezo chip on the surface of a cantilevered beam and use it to excite beam resonances. This scheme has been used extensively in macroscopic-scale, non-contacting electrometer systems. Figure 8.11 shows a diagram of a beam with a thin, surface-mounted piezo element driven by AC voltage to operate in the T-mode. Affixing the chip asymmetrically to one surface of the beam achieves strong actuation. We can readily analyze this configuration using the MDF modeling approach of Section 6.4.

8.4.1 Force couple model

The piezoelectric chip, attached on the surface of the beam and operated in the T-mode, exerts a pair of opposed tangential surface forces directed along the beam axis x. To create an MDF model, we must partition these forces so as to identify force moments, which can then be converted to pairs of normal forces, called *force couples*.

As a starting point, consider Fig. 8.12a, which shows a uniform, x-directed force f_a acting at some point on the top surface of the beam. Because the beam is a linear, elastic medium, we can invoke the superposition principle to add equal and opposite forces f_a and $-f_a$ acting along the neutral line at the same point. Refer to Fig. 8.12b.

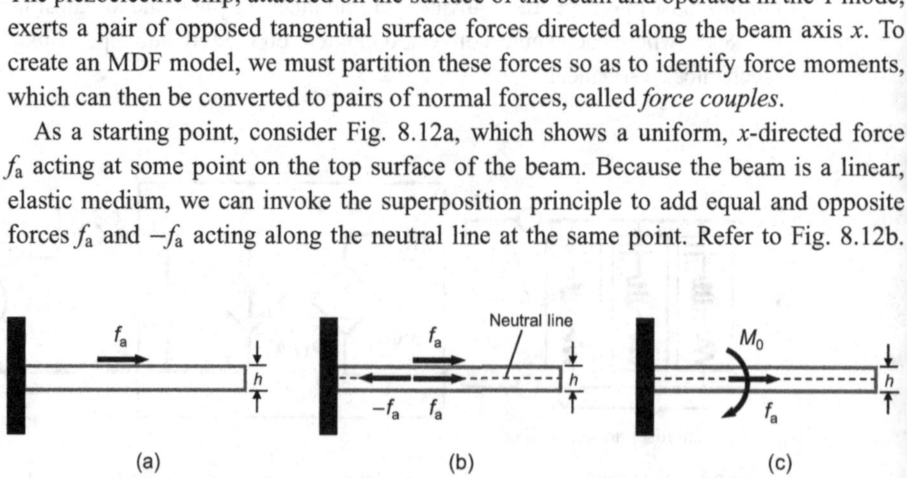

Figure 8.12 Equivalent representation of an axial surface force. (a) An axial force f_a applied on the surface. (b) Equal and opposite axial forces added along the midplane (neutral line). (c) Replacement of the two opposed forces at the surface and the midplane by moment $M_0 = f_a h/2$.

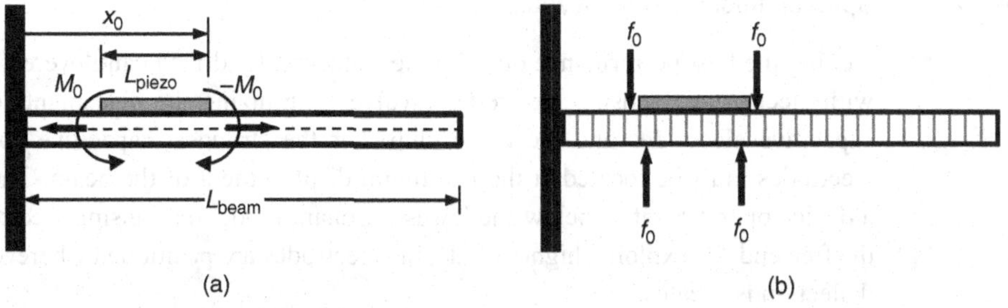

Figure 8.13 Moment and force couple representations for a surface-mounted piezoelectric chip mounted on a cantilevered beam. (a) Axial surface forces exerted by the chip on the beam are represented as a superposition of axial forces acting along the neutral line (which are ignored because the axial stiffness is much higher than the bending stiffness) and two opposed moments. (b) In the n-degrees-of-freedom MDF model, the moments are approximated by pairs of vertical force couples acting at the edges of the chip.

The axial force acting on the surface in the $+x$ direction and the opposed force acting along the neutral line in the $-x$ direction are now combined to form an equivalent moment M_0, as shown in Fig. 8.12c:

$$M_0 = f_a h/2. \tag{8.15}$$

We can neglect the remaining f_a term acting along the neutral line because the axial stiffness of the beam is larger by a factor of $4(L/h)^2$ than the bending stiffness.

The surface-mounted piezo chip actually exerts a pair of oppositely directed surface forces at its two edges, and we can extend this superposition argument by replacing these surface forces by sets of opposed central forces (to be ignored) and opposed moments, $-M_0$ and $+M_0$, acting at $x = x_0$ and $x = x_0 - L_{piezo}$. Refer to Fig. 8.13a. To create the MDF model, the beam is divided into n identical elements of length L/n. Then, as shown in Fig. 8.13b, the two moments acting at the in-board and out-board edges of the chip are replaced by pairs of force couples according to the substitution $M_0 \to (L/n) \times f_0$. The magnitudes of these normal force components are obtained from Eq. (8.15):

$$f_0 = \frac{n}{2}\frac{h}{L} f_a. \tag{8.16}$$

According to the sign convention used here, when $f_0 > 0$, the beam deflects downward. The benefit of formulating the moments in terms of force couples is that these oppositely directed normal force components, acting on adjacent elements, can be implemented directly in an MDF model.

While piezoelectric elements do not respond to DC voltage, they do perform reasonably well down to rather low frequencies. The practical consequence is that the dynamics all the way from low frequencies up to the first beam resonance can be modeled using the quasistatic approximation of Section 6.2.

8.4.2 Optimal placement of piezoelectric element

Getting the best performance out of a piezoelectrically driven cantilevered beam fitted with electrostatic sensing electrodes involves optimizing the placements of both the capacitive electrodes and the piezo chip. For the strongest capacitive coupling, the electrodes must be located at the maximum displacement of the beam. Therefore, for a device operating at or below the lowest resonant mode, the sensing electrode goes at the free end. To exploit a higher mode, the electrodes are positioned wherever the beam deflection is greatest.

Determining the optimum placement of the piezo drive element is complicated because the chip produces a set of opposed moments along its two edges. The following analysis proves that, up to and including the first beam resonance, the mechanical response is maximized when one edge of the chip is located at the fixed end of the beam.

From Appendix D, $M(x)$ is proportional to the beam curvature,

$$M = -YI\frac{\mathrm{d}^2u}{\mathrm{d}x^2}, \tag{8.17}$$

and the distributed force is

$$F = -YI\frac{\mathrm{d}^4u}{\mathrm{d}x^4}. \tag{8.18}$$

As n, the number of elements, increases, f_0, the magnitude of the balanced pairs of opposed normal forces, increases while the distance between them decreases. The product remains constant. In the limit of $n \to \infty$, this distribution becomes a *doublet*, $\delta^{(1)}$, that is, a pair of opposed spatial impulses.[3] For a doublet located at $x = x_0$,

$$F(x) = -M_0\delta^{(1)}(x - x_0). \tag{8.19}$$

Thus,

$$YI\frac{\mathrm{d}^4u}{\mathrm{d}x^4} = -M_0\delta^{(1)}(x - x_0). \tag{8.20}$$

Integrating Eq. (8.20) four times and imposing the four boundary conditions, $u(x=0) = 0$, $\mathrm{d}u/\mathrm{d}x_{x=0} = 0$, $(\mathrm{d}^3u/\mathrm{d}x^3)_{x=L_{\mathrm{beam}}} = 0$, and $(\mathrm{d}^4u/\mathrm{d}x^4)_{x=L_{\mathrm{beam}}} = 0$ to determine the integration constants yields an expression for the beam displacement due to a point moment M_0 located at $x = x_0$:

$$u(x) = \frac{M_0}{2YI}x^2 + \begin{cases} 0, & 0 < x < x_0, \\ -\frac{M_0}{2YI}(x - x_0)^2, & x_0 < x < L_{\mathrm{beam}}. \end{cases} \tag{8.21}$$

The displacement of the free end of the beam, $u_{\mathrm{L}} = u(x = L_{\mathrm{beam}})$, measures the beam response as a function of x_0. From Eq. (8.21),

$$u_{\mathrm{L}} = u(x = L_{\mathrm{beam}}) = \frac{M_0}{YI}\left(L_{\mathrm{beam}}\,x_0 - x_0^2/2\right). \tag{8.22}$$

[3] The doublet is defined by an integral, i.e., $\int_{x_0-}^{x_0+} \delta^{(1)}(x - x_0)\mathrm{d}x \equiv \delta(x - x_0)$.

The net beam displacement induced by the chip is the superposition of the contributions from point moments M_0 and $-M_0$ acting at $x = x_0$ and $x = x_0 - L_{\mathrm{piezo}}$, respectively. The net displacement at the free end of the beam is then

$$u_{\mathrm{Ltot}} = \frac{M_0}{YI}\left(L_{\mathrm{beam}}\,L_{\mathrm{piezo}} + L_{\mathrm{piezo}}^2/2 - L_{\mathrm{piezo}}\,x_0\right). \tag{8.23}$$

Equation (8.23) tells us that the response decreases linearly as x_0 increases. Therefore, the tip displacement is maximized when one edge of the piezoelectric element is located right at $x = 0$, that is, $x_0 = L_{\mathrm{piezo}}$.

8.4.3 Excitation of higher-order resonant modes

Higher-order resonances of a cantilevered beam also can be excited by proper placement of the piezo element [6]. In fact, this selectivity can be realized with any mechanical continuum, including plates or other structures, by proper mounting of piezo elements. The problem of determining optimal chip placement to excite higher-order resonances of continua is considerably more complicated than for the lowest mode. This greater difficulty stems from the fact that, for the higher modes, the quasistatic beam approximation is inapplicable. Instead, the displacement vector must be computed using the equation of motion:

$$\overline{u}_z = [-\omega^2\overline{\overline{m}} - \overline{\overline{a}}^{-1}]^{-1}\overline{f}. \tag{8.24}$$

While analytical solution of Eq. (8.24) is difficult, if not impossible, the MDF approach is well-suited to the task. With relatively straightforward modification, the MATLAB code presented in Section 6.4.5 can be used. The modified code provided in Fig. 8.14 determines the resonant beam response of any specified mode for a piezo chip of fixed length as a function of chip placement x_0.

The program identifies the chip position where a given mode is either maximized or suppressed. Figure 8.15, obtained with the MATLAB code of Fig. 8.14 for $n = 100$ elements, plots beam tip displacement versus chip placement for the first three resonant modes under the assumption that $L_{\mathrm{piezo}} = L_{\mathrm{beam}}/10$. For convenient reference, mode shapes are superimposed on these plots.

These results reveal some interesting behavior. Note that the response of the first mode is not quite linearly related to chip placement as predicted by Eq. (8.23). This discrepancy arises from using the static displacement profile in the analytical model. Even so, maximum amplitude is still achieved with one end of the chip located at the clamped end. A second observation is that the amplitudes for the second and third modes are maximized when the chip is centered near the peak of the displacement curve, where the curvature is largest. Perhaps as intriguing from the standpoint of applications, certain modes can be suppressed by proper chip placement.

```
function piezoPlacementDynamicV2

    g = setGeometry;
   mp = setMatProp;
    n = 100;                           % mesh
    M = eye(n)*mp.rho*g.L*g.h*g.w/n;   % compute mass matrix
    A = zeros(n,n);                    % compute a matrix
    for j = 1:n
        for i = 1:n
            if i>= j
                A(i,j) = g.L^3/(6*mp.E*g.Ix*n^3)*(3*i*j^2-j^3);
            else
                A(i,j) = A(j,i);
            end;
        end;
    end;
    Np = floor(g.Lpiezo/g.L*n);
    ks = 1:(n-Np-1);
    uo = zeros(size(ks));
    w2 = eig(inv(A),M);
   for m = 1:3
        uo(m,:) = zeros(size(ks));
        for k =ks
            f = zeros(n,1);
            f(k)       = -1;                   % moment approx
            f(k+1)     =  1;
            f(k+Np)    =  1;
            f(k+Np+1)  = -1;
            u = (-w2(m)*M+inv(A))^(-1)*f;
            uo(m,k) = abs(u(end));
        end;
        uo(m,:) = uo(m,:)/max(uo(m,:));
        figure; FS = 18;
        [a,h1,h2]=plotyy((ks+(Np+1)/2)/n,uo(m,:),
                        [1:n]/n,u/max(abs(u)));
        % plot decorations omitted
function g = setGeometry
   g.L = 40e-6;            % length in m
   g.h =  2e-6;            % height in m
   g.w =  4e-6;            % width in m
   g.Ix = g.h^3*g.w/12;    % prismatic beam
   g.Lpiezo = g.L/10;
return

function mp = setMatProp
   mp.rho = 2330; % kg/m^3
   mp.E = 150e9;  % GPa
return
```

MDF beam formulation (similar to Fig. 6.10 in Chapter 6).

Comparison of tip displacement values for different piezo actuator placement.

Geometry & material properties functions. Similar to Fig 6.10 in Chapter 6. L_{piezo} is the new parameter defining length of piezo chip.

Figure 8.14 MATLAB code, based on the MDF model from Section 6.4, used to determine cantilevered beam response at resonance as a function of piezo chip placement. The chip length is specified as a fraction of the total beam length. The program finds the first three eigenmodes and then employs the force couple shown in Fig. 8.13b to determine and plot the beam response at each resonant peak.

8.5 Sensing circuits for piezoelectric transducers

Chapter 5 introduced a set of basic inverting amplifier circuits and then showed how to cascade them with capacitive MEMS sensors, particularly three-plate configurations, to build sensor systems. Practical piezoelectric sensors likewise require appropriate signal conditioning and amplification. Piezoelectric elements possess very high, primarily capacitive, output impedance. Typically, the output of piezoelectric sensors is amplified and conditioned by a *charge amplifier*. Kistler developed and commercialized the first practical charge amplifiers for piezoelectrics around 1950. In addition to the original, inverting-operational-amplifier-based designs, there are several distinct realizations for charge amplifiers, including circuits using high-impedance transistors such as JFETs

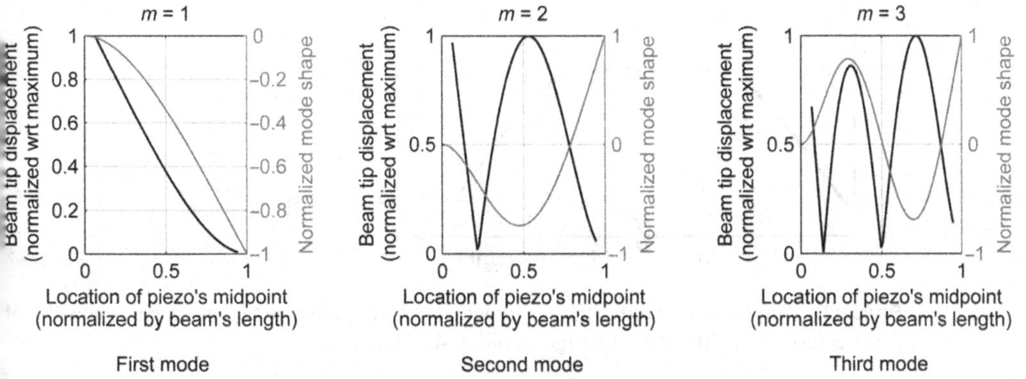

Figure 8.15 Effect of piezoelectric chip placement along the surface of a cantilevered beam as measured by u_{Ltot}, the displacement amplitude at the tip. The chip length is equal to one tenth of the beam length, i.e., $L_{piezo} = 0.1\, L_{beam}$. The first three resonant modes are plotted, with beam displacement curves normalized to the maximum displacement for each mode. Mode shapes are plotted for reference.

and MOSFETs. Here, we restrict attention to the inverting-operational-amplifier-based topology introduced in Chapter 5.

8.5.1 The charge amplifier

The charge amplifier accepts an electrostatic charge input at extremely high impedance and converts this charge to an output voltage at moderate impedance. Thus, the effective units for the transfer function, $\underline{H}_{v/q}$, are volts per coulomb, or farads^{-1}. For inverting-amplifier topologies (see Section 5.3), charge amplification is realized by placing a capacitor in the feedback path. Refer to Fig. 8.16.

For initial simplicity, the sensor itself is represented by a charge source and capacitance C_χ Because C_χ is very small, parasitic capacitance contributions due principally

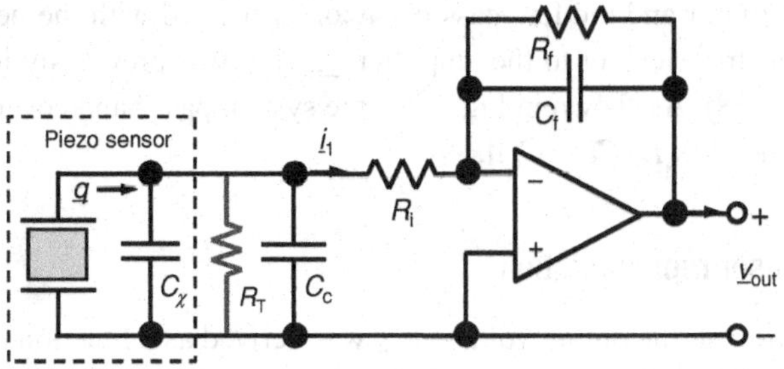

Figure 8.16 Charge amplifier circuit topology using an inverting operational amplifier with capacitive feedback to sense charge output of a piezoelectric sensing element. Note that there is no DC bias. The leakage resistor R_T, shown here as a reminder of low-frequency change dissipation mechanisms, is ignored in the analysis.

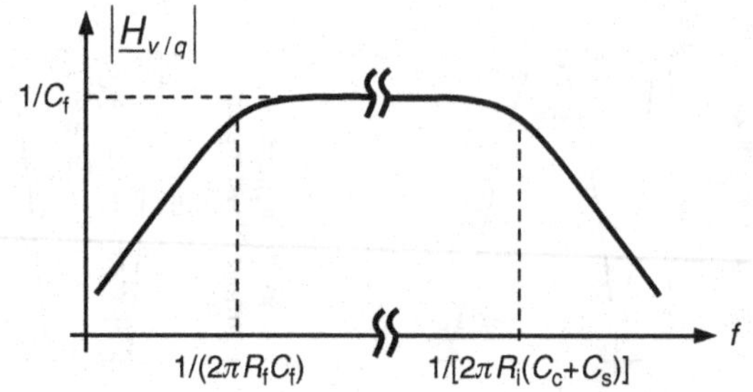

Figure 8.17 Amplitude plot of the transfer function $\underline{H}_{v/q}(j\omega)$ illustrating the useable bandwidth of the operational-amplifier-based charge amplifier shown in Fig. 8.16.

to the cable capacitance C_c cannot be ignored. R_i is the series input resistor to the operational amplifier, while C_f and R_f form the feedback network. The purpose of R_f is to eliminate the effect of biasing currents in the op-amp. As described in Section 5.3.4, this resistor degrades the low-frequency sensitivity of the amplifier, but with little practical consequence because piezo elements do not function at DC anyway. The transfer function for this system is readily derived assuming $e^{j\omega t}$ time-dependence and phasors. Kirchhoff's current law (KCL) at the inverting input of the op-amp gives

$$\underline{i}_1 = -\underline{v}_{out}/R_f - j\omega C_f \, \underline{v}_{out}. \tag{8.25}$$

The large leakage resistor R_T effectively influences only the DC limit, so it is ignored here. Applying KCL a second time to the left of resistor R_i gives

$$j\omega \underline{q} = \underline{i}_1 + j\omega(C_s + C_c)\, R_i \, \underline{i}_1. \tag{8.26}$$

Substituting Eq. (8.25) into Eq. (8.26) and rearranging yields the desired transfer function:

$$\underline{H}_{v/q}(j\omega) = \frac{\underline{v}_{out}}{\underline{q}} = -\frac{1}{C_f} \underbrace{\left(\frac{j\omega R_f C_f}{1 + j\omega R_f C_f}\right)}_{\underline{H}_c(j\omega)} \underbrace{\left(\frac{1}{1 + j\omega R_i (C_s + C_c)}\right)}_{\underline{H}_d(j\omega)}. \tag{8.27}$$

The multiplicative factors $\underline{H}_c(j\omega)$ and $\underline{H}_d(j\omega)$ are, respectively, the high-pass filter behavior of the integrator and the low-pass behavior associated with the network connecting the piezoelectric element to the amplifier. $\underline{H}_c(j\omega)$ was previously identified in Section 5.3.4, Eq. (5.38). As shown in Fig. 8.17, the system pass band covers the range from $1/(2\pi R_f C_f)$ to $1/[2\pi R_i(C_s + C_c)]$.

8.5.2 Two-port piezo sensor representation

In the previous treatment, the output voltage $\underline{v}_{out}$ was derived as a function of charge $\underline{q}$. This approach leads to an illuminating result, but fails to reveal the interplay between electrical and mechanical variables. After all, the input signal to be sensed will most likely be an external mechanical force, displacement, or acceleration. We must account for the external mechanical constraints and can do so by combining the general two-port

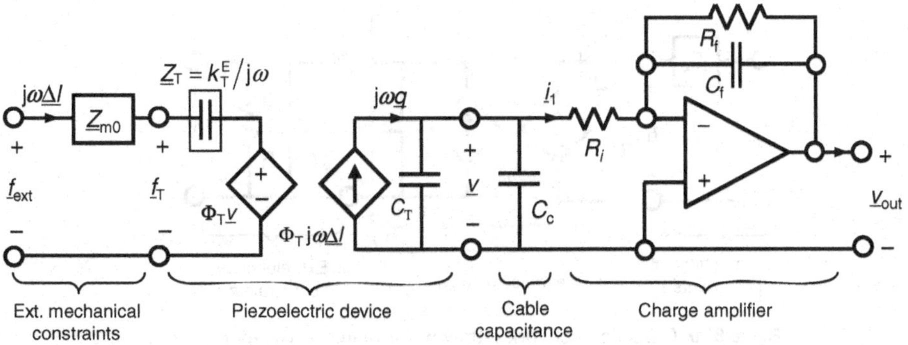

Figure 8.18 Complete cascaded two-port network representation for a piezoelectric sensor with mechanical and electrical constraints. The sensor element is connected to an op-amp realization of a charge amplifier. The leakage resistor R_T is ignored in the analysis.

representation of the piezoelectric transducer from Section 8.3 with the charge amplifier circuit.

Figure 8.18 shows the complete electromechanical network representation for a piezoelectric sensor operated in the T-mode. Here, $C_\chi \to C_T$. On the left-hand side are the external force $\underline{f}_{ext}$ and mechanical impedance $\underline{Z}_{m0}$. The signal conditioning electronics is to the right. For special cases of sensor operation, it is easy to obtain the input/output relation by inspection. For example, if displacement $\underline{\Delta l}$ is the sensed mechanical variable, then, using Eq. (8.27),

$$\underline{v}_{out} = \frac{\Phi_T}{C_f} \underline{H}_c \underline{H}_d \underline{\Delta l}. \tag{8.28}$$

The output voltage is directly proportional to the coupling coefficient Φ_T and inversely related to the feedback capacitor C_f. The output is band-pass limited by the charge amplifier.

A general analytical formulation for this sensor can be derived by performing some matrix manipulations. The starting point is Eq. (8.14), now rewritten using a few convenient definitions to impose consistency with the formulation of capacitive transducers introduced in Chapter 4:

$$\begin{bmatrix} \underline{f}_x \\ \underline{i} \end{bmatrix} = \begin{bmatrix} \underline{Z}_T & \Phi_T \\ -\Phi_T & \underline{Y}_T \end{bmatrix} \begin{bmatrix} j\omega\underline{\Delta l} \\ \underline{v} \end{bmatrix}. \tag{8.29}$$

Here, $\underline{Z}_T = k_T^E/(j\omega)$ and $\underline{Y}_T = j\omega C_T$ are the mechanical impedance and electrical admittance of the transducer, respectively, and Φ_T is defined in Table 8.3. The piezo circuit model appearing in Fig. 8.18 is valid only in the quasistatic regime, where acoustic wave behavior may be ignored. The transmission matrix form for the system results from straightforward manipulation of Eq. (8.29):

$$\begin{bmatrix} \underline{f}_x \\ j\omega\underline{\Delta l} \end{bmatrix} = \frac{1}{\Phi_T} \begin{bmatrix} \Phi_T^2 + \underline{Z}_T\underline{Y}_T & -\underline{Z}_T \\ \underline{Y}_T & -1 \end{bmatrix} \begin{bmatrix} \underline{v} \\ \underline{i} \end{bmatrix}. \tag{8.30}$$

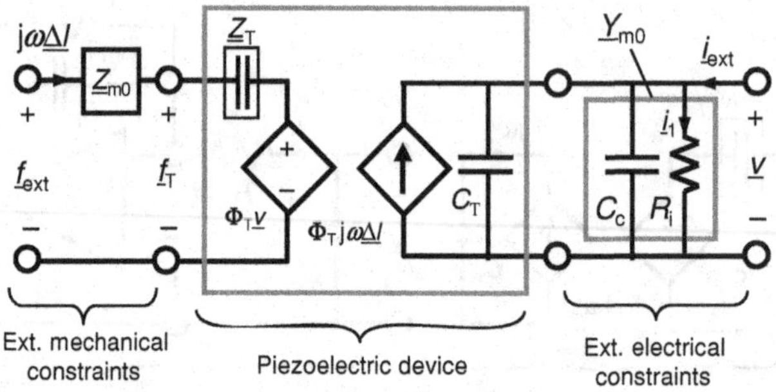

Figure 8.19 Cascaded network redrawn to isolate the two-port piezoelectric transducer from mechanical and electrical constraints. The condition is imposed that the inverting input is at virtual ground. The input voltage to the charge amplifier, $\underline{v}$, corresponds to the same quantity in Fig. 8.18. For normal sensor operation, $\underline{i}_{ext} = 0$.

Note the similarity of this expression to Eq. (4.38). The mechanical and electrical constraints are easily taken into account by a set of substitutions familiar from Section 4.6.3, viz., $\underline{Z}_T \to \underline{Z}_T + \underline{Z}_{m0}$ and $\underline{Y}_T \to \underline{Y}_T + \underline{Y}_{e0}$. Then,

$$\begin{bmatrix} \underline{f}_{ext} \\ j\omega\underline{\Delta l} \end{bmatrix} = \frac{1}{\Phi_T} \begin{bmatrix} \Phi_T^2 + (\underline{Z}_T + \underline{Z}_{m0})(\underline{Y}_T + \underline{Y}_{e0}) & -(\underline{Z}_T + \underline{Z}_{m0}) \\ (\underline{Y}_T + \underline{Y}_{e0}) & -1 \end{bmatrix} \begin{bmatrix} \underline{v} \\ \underline{i}_{ext} \end{bmatrix}. \tag{8.31}$$

Here, $\underline{Z}_{m0}$ is the external lumped parameter mechanical impedance and $\underline{Y}_{e0} \approx 1/R_i + j\omega C_c$. Also, note the appearance of across variables, $\underline{f}_{ext}$ and $\underline{i}_{ext}$, for force and current. The cascaded network representation of Eq. (8.31) appears in Fig. 8.19. Readers should convince themselves that, for sensor operation, $\underline{i}_{ext} \approx 0$.

To validate one limit of Eq. (8.31), we can use the $\underline{i}_{ext} = 0$ condition to obtain

$$\underline{\Delta l} = \frac{\underline{Y}_T + \underline{Y}_{e0}}{j\omega\Phi_T}\underline{v}. \tag{8.32}$$

Then, using

$$\underline{v}_{out} = -\frac{R_f/R_i}{1 + j\omega R_f C_f}\underline{v}, \tag{8.33}$$

the result is

$$\frac{\underline{v}_{out}}{\underline{\Delta l}} = \frac{\Phi_T/C_f}{1 + j\omega R_i(C_f + C_c)} \frac{j\omega R_f C_f}{1 + j\omega R_f C_f}, \tag{8.34}$$

which agrees with Eq. (8.28).

Another transfer function relating external input force to the output voltage of the charge amplifier is also of interest. Assuming $\underline{i}_{ext} = 0$ in Eq. (8.31), one obtains the following relationship for voltage $\underline{v}$ in terms of force $\underline{f}_{ext}$:

$$\underline{v} = \frac{\Phi_T}{\Phi_T^2 + (\underline{Z}_T + \underline{Z}_{m0})(\underline{Y}_T + \underline{Y}_{e0})}\underline{f}_{ext}, \tag{8.35}$$

Then, from Eq. (8.33),

$$\underline{H}_{v/f}(\mathrm{j}\omega) = \frac{\underline{v}_{\mathrm{out}}}{\underline{f}_{\mathrm{ext}}} = -\frac{R_{\mathrm{f}}/R_{\mathrm{i}}}{1+\mathrm{j}\omega R_{\mathrm{f}}C_{\mathrm{f}}}\,\frac{\Phi_{\mathrm{T}}}{\Phi_{\mathrm{T}}^2 + (\underline{Z}_{\mathrm{T}} + \underline{Z}_{\mathrm{m0}})(\underline{Y}_{\mathrm{T}} + \underline{Y}_{\mathrm{e0}})}. \tag{8.36}$$

$\underline{H}_{v/f}$, with units of volts per newton, predicts the frequency-dependent response to an external force of a T-mode piezoelectric sensor connected to an inverting amplifier with capacitive feedback.

8.6 Summary

This chapter presented a quasistatic model for piezoelectric materials and showed that these important materials exhibit intrinsically coupled dielectric and mechanical properties. Piezoelectrics manifest a direct effect, where a mechanical input induces an electrical output, and an indirect effect, where an electrical input produces mechanical motion. There are three distinct piezoelectric phenomena: the L, T, and shear modes. We employed the point forms of the constitutive relations for these modes to develop low frequency, linear, lumped parameter models for the electromechanics of piezoelectric actuators and sensors. These networks take a form similar to those derived for capacitive MEMS transducers. The quasistatic assumption precludes consideration in this text of acoustic wave phenomena, which for MEMS-scale devices would have to operate at frequencies far above the normally recognized limit of micromechanical systems.

Section 8.5 introduced a basic charge amplifier circuit for piezosensors, and showed how it can be incorporated into the electromechanical two-port model to provide complete system-level descriptions for piezo-based sensors. Such amplifiers easily deal with the induced signals and the high impedance, though at the price of frequency-based limits on sensitivity.

A final observation is that piezoelectric elements, long employed as stable frequency standards in electronics applications, are now starting to give way to capacitive MEMS devices. What capacitive devices lack in inherent stability is more than made up for by their lower cost and by the fact that adequate stability can be recovered with feedback control via on-chip electronics. This trend does not mean that piezoelectrics are going away. Impressive progress in new piezoelectric microfabrication technology has been made [1]. Furthermore, piezoelectricity will always possess the advantages of inherent linearity, no voltage bias requirement, and robust actuation and coupling unmatched by capacitive electromechanical devices. One important development is that new MEMS-compatible fabrication processes allow placement of small active elements on resonant, mechanical continua, such as a cantilevered beam and diaphragms. This capability leads to the interesting possibility of using piezo elements to excite or detect specific resonances, while excluding other undesired modes.

Problems

8.1 Consider a mechanically unconstrained L-type, PZT piezoelectric element, 200 μm square and 50 μm thick. Refer to Fig. 8.3 for the geometry.

(a) For an applied voltage of 10 V_{dc}, estimate the change in the thickness Δt of the element, using the ideal model for the L-type piezoelectric effect with material data from Table 8.2.
(b) For the same PZT element and the same applied voltage, what is the net force required to constrain the element to constant thickness?

8.2 Consider a piezoelectric strip of uniaxial PVDF, 500 μm long by 200 μm wide by 100 μm thick. Refer to Fig. 8.4 for the geometry of the T-type device and to Table 8.2 for the properties of PVDF.

(a) What is the capacitance of this device under the constraint of no external mechanical displacement?
(b) Using the ideal model for the T-type piezoelectric effect, calculate the change in the length Δl and the induced (open-circuit) DC voltage when the force exerted on the device is 20 millinewtons. Hint: remember that the open-circuit condition fixes the electric charge to zero.

8.3 Calculate the shear strain angle θ of an initially rectilinear PZT chip that is 300 μm square and 100 μm thick when 20 V_{dc} is applied across the device. Refer to Fig. 8.5 for the geometry and Table 8.2 for the required properties of PZT.

8.4 The objective of this problem is to examine some important distinctions between piezoelectric and electromechanical transducers.

(a) Explain why the mechanical force response of a piezoelectric actuator is linearly related to the applied voltage, while the force of electrical origin varies as the square of the voltage for a capacitive transducer.
(b) Relative to MEMS devices, what are some engineering advantages of piezoelectric devices that derive from their inherent linearity feature?

8.5 Start with the generalized electromechanical network model for a piezoelectric transducer shown in figure (a) on the next page and employ the reflection method presented in Section 2.3.7 to obtain the equivalent circuit specified in figure (b).

(a) Assume an open circuit condition at the electrical port, that is, $\underline{i}_{ext} = 0$, and then find an expression for the equivalent effective mechanical impedance, $\underline{Z}_{m,eff}$, as viewed from the mechanical port.
(b) Now connect the mechanical *RLC* network shown in (b) on the mechanical side to determine the equivalent effective electrical network as viewed from the electrical port. For your answer, draw the circuit and write out expressions for the components in it. Do not attempt to reduce these components to an equivalent admittance.

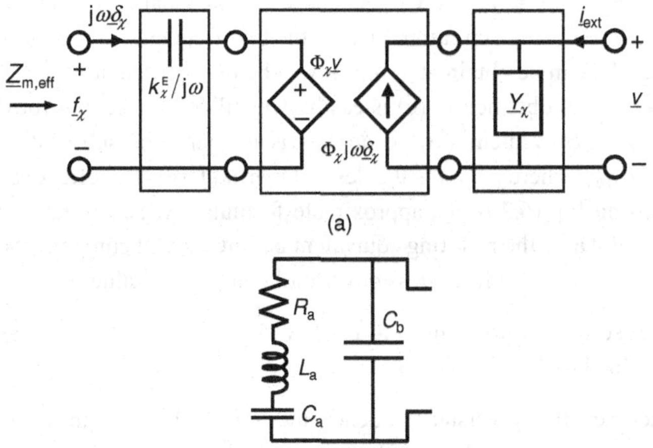

*8.6 A rectangular, surface-mounted T-type piezoelectric chip drives the cantilevered beam, which is shown in side view in the figure below. Both the beam and chip have width w. The beam has density ρ and Young's modulus Y. The piezoelectric chip mounted on the beam has thickness t, coupling coefficient d_{13}, dielectric constant κ^T, and Young's modulus $1/s^E$. Assume $L_{piezo} \ll L_{beam}$, so that the cantilevered beam resonates at a frequency well below the resonance of the piezoelectric chip. In this problem, you are to find an expression for electrical impedance $\underline{Y}_{in}$ that is accurate up to the first beam resonance using the steps outlined.

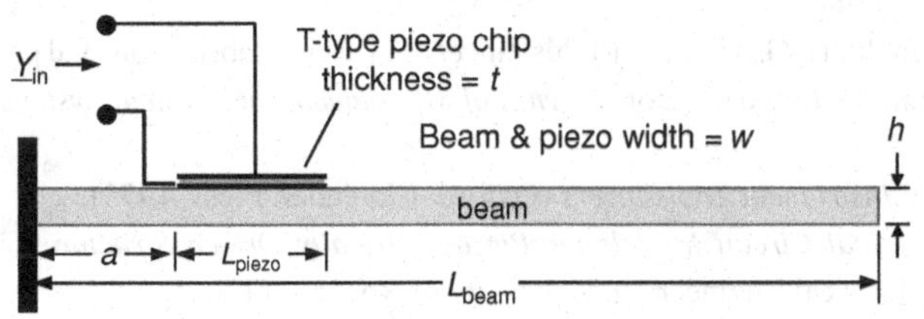

(a) Write the matrix relation for the T-mode piezo chip that expresses $\underline{f}_T$ and $\underline{i}$ in terms of $\underline{v}$ and $j\omega\Delta\underline{l}$. Find expressions for the coefficients of the matrix.

(b) Because the piezo chip is attached to the top surface of the beam, the strain $\varepsilon_x \equiv \partial u_x/\partial x$ must be continuous across the boundary between the chip and the beam. Thus, $\underline{f}_T$ and $\Delta\underline{l}$ are constrained in a linear relationship, which can be derived using $\varepsilon_x(y = h/2)$. Employ Eqs. (D.39) and (D.41) plus an integration to show that, to first order, $\Delta l = 6\, M_0 L_{piezo}/(wh^2Y)$.

(c) Equation (8.15) relates the moment to the x-directed force $\underline{f}_T$. Use this relation to obtain an expression for the low-frequency limit of the mechanical impedance of the beam, $\underline{Z}_{\text{beam}} \equiv \underline{f}_{\text{T}}/\text{j}\omega\Delta\underline{l}$. Then, use the impedance reflection method of Section 2.3.7 to obtain a circuit model for the low-frequency behavior of the device from which $\underline{Y}_{\text{in}}$ can be obtained. (Hint: it will be purely capacitive.)

(d) The circuit model obtained in (c) is easily generalized to account for the effective beam mass. A convenient way to do this is to make the substitution $k_{\text{eq}}/\text{j}\omega \rightarrow k_{\text{eq}}/\text{j}\omega + \text{j}\omega m_{\text{eq}}$, where $\text{j}\omega m_{\text{eq}}$ is the desired inertial term. The effective mass m_{eq} is obtained using Eq. (6.20), the approximate formula for the resonant frequency of the cantilever. Draw the resulting equivalent circuit with all components reflected to the electrical side. Provide expressions for all component values.

8.7 The purpose of this problem is to explore the behavior of the transfer function $\underline{H}_{v/q}(\text{j}\omega)$ as defined by Eq. (8.27).

(a) First sketch separately versus frequency the amplitudes and phases of the terms $\underline{H}_c(\text{j}\omega)$ and $\underline{H}_d(\text{j}\omega)$ from Eq. (8.27). Numerical plots are not needed here; it is only necessary to show the functional behaviors and critical frequencies.

(b) Now use your results from (a) to sketch the phase of $\underline{H}_{v/q}(\text{j}\omega)$ over the range of frequencies $1/[2\pi R_{\text{f}} C_{\text{f}}] \ll f \ll 1/[2\pi R_{\text{i}}(C_{\text{s}} + C_{\text{c}})]$, where the amplitude is approximately constant.

References

1. S. Tadigadapa and K. Mateti, Piezoelectric MEMS sensors: state-of-the-art and perspectives, *Measurement Science & Technology*, 20 (2009), 092001.
2. T. Ikeda, *Fundamentals of Piezoelectricity* (Oxford, UK: Oxford Science Publications, 1996).
3. A. Preumont, *Mechatronics: Dynamics of Electromechanical and Piezoelectric Systems* (Dordrecht, NL: Springer, 2006).
4. H. K. R. Kommepalli, H. G. Yu, C. L. Muhlstein, *et al.*, Design, fabrication, and performance of a piezoelectric uniflex microactuator, *Journal of Microelectromechanical Systems*, 18 (2009), 616–625.
5. H. K. P. Neubert, *Instrument Transducers* (Oxford: Clarendon Press, 1975).
6. N. Nenadic, *Practical Circuit Models for Piezoelectrically Driven Structures*, Ph.D. thesis, Department of Electrical Engineering, University of Rochester (2001).

9 Electromechanics of magnetic MEMS devices

9.1 Preliminaries

For the first century of its existence, the engineering discipline of electromechanics focused almost exclusively on high-current relays, magnetic actuators, and, of course, rotating machines, the latter ranging from fractional horsepower induction motors all the way to gigawatt-rated synchronous alternators [1, 2]. Capacitive transducers never really figured in the discipline because of the focus maintained by the electric power industry on electrical ↔ mechanical power conversion and control. It might seem ironic then that the first eight chapters of this text exclusively concern electrostatic transducers. This emphasis, warranted by the dominance of such devices in the technology of MEMS, testifies to the stern rule of physical scaling in the engineering of useful devices. On the scale of millimeters and below, electrostatic forces hold a considerable advantage.

It would be a mistake, however, to foster the impression that magnetic MEMS will have no role in the future of microsystems technology. They already fill some significant niches in micromechanics and many believe their full potential remains to be tapped. The area-normalized scaling analysis presented in Section A.6 of Appendix A reveals that, based on the present state of materials science and process technology for magnetic materials, the dimensional "break-even" point between capacitive and magnetic MEMS devices is near one millimeter, that is, $\sim 10^3$ microns. This result might seem to suggest that magnetic devices face insurmountable challenges in overtaking capacitive devices in the size range of ~ 100 microns; however, scaling analyses are prone to the biases built into the assumptions used to formulate them, and as a result, may misrepresent the situation.

There are important potential applications for microscale magnetic devices that have no serious competition from capacitive MEMS. Examples include electrical actuators for microvalves and other mechanisms needing strong forces and large strokes. Microfabrication and processing technologies for magnetic materials are steadily advancing. Each new development is likely to create a new niche for magnetic MEMS. Many successful technologies have gotten their start as collections of such niches. Guckel, who did much early work in the field, recognized that magnetic MEMS is "IC friendly" in the sense that the drives and sensing electronics are already available [3]. The reason is that impedances are generally low, so the amplifier bias and leakage currents that plague electronic systems connected to femtofarad capacitances are not a concern. Guckel also pointed out that, while magnetic saturation imposes performance limits, it is a soft failure mechanism, unlikely to damage a device.

9.1.1 Organization and background

This chapter covers the lumped parameter electromechanics of *inductive transducers* and presents a two-port formalism for the linear frequency-dependent behavior of magnetic MEMS actuators and sensors. The organization parallels that of Chapters 3, 4, and 5. The electrical state variables appropriate for inductors, current and magnetic flux, are introduced first. Then the usual conservative system assumption is invoked to obtain lumped parameter models from which forces of electrical origin can be determined. Particular attention is paid to the circuit constraints appropriate for inductive devices. Linearized electromechanical two-port models emerge quite naturally from the analysis.

Magnetic transducers enlist two different types of force. First is the more familiar Lorenz force interaction $\overline{i} \times \overline{B}$, involving the electric current $\overline{i}$ and magnetic field $\overline{B}$ vectors. Second is the ponderomotive force exerted on magnetizable materials by a non-uniform, magnetic field. For magnetic moment $\overline{m}$, it has the form $\mu_0 \overline{m}\, \nabla \overline{H}$, where $\mu_0 = 4\pi \cdot 10^{-7}$ F/m is the permeability of free space and $\overline{H}$ is the magnetic field intensity. There is a clear analogy between this term and the force acting on an electric dipole mentioned in Section 3.1. Just as for capacitive electromechanical systems, it is usually unnecessary to deal directly with field-based formulations for magnetic forces. Rather, we can determine the force of electrical origin from the self- and mutual inductance functions. Even in situations where saturation or other non-linear magnetization effects are dominant, lumped parameter models will often be effective for exploratory analyses.

9.1.2 Note to readers

This chapter is largely self-contained.[1] An advanced student or engineer with a specific interest in magnetic MEMS should be able to learn about inductive electromechanics by starting here; however, the reader should review the basics of magnetoquasistatics and inductor modeling by examining Sections A.4 and A.5 of Appendix A. In addition, a reading of Section 2.3 on basic two-port theory is appropriate. To exploit the analogy between capacitive and inductive transducers, many references are made throughout this chapter to parallel sections in Chapters 3, 4, and 5.

9.2 Lossless electromechanics of magnetic systems

The presentation starts with a definition of the lossless coupling. Just as for capacitive devices, we assume here that all losses – electrical or mechanical in origin – are separated out from the energy transduction mechanism for later treatment as external

[1] The goal of making this chapter self-contained is achieved at the price of some repetition of material from Chapters 3 and 4.

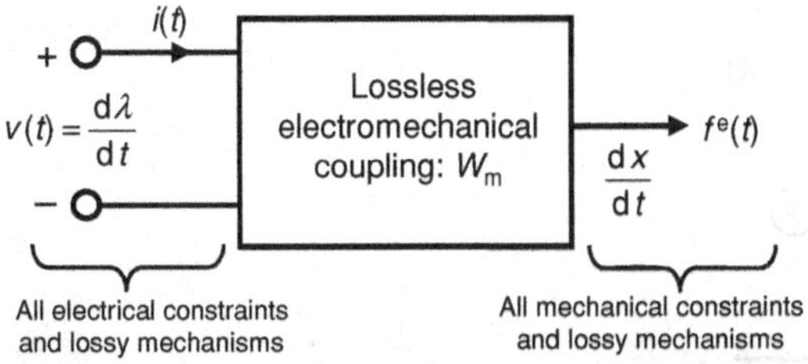

Figure 9.1 Lossless magnetic lumped parameter electromechanical coupling with internal energy storage W_{m}. Note that, in accordance with Faraday's law, the voltage is defined in terms of the magnetic flux linkage $\lambda(t)$.

constraints. This assumption, almost unassailable in the case of capacitive devices with or without dielectric materials, is not so easily defended for magnetic MEMS devices. Generally speaking, ferromagnetic materials are characterized by non-negligible internal loss mechanisms, some frequency-dependent, such as eddy current heating, and some non-linear, such as magnetic hysteresis. We here assume, nevertheless, that such losses will be accounted for as external electrical constraints. When this is untrue, numerical simulations become necessary.

Figure 9.1 shows an electromechanical coupling with one electrical port, characterized by voltage $v(t)$ and current $i(t)$, and one mechanical port, with translational degree of freedom $x(t)$ and associated force of electrical origin $f^{\mathrm{e}}(t)$. By definition, this force arises from the electrical excitation, so must equal zero when the electrical excitation is absent. The coupling itself stores energy W_{m} in magnetic form. System mass, mechanical springs, mechanical damping, voltage or current sources, as well as winding resistance and magnetic losses, are treated as constraints external to the coupling.

Power conservation is expressed in terms of the electrical input, the mechanical output, and the rate at which magnetic energy is stored:

$$vi = f^{\mathrm{e}}\frac{\mathrm{d}x}{\mathrm{d}t} + \frac{\mathrm{d}W_{\mathrm{m}}}{\mathrm{d}t}. \tag{9.1}$$

As defined by Eq. (A.34) of Appendix A, $\lambda(t)$ is the magnetic flux linking the winding of the transducer. According to Faraday's law, $v(t) = \mathrm{d}\lambda/\mathrm{d}t$, so Eq. (9.1) may be rewritten in differential form:

$$\mathrm{d}W_{\mathrm{m}} = -f^{\mathrm{e}}\mathrm{d}x + i\,\mathrm{d}\lambda. \tag{9.2}$$

This equation is the magnetic energy analog to Eq. (3.2).

9.2.1 State variables and conservative systems

The state variables appropriate for the system represented by Eq. (9.2) are x and λ. The force of electrical origin and the electric current are expressed in terms of them, that is,

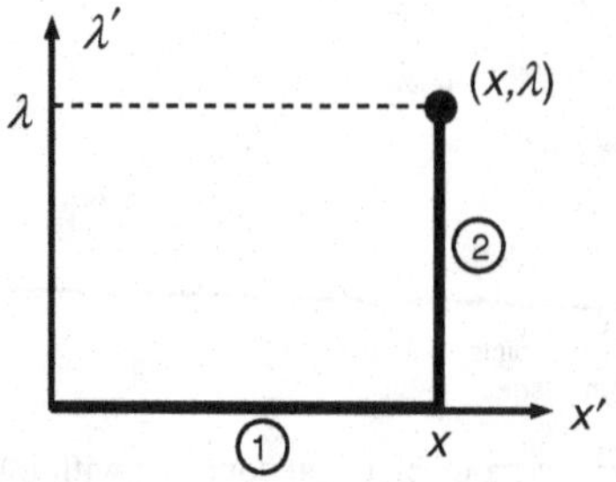

Figure 9.2 Convenient path of integration in (x, λ) state space for the energy function $W_m(x, \lambda)$ of an inductive (magnetic) transducer. The path takes advantage of the assumption that defines the force of electrical origin, namely that $f^e(x', \lambda' = 0) = 0$.

$f^e(x, \lambda)$ and $i(x, \lambda)$, as is the energy function: $W_m(x, \lambda)$. Thus, the key to determining f^e is W_m, the differential of which is

$$dW_m = \left.\frac{\partial W_m}{\partial x}\right|_\lambda dx + \left.\frac{\partial W_m}{\partial \lambda}\right|_x d\lambda. \tag{9.3}$$

Recognizing that, in the absence of external constraints, x and λ are independent, Eqs. (9.2) and (9.3) combine to yield

$$i(x, \lambda) = \left.\frac{\partial W_m}{\partial \lambda}\right|_x \quad \text{and} \quad f^e(x, \lambda) = -\left.\frac{\partial W_m}{\partial x}\right|_\lambda. \tag{9.4}$$

To obtain f^e, we require an expression for the energy.

9.2.2 Evaluation of magnetic energy

Evaluation of W_m involves integrations over the two state variables (x, λ) using Eq. (9.2). Any convenient path in state space can be employed because the system is conservative. Thus, consider the path shown in Fig. 9.2,

$$W_e = \int_{(1)} dW_e + \int_{(2)} dW_e, \tag{9.5}$$

which exploits the property of the force of electrical origin that $f^e(x, \lambda = 0) = 0$.

By integrating first along x' with $\lambda' = 0$, and then along λ' with $x' = x$, the two terms in Eq. (9.5) reduce to a pair of definite integrals:

$$W_m(x, \lambda) = -\int_0^x \underbrace{f^e(x', \lambda' = 0)}_{=0} dx' + \int_0^\lambda i(x, \lambda') d\lambda'. \tag{9.6}$$

The first integral term is zero, so that

$$W_{\mathrm{m}}(x,\lambda)=\int_0^{\lambda} i(x,\lambda')\mathrm{d}\lambda'. \tag{9.7}$$

We may evaluate this integral using the electrical terminal relation. Assuming electrically linear behavior, that is, $\lambda = L(x)i$, where L is inductance,

$$W_{\mathrm{m}}(x,\lambda)=\int_0^{\lambda} \frac{\lambda'}{L(x)}\mathrm{d}\lambda' = \frac{\lambda^2}{2L(x)}. \tag{9.8}$$

9.2.3 Force of electrical origin

The force of electrical origin is obtained by using W_{m} from Eq. (9.8) in Eq. (9.4):

$$f^{\mathrm{e}}(x,\lambda) = -\left.\frac{\partial W_{\mathrm{m}}}{\partial x}\right|_{\lambda} = \frac{\lambda^2}{2L^2(x)}\frac{\mathrm{d}L(x)}{\mathrm{d}x}. \tag{9.9}$$

This partial derivative is performed holding λ constant, but once the force expression is obtained, it is permissible to use the terminal relation $\lambda = L(x)i$ a second time to write

$$f^{\mathrm{e}}(x,i) = \frac{i^2}{2}\frac{\mathrm{d}L(x)}{\mathrm{d}x}. \tag{9.10}$$

9.2.4 Coenergy formulation

An alternate energy function $W'_{\mathrm{m}}(x,i)$, called the *magnetic coenergy*, offers the means to derive Eq. (9.10) directly. This function uses current i as the electrical state variable, a feature which will be found advantageous later in the chapter. The definition of coenergy involves a Legendre transform:

$$W_{\mathrm{m}} + W'_{\mathrm{m}} = \lambda i. \tag{9.11}$$

Thus

$$\mathrm{d}W_{\mathrm{m}} + \mathrm{d}W'_{\mathrm{m}} = \lambda \mathrm{d}i + i\,\mathrm{d}\lambda. \tag{9.12}$$

Combining this expression with Eq. (9.2) yields

$$\mathrm{d}W'_{\mathrm{m}} = f^{\mathrm{e}}\mathrm{d}x + \lambda\mathrm{d}i. \tag{9.13}$$

Equation (9.13), the magnetic analogy to Eq. (3.14) for capacitive transducers, is a statement of *magnetic coenergy conservation*. The state variables are (x, i), so

$$\mathrm{d}W'_{\mathrm{m}} = \left.\frac{\partial W'_{\mathrm{m}}}{\partial x}\right|_{i}\mathrm{d}x + \left.\frac{\partial W'_{\mathrm{m}}}{\partial i}\right|_{x}\mathrm{d}i. \tag{9.14}$$

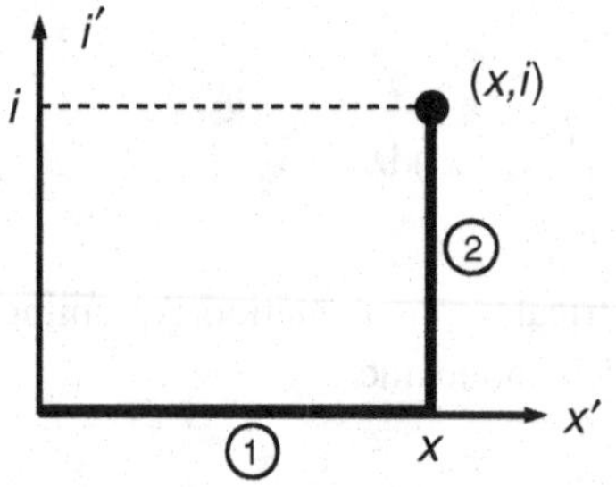

Figure 9.3 Convenient path of integration in (x, i) state space for magnetic coenergy $W'_m(x, i)$. This path takes advantage of the property of the force of electric origin that $f^e(x', i' = 0) = 0$ for an inductive transducer.

Combining Eqs. (9.14) and (9.13) yields

$$\lambda(x, i) = \left.\frac{\partial W'_m}{\partial i}\right|_x \quad \text{and} \quad f^e(x, i) = \left.\frac{\partial W'_m}{\partial x}\right|_i . \tag{9.15}$$

The coenergy is found by integrating Eq. (9.13) in (x, i) state space. Figure 9.3 shows the most convenient integral path:

$$W'_m(x, i) = \int_0^x \underbrace{f^e(x', i' = 0)}_{=0} \mathrm{d}x' + \int_0^i \lambda(x, i')\mathrm{d}i'. \tag{9.16}$$

Just as before, integrating the mechanical variable first takes advantage of the definition that $f^e(x, i = 0) = 0$.

Employing the linear electrical terminal relation, $\lambda = L(x)i$, the coenergy is

$$W'_m(x, i) = \int_0^i \lambda(x, i')\,\mathrm{d}i' = \frac{1}{2}L(x)i^2, \tag{9.17}$$

where $W'_m(x, i)$ takes the familiar form from electric circuit theory for the magnetic energy stored in an inductor. Combining Eq. (9.17) and Eq. (9.15) yields

$$f^e(x, i) = \frac{i^2}{2}\frac{\mathrm{d}L(x)}{\mathrm{d}x}, \tag{9.18}$$

which is identical to Eq. (9.10). For a fixed current, the force of electrical origin depends only on the derivative of inductance, not on its absolute magnitude.

9.2.5 Magnetic non-linearity

For magnetically linear inductors, W_m and W'_m are numerically equal, the only distinction between them being the choice for the electrical state variable: magnetic flux linkage λ or

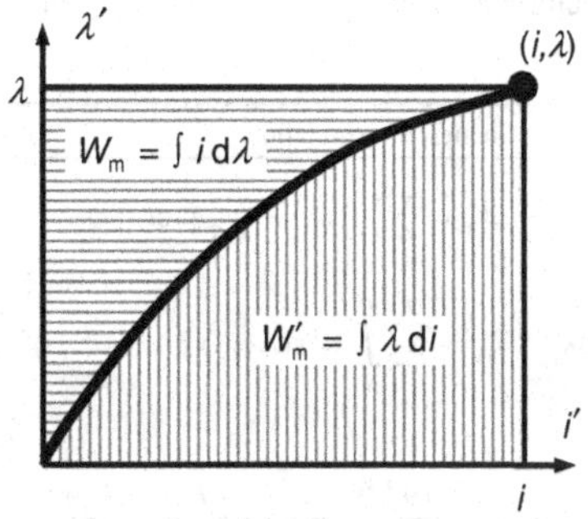

Figure 9.4 Area integrals in (i, λ) space representing magnetic energy and coenergy of a lossless electromechanical transducer for the case of a non-linear inductor made of ferromagnetic material that exhibits saturation. Saturation means that $W'_{\mathrm{m}} > W_{\mathrm{m}}$; nevertheless, a force calculation based on either energy function will yield the same result.

current i, respectively. In reality, however, the iron and nickel-based magnetic materials prevalent in inductors and magnetic transducers exhibit non-linear behavior that can seldom be neglected. Figure 9.4 provides an interpretation of the relationship between W_{m} and W'_{m} for the general case of a saturable ferromagnetic material. Saturation, which limits the magnetic flux density, guarantees that $W'_{\mathrm{m}} > W_{\mathrm{m}}$. Still, even in the presence of non-linearity, Eqs. (9.4) and (9.15) always yield numerically equal expressions for f^{e}.

9.2.6 Multiport magnetic systems

Magnetic MEMS devices can have more than one electrical port and more than one mechanical port. To model such systems, we return to the power conservation principle. Consider a general system with N_{e} electrical and N_{m} mechanical ports:

$$\mathrm{d}W_{\mathrm{m}} = -\sum_{k=1}^{N_{\mathrm{m}}} f_k^{\mathrm{e}} \mathrm{d}x_k + \sum_{k=1}^{N_{\mathrm{e}}} i_k \mathrm{d}\lambda_k. \tag{9.19}$$

Invoking the same arguments as before, now for an enlarged set of state variables, the energy function becomes $W_{\mathrm{m}}(x_1, \ldots, x_{N_{\mathrm{m}}}; \lambda_1, \ldots, \lambda_{N_{\mathrm{e}}})$ and f_k^{e} and i_k may be expressed in terms of appropriate derivatives of it. Likewise, the coenergy function $W'_{\mathrm{m}}(x_1, \ldots, x_{N_{\mathrm{m}}}; i_1, \ldots, i_{N_{\mathrm{e}}})$ can be defined using a Legendre transform, and the variables f_k^{e} and λ_k obtained using it. The derivations are straightforward generalizations. All the important relationships are found in Table 9.1.

One application for these multiport formulations is in modeling transducers with permanent magnets. A systematic way to represent the permanent magnet is to define a virtual port and then use it to account for the magnetization of the material by means of a constant, effective current constraint. Section 9.5 presents an analysis of a permanent magnet transducer.

Table 9.1 Energy and coenergy formulations for a magnetic electromechanical coupling having N_m and N_e mechanical and electrical terminals, respectively. Compare this table with Table 3.1

	Energy formulation	Coenergy formulation
Energy conservation	$dW_m = -\sum_{j=1}^{N_m} f_j^e dx_j + \sum_{j=1}^{N_e} i_j d\lambda_j$	$dW'_m = \sum_{j=1}^{N_m} f_j^e dx_j + \sum_{j=1}^{N_e} \lambda_j di_j$
State space integral for energy function	$W_m = \sum_{j=1}^{N_e} \int_0^{\lambda_j} i_j(x_1, \ldots, x_{N_m}; \lambda_1, \ldots, \lambda_{j-1}, \lambda'_j, 0, \ldots, 0) d\lambda'_j$	$W'_m = \sum_{j=1}^{N_e} \int_0^{i_j} \lambda_j(x_1, \ldots, x_{N_m}; i_1, \ldots, i_{j-1}, i'_j, 0, \ldots, 0) di$
Force of electrical origin	$f_j^e = -\frac{\partial W_m(x_1, \ldots, x_{N_m}; \lambda_1, \ldots, \lambda_{N_e})}{\partial x_j}$	$f_j^e = \frac{\partial W'_m(x_1, \ldots, x_{N_m}; i_1, \ldots, i_{N_e})}{\partial x_j}$
Electrical terminal relations	$i_j = \frac{\partial W_m(x_1, \ldots, x_{N_m}; \lambda_1, \ldots, \lambda_{N_e})}{\partial \lambda_j}$	$\lambda_j = \frac{\partial W'_m(x_1, \ldots, x_{N_m}; i_1, \ldots, i_{N_e})}{\partial i_j}$

For linear inductive network: $\lambda_k = \sum_{j=1}^{N_e} L_{kj}(x_1, x_2, \ldots, x_{N_m}) i_j$, where $k = 1, 2, \ldots, N_e$.

Voltages defined by $v_k = d\lambda_k/dt$.

Legendre transform relating energy and coenergy: $W_m + W'_m \equiv \sum_{j=1}^{N_e} \lambda_j i_j$.

9.3 Basic inductive transducer geometries

Trimmer points out that, based on the physical assumptions commonly accepted for MEMS, magnetic forces scale with volume whereas electrostatic forces scale as area [4]. This rule puts magnetic MEMS at a disadvantage in microdevices, particularly in the case of surface-micromachined structures. The only way to overcome the deficiency is to develop distinctly different geometries and the processes to fabricate them inexpensively. In fact, the most promising magnetic MEMS geometries are *high-aspect-ratio structures*, which differ from surface-micromachined devices in that their height dimensions, normal to the substrate plane, are comparable to the surface dimensions themselves. Because they use comparatively large amounts of magnetic material, high-aspect-ratio structures take advantage of volume-based scaling of the magnetic forces, and they feature the combination of strong force and large stroke.

The less well-developed state of high-aspect-ratio fabrication using magnetic materials still impedes the emergence of new magnetic MEMS technology. In devices larger than a few millimeters, magnetic steels, nickel alloys, sintered ferrites, plus ceramic and rare-earth magnets have been the mainstay of magnetic transducers for many years. Reliance on these materials will certainly carry over to magnetic MEMS. In fact, any such technology not based on them is difficult to envision. LIGA[2] and related electroplating processes hold some promise. Unfortunately, the processing of ferromagnetic materials into structures on the scale of MEMS devices is not inherently compatible with conventional silicon processes. Planar coil structures have been fabricated in stacked

[2] LIGA is an acronym from the German: "Lithographie, Galvanoformung, und Abformung."

configurations on Si wafers, but more work is required to develop this process into an economically viable option. Nanoparticle science may open new possibilities for microscale fabrication of magnetic materials [5].

Figures 9.6a and b reveal microscale inductor designs exemplifying planar and high-aspect-ratio structures. The advantage of the planar spiral coil shown in Fig. 9.6a, fabricated using an enhanced surface process, is that it packs the maximum number of windings into the available chip area. Stacking these coils creates compact, efficient, magnetically coupled components. The flattened solenoidal configuration shown in Fig. 9.6b, fabricated in a number of complex steps, is best classified as a high-aspect-ratio geometry.

Though a variety of devices have been fabricated and tested with encouraging results, no designs can be said to stand out. Even so, one can identify certain generic geometries and then study their general operational characteristics. Among these are *variable-gap* and *variable-area* geometries. Figure 9.7 contains representative inductance versus displacement plots as well as simple magnetic circuit configurations exemplifying them.

Example 9.1 **A high-aspect-ratio magnetic actuator**

Figure 9.5 depicts an actuator tested by Guckel *et al.* [3]. This geometry is a good example of a magnetic circuit that contains the magnetic flux and minimizes undesirable coupling to adjacent circuitry.

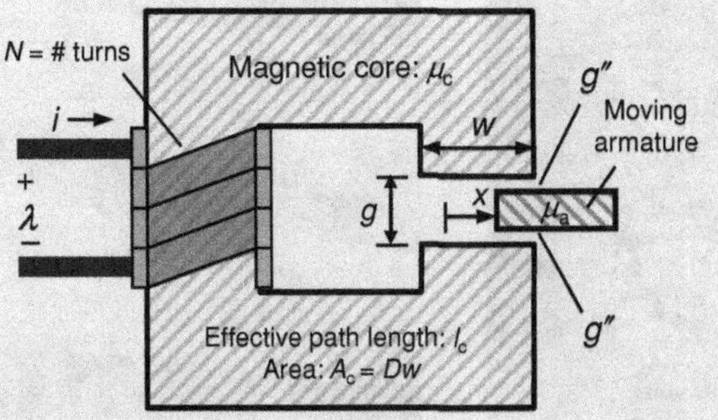

Figure 9.5 A high-aspect-ratio magnetic transducer featuring large stroke, based on a geometry tested by Guckel *et al.* [3]. This device is modeled as a magnetic circuit. The air gaps g'', which are assumed to be small, that is, $g'' \ll g$, provide clearance for motion of the armature.

From the magnetic circuit model of Section A.5.3 in Appendix A, the flux linkage $\lambda \equiv N\Phi$ is related to the magnetomotive force, $mmf \equiv Ni$, and the reluctances:

$$\lambda = N^2 i/[\mathcal{R}_{\text{core}} + \mathcal{R}_{\text{gap}}(x)],$$

where $\mathcal{R}_{\text{core}}(x) = l_c/(\mu_c D\text{w})$. The gap reluctance includes two contributions in parallel:

$$\mathcal{R}_{\text{gap}}(x) = \frac{\mathcal{R}_{\text{air}}(x)\mathcal{R}_{\text{arma}}(x)}{\mathcal{R}_{\text{air}}(x)+\mathcal{R}_{\text{arma}}(x)}.$$

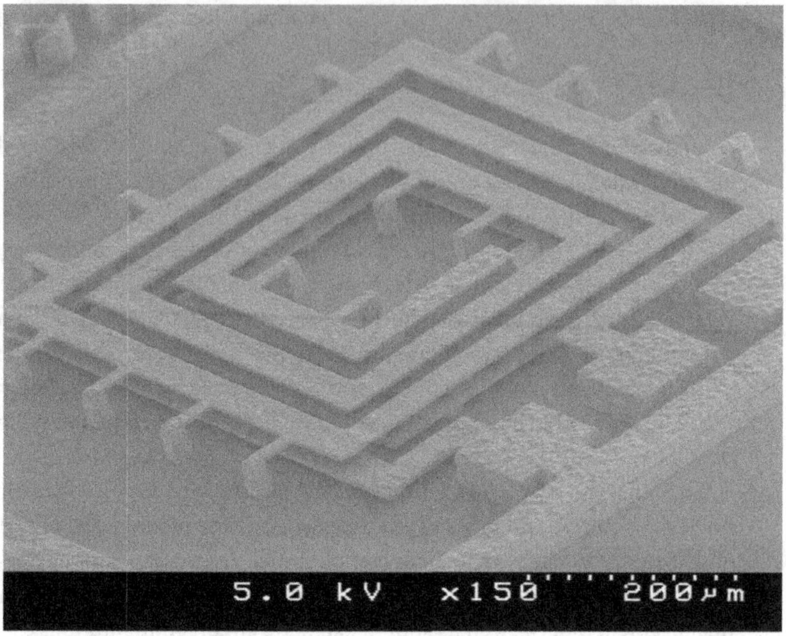

(a)

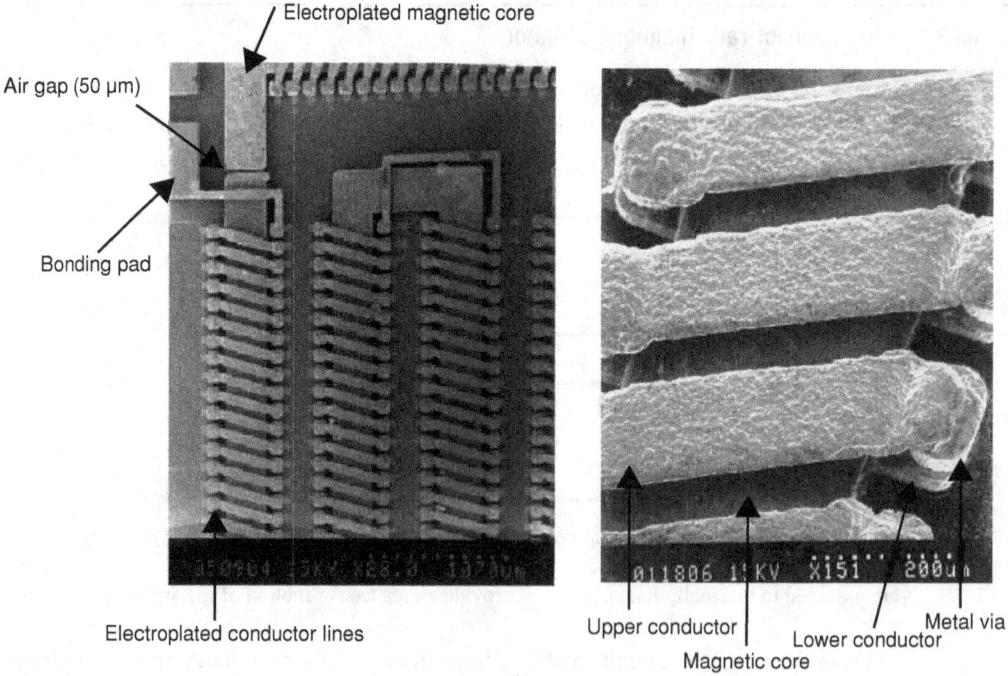

(b)

Figure 9.6 Examples of MEMS-scale inductor geometries. (a) Two-layer, planar spiral coil of thick Cu using an electroplating process and sacrificial photoresistive scaffolding. Spiral coils pack the maximum amount of air inductance per unit area on a chip. When positioned over or under a movable planar sheet of magnetic material, it behaves essentially like a variable-gap inductor [6]. Image provided courtesy of C. D. Meyers, University of Florida. (b) Solenoidal coil geometry with flat Cu windings patterned in two layers connected with vias and enclosing electroplated supermalloy core [7] © IEEE. Magnetic anisotropy, which increases the core permeability, is achieved during fabrication using a permanent magnet and specially designed yoke.

For the air gap on the left-hand side, $\mathcal{R}_{\text{air}}(x) = g/(\mu_0 D x)$. The right-hand side consists of two series terms:

$$\mathcal{R}_{\text{arma}}(x) = \frac{2g''}{\mu_0 D(w-x)} + \frac{g-2g''}{\mu_a D(w-x)}.$$

Using $L(x) = \lambda/i$, the force can be expressed compactly in terms of reluctances:

$$f^e = \frac{i^2}{2}\frac{dL}{dx} = \frac{-N^2 i^2}{2[\mathcal{R}_{\text{core}} + \mathcal{R}_{\text{gap}}(x)]^2}\frac{d\mathcal{R}_{\text{gap}}(x)}{dx}.$$

If $\mu_a, \mu_c \gg \mu_0$ and $g'' \ll g$, the air gaps dominate and the tediously complex force expression reduces to

$$f^e \approx -\mu_0 D N^2 i^2/(4g'').$$

Note that f^e is negative and pulls the magnetizable armature into the gap, confirming the expectation that the force of electrical origin acts to increase inductance.

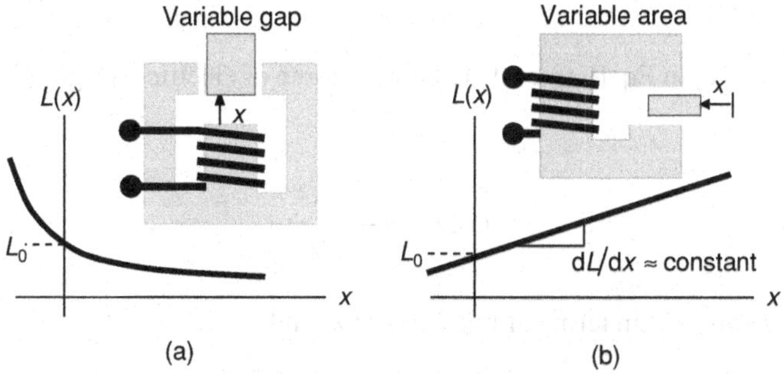

Figure 9.7 Typical dependence of inductance L on mechanical variable for generic magnetic circuit representations of variable-gap and variable-area devices. Refer to Appendix A, Section A.5.3, for a review of magnetic circuits. (a) Variable-gap inductor exhibiting hyperbolic dependence. The slope of $L(x)$ versus x is not constant. This geometry is best suited to short-stroke–high-force applications. (b) Variable-area inductor featuring approximately constant slope over a larger range of motion. This geometry provides relatively lower force at longer stroke.

9.3.1 Variable-gap inductors

The defining attribute of a variable-gap magnetic transducer is a strong dependence of dL/dx on x. Defining the positive direction of x is arbitrary, so the slope can take either sign; what is more important is that, over the range of useful operation, the *second* derivative of $L(x)$ is usually positive, that is,

$$\frac{d^2L}{dx^2} > 0. \tag{9.20}$$

A representative empirical form for the inductance of a variable-gap device is

$$L(x) = \frac{L_0}{1 + x/x_m}, \quad x > 0. \tag{9.21}$$

Consistent with Eq. (9.20), the second derivative of $L(x)$ is positive.

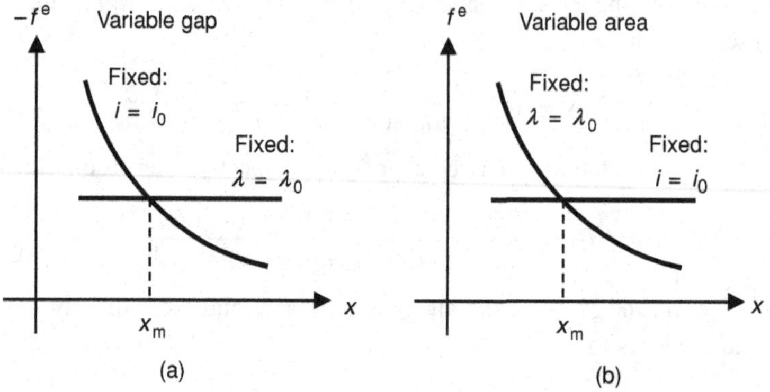

Figure 9.8 Dependence of force f^e on x for generic magnetic transducers at fixed flux and fixed current. The sign of the force is determined by the way x is defined, which is arbitrary. To simplify comparisons, the crossover at x_m is defined by $\lambda_0 = L(x_m)i_0$. (a) Generic variable-gap transducer, such as Eq. (9.21). (b) Generic variable-area transducer, such as Eq. (9.24).

Using Eq. (9.21) in Eq. (9.9) or (9.18) for the force of electrical origin yields, respectively,

$$f^e(x, \lambda) = -\frac{\lambda^2}{2x_m L_0} \tag{9.22}$$

for the force expressed in terms of flux linkage λ, and

$$f^e(x, i) = -\frac{i^2}{2}\frac{L_0/x_m}{(1 + x/x_m)^2} \tag{9.23}$$

when the force is expressed in terms of current i. Refer to Fig. 9.8a.

Example 9.2 **Force magnitude for magnetic actuator**

The inductance in nanohenries (10^{-9} H) of a magnetic actuator is empirically approximated by the following equation:

$$L_{nH}(x_{\mu m}) \approx 50/(1 + 0.01x_{\mu m}),\ x_{\mu m} \geq 0,$$

where $x_{\mu m}$ is in μm (10^{-6} m). It is here convenient to convert this expression to SI units, i.e., henries and meters:

$$L(x) \approx 5.10^{-8}/(1 + 10^4 x).$$

If DC current i_0 passes through the winding, the force, now in newtons, is

$$f^e \approx -2.5\,10^{-4} i_0^2/(1 + 10^4 x)^2.$$

Figure 9.9 plots force versus displacement for $i_0 = 1$ mA. The electrical force acts in the $-x$ direction to increase inductance.

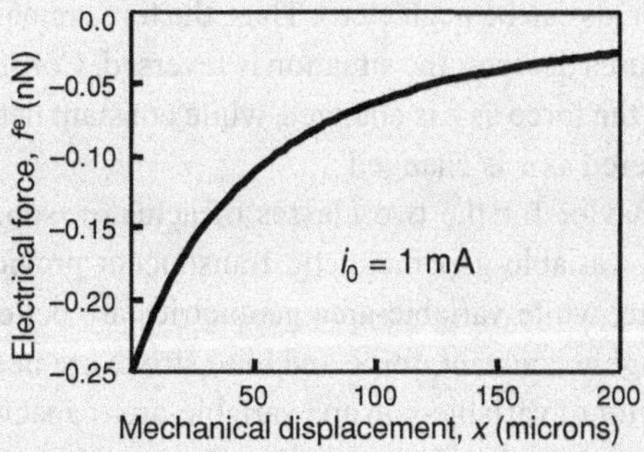

Figure 9.9 Force of electrical origin f^e in nanonewtons plotted versus displacement x in microns for an empirically specified magnetic actuator at a constant DC current excitation of $i_0 = 1$ mA.

9.3.2 Variable-area inductors

As represented in Fig. 9.7b, variable-area inductors are characterized by two conditions: $dL/dx \approx$ constant and $d^2L/dx^2 \approx 0$. Thus, over some range of x,

$$L(x) \approx L_0 (1 + x/x_m), x > 0. \tag{9.24}$$

Using Eq. (9.24) in Eqs. (9.9) and (9.18) for the force yields, respectively,

$$f^e(x, \lambda) = \frac{\lambda^2}{2x_m L_0 (1 + x/x_m)^2} \tag{9.25}$$

and

$$f^e(x, i) = \frac{i^2}{2} \frac{L_0}{x_m}. \tag{9.26}$$

Figure 9.8b shows that f^e is inversely dependent on x when flux is constrained but constant when the current is fixed.

The basic variable-gap and variable-area structures illustrated in Fig. 9.7 do not encompass all realizable magnetic transducers. Practical geometries often blur the distinctions between the ideal variable-gap and variable-area limits.

9.3.3 Nature of magnetic system constraints

The *magnetic circuit* geometries of Figs. 9.7a and b exemplify variable-gap and variable-area devices. For example, when current is held constant for the variable-gap device, the magnetomotive force *mmf* remains fixed because $mmf = Ni$. As the gap becomes larger or smaller, the reluctance changes proportionately and the magnetic flux Φ decreases or increases, respectively. The force changes correspondingly. Thus, the constant-current constraint in a magnetic variable-gap transducer is analogous to fixed voltage in a variable-gap capacitive device.

In contrast, the constant flux linkage constraint is analogous to a constant-charge condition imposed on a capacitive transducer. Referring again to the magnetic circuit model for the variable-gap geometry, changing the gap does not influence the magnetic flux density (as long as fringing fields can be neglected). Thus, the force remains constant as the gap is varied. For variable-area devices, the situation is reversed. Constant current, equivalent to constant *mmf*, fixes the force as x is changed, while constant net flux means that the distribution of flux is altered as x is changed.

Figures 9.8a and b depict behavior for the two classes of actuator, subject to both limiting constraints. In general, variable-gap magnetic transducers provide a strong force effective over a short stroke, while variable-area geometries are better suited for applications where a weaker, roughly constant, force and long stroke are needed. These results are analogous to the behavior of variable-gap and variable-area capacitive devices subject to constant voltage and constant charge constraints, respectively.

The constant-current constraint is readily achieved and easy to understand, but the constant-flux condition is not intuitive. From Faraday's law, $v(t) = d\lambda/dt$, it is evident that the flux can be held constant by imposing a short-circuit condition. This condition, difficult to achieve in ordinary inductive devices owing to winding resistance, is quite literally realized in *superconductive devices*, which trap magnetic flux permanently as long as the superconductor is maintained below its critical temperature.[3]

A further observation to make about magnetic transducers is that, irrespective of any electrical constraint, the force of electrical origin f^e always acts to increase inductance $L(x)$. Capacitive transducers exhibit similar behavior, always acting to increase self-capacitance. Refer to Section 3.4.2.

Example 9.3 **Magnetic transducer with a variable gap**

Figure 9.10 depicts a cantilevered-beam-based microactuator fabricated using a CMOS-compatible process. Ferromagnetic-material-filled vias form a closed magnetic path with only the cantilevered beam exposed. The winding is formed from stacked and bonded Si wafers. The excellent thermal conductivity of crystalline Si prevents overheating.

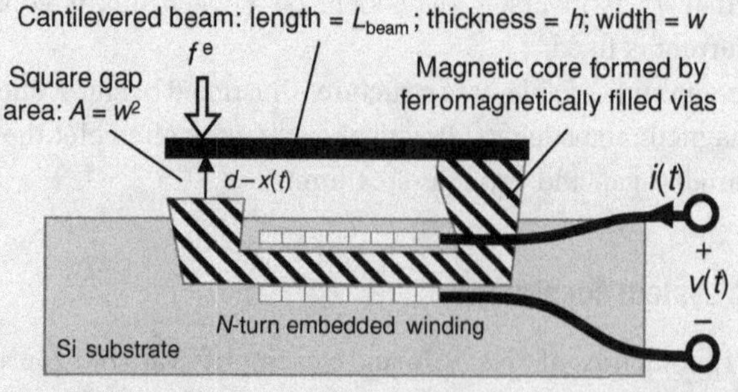

Figure 9.10 Side view of variable-gap magnetic transducer fabricated using a process based on ferromagnetic material-filled vias developed by Sadler [8]. This geometry is a magnetic circuit; the variable air gap is a cantilevered beam. It is assumed that $d \ll w$.

[3] The emergence of practical superconducting MEMS devices awaits major progress in the processing of high-T_c materials.

The magnetic core reluctance is $\mathcal{R}_{core} \approx l_{core}/(\mu_r \mu_0 A_{core})$, where l_{core} and A_{core} are the effective magnetic path length and cross-sectional area, respectively. The air gap is $\mathcal{R}_{gap}(x) = (d - x)/(\mu_0 A)$, where $A = w^2$ is gap area. Using the magnetic circuit model,

$$\Phi = mmf/[\mathcal{R}_{core} + \mathcal{R}_{gap}(x)],$$

where $mmf = Ni$. Then, because $\lambda = N\Phi = Li$,

$$L(x) \equiv \frac{\lambda}{i} = \frac{N^2}{\mathcal{R}_{core} + \mathcal{R}_{gap}(x)},$$

the magnetic force is

$$f^e = \frac{i^2}{2}\frac{dL}{dx} = \frac{N^2 i^2}{2\mu_0 A[\mathcal{R}_{core} + (d - x)/(\mu_0 A)]^2}.$$

If we approximate f^e by a point force acting at distance $w/2$ from the free end, then Eq. (6.8) from Section 6.2 can be used to write equilibrium force balance for the beam:

$$\frac{N^2 i^2}{2\mu_0 A[\mathcal{R}_{core} + (d - x_0)/(\mu_0 A)]^2} \approx \frac{Y_{Si} h^3 w}{4(L_{beam} - w/2)^3} x_0,$$

where Y_{Si} is Young's modulus. Figure 9.11 plots gap $(d - x_0)$ versus current using parameters representative of Sadler's device [8].

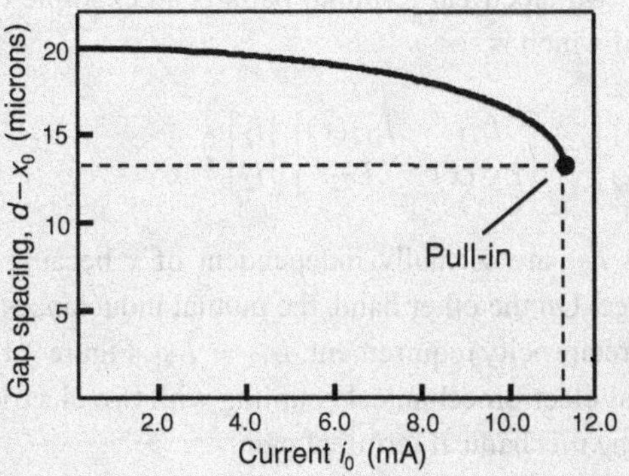

Figure 9.11 Beam deflection at the tip versus current for the actuator shown in Fig. 9.10 using the following parameters. For the Si cantilever: $L_{beam} = 1000$ μm, $w = 100$ μm, $h = 2.0$ μm, $Y_{si} = 1.68\ 10^{11}$ Pa; magnetic core: $d = 20$ μm, $l_{core} = 2000$ μm, $A_{core} = 10^4$ μm^2, $\mu_r = 2000$; winding: $N = 8$. Pull-in of the beam occurs at $i_0 = 11.3$ mA.

9.3.4 A magnetic transducer with two coils

Figure 9.12a shows a magnetic transducer with two planar coils oriented in parallel and sharing a common central axis. The upper coil is free to move up and down along this axis, and the variable spacing between the two coils is $x(t)$. Figure 9.12b is an SEM of a fabricated device that might be used to create this type of MEMS actuator.

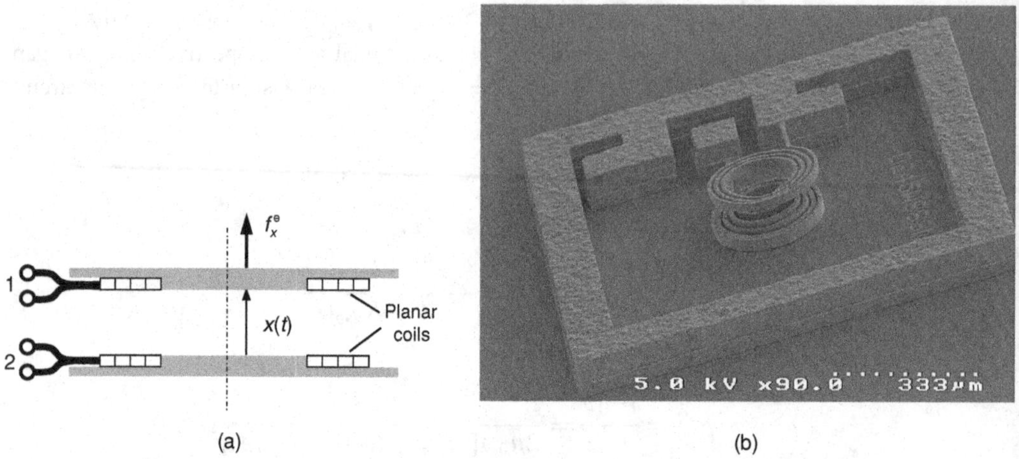

Figure 9.12 Magnetic transducer consisting of two parallel, coupled, planar coils on a common normal axis. (a) Side view of transducer. With no magnetic material in the structure, the self-inductances are virtually independent of displacement x. (b) SEM of inductor consisting of two parallel, circular, spiral coils. The windings in this structure are connected in series but it is evident that the same process could be used to realize the set of two independent coils shown in (a). Image provided courtesy of C. D. Meyer, University of Florida.

This configuration with its two electrical terminal pairs is an example of a coupled circuit, the terminal relation of which is

$$\begin{bmatrix} \lambda_1 \\ \lambda_2 \end{bmatrix} = \begin{bmatrix} L_{11} & L_{12}(x) \\ L_{21}(x) & L_{22} \end{bmatrix} \begin{bmatrix} i_1 \\ i_2 \end{bmatrix}. \tag{9.27}$$

The self-inductances L_{11} and L_{22} are virtually independent of x because there is no magnetic material in this device. On the other hand, the mutual inductances L_{12} and L_{21} must depend on x. To meet the reciprocity requirement, $L_{12} = L_{21}$. Figure 9.13 represents this actuator as an ideal lossless electromechanical coupling with two electrical terminal pairs, one for each coil, and one mechanical terminal pair.

Coenergy is convenient for determining the force of electrical origin here because the terminal relations of Eq. (9.27) suggest i_1 and i_2 for the electrical state variables.

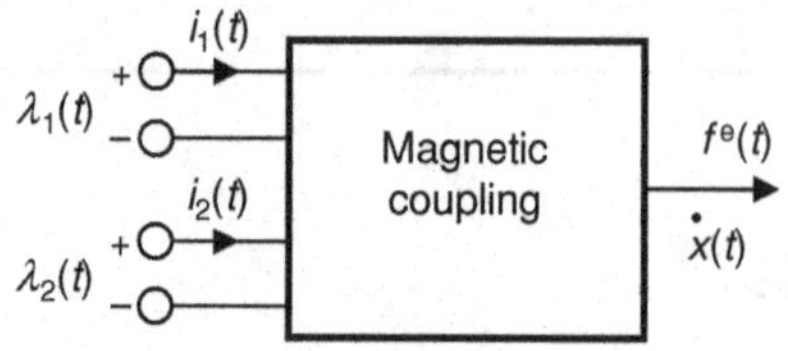

Figure 9.13 Ideal lossless electromechanical representation for the coupled magnetic coil transducer illustrated in Fig. 9.12a.

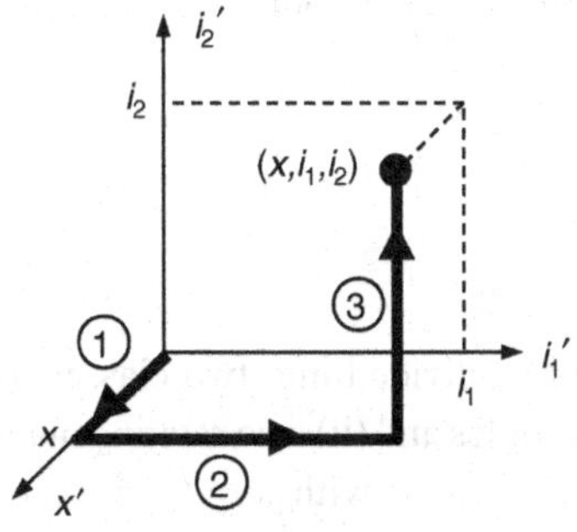

Figure 9.14 Suitable integral path to obtain the coenergy function W'_m using Eq. (9.28) for the multiport magnetic transducer device represented by Fig. 9.13. The first integral term contributes nothing because the force of electrical origin is zero along the first path segment. Because of reciprocity, the order of integration for the currents i_1 and i_2 is immaterial.

Making this choice at the outset obviates the need for a matrix inversion. Referring to the right-hand column of Table 9.1, the conservation relation is

$$dW'_m = f^e dx + \lambda_1 di_1 + \lambda_2 di_2. \tag{9.28}$$

To evaluate W'_m, the mechanical variable x is integrated first to take advantage of the assumption $f^e(x, i_1 = 0, i_2 = 0) = 0$. Figure 9.14 shows a suitable integral path in (x, i_1, i_2) state space. Using this path and the terminal relations from Eq. (9.27) yields

$$W'_m = \frac{1}{2} L_{11} i_1^2 + \frac{1}{2} L_{22} i_2^2 + L_m(x) i_1 i_2, \tag{9.29}$$

where $L_m(x) \equiv L_{12}(x) = L_{21}(x)$. Then, from Eq. (9.15),

$$f^e = i_1 i_2 \frac{dL_m}{dx}. \tag{9.30}$$

Note that the force is proportional to the product of the currents $i_1 i_2$.

More information about the x dependence of the mutual inductance is required to proceed. A simple, empirical expression for this dependence is

$$L_m(x) \approx k_m \sqrt{L_{11} L_{22}}/(1 + \alpha x^2), x \geq 0, \tag{9.31}$$

where α depends on the geometry and $k_m \approx L_m(x = 0)/\sqrt{L_{11} L_{22}}$, the *magnetic coupling coefficient*, is the fraction of magnetic flux from one coil linking the other when $x = 0$. From the definition, it is obvious that $0 < |k_m| < 1$. Though only an approximation, Eq. (9.31) authentically represents the inverse relation of the coupling over at least some range of x. Using it in Eq. (9.30) yields

$$f^e \approx -\frac{2\alpha k_m L_0 i_1 i_2}{(1 + \alpha x^2)^2} x. \tag{9.32}$$

The interaction force between the two coils, dependent on the product of the two currents, can be attractive or repulsive. Such bidirectional capability is advantageous in many transducer applications. The magnitude of the force approaches zero both at $x \to 0$ and

$x \to \infty$. The rule about maximizing self-inductance does not apply here; the interaction is between two coils.

9.4 Rotational magnetic transducers

Rotational magnetic devices for MEMS can be divided into two classes: (i) actuators operating over a finite range of displacement angles and (ii) true rotating machines. Most actuators are high-aspect-ratio assemblies fabricated with *ferrite materials* and they deliver high torque over relatively large angular displacements. Some very interesting centimeter-scale rotating machines have been fabricated and tested [9, 10], and one can anticipate eventual development of relatively inexpensive devices approaching the millimeter barrier. Their torque and speed capabilities will fill a niche that capacitive transducers can never challenge.

9.4.1 Electromechanics of rotating magnetic transducers

The starting point for a lumped parameter formalism of rotating magnetic devices is the conservative coupling. Consider a rotating magnetic transducer with displacement angle $\theta(t)$ and torque of electrical origin τ^{e}. The equation of power conservation is

$$vi = \tau^{\mathrm{e}}\frac{\mathrm{d}\theta}{\mathrm{d}t} + \frac{\mathrm{d}W_{\mathrm{m}}}{\mathrm{d}t}. \tag{9.33}$$

Recasting this expression in terms of differentials yields

$$\mathrm{d}W_{\mathrm{m}} = -\tau^{\mathrm{e}}\mathrm{d}\theta + i\,\mathrm{d}\lambda. \tag{9.34}$$

The independent state variables are θ and λ; current and torque are expressed in terms of them, that is, $i = i(\theta, \lambda)$ and $\tau^{\mathrm{e}}(\theta, \lambda)$. We may express the total differential of $W_{\mathrm{e}} = W_{\mathrm{e}}(\theta, \lambda)$ using partial derivatives with respect to the state variables:

$$\mathrm{d}W_{\mathrm{m}} = \left.\frac{\partial W_{\mathrm{m}}}{\partial \theta}\right|_{\lambda}\mathrm{d}\theta + \left.\frac{\partial W_{\mathrm{m}}}{\partial \lambda}\right|_{\theta}\mathrm{d}\lambda. \tag{9.35}$$

Employing the familiar argument about the independence of θ and λ, Eqs. (9.34) and (9.35) combine to yield

$$i(\theta, \lambda) = \left.\frac{\partial W_{\mathrm{m}}}{\partial \lambda}\right|_{\theta} \quad \text{and} \quad \tau^{\mathrm{e}}(\theta, \lambda) = -\left.\frac{\partial W_{\mathrm{m}}}{\partial \theta}\right|_{\lambda}. \tag{9.36}$$

Integration to obtain W_{m} is accomplished by recognizing that $\tau^{\mathrm{e}}(\theta', \lambda' = 0) = 0$:

$$W_{\mathrm{m}}(\theta, \lambda) = -\int_0^{\theta} \underbrace{\tau^{\mathrm{e}}(\theta', \lambda' = 0)}_{=0}\,\mathrm{d}\theta' + \int_0^{\lambda} i(\theta, \lambda')\,\mathrm{d}\lambda'. \tag{9.37}$$

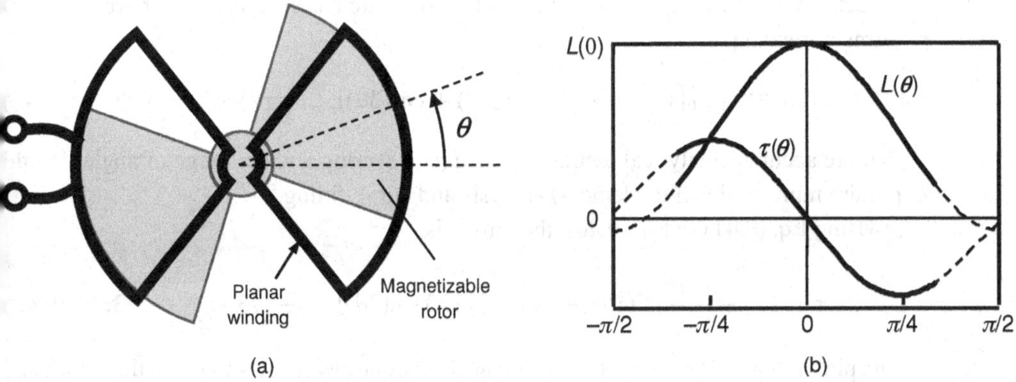

Figure 9.15 A rotational magnetic transducer. (a) The windings consist of a pair of planar coils patterned in the form of two opposed circular sectors. The rotor is a pair of solid pie sections made of magnetically soft material and free to rotate above the coil on the same axis. (b) Inductance and torque as functions of angular displacement θ. The device behaves as a variable-area inductor except close to $\theta = 0$ where the slope of $L(\theta)$ changes from positive to negative. The dashed lines extending toward $\theta = \pm\pi/2$ show that the approximate cosine series expression of Eq. (9.41) starts to lose accuracy near $|\theta| \approx \pi/4$.

For a magnetically linear device, the electrical terminal relation is $\lambda = L(\theta)i$, so

$$W_{\mathrm{m}}(\theta, \lambda) = \int_0^{\lambda} \frac{\lambda'}{L(\theta)} \mathrm{d}\lambda' = \frac{\lambda^2}{2L(\theta)}. \tag{9.38}$$

Combining Eqs. (9.36) and (9.38), the magnetic torque is

$$\tau^{\mathrm{e}} = \frac{\lambda^2}{2L^2} \frac{\partial L}{\partial \theta}. \tag{9.39}$$

We can invoke the electrical terminal relation a second time to express the torque in terms of the current:

$$\tau^{\mathrm{e}} = \frac{i^2}{2} \frac{\partial L}{\partial \theta}. \tag{9.40}$$

Equation (9.40) can also be obtained directly using magnetic coenergy, $W'_{\mathrm{m}}(\theta, i) = L(\theta)i^2/2$, by following the approach in Section 9.2.4.

9.4.2 Rotating magnetic actuator

The transducer shown in Fig. 9.15a consists of a planar coil in the form of two opposed circular sectors with a common axis. Just above the coil is a highly permeable magnetic armature free to rotate about the axis and consisting of a rigidly linked pair of opposed sectors. The armature could be fabricated of FeNi Permalloy® using a LIGA-like process. The inductance $L(\theta)$ is a maximum when the armature completely overlaps the planar winding at $\theta = 0$. Refer to Fig. 9.15b. Over a limited range of angular displacements, the inductance can be roughly approximated by a truncated harmonic series of even (cosine) terms, for example,

$$L(\theta) \approx L_0[1 + \cos(\theta) + \cos(2\theta) + \cos(3\theta)], \quad -\pi/3 < \theta < \pi/3. \qquad (9.41)$$

A more accurate analytical expression for $L(\theta)$ covering a wider range of angles would require numerical (finite-element) analysis and curve-fitting,

Using Eq. (9.41) in Eq. (9.40), the torque is

$$\tau^{e}(\theta) \approx -\frac{L_0 i^2}{2}[\sin(\theta) + 2\sin(2\theta) + 3\sin(3\theta)], \quad -\pi/3 < \theta < \pi/3. \qquad (9.42)$$

The plot in Fig. 9.15b shows that the magnetic torque always acts to move the rotor back toward $\theta = 0$, where inductance in maximized.

Example 9.4 **Torsional magnetic field sensor**

The MEMS-scale magnetic field sensor in Fig. 9.16, based on the classic d'Arsonval movement, uses a piezoresistive element to sense deflection of a DC-current excited planar coil [11]. Using the lumped parameter method to calculate the torque obviates the need to know the geometric details of the coil,

$$\lambda(\theta, i) = Li + BA\sin(\theta).$$

A and L are, respectively, the area and self-inductance of the one-turn coil, and B, the magnetic field to be measured, is assumed to be parallel to the coil when $\theta = 0$.

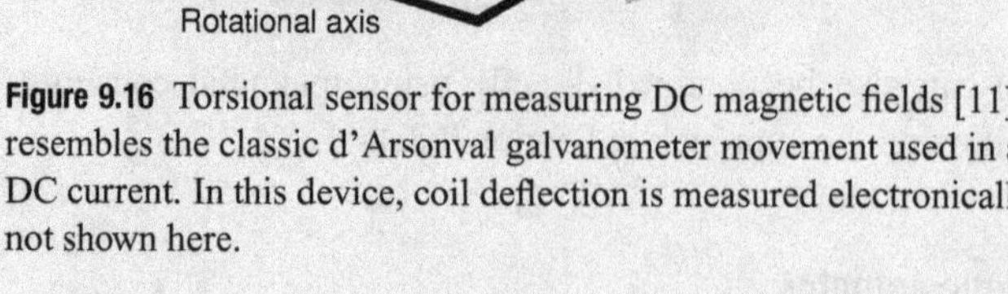

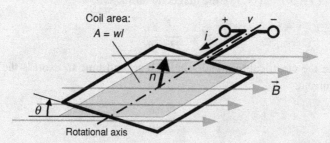

Figure 9.16 Torsional sensor for measuring DC magnetic fields [11]. This device closely resembles the classic d'Arsonval galvanometer movement used in analog meters for measuring DC current. In this device, coil deflection is measured electronically by a piezoresistive element not shown here.

The magnetic coenergy is

$$W'_{\mathrm{m}}(\theta, i) = Li^2/2 + BA\sin(\theta)i.$$

Self-inductance L is independent of angle θ. Thus,

$$\tau^{e} = \left.\frac{\partial W'_{\mathrm{m}}}{\partial\theta}\right|_i = BAi\cos(\theta).$$

Assuming a linear torsional spring with spring constant k_θ, which is relaxed at $\theta = 0$, the torque balance equation is

$$BAi \cos\theta_0 \approx k_\theta \theta_0,$$

which can be solved numerically for θ_0 as a function of i and B.

9.5 Permanent magnet transducers

Advances in microfabrication using permanent magnet materials may one day pave the way for a powerful new class of MEMS devices. Integration of these transducers into working systems will be rapid using off-the-shelf IC electronics. Anticipating such a development, we present here the model of a permanent magnet transducer, employing an approach similar to that used for the electret transducer in Section 3.6. In fact, from the standpoint of energy-based force derivations, transducers using permanent magnets and electrets have much in common. In both cases, the energy stored permanently within the material during processing eliminates the need for external bias.

Figure 9.17 illustrates a movable planar electromagnetic coil oriented parallel to and sharing a common central axis with a permanent magnet in the shape of a thin circular disk. To account for the permanent magnetization, we introduce a *virtual winding* that encloses the magnetized material. This winding, which does not really exist, facilitates the required "thermodynamic" assembly of the system to account for the energy stored in the permanent magnet.[4]

Including the virtual winding, the electrical terminal relation has two terminal pairs and assumes the same canonical form as Eq. (9.27):

$$\begin{bmatrix} \lambda_c \\ \lambda_p \end{bmatrix} = \begin{bmatrix} L_c(x) & L_m(x) \\ L_m(x) & L_p \end{bmatrix} \begin{bmatrix} i_c \\ i_p \end{bmatrix}. \tag{9.43}$$

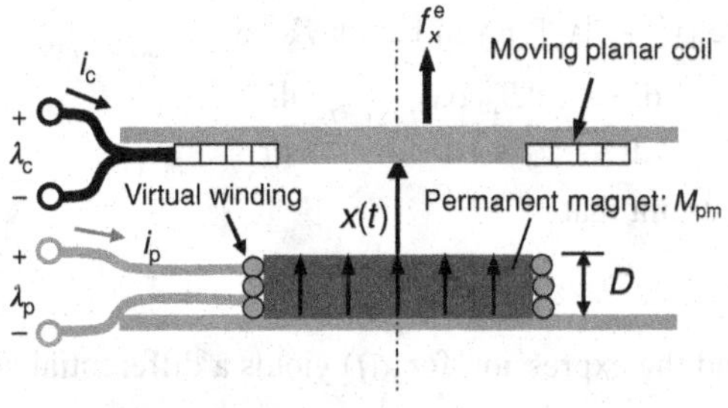

Figure 9.17 Permanent magnetic transducer with a planar coil as the moving element. The coil surrounding the permanent magnet is a virtual winding that accounts for the volume magnetization M_{pm} in the lumped parameter formulation. This winding is analogous to the virtual electrodes depicted in the electret transducer illustrated in Fig. 3.25a.

[4] The apparent current flowing through this virtual winding is called an *Amperian current* [12].

Subscripts "c" and "p" designate, respectively, variables associated with the moving coil and the virtual winding in Fig. 9.17. This device differs from the coupled planar coils described in Section 9.3.4 in that here the self-inductance, L_c, and the mutual inductance, L_m, depend strongly on x. To get started, we adapt Eq. (9.29) to express the coenergy:

$$W'_m = \frac{1}{2}L_c(x)i_c^2 + \frac{1}{2}L_p i_p^2 + L_m(x)i_c i_p. \tag{9.44}$$

The force of electrical origin is then

$$f^e = \frac{\partial W'_m}{\partial x} = i_c i_p \frac{dL_m}{dx} + \frac{i_c^2}{2}\frac{dL_c}{dx}. \tag{9.45}$$

The Amperian current i_p, though itself unmeasurable, is directly related to M_{pm}, the volume magnetization of the permanent magnet. If D, the thickness of the magnet, is small compared with its areal dimensions, then $i_p \approx DM_{pm}$. Further assuming that the permanent magnet is very strong, the mutual coupling term involving the $i_c i_p$ product dominates in Eq. (9.45), so that

$$f^e \approx DM_{pm}\frac{dL_m}{dx}i_c. \tag{9.46}$$

Because M_{pm} is constant, the direction of the force is controlled by the sign of i_c, the current applied to the electromagnet. This inherently strong, bidirectional force is a very desirable attribute for actuators.

Example 9.5 **Equation of motion for magnetic transducer**

The two planar coils in Fig. 9.12a have the following electrical terminal relation:

$$\begin{bmatrix}\lambda_1\\ \lambda_2\end{bmatrix} = \begin{bmatrix}L_1 & L_m(x)\\ L_m(x) & L_2\end{bmatrix}\begin{bmatrix}i_1\\ i_2\end{bmatrix}.$$

Only L_m depends on x. Assume that DC current $i_1 = i_0$ flows through winding 1 and that winding 2, which is movable, is connected to resistor R. Further, assume $x(t) = x_0 + x'\cos(\omega t)$, where $|x'| < |x_0|$. From Faraday's law,

$$v_2 = \frac{d\lambda_2}{dt} = \frac{dL_m}{dx}\frac{dx}{dt}i_0 + L_2\frac{di_2}{dt}.$$

The resistor imposes the constraint that

$$v_2 = -Ri_2.$$

Combining these relations and the expression for $x(t)$ yields a differential equation for $i_2(t)$:

$$L_2\frac{di_2}{dt} + Ri_2 \approx \left.\frac{dL_m}{dx}\right|_{x_0} \omega i_0 x' \sin(\omega t),$$

which is soluble for $i_2(t)$ using phasors. Not surprisingly, the derivative of L_m controls this current.

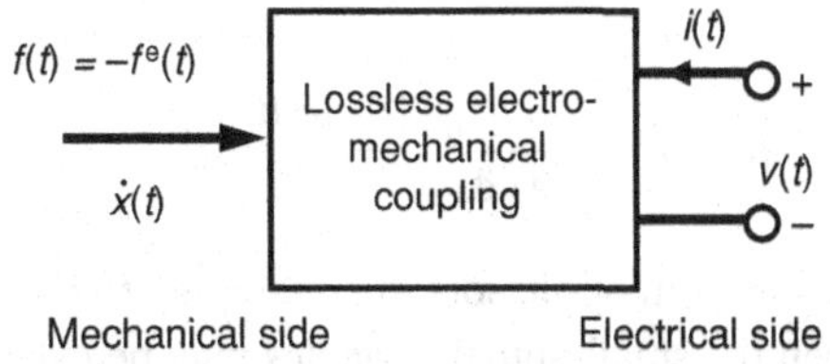

Figure 9.18 Representation for the basic, lossless, electromechanical coupling using terminal variables convenient for two-port analysis: voltage $v(t)$, current $i(t)$, velocity $\dot{x}(t)$, and force, defined as $f(t) = -f^e(t)$.

9.6 Small-signal inductive electromechanics

The fact that most magnetic MEMS devices operate in the linear response regime provides sufficient motivation to develop a linear electromechanical two-port theory for them. Even in the case where large-amplitude motion is of interest and where numerical methods will be needed, much can be learned about such systems by focusing initially on the small-signal dynamics. What follows here parallels development of the two-port theory for capacitive devices found in Section 4.3.

Consider the lossless coupling shown in Fig. 9.18. The variables employed here are: force $f(t) = -f^e(t)$; velocity $\dot{x}(t) = \mathrm{d}x/\mathrm{d}t$, rather than displacement x; voltage, which replaces magnetic flux according to $v(t) = \mathrm{d}\lambda/\mathrm{d}t$; and current $i(t)$. These selections insure consistency with Neubert [13].

The goal of linearization is a set of relations between the small-signal electrical and mechanical variables. The first step is to define equilibrium and perturbation (small-signal) quantities, which we denote here, respectively, by a subscript zero and a prime:

$$\begin{aligned} f(t) &= f_0 + f'(t), && \text{where } |f'(t)/f_0| \ll 1, \\ x(t) &= x_0 + x'(t), && \text{where } |x'(t)/x_0| \ll 1, \\ \lambda(t) &= \lambda_0 + \lambda'(t), && \text{where } |\lambda'(t)/\lambda_0| \ll 1, \\ i(t) &= i_0 + i'(t), && \text{where } |i'(t)/i_0| \ll 1. \end{aligned} \tag{9.47}$$

The formulations based on coenergy $W'_\mathrm{m}(x, i)$ and energy $W_\mathrm{m}(x, \lambda)$ are treated separately to avoid confusion.

9.6.1 M-form transducer matrix based on $W'_\mathrm{m}(x, i)$

In Neubert's M-form transducer formulation, small-signal current i' is the through variable on the electrical side. For this reason, coenergy is used to obtain the M-form transducer matrix for magnetic devices. Expanding Eqs. (9.15) in two Taylor series, then retaining terms to first-order and separating the equilibrium and perturbation quantities yields

$$f_0 = -\frac{\partial W'_\mathrm{m}}{\partial x} \quad \text{and} \quad f' = -\frac{\partial^2 W'_\mathrm{m}}{\partial x^2} x' - \frac{\partial^2 W'_\mathrm{m}}{\partial x \partial i} i' \tag{9.48}$$

and

$$\lambda_0 = -\frac{\partial W'_{\mathrm{m}}}{\partial i} \quad \text{and} \quad \lambda' = \frac{\partial^2 W'_{\mathrm{m}}}{\partial x \partial i} x' + \frac{\partial^2 W'_{\mathrm{m}}}{\partial i^2} i'. \tag{9.49}$$

Letting perturbation variables take the form of phasors, for example, $f'(t) = \mathrm{Re}[\underline{f} e^{j\omega t}]$, the coupled linear relations between the small-signal variables may be expressed using Neubert's M-form transducer matrix, which relates the across variable, $\underline{f}$ and $\underline{v} = j\omega\underline{\lambda}$, to the through variables, $j\omega\underline{x}$ and $\underline{i}$:[5]

$$\begin{bmatrix} \underline{f} \\ \underline{v} \end{bmatrix} = \begin{bmatrix} \underline{Z}_{\mathrm{m}} & \underline{M} \\ -\underline{M}^* & \underline{Z}_{\mathrm{e}} \end{bmatrix} \begin{bmatrix} j\omega\underline{x} \\ \underline{i} \end{bmatrix}. \tag{9.50}$$

The M-form transducer coefficients are

$$\begin{aligned} &\text{mechanical impedance:} && \underline{Z}_{\mathrm{m}} = \frac{-1}{j\omega}\frac{\partial^2 W'_{\mathrm{m}}}{\partial x^2}\bigg|_{x_0, i_0}, \\ &\text{transducer coupling coefficient:} && \underline{M} = -\frac{\partial^2 W'_{\mathrm{m}}}{\partial i \partial x}\bigg|_{x_0, i_0}, \\ &\text{electrical impedance:} && \underline{Z}_{\mathrm{e}} = j\omega \frac{\partial^2 W'_{\mathrm{m}}}{\partial i^2}\bigg|_{x_0, i_0} = j\omega L_0. \end{aligned} \tag{9.51}$$

These coefficients depend on the equilibrium quantities x_0 and i_0. As expected, inductance $L_0 = L(x_0)$ appears in the electrical impedance $\underline{Z}_{\mathrm{e}}$.

9.6.2 N-form transducer matrix based on $W_{\mathrm{m}}(x, \lambda)$

Repeating the linearization exercise upon Eq. (9.4) using the energy function, $W_{\mathrm{m}}(x, \lambda)$, the equilibrium and perturbation components are

$$f_0 = \frac{\partial W_{\mathrm{m}}}{\partial x} \quad \text{and} \quad f' = \frac{\partial^2 W_{\mathrm{m}}}{\partial x^2} x' + \frac{\partial^2 W_{\mathrm{m}}}{\partial x \partial \lambda} \lambda' \tag{9.52}$$

and

$$i_0 = \frac{\partial W_{\mathrm{m}}}{\partial \lambda} \quad \text{and} \quad i' = \frac{\partial^2 W_{\mathrm{m}}}{\partial x \partial \lambda} x' + \frac{\partial^2 W_{\mathrm{m}}}{\partial \lambda^2} \lambda'. \tag{9.53}$$

Again assuming sinusoidal steady-state time-dependence for the small-signal variables, Neubert's *N*-form matrix results:

$$\begin{bmatrix} \underline{f} \\ \underline{i} \end{bmatrix} = \begin{bmatrix} \underline{Z}'_{\mathrm{m}} & \underline{N} \\ -\underline{N}^* & \underline{Y}_{\mathrm{e}} \end{bmatrix} \begin{bmatrix} j\omega\underline{x} \\ \underline{v} \end{bmatrix}. \tag{9.54}$$

[5] The importance of distinguishing through and across variables becomes evident when the equivalent circuits are presented in Section 9.6.3.

Table 9.2 Summary of M- and N-form two-port transducer matrices, circuit models, and coefficients for translational and rotational magnetic transducers. Comparing this table with Table 4.1 for capacitive transducers reveals that the roles of energy and coenergy are reversed with respect to the M- and N-form matrix coefficients

	Translational electromechanical transducer	Rotating electromechanical transducer
M-form transducer matrix	$\begin{bmatrix} \underline{f} \\ \underline{v} \end{bmatrix} = \begin{bmatrix} \underline{Z}_{\mathrm{m}} & \underline{M} \\ -\underline{M}^* & \underline{Z}_{\mathrm{e}} \end{bmatrix} \begin{bmatrix} \mathrm{j}\omega\underline{x} \\ \underline{i} \end{bmatrix}$	$\begin{bmatrix} \underline{\tau} \\ \underline{v} \end{bmatrix} = \begin{bmatrix} \underline{Z}_{\mathrm{m}} & \underline{M} \\ -\underline{M}^* & \underline{Z}_{\mathrm{e}} \end{bmatrix} \begin{bmatrix} \mathrm{j}\omega\underline{\theta} \\ \underline{i} \end{bmatrix}$
Definitions of M-form coefficients	$\underline{Z}_{\mathrm{m}} = \dfrac{-1}{\mathrm{j}\omega} \dfrac{\partial^2 W'_{\mathrm{m}}}{\partial x^2},\ \underline{M} = -\dfrac{\partial^2 W'_{\mathrm{m}}}{\partial i \partial x}$ $\underline{Z}_{\mathrm{e}} = \mathrm{j}\omega \dfrac{\partial^2 W'_{\mathrm{m}}}{\partial i^2} = \mathrm{j}\omega L_0$	$\underline{Z}_{\mathrm{m}} = \dfrac{-1}{\mathrm{j}\omega} \dfrac{\partial^2 W'_{\mathrm{m}}}{\partial \theta^2},\ \underline{M} = -\dfrac{\partial^2 W'_{\mathrm{m}}}{\partial i \partial \theta}$ $\underline{Z}_{\mathrm{e}} = \mathrm{j}\omega \dfrac{\partial^2 W'_{\mathrm{m}}}{\partial i^2} = \mathrm{j}\omega L_0$
	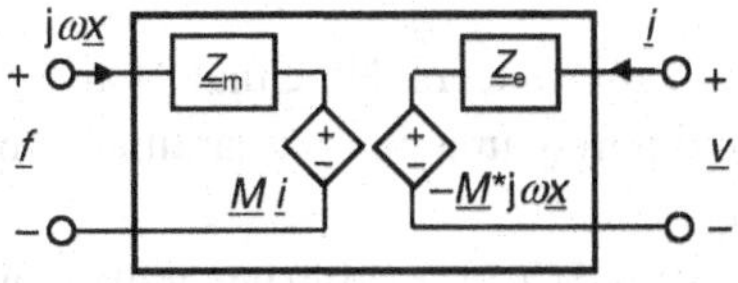	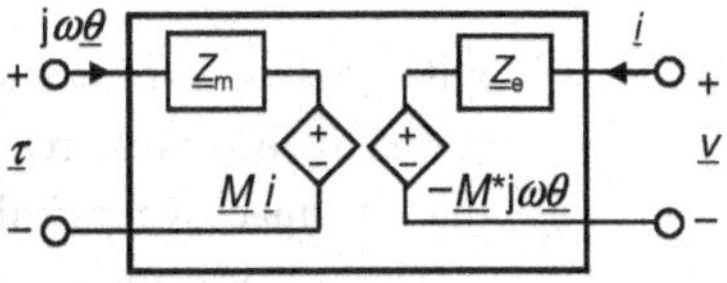
N-form transducer matrix	$\begin{bmatrix} \underline{f} \\ \underline{i} \end{bmatrix} = \begin{bmatrix} \underline{Z}'_{\mathrm{m}} & \underline{N} \\ -\underline{N}^* & \underline{Y}_{\mathrm{e}} \end{bmatrix} \begin{bmatrix} \mathrm{j}\omega\underline{x} \\ \underline{v} \end{bmatrix}$	$\begin{bmatrix} \underline{\tau} \\ \underline{i} \end{bmatrix} = \begin{bmatrix} \underline{Z}'_{\mathrm{m}} & \underline{N} \\ -\underline{N}^* & \underline{Y}_{\mathrm{e}} \end{bmatrix} \begin{bmatrix} \mathrm{j}\omega\underline{\theta} \\ \underline{v} \end{bmatrix}$
Definitions of N-form coefficients	$\underline{Z}'_{\mathrm{m}} = \dfrac{1}{\mathrm{j}\omega} \dfrac{\partial^2 W_{\mathrm{m}}}{\partial x^2},\ \underline{N} = \dfrac{1}{\mathrm{j}\omega} \dfrac{\partial^2 W_{\mathrm{m}}}{\partial \lambda \partial x},$ $\underline{Y}_{\mathrm{e}} = \dfrac{1}{\mathrm{j}\omega} \dfrac{\partial^2 W_{\mathrm{m}}}{\partial \lambda^2} = \dfrac{1}{\mathrm{j}\omega L_0}$	$\underline{Z}'_{\mathrm{m}} = \dfrac{1}{\mathrm{j}\omega} \dfrac{\partial^2 W_{\mathrm{m}}}{\partial \theta^2},\ \underline{N} = \dfrac{1}{\mathrm{j}\omega} \dfrac{\partial^2 W_{\mathrm{m}}}{\partial \lambda \partial \theta},$ $\underline{Y}_{\mathrm{e}} = \dfrac{1}{\mathrm{j}\omega} \dfrac{\partial^2 W_{\mathrm{m}}}{\partial \lambda^2} = \dfrac{1}{\mathrm{j}\omega L_0}$
	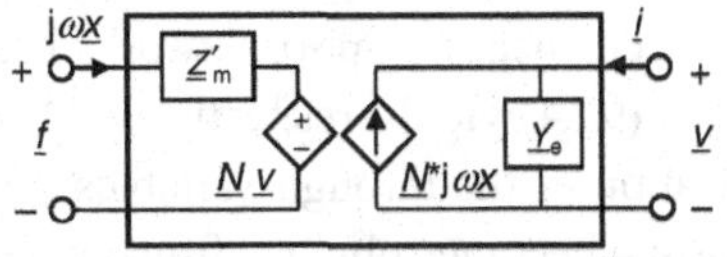	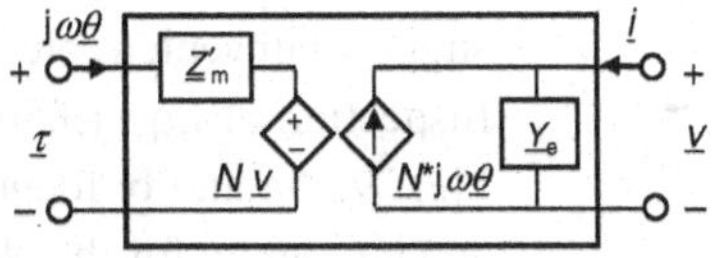
Relationships amongst M- and N-form coefficients	$\underline{Z}'_{\mathrm{m}} = \underline{Z}_{\mathrm{m}} + \lvert\underline{M}\rvert^2/\underline{Z}_{\mathrm{e}},\ \underline{Y}_{\mathrm{e}} = 1/\underline{Z}_{\mathrm{e}},\ \underline{N} = -\underline{M}/\underline{Z}_{\mathrm{e}}$	
Miscellany	$\underline{Z}_{\mathrm{m}}$, $\underline{Z}'_{\mathrm{m}}$, $\underline{N}$, and $\underline{Z}_{\mathrm{e}} = 1/\underline{Y}_{\mathrm{e}}$ are purely imaginary; $\underline{M}$ is purely real.	

This matrix equation is identical in form to Eq. (4.12) for capacitive transducers. The coefficients are now defined in terms of the energy function:

$$
\begin{aligned}
&\text{mechanical impedance:} && \underline{Z}'_{\mathrm{m}} = \frac{1}{\mathrm{j}\omega} \frac{\partial^2 W_{\mathrm{m}}}{\partial x^2}\bigg|_{x_0,\lambda_0}, \\
&\text{transducer coupling coefficient:} && \underline{N} = \frac{1}{\mathrm{j}\omega} \frac{\partial^2 W_{\mathrm{m}}}{\partial \lambda \partial x}\bigg|_{x_0,\lambda_0}, \\
&\text{electrical admittance:} && \underline{Y}_{\mathrm{e}} = \frac{1}{\mathrm{j}\omega} \frac{\partial^2 W_{\mathrm{m}}}{\partial \lambda^2}\bigg|_{x_0,\lambda_0} = \frac{1}{\mathrm{j}\omega L_0}.
\end{aligned}
\qquad (9.55)
$$

The two-port forms of the M- and N-form matrices for translational and rotational magnetic transducers are collected in Table 9.2. The reader should compare this table

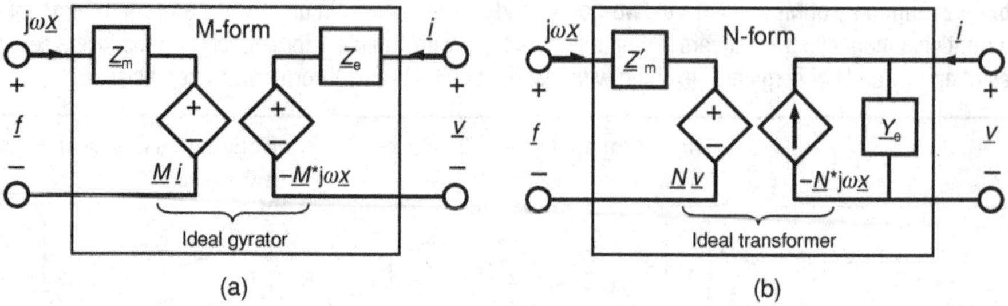

Figure 9.19 Linear two-port network equivalents for magnetic actuators. These equivalent circuits are not unique. (a) The M-form circuit, based on Eqs. (9.50) and (9.51), features two controlled voltage sources (a gyrator). (b) The N-form circuit, based on Eqs. (9.54) and (9.55), uses one controlled voltage source and one controlled current source (a transformer).

with Table 4.1 for capacitive transducers. Note that $\underline{M}$ and $\underline{N}$ for inductive devices are, respectively, purely real and purely imaginary quantities, which is the opposite case to that of capacitive transducers.

The matrices in Eqs. (9.50) and (9.54) readily generalize for multiport networks. Refer back to Table 9.1 for functional definitions of the energy quantities needed to perform the various linearizations. Section 4.3.5 covers $N_e \times N_m$ capacitive transducer networks and its results serve as a guide for generalized magnetic multiport networks.

9.6.3 Linear circuit models for magnetic transducers

The matrix representations for the small-signal AC behavior of magnetic transducers suggest convenient electromechanical two-port networks based on dependent sources. Inspection of Eqs. (9.50) and (9.54) yields, respectively, M- and N-form equivalents of Figs. 9.19a and b. Identification of the through variables – $j\omega\underline{x}$ and $\underline{i}$ for the M-form, and $j\omega\underline{x}$ and $\underline{v}$ for the N-form – becomes obvious from examining these circuits.[6]

Because the left-hand sides of these networks are mechanical ports, it is best now to embrace fully the analogy between linear mechanical and electrical systems. What this means is that, on the mechanical side of the network, velocity, $j\omega\underline{x}$, and force, $\underline{f}$, are treated, respectively, like current and voltage. Kirchhoff's voltage and current laws, KVL and KCL, may then be invoked to account for the external mechanical constraints, Newton's force law and continuity, respectively. A helpful mnemonic to keep this analogy firmly in mind is that $\underline{f}/(j\omega\underline{x})\big|_{\underline{i}=0} \equiv \underline{Z}_m$ and $\underline{f}/(j\omega\underline{x})\big|_{\underline{v}=0} \equiv \underline{Z}'_m$ are called *mechanical impedances*. Section B.2 of Appendix B explains the direct electrical ↔ mechanical analogy that motivates these identifications.

[6] The electromechanical two-ports in Figs. 9.19a and b contain "ideal" transducers identified, respectively, as gyrator and transformer elements. While this result is in apparent harmony with the small-signal M- and N-form models presented for capacitive transducers in Section 4.4, there is a major difference: namely, that for inductive electromechanical transducers, $\underline{M}$ and $\underline{N}$ are, respectively, purely real and purely imaginary, just the reverse of the case for capacitive systems. Thus, any connection with the conventional gyrator and transformer circuit elements introduced in Section 2.3.5 becomes unclear.

9.6.4 External constraints

As models for real sensors and actuators, the networks in Figs. 9.19a and b are incomplete because they do not account for mechanical and electrical constraints external to the electromechanical coupling itself. On the mechanical side, the constraints include the mass of the moving element, any mechanical spring used to position the moving element in proper equilibrium, the damping, and, of course, any externally applied force or torque. On the electrical side, there are the constraints imposed by circuit components, plus external voltage or current drives.

Mechanical constraints are taken into account using Newton's law of force and imposing the continuity condition. As a starting point, assume that the transducer has mass m, linear damping b, a linear mechanical spring k, which exerts no force when $x = x_k$, the electrical force $f(t) = -f^e(t)$, and an external force $f_{ext}(t)$ acting on the mass. Thus,

$$m\frac{d^2x}{dt^2} = -k(x - x_k) - b\frac{dx}{dt} - f(t) + f_{ext}(t). \tag{9.56}$$

One can partition Eq. (9.56) into a static force balance condition relating the equilibrium position x_0 to either i_0 or λ_0,

$$0 = -k(x_0 - x_k) - f_0(x_0, i_0 \text{ or } \lambda_0) + f_{ext,0}, \tag{9.57}$$

and the small-signal (linear) equation of motion,

$$m\frac{d^2x'}{dt^2} = -kx' - b\frac{dx'}{dt} - f'(t) + f'_{ext}(t). \tag{9.58}$$

Imposing exponential ($e^{j\omega t}$) time dependence on the perturbation variables in Eq. (9.58), e.g., $x'(t) = \text{Re}[\underline{x}e^{j\omega t}]$, then yields

$$\underline{f}_{ext} = \underbrace{[j\omega m + b + k/(j\omega)]}_{\underline{Z}_{m0}} j\omega\underline{x} + \underline{f}, \tag{9.59}$$

where $\underline{Z}_{m0} = j\omega m + b + k/(j\omega)$, introduced in Appendix B, Section B.3, is the external mechanical impedance.

Depending on one's choice for the electrical through variable, the circuit constraint may be couched as a Thevenin or a Norton equivalent using either the M- or N-form transducer, respectively:

$$\text{M-form:} \quad \underline{v}_{ext} = \underline{Z}_{e0}\underline{i} + \underline{v} \tag{9.60}$$

and

$$\text{N-form:} \quad \underline{i}_{ext} = \underline{Y}_{e0}\underline{v} + \underline{i}. \tag{9.61}$$

Figures 9.20a and b represent these external mechanical and electrical constraints as electromechanical two-port networks.

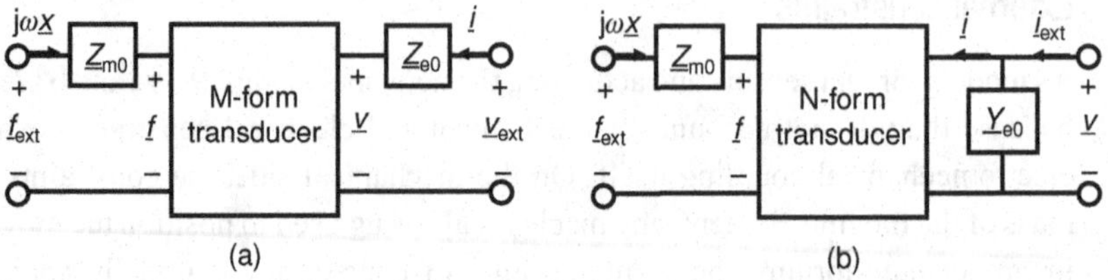

Figure 9.20 Linear two-port transducer networks with external mechanical and electrical constraints. The right-hand (mechanical) side is consistent with the direct mechanical ↔ electrical analogy. Refer to Eq. (9.59) for the definition of $\underline{Z}_{m0}$. (a) The constrained M-form electromechanical network. (b) The constrained N-form electromechanical two-port.

9.6.5 Linear two-port transducers with external constraint

To synthesize complete M- and N-form circuits incorporating all constraints, we only need to insert the two-ports from Figs. 9.19a and b into the networks of Figs. 9.20a and b, respectively. The new networks shown in Figs. 9.21a and b isolate the "ideal" two-port transducers and conveniently group the mechanical impedances on the left and the electrical impedances or admittances on the right.

Inspection of these networks leads to the following matrix relations:

$$\text{M-form:} \quad \begin{bmatrix} \underline{f}_{ext} \\ \underline{v}_{ext} \end{bmatrix} = \begin{bmatrix} \underline{Z}_{m0} + \underline{Z}_m & \underline{M} \\ -\underline{M}^* & \underline{Z}_{e0} + \underline{Z}_e \end{bmatrix} \begin{bmatrix} j\omega\underline{x} \\ \underline{i} \end{bmatrix} \tag{9.62}$$

and

$$\text{N - form:} \quad \begin{bmatrix} \underline{f}_{ext} \\ \underline{i}_{ext} \end{bmatrix} = \begin{bmatrix} \underline{Z}_{m0} + \underline{Z}'_m & \underline{N} \\ -\underline{N}^* & \underline{Y}_{e0} + \underline{Y}_e \end{bmatrix} \begin{bmatrix} j\omega\underline{x} \\ \underline{v} \end{bmatrix}. \tag{9.63}$$

Remember that $\underline{M}$ and $\underline{N}$ are, respectively, pure real and pure imaginary numbers. The canonical similarity of the linear relationships in Eqs. (9.62) and (9.63) highlights the advantage of the direct mechanical ↔ electrical analogy. By presenting the mechanical constraints in terms of simple circuit equations, one becomes more comfortable using voltage and current, respectively, as analogs to force and velocity.

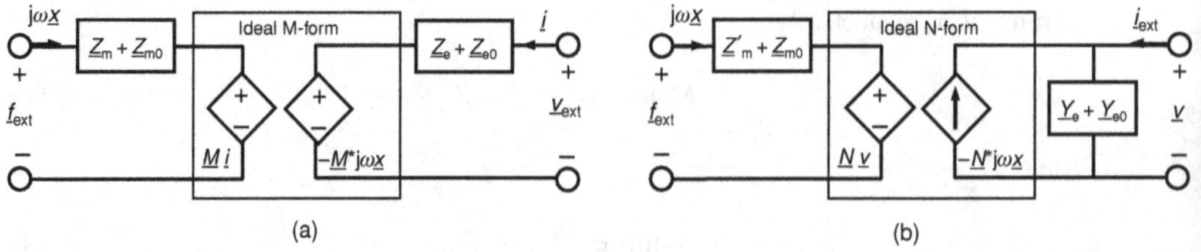

Figure 9.21 Linear two-port magnetic transducer network integrated into external mechanical and electrical constraints represented by the mechanical impedance $\underline{Z}_{m0}$ and either electrical impedance $\underline{Z}_{e0}$ or admittance $\underline{Y}_{e0}$. (a) The M-form network. (b) The N-form network.

9.6.6 Cascade forms

The remaining task is to derive cascade transmission matrix forms for the two-port networks of Figs. 9.21a and b. To do so, one employs simple modifications of the matrix two-port relations from Table 2.3 to express the mechanical variables in terms of the electrical variables.

9.6.6.1 Integrated M-form network

For the M-form network, the cascade form is

$$\begin{bmatrix} \underline{f}_{\text{ext}} \\ j\omega\underline{x} \end{bmatrix} = \underbrace{\begin{bmatrix} 1 & \underline{Z}_{\text{m}} + \underline{Z}_{\text{m0}} \\ 0 & 1 \end{bmatrix}}_{\substack{\text{series-connected} \\ \text{mechanical} \\ \text{impedance}}} \underbrace{\begin{bmatrix} \underline{M} & 0 \\ 0 & -1/\underline{M}^* \end{bmatrix}}_{\substack{\text{ideal M-form} \\ \text{transducer} \\ \text{(gyrator)}}} \underbrace{\begin{bmatrix} 1 & 0 \\ -(\underline{Z}_{\text{e}} + \underline{Z}_{\text{e0}}) & 1 \end{bmatrix}}_{\substack{\text{series-connected} \\ \text{electrical} \\ \text{impedances}}} \begin{bmatrix} \underline{i} \\ \underline{v}_{\text{ext}} \end{bmatrix}. \tag{9.64}$$

The *diagonal matrix* in the middle on the right-hand side of Eq. (9.64) is the *ideal M-form transducer*. It has the form of the gyrator element introduced in Section 2.3.5. With the external mechanical and electrical impedances now folded in, multiplications yield the transmission matrix in Neubert's standard form:

$$\begin{bmatrix} \underline{f}_{\text{ext}} \\ j\omega\underline{x} \end{bmatrix} = \frac{1}{\underline{M}^*} \begin{bmatrix} |\underline{M}|^2 + (\underline{Z}_{\text{m}} + \underline{Z}_{\text{m0}})(\underline{Z}_{\text{e}} + \underline{Z}_{\text{e0}}) & -(\underline{Z}_{\text{m}} + \underline{Z}_{\text{m0}}) \\ (\underline{Z}_{\text{e}} + \underline{Z}_{\text{e0}}) & -1 \end{bmatrix} \begin{bmatrix} \underline{i} \\ \underline{v}_{\text{ext}} \end{bmatrix}. \tag{9.65}$$

9.6.6.2 Integrated N-form network

Using the same approach, the matrix cascade for the N-form network becomes

$$\begin{bmatrix} \underline{f}_{\text{ext}} \\ j\omega\underline{x} \end{bmatrix} = \underbrace{\begin{bmatrix} 1 & \underline{Z}'_{\text{m}} + \underline{Z}_{\text{m0}} \\ 0 & 1 \end{bmatrix}}_{\substack{\text{series-connected} \\ \text{mechanical} \\ \text{impedance}}} \underbrace{\begin{bmatrix} \underline{N} & 0 \\ 0 & -1/\underline{N} \end{bmatrix}}_{\substack{\text{ideal N-form} \\ \text{transducer} \\ \text{(transformer)}}} \underbrace{\begin{bmatrix} 1 & 0 \\ -(\underline{Y}_{\text{e}} + \underline{Y}_{\text{e0}}) & 1 \end{bmatrix}}_{\substack{\text{shunt-connected} \\ \text{electrical} \\ \text{admittances}}} \begin{bmatrix} \underline{v} \\ \underline{i}_{\text{ext}} \end{bmatrix}. \tag{9.66}$$

The embedded diagonal matrix on the right-hand side is the ideal N-form transducer, taking the form of an ideal transformer element. Matrix multiplication yields

$$\begin{bmatrix} \underline{f}_{\text{ext}} \\ j\omega\underline{x} \end{bmatrix} = \frac{1}{\underline{N}^*} \begin{bmatrix} |\underline{N}|^2 + (\underline{Z}'_{\text{m}} + \underline{Z}_{\text{m0}})(\underline{Y}_{\text{e}} + \underline{Y}_{\text{e0}}) & -(\underline{Z}'_{\text{m}} + \underline{Z}_{\text{m0}}) \\ (\underline{Y}_{\text{e}} + \underline{Y}_{\text{e0}}) & -1 \end{bmatrix} \begin{bmatrix} \underline{v} \\ \underline{i}_{\text{ext}} \end{bmatrix}. \tag{9.67}$$

Equations (9.65) and (9.67) are canonically the same as Eqs. (4.47) and (4.48) for capacitive transducers. Only the coefficient definitions of Eqs. (9.51) and (9.55) distinguish them.

9.7 Two-port models for magnetic MEMS

Equations (9.65) and (9.67), or their two-port network equivalents, can be used to derive predictive models for the small-signal frequency-dependent behavior of magnetic

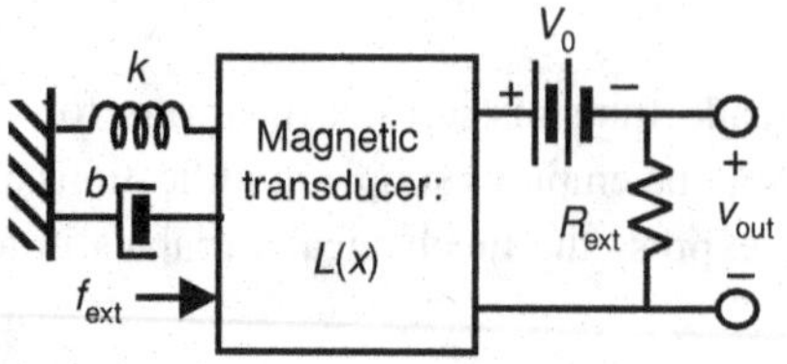

Figure 9.22 A generalized current-biased magnetic actuator. Current i_0 is controlled by the series combination of the external resistor R_{ext} and the winding resistance R_w. The mechanical constraints include mass m, a linear spring of spring constant k, mechanical damping b, and an externally applied force f_{ext}.

transducers and actuators. This section presents examples of such systems under various types of electrical and mechanical constraints.

9.7.1 Current-biased magnetic transducers

One way to achieve strong coupling interactions in magnetic MEMS transducers is to use DC bias current. To investigate the influence of bias, consider a transducer having inductance $L(x)$ and series winding resistance R_w, with a DC voltage V_0 connected to the device through an external resistor R_{ext}. See Fig. 9.22. R_{ext} provides the means to monitor current. The equilibrium (bias) current for this arrangement is $i_0 = V_0/R_0$, where $R_0 = R_{ext} + R_w$.

If the spring attached to the moving element is in its relaxed state at $x = x_k$ and if no zero-order external force acts on the transducer, that is, $f_{ext,0} = 0$, then mechanical equilibrium is

$$-k(x_0 - x_k) + \frac{i_0^2}{2}\frac{\mathrm{d}L}{\mathrm{d}x}\bigg|_{x_0} = 0. \tag{9.68}$$

Once $L(x)$ is specified, analytically or numerically, Eq. (9.68) may be solved for the equilibrium relationship between x_0 and i_0.

The net output voltage $v_{out}(t)$ monitored across R_{ext} includes a DC offset, $-R_{ext}V_0/(R_0)$, and the small-signal voltage, $v'_{out}(t) = -R_{ext}i'(t)$. Using phasor quantities, we turn to the M-form circuit, Fig. 9.21a, which is more convenient here. Accordingly, $\underline{Z}_{e0} = R_0$ and $\underline{v}_{ext} = 0$. From Eq. (9.65),

$$\underline{f}_{ext} = \frac{|\underline{M}|^2 + (\underline{Z}_m + \underline{Z}_{m0})(\underline{Z}_e + \underline{Z}_{e0})}{\underline{M}^*}\underline{i}. \tag{9.69}$$

Using the phasor expression for the small-signal current, $\underline{i} = -\underline{v}_{out}/R_{ext}$, the transfer function relating output voltage to input force is

$$\underline{H}_{v/f}(j\omega) = \frac{\underline{v}_{out}}{\underline{f}_{ext}} = \frac{-R_{ext}\underline{M}^*}{|\underline{M}|^2 + (\underline{Z}_m + \underline{Z}_{m0})(\underline{Z}_e + \underline{Z}_{e0})}. \tag{9.70}$$

By the convention adopted in Chapter 4, the subscript of $\underline{H}$ identifies output and input quantities. Thus, "*v/f*" means voltage output for force input. Using the expressions for mechanical impedance $\underline{Z}_\mathrm{m}$ and coupling coefficient $\underline{M}$ from Table 9.2, the transfer function reduces to the more readily interpretable form:

$$\underline{H}_{v/f}(\mathrm{j}\omega) = \left(\frac{R_\mathrm{ext}}{R_0}\right) \frac{-V_0 \dfrac{\mathrm{d}L}{\mathrm{d}x}}{\left(\dfrac{V_0}{R_0}\dfrac{\mathrm{d}L}{\mathrm{d}x}\right)^2 + [\mathrm{j}\omega m + b + k_\mathrm{eff}/(\mathrm{j}\omega)]\,(\mathrm{j}\omega L_0 + R_0)}, \tag{9.71}$$

where $k_\mathrm{eff} = k - \frac{1}{2}(V_0/R_0)^2 \mathrm{d}^2L/\mathrm{d}x^2$ is an effective mechanical spring constant. One can investigate the frequency dependence of $\underline{H}_{v/f}(\mathrm{j}\omega)$ systematically by contrasting the critical frequency of the LR circuit, $\omega_\mathrm{crit} = R_0/L_0$, with the intended operating frequency range, which is typically centered at the mechanical resonance: $\omega_0' = \sqrt{k_\mathrm{eff}/m}$.

9.7.1.1 Low-frequency limit: $\omega_0' \ll \omega_\mathrm{crit}$

If the mechanical resonance is well below ω_crit, an effective constant current constraint is enforced and the transfer function takes the form of a second-order resonator:

$$\underline{H}_{v/f}(\mathrm{j}\omega) \approx \frac{-\mathrm{j}\omega \dfrac{R_\mathrm{ext} V_0}{R_0^2 k_\mathrm{eff}} \dfrac{\mathrm{d}L}{\mathrm{d}x}}{1 - (\omega/\omega_0)^2 + \mathrm{j}2\xi_\mathrm{eff}(\omega/\omega_0')}, \quad \omega_0' \ll \omega_\mathrm{crit}. \tag{9.72}$$

When $\mathrm{d}^2L/\mathrm{d}x^2 > 0$, the magnetic coupling has the effect of decreasing k_eff, the stiffness of the system. If k_eff goes negative, the equilibrium becomes unstable. Such instability is considered in Section 9.8. The denominator of $\underline{H}_{v/f}(\mathrm{j}\omega)$ contains an effective dissipation ratio,

$$\zeta_\mathrm{eff} = b_\mathrm{eff}/(2\sqrt{m k_\mathrm{eff}}), \tag{9.73}$$

where

$$b_\mathrm{eff} = b + \frac{V_0^2}{R_0^3}\left(\frac{\mathrm{d}L}{\mathrm{d}x}\right)^2. \tag{9.74}$$

Equation (9.74) teaches that system damping is controlled through the bias voltage V_0. The physical origin of this effect is the energy dissipated via ohmic heating of the resistors induced by the motion.[7] Such adjustable damping, called *dynamic braking*, provides a capability to control second-order resonators. For example, in the transient response to a mechanical step input, the fastest approach to the new equilibrium *without overshoot* results when $\zeta_\mathrm{eff} = 1.0$. This is the well-known *critical damping* condition and, as long as the inherent mechanical damping is not already too high, it is attainable by adjusting voltage V_0. Details of under-damped and over-damped second-order systems are found in Section B.1.5 of Appendix B.

[7] Adjustable damping is virtually unrealizable in passive capacitive MEMS devices because of their very high resistances and very low currents.

9.7.1.2 High-frequency limit: $\omega_0'' \gg \omega_{\text{crit}}$

In the high-frequency limit, the transfer function reduces to

$$\underline{H}_{v/f}(j\omega) \approx \frac{-\dfrac{R_{\text{ext}}V_0}{R_0 k''_{\text{eff}} L_0}\dfrac{dL}{dx}}{1-(\omega/\omega_0'')^2 + j2\zeta''_{\text{eff}}(\omega/\omega_0'')}, \quad \omega_0'' \gg \omega_{\text{crit}}, \tag{9.75}$$

where $\omega_0'' = \sqrt{k''_{\text{eff}}/m}$ and $k''_{\text{eff}} = k - (V_0/R_0)^2[(d^2L/dx^2)/2 - (dL/dx)^2/L_0]$. This case corresponds to an effectively constant magnetic flux constraint. As before, the effective spring constant and thus the resonant frequency can be adjusted by changing the bias. On the other hand,

$$\zeta''_{\text{eff}} = b\Big/\left(2\sqrt{mk''_{\text{eff}}}\right). \tag{9.76}$$

Because the damping coefficient b is independent of bias voltage, adjustment of Q to achieve critical damping can be realized only through the rather weak dependence of k''_{eff} on V_0.

9.7.2 Variable-gap and variable-area transducers

One can gain physical insight about current-biased magnetic transducer behavior by separately considering the variable-gap and variable-area devices defined by Eqs. (9.21) and (9.24). Table 9.3, which collects effective spring constant and damping coefficient expressions for the low- and high-frequency limits, shows that variable-gap and variable-area devices behave quite differently. For example, when $\omega_0'' \ll \omega_{\text{crit}}$, magnetic coupling decreases the effective spring constant of the variable-gap transducer, but has no effect at all on the variable-area device. On the other hand, when $\omega_0'' \gg \omega_{\text{crit}}$, k_{eff} is fixed for the variable-gap device but increases for the variable-area structure.

Current bias increases the effective damping for both variable-gap and variable-area devices but only in the low-frequency limit, that is, $\omega_0' \ll \omega_{\text{crit}}$. The ability to adjust both the frequency and the damping of a mechanical resonator via DC voltage control would be a potentially important advance for MEMS technology.

9.7.3 A magnetic MEMS resonator

The device shown in Fig. 9.23a serves to model a practical magnetic MEMS resonator. Driven at its resonance, such mechanisms are used in gyroscopes and many other microdevices. To develop an electromechanical two-port, assume that the planar coil is DC current-biased by voltage V_0. The bias current is $i_0 = V_0/R_0$, where $R_0 = R_{\text{ext}} + R_w$, R_{ext} is an external resistor, and R_w is the winding resistance. The inductance $L(x)$, expected to have the general form plotted in Fig. 9.23b, satisfies the conditions $L(x \to \infty) \ll L(x = 0)$ and $dL/dx\,|_{x=0} = 0$. In a practical design, $L(x)$ would be

Table 9.3 Effective spring constant and damping coefficient for current-biased variable-gap and variable-area magnetic transducers. Equilibrium is assumed to exist at x_0 and $i_0 = V_0/R_0$. The effect of magnetic coupling on the effective spring constant for variable-gap and variable-area devices is similar to capacitive devices, but the effect on damping is unique to magnetic MEMS transducers

	Generic variable-gap transducer	Generic variable-area transducer
Inductance formula	$L(x) = \dfrac{L_0}{1 + x/x_m}$, so $\dfrac{dL}{dx} = -\dfrac{L_0/x_m}{(1 + x/x_m)^2}$, $\dfrac{d^2L}{dx^2} = \dfrac{2L_0/x_m^2}{(1 + x/x_m)^3}$	$L(x) \approx L_0(1 + x/x_m)$, so $\dfrac{dL}{dx} = \dfrac{L_0}{x_m}$, $\dfrac{d^2L}{dx^2} = 0$
Low frequency limit: $\omega_0' \ll \omega_{crit}$	Spring constant decreases when V_0 is increased (instability possible); damping is increased $k_{eff} = k - \left(\dfrac{V_0}{R_0}\right)^2 \left[\dfrac{L_0/x_m^2}{(1 + x_0/x_m)^3}\right]$, $b_{eff} = b + \dfrac{V_0^2}{R_0^3}\left[\dfrac{L_0^2/x_m^2}{(1 + x_0/x_m)^4}\right]$	Spring constant is unaffected by voltage; damping is increased $k_{eff} = k$, $b_{eff} = b + \dfrac{V_0^2}{R_0^3}\left[\dfrac{L_0^2}{x_m^2}\right]$
High frequency limit: $\omega_0'' \gg \omega_{crit}$	No effects on mechanics: $k_{eff}' = k$, $b_{eff}' = b$	Spring constant is increased but damping is unaffected: $k_{eff} = k + \left(\dfrac{V_0}{R_0}\right)^2 \left[\dfrac{L_0}{x_m^2}\right]$, $b_{eff}'' = b$

extracted from a finite-element computation but here we employ a simple empirical equation:

$$L(x) \approx L_r e^{-\alpha x^2}. \tag{9.77}$$

With no force exerted externally, that is, $f_{ext,0} = 0$, the mechanical equilibrium position x_0 is determined by balancing the spring force against the equilibrium component of the

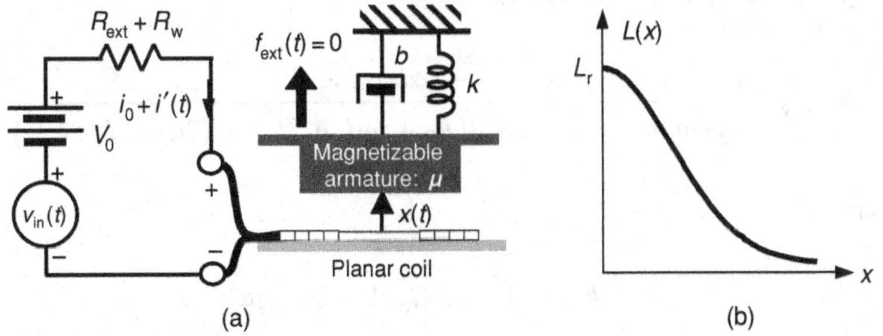

Figure 9.23 A current-biased magnetic transducer. (a) Schematic showing circuit connections of stationary planar coil patterned in the substrate below a moving, magnetically soft armature element. The armature has mass m, mechanical damping b, and a linear spring force of constant k. The spring exerts no force when $x = x_k$. (b) Plot of inductance L versus x using empirical Eq. (9.77).

force of electrical origin:

$$-k(x_0 - x_k) + \frac{i_0^2}{2}\frac{\mathrm{d}L}{\mathrm{d}x}\bigg|_{x_0} = 0. \tag{9.78}$$

Using Eq. (9.77),

$$k(x_0 - x_k) + \alpha L_r i_0^2 x_0 \mathrm{e}^{-\alpha x_0^2} = 0, \tag{9.79}$$

which can be solved numerically for x_0 versus i_0.

To investigate the small-signal dynamics, the M-form transducer matrix is a good choice. Assume the device is driven by a sinusoidal voltage source: $v_{\mathrm{in}}(t) = \mathrm{Re}[\underline{v}_{\mathrm{ext}} e^{\mathrm{j}\omega t}]$. The most convenient way to proceed is to use the condition $\underline{f}_{\mathrm{ext}} = 0$ in the matrix Eq. (9.62) to obtain a transfer function relating output velocity $\mathrm{j}\omega\underline{x}$ to input voltage $\underline{v}_{\mathrm{ext}}$. Solving for the velocity,

$$\underline{v}_{\mathrm{ext}} = -\left[\frac{(\underline{Z}_{\mathrm{m0}} + \underline{Z}_{\mathrm{m}})(\underline{Z}_{\mathrm{e0}} + \underline{Z}_{\mathrm{e}})}{\underline{M}} + \underline{M}^*\right]\mathrm{j}\omega\underline{x}. \tag{9.80}$$

So the relevant transfer function is

$$\underline{H}_{\mathrm{j}\omega x/v} = \frac{\mathrm{j}\omega\underline{x}}{\underline{v}_{\mathrm{ext}}} = \frac{-\underline{M}}{(\underline{Z}_{\mathrm{m0}} + \underline{Z}_{\mathrm{m}})(\underline{Z}_{\mathrm{e0}} + \underline{Z}_{\mathrm{e}}) + |\underline{M}|^2}. \tag{9.81}$$

Making some substitutions yields

$$\underline{H}_{\mathrm{j}\omega x/v} = \frac{i_0 \mathrm{d}L/\mathrm{d}x}{[\mathrm{j}\omega m + b + k_{\mathrm{eff}}/(\mathrm{j}\omega)](R_0 + \mathrm{j}\omega L(x_0)) + (i_0 \mathrm{d}L/\mathrm{d}x)^2}. \tag{9.82}$$

Using Eq. (9.77) for inductance $L(x)$ gives

$$\underline{H}_{\mathrm{j}\omega x/v} = \frac{-2\alpha i_0 x_0 L_r \mathrm{e}^{-\alpha x_0^2}}{[\mathrm{j}\omega m + b + k_{\mathrm{eff}}/(\mathrm{j}\omega)]\left(R_0 + \mathrm{j}\omega L_r \mathrm{e}^{-\alpha x_0^2}\right) + \left(2\alpha i_0 x_0 L_r \mathrm{e}^{-\alpha x_0^2}\right)^2}, \tag{9.83}$$

where

$$k_{\mathrm{eff}} = k - \alpha L_r i_0^2 \left(1 - 2\alpha x_0^2\right)\mathrm{e}^{-\alpha x_0^2}. \tag{9.84}$$

The effect of the magnetic coupling on k_{eff} and thus the resonant frequency diminishes rapidly as the equilibrium location of the magnetic armature moves away from the coil. Because this device is current-biased, the low- and high-frequency limits introduced in Section 9.7.1 can again be used to explore the electromechanical behavior. This exercise is left to the reader.

Example 9.6 **Dynamics of torsional sensor**

We can use the small-signal formalism to study the response to spurious mechanical excitations of the field meter already presented in Example 9.4. Consider the DC-biased system shown in Fig. 9.24.

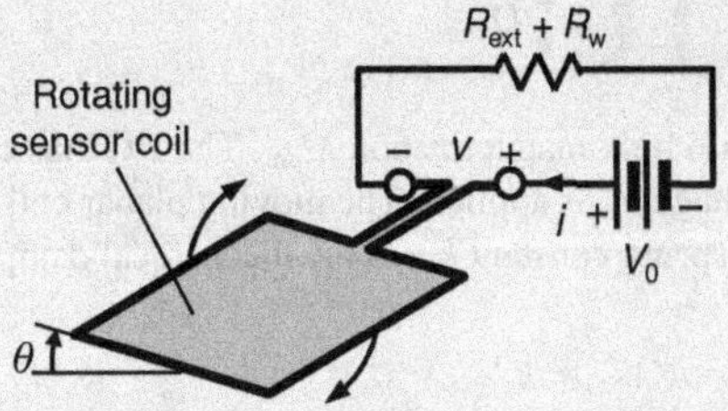

Figure 9.24 DC bias arrangement of torsional magnetic field meter subject to spurious torque $\tau'(t)$. Neither the mechanical torsion spring nor the piezoresistive element that detects the deflection is shown.

Starting with the rotational transducer matrix found in Table 9.2 and then incorporating the external constraints gives

$$\begin{bmatrix} \underline{\tau}_{ext} \\ \underline{v}_{ext} \end{bmatrix} = \begin{bmatrix} \underline{Z}_m + \underline{Z}_{m0} & \underline{M} \\ -\underline{M}^* & \underline{Z}_e + \underline{Z}_{e0} \end{bmatrix} \begin{bmatrix} j\omega\underline{\theta} \\ \underline{i} \end{bmatrix},$$

where $\underline{Z}_{m0} = j\omega I + b_\theta + k_\theta/(j\omega)$, I is the coil's moment of inertia, b_θ is the torsional damping coefficient, and k_θ is the torsional spring constant. Using the coenergy defined in Example 9.4, $\underline{Z}_m \approx 0$, $\underline{M} = BA$, and $\underline{Z}_e = j\omega L_0$. Setting $\underline{v}_{ext} = 0$ and manipulating the above matrix equation yields a transfer function relating the coil response to an external torque input, which might represent vibrational noise;

$$\underline{H}_{j\omega\theta/\tau} = \frac{j\omega\underline{\theta}}{\underline{\tau}_{ext}} = \frac{(j\omega L_0 + R_{ext} + R_w)}{[j\omega I + b_\theta + k_\theta/(j\omega)](j\omega L_0 + R_{ext} + R_w) + (BA)^2}.$$

Typically, $\omega_{LR} \ll \omega_0$, where $\omega_{LR} = (R_{ext} + R_w)/L_0$ and $\omega_0 = (k_\theta/I)^{1/2}$. Then, low and high frequency limits of $\underline{H}_{j\omega\theta/\tau}$ can be specified:

$$\underline{H}_{j\omega\theta/\tau} = \begin{cases} \dfrac{1}{j\omega I + [b_\theta + (BA)^2/(R_{ext} + R_w)] + k_\theta/(j\omega)}, & \omega \ll \omega_{LR}, \\ \dfrac{1}{j\omega I + b_\theta + [k_\theta + (BA)^2/L_0]/(j\omega)}, & \omega \gg \omega_{LR}. \end{cases}$$

For mechanical disturbances at low frequency, magnetically coupled dynamic braking is evident while, for high frequency, the coupling increases the effective spring constant, thus raising the apparent resonant frequency.

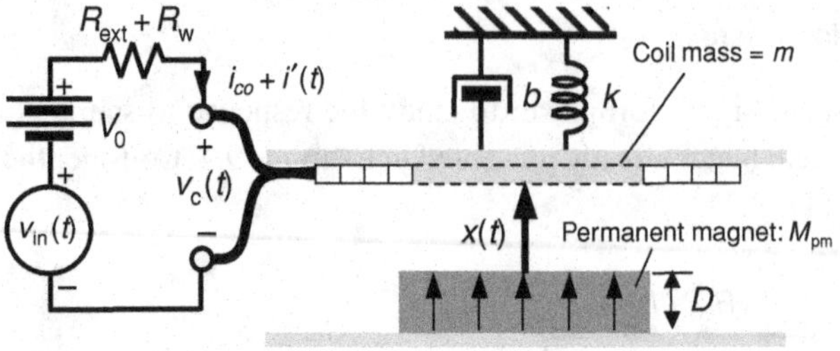

Figure 9.25 Permanent magnet actuator with bulk magnetization M_{pm}. The DC bias current is $i_{co} = V_0/(R_w + R_{ext})$ and a small AC voltage v_{in} is applied. The moving planar coil of mass m is linked to a linear restoring spring having spring constant k and the mechanical damping is b.

9.7.4 A permanent magnet actuator

The small-signal behavior of permanent magnet-based transducers can be described readily using the M-form matrix. Consider the actuator shown in Fig. 9.25, which resembles a geometry considered in Section 9.5. As always, we start by defining the equilibrium. Assume that the mechanical restoring spring is relaxed at $x = x_k$ and, further, that external DC voltage V_0 establishes the bias current $i_{co} = V_0/(R_w + R_{ext})$ in the moving coil. Using Eq. (9.46) for the force of electrical origin to write the equation for the mechanical equilibrium gives

$$k(x_0 - x_k) + DM_{pm}\frac{dL_m}{dx}i_{co} \approx 0. \tag{9.85}$$

With knowledge of $L_m(x)$, this equation can be solved for x_0 in terms of i_{co}.

The best way to develop a small-signal model is to start with the coenergy. Using Eq. (9.44) and ignoring self-inductance terms, which will be small,

$$W'_m \approx L_m(x)i_{pm}i_c = L_m(x)DM_{pm}i_c. \tag{9.86}$$

Recall that $i_{pm} = DM_{pm}$ is the Amperian current already introduced in Section 9.5. Knowing W'_m, it is easy to formulate the desired two-port, M-form transducer matrix:

$$\begin{bmatrix} \underline{f}_{ext} \\ \underline{v}_{ext} \end{bmatrix} = \begin{bmatrix} \underline{Z}_m + \underline{Z}_{m0} & \underline{M} \\ -\underline{M}^* & \underline{Z}_e + \underline{Z}_{e0} \end{bmatrix} \begin{bmatrix} j\omega\underline{x} \\ \underline{i} \end{bmatrix}, \tag{9.87}$$

where $\underline{Z}_{m0} = j\omega m + b + k/(j\omega)$, $\underline{Z}_{e0} = R_{ext} + R_w$, and

$$\begin{aligned} \underline{Z}_m &= -\frac{DM_{pm}i_{co}}{j\omega}\left.\frac{d^2L_m}{dx^2}\right|_{x_0}, \\ \underline{M} &= -DM_{pm}\left.\frac{dL_m}{dx}\right|_{x_0}, \\ \underline{Z}_e &= j\omega L_c(x_0). \end{aligned} \tag{9.88}$$

The sign of the coupling coefficient $\underline{M}(x_0)$ is determined by the sign of the permanent magnetization $M_{\rm pm}$. Letting $\underline{v}_{\rm ext} = \underline{x}_{\rm in}$ and $\underline{f}_{\rm ext} = 0$, and then rearranging gives

$$\underline{H}_{{\rm j}\omega x/v} \equiv \frac{{\rm j}\omega\underline{x}}{\underline{v}_{\rm in}} = \frac{-DM_{\rm pm}{\rm d}L_{\rm m}/{\rm d}x}{[{\rm j}\omega m + b + k_{\rm eff}/({\rm j}\omega)]\,[R_0 + {\rm j}\omega L_{\rm c}(x_0)] + \left[DM_{\rm pm}{\rm d}L_{\rm m}/{\rm d}x\right]^2}, \tag{9.89}$$

where the effective spring constant is

$$k_{\rm eff} = k - \frac{DM_{\rm pm}i_{\rm co}}{2}\frac{{\rm d}^2L_{\rm m}}{{\rm d}x^2}. \tag{9.90}$$

$\underline{H}_{{\rm j}\omega x/v}$ has almost the same form as Eq. (9.82), the transfer function for the magnetically coupled resonator. The only difference is that the electromechanical coupling mechanism is via mutual inductance rather than self-inductance. Providing a voltage source adds to system cost, but also provides functionality to tune the resonant frequency *upward* or *downward*.

9.8 Stability of magnetic transducers

Depending on the electrical constraints, some magnetic transducer configurations exhibit electromechanical instability. To avoid such undesirable behavior, means must exist to predict the conditions that lead to it. System stability is always defined with respect to a mechanical equilibrium. For a stable second-order system, a small perturbation produces oscillatory motion that decays in time. On the other hand, an unstable system responds to the same perturbation by unbounded growth. If this growth is purely exponential, then it is called *absolute instability*.[8] The material non-linearity of a magnetic MEMS device may affect the time-dependent character of this growth, but usually does not influence either the threshold or initially exponential evolution of instability.

9.8.1 Use of small-signal analysis

Because it is always defined with respect to an equilibrium, instability can be analyzed using the small-signal analysis method introduced in Section 9.6. Consider a linear magnetic actuator with one degree of freedom and inductance $L(x)$. The starting point is Eq. (9.59), the small-signal, electromechanically coupled equation of motion. Using the matrix expressions from Eq. (9.50) and (9.54), one obtains, respectively,

$$\text{M-form:}\quad \underline{f}_{\rm ext} = (sm + b + k/s + \underline{Z}_{\rm m})\,s\underline{x} + \underline{M}\,\underline{i} \tag{9.91}$$

and

$$\text{N-form:}\quad \underline{f}_{\rm ext} = \left(sm + b + k/s + Z'_{\rm m}\right)s\underline{x} + \underline{N}\,\underline{v}. \tag{9.92}$$

[8] Other forms of instability, such as *overstability*, are not considered here.

For convenience in identifying marginal stability conditions, we have substituted $j\omega \rightarrow s$, where s is complex frequency. The assumed time dependence of perturbations from equilibrium then becomes e^{st} and the condition for instability is $\text{Re}[s] > 0$.

A simple way to expose the stability condition for an electromechanical system is to define the transfer function relating displacement $\underline{x}$ to external force $\underline{f}_{\text{ext}}$ with the constraint of no external electrical input. For the M- and N-forms given, this means $\underline{i} = 0$ and $\underline{v} = 0$, respectively, in which case Eqs. (9.91) and (9.92) collapse to a common form for the desired transfer function:

$$\underline{H}_{x/f}(s) \equiv \frac{\underline{x}}{\underline{f}_{\text{ext}}} = \frac{1}{ms^2 + bs + k_{\text{eff}}}. \tag{9.93}$$

The effective spring constant k_{eff}, which takes different forms depending on the electrical constraint, is the key to identifying the stability threshold. For constant current ($\underline{i} = 0$):

$$\begin{aligned} k_{\text{eff}} &= k_i = k - \partial^2 W'_{\text{m}}/\partial x^2 \\ &= k - \frac{i_0^2}{2}\frac{\mathrm{d}^2 L}{\mathrm{d}x^2} \end{aligned} \tag{9.94}$$

and for constant flux ($\underline{v} = 0$):

$$\begin{aligned} k_{\text{eff}} &= k_\lambda = k + \partial^2 W_{\text{m}}/\partial x^2 \\ &= k + \frac{\lambda_0^2}{2L^2}\left[\frac{2}{L}\left(\frac{\mathrm{d}L}{\mathrm{d}x}\right)^2 - \frac{\mathrm{d}^2 L}{\mathrm{d}x^2}\right]. \end{aligned} \tag{9.95}$$

9.8.2 Constant current and constant flux limits

The denominator of $\underline{H}_{x/f}(s)$ is quadratic in s and the roots of this quadratic are the natural frequencies or poles of the system. These poles, s_1 and s_2, determine the time-dependent evolution of any initial perturbation:

$$s_{1,2} = -\frac{b}{2m} \pm \sqrt{\frac{b^2}{4m^2} - \frac{k_{\text{eff}}}{m}}. \tag{9.96}$$

As long as $k_{\text{eff}} > 0$, then $\text{Re}(s_1) < 0$ and $\text{Re}(s_2) < 0$, signifying stability. On the other hand, if $k_{\text{eff}} < 0$, then $\text{Re}(s_1) > 0$, in which case, an initial perturbation will grow exponentially as $e^{s_1 t}$. The marginal condition separating stable and unstable behavior is $k_{\text{eff}} = 0$.

9.8.3 General stability criteria

General stability conditions for the inductance function $L(x)$ emerge from examination of the effective spring constant expressions of Eqs. (9.94) and (9.95). *Sufficient conditions* for stability are

$$\text{constant current:} \quad \mathrm{d}^2 L/\mathrm{d}x^2 \leq 0 \tag{9.97}$$

Table 9.4 Stability of generic variable-gap and variable-area magnetic transducers defined by Eq. (9.99) under constant current (i_0) and (short-circuit) constant flux (λ_0) constraints. Equilibrium is assumed to exist at $x = x_0$. Only with great care should these results be applied to other transducer geometries

	Generic variable-gap inductor	Generic variable-area inductor
Electrical constraint	$L(x) = L_0/(1+x/x_m)$, so $\frac{d^2L}{dx^2} > 0$	$L(x) = L_0(1+x/x_m)$, so $\frac{d^2L}{dx^2} = 0$
Constant current: $i = i_0$, $\underline{i} = 0$	$k_{eff} = k - L_0 i_0^2/[x_m(1+x_0/x_m)^3]$, pull-in instability if $\lvert i_0\rvert > \sqrt{k x_m (1+x_0/x_m)^3/L_0}$	$k_{eff} = k$, always stable no effect on resonant frequency
Constant flux: $\lambda = \lambda_0$, $\underline{v} = 0$	$k_{eff} = k$, always stable; no effect on resonant frequency	$k_{eff} = k + \lambda_0^2/\left[L_0 x_m^2 (1+x_0/x_m)^3\right]$, always stable; magnetic coupling increases resonant frequency

and

$$\text{constant flux:} \quad \frac{d^2L}{dx^2} \le \frac{2}{L_0}\left(\frac{dL}{dx}\right)^2. \tag{9.98}$$

These conditions, dependent only on the device geometry, are similar to those identified for capacitance in Section 4.8.2.

9.8.4 Variable-gap and variable-area devices

To explore instability conditions further, consider the generic variable-gap and variable-area models previously introduced in Sections 9.3.1 and 9.3.2:

$$L(x) = \begin{cases} L_0/(1+x/x_m), & \text{variable gap,} \\ L_0(1+x/x_m), & \text{variable area.} \end{cases} \tag{9.99}$$

Table 9.4 summarizes the results for these generic expressions. As predicted, only the variable-gap device can be unstable, and then only if the current exceeds a threshold. Variable-area inductive transducers never exhibit instability. The criteria here, while instructive, are far from inclusive or general. For all but clearly identifiable variable-gap and variable-area geometries, it is best to investigate stability by starting with the small-signal formulation, deriving an effective spring constant, and then looking for the instability threshold: $k_{eff} = 0$.

9.9 Magnetic MEMS sensors

Macroscale magnetic sensors have been in use for many years, but the considerable knowledge and experience gained with this technology have not yet migrated down to microscale devices. When it does, fully integrated signal amplification and conditioning will be needed to exploit the capabilities of magnetic MEMS. Modern power

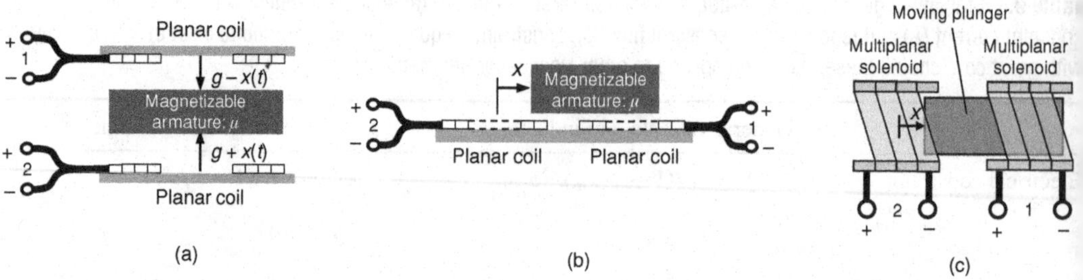

Figure 9.26 Conceptual geometries of three-terminal inductor designs for current half-bridge sensors, all featuring identical windings and a moving armature of magnetically soft material. At mechanical equilibrium, $\mathrm{d}L_1/\mathrm{d}x = -\mathrm{d}L_2/\mathrm{d}x$. (a) Coaxial planar coils with magnetizable armature between them forming a variable-gap device. (b) Side-by-side planar coils forming a variable-area device. (c) Coaxially mounted flat solenoidal coils with moving plunger.

electronics may play a role here. This section introduces circuit topologies for two important three-terminal magnetic sensors: the *current bridge* and the *linear variable differential transformer*. These schemes use DC and AC bias current, respectively. Emphasis is placed on general principles; details of the electronics are not addressed. Readers with knowledge of electronics will recognize that the moderate impedances and currents characterizing magnetic sensors assure compatibility with common IC circuits.

The more promising applications for micromechanical magnetic devices may be as actuators because of their strong forces and large ranges of motion achievable by high-aspect-ratio geometries. There is, however, serious interest in magnetic sensors, too. Many possible applications exist, one example being their combination with actuators to create a new generation of smart electromechanical devices based on feedback control.

9.9.1 DC biased current-bridge sensor

The current-bridge sensor is in many respects similar to the capacitive half-bridge covered in Section 5.4. Figure 9.26 illustrates several magnetic sensor geometries for which the current-bridge topology is well-suited. MEMS realizations for these geometries must be high-aspect-ratio structures fabricated with ferromagnetic materials and featuring pairs of symmetrical windings. A magnetically soft moving armature provides the essential inductance variation; when the armature is displaced from its equilibrium, one inductance increases and the other decreases. To achieve good linearity, the two inductors must be antisymmetric in x, that is, $L_1(x) = L_2(-x)$, as shown in Fig. 9.27. If we assume equilibrium at $x = 0$, then $L_1(x=0) = L_2(x=0) \equiv L_0$ and $\mathrm{d}L_1/\mathrm{d}x|_{x=0} \approx -\mathrm{d}L_2/\mathrm{d}x|_{x=0}$. It is this latter condition that suppresses non-linearity effects. The Taylor series expansion of the inductance function is

$$L_{1,2}(x') = L_0 \pm \frac{\mathrm{d}L}{\mathrm{d}x}x' + \frac{1}{2}\frac{\mathrm{d}^2L}{\mathrm{d}x^2}(x')^2 + \ldots, \tag{9.100}$$

where $L(x) \equiv L_1(x) = L_2(-x)$ and $x'(t)$ is the perturbation displacement. Mutual inductance is ignored in the model.

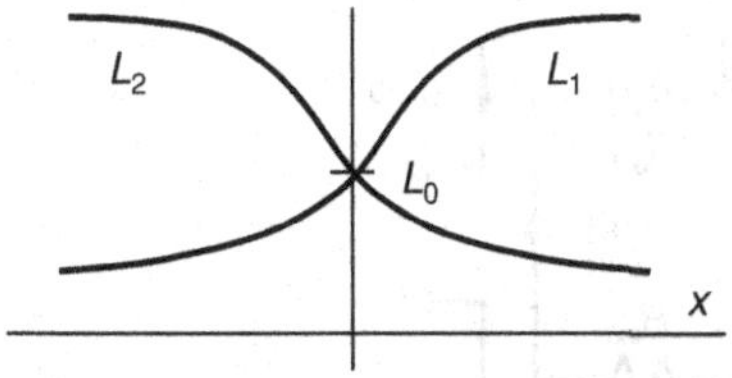

Figure 9.27 Antisymmetric spatial dependences of two matched inductive sensors, $L_1 = L(x)$ and $L_2 = L(-x)$, where $x = 0$ is the assumed equilibrium.

Current-bridge sensors are somewhat similar to the three-plate capacitive devices introduced in Section 4.7.5. Both employ DC bias – current and voltage for inductive and capacitive sensors, respectively. Furthermore, both derive inherent bidirectional sensitivity and good response linearity from paired antisymmetric sensing elements. The difference is that, while the capacitive half-bridge shuttles charge between two series-connected capacitors, the current bridge divides the current between shunt-connected inductors, L_1 and L_2. This current redistribution is reflected in antisymmetric voltage changes across series resistors.[9]

The basic topology for the current bridge is provided in Fig. 9.28. Identical resistors in series with L_1 and L_2 divide the equilibrium current equally between the two windings. The net currents flowing in the two branches are $i_1(t) = I_0/2 + i_1'(t)$ and $i_2(t) = I_0/2 + i_2'(t)$. KCL guarantees that $i_1' = -i_2'$. This perturbation current, effectively circulating around the loop formed by the inductors, generates antisymmetric voltage perturbations across the paired external resistors R_{ext}. A high-impedance *instrumentation amplifier* detects the signal. As shown, the winding resistances, R_{w}, inevitable and virtually never negligible in a MEMS-scale inductive sensor, reduces the output voltage by a resistive voltage divider term.

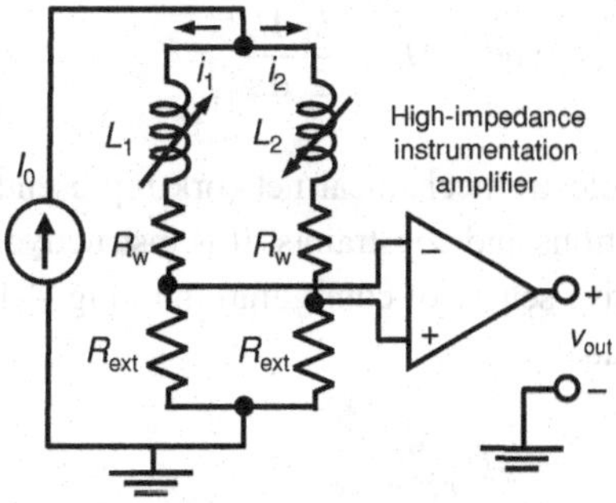

Figure 9.28 Circuit topology of current-bridge sensor using a three-terminal inductive element with constant DC current bias I_0. Balancing the resistors assures that the current is divided equally between the two inductors. This differential circuit avoids DC offset and employs a high-impedance instrumentation amplifier for signal amplification.

9 Resistors are not required in the capacitive half-bridge because the charge redistribution leads directly to a detectable voltage.

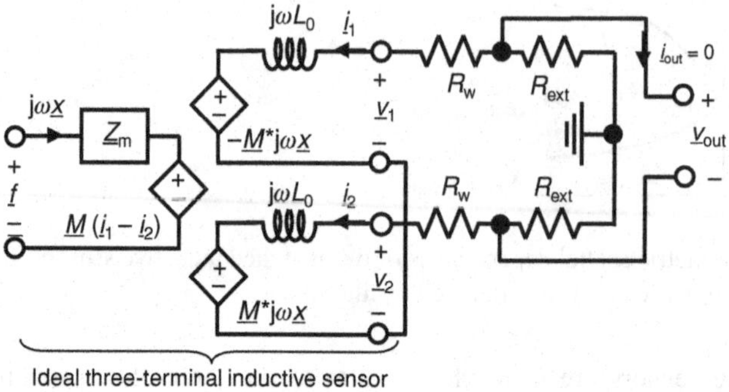

Figure 9.29 Equivalent electromechanical network for the current-bridge sensor shown in Fig. 9.28 and modeled by the matrix in Eq. (9.101). The antisymmetry of the ganged inductors, L_1 and L_2, makes further simplification possible.

The M-form electromechanical matrix is convenient here because of the current constraint. Thus, the coenergy $W'_{\mathrm{m}}(x, i_1, i_2) = \frac{1}{2}L_1(x)i_1^2 + \frac{1}{2}L_2(x)i_2^2$ is used. With x, i_1, and i_2 as state variables, the general matrix representation for the electromechanical three-port becomes

$$\begin{bmatrix} \underline{f} \\ \underline{v}_1 \\ \underline{v}_2 \end{bmatrix} = \begin{bmatrix} \underline{Z}_{\mathrm{m}} & \underline{M} & -\underline{M} \\ -\underline{M}^* & \mathrm{j}\omega L_0 & 0 \\ \underline{M}^* & 0 & \mathrm{j}\omega L_0 \end{bmatrix} \begin{bmatrix} \mathrm{j}\omega \underline{x} \\ \underline{i}_1 \\ \underline{i}_2 \end{bmatrix}. \tag{9.101}$$

The matrix takes this form as long as (i) the DC bias current divides itself equally between the two windings and (ii) the inductances are antisymmetric. The coefficients are obtained from straightforward multiport generalizations of Table 9.2:

$$\underline{Z}_{\mathrm{m}} = \frac{-1}{\mathrm{j}\omega}\frac{I_0^2}{4}\frac{\mathrm{d}^2L}{\mathrm{d}x^2} \quad \text{and} \quad \underline{M} = \frac{I_0}{2}\frac{\mathrm{d}L}{\mathrm{d}x}\bigg|_{x_0}. \tag{9.102}$$

Figure 9.29 introduces an equivalent electromechanical network representation incorporating the external electrical connections and constraints. It is instructive to compare this circuit with the three-plate capacitive sensor of configuration in Fig. 4.17.

Simplification is possible here because

$$\underline{i}_2 = -\underline{i}_1. \tag{9.103}$$

Then, incorporating the mechanical constraints, the network reduces to the two-port revealed in Fig. 9.30. In this figure, $\underline{f}_{\mathrm{ext}}$ and $\underline{Z}_{\mathrm{m0}}$ are the externally applied force and mechanical impedance, respectively. Inspection of Fig. 9.30 yields a convenient 2×2 electromechanical matrix:

$$\begin{bmatrix} \underline{f}_{\mathrm{ext}} \\ \underline{v}_2 - \underline{v}_1 \end{bmatrix} = \begin{bmatrix} \underline{Z}_{\mathrm{m}} + \underline{Z}_{\mathrm{m0}} & -2\underline{M} \\ 2\underline{M}^* & 2\mathrm{j}\omega L_0 \end{bmatrix} \begin{bmatrix} \mathrm{j}\omega \underline{x} \\ \underline{i} \end{bmatrix}, \tag{9.104}$$

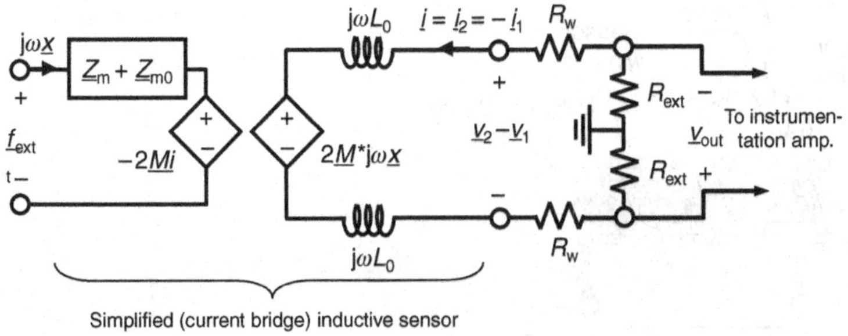

Figure 9.30 Simplified (two-port) transducer model for the current-bridge sensor of Fig. 9.28. The bipolar differential output voltage connects directly to an instrumentation amplifier, which draws virtually no current.

where $\underline{i} = \underline{i}_2 = -\underline{i}_1$ and, for convenience, the differential voltage $(\underline{v}_2 - \underline{v}_1)$ has been introduced. The external constraint imposed by the net resistance upon the small-signal AC circuit is

$$\underline{v}_2 - \underline{v}_1 = -2\,(R_{\text{ext}} + R_{\text{w}})\,\underline{i}. \tag{9.105}$$

The winding resistance attenuates the output voltage, so

$$\underline{v}_{\text{out}} = \left(\frac{R_{\text{ext}}}{R_{\text{ext}} + R_{\text{w}}}\right)(\underline{v}_1 - \underline{v}_2). \tag{9.106}$$

Using Eqs. (9.105) and (9.106) in Eq. (9.104) yields a transfer function relating the output voltage to sensor velocity:

$$\underline{H}_{v/\text{j}\omega x}(\text{j}\omega) \equiv \frac{\underline{v}_{\text{out}}}{\text{j}\omega\underline{x}} = \frac{2R_{\text{ext}}\underline{M}^*}{\text{j}\omega L_0 + R_{\text{ext}} + R_{\text{w}}}. \tag{9.107}$$

This result, easily confirmed by direct examination of Fig. 9.30, reveals that the winding resistance R_{w} reduces the output signal and raises the frequency of the dominant system pole. Depending on whether the sensor is to detect displacement or velocity, this frequency shift may or may not be a problem.

9.9.2 Linear variable differential transformer sensor

Figure 9.31 illustrates a simple linear variable differential transformer (LVDT). This configuration, common in macroscale magnetic sensors, can be implemented in MEMS. The middle coil is excited by an AC signal and the differential voltage across the series connection of the two sensing coils is monitored. The operational principle is simple. When the magnetically soft armature is displaced from $x = 0$, the mutual coupling between the central coil and the sensing coils becomes unbalanced, leading to a difference signal $v_{\text{out}}(t)$.

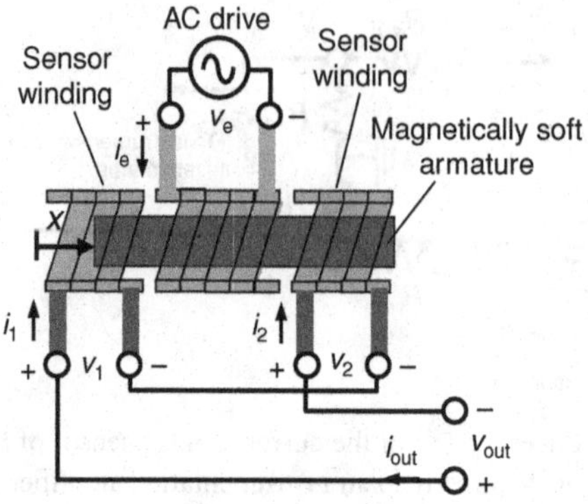

Figure 9.31 Conceptual geometry for a linear variable differential transformer (LVDT) showing electrical connections. All coils are stationary, and the middle coil is excited by AC. Voltages induced in the paired outer coils 1 and 2 change their relative magnitudes as the armature moves back and forth. The differential output can be connected directly to a high-impedance instrumentation amplifier.

The linear electrical terminal relation for this device is

$$\begin{bmatrix} \lambda_1 \\ \lambda_2 \\ \lambda_e \end{bmatrix} = \begin{bmatrix} L_1(x) & 0 & L_{m1}(x) \\ 0 & L_2(x) & L_{m2}(x) \\ L_{m1}(x) & L_{m2}(x) & L_e \end{bmatrix} \begin{bmatrix} i_1 \\ i_2 \\ i_e \end{bmatrix}. \tag{9.108}$$

Equation (9.108) ignores magnetic coupling between the two sensor windings and, further, assumes that the self-inductance of the drive coil L_e does not depend on x. To meet the antisymmetry requirement, $dL_1/dx = -dL_2/dx$. More importantly, $L_{m1}(x) = L_{m2}(-x)$, as shown in Fig. 9.32, so that $dL_{m1}/dx = -dL_{m2}/dx$. This property of the mutual inductance guarantees that $v_1(t) = -v_2(t)$ and is the basis of operation of the linear variable differential transformer.

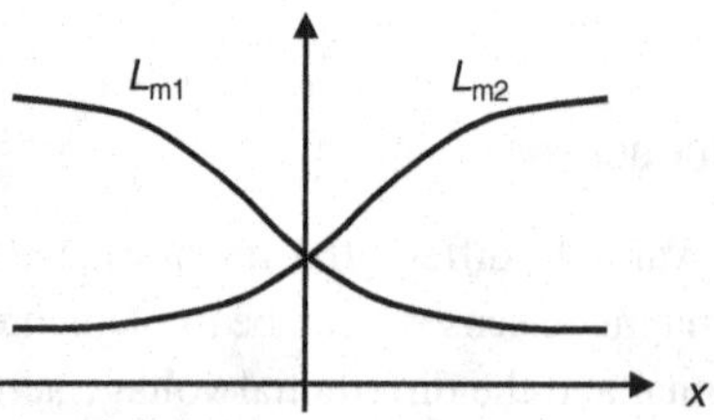

Figure 9.32 The antisymmetric x dependence of mutual inductances L_{m1} and L_{m2} for a linear variable differential transformer (LVDT) type of magnetic sensor, as exemplified by Fig. 9.31. Compare the x-dependences of these mutual inductances with the self-inductances for the current bridge illustrated in Fig. 9.27.

The AC bias means that the small-signal electromechanical transducer formalism cannot be used here. A new linearization is required. To first order in $x(t)$, the time-dependent sensor winding voltages are

$$v_1 = \frac{d\lambda_1}{dt} = L_1\frac{di_1}{dt} + i_1\frac{dL_1}{dx}\frac{dx}{dt} + \left(L_{m1} + \frac{dL_{m1}}{dx}x\right)\frac{di_e}{dt} + i_e\frac{dL_{m1}}{dx}\frac{dx}{dt},$$
$$v_2 = \frac{d\lambda_2}{dt} = L_2\frac{di_2}{dt} + i_2\frac{dL_2}{dx}\frac{dx}{dt} + \left(L_{m2} + \frac{dL_{m2}}{dx}x\right)\frac{di_e}{dt} + i_e\frac{dL_{m2}}{dx}\frac{dx}{dt}. \quad (9.109)$$

Defining $L_0 \equiv L_1(x=0) = L_2(x=0)$ and $L_m(x) \equiv L_{m1}(x) = L_{m2}(-x)$, and recognizing $i \equiv i_1 = -i_2$, the output voltage is

$$v_{out} = v_1 - v_2 = \underbrace{2L_0\frac{di}{dt}}_{\text{self term}} + \underbrace{2\frac{dL_m}{dx}\left[x\frac{di_e}{dt} + i_e\frac{dx}{dt}\right]}_{\text{mutual coupling term}}. \quad (9.110)$$

If high-impedance amplification is used to detect v_{out}, the sensor coil current i will be so small that the self-inductance term in Eq. (9.110) can be neglected. Then, after regrouping, the output voltage is

$$v_{out}(t) \approx 2\frac{dL_m}{dx}\frac{d[xi_e]}{dt}. \quad (9.111)$$

As expected, this signal is proportional to the spatial derivative of mutual inductance; however, the dependence on the displacement involves the time derivative of the $x(t)i_e(t)$ product. Assuming sinusoidal current excitation,

$$i_e(t) = I_e \cos(\omega_e t), \quad (9.112)$$

the output voltage is

$$v_{out}(t) \approx 2I_e\frac{dL_m}{dx}\left[\cos(\omega_e t)\frac{dx}{dt} - \omega_e \sin(\omega_e t)\, x\right]. \quad (9.113)$$

This equation reveals that the LVDT sensor is actually a modulation scheme with the applied current $i_e(t)$ serving as a carrier signal. Good replication of the detected signal requires that the carrier frequency ω_e be much larger than the highest expected frequency component of $x(t)$. If this condition is maintained, then the first term on the right-hand side of Eq. (9.113) can be ignored and the output voltage $v_{out}(t)$ reduces to a double-sideband suppressed carrier amplitude-modulated signal:

$$v_{out} \approx -2\omega_e I_e\frac{dL_m}{dx}\sin(\omega_e t)\, x(t). \quad (9.114)$$

This signal has the same canonical form as Eq. (5.65), the output voltage of the AC-excited, capacitive half-bridge. Synchronous demodulation, described in Section 5.5.6, can be used to extract a signal replicating $x(t)$.

The LVDT scheme is not amenable to the electromechanical transducer matrix formalism. Instead, Eq. (9.114) has been derived from a linearization based on Faraday's law. While this approach provides a relationship between output voltage $v_{out}(t)$ and displacement $x(t)$, it yields no information about the actual electromechanics of the system. For example, if the input signal is actually an external mechanical force, then a linearization of the dynamic equation of motion is needed to relate $x(t)$ to this force. This exercise leads to a linear differential equation with a time-varying coefficient. If the carrier frequency is sufficiently high, then simplification is possible.

9.10 Summary

This chapter, following the general outline of Chapters 3 through 5 for capacitive MEMS, has presented a concise treatment of the basic electromechanics of magnetic MEMS. We started out by defining the lossless electromechanical coupling using the standard method of virtual work. The energy and coenergy functions provide means to determine the force of electrical origin, based entirely on knowledge of inductance versus mechanical displacement, that is, $L(x)$, and either current or magnetic flux. Linearization was then employed to develop a general linear multiport model for the frequency-dependent behavior of electromechanical transducers. This approach yielded electromechanical transfer functions and also stability criteria. The last topic covered was magnetic sensors. Two simple topologies for differential sensing schemes, the inductive current bridge and the linear variable differential transformer, were presented. Very clear similarities of these networks and their frequency-dependent functionality to the capacitive sensor schemes presented in Chapter 5 were uncovered.

The technology of magnetic MEMS is still in its infancy, a fact that explains the occasionally somewhat speculative tone evident in this chapter. Before magnetic MEMS technology can become feasible and economically competitive, significant advances in the fabrication of high-aspect-ratio structures formed from both magnetically soft ferromagnetic and permanent magnetic materials will have to be made. The impetus for such development will probably arise from a demand for actuators providing larger forces and longer mechanical strokes than capacitive MEMS can offer. One may anticipate that such devices will have physical dimensions generally larger than capacitive counterparts, though at this stage it is difficult to predict just what this size range will be. In any case, it is safe to assume that (i) smaller MEMS devices will probably always be capacitive, while (ii) a new class will emerge, consisting of somewhat larger and more robust inductive electromechanical devices.

Problems

9.1 The square planar coil shown on the next page, which has $N = 3$ turns, is patterned on a substrate in copper of thickness $h = 0.6$ μm and resistivity $\rho = 1.8 \cdot 10^{-7}$ Ω-m. The defined dimensions are $d_{out} = 2$ mm, $g = 3$ μm, and $w = 60$ μm. Calculate the inductance and resistance of this coil using the semi-empirical formulas:

$R \approx 4N\rho d_{avg}/(wh)$ and $L \approx 2.34\mu_0 N^2 d_{avg}/[1 + 2.75F_f]$, where $d_{avg} = (d_{in} + d_{out})/2$ and $F_f = (d_{out} - d_{in})/(d_{out} + d_{in})$.

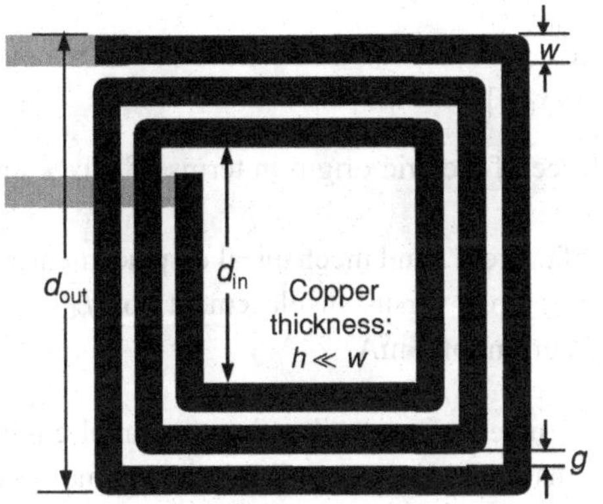

9.2 Over its useable range of motion, the dependence of the inductance of a magnetic actuator on the mechanical variable is $L(x) = L_0 x_0/x$, where $x > 0$. Find the time-dependent voltage $v(t)$ for the following sets of conditions:

(a) $i(t) = I_0 e^{-t/\tau}$ and $x(t)$ is constant and equal to x_0;
(b) $i(t) = I_0$ (constant) and $x(t) = x_0 - u_0 t$;
(c) $i(t) = I_0 \cos(\omega t)$ and $x(t) = x_0(1 + \alpha e^{-t/\tau})$.

9.3 For the inductor $L(x) = L_0 x_0/x$, where $x > 0$, find time-dependent expressions for the force of electrical origin $f^e(t)$ subject to each of the conditions (a), (b), and (c) described in Problem 9.2.

9.4 For the magnetic actuator described in Problem 9.2, assume that a steady DC current I_0 is applied and the mechanical variable is held at $x = x_0$. The device is then short-circuited to trap the magnetic flux. Assume that the winding resistance is zero. With the current i no longer externally constrained, what is the algebraic relationship between i and the mechanical variable x?

9.5 A magnetic actuator is driven at a constant DC current $i = 10$ mA. Over some interval of time the velocity of the moving element is constant at $dx/dt = 10$ μm/ms and the mechanical power is constant: $p_m(t) = 100$ μW. Selecting units convenient for your answers, what are the mechanical force $f^e(t)$ and the terminal voltage $v(t)$ in this same time interval?

9.6 The goal of this problem is to become familiar with practical units of magnetic MEMS devices. A magnetic MEMS device has a measured inductance of $L_{nH} = 1000 + 100x_{\mu m}$, where L_{nH} is in nanohenries and $x_{\mu m}$ is in microns.

(a) What is the magnitude of the static force of electrical origin f^e in newtons for DC current $i = 100$ mA and $x_{\mu m} = 5$ μm?

(b) If $x_{\mu m}(t) = 5\cos(2\pi \cdot 10^3 t)$, with t in seconds, and $i = 100$ mA, obtain expressions for the voltage $v(t)$ and the stored, magnetostatic coenergy $W_e'(t)$. Choose units appropriate for each quantity.

9.7 Over the displacement range $0 < x < 0.9x_0$, the inductance of a magnetic actuator is found to be $L(x) \approx L_0/[1 - (x/x_0)^2]$.

(a) Find an expression for the force of electric origin in terms of flux λ and displacement x.
(b) Now find the force in terms of current i and mechanical displacement x.
(c) Using suitable units, plot the force versus displacement for $L_0 = 10$ nH and $x_0 = 100$ μm at constant DC current of 5 mA.

9.8 A magnetic actuator is constructed with a coil and two parallel circular disks of very high permeability magnetic material. The upper disk is free to move up and down with its face remaining parallel to the lower disk. A mechanical stop prevents the two faces from actually touching, so that the gap is $x(t) + g_0$, where $x(t) \geq 0$ and $g_0 > 0$. Assume the device reluctance to be $\mathcal{R}(x) \approx (x + g_0)/\mu_0 A$.

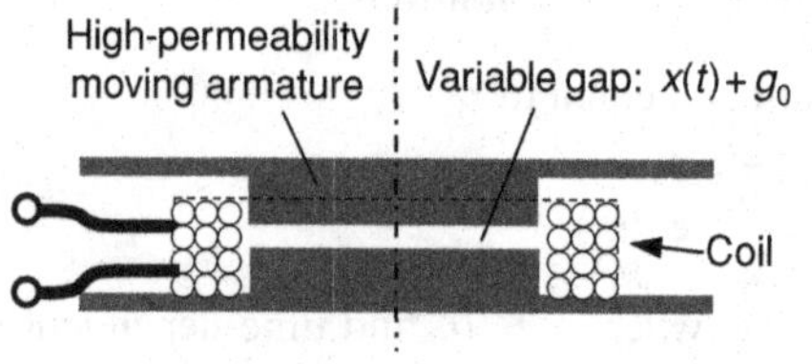

(a) Using energy methods, find the force of electrical origin in terms of both (x, i) and (x, λ).
(b) Sketch the x-dependence of this force when the magnetic flux λ is held constant. Explain this result.
(c) Repeat (b) for the case of constant current i.

9.9 An implicit empirical expression for the electrical terminal relation of an electrically non-linear actuator as it approaches the magnetic saturation limit is $i = I_0[(\lambda/\lambda_0) + 2(\lambda/\lambda_0)^3]/[1 + (x/x_0)^2]$.

(a) Which energy function, energy or coenergy, is the more convenient choice to use for determining the force of electrical origin? Why?
(b) Use your choice of an energy function to obtain an expression for this force.

9.10 The force of electrical origin for a magnetic MEMS device has the following dependence on current i and the mechanical variable x: $f^e(x, i) = L_0 i^2 x/(a^2 + x^2)$, where $0 < x < a$. Within a constant factor, find an expression for the inductance $L(x)$.

9.11 Measured inductance values versus displacement are tabulated for a magnetic MEMS device in the table:

Mechanical displacement x (microns)	Inductance $L(x)$ (microhenries)
0	100
5	110
10	120
15	129
20	139
25	149
30	158
35	165
40	171
45	176
50	179
55	180
60	181

(a) Employ polynomial curve fitting to obtain an algebraic expression for the inductance $L(x)$.
(b) In a single graph, plot the inductance data points and the fitted curve versus x.
(c) Use the fitted curve to calculate the force f^e. Plot this force versus displacement x for a DC current of 20 mA. Use convenient units.
(d) Investigate how the number of terms used for the polynomial curve fit influences the force estimate. Discuss the limits imposed by polynomial data fits on the accuracy of modeling MEMS devices.

MATLAB functions **polyfit** and **polyder** can be used, respectively, to obtain a fitted curve for the inductance data and then to calculate the derivative for the force of electrical origin. Refer to: www.mathworks.com/help/techdoc/ref/polyfit.html and www.mathworks.com/help/techdoc/ref/polyder.html.

9.12 A linear magnetic actuator with two windings and two mechanical degrees of freedom has the following terminal relations:

$$\lambda_1 = L_1 i_1 + M(x, y) i_2,$$
$$\lambda_2 = M(x, y) i_1 + L_2 i_2.$$

Here, L_1 and L_2 are constants and the mutual inductance M is

$$M(x, y) = M_0 \sinh(-\alpha x) \sin(\beta y).$$

(a) Using magnetic coenergy, find expressions for the x- and y-directed forces of electrical origin in terms of i_1, i_2, x, and y.
(b) Use a Legendre transform to define a suitable new energy function and use it to derive an expression for the y-directed force of electrical origin as a function of λ_1, i_2, x, and y.

9.13 The coupled, non-linear magnetic actuator depicted in the figure has the following electrical terminal relations:

$$\lambda_a = Ax_1 \tanh(i_a/i_0) + Bx_1x_2(i_b/i_0),$$
$$\lambda_b = Bx_1x_2(i_a/i_0) + Cx_2 \tanh(i_b/i_0).$$

Find expressions for the forces of electrical origin, f_1^e and f_2^e, in terms of the position and current variables.

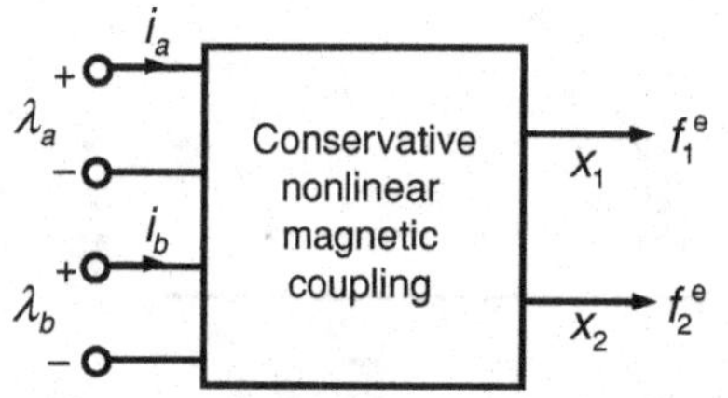

***9.14** A movable bar of high-permeability magnetic material having length D and width w rests above and is parallel to a uniform array of parallel conductors of pitch w. The conductors are embedded in a substrate of the same high-permeability material. A non-magnetic layer of thickness $g \ll w$ covers the array. Over the range of motion $-w/2 < x < w/2$, the device can be modeled as a magnetic circuit having reluctance dominated by the non-magnetic layer: $\mathcal{R}(x) \approx 2wg/[\mu_0 D(w^2 - x^2)]$.

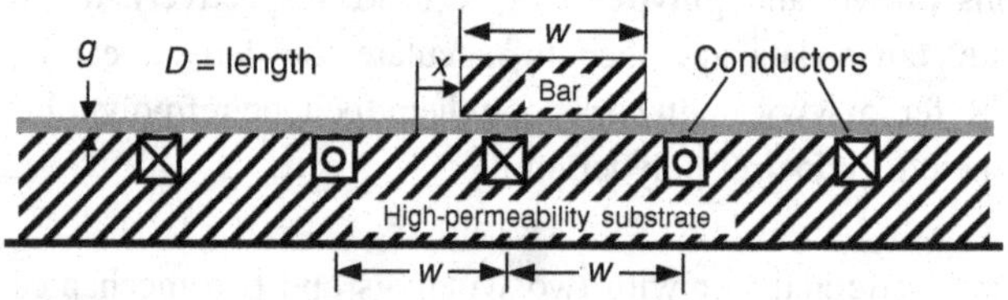

(a) Ignore fringing fields to derive the above expression for $\mathcal{R}(x)$.
(b) Find an expression for the x-dependent portion of the inductance.
(c) Obtain an expression for the force of electrical origin acting on the bar in terms of the current.

Hint: refer to the review of magnetic circuits found in Section A.5.3 of Appendix A.

9.15 The terminal relation of many simple, linear, magnetic transducers can be expressed in terms of the number of turns N and the reluctance $\mathcal{R}(x)$ as $\lambda = N^2 i/\mathcal{R}(x)$.

(a) Use this expression to write the energy function and the N-form transducer matrix coefficients in terms of the reluctance $\mathcal{R}(x)$.
(b) Now use this expression to obtain the coenergy expression and the M-form transducer matrix coefficients.

*9.16 The moving armature of the magnetic actuator described in Problem 9.8 has mass m. It is linked to a linear spring of spring constant k that is relaxed at $x = x_k$, and a mechanical damper with coefficient b. The winding has DC current $i = I_0$.

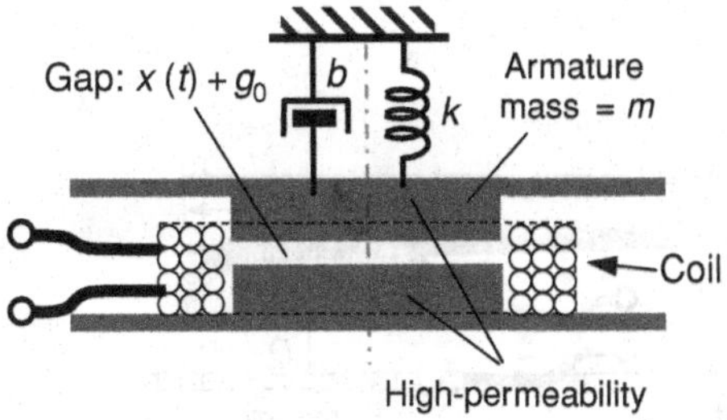

(a) Write the general (non-linear) equation of motion for this device.
(b) From (a) extract but do not solve the static equilibrium equation.
(c) Obtain a linearized equation of motion for small mechanical perturbations, assuming the current is fixed at $i = I_0$.
(d) Can this system exhibit instability?

9.17 A simple permanent magnet transducer is illustrated in Fig. 9.17. Using Eq. (9.43) for the electric terminal relation of this device, find the general M-form matrix in terms of the defined inductances. Then, taking advantage of the fact that the fictitious (Amperian) current i_p representing the permanent magnetization is a constant with no perturbation component, develop a linear electromechanical two-port model for the transducer.

9.18 Starting with the non-linear terminal relation provided for Problem 9.9, derive the N-form transducer matrix and its coefficients for the small-signal two-port electromechanical network.

*9.19 The actuator described in Problem 9.8, and then elaborated in Problem 9.16, is now connected to a DC voltage source V_0 through resistor R as shown in the figure.

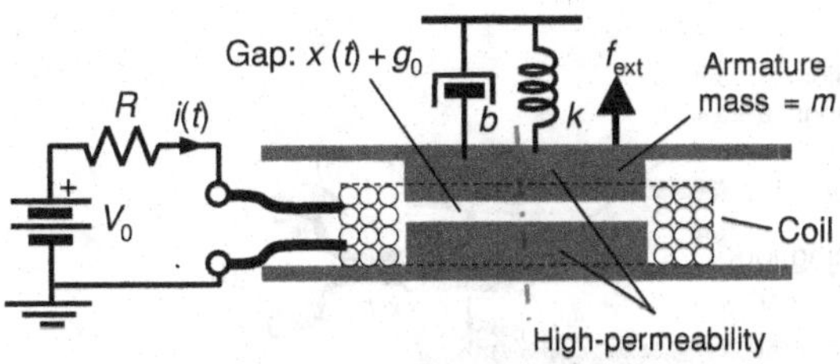

(a) Use the M-form transducer formation and the external constraints to develop a complete matrix model to predict the small-signal behavior of this device, making sure to define all necessary matrix coefficients.
(b) Find an expression for the system function relating input force $\underline{f}_{ext}$ to the small-signal voltage across the resistor R, that is, $\underline{H}(j\omega) = R\underline{i}/\underline{f}_{ext}$.
(c) Can this device be unstable? If yes, what is the threshold condition on current I_0?

***9.20** A magnetic MEMS device is fabricated with a square $a \times a$ slab of magnetizable material attached at the end of a cantilever. The beam length, width, and thickness are L, a, and h. Assume that the moving magnetic slab has mass m_{slab} and is small enough compared with the beam length, that is, $a \ll L$, to justify use of a lumped parameter model for the beam.

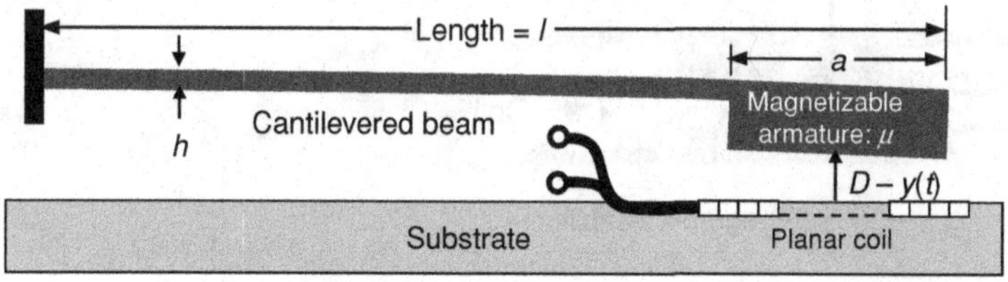

(a) Propose a reasonable analytical function for the device inductance. Defend your choice but also point out what its limitations might be.
(b) Use this inductance to obtain an equation relating the static deflection y_0 to DC current I_0. Under the assumption that $y_0 \ll D$, solve this equation to obtain an approximation for y_0.
(c) Obtain an approximate expression for the resonant frequency as a function of I_0 and other parameters.
(d) Find expressions for the threshold current and the marginal stable value of y_0 at which pull-in occurs.

Hint: refer to Section 6.2 concerning the point-loaded cantilevered beam.

***9.21** A coil of N turns wound on a very high permeability core of area A is positioned symmetrically between the tines of a tuning fork, which is also of high magnetic permeability. The device is designed to operate near or below the fundamental resonance of the tuning fork. In this fundamental mode, the two tines alternately bend in and out together. Thus, the two air gaps can be approximated by $g + x(t)$, where $|x(t)| < g$.

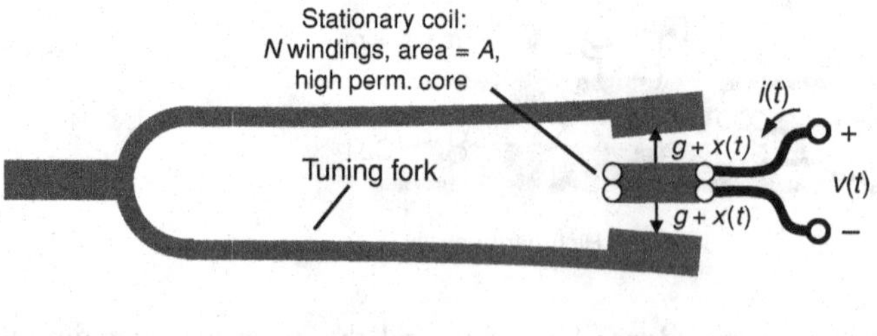

(a) Use a magnetic circuit model that neglects the reluctance of the tuning fork and the coil core compared with the air gaps to obtain an approximate expression for inductance $L(x)$.
(b) Now let I_0 be the DC current bias applied to the coil. Using m_{eq} and k_{eq} as effective mass and mechanical spring constant values for the tuning fork at and below the

first resonance, respectively, employ either the M- or N-form model and to find an expression for the small-signal electrical impedance of this transducer.

***9.22** Two identical planar coils are mounted parallel on a common axis at spacing $2d$. Between them is a magnetic plate, also parallel, that moves up and down. This plate magnetically shields the coils from each other and, thus, only influences their self-inductances: $L_1(x) = L_0 e^{+\alpha x}$ and $L_2(x) = L_0 e^{-\alpha x}$. The internal winding resistance of each coil is R_w. The device is connected in a current-bridge configuration. To provide the bias current, identical external resistors R_{ext} connect the coils to a common DC voltage supply V_0. Assume that the plate has mass m, effective linear damping constant b, and is held in position at $x = 0$ by a linear spring having spring constant k.

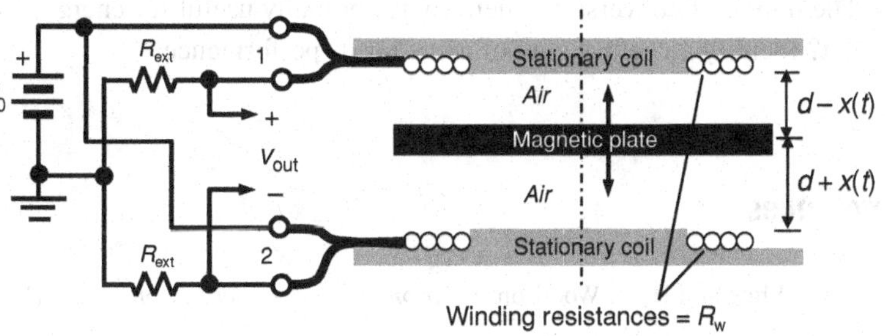

(a) Find the equilibrium conditions on current and $x(t)$ for this system with no external applied mechanical force.
(b) Write out the 3 × 3 M-form transducer matrix for this magnetic sensor and provide expressions for all the coefficients.
(c) Obtain an expression for $\underline{H}_{v/x}(j\omega) \equiv \underline{v}_{out}/\underline{x}$, the small-signal transfer function.

***9.23** The energy harvester shown below is constructed using a permanent magnet disk and a planar coil. The permanent magnet is attached to a membrane and is free to move up and down in response to vibration of the surface upon which the device is mounted. The planar coil detects this motion as an electrical signal and converts some of it to electric energy. The object of this open-ended problem is to develop a model for the device and then estimate the power that can be harvested.

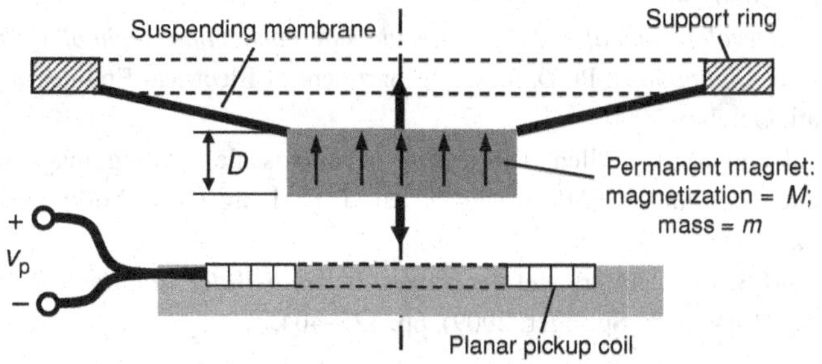

(a) Use a simple mass-and-spring representation for the movable mass and a likewise simple electric circuit model for the pick-up coil to obtain an expression for the frequency-dependent open-circuit voltage induced when the device is subject to sinusoidal acceleration $a(t) = A\cos(\omega t)$. Use Q for the resonator quality factor.

(b) In part (a), you defined a number of parameters, including a spring constant and a self-inductance. You may refer to the text or to other available resources to develop a model from which estimates for these parameters can be obtained.

(c) Assuming a resistor R attached to the electrical terminals as the load, find an expression for the electrical power delivered to this load. Do not neglect either self-inductance or winding resistance of the coil.

(d) Comment on the frequency dependence of the deliverable power.

(e) The notion of conversion efficiency is not really useful for energy harvesters. Why not? Identify a useful figure of merit for its performance.

References

1. D. C. White and H. H. Woodson, *Electromechanical Energy Conversion* (New York: Wiley, 1959).
2. H. H. Woodson and J. R. Melcher, *Electromechanical Dynamics. Part I* (New York: Wiley, 1968).
3. H. Guckel, T. Earles, J. Klein, J. D. Zook, and T. Ohnstein, Electromagnetic linear actuators with inductive position sensing. *Sensors and Actuators A*, 53 (1996), 386–391.
4. W. S. N. Trimmer, Microrobots and micromechanical systems, *Sensors and Actuators*, 19 (1989), 267–287.
5. O. Cugat, J. Delamare, and G. Reyne, Magnetic microsystems: MAG-MEMS. In *Magnetic Nanostructures in Modern Technology*, ed. B. Azzerboni, G. Asti, L. Pareti, and M. Ghidini (Dordrecht, Netherlands: Springer, 2008), pp. 105–125.
6. C. D. Meyer, S. S. Bedair, B. C. Morgan, and D. P. Arnold, High-inductance-density, air-core, power inductors and transformers designed for operation at 100–500 MHz, *IEEE Transactions on Magnetics*, 46:6 (2010), 2236–2239.
7. J. Y. Park and M. G. Allen, Integrated electroplated micromachined magnetic devices using low temperature fabrication processes, *IEEE Transactions on Electronics Packaging Manufacturing*, 23 (2000), 48–55.
8. D. J. Sadler, *Development of a New Magnetic Interconnection Technology for Magnetic MEMS Device Applications*. Ph.D. thesis, Department of Electrical Engineering, University of Cincinnati, October, 2000.
9. D. P. Arnold and M. G. Allen, Fabrication of microscale rotating magnetic machines. In *Multi-Wafer Rotating MEMS Machines*, ed. J. H. Lang (New York: Springer, 2009), pp. 157–190.
10. J. H. Lang and S. Das, Motors and generators. In *Multi-Wafer Rotating MEMS Machines*, ed. J. H. Lang (New York: Springer, 2009), pp. 325–404.

11. B. Eyre, L. Miller, and K. S. J. Pister, MEMS magnetic sensor in standard CMOS, *Science Closure and Enabling Technologies for Constellation Class Missions* (University of California at Berkeley, 1998), pp. 99–102.
12. N. H. Frank, *Introduction to Electricity and Magnetism*, 2nd edn (New York: McGraw-Hill, 1950), Chapter 14, Section 2, pp. 257–260.
13. H. K. P. Neubert, *Instrument Transducers* (Oxford: Clarendon Press, 1975).

A Review of quasistatic electromagnetics

A.1 Integral equations of electromagnetism

This appendix provides a concise summary of the laws of electromagnetics and then introduces the electroquasistatic and magnetoquasistatic approximations for the two physical regimes that rule the electromechanics of MEMS devices. The electroquasistatic approximation is the basis of the capacitive circuit representations introduced in Chapter 2 and employed throughout the text to model electromechanical transducers. The magnetoquasistatic regime is the starting point for magnetic devices, which are covered in Chapter 9. We use SI units exclusively and the nomenclature is chosen to maximize consistency with the literature while still providing essential flexibility for electromechanical modeling.

A good strategy for developing reduced-order circuit models for electromechanical devices is to minimize the use of vector fields. An efficient way to take advantage of this strategy is to introduce the field theory of Maxwell using vectors and the integral laws of electromagnetism first and then use these laws to develop lumped parameter models for electromechanical devices. The integral laws are

$$\text{Gauss's law:} \quad \oint_S \overline{D} \cdot d\overline{A} = \oint_V \rho_{\text{free}} \mathrm{d}V, \tag{A.1}$$

$$\text{Solenoidal law:} \quad \oint_S \overline{B} \cdot \mathrm{d}\overline{A} = 0, \tag{A.2}$$

$$\text{Faraday's law:} \quad \oint_c \overline{E} \cdot \mathrm{d}\overline{l} = -\frac{\mathrm{d}}{\mathrm{d}t} \int_{S_c} \overline{B} \cdot \mathrm{d}\overline{A}, \tag{A.3}$$

$$\text{Ampère's law:} \quad \oint_c \overline{H} \cdot \mathrm{d}\overline{l} = \int_{S_c} \overline{J} \cdot \mathrm{d}\overline{A} + \frac{\mathrm{d}}{\mathrm{d}t} \int_{S_c} \overline{D} \cdot \mathrm{d}\overline{A}. \tag{A.4}$$

$\overline{D}$ and $\overline{E}$ are the electric flux (in coulombs/m^2) and electric field (volts/m) vectors and $\overline{B}$ and $\overline{H}$ are the magnetic flux (teslas) and intensity (amperes/meter) vectors, all respectively. $\overline{J}$ is the free electric current density vector (amperes/m^2) and ρ_{free} is the free electric volume charge density (coulombs/ m^3). The key to gaining insights about these equations is to understand the relationships between the various contours, surfaces, and volumes that define the integrals. First, the choices for surfaces S and contours c appearing on the left-hand sides of these equations are completely arbitrary. Second, the closed surface S in Eq. (A.1) defines the volume V. Third, the closed contour c in Eqs. (A.3) and (A.4) defines the outer edge of the open area S_c.

Table A.1 Summary of the field variables and other relevant quantities, plus units, used in electroquasistatics and magnetoquasistatics

Symbol	Field variable name	SI units
$\overline{H}$	Magnetic field intensity	A/m
$\overline{J}$	Free current density	A/m^2
$\overline{K}_f$	Free surface current density	A/m
$\overline{B}$	Magnetic flux density	Tesla (webers/m^2)
$\overline{M}$	Magnetization density	A/m
$\overline{E}$	Electric field intensity	V/m
$\overline{D}$	Electric displacement	C/m^2
ρ_f	Free charge density	C/m^3
σ_f	Free surface charge density	C/m^2
$\overline{P}$	Polarization density	C/m^2
μ_0	Free space permeability	$4\pi \times 10^{-7}$ H/m
ε_0	Free space permittivity	8.854×10^{-12} F/m
ϕ	Electrostatic potential	V

The laws of electromagnetism must be supplemented by material properties. The most familiar linear isotropic forms for these constitutive relations are

$$\overline{D} = \varepsilon\overline{E},\ \overline{B} = \mu\overline{H},\ \overline{J} = \sigma\overline{E}, \tag{A.5}$$

where ε is dielectric permittivity (in farads/m), μ is magnetic permeability (henries/m), and σ is ohmic electrical conductivity (siemens/m). These approximations provide adequate descriptions of many practical materials. Table A.1 summarizes some terminology used in this section.

Electromagnetic energy is an important consideration in electromechanics, and Maxwell's equations are consistent with energy conservation. For a linear medium with neither dielectric nor magnetic loss, the integral form for energy conservation is

$$\oint_S \overline{E} \times \overline{H} \cdot \mathrm{d}\overline{A} + \frac{\partial}{\partial t}\left[\int_V \underbrace{\tfrac{1}{2}\varepsilon|\overline{E}|^2\,\mathrm{d}V}_{u_\mathrm{e}} + \int_V \underbrace{\tfrac{1}{2}\mu|\overline{H}|^2\,\mathrm{d}V}_{u_\mathrm{m}}\right] = -\sigma\int_V |\overline{E}|^2\mathrm{d}V. \tag{A.6}$$

The so-called Poynting vector, $\overline{E} \times \overline{H}$, is interpreted as power flux (watts/m^2), while $u_\mathrm{e} = \frac{1}{2}\varepsilon|\overline{E}|^2$ and $u_\mathrm{m} = \frac{1}{2}\mu|\overline{H}|^2$ are, respectively, the volume densities of electric and magnetic energy (joules/m^3). The term on the right-hand side is Joule heating, that is, power converted irreversibly from electrical form to heat. In Eq. (A.6), just as before, V is the volume defined by the closed surface S.

Equations (A.1) through (A.4) and (A.6), while not compact like the point forms of Maxwell's equations and Poynting's theorem, are more naturally suited to our objective, which is to model MEMS devices as circuit elements. These forms make it easy to extract lumped parameter variables, such as voltage and current from appropriately executed integrals. The next two sections present integral formulations for the two special cases of (i) electroquasistatics, describing capacitive devices, and (ii) magnetoquasistatics for inductive devices.

A.2 Electroquasistatics

The electroquasistatic approximation neglects magnetic fields and induction, thereby considerably simplifying the equations. This approximation is well-justified for capacitive devices having physical dimensions typical for MEMS as long as they are operated at frequencies below $\sim 10^{10}$ Hz. Setting all magnetic terms in Eqs. (A.1) through (A.4) to zero and then rearranging, the equations for capacitive systems become

$$\text{Gauss's law:} \quad \oint_S \overline{D} \cdot \mathrm{d}\overline{A} = q_{\text{enclosed by } S}, \tag{A.7}$$

$$\text{Conservative } \overline{E} \text{ field:} \quad \oint_c \overline{E} \cdot \mathrm{d}\overline{l} = 0 \quad \text{or} \quad \overline{E} = -\nabla\phi, \tag{A.8}$$

$$\text{Charge conservation:} \quad \oint_S \overline{J} \cdot \mathrm{d}\overline{A} = -\frac{\mathrm{d}}{\mathrm{d}t} q_{\text{enclosed by } S}, \tag{A.9}$$

where $q_{\text{enclosed by S}}$ is the total free electric charge (coulombs) enclosed within the volume V, which is defined in turn by the closed surface S,

$$q_{\text{enclosed by } S} = \int_V \rho_{\text{free}}\, \mathrm{d}V; \tag{A.10}$$

ϕ is the scalar electrostatic potential (volts). Equation (A.10) includes all volume, surface, line, and point charges residing within V. The contour c in Eq. (A.8) may be any path that closes on itself; an electric field $\overline{E}$ satisfying this relation is called a conservative field. The work done moving an electrical test charge from one point to another in a conservative field is independent of the path taken.

It is sometimes convenient to express the linear dielectric constitutive relation using the dimensionless *dielectric constant*: $\kappa \equiv \varepsilon/\varepsilon_0$. Then, for an ohmically conductive dielectric,

$$\overline{D} = \kappa\varepsilon_0 \overline{E} \quad \text{and} \quad \overline{J} = \sigma \overline{E}, \tag{A.11}$$

where $\varepsilon_0 = 8.854 \cdot 10^{-12}$ F/m is the permittivity of free space.

In the electroquasistatic limit, the volume density of stored electrical energy is

$$u_{\mathrm{e}} = \tfrac{1}{2}\varepsilon|\overline{E}|^2. \tag{A.12}$$

A.3 Capacitive devices

The laws of electroquasistatics govern capacitive devices. Lumped parameter capacitors consist of a minimum of two electrical conductors (electrodes) connected to external circuitry. Figure A.1 depicts a generalized capacitor. When voltage is applied, electric

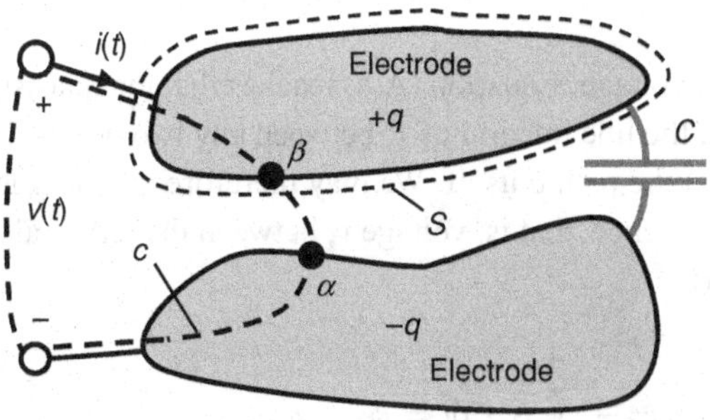

Figure A.1 Generalized capacitor with voltage, current, and charge conventionally defined. Note the closed surface S in Eq. (A.10) and the line integral c in Eq. (A.8). α and β mark the start and end points for the portion of c between the negative and positive electrodes.

charge is stored upon the surfaces of these conductors. In a normal capacitor, there is no free electrical charge in the space between the electrodes. Then, according to Gauss's law, Eq. (A.7), all lines of electric flux $\overline{D}$ starting on the positive electrode must terminate on the negative electrode. This means, in turn, that the two electrodes contain equal (in magnitude) but opposite (in sign) quantities of charge, that is, $+q$ and $-q$. A "charged" capacitor in fact has *no net charge* at all, but rather *separated* charge. The familiar energy quantity $Cv^2/2$ (in joules) is the work required to separate these charges.

A sign convention is necessary here; we define q to be the charge residing on the electrode with positive voltage polarity. Thus, q and v always have the same sign. An opposite charge $-q$ resides on the other, negative polarity, electrode. The voltage $v(t)$ measures only the *difference* in the electrostatic potential ϕ between the + and − terminals, that is, $v \equiv \phi_+ - \phi_-$.

A.3.1 Definition of capacitance

According to the lumped parameter model for the capacitor of Fig. A.1, electric current is confined to flow in the wires connected to the electrodes. Thus, Eq. (A.9) for charge conservation becomes

$$i(t) = \mathrm{d}q/\mathrm{d}t. \tag{A.13}$$

If the region between the electrodes is free space or any electrically linear dielectric material, then $q(t)$, the electric charge on the positive electrode, must be linearly related to the voltage, $v(t)$. The proportionality constant of this linear relationship is the capacitance C:

$$q = Cv. \tag{A.14}$$

Combining Eqs. (A.13) and (A.14), and recognizing that the capacitance of a transducer will depend on some mechanical variable, $x(t)$, we obtain Eq. (1.4):

$$i(t) = \frac{\mathrm{d}q}{\mathrm{d}t} = C(x)\frac{\mathrm{d}v}{\mathrm{d}t} + \frac{\mathrm{d}C}{\mathrm{d}x}v\frac{\mathrm{d}x}{\mathrm{d}t}. \tag{1.4}$$

A.3.2 Definition of voltage

More needs to be said about voltage. Equation (A.8) teaches that the quasistatic electric field is conservative, that is, the line integral of $\overline{E}$ between any two points depends only on the end points and not on the path c itself. We may therefore unambiguously define an electrostatic potential difference, that is, voltage v, between the terminals in terms of a line integral of this vector:

$$v = -\int_{-}^{+} \overline{E} \cdot d\overline{l} = \phi_+ - \phi_-. \tag{A.15}$$

A general path for the integral from the negative to the positive terminal is depicted in Fig. A.1. If the connecting wires and the electrodes are perfectly conducting, usually an excellent approximation for a capacitor, then the electric field is zero inside them, and the only non-zero portion of the line integral is the gap between the electrodes. Thus,

$$v = -\int_{\alpha}^{\beta} \overline{E} \cdot d\overline{l}, \tag{A.16}$$

where α and β are defined in Fig. A.1. Equation (A.16), the definition of voltage, will demand re-examination for the case of magnetostatics, where the electric field is no longer conservative.

A.3.3 Parallel-plate capacitor

The geometries of practical capacitive MEMS devices are often sufficiently complex to necessitate numerical solution to obtain capacitance versus displacement. Virtually all modern finite-element modeling (FEM) and boundary-element modeling (BEM) software packages have built-in capacitance calculation tools. Fortunately, the newcomer to MEMS can study the basics of capacitive transducers without immediately resorting to FEM analyses. In fact, we can explore virtually all the important behavior of capacitive transducers by considering two distinct capacitive transducer types: variable-gap and variable-area geometries.

The starting point for both these types is the so-called parallel-plate capacitor shown in Fig. A.2. In the parallel-plate capacitor, the spacing of the parallel electrodes d is very small compared with the areal dimensions. The consequence is that the influence on capacitance C of the non-uniform (fringing) electric field around the periphery becomes insignificant. Delving into electromechanics, we will encounter some apparent paradoxes arising from this approximation; however, we will also confirm the extraordinary robustness of the parallel plate/uniform electric field approximation.

It is a simple matter to use the electroquasistatic equations to derive the capacitance of the parallel-plate device in Fig. A.2. Let us assume that the electrodes are in free space with no free electric volume charge present. Employing the uniform electric field

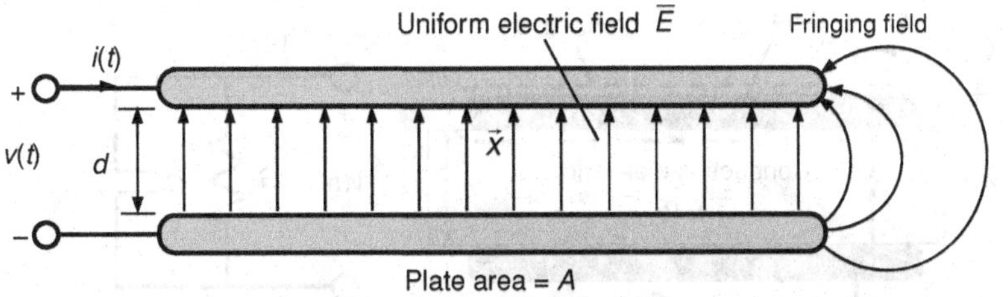

Figure A.2 The basic parallel-plate capacitor of area A and spacing d. If the spacing is small compared with the areal dimensions, the fringing field contribution to capacitance becomes insignificant. In this figure, the electric field lines are defined with respect to the x direction.

approximation and entirely neglecting the fringing fields, Eq. (A.7) gives

$$\overline{E} = -(q/(\varepsilon_0 A))\overline{i}_x, \tag{A.17}$$

where $\overline{i}_x$ is the x-directed unit vector. The negative sign arises because the electric field actually points from positive to negative charge. Using Eq. (A.17) in Eq. (A.16) yields

$$v = -\int_{\alpha}^{\beta} (-q/(\varepsilon_0 A))\,\mathrm{d}x = (q/(\varepsilon_0 A))d. \tag{A.18}$$

Then from Eq. (A.14), the capacitance is

$$C \approx \varepsilon_0 A/d. \tag{A.19}$$

This familiar expression is accurate as long as neglect of the fringing fields is justified. In some MEMS devices, this assumption does not apply; nevertheless, the critical electromechanical behavior is usually controlled by the parallel-plate (uniform field) portion of capacitance.

If the space between the plates is filled with a linear dielectric as defined by Eq. (A.11), then

$$C \approx \kappa\varepsilon_0 A/d. \tag{A.20}$$

The cases of particular interest in MEMS devices – variable-gap and variable-area capacitive transducers – result from allowing, respectively, electrode spacing or area to become dependent on a mechanical variable. Rotating capacitive transducers are often reducible to variable-gap or variable-area devices.

It should be obvious that there is nothing to preclude both d and A from being variables, in which case, the device has more than one mechanical degree of freedom. This situation might arise by intent or, if some mode of motion unrecognized during the design phase emerges, accidentally. Sometimes these motions trace back to fabrication defects. Performance degradation of MEMS devices resulting from unanticipated mechanics is fairly common.

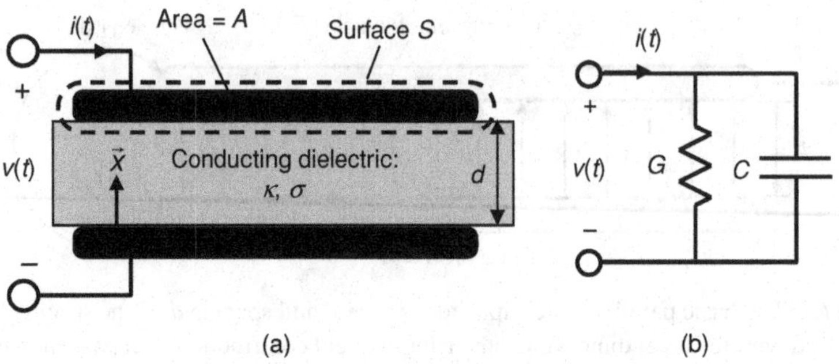

Figure A.3 Parallel-plate geometry. (a) Electrodes are separated by a slab of ohmically conducting dielectric material. (b) The equivalent electric circuit model consists of a capacitor C in parallel with a linear conductance G.

A.3.4 Electrical conductivity

We can learn more about circuit models using the parallel-plate geometry. Assume that the space between the plates is filled with a linear, conductive, dielectric so that $\overline{D} = \varepsilon\overline{E}$ and $\overline{J} = \sigma\overline{E}$. To proceed, it is convenient to let the voltage depend on time, that is, $v(t)$. We then modify Eq. (A.13) to account for current flowing through the ohmic conductive medium between the plates. Let the surface S shown in Fig. A.3a completely enclose the upper electrode and then invoke Eq. (A.9) to get

$$i = -A\sigma E_x + \mathrm{d}q/\mathrm{d}t. \tag{A.21}$$

Using $E_x = -v/d$ for the uniform electric field along with the expression for the charge q in terms of capacitance from Eq. (A.20) yields

$$i = Gv + C\frac{\mathrm{d}v}{\mathrm{d}t}. \tag{A.22}$$

$G \approx \sigma A/d$, the electrical conductance, is the reciprocal of resistance R: $G = 1/R$. The equivalent circuit for the device described by Eq. (A.22) is the parallel RC network in Fig. A.3b. For clarity in the above derivation, we temporarily suppressed the time dependence of C to get Eq. (A.22). In general, however, for a MEMS device, both G and C will depend on time through their functional dependence on a mechanical variable.

The circuit model of Fig. A.3b is quite general. In fact, the parallel RC model is valid for any geometry having a homogeneous lossy dielectric medium, not just the parallel-plate capacitor. Furthermore, as long as the material is homogeneous, the quotient of C and G depends only on intrinsic electrical properties and is independent of geometry:

$$C/G = \varepsilon/\sigma. \tag{A.23}$$

$\tau = \varepsilon/\sigma$, which has units of seconds, is called the charge relaxation time [1].

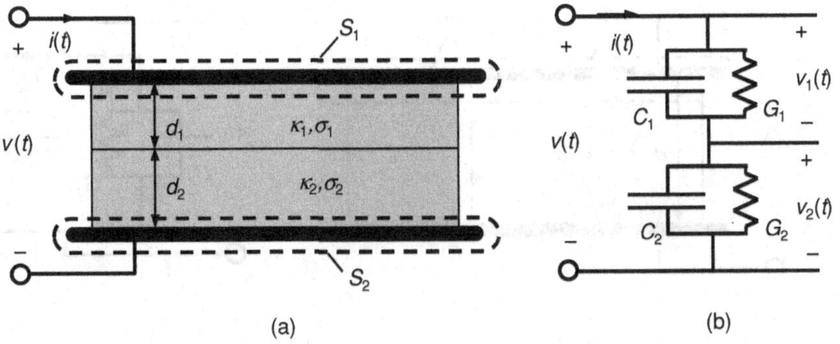

Figure A.4 The two-layer capacitor geometry, also known as Maxwell's capacitor. (a) Side view, showing geometrical dimensions and definitions for surfaces S_1 and S_2. (b) AC circuit model as the series connection of two parallel RC networks.

A.3.5 Circuit models

We can use the electroquasistatic integral relations to formulate circuit models for more complex capacitive geometries by identifying two generic cases. First, consider the two-layer capacitor in Fig. A.4a, consisting of two conductive dielectric slabs of thicknesses d_1 and d_2, dielectric constants κ_1 and κ_2, and electrical conductivities σ_1 and σ_2. Invoke the uniform field approximation in both layers to define voltage drops across the two layers: $v_1(t) = -d_1 E_{x,1}$ and $v_2(t) = -d_2 E_{x,2}$. Then, from Eq. (A.8) and the definition of voltage,

$$v(t) = v_1(t) + v_2(t). \tag{A.24}$$

Performing the surface integrals specified by Eqs. (A.7) and (A.9), once each for the closed surfaces S_1 and S_2 that enclose the top and bottom electrodes, gives, after some manipulation, two equations for $i(t)$:

$$i(t) = G_1 v_1 + C_1 \frac{dv_1}{dt} \quad \text{and} \quad i(t) = G_2 v_2 + C_2 \frac{dv_2}{dt}, \tag{A.25}$$

where

$$C_1 = \varepsilon_1 A/d_1, \quad G_1 = \sigma_1 A/d_1, \quad C_2 = \varepsilon_2 A/d_2, \quad G_2 = \sigma_2 A/d_2. \tag{A.26}$$

These results suggest the equivalent circuit of Fig. A.4b, the series combination of two parallel RC networks.

Figure A.5a shows a second generic geometry consisting of two different dielectric media arranged side by side to fill the space between the parallel plates. Again, we invoke the uniform field approximation, i.e., $v(t) = -dE_x$, and then apply Eqs. (A.7) and (A.9) over the closed surface S. Using Eq. (A.13),

$$i(t) = (G_a + G_b)v + (C_a + C_b)\frac{dv}{dt}, \tag{A.27}$$

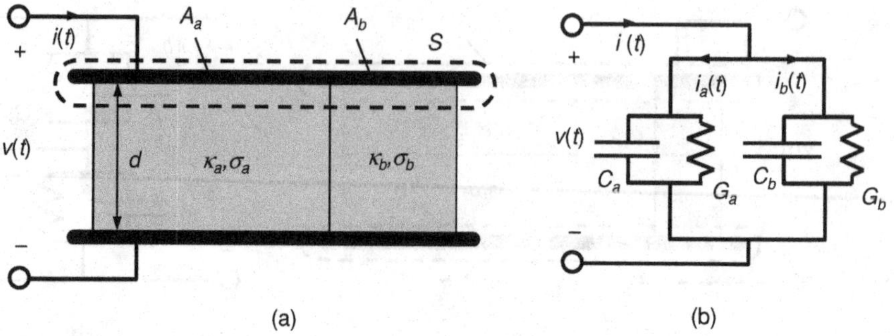

Figure A.5 The parallel-media capacitive geometry. (a) Side view showing dimensions and the definition for the closed surface S. (b) Circuit model as the shunt connection of two parallel RC networks.

where

$$C_a = \varepsilon_a A_a/d, \quad G_a = \sigma_a A_a/d, \quad C_b = \varepsilon_b A_b/d, \quad G_b = \sigma_b A_b/d. \tag{A.28}$$

Identifying

$$i_a = G_a v + C_a \frac{dv}{dt} \quad \text{and} \quad i_b = G_b v + C_b \frac{dv}{dt}, \tag{A.29}$$

and, using $i(t) = i_a(t) + i_b(t)$, we identify the equivalent circuit to be the parallel connection of two RC networks, as shown in Fig. A.5b.

A.3.6 Inspection methods for MEMS modeling

The parallel RC networks derived previously for various parallel-plate configurations serve as building blocks from which equivalent circuits for more complex devices can be constructed by inspection. This technique saves the time and effort required to apply the integral electroquasistatics laws, Eqs. (A.7), (A.8), and (A.9). Even when the uniform field approximation itself is not justified, this method is often applicable.

Inspection is a convenient way to create reduced-order lumped parameter models for MEMS devices. Because the conductivities of dielectrics in MEMS devices are usually low, conductances are ordinarily negligible, so these circuits will usually consist only of capacitors in parallel or in series, or both.

A.4 Magnetoquasistatics

In the magnetoquasistatic approximation, the accumulation of net electric charge is assumed to be negligible, making magnetic induction the sole source for the electric field. This approximation is well-justified for magnetic MEMS devices of typical physical dimensions for frequencies below $\sim 10^{10}$ Hz. Imposing this approximation on Eqs. (A.1)

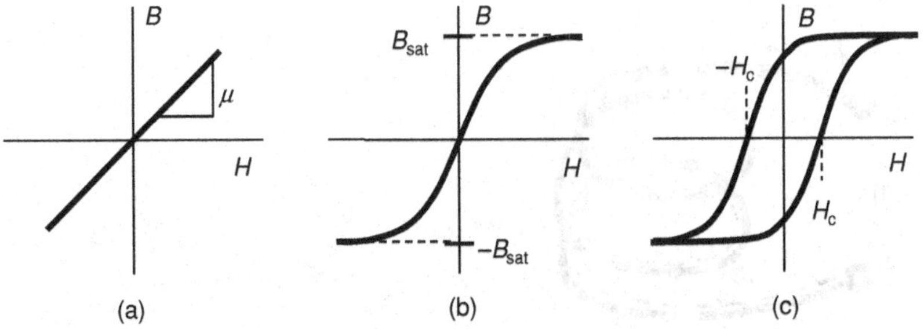

Figure A.6 Representative $B(H)$ curves for linear and typical non-linear (ferrous) media. (a) Ideal linear medium. (b) Medium exhibiting saturation. (c) Medium exhibiting hysteresis and saturation.

through (A.4) yields the following integral forms governing inductive systems:

$$\text{Solenoidal } \overline{B} \text{ field:} \quad \oint_S \overline{B} \cdot \mathrm{d}\overline{A} = 0, \tag{A.30}$$

$$\text{Ampère's circuital law:} \quad \oint_c \overline{H} \cdot d\overline{l} = i_{\text{enclosed by } c}, \tag{A.31}$$

$$\text{Faraday's law:} \quad \oint_c \overline{E} \cdot \mathrm{d}\overline{l} = -\frac{\mathrm{d}}{\mathrm{d}t} \lambda_{\text{linked by } c}, \tag{A.32}$$

$$\text{Current continuity:} \quad \oint_S \overline{J} \cdot \mathrm{d}\overline{A} = 0. \tag{A.33}$$

Equation (A.32) uses the flux linkage (in webers) linked by closed contour c:

$$\lambda_{\text{linked by } c} = \int_{S_c} \overline{B} \cdot \mathrm{d}\overline{A}. \tag{A.34}$$

For a winding with N turns, this integral links the magnetic flux N times. The current in Eq. (A.31) is defined by an area integral of current density:

$$i_{\text{enclosed by } c} = \int_{S_c} \overline{J} \cdot \mathrm{d}\overline{A}. \tag{A.35}$$

Here S_c is any surface, the outer edge of which is defined by contour c for the line integral in Eq. (A.31).

The linear, constitutive law for magnetic media is often written $\overline{B} = \mu_r \mu_0 \overline{H}$. Here, $\mu_r = \mu/\mu_0$, a unitless quantity, is the relative permeability and $\mu_0 = 4\pi \cdot 10^7$ H/m is the permeability of free space, a physical constant. Another often-used quantity is the magnetic susceptibility, $\chi_m \equiv \mu_r - 1$.

From Eq. (A.6), the magnetic energy storage density is

$$u_m = \tfrac{1}{2}\mu|\overline{H}|^2. \tag{A.36}$$

Ferromagnetic materials seldom exhibit the linear behavior illustrated in the $\overline{B}(\overline{H})$ curve of Fig. A.6a. There are significant saturation and hysteresis effects, as illustrated in Figs. A.6b and c. Saturation is the limit reached when all the *magnetic domains* within a

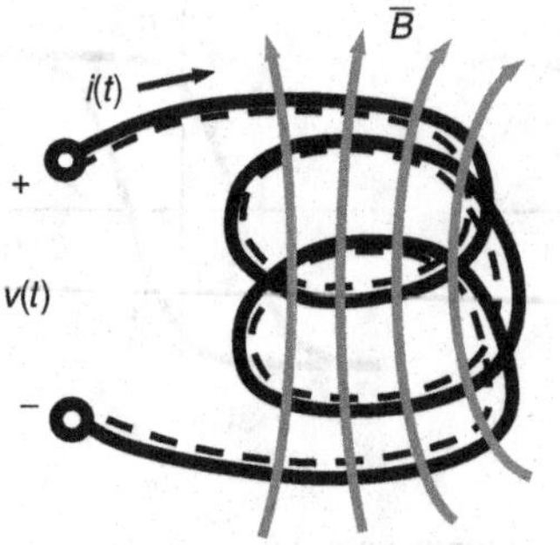

Figure A.7 A generalized inductive device having a current winding of N turns. Some typical lines of the magnetic flux density $\overline{B}$ are shown. If the magnetic flux is Φ, then the flux linkage as defined by Eq. (A.34) will be $\lambda = N\Phi$. The dashed line shows the contour integral path through the winding used in Eq. (A.38).

ferromagnetic material have become aligned. Once this state is reached, a material cannot be further magnetized, thus limiting $|\overline{B}| \leq B_{sat}$, as shown in Fig. A.6b. Hysteresis is an energy-loss mechanism resulting from the tendency of the domains, once aligned, to stay that way and resist reorientation until the magnetic intensity $\overline{H}$ is actually reversed. Finite energy is required to demagnetize hysteretic materials. This energy, irreversibly lost as heat, is equal to the area enclosed by the loop in Fig. A.6c.

A.5 Inductive (magnetic field) devices

Figure A.7 depicts an inductive device consisting of a wire formed into a coil. The coil of an inductor is usually wrapped around a magnetizable material, but we assume here that the winding is in free space, so that $\overline{B} = \mu_0 \overline{H}$. Taking this linear behavior into account in Eqs. (A.31) and (A.34) guarantees that the magnetic flux $\lambda(t)$ will be a linear function of the electric current $i(t)$:

$$\lambda(t) = Li(t). \tag{A.37}$$

The proportionality constant L is called the inductance (and it is measured in henries). In general, evaluation of inductance requires solution for the magnetic flux density vector $\overline{B}$, followed by an integration to obtain λ. For all but the simplest geometries, numerical methods are unavoidable.

A.5.1 Voltage definition in magnetic devices

Because the electric field in the presence of a time-varying magnetic field is not conservative, we must reconsider the definition of voltage. Starting from Faraday's law, Eq. (A.32), we may write

$$v = -\int_{-}^{+} \overline{E}_{wire} \cdot d\overline{l} + d\lambda/dt, \tag{A.38}$$

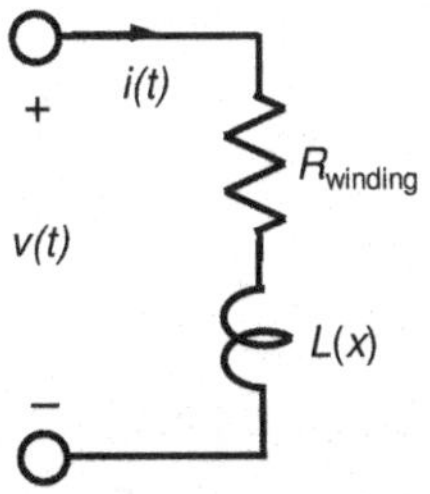

Figure A.8 The equivalent series RL circuit for the generalized inductor shown in Fig. A.7. The assumptions needed for this model are that any magnetic materials in the device are linear and that the winding consists of wire having finite electrical conductivity. The resistance R_{winding} can be computed with reasonably accuracy from Eq. (A.41), as long as the cross-section of the wire varies slowly.

where $\overline{E}_{\text{wire}}$ is the electric field inside the wire used to form the winding. If the wire is perfectly conducting, then $\overline{E}_{\text{wire}} = 0$. In that case, using Eqs. (A.37) and (A.38) yields the following expression relating voltage and current:

$$v = L\frac{\mathrm{d}i}{\mathrm{d}t} + i\frac{\mathrm{d}L}{\mathrm{d}x}\frac{\mathrm{d}x}{\mathrm{d}t}. \tag{A.39}$$

Equation (A.39) recognizes that the inductance of a magnetic transducer can be a function of some mechanical variable, $x(t)$. The first term is familiar from circuit theory, while the second term, proportional to $\mathrm{d}L/\mathrm{d}x$, is the motional (or speed) voltage. Compare this second term with the motional current in Eq. (1.4), particularly noting their respective dependencies on $\mathrm{d}L/\mathrm{d}x$ and $\mathrm{d}C/\mathrm{d}x$.

A.5.2 Case of resistive winding

An important generalization for real inductors results from recognizing that the wire of the winding has finite electrical conductivity σ. Then, using Ohm's law $\overline{J} = \sigma\overline{E}$, the non-zero electric field in the wire is $\overline{E}_{\text{wire}} \approx -i/(\sigma\, A_x)\,\vec{u}_l$, where $\mathrm{d}\overline{l} = \vec{u}_l\,\mathrm{d}l$, $\vec{u}_l$ is a unit vector parallel to the direction of current flow in the wire, and A_x is the cross-sectional area of the wire. Equation (A.38) becomes

$$v = R_{\text{winding}} i + \mathrm{d}\lambda/\mathrm{d}t, \tag{A.40}$$

where R_{winding} is the winding resistance. If A_x varies at most slowly along the wire length,

$$R_{\text{winding}} \approx \int_{-}^{+} \frac{\mathrm{d}l}{\sigma A_x}. \tag{A.41}$$

The new voltage expression becomes

$$v = R_{\text{winding}}\, i + L\frac{\mathrm{d}i}{\mathrm{d}t} + i\frac{\mathrm{d}L}{\mathrm{d}x}\frac{\mathrm{d}x}{\mathrm{d}t}. \tag{A.42}$$

Figure A.8 shows the familiar series RL circuit model.

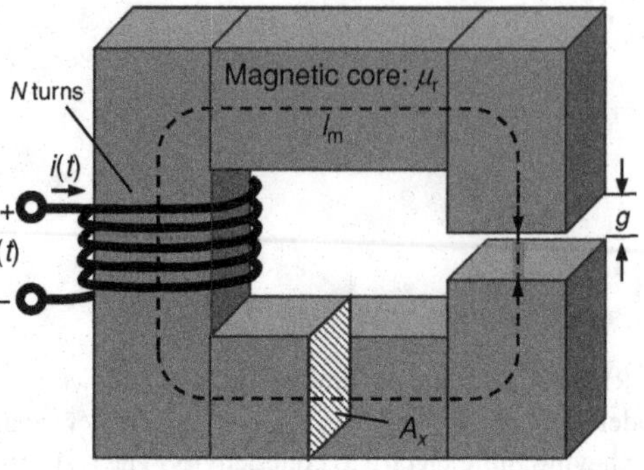

Figure A.9 A simple example of an inductive device readily modeled as a magnetic circuit. The basic assumption is that the magnetic flux Φ is contained inside the core. N is the number of windings. The length l_m and cross-section A_x defined in the figure are effective values.

A.5.3 Magnetic circuits

Magnetic circuits are a special class of geometries that serve as practical models for many inductors, transformers, and magnetic actuators; however, until recently, limitations in microfabrication technology have impeded development of MEMS devices in such configurations. This situation is changing as advances are made in fabricating high-aspect-ratio microstructures with ferromagnetic materials.

A magnetic circuit consists of a re-entrant magnetic core, which directs and confines the flux within a path linked by one or more windings. The device shown in Fig. A.9 features a magnetizable core of relative permeability μ_r, average cross-section A_x, and effective length l_m, with an air gap of length g.

If the core permeability is high, that is, $\mu_r \gg 1$, the magnetic field lines are largely contained within it. Then, the solenoidal nature of the magnetic flux density $\overline{B}$, that is, Eq. (A.30), suggests a useful analogy between magnetic flux,

$$\Phi = \int_{A_x} \overline{B} \cdot \mathrm{d}\overline{A}, \tag{A.43}$$

and electric current. In particular, the magnetic flux circulating around the core exhibits continuity, obeying a KCL-like rule. Thus, $\Phi \approx$ constant at any cross-section, irrespective of the geometry of the core.

Using a closed path c inside the core for the line integral in Eq. (A.31), Ampère's circuital law gives

$$Ni = \int_{\text{gap}} \overline{H} \cdot \mathrm{d}\overline{l} + \int_{\text{core}} \overline{H} \cdot \mathrm{d}\overline{l}. \tag{A.44}$$

Assuming that the magnetic fields in the gap H_{gap} and the core H_{core} are uniform, the integral terms in Eq. (A.44) may be approximated as

$$Ni = H_{\text{gap}}\, g + H_{\text{core}}\, l_m, \tag{A.45}$$

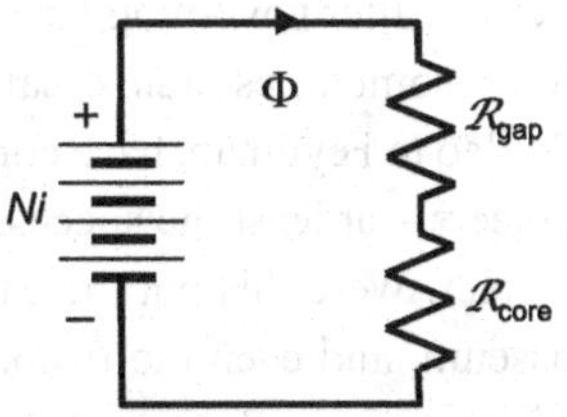

Figure A.10 A magnetic circuit model for the linear magnetic device shown in Fig. A.9. Here, $\mathcal{R}_{gap}$ and $\mathcal{R}_{core}$ are reluctances for the air gap and the magnetic core, respectively. Magnetomotive force, *Ni*, and magnetic flux, Φ, are analogous, respectively, to voltage and current of a DC electric circuit.

where l_m is an average path length within the core. Because the flux Φ must be constant at any cross-section, $H_{gap} = \Phi/\mu_0 A_x$ and $H_{gap} = \Phi/\mu A_x$. Using these expressions in Eq. (A.45) gives

$$Ni = (\mathcal{R}_{gap} + \mathcal{R}_{core})\Phi, \tag{A.46}$$

where $\mathcal{R}_{gap} = g/\mu_0 A_x$ and $\mathcal{R}_{core} = l_m/\mu A_x$ are the *reluctances* of the air gap and the magnetic core, respectively. Notice that the reluctance expressions are directly proportional to path length and inversely proportional to area, just like resistance. The product *Ni*, interpreted as *magnetomotive force* (mmf), is analogous to voltage, just as the magnetic flux Φ is analogous to electric current. Looked at in this way, Eq. (A.46) is a direct analogy to Ohm's law. Figure A.10 shows the magnetic circuit equivalent to the device in Fig. A.9.

The magnetic flux linkage defined by Eq. (A.34) is directly proportional to *N*, the number of windings, that is, $\lambda = N\Phi$. Accordingly,

$$\lambda = \frac{N^2 i}{\mathcal{R}_{gap} + \mathcal{R}_{core}}, \tag{A.47}$$

so the inductance for the magnetic device of Fig. A.9 is

$$L = \frac{\mu_0 N^2 A_x}{g + l_m/\mu_r}. \tag{A.48}$$

Inductance is directly proportional to flux area and the number of turns squared, and inversely proportional to a permeability-weighted sum of the lengths of the sections of the re-entrant magnetic path. Even in more complex magnetic circuit devices, the same basic form is retained for inductance expressions.

A.6 Energy considerations and scaling

Bulky magnetic machines characterized the first hundred years of electrotechnology. So how can it be that capacitive devices are so dominant in the MEMS world? This is a scaling question, and the key to answering it lies in comparing how magnetic and electric systems behave as size is reduced. Serious attention to scaling helps us to understand the capabilities and the limits of MEMS technology. When it comes to physical phenomena, changing the dimensional scale of observation does not alter governing laws, but it does rearrange the relative importance of different effects, sometimes dramatically. Over the centuries, many natural scientists, ranging from Galileo to Feynman, have considered the implications of scaling in our natural world. Engineers – at least the successful ones! – have certainly understood the essentials of the subject. Were this not the case, marvels such as the Egyptian pyramids, the Roman Colosseum, and even the Brooklyn Bridge would not still be standing today. It might be fair to assume that many constructions no longer standing and now long forgotten probably failed because someone did not account for scaling. The second half of the twentieth century witnessed a fascinating redirection of our interest in physical scaling – from making things ever larger to making them ever smaller, first down to the scale of microns and now pushing to nanometers.[1]

A.6.1 Energy-based comparisons

The conventional approach to assessing the relative merits of capacitive and inductive MEMS is to compare their attainable volume energy storage densities: u_e and u_m. The rationale is that these quantities should accurately predict usable energy conversion capability. Equations (A.12) and (A.36) provide definitions for these volumetric energy densities in J/m^3 as functions of electric and magnetic field magnitudes. The challenge is to identify field values that are realistically attainable in microdevices.

On the microscale, the upper limit for the electric field E_{max} is usually imposed by one of two electrical breakdown mechanisms: Townsend (spark) discharge or field emission. Device dimensions, usually the length of an air gap, and ambient pressure determine which one governs. For air gaps on the scale of $\sim$2 to $\sim$10 μm at atmospheric pressure, the so-called *minimum sparking potential*, $V_{min} = 327$ volts controls breakdown, in which case, $E_{max} \approx V_{min}/d$. A common estimate for B_{max} is the saturation flux density of typical ferromagnetic materials, namely, $B_{max} \approx B_{sat} \sim 1$ T.

Table A.2 summarizes computations of u_e and u_m for these extreme upper limits on the electric and magnetic fields. The electroquasistatic energy density is approximately fifty times larger than the magnetostatic value. But is this really correct? For one thing, the minimum sparking potential, 327 volts, is far higher than the operating voltage of any conventional electronic system to be connected to our capacitive MEMS device. A more realistic upper limit might be 5 volts, in which case u_e drops by nearly four orders of magnitude. As a result, magnetic systems seem to gain the advantage based on energy density. But the choice of $B_{max} \approx 1$ T for the maximum attainable flux density is also questionable. In fact, major breakthroughs will be required in fabricating patterned ferromagnetic structures before the 1 T limit can be attained.

A big concern about energy-density-based comparisons is that they make the inherently questionable assumption that the two hypothetical devices being compared have the same volume. This assumption simply does not bear up to scrutiny if one considers

1 In 1989, William Trimmer introduced a scaling analysis methodology, the essential feature of which is a simple notational scheme that facilitates examination of how physical effects change in their relative importance when dimensions are altered subject to realistic constraints [2]. We do not employ Trimmer's analysis in this text, but strongly recommend its study to any serious student of microelectromechanical systems.

Table A.2 Useful expressions for calculating energy densities and typical parameter values associated with electric and magnetic fields

	Electroquasistatic (capacitive) systems	Magnetostatic (inductive) systems
Energy density (J/m^2)	$\kappa\varepsilon_0 E^2/2$	$B^2/2\mu_r\mu_0$
Physical mechanism of limits imposed on fields	Minimum sparking potential: $V_{min} = 327$ volts	Ferromagnetic saturation
Maximum field values	$E_{max} = V_{min}/d$, assuming $d = 5\,\mu$m: $E_{max} \sim 6.5 \cdot 10^7$ V/m	$B_{max} \sim 1$ T
Maximum energy densities	Assuming $\kappa = 1$: $u_e \sim 1.9 \cdot 10^4$ J/m^3	Assuming $\mu_r = 1000$: $u_m \sim 4 \cdot 10^2$ J/m^3

the very different topologies of capacitors and inductors, especially on the microscopic scale. Based on the issues raised here, it seems advisable not to rely too heavily on computations of u_e and u_m when comparing capacitive and inductive MEMS capabilities.

A.6.2 Area-based energy comparison

An alternative approach is to compare the net stored energy of capacitive and inductive devices scaled to micromechanical systems dimensions and taking up identical *areas* on a chip. The compelling argument favoring this approach is that chip area is very expensive real estate coveted by each design engineer working on a new microdevice. This comparison makes no assumption at all about the respective working volumes of the two devices, and indeed allows for the high-aspect-ratio structures favored in magnetic MEMS technology.

The hypothetical capacitive device is a parallel-plate geometry fabricated using surface micromachining technology. The inductor is the planar coil with N turns illustrated in Fig. A.11. Table A.3 includes approximate, analytical expressions for C and

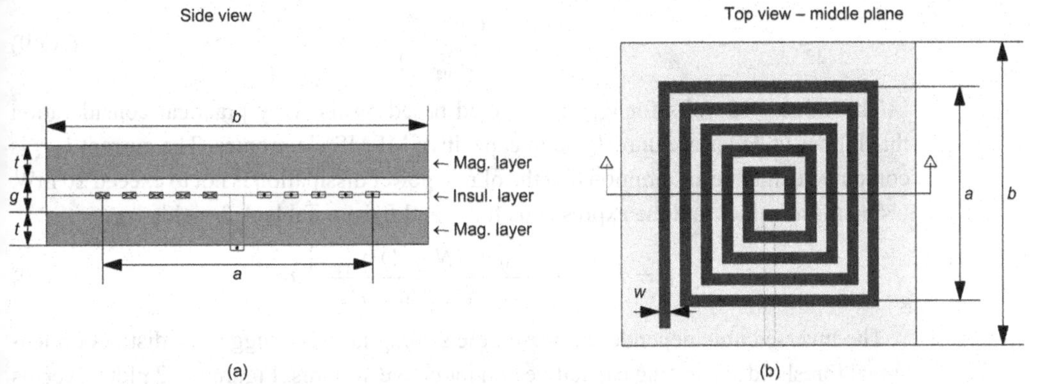

Figure A.11 Geometry of a planar inductor coil. (a) Side view showing the layered structure with conductors and magnetic material. (b) Top view showing the spiral coil design.

Table A.3 Useful expressions and typical parameter values for circuit energy storage calculations and MEMS scaling analysis. All dimensions are referenced to fixed area $A = 1\ \text{mm} \times 1\ \text{mm}$. It is assumed that the scale factor α is applied to all dimensions, except for h, the thickness of the copper traces that form the planar coil

Quantity	Capacitive systems	Inductive systems
Stored energy (J)	$\frac{1}{2}Cv^2$	$\frac{1}{2}Li^2$
Circuit parameters	Capacitance: [a] $C = \kappa\varepsilon_0 A/d$	Inductance:[b] $L = \dfrac{\mu_0 N^2 a^2}{\left[2 + \dfrac{2a^2}{2b^2 - a^2}\right] g + \dfrac{\sqrt{2}\,a\,b}{8\,\mu_r\,t}}$ Resistance: $R = 2\rho(N+1)a/(wh)$
Material properties	Dielectric constant: $\kappa = 1$	Cu resistivity: $\rho = 1.7 \cdot 10^{-7}$ ohm-m Ferromagnetic element: $\mu_r = 1000$
Dimensions of reference element	$A = b^2;\ d = 2 \cdot 10^{-6}\,\text{m}$	$N = 10$, $b = 1\ \text{mm},\ a = b/\sqrt{2},\ t = 0.3\ \mu\text{m},$ $g = 1\ \mu\text{m},\ w = 30\ \mu\text{m},\ h = 0.3\ \mu\text{m}$
Magnitude limits	$v_{max} = 5$ V[c]	$i_{max} = \sqrt{P_{max}/R}$[d] $P_{max} = 20$ mW

[a] Capacitor is assumed to be fabricated using surface-fabrication technology.
[b] Refer to Fig. A.11 for geometry of the planar coil.
[c] Voltage v_{max} is chosen to be compatible with CMOS circuitry.
[d] Current i_{max} is limited by the maximum allowed Joule heat dissipation of the resistor.

L, along with dimensional information for both devices. Note that the dimensional values appearing in this table are referenced to devices covering a fixed chip area $A = 1\ \text{mm} \times 1\ \text{mm} = 10^{-6}\ \text{m}^2$.

To facilitate the analysis, we introduce the scaling factor α, which serves as the multiplier for all physical dimensions provided in the table, that is, $a \to \alpha a, t \to \alpha t, d \to \alpha d, \ldots$, except for h, the thickness of the Cu traces of the planar coil. Thus, for $\alpha = 0.5$ or $\alpha = 2$, all scalable device dimensions are either halved or doubled, respectively. We now define the ratio r of stored energies for the hypothetical capacitor C and inductor L, each of which takes up area A on the chip:

$$r = \frac{Cv_{max}^2/2}{Li_{max}^2/2}. \tag{A.49}$$

The value of 5 volts for v_{max} is selected based on the very practical consideration that higher voltages are unrealistic in capacitive MEMS electronics. The current i_{max} is computed using the assumption that the ohmic power dissipation is not to exceed 20 mW. Using these values and the expressions for C and L from Table A.3 yields

$$r = \left[0.354 \frac{\rho\kappa\varepsilon_0}{\mu_r\mu_0} \frac{b^3(N+1)}{dhtwN^2} \frac{v_{max}^2}{P_{max}}\right] \alpha^{-3}. \tag{A.50}$$

The inverse cubic dependence of r on the scaling factor α suggests a distinct dimensional threshold separating capacitive and inductive systems. Figure A.12 plots r versus α over the range $0.1 \le \alpha \le 10$. The $r = 1$ break-even condition occurs near $\alpha \approx 1.4$,

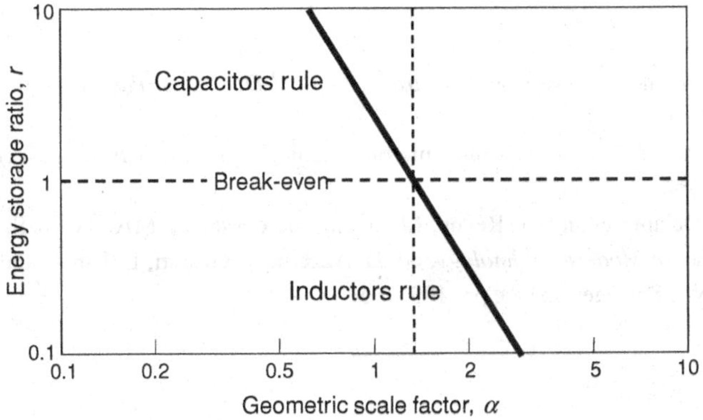

Figure A.12 Ratio of stored energy for parallel-plate capacitor and planar spiral inductor under the assumption that the two devices take up identical area A on a chip. See Table A.3 for detailed specifications. For parameters chosen, the capacitor does better for $A < \sim 1\ \text{mm}^2$, while the inductor wins out for larger sizes.

that is, an area of $A \approx 2\ \text{mm}^2$. Capacitive and inductive MEMS devices are favored at, respectively, smaller and larger sizes.

Cugat *et al.* [3] presented another scaling analysis that leads to still greater optimism about the prospects for magnetic MEMS. The essential tenet of their approach is that determining the maximum allowed current density in a planar coil must account for heat dissipation through the surface. Thus, reducing the scale factor α actually boosts the permissible current density. Values from 10^3 to 10^5 A/mm^2 are claimed. For reference, the uniform current density for the spiral coil used in this analysis predicts a current density of $\sim 10^3$ A/mm^2.

A.6.3 Summary

The primary goal of the scaling analysis presented in Section A.6.2 is to raise more questions about the conventional view of capacitive versus inductive MEMS. The flaw of our analysis, and probably all the others, is that it fails to consider comparative transducer functionality. For example, no consideration is given to such specifics as frequency response or the electromechanical force versus stroke. Recourse to Trimmer's method might provide the means to deal with these performance issues. Were we to employ it, we would probably discover that these two microelectromechanical technologies will never even be suited for similar applications. The significant distinctions between the geometries of practical capacitive and inductive devices and also the near impossibility of predicting exactly where the fabrication technology of magnetic materials will be in ten years make it a very risky proposition to issue pronouncements about the relative merits of capacitive and magnetic MEMS. At the writing of this text, magnetic microdevices claim only a small share of the MEMS market. Still, the promise of this technology suggests that students invest time for study of Chapter 9 in order to be ready for its emergence.

References

1. H. H. Woodson and J. R. Melcher, *Electromechanical Dynamics. Part II*, (New York: Wiley, 1968), Section 7.2.
2. W. S. N. Trimmer, Microrobots and micromechanical systems. *Sensors and Actuators*, 19 (1989), 267–287.
3. O. Cugat, J. Delamare, and G. Reyne, Magnetic microsystems: MAG-MEMS. In *Magnetic Nanostructures in Modern Technology*, ed. B. Azzerboni, G. Asti, L. Pareti, and M. Ghidini. (Dordrecht, NL: Springer, 2008), pp. 105–125.

B Review of mechanical resonators

This appendix reviews the essentials of lumped parameter mechanical systems relevant to microelectromechanical sensors and actuators. Our focus here is mechanical resonators. This choice is based on the reality that, in the MEMS world, stictive phenomena such as van der Waals forces or capillarity are always lurking. Like a tree snagging a kite from the air, these forces can fasten upon a moving element and affix it permanently to an adjacent surface. For this reason, resonance is usually the best way to exploit large motions reliably in a MEMS device. We start with a simple spring-mass system with one degree of freedom and Newton's second law in scalar form. All the main concepts are derived from the single-degree-of-freedom linear model, and then generalized to multiple-degree-of-freedom models and continuous systems using modal analysis.

The approach taken in this appendix is tailored to students having some familiarity with Laplace transforms, from the perspective of either resonant electric circuits or mechanical resonators. The starting point is the differential equation of motion; simple Laplace techniques are introduced so that solutions to the equation can be obtained and the important properties of this solution can be studied.

B.1 Degrees of freedom (DOF): spring-mass system

Consider the linear spring-mass system shown in Fig. B.1a. It is characterized by three lumped parameters: the mass m [kg], the stiffness of the spring k [N/m], and the viscous damping b [N s/m]. The mass is constrained to move along a straight line and its position with respect to some reference point is denoted by $x(t)$. The external force f_{ext} is applied along the x direction, thus allowing us to replace vector variables, such as force and displacement, by scalars. While this one-dimensional (1-D) idealization may seem overly simplistic for describing practical situations, the approach provides very useful first-order approximations for many MEMS devices.

The essential dynamics of this 1-D system are lumped into three types of attribute: inertial properties (kinetic energy storage) are lumped into the mass m; elastic properties (potential energy storage) are lumped into stiffness k; and motion-retarding properties (mechanical energy dissipated as heat) are lumped into the damping b.[1] Often, both

[1] In MEMS devices, the assumption of linearity is most questionable for the damping term. Some of the most important damping mechanisms, e.g., Coulomb damping, are simply not linear.

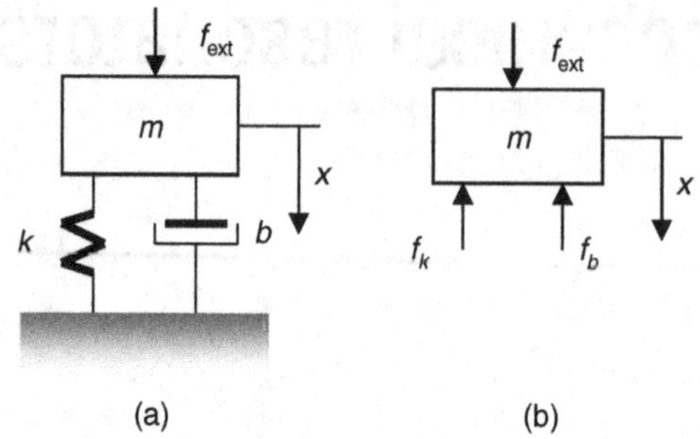

Figure B.1 The basic one-dimensional spring-mass system. (a) A symbolic sketch of the damped mass-spring resonator. (b) The free-body diagram.

for convenience and to emphasize that these parameters result from a process aimed at model simplification, we shall refer to them as *equivalent mass*, *equivalent stiffness*, and *equivalent damping*. We will be able to take considerable advantage of these equivalent values when modeling dynamic continua, such as bending beams and plates.

Restricting the motion to one coordinate means that the system has one degree of freedom (DOF). Such systems are usually referred to as single-degree-of-freedom (SDF) systems. Both translational motion in one direction and shaft rotation about a fixed axis are examples of mechanical systems with a single DOF. Furthermore, a material point constrained to move along a line, straight or curved, also has one DOF. Many MEMS continua with one dominant mode of motion can be approximated quite satisfactorily as SDF systems.

B.2 Newton's law: free-body diagrams

Figure B.1b shows a *free-body diagram* representation of an SDF. Newton's second law may be stated

$$m\ddot{x} = \sum f = f_{\text{ext}} - f_k - f_b, \tag{B.1}$$

where f_{ext} is the external force, f_b is the retarding force of the damper, and f_k is the restoring force of the spring. Note that the sign chosen for f_{ext} is opposite to f_b and f_k. The constitutive relations for a linear spring and a linear damper are $f_k = kx$ and $f_b = b\dot{x}$, respectively, where both k and b are positive. Introducing these relations into Eq. (B.1) results in the *equation of motion*,

$$m\ddot{x} + b\dot{x} + kx = f_{\text{ext}}, \tag{B.2}$$

which is a second-order, linear, ordinary differential equation. Section B.4 addresses the subject of system parameter measurement. Often, the equivalent mass and equivalent stiffness of MEMS devices must be approximated analytically. Procedures for doing so are presented in Section B.4. Equivalent damping, usually very difficult to quantify, is often determined experimentally.

B.2.1 Solving the equation of motion for an SDF

The equation of motion is solved using unilateral Laplace transforms. After presenting the formal solution for Eq. (B.2), we will provide some physical interpretation for it. The Laplace transform is defined by

$$\underline{y}(s) = \mathcal{L}\{y(t)\} = \int_{0-}^{\infty} y(t)\mathrm{e}^{-st}\mathrm{d}t, \tag{B.3}$$

and the inverse transform is

$$y(t) = \mathcal{L}^{-1}\{\underline{y}(s)\} = \frac{1}{2\pi j}\int_{\gamma-j\infty}^{\gamma+j\infty} Y(s)\mathrm{e}^{st}\mathrm{d}s. \tag{B.4}$$

$y(t)$ and $Y(s)$ form what is called a transform pair, signified by the notation $y(t) \leftrightarrow \underline{y}(s)$. In these definitions, the complex quantity s is regarded as a complex frequency. The Laplace transform of the time derivative of $y(t)$ may be derived directly from Eq. (B.3) using partial integration:

$$\mathcal{L}\{\dot{y}(t)\} = s\underline{y}(s) - y(0^-). \tag{B.5}$$

Note that L$\{\dot{y}(t)\}$ depends on the initial condition $y(0^-)$. The rule of Eq. (B.5) can be applied multiple times for higher-order derivatives. For example, the Laplace transform of the second derivative is

$$\mathcal{L}\{\ddot{y}(t)\} = s^2\underline{y}(s) - sy(0^-) - \dot{y}(0^-). \tag{B.6}$$

Note that this transform carries along information about the values of y and its first derivative at $t = 0$. For more details on Laplace transforms, refer to a textbook on signals and systems, e.g. Papoulis [1].

Applying Eqs. (B.5) and (B.6) to Eq. (B.2) gives, after some rearrangement,

$$\underbrace{(ms^2 + bs + k)}_{\substack{\text{characteristic}\\ \text{polynomial } \lambda(s)}}\underline{x}(s) = \underline{f}_{\text{ext}}(s) + \underbrace{m\dot{x}(0^-) + (ms + b)x(0^-)}_{\text{initial conditions}}, \tag{B.7}$$

where $\underline{x}(s)$ and $\underline{f}_{\text{ext}}(s)$ are the Laplace transforms of $x(t)$ and $f_{\text{ext}}(t)$, respectively. Solving Eq. (B.7) for $\underline{x}(s)$, we obtain

$$\underline{x}(s) = \frac{\underline{f}_{\text{ext}}(s)}{ms^2 + bs + k} + \frac{m\dot{x}(0^-)}{ms^2 + bs + k} + \frac{(ms + b)\,x\,(0^-)}{ms^2 + bs + k}. \tag{B.8}$$

Equation (B.8) reveals that the solution of a linear differential equation is the sum of the forced (driven) and the homogenous responses, the latter consisting of the sum of transient terms dependent on the initial conditions and manifesting the natural frequencies of the system. The time domain solution is obtained by taking the inverse Laplace transform of the right-hand side of Eq. (B.8):

$$x(t) = \underbrace{\mathcal{L}^{-1}\left\{\frac{\underline{f}_{\text{ext}}(s)}{ms^2 + bs + k}\right\}}_{\text{response due to forcing}} + \underbrace{\mathcal{L}^{-1}\left\{\frac{m\dot{x}(0^-)}{ms^2 + bs + k} + \frac{(ms + b)x(0^-)}{ms^2 + bs + k}\right\}}_{\text{response due to initial conditions}}. \tag{B.9}$$

Note that the driven response, the first term in Eq. (B.9), does not depend on the initial conditions. Also, for resonators and other second-order systems, the homogeneous response will, in general, have two components. For a passive system, both of these terms decay in time.

First, consider the driven response defined by

$$\underline{x}_{\text{driven}}(s) = \frac{1}{ms^2 + bs + k}\underline{f}_{\text{ext}}(s). \tag{B.10}$$

We make no assumption about the excitation; Eq. (B.10) is valid for any $f_{\text{ext}}(t)$. In the complex frequency domain, the transfer function is the ratio of the output to the input when $x(0^-) = \dot{x}(0^-) = 0$:

$$\underline{H}(\underline{s}) = \left.\frac{\underline{x}_{\text{driven}}(\underline{s})}{\underline{f}_{\text{ext}}(\underline{s})}\right|_{x(0^-)=0,\dot{x}(0^-)=0} = \frac{1}{m\underline{s}^2 + b\underline{s} + k}. \tag{B.11}$$

$H(\underline{s})$ provides a complete description of the response for our resonator (and, more generally, does the same for any linear system). The inverse transform of $H(\underline{s})$, known as the impulse response $h(t)$, is an entirely equivalent way to characterize a system:

$$h(t) = \mathcal{L}^{-1}\{\underline{H}(\underline{s})\}. \tag{B.12}$$

B.2.2 Poles of the transfer function

In general, $\underline{H}(\underline{s})$ will be in the form of a fractional polynomial expression. The roots of the numerator are the zeros and those of the denominator are the poles. Both zeros and poles are, in general, complex numbers. The characteristic equation from which the poles are obtained results from setting the denominator of $\underline{H}(\underline{s})$ to zero:

$$\lambda(\underline{s}) = m\underline{s}^2 + b\underline{s} + k = 0. \tag{B.13}$$

The *characteristic polynomial* $\lambda(\underline{s})$ was first introduced in Eq. (B.7). Equation (B.13) is a quadratic with two roots, $\underline{s}_1$ and $\underline{s}_2$:

$$\underline{s}_{1,2} = -\frac{b}{2m} \pm \sqrt{\left(\frac{b}{2m}\right)^2 - \frac{k}{m}}. \tag{B.14}$$

For a passive system, $b > 0$; thus, the real parts of these two roots are always negative. Based on the parameters b, m, and k, one may distinguish four cases:

$b = 0$ undamped case: imaginary complex-conjugate poles;
$b^2 < 4km$ under-damped case: complex-conjugate poles;
$b^2 = 4km$ critically damped case: negative, double real poles;
$b^2 > 4km$ over-damped case: real, negative, unequal poles.

Although the undamped case is physically non-realizable, because some energy loss mechanism is always present, this limit remains a useful idealization in modeling, for example, in the case of mechanical systems having more than one degree of freedom. In the limit of weakly damped systems, the roots may be approximated by

$$\underline{s}_{1,2} = \pm j\omega_0, \tag{B.15}$$

where ω_0 is the natural frequency,

$$\omega_0 = \sqrt{\frac{k}{m}}. \tag{B.16}$$

In the general case, where loss may not be ignored, the roots defined by Eq. (B.14) are conveniently expressed as

$$\underline{s}_{1,2} = -\zeta\omega_0 \pm \underbrace{j\omega_0\sqrt{1-\zeta^2}}_{=\omega_d}, \tag{B.17}$$

where ζ and ω_d are, respectively, the damping coefficient and the damped natural frequency:

$$\zeta = \frac{1}{2}\frac{b}{\sqrt{mk}} \quad \text{and} \quad \omega_d = \omega_0\sqrt{1-\zeta^2}. \tag{B.18}$$

The definitions of ω_0 and ζ provide a very convenient and general way to express the equation of motion of a second-order system in the time domain:

$$\ddot{x} + 2\zeta\omega_0\dot{x} + \omega_0^2 = f_{ext}/m. \tag{B.19}$$

More information about ω_0 and ω_d is provided in the next section, when we examine the transfer function in the sinusoidal steady-state frequency domain.

There are important exceptions, but many MEMS resonators may be described adequately as under-damped SDFs, that is, $\zeta < 1$, in which case, the roots $\underline{s}_1$ and $\underline{s}_2$ are a complex conjugate pole pair. For a critically damped system, $\zeta = 1$, so that the poles are identical and purely real: $\underline{s}_1 = \underline{s}_2 = -\zeta\omega_0$. Critically damped systems have the attribute of the fastest response to a step input. Over-damped systems, defined by the condition $\zeta > 1$, have two unequal real poles given by

$$\underline{s}_{1,2} = -\zeta\omega_0 \pm \omega_0\sqrt{\zeta^2 - 1}. \tag{B.20}$$

Figure B.2 plots the root locus in the complex plane using the damping coefficient ζ as the varied parameter.

B.2.3 Steady-state response to periodic forcing: the transfer function

If the forcing function is a sinusoid, we may employ the complex exponential form

$$f_{ext}(t) = \text{Re}[\underline{f}_{ext} e^{j\omega t}]. \tag{B.21}$$

For algebraic convenience, let $f_{ext}(t) \to \underline{f}_{ext} e^{j\omega t}$. After doing all the manipulations necessary to get a solution, we may invoke the superposition principle by taking the real part of the solution to revert to the time domain. The Laplace transform of the complex exponential forcing function is

$$\underline{f}_{ext}(\underline{s}) = \mathcal{L}\{F_0 e^{j\omega t}\} = \frac{F_0}{\underline{s} - j\omega}. \tag{B.22}$$

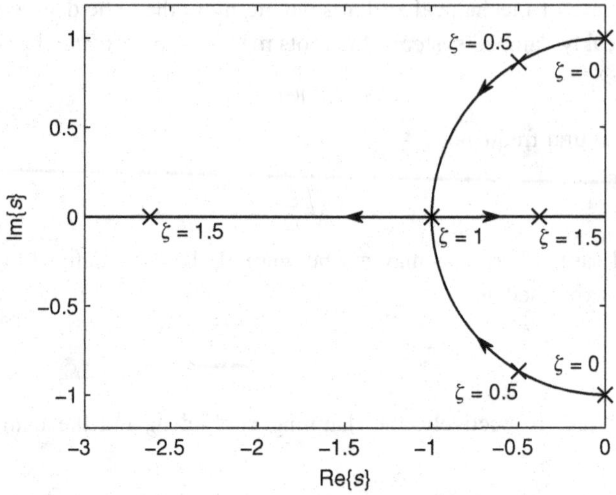

Figure B.2 Loci of the complex conjugate pole pair in the s-plane for a second-order resonator as the damping coefficient ζ ranges from 0 to ∞.

Using Eq. (B.22) in Eq. (B.10) gives

$$\underline{x}(\underline{s}) = \frac{F_0}{(\underline{s} - j\omega)\left(m\underline{s}^2 + b\underline{s} + k\right)} = \underbrace{\frac{\underline{c}_1}{\underline{s} - j\omega}}_{\substack{\text{forced term:} \\ \underline{x}_{ss}(\underline{s})}} + \underbrace{\frac{\underline{c}_2}{m\underline{s}^2 + b\underline{s} + k}}_{\substack{\text{transient term:} \\ \underline{x}_{tr}(\underline{s})}}, \tag{B.23}$$

where $\underline{c}_1$ and $\underline{c}_2$ are complex coefficients arising from the partial fraction expansion of $F_0/(\underline{s} - j\omega)(m\underline{s}^2 + b\underline{s} + k)$. Like Eq. (B.8), Eq. (B.23) partitions $\underline{x}(\underline{s})$ into forced and transient response components, respectively, $\underline{x}_{ss}(\underline{s})$ and $\underline{x}_{tr}(\underline{s})$. The forced response is itself a sinusoidal steady-state term. The net transient response is the sum of $\underline{x}_{tr}(\underline{s})$ and terms due to non-zero initial conditions, that is, non-zero $x(0^-)$ and $\dot{x}(0^-)$. Comparing $\underline{x}_{ss}(\underline{s})$ with Eq. (B.22), we see that the response exhibits the same complex exponential form as the excitation:

$$x_{ss}(t) = \mathcal{L}^{-1}\{\underline{x}_{xs}(\underline{s})\} = \underline{c}_1 e^{j\omega t}, \tag{B.24}$$

where

$$\underline{c}_1 = \left.\frac{F_0}{m\underline{s}^2 + b\underline{s} + k}\right|_{\underline{s}\to j\omega} = \underbrace{\left(\frac{1}{-m\omega^2 + b(j\omega) + k}\right)}_{\underline{H}(j\omega)} F_0 \tag{B.25}$$

$\underline{H}(j\omega)$ is the transfer function for the sinusoidal steady-state case: $\underline{s} = j\omega$. Substituting Eq. (B.25) into Eq. (B.24) yields

$$x_{ss}(t) = \underline{H}(j\omega) F_0 e^{j\omega t} = A(\omega) F_0 e^{j[\omega t + \phi(\omega)]}, \tag{B.26}$$

where $A(\omega)$ and $\phi(\omega)$ are the magnitude and phase of $\underline{H}(\mathrm{j}\omega)$, respectively:

$$A(\omega) = |\underline{H}(\mathrm{j}\omega)| \tag{B.27}$$

and

$$\phi(\omega) = \angle\underline{H}(\mathrm{j}\omega) = \arctan\left(\frac{\mathrm{Im}[\underline{H}(\mathrm{j}\omega)]}{\mathrm{Re}[\underline{H}(\mathrm{j}\omega)]}\right). \tag{B.28}$$

So far, we assumed the excitation to be in the form of a complex exponential. Now, with all the manipulations completed, we can obtain the steady-state response in the time domain by taking the real part of Eq. (B.26):

$$f(t) \rightarrow \mathrm{Re}\lfloor F_0 \mathrm{e}^{\mathrm{j}\omega t}\rfloor = A(\omega) F_0 \cos\lfloor \omega t + \phi(\omega)\rfloor. \tag{B.29}$$

By normalizing the frequency, that is, $\omega_{\mathrm{r}} \equiv \omega/\omega_0$, the transfer function for the resonator takes on a form that offers insight into its driven behavior:

$$\underline{H}(\mathrm{j}\omega) = \frac{1/k}{(1-\omega_{\mathrm{r}}^2) + \mathrm{j}2\zeta\omega_{\mathrm{r}}}. \tag{B.30}$$

Equation (B.30) shows that the response at resonance, that is, $\omega = \omega_0$, is limited only by the damping. Thc magnitude and phase of $\underline{H}(\mathrm{j}\omega)$ are

$$A(\omega) = |\underline{H}(\mathrm{j}\omega)| = \frac{1/k}{\sqrt{(1-\omega_{\mathrm{r}}^2)^2 + (2\zeta\omega_{\mathrm{r}})^2}} \tag{B.31}$$

and

$$\phi(\omega) = \angle\underline{H}(\mathrm{j}\omega) = -\arctan\left(\frac{2\zeta\omega_r}{1-\omega_r^2}\right). \tag{B.32}$$

Figure B.3 shows the amplitude and phase responses for a second-order resonator for several values of the damping factor ζ. As frequency is reduced, the limit of the normalized amplitude response approaches one, because in the limit of zero frequency, the displacement equals the static displacement $x_0 = F_0/k$. As expected, the phase response ϕ approaches zero as frequency goes to zero. At resonance, $\omega_{\mathrm{r}} = 1$, the phase delay is 90°. For $\omega_{\mathrm{r}} \gg 1$, $\phi \rightarrow 180°$.

The transfer function first presented in Eq. (B.11) defines the input as the force and the output as the mechanical displacement, but there are other, very useful alternatives. One in particular, used for accelerometers, treats acceleration as the output:[2]

$$\underline{H}_{\mathrm{a}}(\mathrm{j}\omega) = \frac{\omega^2 \underline{X}(\mathrm{j}\omega)}{\underline{f}(\mathrm{j}\omega)} = \frac{-\omega_{\mathrm{r}}^2/m}{(1-\omega_{\mathrm{r}}^2) + \mathrm{j}2\zeta\omega_{\mathrm{r}}}. \tag{B.33}$$

The amplitude and phase for this complex function, respectively, $|\underline{H}_{\mathrm{a}}(\mathrm{j}\omega)|$ and $\angle\underline{H}_{\mathrm{a}}(\mathrm{j}\omega)$, are plotted for different values of the damping factor ζ in Fig. B.4. Note that the normalized acceleration asymptotically approaches one for $\omega \gg \omega_0$. It is instructive to compare this plot with the displacement transfer function plotted in Fig. B.3.

[2] The transfer function defined in Eq. (B.30) is commonly employed for seismic instruments and other sensors that measure displacement, while the form used in Eq. (B.33) is suited for accelerometers.

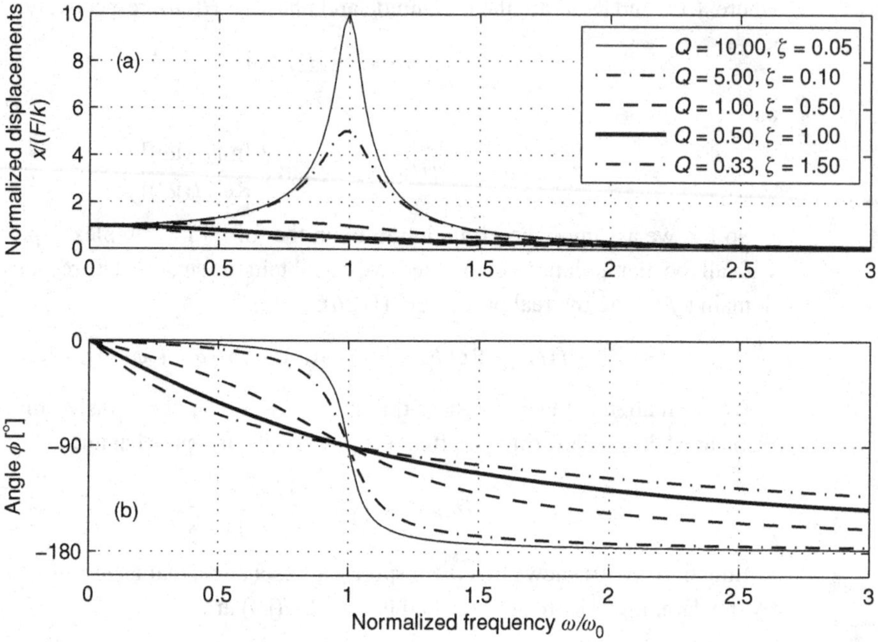

Figure B.3 The transfer function $\underline{H}$, that is, displacement vs. force, for a second-order SDF resonator system. (a) Amplitude response. (b) Phase response.

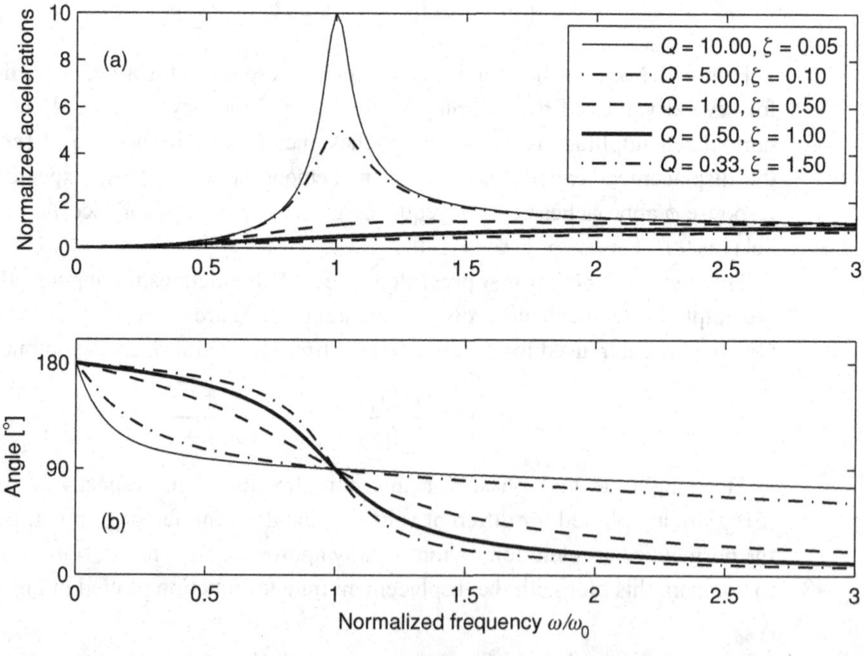

Figure B.4 The transfer function $\underline{H}_a$, that is, acceleration vs. force, for a resonator (SDF) system. Compare this plot with Fig. B.3. (a) Amplitude response. (b) Phase response.

B.2.4 The quality factor

Instead of the damping factor ζ, another quantity more familiar in circuit theory called the quality factor Q may be used:

$$Q = \frac{1}{2\zeta} = \frac{\sqrt{km}}{b}. \tag{B.34}$$

Valuable physical insight about the quality factor of a resonator arises from its definition as the ratio of average stored energy to average energy dissipated per cycle. For a mechanical resonator, the system energy constantly changes back and forth between elastic and kinetic forms, so we may define Q in either of two ways:

$$Q = \frac{\frac{k}{2}\langle \dot{x}^2 \rangle / \omega_0}{\frac{b}{2}\langle \dot{x}^2 \rangle} = \frac{\frac{m}{2}\langle \dot{x}^2 \rangle \omega_0}{\frac{b}{2}\langle \dot{x}^2 \rangle} = \frac{\sqrt{km}}{b}. \tag{B.35}$$

The quality factor is sometimes referred to as *the magnification factor* because $A(\omega = \omega_0) = Q/k = QA(\omega = 0)$, as can be shown by introducing Eq. (B.34) into Eq. (B.31).

B.2.5 State-space representation and phase plots

State variables provide a convenient alternative formalism to describe the dynamics of an SDF. The state space form has certain advantages, one being that it is better suited when the equation of motion must be solved by numerical integration, for example, if non-linearities are present. To use the state space formalism for the resonator, one starts with definitions for two state variables, usually displacement, $q_1 = x(t)$ and velocity, $q_2 = \mathrm{d}x/\mathrm{d}t$. These definitions are used in Eq. (B.2) to obtain two coupled, first-order, differential equations:

$$\dot{q}_1 = q_2, \tag{B.36a}$$

$$\dot{q}_2 = -\frac{k}{m}q_1 - \frac{b}{m}q_2 + \frac{f_{\text{ext}}}{m}. \tag{B.36b}$$

A good way to interpret the significance of the state variables is to recognize their relationship to the energy-storing mechanisms of the resonator. In the present case, displacement (q_1) is a measure of the potential energy stored in the spring while velocity (q_2) measures the kinetic energy of the mass. Rewriting Eqs. (B.36a) and (B.36b) in matrix form yields

$$\begin{bmatrix} \dot{q}_1 \\ \dot{q}_2 \end{bmatrix} = \underbrace{\begin{bmatrix} 0 & 1 \\ -k/m & -b/m \end{bmatrix}}_{\overline{\overline{a}}} \begin{bmatrix} q_1 \\ q_2 \end{bmatrix} + \begin{bmatrix} 0 \\ 1/m \end{bmatrix} f_{\text{ext}}, \tag{B.37}$$

where $\overline{\overline{a}}$, as identified in Eq. (B.37), is called the system matrix. Note that the characteristic polynomial can be obtained here from $\overline{\overline{a}}$: $\lambda(s) = \det(s\,\overline{\overline{1}} - \overline{\overline{a}})$, where $\overline{\overline{1}}$ is the identity matrix of the same order as $\overline{\overline{a}}$. As before, the roots of $\lambda(s) = 0$ are the system poles.

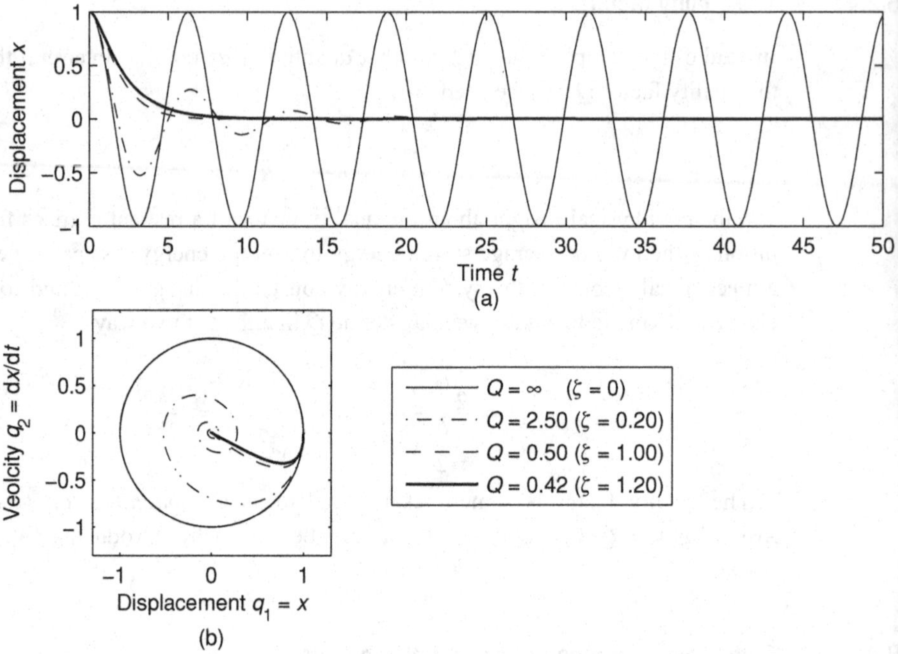

Figure B.5 The response of a second-order mechanical resonator with an initial displacement but no other external forcing for selected values of damping factor ζ. (a) Time-domain plot of $x(t)$ versus t. (b) Phase space trajectory: $q_1(t)$ versus $q_2(t)$.

Figures B.5a and B.5b show the response in the time domain and in phase space, respectively, for a mechanical resonator that has been displaced from its equilibrium position for various values of the damping factor ζ: $\zeta = 0$ (undamped), $\zeta = 0.2$ (under-damped), $\zeta = 1.0$ (critically damped), and $\zeta = 1.2$ (over-damped). In the undamped case, there is no dissipation, so energy flows back and forth forever between potential and kinetic forms. When the potential energy reaches its peak, the velocity and therefore the kinetic energy are zero. When the kinetic energy achieves its maximum, the displacement and the potential energy stored in the spring are zero.

B.3 Electromechanical analogies: mechanical impedance and admittance

If we choose velocity $\dot{x}$ instead of displacement x as the dependent variable, Eq. (B.2) takes the form of an integro-differential equation:

$$m\frac{\mathrm{d}}{\mathrm{d}t}(\dot{x}) + b\dot{x} + \int \dot{x}\,\mathrm{d}s = f_{\mathrm{ext}}. \tag{B.38}$$

In complex (phasor) notation, Eq. (B.38) becomes

$$\left(\mathrm{j}\omega m + b + \frac{1}{\mathrm{j}\omega(1/k)}\right)(\mathrm{j}\omega \underline{x}) = \underline{f}_{\mathrm{ext}}. \tag{B.39}$$

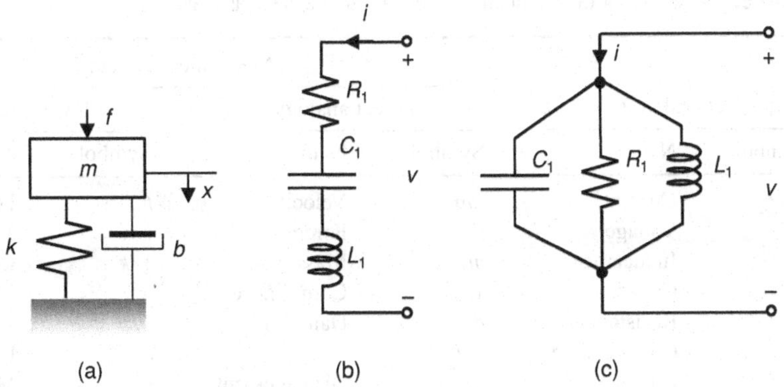

Figure B.6 Electrical analogies for mechanical resonators. (a) The mechanical SDF system. (b) Serial *RLC* circuit. (c) Parallel *RLC* circuit.

A very powerful interpretation of Eq. (B.39) emerges if we seek analogies between the resonator shown in Fig. B.6a and the RLC circuits shown in Figs. B.6b and B.6c. In phasor form, the $v - i$ relationship for the series RLC circuit of Fig. B.6b is

$$\underbrace{\left(\mathrm{j}\omega L + R + \frac{1}{\mathrm{j}\omega C}\right)}_{\underline{Z}_e} \underline{i} = \underline{v}. \tag{B.40}$$

The equivalent relation for the parallel RLC circuit shown in Fig. B.6c is

$$\underbrace{\left(\mathrm{j}\omega C + \frac{1}{R} + \frac{1}{\mathrm{j}\omega L}\right)}_{\underline{Y}_e} \underline{v} = \underline{i}. \tag{B.41}$$

Equations (B.39), (B.40), and (B.41) are canonically identical, a direct consequence of the essential similarity of second-order electrical and mechanical resonance. The two systems are entirely analogous. Both are characterized by sinusoidal transients that decay in time. The roles of potential and kinetic energy in the mechanical resonator are played by capacitive and inductive energy storage in the electrical resonator. Mechanical damping is replaced by electrical resistance. The choice of a mechanical interpretation for an RLC circuit or an electrical interpretation for the mechanical resonator depends on the use for which the analogy is to be applied. For electromechanical system modeling, electrical analogs for the mechanical resonators have the greater utility, simply because MEMS devices are always interfaced to electronic circuitry. We will find that these electrical analogs have equal utility in actuator and sensor systems.

Further examination of Eqs. (B.40) and (B.41) indicates that, depending on whether we select voltage ($\underline{v}$) as the analog to force ($\underline{f}_{\mathrm{ext}}$) and current ($\underline{i}$) as the analog to velocity ($\mathrm{j}\omega\underline{x}$), or the reverse, then, respectively, the series or parallel RLC network is appropriate. The first of these is known as the *direct analogy* and the second is the *indirect analogy*.

Table B.1 Variables for the direct and indirect electromechanical analogies

Electrical system		Mechanical system: Direct analogy		Mechanical system: Inverse analogy	
Symbol	Name	Symbol	Name	Symbol	Name
i	Current	$j\omega x$	Velocity	f	Force
v	Voltage	f	Force	$j\omega x$	Velocity
L	Inductance	m	Mass	$1/k$	Compliance
C	Capacitance	$1/k$	Compliance	m	Mass
R	Resistance	b	Damping	$1/b$	–
G	Conductance	$1/b$	–	b	Damping
Q	Charge	x	Displacement	p	Momentum
λ	Flux linkage	p	Momentum	x	Displacement

B.3.1 The direct analogy

If we state that force is analogous to voltage and that velocity is analogous to current, then the ratio of force to velocity is called a mechanical impedance:

$$\underline{Z}_{\text{m0}} = \frac{\underline{f}_{\text{ext}}}{j\omega}\underline{x} = j\omega m + b + \frac{k}{j\omega} \tag{B.42}$$

According to this scheme, the "impedances" due to inertia, damping, and the spring force are additive, that is, series-connected as shown in Fig. B.6b.

B.3.2 The inverse analogy

Alternately, if we identify velocity and force to be is analogous, respectively, to voltage and current, the mechanical admittance is

$$\underline{Y}_{\text{m0}} = \frac{\underline{f}_{\text{ext}}}{j\omega \underline{X}} = j\omega m + b + \frac{k}{j\omega}. \tag{B.43}$$

In this case, inertia, damping, and the restoring force of the spring are in effect admittances connected in parallel, as shown in Fig. B.6c.

In Chapter 4, the concepts of mechanical impedance and admittance are used extensively to facilitate the construction of models accounting for the mechanical constraints imposed on electromechanical two-port network. Care must be taken in using these mechanical quantities because their units do not reflect the inverse relationship of electrical impedance and admittance. Table B.1 contains a summary of the important equivalencies employed for the direct and inverse analogies. Note that the terms $\underline{Z}_{\text{m0}}$ and $\underline{Y}_{\text{m0}}$ are chosen for their consistency with usage in Chapter 4.

Electromechanical analogies have a long history [2, 3]. Generally, the direct analogy is more popular, but some texts, e.g., Woodson and Melcher [4], prefer the inverse analogy. Neubert [5] prefers the matrix approach but commits to neither direct nor inverse analogy. However, $\underline{Z}_{\mathrm{m}}$, defined in his transducer matrix, and used throughout this book, implies the direct analogy.[3] This book employs the direct analogy.

B.4 Experimental determination of resonator parameters

In practice, the parameters of a fabricated MEMS resonator are often measured experimentally. There are basically two methods for doing so. In the first method, one observes the decaying transient in time, either after the driving force has been removed or after the system has been subjected to a mechanical stimulus giving it an initial displacement or velocity. This approach leads to an estimate for the damping factor ζ. The second experimental method involves measuring the frequency response: the resonator is driven with a sinusoidal force of constant amplitude while the frequency is swept over the range of interest. In this case, damping is accessed through the quality factor Q and Eq. (B.34).

B.4.1 Transient response method

There are several ways to initiate transients on a MEMS device. The simplest is to turn on the periodic forcing for enough time to achieve steady-state motion and then turn it off. Depending on the timing of the turn-off, which can be controlled accurately, initial conditions ranging from $x(0) = 0$ to $\dot{x} = 0$ can be imposed. A seemingly crude yet often effective method is to strike the device a sharp blow or to rap it against a table or bench. Obviously, initial conditions are hard to control with such techniques; however, with due care taken to avoid damaging the device, this method is in fact simple and time-saving. Another method that works for larger devices like cantilevered beams is to displace the moving element statically with a small micromanipulator and then to release it and observe the transient. The condition achieved here is $x(0) = x_0$ and $\dot{x} = 0$.

Consider an under-damped resonator: $\zeta < 1$. The response to the non-zero initial conditions is the inverse Laplace transform of the last two terms grouped together in Eq. (B.9). Equivalently, one may note that the response is a linear combination of $\exp(\underline{s}_1 t)$ and $\exp(\underline{s}_2 t)$, where $\underline{s}_1$ and $\underline{s}_2$ are roots, as defined by Eq. (B.17). The general form of the solution is

$$x(t) = A\mathrm{e}^{-\zeta\omega_0 t}\cos(\omega_{\mathrm{d}}t) + B\mathrm{e}^{-\zeta\omega_0 t}\sin(\varepsilon_{\mathrm{d}}t), \tag{B.44}$$

where A and B are unknown constants that depend on the initial conditions. For example, if $x(0^-) = 0$, then $A = 0$ and $B = \dot{x}_0/\varepsilon_{\mathrm{d}}$, then

$$x(t) = \frac{\dot{x}_0}{\omega_{\mathrm{d}}}\mathrm{e}^{-\zeta\omega_0 t}\sin(\omega_{\mathrm{d}}t). \tag{B.45}$$

The response due to non-zero velocity and zero displacement as initial conditions is simply the scaled impulse response of the mechanical resonator.

[3] $\underline{Z}_{\mathrm{m}}$ is the ratio between force $\underline{f}$ and velocity $\mathrm{j}\omega\underline{x}$.

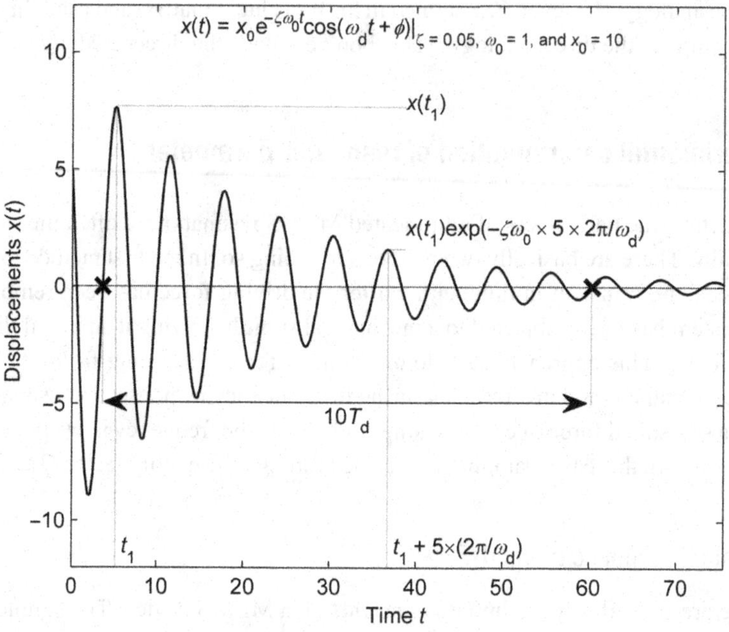

Figure B.7 Quantitative characterization of damped oscillations: (i) measurement of damping factor by means of the logarithmic decrement method and (ii) oscillation period measured using zero crossings.

For the more general case of non-zero displacement and velocity, solution for A and B followed by some manipulation yields

$$x(t) = \underbrace{\sqrt{x_0^2 + \left(\frac{\dot{x}_0 + \zeta\varepsilon_0 x_0}{\omega_d}\right)^2}}_{x_0} \mathrm{e}^{-\zeta\omega_0 t} \cos\left[\underbrace{\omega_d t - \arctan\left(\frac{\dot{x}_0 + \zeta\omega_0 x_0}{\omega_d x_0}\right)}_{\phi}\right]. \tag{B.46}$$

Figure B.7 shows an example of such damped oscillations. Good accuracy can usually be achieved by measuring two peaks n periods apart, and then taking the logarithm of their ratio, that is, $\delta_n \equiv \ln[x(t_0)/x(t_0 + nT_\mathrm{d})]$, where T_d is the period of oscillation and δ_n is called the *logarithmic decrement*. Using the definition of ω_d, it may be written as

$$\delta_n = \frac{2\pi n\zeta}{\sqrt{1-\zeta^2}}. \tag{B.47}$$

In the limit of small damping, $\zeta \approx \delta_n/(2\pi n)$. For large damping, it is more convenient to express ζ in terms of the logarithmic decrement δ_n:

$$\zeta = \frac{\delta_n}{\sqrt{\delta_\mathrm{n}^2 + (2\pi n)^2}}. \tag{B.48}$$

Note that the logarithmic decrement method is not useful when $\zeta > 1$.

The resonant frequency ω_d is determined by measuring the period of the damped oscillation, $T_d = 2\pi/\omega_d$. T_d is most easily measured by looking at zero crossings; the time difference is measured and then divided by the number of periods. With estimates for ζ and T_d, the natural frequency is

$$\omega_0 = \frac{2\pi}{T_d\sqrt{1-\zeta^2}}. \tag{B.49}$$

Such dynamic measurements can never provide direct estimates for k and m.

An alternate way to measure ζ and T_d is to fit measured waveform data to the functional form of Eq. (B.45) by a *least-squares* method or some other error minimization algorithm:

$$\text{error} = \sqrt{\sum_{i=1}^{N}\left(x\,(i) - \frac{\dot{x}_0}{\omega_d}\mathrm{e}^{-\zeta\omega_0 t}\,\sin(\omega_d t_i)\right)^2}. \tag{B.50}$$

This approach has the advantages that it exploits all the available data and is easy to automate. The disadvantage is that it is more susceptible to errors due to noisy data. Also, the initial condition on velocity, which does not affect the resonator properties, must be treated as a fitting parameter.

The data-fitting method actually finds its best application for over-damped and for critically damped resonators. For over-damped devices, Eq. (B.46) is replaced by

$$x(t) = A\mathrm{e}^{-\zeta\omega_0 t}\cosh(\sqrt{\zeta^2-1}\,\omega_0 t) + B\mathrm{e}^{-\zeta\omega_0 t}\sinh(\sqrt{\zeta^2-1}\,\omega_0 t), \tag{B.51}$$

while for a critically damped resonator, the response takes the form

$$x(t) = A\mathrm{e}^{-\zeta\omega_0 t} + Bt\mathrm{e}^{-\zeta\omega_0 t}. \tag{B.52}$$

B.4.2 Frequency sweep method

Implementation of the frequency response method involves examination of the magnitude of the response as the frequency is swept through the range of interest. Consider the under-damped case. From Eq. (B.32), $|\underline{H}(\mathrm{j}\omega_0)| = Q$. The simplest way to measure Q is to take the ratio of the (peak) magnitude of motion at resonance and at DC, that is, $Q = |x(t)|_{\omega=\omega_0}/|x(t)|_{\omega=0}$. The natural frequency is identified as the frequency of the peak. More careful analysis shows that the magnitude achieves its maximum slightly below resonance, at the damped natural frequency:

$$\frac{\partial|H(\omega)|}{\partial\omega} = 0 \Rightarrow \omega = \omega_0\sqrt{1-\frac{1}{4Q^2}} = \omega\sqrt{1-\zeta^2} = \omega_d. \tag{B.53}$$

Also, the peak itself is slightly higher than QF_0/k:

$$|\underline{H}(\omega_d)| = Q\frac{1}{\sqrt{1-3/(16Q^4)}} \approx Q\left(1+\frac{3}{32Q^4}\right). \tag{B.54}$$

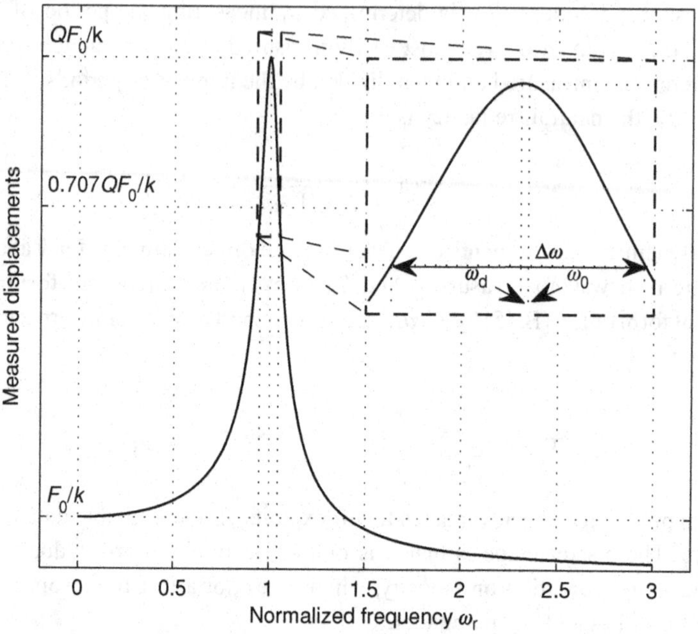

Figure B.8 Parameter estimation from the magnitude response; ω_0 corresponds to the peak of the magnitude response. The bandwidth $\Delta\omega$ is the difference between frequencies at the half-power points. Note that $Q = \omega_0/\Delta\omega$.

The error consequent to neglecting this correction is usually small; for $Q \geq 5$, it is usually less than 1%.

In some MEMS resonators, DC measurements cannot be easily made. In this case, Q is estimated by bandwidth measurements. The half-power points are the two frequencies on either side of the resonance where the magnitude of the response is equal to $1/\sqrt{2} \sim 0.707$ of the peak value. The difference between these two frequencies, $\Delta\omega$, is called the *bandwidth*. Refer to Fig. B.8. The quality factor Q defined before in terms of the damping factor, Eq. (B.34), can be expressed as the ratio of the bandwidth and the resonance frequency:

$$Q = \frac{\omega_0}{\Delta\omega}. \tag{B.55}$$

If the half-power points cannot be obtained directly, it is still possible to employ curve fitting. In fact, curve fitting is a good choice for over-damped and critically damped systems. Care must be exercised when dealing with relatively closely spaced resonances.

During frequency sweep measurements, it is important to verify that the peak occurs at the same frequency regardless of the direction of the sweep (from low to high, or from high to low frequency). If this test fails, the system is operating outside of the linear regime.

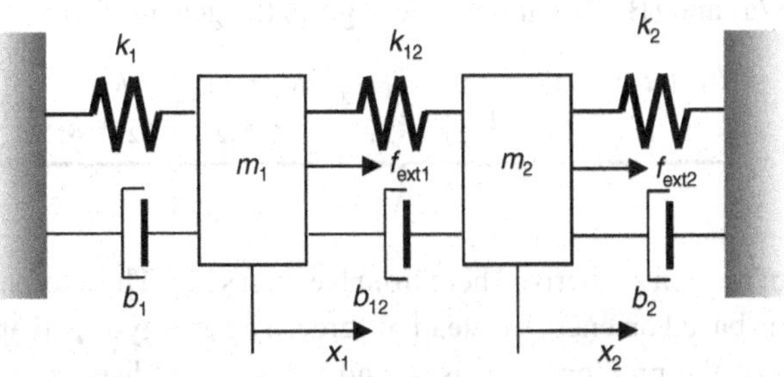

Figure B.9 Coupled mechanical resonators exemplifying a two-degrees-of-freedom mechanical device having two masses coupled by springs and dampers.

B.5 Multiple DOF systems

Though many MEMS devices can be classified as SDF systems, a significant fraction are not. The most important example is the gyroscope, which usually features two degrees of freedom. A different sort of example is the cantilevered beam. The cantilevered beam is, of course, a continuum with effectively $n \to \infty$ DOFs. Nevertheless, perfectly adequate models for beams can be constructed with a finite number of degrees of freedom. These examples provide sufficient motivation for the consideration of MDF systems.

In general, the dynamics of MDF systems is difficult to describe. Newton was successful in solving the two-body problem, but scientists and mathematicians vainly sought further generality for about 200 years until Poincare proved that no closed-form solution exists for the three-body problem. Fortunately, in the MEMS world there is no true free body motion. Furthermore, with the allowed motion generally quite restricted, we may confidently generalize from SDF to MDF systems in a straightforward manner.

B.5.1 Equations of motion

While an SDF system is described by one scalar equation of motion, such as Eq. (B.2), an MDF model with n degrees of freedom requires n such equations. For convenience in describing these systems, we employ matrices. The general form of the equation of motion for an MDF system with n degrees of freedom is

$$\overline{\overline{m}}\,\ddot{\overline{x}} + \overline{\overline{b}}\dot{\overline{x}} + \overline{\overline{k}}\overline{x} = \overline{f}_{\text{ext}}, \tag{B.56}$$

where $\overline{\overline{m}}$, $\overline{\overline{b}}$, and $\overline{\overline{k}}$ are $n \times n$ matrices characterizing mass, damping, and stiffness, respectively. Note that the number of system parameters increases geometrically with n. Displacement $\overline{x}$ and force $\overline{f}_{\text{ext}}$ vectors are represented as $n \times 1$ column matrices.

Consider first the coupled resonator with two masses, m_1 and m_2, shown in Fig. B.9. The coupled linear equations of motion are

$$m_1\ddot{x}_1 = f_{\text{ext1}} - k_1x_1 - k_{12}(x_1 - x_2) - b_1\dot{x}_1 - b_{12}(\dot{x}_1 - \dot{x}_2), \tag{B.57a}$$

$$m_2\ddot{x}_2 = f_{\text{ext2}} - k_2x_2 - k_{12}(x_2 - x_1) - b_2\dot{x}_2 - b_{12}(\dot{x}_2 - \dot{x}_1). \tag{B.57b}$$

Casting Eqs. (B.57a) and (B.57b) into matrices gives the general form of Eq. (B.56):

$$\underbrace{\begin{bmatrix} m_1 & 0 \\ 0 & m_2 \end{bmatrix}}_{\overline{\overline{m}}} \underbrace{\begin{bmatrix} \ddot{x}_1 \\ \ddot{x}_2 \end{bmatrix}}_{\ddot{\overline{x}}} + \underbrace{\begin{bmatrix} b_1 + b_{12} & -b_{12} \\ -b_{12} & b_2 + b_{12} \end{bmatrix}}_{\overline{\overline{b}}} \underbrace{\begin{bmatrix} \dot{x}_1 \\ \dot{x}_2 \end{bmatrix}}_{\dot{\overline{x}}} + \underbrace{\begin{bmatrix} k_1 + k_{12} & -k_{12} \\ -k_{12} & k_2 + k_{12} \end{bmatrix}}_{\overline{\overline{k}}} \underbrace{\begin{bmatrix} x_1 \\ x_2 \end{bmatrix}}_{\overline{x}}. \tag{B.58}$$

The equations of motion are derived here from Newton's law. The alternative approach, not covered here, is based on energy instead of force. The energy-based approach scales nicely as the size of the problem increases, and it has some distinct advantages. The reader is referred to Crandall [6] or Moon [7] for an introduction to energy methods.

B.5.2 Natural frequencies and mode shapes

To find the dynamic modes and their natural frequencies, it is beneficial to resort to the limit of zero damping. Consider then the zero damping limit of the homogenous equation of motion obtained by setting $\overline{f}_{\text{ext}} = \overline{0}$ in Eq. (B.56). Equation (B.58) reduces to

$$\overline{\overline{m}}\,\ddot{\overline{x}} + \overline{\overline{k}}\overline{x} = \overline{0}. \tag{B.59}$$

Assume that all degrees of freedom exhibit sinusoidal motion at a single common frequency, that is, $\overline{x}(t) = \overline{x}_0 \cos(\omega t)$. Then, Eq. (B.59) becomes

$$[-\omega^2 \overline{\overline{m}} + \overline{\overline{k}}] = \overline{0}. \tag{B.60}$$

A non-trivial solution to Eq. (B.60) exists only when the determinant of the matrix is zero:

$$\lambda(\mathrm{j}\omega) = \det([-\omega^2 \overline{\overline{m}} + \overline{\overline{k}}]) = \overline{0}. \tag{B.61}$$

Equation (B.61), the characteristic equation of an MDF system, is an nth order polynomial in ω^2. The squares of these natural frequencies, ω_i^2, are the eigenvalues. For zero damping, all eigenvalues are real.[4] Each eigenvalue has its own corresponding eigenvector $\overline{\phi}_i$.

For vibrating systems, the eigenvalue–eigenvector pair has a nice physical interpretation: the eigenvalue is the square of the frequency at which the resonator oscillates and the eigenvector contains information about the shape of the oscillations in space. Eigenvectors are also referred to as the natural modes or mode shapes. The mode shapes are obtained by substituting the eigenvalues into Eq. (B.59) and then solving for $\overline{x}$. Mode shapes are only known within a multiplicative constant: if $\overline{x}$ satisfies Eq. (B.60), then so does $\alpha\overline{x}$, where α is any real number.

Let us examine how eigenanalysis is used to decouple n coupled resonators. Eigenanalysis provides a framework to realize mathematically the intuitive notion of representing a complex mechanical system by a set of n independent resonators. The method is based on the approximation that near each resonance even a complex system tends

[4] A non-zero damping matrix gives rise to complex natural frequencies.

to behave like a simple SDF, with all mass elements oscillating in accordance with the mode shape or eigenvalue.

To start, we compute all n eigenvectors and arrange them into the modal matrix $\overline{\overline{\Phi}}$ as follows:

$$\overline{\overline{\Phi}} = [\overline{\phi}_1 \overline{\phi}_2 \ldots \overline{\phi}_n]. \tag{B.62}$$

Next, we pre-multiply the equation of motion by the transpose of the modal matrix $\overline{\overline{\Phi}}^{\mathrm{T}}$:

$$\overline{\overline{\Phi}}^{\mathrm{T}} \overline{\overline{m}}\, \ddot{\overline{x}} + \overline{\overline{\Phi}}^{\mathrm{T}} \overline{\overline{k}} \overline{x} = \overline{\overline{\Phi}}^{\mathrm{T}} \overline{f}_{\mathrm{ext}}. \tag{B.63}$$

Finally, we introduce modal displacements $\overline{\eta}$ via $\overline{x} = \overline{\overline{\Phi}} \overline{\eta}$ into Eq. (B.63):

$$\overline{\overline{\Phi}}^{\mathrm{T}} \overline{\overline{m}}\, \overline{\overline{\Phi}} \ddot{\overline{\eta}} + \overline{\overline{\Phi}}^{\mathrm{T}} \overline{\overline{k}}\, \overline{\overline{\Phi}} \overline{\eta} = \overline{\overline{\Phi}}\, \overline{f}_{\mathrm{ext}}. \tag{B.64}$$

$\overline{\overline{\Phi}}$ transforms $\overline{\overline{m}}$ and $\overline{\overline{k}}$ into diagonal matrices. The modal mass matrix $\overline{\overline{m}}$ is

$$\overline{\overline{m}} = \overline{\overline{\Phi}}^{\mathrm{T}}\, \overline{\overline{m}}\, \overline{\overline{\Phi}} = \begin{bmatrix} M_1 & 0 & \ldots & 0 \\ 0 & M_2 & \ldots & 0 \\ \ldots & \ldots & \ldots & \ldots \\ 0 & 0 & \ldots & M_n \end{bmatrix} \tag{B.65}$$

and the modal stiffness matrix $\overline{\overline{K}}$ is

$$\overline{\overline{K}} = \overline{\overline{\Phi}}^{\mathrm{T}}\, \overline{\overline{k}}\, \overline{\overline{\Phi}} = \begin{bmatrix} K_1 & 0 & \ldots & 0 \\ 0 & K_2 & \ldots & 0 \\ \ldots & \ldots & \ldots & \ldots \\ 0 & 0 & \ldots & K_n \end{bmatrix}. \tag{B.66}$$

The off-diagonal elements of $\overline{\overline{M}}$ and $\overline{\overline{K}}$ are zero because the mode shapes are orthogonal with respect to the mass and stiffness matrices $\overline{\phi}_i^{\mathrm{T}} \overline{\overline{m}} \phi_j = 0$ and $\phi_i^{\mathrm{T}} \overline{\overline{k}} \phi_j = 0$ for $i \neq j$.

As stated, the eigenvectors are only known within a multiplicative constant, and so can be normalized in any of several ways. One approach is to choose the multiplicative constant, so that

$$\phi_i^{\mathrm{T}} \overline{\overline{m}} \phi_i = 1. \tag{B.67}$$

With this normalization, the $\overline{\overline{M}}$ matrix transforms into the identity matrix $\overline{\overline{I}}$,

$$\overline{\overline{M}} = \Phi^{\mathrm{T}} \overline{\overline{m}} \Phi = \overline{\overline{1}}, \tag{B.68}$$

and $\overline{\overline{k}}$ transforms into a diagonal matrix, with the diagonal elements equal to $\omega_i{}^2$,

$$\overline{\overline{K}} = \overline{\overline{\Phi}}^{\mathrm{T}} \overline{\overline{k}} \overline{\overline{\Phi}} = \begin{bmatrix} \omega_1^2 & 0 & \cdots & 0 \\ 0 & \omega_2^2 & \cdots & 0 \\ \ldots & \ldots & \ldots & \ldots \\ 0 & 0 & \cdots & \omega_n^2 \end{bmatrix}. \tag{B.69}$$

B.5.3 Example of a system with two degrees of freedom

As an illustrative example, consider the system with two DOFs described by Eq. (B.58) but with zero damping: $b_1 = b_{12} = b_2 = 0$. According to Eq. (B.61), we have

$$\begin{vmatrix} -m_1\omega^2 - k_1 - k_{12} & k_{12} \\ k_{12} & -m_2\omega^2 - k_2 - k_{12} \end{vmatrix}, \tag{B.70}$$

the roots of which are[5]

$$\omega_{1,2}^2 = \frac{m_1 k_2 + m_2 k_1 \mp \sqrt{(m_1 k_2 + m_2 k_1)^2 - 4 m_1 m_2 \left(k_1 k_2 + k_{12}^2\right)}}{2 m_1 m_2}. \tag{B.71}$$

To simplify the algebra for the example, we introduce some additional simplifying assumptions: $m_1 = m_2 = m$ and $k_1 = k_2 = 2k_{12} = k$. Then, Eq. (B.71) simplifies to

$$\omega_{1,2}^2 = \begin{cases} k/(2m), \\ 3k/(2m). \end{cases} \tag{B.72}$$

After normalization to the mass matrix, the eigenvectors become

$$\overline{\phi}_1 = \frac{1}{\sqrt{2m}} \begin{bmatrix} 1 \\ 1 \end{bmatrix}, \tag{B.73a}$$

$$\overline{\phi}_2 = \frac{1}{\sqrt{2m}} \begin{bmatrix} 1 \\ -1 \end{bmatrix}. \tag{B.73b}$$

Finally, the governing equation,

$$\begin{bmatrix} m & 0 \\ 0 & m \end{bmatrix} \ddot{\overline{x}} \begin{bmatrix} k & -k/2 \\ -k/2 & k \end{bmatrix} \overline{x} = \begin{bmatrix} f_{\text{ext1}} \\ f_{\text{ext2}} \end{bmatrix}, \tag{B.74}$$

becomes, after similarity transformation,

$$\begin{bmatrix} 1 & 0 \\ 0 & 1 \end{bmatrix} \ddot{\overline{\eta}} + \begin{bmatrix} k/(2m) & 0 \\ 0 & 3k/(2m) \end{bmatrix} \overline{\eta} = \frac{1}{\sqrt{2m}} \begin{bmatrix} f_{\text{ext1}} & +f_{\text{ext2}} \\ f_{\text{ext1}} & -f_{\text{ext2}} \end{bmatrix}. \tag{B.75}$$

In the matrix equation Eq. (B.75), the two modes are decoupled. Figure B.11 shows a schematic representation of this transformation.

Figure B.10 shows the two mode shapes. The mode shapes revealed in Fig. B.10 are, respectively, symmetric and antisymmetric because of the assumptions made about masses and spring constants. At the resonances, m_1 and m_2 move the same amount, either in phase or out of phase. In the more general case of $m_1 \neq m_2$ or $k_1 \neq k_2$, the masses will not move the same amount, but they still move in the same direction for the

[5] Because it is customary to order natural frequencies in ascending order, we use −/+ instead of +/−.

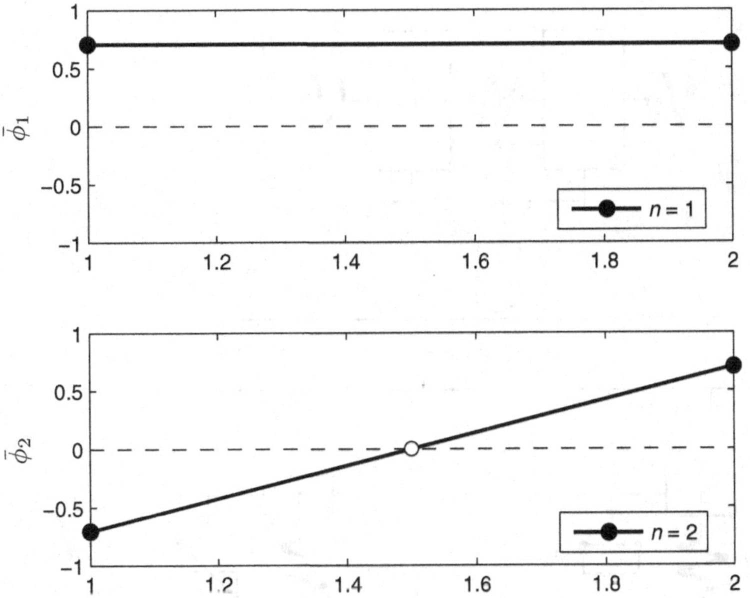

Figure B.10 The mode shapes of a symmetrical system with two degrees of freedom.

first mode and in the opposite directions (out-of-phase) for the second mode. This result becomes evident in writing out the mode shape expressions:[6]

$$\bar{\phi}_1 = \begin{bmatrix} \dfrac{k_2 m_1 - k_1 m_2 + \sqrt{(m_1 k_2 + m_2 k_1)^2 - 4 m_1 m_2 (k_1 k_2 - k_{12}^2)}}{4 k_1 m_1} \\ 1 \end{bmatrix}, \quad \text{(B.76)}$$

$$\bar{\phi}_2 = \begin{bmatrix} \dfrac{k_2 m_1 - k_1 m_2 - \sqrt{(m_1 k_2 + m_2 k_1)^2 - 4 m_1 m_2 (k_1 k_2 - k_{12}^2)}}{4 k_1 m_1} \\ 1 \end{bmatrix}. \quad \text{(B.77)}$$

In the second mode, there is a location in the mode shape curve that does not move. This point is referred as the node of the resonance. In general, the nth mode will have $n - 1$ nodes.

While the algebraic expressions quickly become messy, as seen in Eqs. (B.76) and (B.77), numerical methods can handle high-order matrices with no difficulty.

Figure B.12 exemplifies a more complex system featuring $n = 4$ degrees of freedom.

The mode shapes plotted in Fig. B.13 reveal that the number of nodes is one less than the number of the mode itself.

[6] Here, the mode shapes have not been normalized to the mass matrix, to avoid lengthy expressions. The reader is invited to visit the text website to learn how solutions can be obtained using symbolic manipulation and numerical calculations.

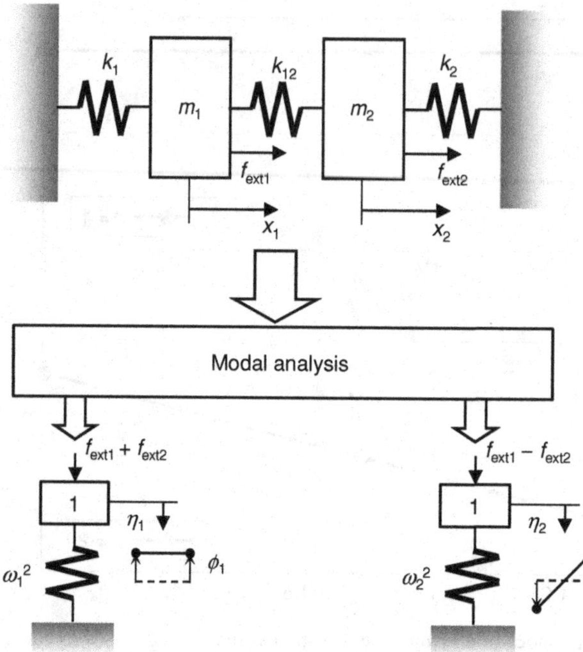

Figure B.11 Modal analysis visualization. Modal analysis is a process that converts a coupled oscillator into a series of decoupled oscillators, two in this case. With our choice of normalization of the mode shapes (eigenvectors), the masses of the decoupled resonators are equal to one, and the spring constants are equal to the squares of the corresponding natural frequencies. It is important to keep in mind that the generalized coordinates η_i are multiplying mode shapes $\overline{\phi}_i$.

B.5.4 Approximations for continuous systems

The cantilever beam is an important example of a mechanical continuum that can be modeled as a system with a finite number of DOFs. See Fig. B.14. For example, if interest is restricted to the motion at the free end of a homogeneous beam at frequencies up to the fundamental frequency, an SDF model may suffice and more complicated models may not be required. Appendix D shows that, with this restriction, one can estimate the stiffness from the static model (see Eq. (D.52)) and then approximate the equivalent mass as $3/8$ of the total beam mass (see Eq. (D.61)). Modeling the equivalent damping is, of course, far more complicated; typically, empirical values for damping are obtained from experiments. The important idea is that, for a large class of important devices, the dynamic complexity of many continua can be reduced to systems as simple as the resonator shown in Fig. B.1.

Continua usually possess a countably infinite number of modal resonances. For example, Fig. B.14b shows the second mode for the cantilevered beam. In cases where higher-order resonant dynamics are important, the SDF modeling approximation can still be taken advantage of, as long as the system is operated close to resonance.

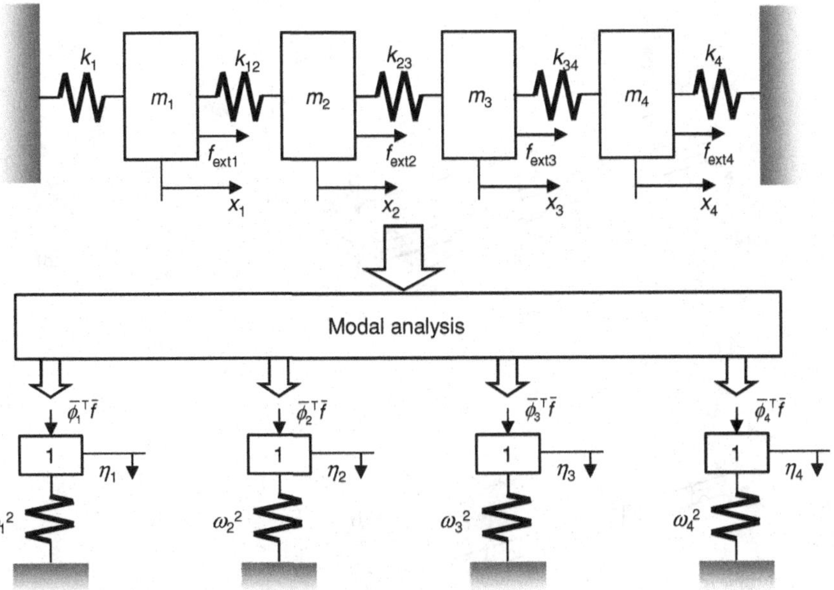

Figure B.12 Transformation of fourth-order coupled mechanical systems into four decoupled oscillators described by η_1, η_2, η_3, and η_4.

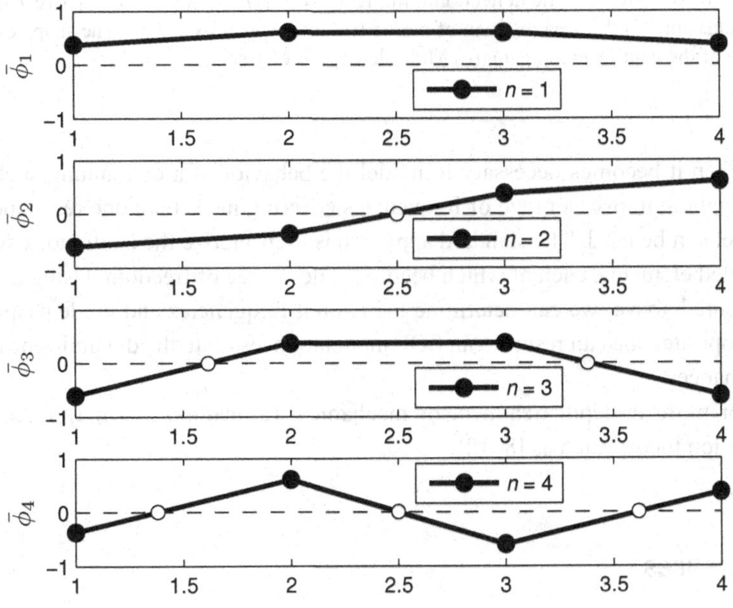

Figure B.13 Mode shapes of a fourth-order system. Note that the mode shapes reveal, respectively, 0, 1, 2, and 3 nodes for modes $n = 1$ through 4. For simplicity, it is assumed here that all the masses and all the spring constants are identical.

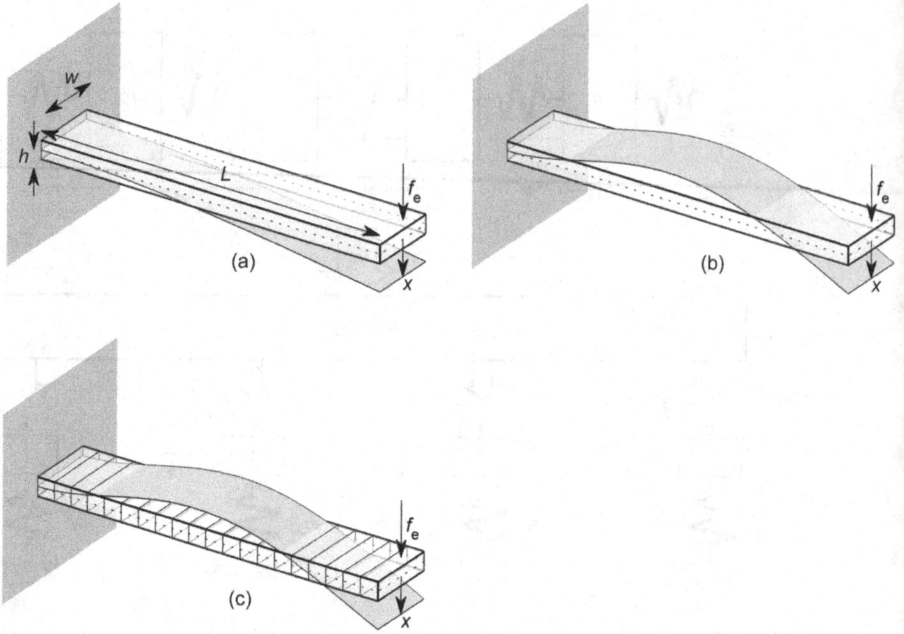

Figure B.14 A cantilevered beam of length L, width w, and thickness h. Near each resonance, the system behaves like a second-order system. (a) For the first mode, the equivalent mass is $m_{eq} = (3/8)\rho Lwt$, where ρ is the mass density of the beam, and the equivalent stiffness, determined from the static deflection limit, is $k_{eq} = 3YI/L^3 = \frac{1}{4}Ywh^3/L^3$, where Y is the Young modulus and I is the area moment of inertia (refer to Appendix D). (b) The shape of the second mode of the cantilevered beam. (c) Multiple-degree-of-freedom model.

When it becomes necessary to model the behavior of a continuum, such as a beam or diaphragm, over a range of frequencies covering more than one resonance, an MDF model can be used. The standard approach is to discretize the continuous system into n coupled elements, each of which becomes one degree of freedom. Using the formalism presented above, we can determine the natural frequencies and mode shapes. Then, as appropriate, one can resort to an SDF model to investigate the dynamics near individual resonances.

For more in-depth treatment of mechanical resonators, one may refer to texts on vibration theory such as [8–12].

References

1. A. Papoulis, *Circuits and Systems: A Modern Approach* (New York: Holt, Rinehart, and Winston, 1980).
2. F. A. Firestone, A new analogy between mechanical and electrical systems, *Journal of the Acoustical Society of America*, 4 (1933), 249.

3. W. P. Mason, *Electromechanical Transducers and Wave Filters* (New York: Van Nostrand Reinhold, 1946).
4. H. H. Woodson and J. R. Melcher, *Electromechanical Dynamics* (New York: Wiley, 1968).
5. H. K. P. Neubert, *Instrument Transducers: An Introduction to Their Performance and Design*, 2nd edn. (Oxford: Clarendon Press, 1975).
6. S. H. Crandall, *Dynamics of Mechanical and Electromechanical Systems* (New York: McGraw-Hill, 1968).
7. F. C. Moon, *Applied Dynamics: With Applications to Multibody and Mechatronic Systems* (New York: Wiley, 1998).
8. R. R. Craig, *Structural Dynamics: An Introduction to Computer Methods* (New York: Wiley, 1981).
9. A. D. Dimarogonas, *Vibration for Engineers*, 2nd edn (Upper Saddle River, NJ: Prentice Hall, 1996).
10. L. Meirovitch, *Analytical Methods in Vibrations* (New York: Macmillan, 1967).
11. L. Meirovitch, *Principles and Techniques of Vibrations* (Upper Saddle River, NJ: Prentice Hall, 1997).
12. A. A. Shabana, *Vibration of Discrete and Continuous Systems*, 2nd edn (New York: Springer, 1997).

C Micromachining

The field of microsystems has an interesting history. In the late 1950s, the physicist Richard P. Feynman [1] delivered a presentation during which he posed two technological challenges designed to stimulate development of new ways to design and build miniature machines. The first challenge was to write a page of a book on a surface 25 000 times smaller in linear dimensions than the original, to be read by an electron microscope. The second prize was to build a fully functional, miniature, electric motor with dimensions less than 1/64 of an inch. Surprisingly, the second prize was claimed first in 1960, using conventional fabrication methods. On the other hand, the technology required to accomplish the first challenge did not emerge until 1987. One early MEMS device, the resonant gate transistor shown in Fig. C.1, was demonstrated in 1967 [2], but the field did not really begin to emerge till the 1980s and 1990s. Today there is a vast array of methods and processes for fabricating micromechanical devices. This appendix offers a concise introduction to the fundamental concepts of microfabrication and to the ways it is used to create MEMS devices.

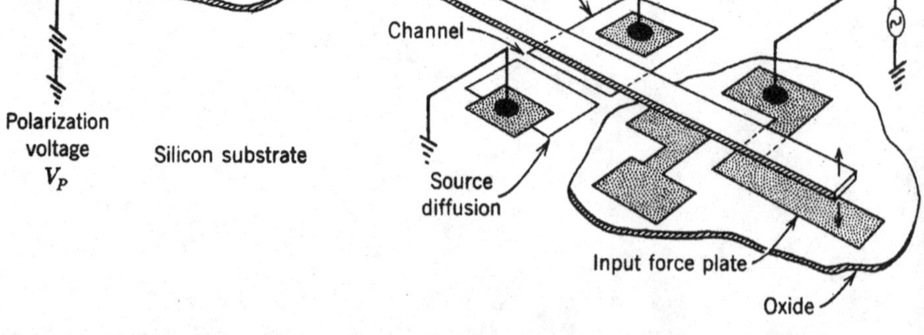

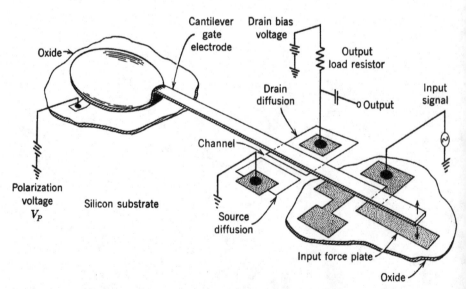

Figure C.1 A very early MEMS device – the resonant gate transistor [2], © IEEE. Used with permission. This device incorporates a cantilevered beam as the mechanical element.

Most of the fabrication processes emerged from the revolution in the microelectronics industry. Those adapted or specifically developed for MEMS are referred to collectively as *micromachining*. These batch processes have the considerable advantage that they can deliver large numbers of device replicas on a wafer simultaneously.

The fact that batch processing of MEMS devices was quickly achieved led to somewhat unrealistic expectations based on comparisons with the semiconductor industry. While serving to generate excitement in the beginning, these overly optimistic expectations actually had an inadvertently negative impact on MEMS development. The challenge, initially unrecognized, was that, unlike microelectronics, MEMS technologies are expected to operate over a wide range of physical environments and conditions. In addition, compared with VLSI, which is based on a very small set of building blocks – viz., transistors, diodes, and capacitors – MEMS devices must be virtually custom designed and fabricated. In this respect, only analog electronic circuits are even close to MEMS, and, in fact, many are embedded on a chip with customized analog circuitry. In any case, despite initial hype and early disappointments, the field has reached maturity and the principal design and development processes are now well established.

Because the art and science of micromachining is rich and dynamic, proper coverage is possible only in reference volumes [3–5]. This short appendix keeps the discussion at a basic level, introducing essential concepts and vocabulary without getting into detail. The goal here is to motivate further investigation by students who recognize the need to learn about and compare the capabilities of specific processes. After reading this appendix, you will not even be considered *dangerous*, but we hope that you will have become curious.

There are two largely distinct families of MEMS processes: *bulk* and *surface* micromachining. In bulk micromachining operations, devices are fashioned from the wafer structure itself by selective etching of the material. See Fig. C.2a. In surface micromachining, on the other hand, the wafer serves merely as a platform upon which

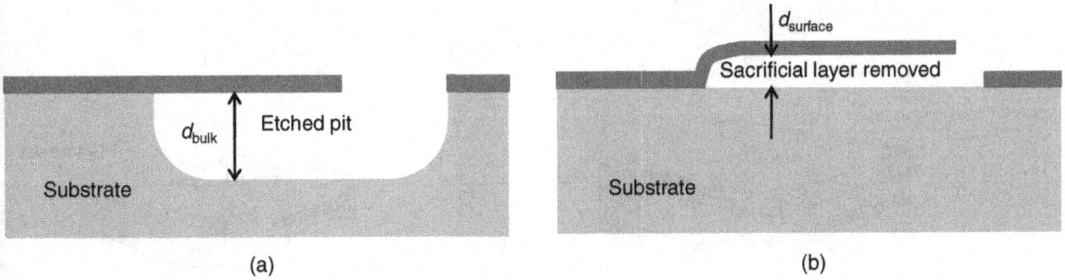

Figure C.2 Bulk- and surface-micromachined structures. The drawings are not to scale. The vertical dimensions in bulk micromachining are much larger than those in surface-micromachined structures, i.e., $d_{bulk} \gg d_{surface}$. (a) Bulk micromachined beam formed from the substrate material. (b) Surface-micromachined beam patterned from deposited layer on top of the wafer. This figure is adapted from [3].

the device is built, usually by depositing and then patterning material into desired structures, as shown in Fig. C.2b.

C.1 Planar processes

The majority of commercially available MEMS devices are created using planar processes, where the desired pattern is transferred from a mask to the substrate and material is added by deposition and then selectively removed by etching. The transfer of the pattern from the mask to the substrate is called *photolithography*. Figure C.3 illustrates the transfer of a simple rectangular shape from mask to wafer. The substrate is covered by a uniform 1–10 μm layer of photoresist, typically applied by *spin coating*, and then ultraviolet (UV) light is shone through a mask, which is usually made of quartz glass with a patterned film of chromium. The UV light alters the chemical properties of the photoresist, making it either more soluble (positive photoresist) or less soluble (negative photoresist). The more soluble material is then removed to replicate the pattern on the substrate. The photolithographic process depicted here, called *contact* or *proximity lithography*, is simple but offers relatively low resolution. For higher resolution, a larger mask is used and optical projection lithography, not shown in the figure, is employed to reduce the dimensions.

Masks, which specify the shape of the mechanical structure, electrodes, planar electrical traces, and insulation (oxide), are central to MEMS design and processing.

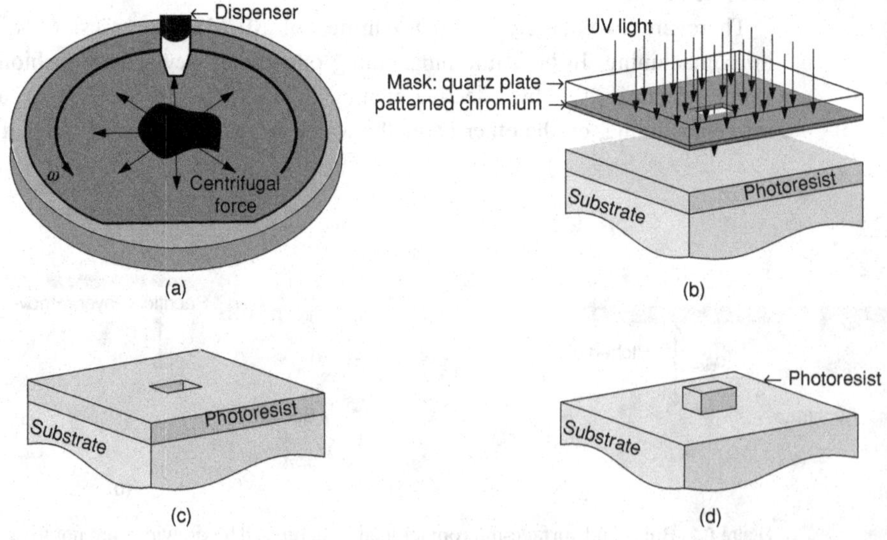

Figure C.3 Basics of contact photolithography. (a) Deposition of photoresist using spin-coating. (b) Photoresist exposure to UV through the mask. (c) After the etch step for positive photoresist. (d) After the etch step for negative photoresist. The choice between positive and negative photoresist processes depends on the structure to be fabricated and the materials to be used.

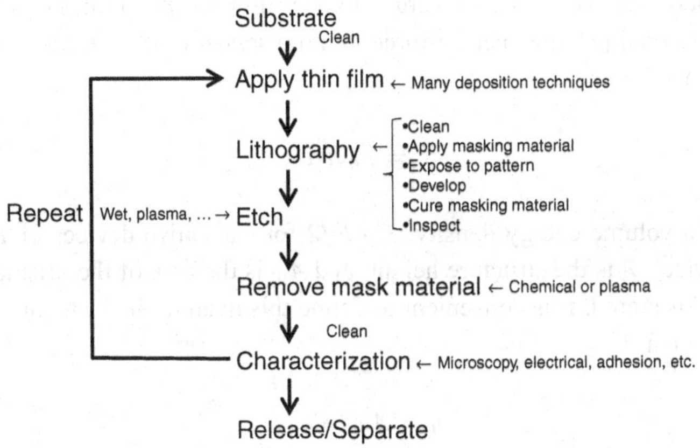

Figure C.4 Generic micromachining process flow chart, adapted from Jackson [5].

Usually, a sequence of lithographic steps involving several masks is employed in surface micromachining. Figure C.4 is a micromachining flow chart based on Jackson [5] that lays out the manufacturing steps employed for MEMS devices. Tools used in MEMS include thin-film deposition (*chemical vapor deposition*, *plasma-enhanced chemical vapor deposition*, *sputtering*, and *evaporation*); etching (*wet chemical isotropic*, *wet chemical anisotropic*, *plasma etching*, *reactive ion etching*, and *deep reactive ion etching*); machining (*drilling*, *milling*, and *electric discharge machining*); bonding (*fusion bonding*, *anodic bonding*, and *adhesives*); and annealing [3].

Typically, a microfabrication process uses one tool to accomplish each manufacturing step. Examples of the sequence of steps for one surface micromachining process and one bulk micromachining process are examined in Section C.4.2.3 (see Fig. C.18) and Section C.5 (see Fig. C.20), respectively.

C.2 Characteristics of MEMS structures

Understanding the capabilities and limitations of available micromachining processes is crucial to MEMS design. Because they are created from planar masks, 3-D structures have the effective form of extruded 2-D geometries. Thus, neither a sphere nor a torus can be fabricated readily. Also, 3-D structures usually feature dimensions that are much larger in the plane of the wafer (x and y) than in the normal (z) direction. The properties of pseudo 2-D structures differ markedly from the ponderable 3-D shapes of our everyday experience. For example, for flat structures with $z \ll x$ and $z \ll y$, the perpendicular theorem for moments of inertia can be used: $I_z \approx I_x + I_y$ (see Section 7.4.5.2 for more details).

It is instructive to consider the energy storage in prismatic volumes. Following the argument of Guckel [6], the energy stored in a transducer (with extruded z dimension) can be written as

$$W = \rho_W h A_W, \tag{C.1}$$

where ρ_W is a volume energy density: $\bar{D} \cdot \bar{E}/2$ for capacitive devices or $\bar{B} \cdot \bar{H}/2$ for magnetic devices; h is the structure height; and A_W is the area of the prismatic volume where energy is stored. It is convenient to define this quantity in terms of a fraction of the total chip area A_{chip}, that is,

$$\eta = A_W / A_{\text{chip}}. \tag{C.2}$$

Inserting Eq. (C.2) into Eq. (C.1) and rearranging yields a figure of merit for prismatic transducers:

$$\frac{W}{A_{\text{chip}}} = \rho_W \eta h. \tag{C.3}$$

Because W is directly proportional to h, microfabrication processes featuring high-aspect-ratio structures yield intrinsically more sensitive sensors and more robust actuators.

In the macroscopic world, large-amplitude motion is often achieved by exploiting friction (rolling, walking). In the microscopic world, on the other hand, the large surface areas of microstructures lead to strong stictive forces that can permanently disable small moving parts if they come into contact. To avoid these stictive forces, large-amplitude motion in MEMS is achieved using mechanical resonance of the suspended mechanical structures. For example, while typical macroscale gyros employ a spinning element, their MEMS equivalents use vibration.

If a micromechanical device can be designed and fabricated, then, thanks to the microelectronics heritage inherent in available fabrication processes, it can often be turned into a real MEMS device with relative ease. When a MEMS version of an existing macroscale device is the goal, it is usually best to prepare a clean-sheet design based on the process constraints, rather than to attempt a miniaturization of the original device. Although there are many processing alternatives, MEMS designers may find themselves limited to the toolset in which their organization has invested.

Finally, MEMS devices, particularly sensors, are designed to interact with their environment, so the packaging significantly influences performance. Because packaging is central both to process integration and system-level design, many decisions come into play, for example, whether to fabricate the micromechanical device and the electronics on the same chip or on separate chips with wire-bonded connections. Another decision involves the choice of materials for the enclosure. Detailed case studies of packaging have been published [7].

C.3 Materials for MEMS

The range of materials used in MEMS is wider than that of the semiconductor industry. Even so, silicon (Si) is currently the most important material in MEMS. It is used as a mechanical structural material and for electrodes. In 1982, Petersen published a detailed description of mechanical and electrical properties of Si [8]. An online table of the Si properties is available [9].

Silicon dioxide, SiO_2, which forms naturally on silicon, serves as an excellent insulating material. Other conductive materials, usually used for traces and vias, include metals, such as aluminum and copper.

C.3.1 Wafer materials

The choice of wafer material is dictated by selections made long ago by the semiconductor industry. For application in MEMS, the crystalline semiconductors, principally Si, Ge, and GaAs, have certain advantages [3]: (i) they are very well characterized; (ii) their processing technology is mature; (iii) they have useful inherent anisotropy; (iv) they are relatively easy to integrate with electronics. In addition, the native oxide of silicon, an excellent dielectric, can be grown in a controlled way on an Si surface. For piezoelectric transducer devices, quartz is an attractive substrate material.

Single-crystal Si wafers are manufactured using either the floating zone (FZ) or Czochralski (CZ) method. In the CZ process, which is more commonly employed, high-purity Si is melted in a quartz crucible. Dopants are added to the melt in specified amounts to make the silicon conductive (*p*- or *n*-type). Then a precisely oriented crystal seed is immersed into the crucible and slowly pulled upwards while simultaneously being rotated. See Fig. C.5a. The ingot or boule is cooled and later sliced at the correct angle to produce oriented wafers. See Fig. C.5b. In MEMS, the wafer may serve as the principal structural material for bulk micromachining or as a substrate for surface micromachining processes.

Silicon-on-insulator (SOI) wafers, now coming into use due to their distinct advantages in certain MEMS designs (see Section C.6.1), require considerable additional processing. Silicon-on-insulator wafers are sandwich-like structures with two layers of single-crystal silicon separated by an insulating layer, usually SiO_2. There are several methods for manufacturing these wafers, including oxygen implantation, wafer bonding, and a seed method. *Separation by implantation of oxygen* (SIMOX) employs oxygen ion beam implantation followed by high-temperature annealing to form a buried SiO_2 layer. The wafer bonding method is based on bonding of a Si wafer on the top of another, previously oxidized Si wafer and subsequently etching away most of the top wafer. The seed method grows a Si layer upon a previously oxidized wafer. Details on manufacturing of SOI wafers is found in Celler and Cristoloveanu [10].

C.3.2 Other materials

Electromechanical devices consist of the following basic elements: a mechanical structure, electrodes, and insulators. The electrodes are often formed from the structural

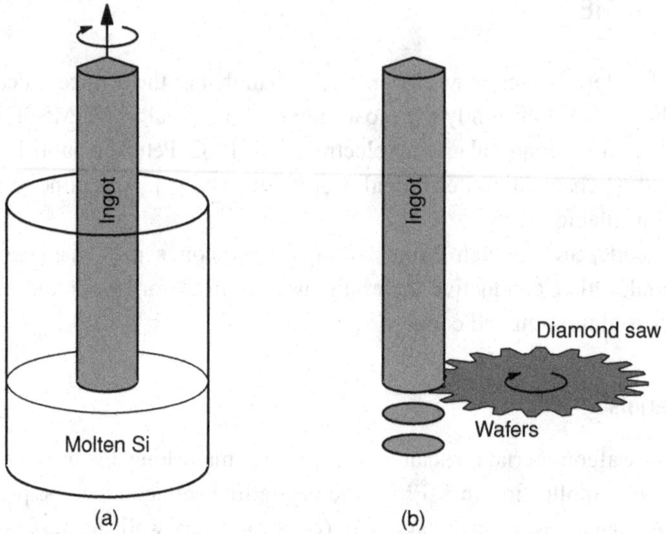

Figure C.5 The manufacture of Si wafers involves two steps. (a) The CZ process for producing the single-crystal Si ingot or boule. (b) Cutting with a diamond saw, which slices the ingot into the desired wafers.

material, but they are connected to the terminals using sputtered metal. Most often, this metal is aluminum (Al), but copper (Cu) is sometimes employed to take advantage of its superior thermal properties. Other materials for such use include gold (Au), nickel (Ni), titanium (Ti), zinc oxide (ZnO), gallium arsenide (GaAs), and cadmium sulfide (CdS). Metals, most commonly employed for wire traces to connect the electrodes of MEMS devices, can serve as structural materials in surface micromachining or LIGA processing.

C.3.3 Crystallographic properties of Si

Solid silicon can be found in three states: amorphous, polycrystalline (polysilicon), and single-crystal, as sketched in Fig. C.6. In the amorphous state sketched in Fig. C.6a, Si atoms are arranged in a random fashion. Polycrystalline silicon is a collection of randomly oriented subdomains, within each of which the atoms are in crystalline arrangement. Single-crystal Si has the same regular structure throughout its volume. Amorphous silicon is not commonly used in MEMS. Polysilicon is employed

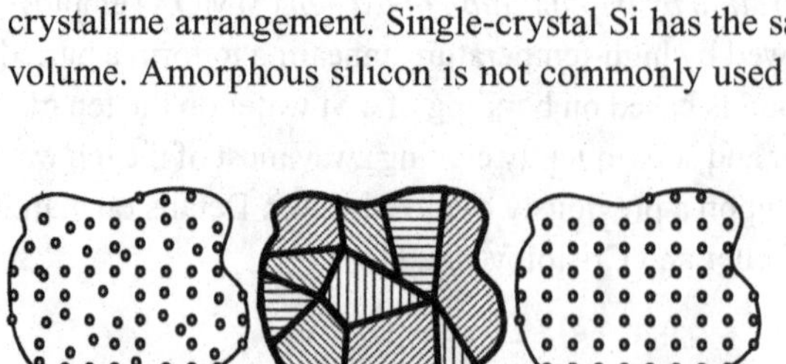

Figure C.6 The three states of solid Si. (a) Amorphous. (b) Polycrystalline. (c) Single-crystal.

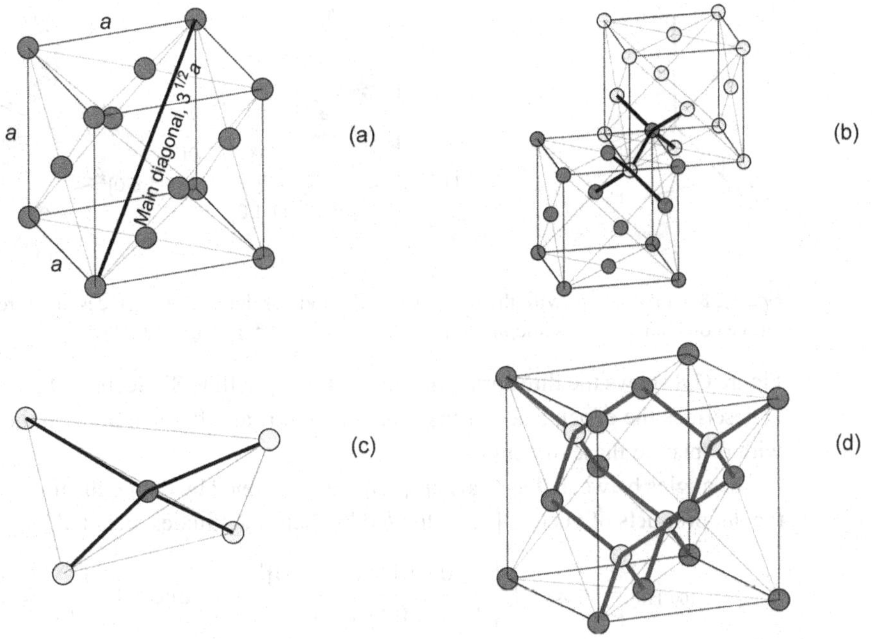

Figure C.7 Diamond crystalline structure of Si (lattice): (a) Basic cube with atoms at vertices and at the middle of each face (a is the length of each face). A main diagonal is indicated. (b) Two intersecting cubes with one overlapping the other by a quarter of the length of the diagonal ($(a\sqrt{3})/4$). The atoms at the cube vertex are connected to the atoms on the three nearest faces and the nearest vertex of the adjacent cube. (c) Isolated tetrahedron of four Si atoms from the upper cube with an atom at the center. (d) Basic diamond cube showing four atoms (lighter color) from neighboring cubes in its interior.

as a structural material for surface micromachining. Single-crystal Si is employed as the structural material in bulk micromachining and SOI structures. The remainder of this section is devoted to single-crystal Si.

The arrangement of atoms in a crystal is called the lattice. Crystalline silicon has a *diamond structure*.[1] To visualize the crowded Si lattice, it is convenient to assemble this structure step by step. Accordingly, Fig. C.7a shows the basic building element – a cube with an atom at each vertex and at the center of each face. To build larger crystals, basic cubes are intersected along the main diagonal at a periodic distance of $(3a\sqrt{3})/4$. See Fig. C.7b. An atom at the vertex of a cube is connected to the nearest vertex of the neighboring cube and the three nearest faces of the neighboring cube. These four atoms form a tetrahedron, as shown in Fig. C.7c. Figure C.7d highlights the tetrahedron with the basic diamond structure and shows the crystalline bonds of the neighboring cubes. These crystalline bonds give rise to the anisotropic properties of Si.

The *Miller index* system is used to identify the principal planes and directions that characterize crystals. In a cubic crystal, the [*hkl*] direction is normal to the (*hkl*) plane.

[1] The MATLAB™ routine used to generate this figure is available for download from the book's website.

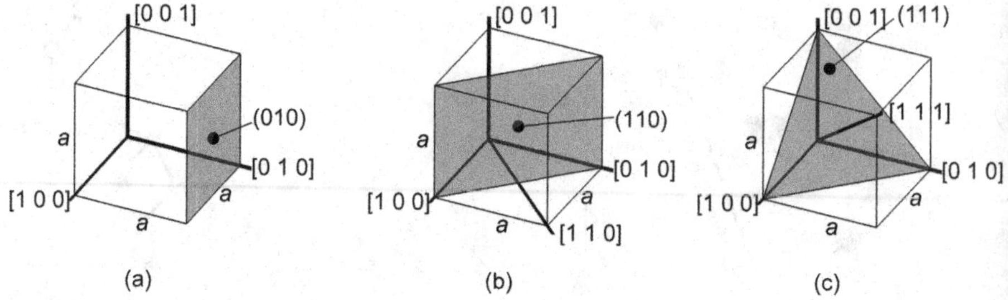

Figure C.8 Cubic lattice with the principal Miller indices defined by surfaces and directions. The lattice constant a is the same in all three dimensions, and all angles are 90°.

Figure C.8 shows the three principal planes for crystalline Si. Section D.11 of Appendix D describes the dependence of the Young modulus and Poisson ratio on the orientation with respect to these directions.

The angles between these various planes are obtained by taking the inverse cosines of the dot products of vectors [. . .] divided by their magnitudes, denoted by $\| \dots \|$:

$$\alpha_{[100,110]} = \arccos\left(\frac{[1\ \ 0\ \ 0]\cdot[1\ \ 1\ \ 0]^{\mathrm{T}}}{\|1\ \ 0\ \ 0\|\,\|1\ \ 1\ \ 0\|}\right) = \arccos\left(\frac{1}{\sqrt{2}}\right) = 45^\circ, \quad \text{(C.4)}$$

$$\alpha_{[100,111]} = \arccos\left(\frac{[1\ \ 0\ \ 0]\cdot[1\ \ 1\ \ 1]^{\mathrm{T}}}{\|1\ \ 0\ \ 0\|\,\|1\ \ 1\ \ 1\|}\right) = \arccos\left(\frac{1}{\sqrt{3}}\right) = 54.74^\circ, \quad \text{(C.5)}$$

$$\alpha_{[110,111]} = \arccos\left(\frac{[1\ \ 1\ \ 0]\cdot[1\ \ 1\ \ 1]^{\mathrm{T}}}{\|1\ \ 1\ \ 0\|\,\|1\ \ 1\ \ 1\|}\right) = \arccos\left(\frac{2}{\sqrt{6}}\right) = 35.26^\circ. \quad \text{(C.6)}$$

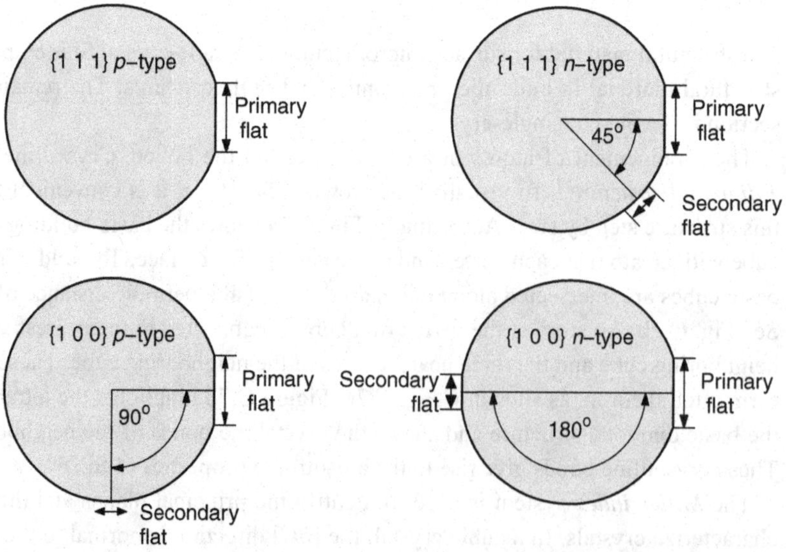

Figure C.9 Primary and secondary wafer flats for "six-inch" and smaller wafers of p- and n-type Si.

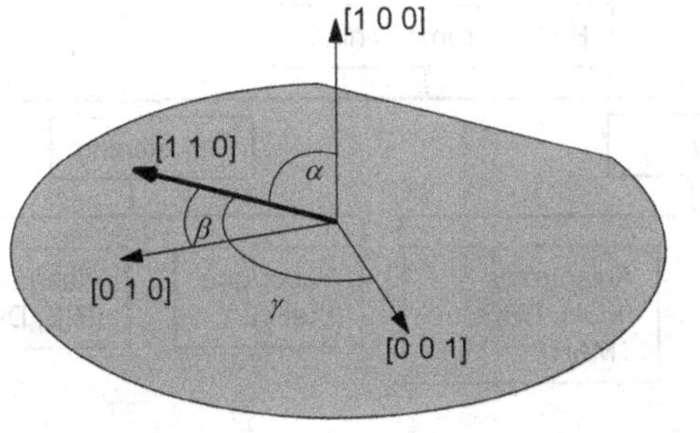

Figure C.10 Definition of direction cosines of vector [110] on a (100) wafer shown in perspective view. The vector [100] is perpendicular to the plane of the wafer and [110] is parallel to the wafer flat. The direction cosines of [110] are $l = \cos(\alpha)$, $m = \cos(\beta)$, and $n = \cos(\gamma)$, where $\alpha = 90°$, $\beta = 45°$, and $\gamma = 135°$.

Because etch rates depend strongly on orientation, wafers are specified with respect to the crystallographic axes. To facilitate wafer processing, these axes are defined by flats. The primary flat reveals the crystalline plane of high symmetry (usually 110) and the position of the secondary flat with respect to the primary flat identifies the wafer orientation and doping. The convention for "six-inch" wafers is shown in Fig. C.9.[2]

The flats are used in "six-inch" and smaller wafers. Larger wafers ("eight-inch" (200 mm) and "twelve-inch" (300 mm)) do not employ the flats. Instead, the orientation is marked by a small notch and the information on doping is not marked directly on the wafer.

A particular direction in a crystal is specified using direction cosines $l = \cos(\alpha)$, $m = \cos(\beta)$, and $n = \cos(\gamma)$. For example, consider the (100) wafer with its flat parallel to [110] direction, as shown in Fig. C.10. Direction [100] is perpendicular to the plane of the wafer, and direction [110] is parallel to the wafer's flat. For the vector [110], the angles are $\alpha = 90°$, $\beta = 45°$, and $\gamma = 135°$ and the direction cosines are $l = 0$, $m = 1/\sqrt{2}$, and $n = -1/\sqrt{2}$.

C.4 Bulk micromachining

In bulk micromachining, the structure is formed from the wafer material by etching into the substrate itself, which is typically single-crystal silicon. Bulk micromachining may be thought of as subtractive processing because the masks determine material to be removed. There are two main categories of etching in bulk micromachining: wet processing, which employs aqueous chemistries, and dry processing, which relies on vapor or plasma interactions. Wet processing is further subdivided into anisotropic and isotropic etching. Classification of the principal etching processes used in bulk

[2] The actual diameter of "six-inch" wafers is 150 mm, which equals 5.9 inches.

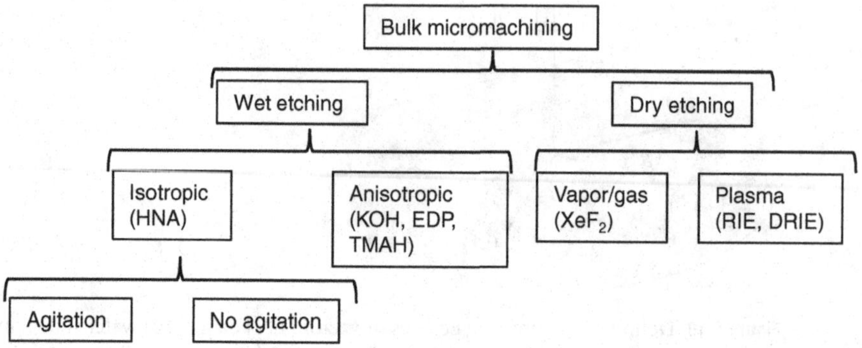

Figure C.11 Very basic classification of bulk micromachining processes.

micromachining is shown in Fig. C.11. Kovacs *et al.* [11] provide a comprehensive list with detailed descriptions of different processes.

C.4.1 Wet chemical etching

Wet chemical etching yields robust devices with relatively simple geometries. Some early examples of successful fabrications based on this process are the nozzles for inkjet printers and piezoresistive pressure sensor elements. Wet etch patterning is controlled by photolithographic masks and by implanted boron, which serves as an etch stop. In general, submicron features are not attainable with wet etching because the chemistry is difficult to control on that scale [12].

C.4.1.1 Anisotropic wet etching

Bulk micromachining of single-crystal Si originally started with anisotropic chemical etching using potassium hydroxide (KOH) [8]. The etch rates are, respectively, ~400 and ~600 times faster in the (100) and (110) directions than in the (111) direction [11]. Figure C.12 illustrates typical pits for (100) and (111) wafers. Undercutting, as illustrated in Fig. C.12c, occurs when a photolithographic mask is not correctly aligned with the crystallographic axes. Refer to Bean [13] for a detailed treatment of anisotropic etching of silicon.

The SEM of Fig. C.13 illustrates a pyramidal pit etched out using a square mask and KOH etchant.

Other chemicals commonly used for anisotropic etching include tetramethyl ammonium hydroxide (TMAH) and ethylenediamine-pyrocatechol (EDP). These etchants are selective; in particular, they do not etch aluminum. This desirable property makes them compatible with CMOS technology, making it possible to integrate transducers and electronics on the same chip. Table C.1 offers a comparison of these three chemistries. Notice that TMAH and EDP are less selective than KOH. Also, EDP is quite hazardous.

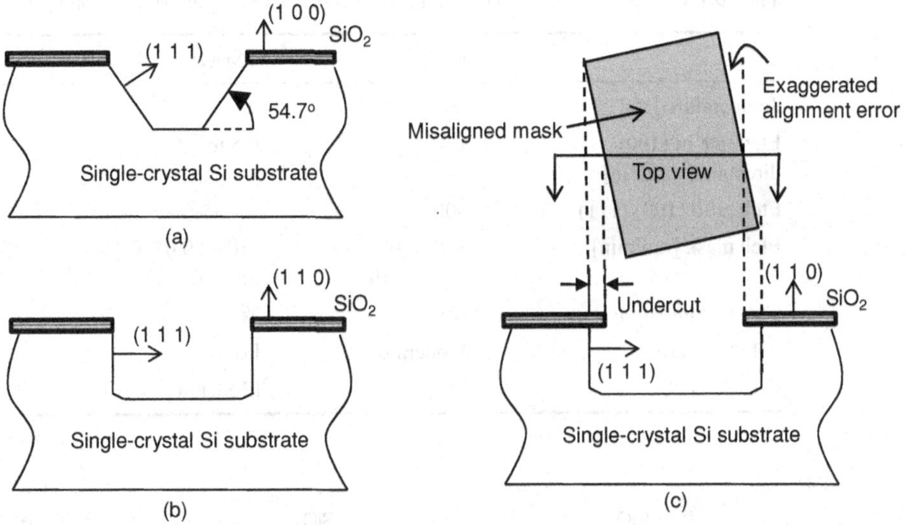

Figure C.12 Typical cavities achieved with anisotropic etching of crystalline Si using KOH. (a) For (100) wafer, the angle between any sidewall and the floor is are 54.7°. (b) For (110) wafer, the angle between any sidewall and the floor is 90°. (c) Undercutting phenomenon caused by mask misalignment as shown above the pit.

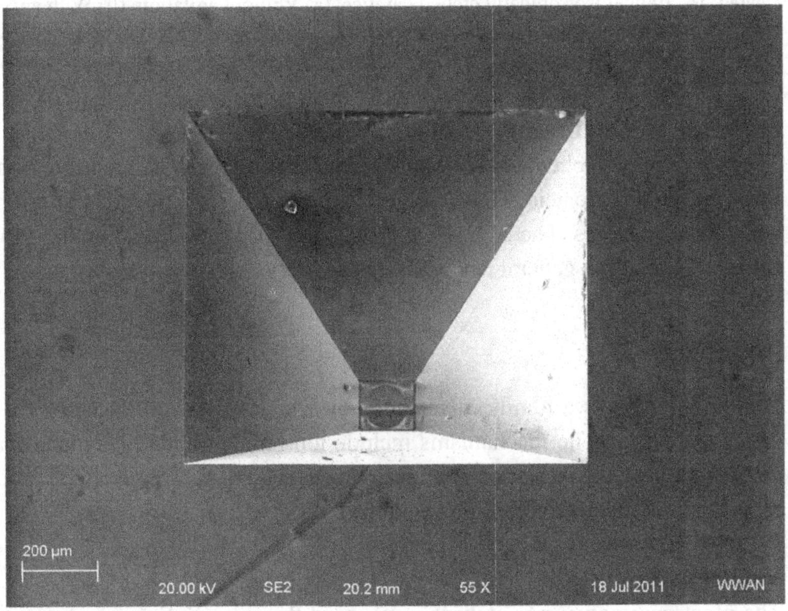

Figure C.13 Scanning electron micrograph (SEM) of a pyramidal pit etched in a (100) Si wafer. The image is credited to Weiqiang Wang of University of Rochester. Used with permission. In this example, a boron-doped layer on the back side of the wafer has stopped the etch process.

Table C.1 Comparison of three most common chemistries for anisotropic Si etching [11, 14]

	KOH	TMAH	EDP
Temperature [°C]	80	90	115
Etch rate in (100) direction [μm/min]	1.4	0.5 to 1.5	0.75
Etch ratio (100)/(111)	400	10–35	35
Etch mask [nm/min]	Si_3N_4: 0 SiO_2: <10	Si_3N_4: 0.05–0.25 SiO_2: 0	Si_3N_4: 0.1 SiO_2: 0.2
CMOS compatible	No	Yes	Yes
Safety hazards	Moderate	Low	High
Disposal	Easy	Moderate	Difficult

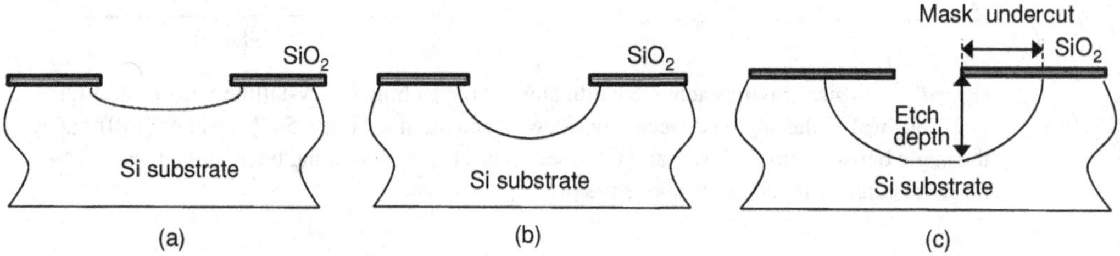

Figure C.14 Typical isotropically etched cavities. (a) Without agitation. (b) With agitation. (c) The undercutting phenomenon (exaggerated). Isotropic etching always undercuts silicon under the mask if the etch process is allowed to proceed for too long.

C.4.1.2 Isotropic wet etching

The most common chemical for wet isotropic etching is HNA, a mixture of hydrofluoric acid (HF) and nitric acid (HNO). Figure C.14 shows cross-sections of typical isotropically etched cavities. The etch rate is as fast as 3 μm/min. Half-circular profiles can be achieved, though in general cavity shapes are very sensitive to agitation.

C.4.2 Dry etching

Dry etching covers a family of methods by which wafers are etched by a gas, vapor, or plasma phase. The mechanisms include ion bombardment, chemical reaction, or a combination of physical and chemical mechanisms [4]. Dry etching produces high-aspect-ratio structures with high resolution.

C.4.2.1 Vapor etching

The most common vapor etchant is xenon difluoride (XeF_2), which is employed for post-processing of CMOS wafers because it attacks neither aluminum nor SiO_2. For example, a MEMS inductor can be formed after the CMOS components are fabricated. Etching with XeF_2 typically leaves a rough Si surface, but the advantage of this method is that it can sometimes achieve results similar to plasma etching without the need for

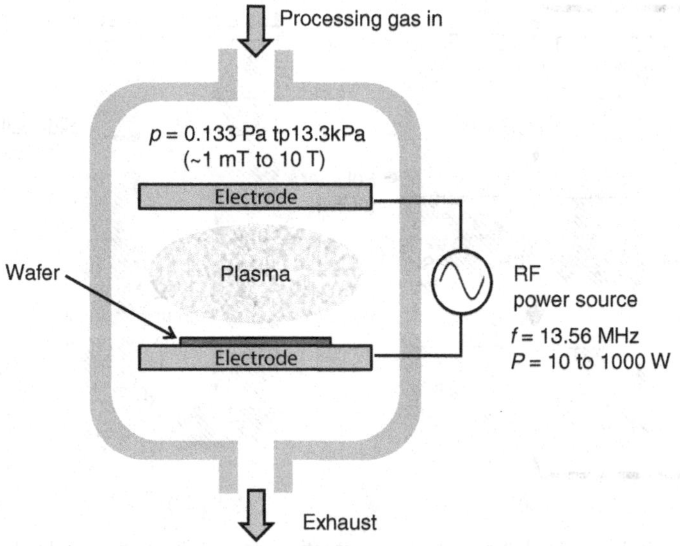

Figure C.15 Simplified sketch of plasma etching tool. The wafer to be etched is placed on one of the RF electrodes inside the vacuum–plasma chamber. This figure is adapted from Srinivasan [14].

expensive equipment [3]. Because of its selectivity, XeF_2 is widely used in surface micromachining, for example in combination with polysilicon as a sacrificial layer [7]. Read more about this process in Section C.5.

C.4.2.2 Plasma etching

Coburn and Winters [15] showed that the etch rate of Si can be increased by an order of magnitude using an ion beam. A radio-frequency (RF) power source produces a glow discharge plasma in a vacuum chamber, as illustrated in Fig. C.15. The plasma produces abundant ions, which, when directed at the wafer surface, remove the material by direct bombardment or chemical reaction. Figure C.15 is very much simplified. In practice, electrode shape, wafer placement, the number of wafers, and the number of electrodes influence the process [3, 4].

The relative importance of the bombardment and chemical mechanisms is closely related to pressure. At lower pressures, there is less scattering of ions and the bombardment mechanism dominates. This type of etching is directional, less selective, and generally achieves lower etch rates [4]. If the chamber pressure is increased, chemical mechanisms grow in importance, leading to better sensitivity and higher etch rates. As in wet etching, dry chemical etching can be either isotropic or anisotropic. *Reactive ion etching* (RIE) strikes a balance between physical sputtering and chemical reactions, resulting in rapid and highly directional etching. Other important dry processes include ion beam etching (IBE), chemically assisted ion beam etching (CAIBE), reactive ion beam etching (RIBE), and magnetically enhanced ion etching

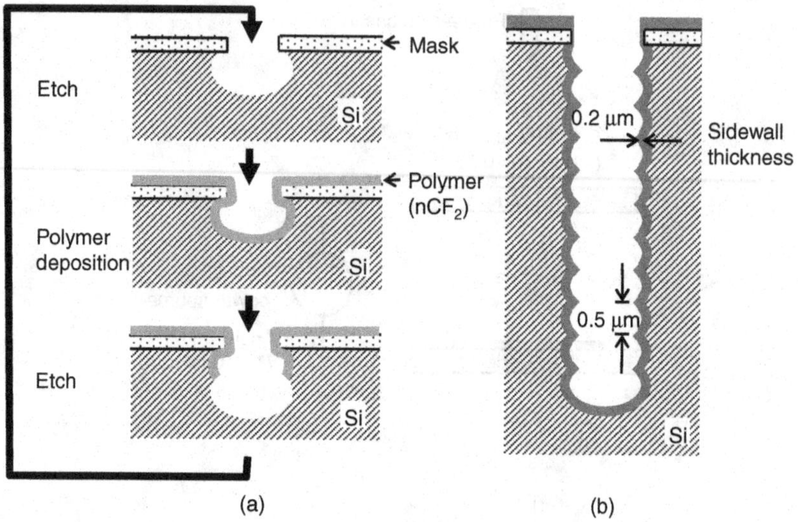

Figure C.16 The Bosch (DRIE) process. (a) Alternating reactive ion etching and polymer deposition steps. (b) Resulting deep vertical etch. This sketch is not to scale and the scalloping of the sidewalls as depicted here is greatly exaggerated.

(MIE). Detailed information about these processes is available in the reference literature [3–5].

A particularly important dry etch process is *deep reactive ion etching* (DRIE), developed by researchers at Bosch [16]. In this process, reactive ion etching is alternated with polymer deposition, as illustrated in Fig. C.16a. A sulfur hexafluoride (SF_6) plasma is used for the reactive ion etching and an octafluorocyclobutane (C_4F_8) plasma for the polymer deposition. The resulting process is independent of the crystallographic orientation of silicon. See Fig. C.16b. Very high aspect ratios (>200) have been achieved at etch rates of 1–2 μm/min. This rate compares favorably with wet etching processes. Refer to Table C.1. Deep reactive ion etching can produce very deep etches, including through-wafer machining. The etch rate is geometry-dependent, and small features etch slower than large ones. Consequently, the achievable depth of pits and holes is directly proportional to feature size.

Figure C.17 shows an SEM of electrodes carved out of silicon using a Bosch process. Note that the scallops are less than 0.5 μm and that the walls are close to being perfectly vertical.

C.4.2.3 Process integration example

Shaw *et al.* [12, 17] described an integrated dry process, *single-crystal silicon reactive etch and metal* (SCREAM I), where several processes are integrated to produce useful microstructures ranging from cantilevers to accelerometers [18]. The success of this scheme depends on a clever combination of anisotropic and isotropic dry etch steps.

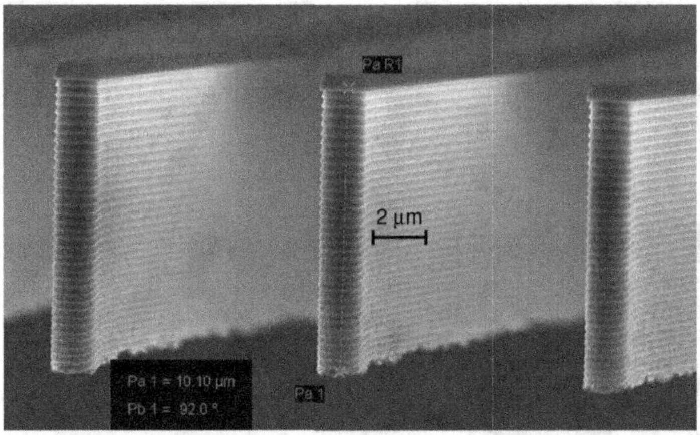

Figure C.17 An image of released beams produced by a DRIE process. The scallops are visible. The image is credited to Mollie Devoe of Kionix Inc. Used with permission.

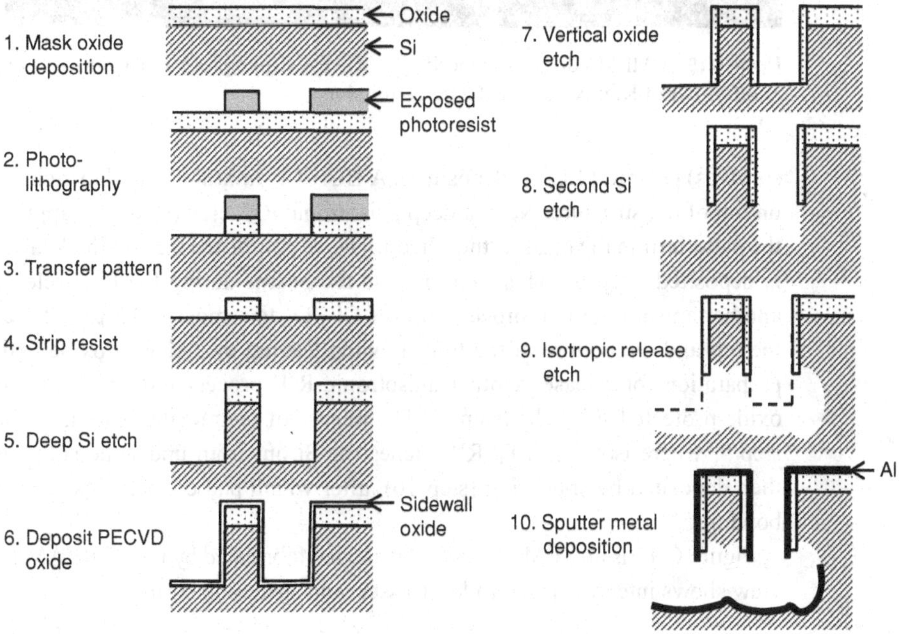

Figure C.18 Sketch of the steps of the SCREAM process adapted from [12].

Figure C.18 illustrates the individual process steps. The first step lays down a 1–2 μm thick oxide mask using plasma-enhanced chemical vapor deposition (PECVD). In step 2, photolithography is used to define the pattern. The transfer of the pattern from the photoresist to the oxide is achieved using magnetic-field-assisted CHF_3 ion etching (MIE) in step 3. Step 4 strips the resist using O_2 plasma etch. The photoresist mask can be retained for additional masking in deeper etches and then removed (if any photoresists

Figure C.19 A MEMS device produced by a SCREAM-like process. The image is credited to Scott Miller of Kionix Inc. Used with permission.

survives) prior to PECVD deposition. A BCl_3/Cl_2 mixture is used to etch the aluminum on top of the structure. Next, a deep anisotropic RIE etch (step 5) is applied to transfer the pattern from the mask to the silicon. Then a conformal 0.3 μm thick layer of PECVD is deposited (step 6). This step protects the beams during the later release. Step 7 is another anisotropic RIE process. It uses CF_4/O_2 to remove ~0.3 μm of the oxide from the top and the bottom of the trench, while leaving the sidewall oxide undisturbed. In preparation for release, another anisotropic RIE process removes 3–5 μm below the oxide-protected sidewalls (step 8). The release of the moving parts is accomplished in step 9, where isotropic SF_6 RIE etches the Si out from under the sidewalls. Metal is then deposited by sputtering (step 10), after which pad connections are made by wire bonding.

Figure C.19 is an SEM of a complex device fabricated by the SCREAM process. This view shows integrated electrodes, trusses, and stiffening struts.

C.5 Surface micromachining

In surface micromachining, the structure is built upon the wafer in a layer-by-layer fashion by depositing thin films of various materials and subsequently selectively patterning and removing material. The features in the wafer plane (xy) are defined by dry-etch processes and the structure is usually released by a wet-etch process. These processes are considered additive, as they add structure on the top of a wafer (as opposed to

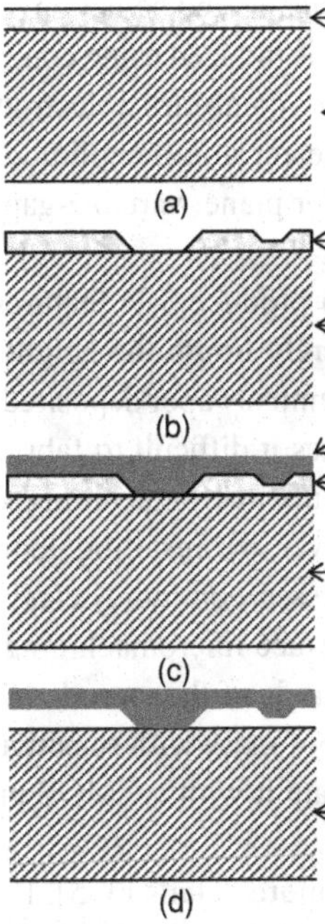

Figure C.20 Typical process steps in surface micromachining. (a) Deposition of sacrificial layer. (b) Photolithographic patterning and etching of the sacrificial layer. (c) Deposition of polysilicon layer. (d) Removal of the sacrificial scaffolding layer to release the part.

bulk processes, which excavate into the wafer). The dominant structural material in surface micromachining is polysilicon, though sometimes metals are used as the structural materials with polysilicon serving as the scaffolding. An important application featuring aluminum (for its reflectivity) as structural material and polysilicon as scaffolding is the digital micromirror chip of Texas Instruments. See Example 3.5 and, for more details, Senturia [7].

The principal steps for a surface-micromachining process are outlined in Fig. C.20. Here, a polysilicon structure is first formed on top of a Si wafer and then released by removing the scaffolding.

Surface micromachining is a very common fabrication technology suitable for high-volume manufacturing. Consequently, surface-micromachined devices lend themselves to integration with combinations of actuating and sensing electronics on the same chip. Prior to introduction of dry bulk micromachining in the mid-1990s, high-resolution structures could only be made using surface micromachining. Rather intricate structures were designed, built, and demonstrated in the 1980s, such as electrostatic actuators (including comb drives), resonators, and micromotors. A rich collection of pioneering papers describing these early devices that shaped the later development in MEMS is found in Trimmer [19], including the very first MEMS device – the resonant gate transistor by Nathanson [2] of Fig. C.1. Surface micromachining made possible many commercial devices, such as air-bag accelerometers (Analog Devices) and projection displays (Texas Instruments).

Surface-micromachined devices enable versatile sensing and actuation schemes. For electrostatic actuators, motion in the plane of the wafer is induced using comb electrodes. If large-amplitude motion is called for, then variable-area comb-drive electrodes are used. On the other hand, for the high sensitivity needed to detect small-amplitude motion and for devices based on motion perpendicular to the wafer plane, variable-gap geometries are a better choice.

Generally, surface micromachining can be used to produce 3-D features. However, the nature of the deposition processes involved severely limits the height in surface-micromachined parts. For instance, the thickness of chemical vapor deposited polysilicon is only a few microns [4]. This height limitation makes it difficult to fabricate moving elements having sufficient mass for inertial sensing. Bulk micromachining, on the other hand, features thicker, more massive structures; however, the resulting geometry is not truly 3-D, but extruded 2-D. Reactive ion etching processes, commonly classified as bulk micromachining processes, can be combined with surface micromachining to produce thicker structures. The geometry of these thicker layers is of the extruded 2-D type. In an important example, RIE is employed to carve a proof mass with sensitive electrodes out of silicon-on-insulator (SOI) materials and integrate it with a CMOS process [20]. Section C.6.1 provides more details on SOI.

Refer to the standard texts on microfabrication for more details [3–5]. Bustillo *et al.* [21] provide an excellent review of the state of surface micromachining at the end of the twentieth century.

C.6 Potpourri of other processes and materials

As explained earlier in this appendix, micromachining technology is very rich and diverse. This final section briefly describes a few other microfabrication processes that further illustrate this diversity.

C.6.1 Silicon-on-insulator wafers

Silicon-on-insulator wafers, already introduced in Section C.3.1, feature a layer of single-crystal Si on top of an insulator, typically SiO_2. This technology, introduced in the 1970s for applications such as spacecraft electronics (which require immunity to cosmic alpha rays), has more recently enabled denser integration of integrated circuits [10]. In MEMS, silicon-on-insulator (SOI) micromachining employs dry bulk micromachining processes (DRIE) to create high-aspect-ratio structures. Although DRIE etching of SOI structures is classified as bulk micromachining, it can be regarded as a type of surface micromachining [21]. Brosnihan *et al.* [20] fabricated capacitive sensors based on SOI wafers, and demonstrated 45 μm deep single-crystal Si beams and tested their compatibility with CMOS.

Milanovic [22], using front-side and back-side etching of SOI wafers, fabricated interdigitated beams, which are capable of upward and downward actuation and also bidirectional rotation. Figure C.21a illustrates an application in a MEMS mirror device.

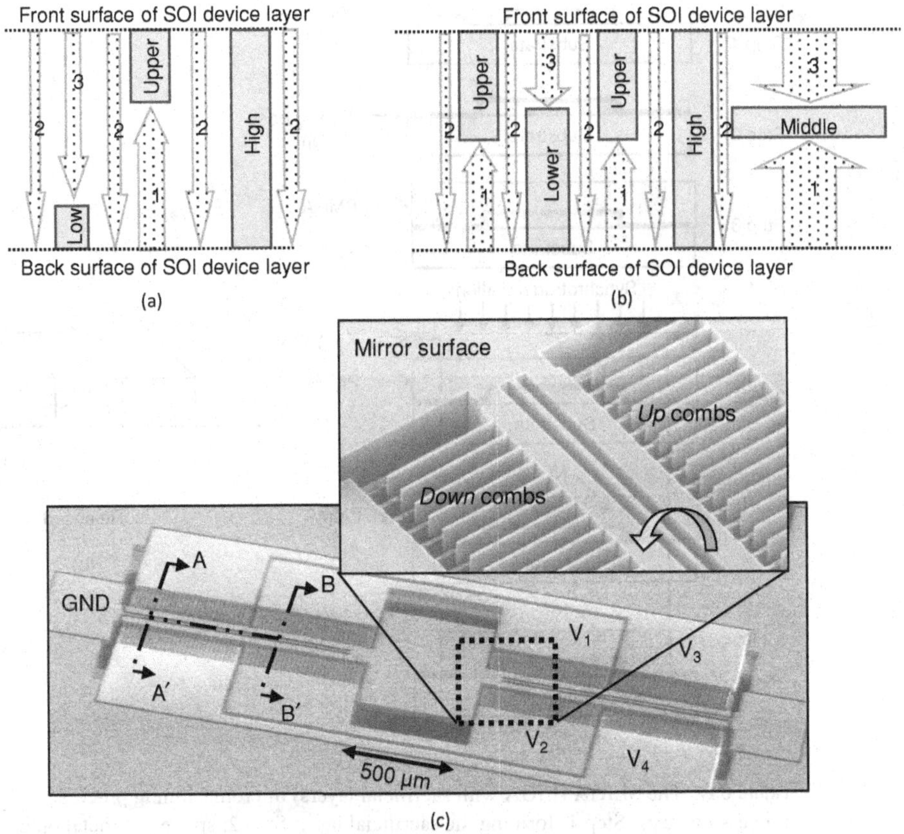

Figure C.21 Manufacturing of beams of different heights from an SOI wafer [22]. (a) When the lower and upper beams are both shorter than one-half of the high beam, three beam heights can be produced. (b) When the lower and upper beams are higher than one-half of the high beam, there is an overlap and four beam heights can be created. (c) SEM micrograph of a micromirror with four isolated comb drive sets. The images are credited to Veljko Milanovic. Copyright © IEEE. Used with permission.

One half of the variable-area comb drive (the upper beams) is produced on the front side of the SOI wafer, and the other half (the lower beams) on the back side. With no applied voltage, the electrodes do not overlap and the vertical distance between the upper and lower beams is equal to the insulator thickness. An SEM image of the interdigitated electrode structure is shown in Fig. C.21b.

C.6.2 LIGA

The LIGA process was developed and introduced by W. Ehrfeld *et al.* at the Karlsruhe Nuclear Research Laboratory in the 1980s [23]. LIGA, which stands for lithography, electroplating, and molding (in German, *Lithographie, Galvanoformung, und Abformung*), comprises a diverse set of micromachining processes for producing non-silicon (metallic) structures with very high aspect ratio, as high as ~100 to 1.

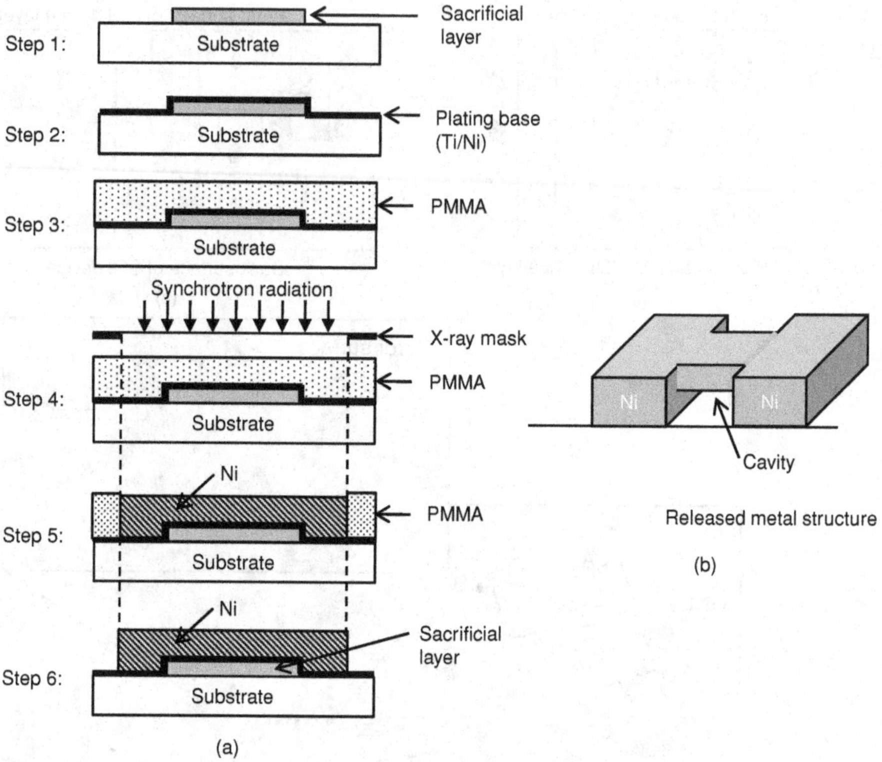

Figure C.22 The SLIGA (LIGA with sacrificial layers) micromachining process. (a) Sequence of processing steps. Step 1: forming the sacrificial layer. Step 2: sputtering metal on base plate. Step 3: PMMA cast and anneal. Step 4: X-ray lithography. Step 5: Ni electroplating. Step 6: Removal of PMMA and plating base in preparation for removal of scaffolding. (b) Free-standing metal part. Adapted from [6].

Polymethylmethacrylate (PMMA) serves as the resist for the lithography. First, a thick layer (up to 500 μm) of this resist is formed on the wafer. Such thicknesses cannot be achieved by spin coating, so other methods, such as *in-situ* polymerization and cast-and-anneal processes, are used. The critical property of PMMA is its sensitivity to X-rays. When the resist layer is exposed through a patterned mask to a beam of high-energy X-rays from a synchrotron, the exposed material becomes soluble and the patterned mask is transferred to the PMMA. The exposed material is removed using a chemical solvent. This particular technique is called *deep X-ray lithography* (DXRL, or sometimes DXL). Next, metal is electrodeposited onto the photoresist mold. Removing the remainder of the PMMA scaffolding then produces a free-standing metal structure. This metal structure may itself be the final product or it may be used as a mold for making small, precision, plastic parts.

Figure C.22 illustrates a LIGA-based process known as *surface micromachining and LIGA* (SLIGA) [6]. This process starts with the creation of a patterned sacrificial layer, which serves as the scaffolding for the resulting free-standing structure. After this layer

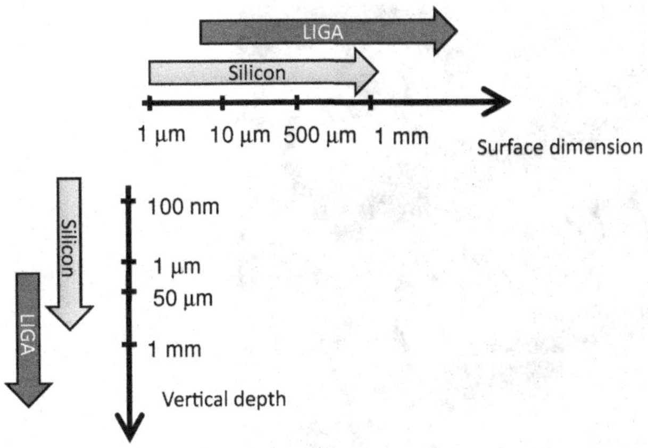

Figure C.23 Comparison of the achievable device dimensions for silicon-based and LIGA devices, adapted from Rasmussen *et al.* [24].

is formed on the wafer, a conformal layer of metal (titanium or nickel) is sputtered uniformly on the wafer. In the third step, PMMA is laid down as a resist layer. Next, in the fourth step, X-ray lithography is used to pattern the resist. The fifth step is electroplating of nickel. The last two steps remove PMMA and the scaffolding to release the free-standing metal structure.

The LIGA process produces very nearly vertical sidewalls with run-outs less than 0.1 μm per 100 μm of structure height [6]. The chief attraction of LIGA is that it can produce precision high-aspect-ratio structures on a scale larger than that achievable with silicon-based processes. Refer to Fig. C.23 for a dimensional comparison of Si- and LIGA-based processing capabilities. The disadvantage of LIGA is that, compared with other MEMS processes, it is expensive and requires access to a synchrotron for the X-ray lithography. Furthermore, sensing and actuating electronics cannot be packaged on the same chip with the mechanical device.

In 2004, the *International LIGA Interest Group* was formed to bring together major researchers, practitioners, manufacturers, and users to solve technical problems and accelerate commercialization of this unique process [5].

C.6.3 SU-8 and UV LIGA

Introduction of SU-8, an epoxy-based negative resist material, made it possible to extend the LIGA concept to UV lithography. Such processes, referred to as UV LIGA, cannot match the resolution of X-ray LIGA, but they are very inexpensive and easy to implement [25]. SU-8 was developed at IBM and first applied to MEMS in the 1990s [26]. Unlike PMMA, SU-8 can be laid down using spin coating in a wide range of thicknesses: ~750 nm to ~450 μm. It is compatible with electroplating, so can be used to form high-aspect-ratio metal structures. In UV LIGA, the structural material can be either a metal or the SU-8 itself. Dai *et al.* fabricated and tested the high-aspect-ratio,

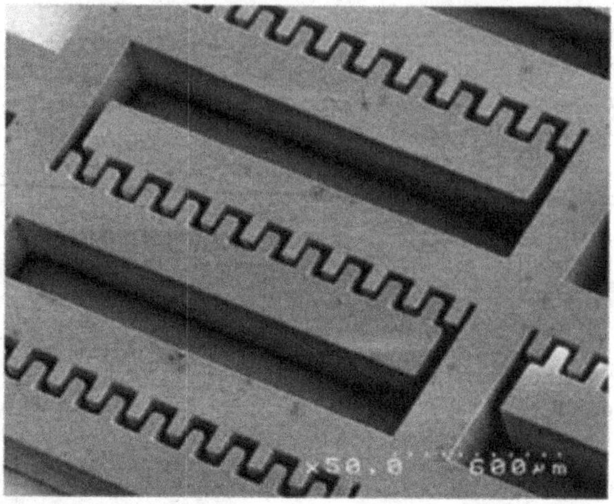

Figure C.24 SEM image of a high-aspect-ratio, variable-area, comb-drive microactuator fabricated using a UV-LIGA process. In this device, SU-8 serves as the structural material [27]. IEEE©. Used with permission.

variable-area comb-drive microactuator shown in Fig. C.24 using SU-8 as the structural material [27].

C.7 Conclusion

This short appendix has introduced some of the fundamental concepts and processing vocabulary for fabricating microelectromechanical sensors and actuators. Coverage has emphasized Si micromachining. The two principal families of processes have been introduced: bulk and surface micromachining. Bulk-micromachined structures, excavated into a silicon wafer, feature deep structures and extruded 2-D geometries. Surface micromachining has greater flexibility with respect to achievable geometries and excellent compatibility with CMOS electronics, but is limited to thin low-aspect geometries. Finally, to exemplify the diverse evolving landscape of microfabrication, brief coverage of SOI, X-ray LIGA, and UV LIGA processes was provided.

References

1. R. P. Feynman, There's plenty of room at the bottom [data storage]. *Journal of Microelectromechanical Systems*, 1 (1992), 60–66.
2. H. C. Nathanson, W. E. Newell, R. A. Wickstrom, and J. R. Davis, Jr., The resonant gate transistor. *IEEE Transactions on Electron Devices*, 14 (1967), 117–133.
3. G. T. A. Kovacs, *Micromachined Transducers Sourcebook* (Boston, MA: WCB, 1998).
4. M. J. Madou, *Fundamentals of Microfabrication: The Science of Miniaturization*, 2nd edn (Boca Raton, FL: CRC Press, 2002).

5. M. J. Jackson, *Microfabrication and Nanomanufacturing* (Boca Raton, FL: CRC/Taylor & Francis, 2006).
6. H. Guckel, High-aspect-ratio micromachining via deep X-ray lithography. *Proceedings of the IEEE*, 86 (1998), 1586–1593.
7. S. D. Senturia, *Microsystem Design* (Boston, MA: Kluwer Academic Publishers, 2001).
8. K. E. Petersen, Silicon as a mechanical material. *Proceedings of the IEEE*, 70 (1982), 420–457.
9. K. Pister, *Properties of Silicon*. www-bsac.eecs.berkeley.edu/~pister/crystal.pdf.
10. G. K. Celler and S. Cristoloveanu, Frontiers of silicon-on-insulator. *Journal of Applied Physics*, 93 (2003), 4955.
11. G. T. A. Kovacs, N. I. Maluf, and K. E. Petersen, Bulk micromachining of silicon. *Proceedings of the IEEE*, 86 (1998), 1536–1551.
12. K. A. Shaw, Z. L. Zhang, and N. C. MacDonald, SCREAM I: a single mask, single-crystal silicon, reactive ion etching process for microelectromechanical structures. *Sensors and Actuators A*, 40 (1994) 63–70.
13. K. E. Bean, Anisotropic etching of silicon. *IEEE Transactions on Electron Devices*, 25 (1978), 1185–1193.
14. T. Srinivasan, *MEMS Fabrication I: Process Flows and Bulk Micromachining*. www-bsac.eecs.berkeley.edu/projects/ee245/Lectures/lecturepdfs/Lecture2.BulkMicromachining.pdf.
15. J. W. Coburn and H. F. Winters, Ion and electron assisted gas surface chemistry – an important effect in plasma etching. *Journal of Applied Physics*, 50 (1979), 3189–3196.
16. F. Lärmer and A. Schilp, *Verfahren zum anisotropen Ätzen von Silicium*. German Patent, DE4241045. 1994.
17. K. A. Shaw and N. C. MacDonald, Integrating SCREAM micromachined devices with integrated circuits. In *IEEE, The Ninth Annual International Workshop on Micro Electro Mechanical Systems, MEMS '96. 'An Investigation of Micro Structures, Sensors, Actuators, Machines and Systems'*. 1996, pp. 44–48.
18. K. A. Shaw, S. G. Adams, and N. C. MacDonald, *Microelectromechanical Lateral Accelerometer*. US Patent US5,563,343, 1996.
19. W. Trimmer and Institute of Electrical and Electronics Engineers, *Micromechanics and MEMS: Classic and Seminal Papers to 1990* (New York: IEEE Press, 1997).
20. T. J. Brosnihan, J. M. Bustillo, A. P. Pisano, and R. T. Howe, Embedded interconnect and electrical isolation for high-aspect-ratio, SOI inertial instruments. In *International Conference on Solid State Sensors and Actuators. TRANSDUCERS '97 Chicago*. Vol. 1 (1997), pp. 637–640.
21. J. M. Bustillo, R. T. Howe, and R. S. Muller, Surface micromachining for microelectromechanical systems. *Proceedings of the IEEE*, 86 (1998), 1552–1574.
22. V. Milanovic, Multilevel beam SOI-MEMS fabrication and applications. *Journal of Microelectromechanical Systems*, 13 (2004), 19–30.
23. E. W. Becker, W. Ehrfeld, P. Hagmann, A. Maner, and D. Münchmeyer, Fabrication of microstructures with high aspect ratios and great structural heights by synchrotron radiation lithography, galvanoforming, and plastic moulding (LIGA process). *Microelectronic Engineering*, 4 (1986), 35–56.
24. J. Rasmussen, W. Bonivert, and J. Krafcik, *High Aspect Ratio Metal MEMS (LIGA) Technologies for Rugged, Low-Cost Firetrain and Control Components*. Presented at the NDIA 47th Annual Fuze Conference, 2003.

25. U. Wallrabe and V. Salle, LIGA technology for R&D and industrial applications. In *MEMS: A Practical Guide to Design, Analysis, and Applications*, ed. J. G. Korvink and O. Paul (Berlin: Springer, 2006), pp. 853–899.
26. H. Lorenz, M. Despont, N. Fahrni, *et al.*, SU-8: a low-cost negative resist for MEMS. *Journal of Micromechanics and Microengineering*, 7 (1997), 121.
27. W. Dai, K. Lian, and W. Wang, Design and fabrication of a SU-8 based electrostatic microactuator. *Microsystem Technologies*, 13 (2007), 271–277.

D A brief review of solid mechanics

This appendix reviews some basic elements of solid mechanics. The emergence of MEMS technology has revitalized interest in this long-established field because the very small size of MEMS structures introduces scaling issues not examined in the past. The review starts with definitions for stress and strain, and then introduces Hooke's law, the linear relationship between them. After a qualitative discussion of force balance, material properties, and the principle of continuity of deformation, we then consider axial, torsional, and bending deformation of simple one-dimensional elements, namely, bars, rods, and beams, respectively. One two-dimensional system of importance to MEMS devices, the deforming plate, is also considered. Finally, we cover the anisotropic, elastic properties of crystalline silicon. Material non-linearity is not considered.

D.1 Stress

Consider the static deformation of a bar having uniform cross-section shown in Fig. D.1a. The bar is anchored at one end and acted upon by an axial force f at the other end. The average stress σ_{av} [Pa] inside the bar is the ratio of the force and the area A:

$$\sigma_{\text{av}} = \frac{f}{A}. \tag{D.1}$$

From equilibrium considerations, Eq. (D.1) applies for any cross-section perpendicular to the axis of the bar.

In Fig. D.1a, the force and normal of the defined area A are parallel. A more general definition for stress emerges if we allow the area element to have arbitrary orientation. Let the angle between its normal and the force, still acting axially, be equal to α, as depicted in Fig. D.1b. Because the equilibrium cannot be influenced by the orientation of the cross-section, the stress acting on the area must consist of components acting both parallel and perpendicular to the normal. The normal component is called the *normal stress*, while components perpendicular to the normal are *shear stresses*.

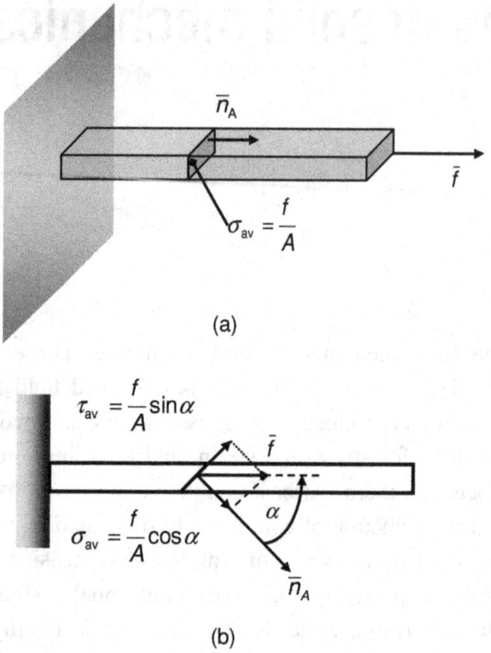

Figure D.1 A bar subjected to axial stress, fixed at one end and placed in tensile or compressive stress by the axial force $\overline{f}$ at the other end. (a) σ_{av} is the average stress at any cross-section of the bar perpendicular to its axis. The area normal $\overline{n}_A$ is parallel to the force $\overline{f}$. (b) The stress on an arbitrary plane at angle α to the axial force $\overline{f}$ has normal and shear components.

Although Eq. (D.1) is a scalar, the force $\overline{f}$ and the normal $\overline{n}_A$ are both vectors. Therefore, generalization is required. We can accommodate the linear relationship between $\overline{f}$ and $A\overline{n}_A$ using a matrix representation for the stress as follows:

$$\begin{bmatrix} f_x \\ f_y \\ f_z \end{bmatrix} = \underbrace{\begin{bmatrix} \sigma_x & \tau_{xy} & \tau_{xz} \\ \tau_{yx} & \sigma_y & \tau_{yz} \\ \tau_{zx} & \tau_{zy} & \sigma_z \end{bmatrix}}_{\overline{\overline{\sigma}}} \begin{bmatrix} n_{Ax} \\ n_{Ay} \\ n_{Az} \end{bmatrix} A, \tag{D.2}$$

where n_{Ax}, n_{Ay}, and n_{Az} are the direction cosines of the unit normal. Equation (D.2) identifies the standard formulation for the *stress tensor* $\overline{\overline{\sigma}}$, sometimes referred to as the *state of stress*. Normal stresses are denoted by σ and shear stresses by τ.

The stress tensor, providing a point description of the state of stress, is a field variable. The formal mathematical definition specifies the stress components acting on face A_n due to force F_m in the limit of infinitesimal area elements, that is,

$$\sigma_m = \lim_{\Delta A_m \to 0} \frac{\Delta F_m}{\Delta A_m}, \quad m = x, y, z, \quad \text{and} \quad \tau_{mn} = \lim_{\Delta A_n \to 0} \frac{\Delta F_m}{\Delta A_n}, \quad m \neq n. \tag{D.3}$$

Figure D.2 provides a physical interpretation of the stress tensor. The differential volume element, $dxdydz$, has six sides, each having an associated normal. The normals

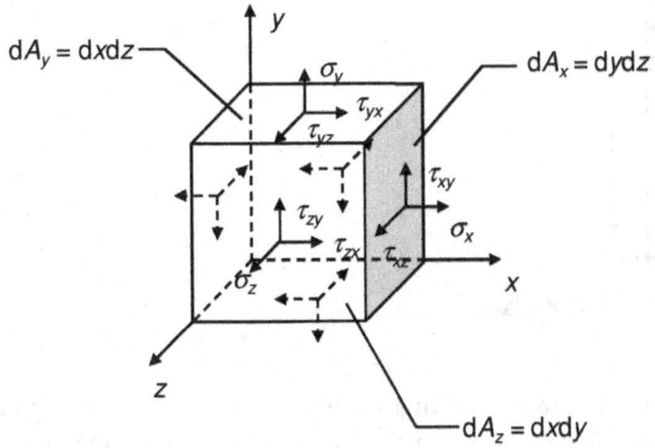

Figure D.2 State of stress of an infinitesimal rectangular volume element with rectilinear edges dx, dy, and dz.

of all positive faces coincide with the axes, while the normals of the negative faces are in the opposite directions. One normal and two shear stresses act upon each area element. Note the sign convention: when $\sigma_m > 0$, the material in the box is in tensile stress along that axis, while for $\sigma_m < 0$, the box is in compressive stress. τ_{mn} is the shear stress acting on the nth face in the m direction.

To meet the requirement that each infinitesimal volume element remain in a state of equilibrium, the stress tensor must be symmetric, i.e.,

$$\tau_{mn} = \tau_{nm}. \tag{D.4}$$

This condition guarantees that the vector sum of forces is zero and that the sum of moments about each axis is zero. The symmetry condition is easily demonstrated for special cases. Consider, for example, the state of pure shear: $\sigma_x = \sigma_y = \sigma_z = 0$. Equilibrium of moments about the center of the cube in the direction of the z-axis requires

$$\tau_{xy}\underbrace{\mathrm{d}z\mathrm{d}y}_{dAx}\underbrace{\frac{\mathrm{d}x}{2}}_{\text{moment arm}} + \tau_{xy}\underbrace{\mathrm{d}z\mathrm{d}y}_{dAx}\underbrace{\frac{\mathrm{d}x}{2}}_{\text{moment arm}} - \tau_{yx}\underbrace{\mathrm{d}z\mathrm{d}x}_{dAy}\underbrace{\frac{\mathrm{d}x}{2}}_{\text{moment arm}} - \tau_{yx}\underbrace{\mathrm{d}z\mathrm{d}x}_{dAy}\underbrace{\frac{\mathrm{d}x}{2}}_{\text{moment arm}} = 0. \tag{D.5}$$

Because dx and dy are independent, Eq. (D.5) reduces to $\tau_{xy} = \tau_{yx}$. Using the same argument along the other axes yields $\tau_{xz} = \tau_{zx}$ and $\tau_{yz} = \tau_{zy}$.

Now consider equilibrium of the translational forces on the infinitesimal volume element dxdydz in the presence of a body force $\overline{F}$. Setting the sum of forces to zero along each axis and rearranging yields

$$\frac{\partial \sigma_x}{\partial x} + \frac{\partial \tau_{yx}}{\partial y} + \frac{\partial \tau_{zx}}{\partial z} + F_x = 0, \tag{D.6}$$

$$\frac{\partial \tau_{xy}}{\partial x} + \frac{\partial \sigma_y}{\partial y} + \frac{\partial \tau_{zy}}{\partial z} + F_y = 0, \tag{D.7}$$

$$\frac{\partial \tau_{xz}}{\partial x} + \frac{\partial \tau_{yz}}{\partial y} + \frac{\partial \sigma_z}{\partial z} + F_z = 0. \tag{D.8}$$

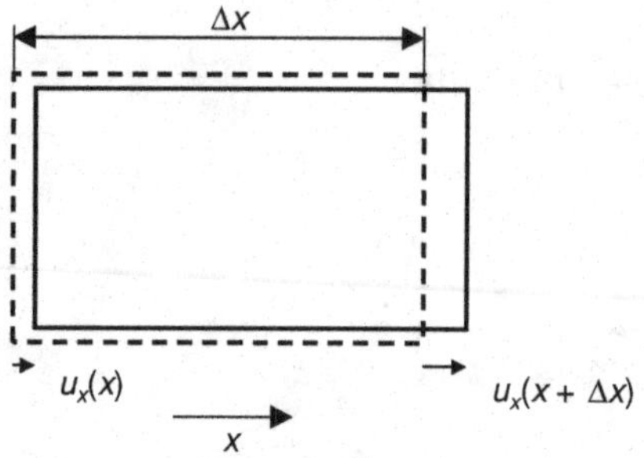

Figure D.3 One-dimensional normal strain deformation of a differential volume element. The dashed rectangle represents the undeformed element; the solid rectangle is the same element after deformation.

Previously, we noted that rotation of the coordinate system alters the stress components. The conclusion of this observation is that the stress tensor must exhibit continuity, that is, the continuous transformation during coordinate system rotation. If $\overline{\overline{\sigma}}$ and $\overline{\overline{\sigma}}_{\mathrm{r}}$ are the stress tensors in the original and rotated coordinate systems, then

$$\overline{\overline{\sigma}}_{\mathrm{r}} = \overline{\overline{R}}\sigma\overline{\overline{R}}^{\mathrm{T}}, \tag{D.9}$$

where, for example, the matrix for a rotation of angle θ about the z-axis is

$$\overline{\overline{R}}_{z,\theta} = \begin{bmatrix} \cos(\theta) & -\sin(\theta) & 0 \\ \sin(\theta) & \cos(\theta) & 0 \\ 0 & 0 & 1 \end{bmatrix}. \tag{D.10}$$

A $\pi/4$ radian rotation of the coordinate system about the z-axis changes a state of pure shear stress in the xy-plane into a state of pure normal stress in the same plane:

$$\begin{bmatrix} 0 & \tau_{xy} & 0 \\ \tau_{xy} & 0 & 0 \\ 0 & 0 & 0 \end{bmatrix} \xrightarrow{\overline{\overline{R}}_{z,\pi/4}} \begin{bmatrix} \tau_{xy} & 0 & 0 \\ 0 & -\tau_{xy} & 0 \\ 0 & 0 & 0 \end{bmatrix}. \tag{D.11}$$

D.2 Strain

Strain describes the deformation of a solid in response to stress. While strain is a tensor quantity, we first examine the case of simple axial deformation to introduce the concept. Consider an infinitesimal volume element of length Δx embedded in a bar that is subjected to tensile stress, as depicted in Fig. D.3. One can describe the x-directed lengthening by first defining the point-by-point displacement $u(x)$ along the axis of the bar. The normal strain is related to the difference in the displacement at two closely spaced points, that is,

$$\varepsilon_x(x) = \lim_{\Delta x \to 0}\left(\frac{u_x(x+\Delta x) - u_x(x)}{\Delta x}\right) = \frac{\partial u_x}{\partial x}. \tag{D.12}$$

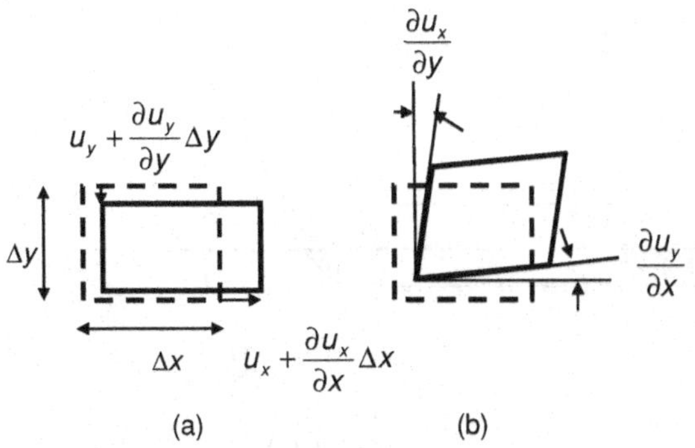

Figure D.4 Strain deformation for a two-dimensional element. The dashed figures represent the undeformed element and solid shapes represent it after deformation. (a) Normal strain. (b) Shear strain. The total angle change for small deformations is $\partial u_y/\partial x + \partial u_x/\partial y$.

The defining expressions for the other normal (dilatational) strain components, ε_y and ε_z, are similar in form to Eq. (D.12).

Solids will also deform under shear stresses. In the limit of small angle changes, the *shear strain* is[1]

$$\gamma_{mn} = \gamma_{nm} = \frac{\partial u_n}{\partial x_m} + \frac{\partial u_m}{\partial x_m}, \qquad m \neq n. \tag{D.13}$$

Note that, from purely geometrical considerations, shear strain components must be symmetric. Figure D.4 depicts two-dimensional rectangular elements in pure normal and pure shear strain.

For the three-dimensional rectangular volume element of Fig. D.2, the *strain tensor*, also called the *state of strain*, is

$$\overline{\overline{e}} = \begin{bmatrix} \varepsilon_x & 1/2\gamma_{xy} & 1/2\gamma_{xz} \\ 1/2\gamma_{xy} & \varepsilon_y & 1/2\gamma_{yz} \\ 1/2\gamma_{xz} & 1/2\gamma_{yz} & \varepsilon_z \end{bmatrix}. \tag{D.14}$$

If the coordinate system is rotated, the strain tensor transforms in the same fashion as the stress tensor:

$$\overline{\overline{e}}_r = \overline{\overline{R}}\,\overline{\overline{e}}\,\overline{\overline{R}}^{\mathrm{T}}. \tag{D.15}$$

It is informative to examine the transformation of strain for the same $\pi/4$ rotation about the z-axis considered in Section D.1 for the stress tensor:

$$\begin{bmatrix} 0 & \gamma_{xy}/2 & 0 \\ \gamma_{xy}/2 & 0 & 0 \\ 0 & 0 & 0 \end{bmatrix} \xrightarrow{\overline{\overline{R}}_{z,\pi/4}} \begin{bmatrix} \gamma_{xy}/2 & 0 & 0 \\ 0 & -\gamma_{xy}/2 & 0 \\ 0 & 0 & 0 \end{bmatrix}. \tag{D.16}$$

Like the stress tensor, pure shear strain transforms into pure normal strain in the xy-plane. This result is purely a consequence of geometry.

1 Equation (D.13) defines *engineering shear strain*, which is used in this text. Mathematical shear strain is defined as ½ of the engineering shear strain, that is, $\varepsilon_{xy} = \gamma_{xy}/2$.

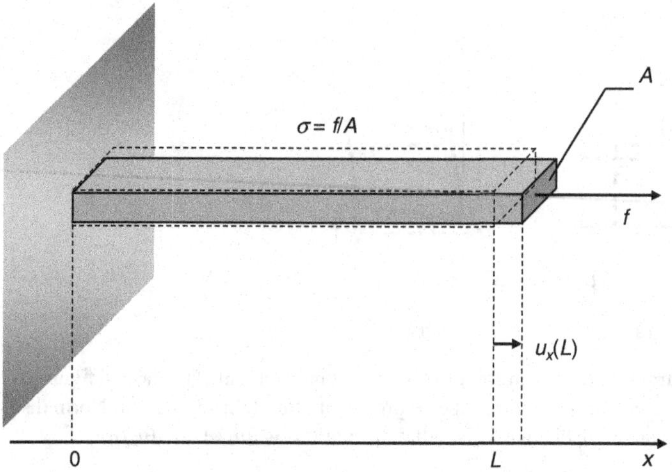

Figure D.5 A bar in tension is the simplest elasticity problem. In general, elongation in the x direction is accompanied by contraction in the other two directions, a phenomenon referred to as the Poisson effect.

D.3 Hooke's law

Stress and strain are related through the elastic constitutive properties of the solid material. To a good approximation, $\overline{\overline{\sigma}}$ and $\overline{\overline{\varepsilon}}$ are linearly related for many materials. The coefficients of this linear relationship are the *elastic constants*.

To explore the implications of a linear constitutive law, one can again refer initially to the simple case of axial deformation, as depicted in Fig. D.5. Stress and strain are given by Eqs (D.1) and (D.12), respectively. The normal stresses and strains are related through Young modulus Y:[2]

$$\sigma = Y\varepsilon, \tag{D.17}$$

where Eq. (D.17) is known as Hooke's law. Because strain is non-dimensional, Y has the same units as stress [Pa]. Section D.5 will reveal an analogy of Hooke's law with a linear, mechanical spring.

Another linear stress–strain relation may be identified for the case of pure shear:

$$\tau = G\gamma, \tag{D.18}$$

where G is the shear modulus.

To achieve general tensor forms for Eqs. (D.17) and (D.18), we must consider the mutual dependences of the strain elements. From mass conservation, one expects deformation of the material in the direction of an applied force to be accompanied by deformations in the perpendicular directions. For example, Fig. D.5 suggests lateral contractions

2 In this text, we use Y for Young modulus instead of the more commonly used E, to avoid confusion with the electric field vector in Chapter 8.

of an elongated bar under tensile stress. The Poisson ratio ν, a non-dimensional quantity, relates the magnitudes of the axial and lateral strains,

$$\nu = -\frac{\varepsilon_y}{\varepsilon_x} = -\frac{\varepsilon_z}{\varepsilon_x}, \tag{D.19}$$

and is a non-dimensional material property. For most materials $\nu \approx 0.3$; for crystalline silicon, the most important MEMS material, ν depends on direction (see Section D.12). The material constants Y, G, and ν are not independent. Their relationship can be obtained by considering the $\pi/4$ rotational transformations performed in Sections D.2 and D.3, that is, Eqs. (D.11) and (D.16). For the unrotated system,

$$\tau_{xy} = G\gamma_{xy}. \tag{D.20}$$

On rotation, τ_{xy} and γ_{xy} become normal stress and normal strain, related by

$$\gamma_{xy}/2 = \underbrace{\frac{1}{Y}\tau_{xy}}_{\text{uniaxial part}} + \underbrace{\frac{\nu}{Y}\tau_{xy}}_{\text{Poisson part}} = \frac{(1+\nu)}{Y}\tau_{xy}. \tag{D.21}$$

Thus,

$$G = \frac{Y}{2(1+\nu)}. \tag{D.22}$$

Stress and strain are second-order tensors. In general, to relate two second-order tensors, one needs a fourth-order tensor; however, because of their symmetry, only six independent quantities are needed to specify stress and strain. Therefore, fourth-order tensors can be avoided altogether by defining stress and strain as vectors [1]:

$$\overline{\sigma} = [\sigma_x \quad \sigma_y \quad \sigma_z \quad \tau_{xy}\, \tau_{yz} \quad \tau_{zx}]^{\mathrm{T}}, \tag{D.23}$$

$$\overline{e} = [\varepsilon_x \quad \varepsilon_y \quad \varepsilon_z \quad \gamma_{xy}\, \gamma_{yz} \quad \gamma_{zx}]^{\mathrm{T}}. \tag{D.24}$$

Then, for a linear elastic material,

$$\overline{\sigma} = \overline{\overline{C}}\,\overline{e}, \tag{D.25}$$

where $\overline{\overline{C}}$ is a symmetric 6×6 material stiffness matrix with 21 independent constants. For the important case of isotropic homogenous materials, $\overline{\overline{C}}$ simplifies because the matrix elements reduce to functions of only two constants, Young modulus Y and

Poisson ratio ν:

$$\overline{\overline{C}} = \frac{Y(1-\nu)}{(1+\nu)(1-2\nu)} \begin{bmatrix} 1 & \frac{\nu}{(1-\nu)} & \frac{\nu}{(1-\nu)} & 0 & 0 & 0 \\ \frac{\nu}{(1-\nu)} & 1 & \frac{\nu}{(1-\nu)} & 0 & 0 & 0 \\ \frac{\nu}{(1-\nu)} & \frac{\nu}{(1-\nu)} & 1 & 0 & 0 & 0 \\ 0 & 0 & 0 & \frac{(1-2\nu)}{2(1-\nu)} & 0 & 0 \\ 0 & 0 & 0 & 0 & \frac{(1-2\nu)}{2(1-\nu)} & 0 \\ 0 & 0 & 0 & 0 & 0 & \frac{(1-2\nu)}{2(1-\nu)} \end{bmatrix}. \tag{D.26}$$

Note that the fourth diagonal element relating γ_{xy} to τ_{xy} reduces to

$$\frac{\tau_{xy}}{\gamma_{xy}} = \frac{Y}{2(1+\nu)}, \tag{D.27}$$

which is consistent with Eq. (D.22).

D.4 General problem of linear elasticity

In general, linear elasticity problems concern solutions for the stresses and displacements within solids, subject to external body forces and applied stresses. In addition, there are boundaries with either applied forces or physical constraints. Within the domain of such problems, there are 15 unknowns:

(a) Six components of the stress vector: $(\sigma_x, \sigma_y, \sigma_z, \tau_{xy}, \tau_{yz}, \tau_{zx})$;
(b) Six components of the strain vector: $(\varepsilon_x, \varepsilon_y, \varepsilon_z, \gamma_{xy}, \gamma_{yz}, \gamma_{zx})$;
(c) Three displacement components: (u_x, u_y, u_z).

To solve for these unknowns there are 15 equations, including:

(a) Three equilibrium equations, Eqs. (D.6) through (D.8);
(b) Six strain–displacement relationships: Eqs. (D.12) and (D.13);
(c) Six stress–strain relationship: Eq. (D.25).

The relations between strain and displacement are not independent. The continuity condition imposed on the displacements establishes some relationships between them and the strains, namely, six partial differential equations relating strains to displacement (see, e.g., [2]).

It is clear that solid mechanics problems involve a daunting complexity of coupled partial differential equations, the solution of which is seldom straightforward. Fortunately, the majority of MEMS geometries are simple enough to facilitate approximations and then solutions that are intuitive and relatively easy to interpret. In the next three sections of this appendix, we introduce and examine the most important examples of these: bars placed in tension, rods placed in torsion, and bent beams.

D.5 Deformation of axially loaded bar

The problem of an axially loaded bar serves well in a review of solid mechanics appropriate to MEMS devices. From the definition of strain

$$\varepsilon_x = \mathrm{d}u_x/\mathrm{d}x. \tag{D.28}$$

Combining the equilibrium equation, Eq. (D.1), with Hooke's law, Eq. (D.17), and Eq. (D.28), using the fixed-wall boundary condition, $u_x(0) = 0$, and then integrating gives

$$u_x(x) = \int_0^x \frac{f}{AY}\mathrm{d}x = \frac{f}{AY}x. \tag{D.29}$$

The displacement increases linearly from zero at the fixed end to a maximum at the tip of the bar, $x = L$. Expressing the force in terms of the displacement at the tip of the bar gives

$$f = \frac{AY}{L}u_x(L), \tag{D.30}$$

which has the same form as the force–displacement relation of a linear mechanical spring, $f = k_{\mathrm{eq}}u_x$, where

$$k_{\mathrm{eq}} = AY/L \tag{D.31}$$

is an equivalent spring constant. In more complex geometries, such as a tapered beam, where the cross-section A is a function of x, evaluation of the integral in Eq. (D.29) is more involved. Yet, as long as the elastic solid is linear, an equivalent spring constant can always be defined.

D.6 Torsion of a rod

Simple elements subjected to torsion are called *rods*. This terminology distinguishes them from *bars*, which experience uniaxial deformation, and *beams*, which are laterally deflected. Although a proper mathematical treatment of rods of arbitrary cross-section is far beyond our scope here, elements with a circular cross-section are readily analyzed. While the circular cross-section is not common in MEMS structures, there are examples of devices with torsional mechanical elements. These practical geometries are usually modeled using a simple correction factor to the circular cross-section case.

For the circular rod illustrated in Fig. D.6(a), the angle is a linear function of torque [3],

$$\phi = \frac{TL}{GI_p}, \tag{D.32}$$

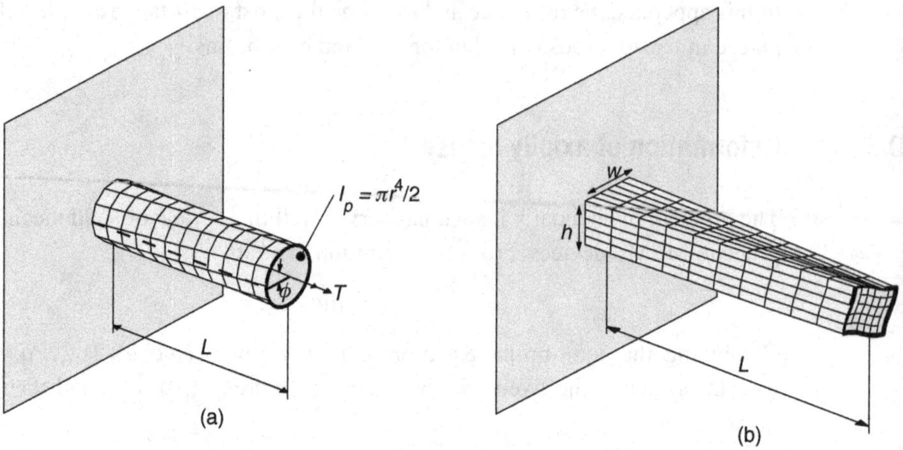

Figure D.6 Rods under torsional stress. (a) Torsion applied to a uniform circular rod. Each infinitesimal elemental disk undergoes rotation with no deformation in the axial direction. (b) Torsion applied to a prismatic element. In this case, the distortion is far more complex, resulting in warping.

where ϕ is the twist angle at the end of the rod, T is the externally applied torque, L is the rod length, G is the shear modulus from Eq. (D.18), and I_p is the *moment of inertia* defined about the central axis,

$$I_p = \int_A r^2 \mathrm{d}A. \tag{D.33}$$

Note that Eq. (D.32) is analogous in form to Eq. (D.30), with φ, T, G, and I_p replacing u_x, F, Y, and A, respectively.

What complicates the situation for rods with non-circular cross-sections under torsion is that deformations are not restricted to the cross-sectional plane. This effect, depicted in Fig. D.6b, is called warping. In general, warping is non-negligible and greatly complicates the distribution of stress and strain.

For rectangular cross-sectional rods elements typical of MEMS devices, Eq. (D.32) is replaced by the approximate relation [4]

$$\phi = TL/(G\beta wh^3), \tag{D.34}$$

where w and h are the width and thickness of the rod, and $L \gg w > h$. The coefficient β depends monotonically on the ratio w/h. For a square cross-section, that is, $w/h = 1$, $\beta \approx 0.14$, while in the limit of $w/h \to \infty$, $\beta = 0.333$.

D.7 Bending of beams

A beam is a structural element that bends in response to a force applied perpendicular to the axis. There are three ways to impart bending deformation to a beam: concentrated load f_y, distributed load F_y, and moment M.

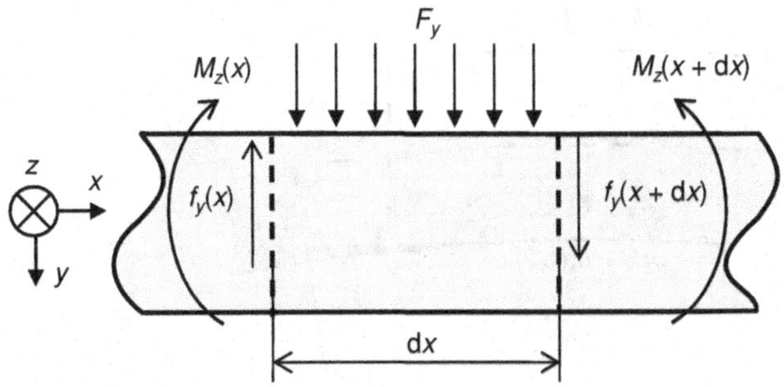

Figure D.7 Differential beam element in equilibrium. The sum of all forces acting in the y direction must be zero. The sum of moments about the z-axis must also be zero.

D.7.1 Force and moment equilibria

Figure D.7 shows a differential element subjected to moments about the z-axis, M_z, loads along the y-axis, f_y, and distributed loads per unit length, F_y. The differential equation for bending arises from the equilibrium condition applied to a beam element. The vector sum of forces acting in the y direction on an infinitesimal element,

$$f_y(x + \mathrm{d}x) - f_y(x) + F_y \mathrm{d}x = 0, \tag{D.35}$$

is, in the limit $\mathrm{d}x \to 0$,

$$\mathrm{d}f_y/\mathrm{d}x = -F_y. \tag{D.36}$$

The other equilibrium condition involves a summation of moment terms:

$$f_y(x + \mathrm{d}x)\mathrm{d}x/2 + f_y(x)\mathrm{d}x/2 + M_z(x) - M_z(x + \mathrm{d}x) = 0, \tag{D.37}$$

which in the $\mathrm{d}x \to 0$ limit yields

$$\mathrm{d}M_z/\mathrm{d}x = f_y. \tag{D.38}$$

D.7.2 Beam in pure bending

Figure D.8 depicts a beam element in a state of *pure bending*. The portion of the beam above the *midplane* elongates and is under tension, while the lower portion contracts and endures compression. Continuity of deformation then requires the existence of a surface inside the beam, which neither elongates nor contracts. This locus, denoted in the figure by the dashed line, is the *neutral line*. The beam's radius of curvature ρ is measured from the undeflected midplane to the neutral line; because there is no elongation at the midplane, $\rho \mathrm{d}\theta$ remains constant. The deformation varies linearly from the neutral line and attains its maxima and minima at the top and bottom surfaces, respectively. If, as depicted in Fig. D.8, the beam happens to be symmetric, the neutral line is the line of symmetry.

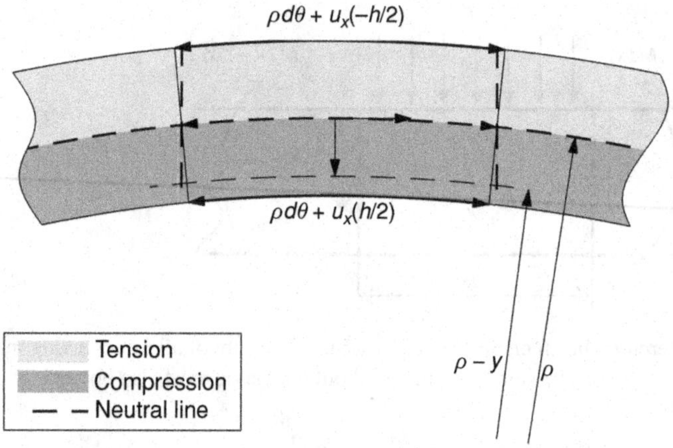

Figure D.8 Beam element in pure bending, with bending exaggerated for clarity. Each plane surface of the unbent beam undergoes rotation about the neutral axis, but no other form of distortion affects these planes.

The strain in the x direction is obtained using the definitions in Fig. D.8 directly from the following geometric relations:

$$\left.\begin{aligned} \rho \mathrm{d}\theta &= \mathrm{d}x \\ \mathrm{d}x + u_x &= (\rho - y)\mathrm{d}\theta \\ \varepsilon_x &= \frac{u_x}{\mathrm{d}x} \end{aligned}\right\} \Rightarrow \varepsilon_x = -\frac{y}{\rho}. \tag{D.39}$$

Then, from Hooke's law,

$$\sigma_x = -Yy/\rho. \tag{D.40}$$

The total moment due to elastic forces is obtained by integration of the product of the elemental force $\sigma_x \mathrm{d}A$ and the associated moment arm y:

$$M = \int_A y \underbrace{\sigma_x \mathrm{d}A}_{\mathrm{d}F} = -\frac{Y}{\rho} \underbrace{\int_A y^2 \mathrm{d}A}_{I_x} = -\frac{YI_x}{\rho}. \tag{D.41}$$

The *area moment of inertia* I_x, also called the second moment of inertia, plays a role in bending similar to that of area A in uniaxial deformation. For a beam with rectangular cross-section, having width w and thickness h,

$$I_x = \int_{-h/2}^{h/2} y^2 \underbrace{w\mathrm{d}y}_{\mathrm{d}A} = \frac{1}{12} wh^3. \tag{D.42}$$

The relationship between moment M and stress σ_x is

$$\sigma_x = My/I_x, \tag{D.43}$$

where $\sigma_x = 0$ along the neutral line, depends linearly on y, and is inversely proportional to I_x. Equation (D.41) relates the moment M to the *curvature*, $1/\rho$, which is

$$\frac{1}{\rho} = \frac{\mathrm{d}^2u_y/\mathrm{d}x^2}{[1+(\mathrm{d}u_y/\mathrm{d}x)^2]^{3/2}} \approx \frac{\mathrm{d}^2u_y}{\mathrm{d}x^2}, \tag{D.44}$$

where u_y denotes transverse displacement of the neutral line with respect to the reference plane. Inserting Eq. (D.44) in Eq. (D.41) yields a differential equation relating small transverse (bending) displacements to the moment:

$$M = -YI\frac{\mathrm{d}^2u_y}{\mathrm{d}x^2}. \tag{D.45}$$

Equation (D.45) is actually the most general form of the partial differential equation for a bending beam, because shear and moment loading can be expressed mathematically using spatial impulse and doublet functions, respectively. Two other types of load – point and distributed shear – are easily related to transverse displacement by combining Eq. (D.45) with Eqs. (D.38) and (D.36), respectively:

$$f_y = -\frac{\mathrm{d}}{\mathrm{d}x}\left(YI_x\frac{\mathrm{d}^2u_y}{\mathrm{d}x^2}\right) \tag{D.46}$$

and

$$F_y = \frac{\mathrm{d}^2}{\mathrm{d}x^2}\left(YI_x\frac{\mathrm{d}^2u_y}{\mathrm{d}x^2}\right). \tag{D.47}$$

The beams designed into many MEMS devices are uniform, meaning that the product YI_x does not depend on x and can be moved outside the derivative:

$$F_y = YI_x\frac{\mathrm{d}^4u_y}{\mathrm{d}x^4}. \tag{D.48}$$

This equation holds for the long slender beams. See Section D.10 and Eq. (D.84) for the case of short wide beams.

D.7.3 Boundary conditions for bending beams

In general, beam bending is a fourth-order boundary value problem requiring four boundary conditions to achieve a solution. There are two types of boundary condition. *Natural* (also known as *geometric* or *Neumann*) *boundary conditions* constrain either the displacement itself or the slope, or both, usually at the end of a beam. The other type, *force boundary conditions*, account for externally applied forces or moments and can be applied anywhere along the length of a beam.

Table D.1 summarizes the mathematical formulations for the most commonly recognized boundary conditions. Note that the fixed (clamped) condition is purely geometric. The free-end conditions are of the force type. The pinned and guided boundary conditions are mixed, that is, they consist of one of each type.

Table D.1 The common boundary conditions relevant to bending beam problems

Boundary condition type	Geometric	Force
Fixed[a] 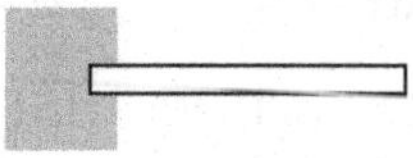	$u_y = 0$; $\mathrm{d}u_y/\mathrm{d}x = 0$	None
Free 	None	$YI\mathrm{d}^2u_y/\mathrm{d}x^2 = M$ and $YI\mathrm{d}^3u_y/\mathrm{d}x^3 = f_y$
Pinned 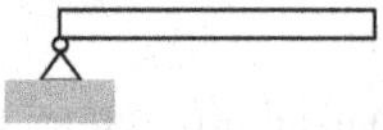	$u_y = 0$	$YI\mathrm{d}^2u_y/\mathrm{d}x^2 = M$
Guided 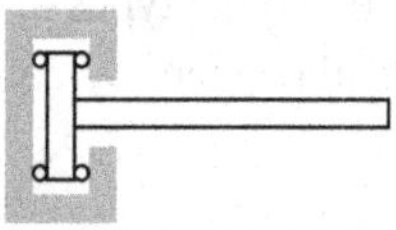	$\mathrm{d}u_y/\mathrm{d}x = 0$	$YI\mathrm{d}^3u_y/\mathrm{d}x^3 = f_y$

[a] This boundary condition is also known as *clamped* and *built in*.

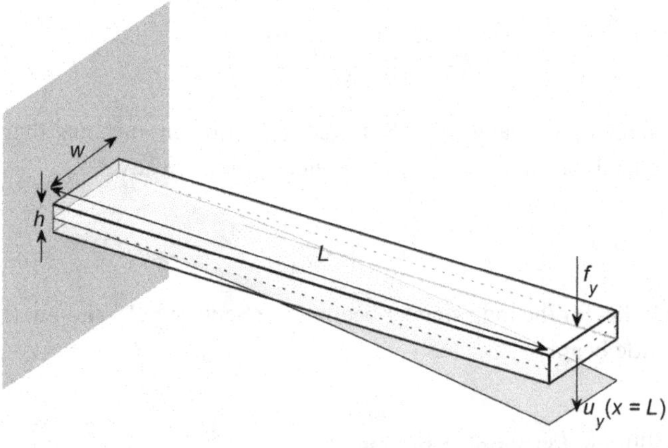

Figure D.9 A cantilevered beam deflected by a transverse point force f_y applied at the free end, $x = L$. In this sketch, only the deflection of the neutral line is depicted.

D.7.4 Cantilevered beam

Probably because it is easiest to fabricate, the cantilevered beam shown in Fig. D.9 is the most commonly used mechanical element found in MEMS systems. In the cantilevered beam, one end is fixed and the other end is free. An easy way to solve for the transverse displacement $u_y(x)$ of the cantilevered beam is to assume a polynomial solution form. Because there are four boundary conditions, a third-order polynomial is required:

$$u_y = a_0 + a_1x + a_2x^2 + a_3x^3. \tag{D.49}$$

The zero displacement and zero slope conditions at the fixed end, $x = 0$, are met if $a_0 = a_1 = 0$. The zero moment boundary condition, $\mathrm{d}^2u_y/\mathrm{d}x^2 = 0$, applied at $x = L$, provides a relationship between a_2 and a_3. Finally, using the force boundary condition of Eq. (D.46) yields

$$u_y = \frac{f_y}{6YI_x}(3Lx^2 - x^3). \tag{D.50}$$

As expected, the displacement increases monotonically, with its maximum at the free end, $x = L$:

$$u_y(L) = f_y L^3/(3YI_x). \tag{D.51}$$

The equivalent stiffness of the cantilevered beam acted upon by a force of its free end is the ratio of the force f_y and displacement $u_y(L)$:

$$k_{\text{cant}} = 3\frac{YI_x}{L^3} = \frac{1}{4}Yw\left(\frac{h}{L}\right)^3. \tag{D.52}$$

D.7.5 Stiffness of a guided beam

The guided beam has one end fixed and the other constrained, such that its slope remains zero in the displaced state. Macroscopic beams can be formed with this constraint, as sketched in the third row of Table D.1. Although a guided beam may not seem to be a realizable MEMS structure, its boundary conditions are used to analyze a very important MEMS structure called the *folded beam*, which is considered Section 7.3. The boundary conditions for this configuration are

$$\begin{aligned} u(x=0) &= 0, \\ \frac{du}{dx}(x=0) &= 0, \\ \frac{du}{dx}(x=L) &= 0, \\ EI\frac{d^3u}{dx^3}(x=L) &= -f_y. \end{aligned} \tag{D.53}$$

Applying the boundary conditions of Eq. (D.53) to Eq. (D.49) and solving for the four coefficients yields

$$u_y(x) = \frac{f}{12YI_x}(3Lx^2 - 2x^3). \tag{D.54}$$

The equivalent stiffness at the tip of the beam $x = L$ is

$$k_{\text{guided}} = \frac{12YI_x}{L^3} = Yw\left(\frac{h}{L}\right)^3 = 4k_{\text{cant}}. \tag{D.55}$$

D.7.6 Stiffness coefficients for bars, rods, and beams

Table D.2 summarizes the effective stiffness coefficients for a prismatic element of length L, width w, and thickness h, subjected to uniaxial, torsional, and bending stresses. Note

Table D.2 Effective stiffness of a prismatic element of length L, width w, and thickness h, subjected to various types of stress

Type of deformation	Uniaxial	Torsional	Bending
Effective stiffness coefficient	$k_{ux} = \dfrac{F}{u_x} = \dfrac{Ywh}{L}$	$k_\phi = \dfrac{T}{\phi} = \dfrac{Yw\beta h^3}{2(1+\nu)L}$	$k_b = \dfrac{3YI}{L^3} = \dfrac{Ywh^3}{4L^3} = \dfrac{1}{4}k_{ux}(h/L)^2$

that the bending stiffness k_b is directly related to the uniaxial stiffness k_{ux}. In particular, k_b is smaller than k_{ux} by a factor of $4(L/h)^2$. Because the units of k_f [N-m] are different from those of k_b and k_{ux} [N/m], direct comparison of the torsional stiffness expression with uniaxial or bending is not meaningful, though the analogy is evident.

D.8 Mechanical resonances of a prismatic element

Many MEMS structures beneficially exploit mechanical resonance, so there is strong motivation to investigate the phenomenon in continua such as bars, rods, and beams. For MEMS, the most important example is the resonant bending beam. In general, dynamic behavior is very complex, but, by sacrificing some generality, we can take advantage of a simplification offering sufficient accuracy for most MEMS applications.

Because MEMS devices are often operated close to their lowest mechanical resonance, it is adequate to fall back on an analogy to the mass–spring mechanical resonator presented in Appendix B. The natural frequency of the mass–spring resonator is $\omega_0 = \sqrt{k/m}$, where k and m are, respectively, the spring constant and mass. For convenience, damping is neglected here, though it can be added later. To model mechanical continua near resonance, one approach is to identify equivalent spring constant and mass values, k_eq and m_eq, for the structure. At least for the lowest resonant mode of a prismatic element, it is often sufficient to use effective stiffness values derived from static analyses to estimate k_eq. See Table D.2. We then employ an energy argument and knowledge of the static deflection profile to estimate m_eq. Finally, $\omega_0 = \sqrt{k_\mathrm{eq}/m_\mathrm{eq}}$ is used to estimate the resonant frequency. At least for the more well-studied geometries, like the cantilevered beam, estimates obtained in this way compare within $\sim$10% of values obtained from analytical or numerical computation.

D.8.1 Resonance of uniaxial bar

It might be tempting to use the total mass of the bar to estimate m_eq, but this crude approximation ignores the fact that, at resonance, not all portions of the bar have the same motional amplitude. In fact, if the goal is to estimate the equivalent mass with respect to motion at the free end, then the total mass grossly overestimates m_eq. A better approach acknowledges the x-dependence of the amplitude of the motion. From consideration of the multiple-degree-of-freedom systems introduced in Appendix B, we know that at resonance the entire structure moves according to the mode shape and in phase. Using the static bar deformation to approximate the mode shape of the lowest resonance, then, based on a kinetic energy argument referenced to the free end,

$$m_{\mathrm{eq}} \approx \frac{\int_0^L u_x^2(x)\rho A(x)\mathrm{d}x}{u_x^2(L)}. \tag{D.56}$$

Using (D.29),

$$m_{\mathrm{eq}} \approx \frac{\int_0^L u_x^2(L)(x/L)^2 \rho wh\mathrm{d}x}{u_x^2(L)} = m_{\mathrm{bar}}/3, \tag{D.57}$$

where $m_{\mathrm{bar}} = \rho whL$ is the total mass. The estimate for the resonance of the first mode is then

$$\omega_{1,ux} \approx \frac{1}{L}\sqrt{\frac{3Y}{\rho}}. \tag{D.58}$$

The exact solution for the purely axial vibrational resonances of a bar with one end fixed and the other free is [5]

$$\omega_{k,ux} = \frac{(2k-1)\pi}{2L}\sqrt{\frac{Y}{\rho}}, \quad \text{for } k = 1, 2, \ldots \tag{D.59}$$

Comparing Eqs. (D.58) and (D.59) for $k = 1$, the discrepancy for the resonant frequency is ~10%.

D.8.2 Bending beam resonance

The same exercise can be repeated for the lowest resonant mode of the cantilevered beam. The first step is to evaluate the equivalent mass from an integral along the length of the beam:

$$m_{\mathrm{eq}} \approx \frac{\int_0^L u_y^2(x)\rho A\mathrm{d}x}{u_y^2(L)}. \tag{D.60}$$

Using Eq. (D.50) for the static deflection,

$$m_{\mathrm{eq}} \approx \int_0^L \left[\frac{3x^2}{2L^2} - \frac{x^3}{2L^3}\right]^2 \rho wh\mathrm{d}x = \frac{3}{8}m_{\mathrm{beam}}, \tag{D.61}$$

where $m_{\mathrm{beam}} = \rho whL$. The estimate for the resonance frequency is then

$$\omega_{1,uy} \approx \frac{0.91h}{L^2}\sqrt{\frac{Y}{\rho}}. \tag{D.62}$$

The exact analytical solution for the lowest beam resonance with one end fixed and the other end free is [5]

$$\omega_{1,uy} = \frac{1.018h}{L^2}\sqrt{\frac{Y}{\rho}}. \tag{D.63}$$

The discrepancy between Eqs. (D.62) and (D.63) is ~11%.

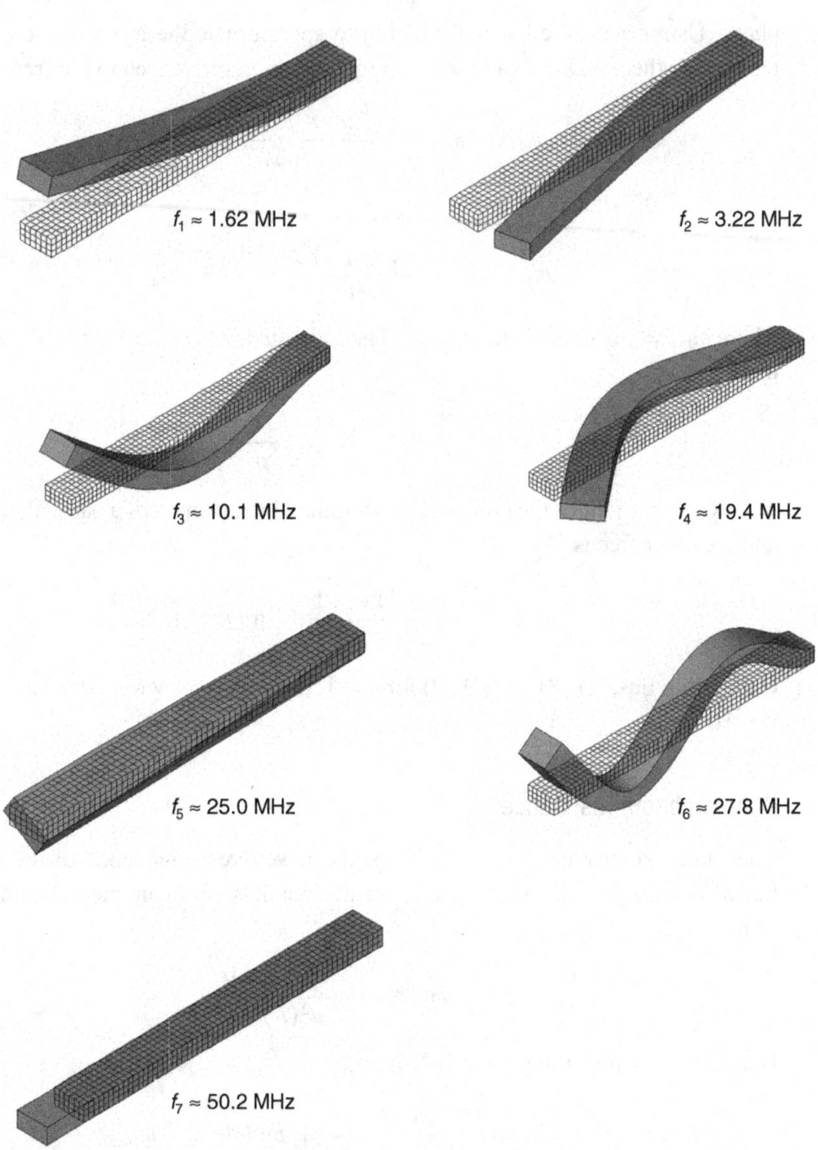

Figure D.10 The first seven resonant modes and their natural frequencies of a silicon beam (length $L = 40$ μm, width $w = 4$ μm, and thickness $h = 2$ μm), obtained using finite-element analysis in ANSYS. Bending modes in any one DOF are widely separated, but w/h determines modal separations for motion in different directions.

D.8.3 More about beam resonances

Because the method described previously for estimating beam resonances breaks down for the higher modes, it is more effective to turn to finite-element analysis (FEA). Doing so reveals just how complex the resonant behavior is, even for a simple prismatic beam. Figure D.10 illustrates the mode shapes and enumerates the resonant frequencies of the first seven modes for a prismatic element of rectangular cross-section with length

$L = 40\ \mu\text{m}$, width $w = 4\ \mu\text{m}$, and height $h = 2\ \mu\text{m}$ clamped at the end. The material properties used are those of crystalline silicon, for which $Y = 150$ GPa,[3] $\nu = 0.17$, and $\rho = 2330\ \text{kg/m}^3$.

Of the first seven, five are bending modes, one is torsional, and the mode at the highest frequency is uniaxial deformation. The lowest resonance is the bending mode along the direction of the thinnest beam dimension. The other modes follow in a frequency-based order that depends strongly on the relative values of the length L, thickness h, and width w. In general, prismatic structures are most compliant to bending, less so to twisting (torsion), and least so to uniaxial deformation. The lesson of Fig. D.10 is that an important aspect of MEMS design is to be aware of unwanted modes that can affect device performance.

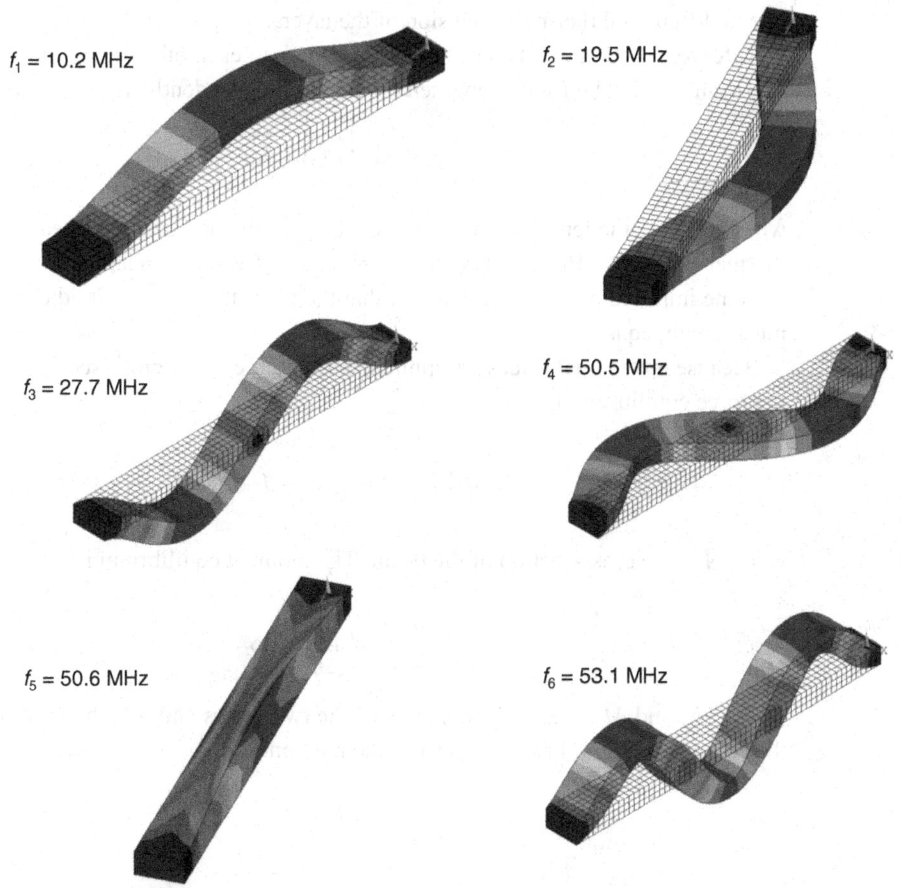

Fig. D.11 The first six modes of a doubly clamped Si beam of length $L = 40\ \mu\text{m}$, width $w = 4\ \mu\text{m}$, and thickness $h = 2\ \mu\text{m}$, obtained using FEA. The first three bending modes in the y direction are 1, 3, and 6. Modes 2 and 4 are the first two bending in the x direction. Mode 5 is the first twisting mode.

3 This is an average value of Young modulus Y. See Section D.12 for more details on the effect of the crystalline anisotropy of Si.

The first six resonances of the same beam with the same dimensions used to obtain Fig. D.10, but now clamped at both ends, are shown in Fig. D.11. Though the frequencies are much higher, these modes are essentially the same as those for the cantilevered beam.

D.9 Temperature-induced bending of a composite beam

Many MEMS structures are composites, consisting of two or more materials. For example, a silicon beam can have deposited oxides and metals (typically aluminum and sometimes copper). These depositions are made at elevated temperatures. As the composite structures cool, the dissimilar properties of the materials impart bending moments to the structure. This section examines an idealized two-layer beam subjected to bending due to differential thermal expansion of the layers.

Refer to Fig. D.12. If the layers did not adhere to each other along the interface, the two beams would be free to lengthen or shorten independently according to

$$L(T) = L(T_0) + \alpha T, \tag{D.64}$$

where $L(T_0)$ is the length at the reference temperature and α is the linear coefficient of thermal expansion. Here the two layers are modeled as independent homogenous beams with the imposed geometric constraint that their lengths along their adjoining interface must remain equal.

Because no external forces are applied, the net force at any cross-section is zero. Thus, the force equilibrium is

$$\int_A \sigma \mathrm{d}A = 0 \rightarrow f_{\mathrm{Si}} = f_{\mathrm{SiO_2}} = f, \tag{D.65}$$

where A is the cross-section of the beam. The moment equilibrium is

$$f\frac{h}{2} = M_{\mathrm{Si}} + M_{\mathrm{SiO_2}}, \tag{D.66}$$

where M_{Si} and $M_{\mathrm{SiO_2}}$ are the moments of the two layers and h is the total thickness of the composite beam. From Eq. (D.41), the moments are

$$M_{\mathrm{SiO_2}} = \frac{Y_{\mathrm{SiO_2}} I_{\mathrm{SiO_2}}}{\rho}, \tag{D.67}$$

$$M_{\mathrm{Si}} = \frac{Y_{\mathrm{Si}} I_{\mathrm{Si}}}{\rho}, \tag{D.68}$$

where ρ, I_{Si}, and $I_{\mathrm{SiO_2}}$ are the radius of curvature of the beam, the area moment of inertia of the silicon layer, and the area moment of inertia of the silicon dioxide layer, respectively. Introducing the moment equations, Eqs. (D.67) and (D.68), into the moment

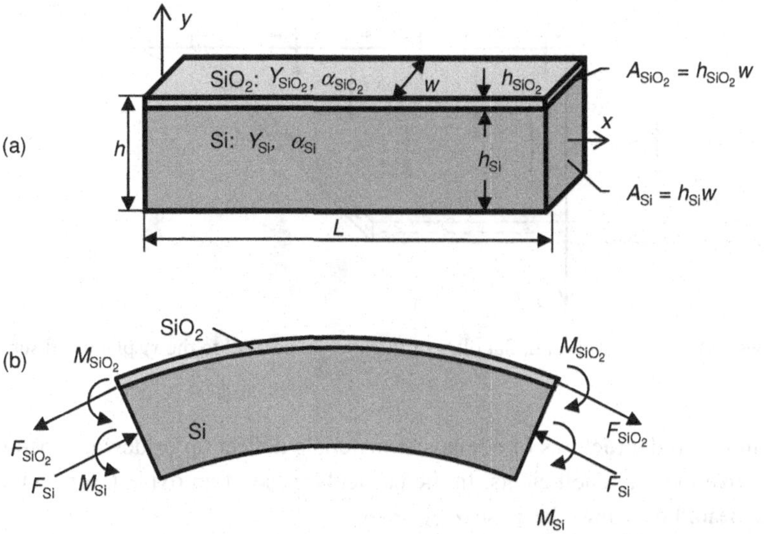

Figure D.12 Idealized geometry of Si beam with oxide layer on one side. (a) Geometry of undeformed beam: (b) Deformed beam with free-body forces and moments. SiO_2 has a smaller temperature coefficient than Si and, therefore, contracts less than Si as it cools.

equilibrium, Eq. (D.66), yields

$$f = \frac{2}{h}\left(\frac{Y_{SiO_2} I_{SiO_2}}{\rho} + \frac{Y_{Si} I_{Si}}{\rho}\right). \tag{D.69}$$

Since the lengths of the two layers were equal during elevated temperature processing and are constrained to remain equal at the interface, the strains at the interface must be equal:

$$\alpha_{SiO_2}\Delta T + \frac{F}{Y_{SiO_2}A_{SiO_2}} + \frac{h_{SiO_2}}{2\rho} = \alpha_{Si}\Delta T - \frac{F}{Y_{Si}A_{Si}} - \frac{h_{Si}}{2\rho}, \tag{D.70}$$

where ΔT is the temperature difference between the processing and the room temperatures. Solving Eq. (D.70) for the curvature $1/\rho$, after plugging in Eq. (D.69), yields

$$\frac{1}{\rho} = \frac{(\alpha_{SiO_2} - \alpha_{Si})\Delta T}{\frac{h}{2} + \frac{2(Y_{SiO_2}I_{SiO_2} + Y_{Si}I_{Si})}{h}\left(\frac{1}{Y_{SiO_2}h_{SiO_2}} + \frac{1}{Y_{Si}h_{Si}}\right)} \tag{D.71}$$

Equation (D.71) shows that curvature is proportional to temperature difference ΔT and the difference in the linear coefficients of thermal expansion $\alpha_{SiO_2} - \alpha_{Si}$.

Once the curvature is known, the displacement can be calculated. For example, to evaluate the displacement at the tip of a cantilevered beam, it is sufficient to integrate the curvature twice (see Eq. (D.44)) and then evaluate it at $x = L$. Thus

$$u(L) = \frac{L^2}{2\rho}. \tag{D.72}$$

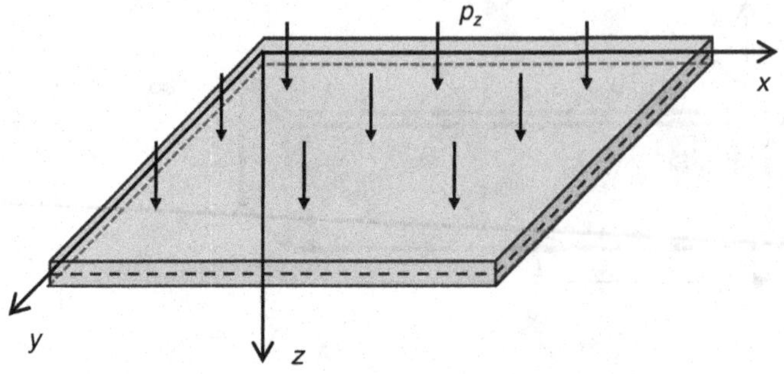

Figure D.13 A uniform, thin, flat plate of thickness h parallel to the xy plane and subject to uniform pressure p_z.

Cantilevered structures of composite materials deflect up or down, depending on their relative thermal coefficients. In the particular case of an oxide layer on top of silicon, the beam bows down because $\alpha_{SiO_2} < \alpha_{Si}$.

For more information on this topic, refer to Timoshenko [6] and Chu *et al.* [7].

D.10 Deformation of thin flat plates

Rigorous analysis of deforming plates is far beyond our scope here. Instead, we pursue the limited objective of deriving the governing equation for static deformations of a thin uniform plate subjected to a distributed normal force. The result is a generalization of the beam equation, Eq. (D.47). The development presented here is based on Saada [8]. More detailed coverage is found in Timoshenko and Woinowsky-Krieger [9].

Figure D.13 depicts a thin, flat, uniform plate parallel to the xy plane. The thickness h is assumed to be much smaller than any plate dimensions in the xy plane. The model, referred to as Kirchhoff's theory, is based on the following assumptions:

- The midplane is assumed to be in an unstrained state, similar to the neutral line of the bending beam, as discussed in Section D.7.
- The normal strain in the z direction is assumed to be negligible, that is, $\varepsilon_z \approx 0$. Thus, from Hooke's law, Eq. (D.17), and the Poisson ratio, Eq. (D.19),

$$\varepsilon_x = (\sigma_x - \nu\sigma_y)/Y,$$
$$\varepsilon_y = (\sigma_y - \nu\sigma_x)/Y. \tag{D.73}$$

- Only γ_{xy} strains are considered; the normals to the neutral plane remain normal to the neutral plane during bending. In other words, out-of-plane shears are neglected:

$$\gamma_{yz} = \frac{\partial u_y}{\partial z} + \frac{\partial u_z}{\partial y} \approx 0,$$
$$\gamma_{zx} = \frac{\partial u_x}{\partial z} + \frac{\partial u_z}{\partial x} \approx 0. \tag{D.74}$$

The normal displacement u_z is a function of x and y,

$$u_z = u_z(x, y), \tag{D.75}$$

while the in-plane displacements are approximately

$$u_x \approx -z\frac{\partial u_z}{\partial x} \quad \text{and} \quad u_y \approx -z\frac{\partial u_z}{\partial y}. \tag{D.76}$$

Combining Eqs. (D.73) and (D.76) gives

$$\begin{aligned} \varepsilon_x &= -z\frac{\partial^2 u_z}{\partial x^2} = \frac{1}{Y}(\sigma_x - \nu\sigma_y), \\ \varepsilon_y &= -z\frac{\partial^2 u_z}{\partial y^2} = \frac{1}{Y}(\sigma_y - \nu\sigma_x), \\ \gamma_{xy} &= \frac{1}{G}\tau_{xy}. \end{aligned} \tag{D.77}$$

Solving Eq. (D.77) for stresses then gives

$$\begin{aligned} \sigma_x &= \frac{Y}{1-\nu^2}(\varepsilon_x + \nu\varepsilon_y) = -\frac{Y}{1-\nu^2}z\left(\frac{\partial^2 u_z}{\partial x^2} + \nu\frac{\partial^2 u_z}{\partial y^2}\right), \\ \sigma_y &= \frac{Y}{1-\nu^2}(\varepsilon_y + \nu\varepsilon_x) = -\frac{Y}{1-\nu^2}z\left(\frac{\partial^2 u_z}{\partial y^2} + \nu\frac{\partial^2 u_z}{\partial x^2}\right), \\ \tau_{xy} &= G\gamma_{xy} = -\frac{Y(1-\nu)}{1-\nu^2}z\frac{\partial^2 u_z}{\partial x \partial y}. \end{aligned} \tag{D.78}$$

Just as with beams, the stresses depend linearly on distance from the neutral plane. Following the beam derivation, we identify two radii of curvature:

$$\begin{aligned} \frac{1}{\rho_x} &\approx \frac{\partial^2 u_z}{\partial x^2}, \\ \frac{1}{\rho_y} &\approx \frac{\partial^2 u_z}{\partial y^2}. \end{aligned} \tag{D.79}$$

The simplifying assumptions that ignored σ_z, γ_{yz}, and γ_{zx} make it impossible to determine σ_z, τ_{yz}, and τ_{zx} from Hooke's law, but we can use the equilibrium conditions, Eqs. (D.6) to (D.8), by assuming no body forces ($F_x = F_y = F_z = 0$). Then, from Fig. D.13, the boundary conditions are

$$\begin{aligned} &\tau_{xz} = \tau_{yz} = 0, z = \pm h/2, \\ &\sigma_z = 0, z = h/2, \\ &\sigma_z = -p_z, z = -h/2. \end{aligned} \tag{D.80}$$

Combining Eq. (D.78) with the equilibrium relations, Eqs. (D.6) to (D.8), and integrating using the boundary conditions (D.80) yields

$$\begin{aligned}
\tau_{zx} = \tau_{xz} &= -\int\left(\frac{\partial\sigma_x}{\partial x} + \frac{\partial\tau_{yx}}{\partial y}\right)\mathrm{d}z = -\frac{Y}{2(1-\nu^2)}\left(\frac{h^2}{4} - z^2\right)\frac{\partial}{\partial x}(\nabla^2 u_z),\\
\tau_{zy} = \tau_{yz} &= -\int\left(\frac{\partial\sigma_y}{\partial y} + \frac{\partial\tau_{xy}}{\partial x}\right)\mathrm{d}z = -\frac{Y}{2(1-\nu^2)}\left(\frac{h^2}{4} - z^2\right)\frac{\partial}{\partial y}(\nabla^2 u_z),\\
\sigma_z &= -\int\left(\frac{\partial\tau_{xz}}{\partial x} + \frac{\partial\tau_{yz}}{\partial y}\right)\mathrm{d}z = -\frac{Y}{2(1-\nu^2)}\left(\frac{h^3}{12} - \frac{h^2 z}{4} + \frac{z^3}{3}\right)\nabla^4 u_z,
\end{aligned} \tag{D.81}$$

where ∇^2 is the Laplacian differential operator. In Cartesian coordinates,

$$\nabla^2\varphi = \frac{\partial^2\varphi}{\partial x^2} + \frac{\partial^2\varphi}{\partial y^2} + \frac{\partial^2\varphi}{\partial z^2}. \tag{D.82}$$

Finally, equating the expressions for σ_z from Eqs. (D.80) and (D.81) yields Lagrange's equation,

$$\rho_z(x, y) = \underbrace{\frac{Yh^3}{12(1-\nu^2)}}_{D}\nabla^4 u_z. \tag{D.83}$$

The coefficient D defined in Eq. (D.83) is the *flexural rigidity*. This term is similar to YI_x, as identified in Eq. (D.48). It is instructive to compare Eq. (D.83) and Eq. (D.48) more closely. Consider a narrow plate of width w. When $\partial u_z/\partial y = 0$, multiplying both sides of Eq. (D.83) by the width w yields

$$\underbrace{\rho_z(x, y)w}_{F_z} = \underbrace{\frac{Yh^3w}{12}}_{I_x}\frac{1}{(1-\nu^2)}\frac{\mathrm{d}^4 u_z}{\mathrm{d}x^4}, \tag{D.84}$$

where F_z is the force per unit length, as before. Notice that the last equation has an additional multiplier, $1/(1-\nu^2)$, compared with the beam equation. Since, for most materials, $\nu \approx 0.3$, the disagreement between the two expressions is less than 10%. Equation (D.84) is used for short wide beams and $Y/(1-\nu^2)$ is referred to as the Young modulus for wide beams.

D.11 Circular plates

The uniform circular plate, clamped around its edges, is an important MEMS structure. Section 6.6 presents a model for a capacitive transducer based on this geometry. The result of Section D.10 can be used to relate displacement $u(r)$ to a uniformly applied pressure p_z.

For the case of azimuthal symmetry, the Laplacian reduces to

$$\nabla^2 u_z = \frac{1}{r}\frac{\mathrm{d}}{\mathrm{d}r}\left(r\frac{\mathrm{d}u_z}{\mathrm{d}r}\right). \tag{D.85}$$

Inserting the last equation into Eq. (D.83) yields

$$\rho_z = D\frac{1}{r}\frac{\mathrm{d}}{\mathrm{d}r}\left(r\frac{\mathrm{d}}{\mathrm{d}r}\left(\frac{1}{r}\frac{\mathrm{d}}{\mathrm{d}r}\left(r\frac{\mathrm{d}u_z}{\mathrm{d}r}\right)\right)\right). \tag{D.86}$$

For uniform pressure p_z and structural rigidity D, Eq. (D.86) can be integrated directly:

$$u_z(r) = \frac{p_z r^4}{64D} + C_1 \ln(r) + C_2 r^2 + C_3 r^2 \ln(r) + C_4. \tag{D.87}$$

Since only finite values of the displacement $u_z(r)$, the slope $\mathrm{d}u_z(r)/\mathrm{d}r$, and the curvature are allowed at $r = 0$, C_1 and C_3 must be zero. The remaining boundary conditions imposed at $r = r_a$ by the clamped edge,

$$u_z(r_a) = 0 \quad \text{and} \quad \frac{\mathrm{d}u_z}{\mathrm{d}r}(r = r_a) = 0, \tag{D.88}$$

can be used to determine C_2 and C_4. The result is

$$u_z(r) = \frac{r_a^4}{64D}\left[1 - \left(\frac{r}{r_a}\right)^2\right]^2 p_z. \tag{D.89}$$

D.12 Elastic properties of crystalline silicon

Silicon, the most important structural material in MEMS, is significantly anisotropic. This section briefly examines the impact of silicon anisotropy on MEMS devices, using the approach of Wortman and Evans [10]. More details on the physical anisotropy of crystalline materials are found in Nye [11], and a comprehensive treatment of the anisotropy of Si as applied to MEMS is found in Hopcroft *et al.* [12].

The elastic matrix $\overline{\overline{C}}$ for Si is sparse and has only three independent coefficients:

$$\overline{\overline{C}} = \begin{bmatrix} C_{11} & C_{12} & C_{12} & 0 & 0 & 0 \\ C_{12} & C_{11} & C_{12} & 0 & 0 & 0 \\ C_{12} & C_{12} & C_{11} & 0 & 0 & 0 \\ 0 & 0 & 0 & C_{44} & 0 & 0 \\ 0 & 0 & 0 & 0 & C_{44} & 0 \\ 0 & 0 & 0 & 0 & 0 & C_{44} \end{bmatrix}, \tag{D.90}$$

where $C_{11} = 165.7$ GPa, $C_{12} = 63.9$ GPa, and $C_{44} = 79.6$ GPa.

The compliance matrix, defined as the inverse of the elastic matrix, has the same form as the elastic matrix:

$$\overline{\overline{S}} = \overline{\overline{C}}^{-1}, \tag{D.91}$$

where $S_{11} = 7.68 \times 10^{-12}$ m^2/N, $S_{12} = -2.14 \times 10^{-12}$ m^2/N, and $S_{44} = 12.6 \times 10^{-12}$ m^2/N.

By convention, processed Si wafers are defined by the crystal orientation of the wafer plane. Most of these, used in microelectronic and MEMS fabrication, are of the (100) type, shown in Fig. D.14. This section is limited to this wafer type.

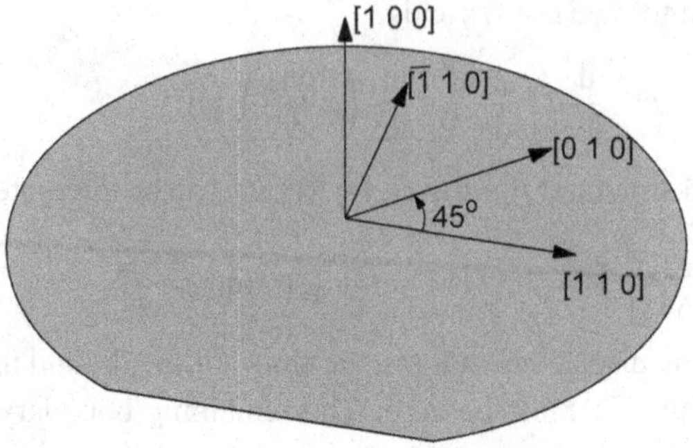

Figure D.14 A perspective view of a standard (100) Si wafer with various crystalline directions indicated. The periodicity of the material properties is 90°, i.e., the properties are the same along the x-axis and the y-axis, but differ significantly in between.

To consider anisotropy, a generalization of the definitions of Young modulus and Poisson ratio is needed. The Young modulus in the direction i is defined as

$$Y_i = \frac{\sigma_i}{\varepsilon_i}. \tag{D.92}$$

From Eqs. (D.25), (D.91), and (D.92), we have

$$Y_i = \frac{1}{S_{ii}}. \tag{D.93}$$

The Poisson ratio between axes i and j is defined as

$$\nu_{ij} = -\frac{\varepsilon_j}{\varepsilon_i}. \tag{D.94}$$

Then, from Eqs. (D.25), (D.91), and (D.94),

$$\nu_{ij} = -\frac{S_{ij}}{S_{ii}}. \tag{D.95}$$

Consider how the Young modulus and Poisson ratio vary in different directions. One way to obtain the Young modulus about an arbitrary axis r is to rotate the compliance tensor so that the x-axis (1-axis) coincides with the axis r; compute the "11" component of the rotated fourth-order tensor S_{r11}; and then use Eq. (D.93) to find Y_r. The details of this transformation are beyond of the scope of this book. For our purposes, it suffices to present the final results. If l, m, and n are the direction cosines between the direction of interest and the cubic axes,[4]

$$l = \cos(\alpha), \quad m = \cos(\beta), \quad n = \cos(\gamma), \tag{D.96}$$

then the Young modulus in this direction is given by

$$Y = \left[S_{11} - 2\left[(S_{11} - S_{12}) - \frac{1}{2}S_{44}\right]\left(l^2m^2 + m^2n^2 + l^2n^2\right)\right]^{-1}. \tag{D.97}$$

4 More details on crystalline axes are given in Section C.3.3.

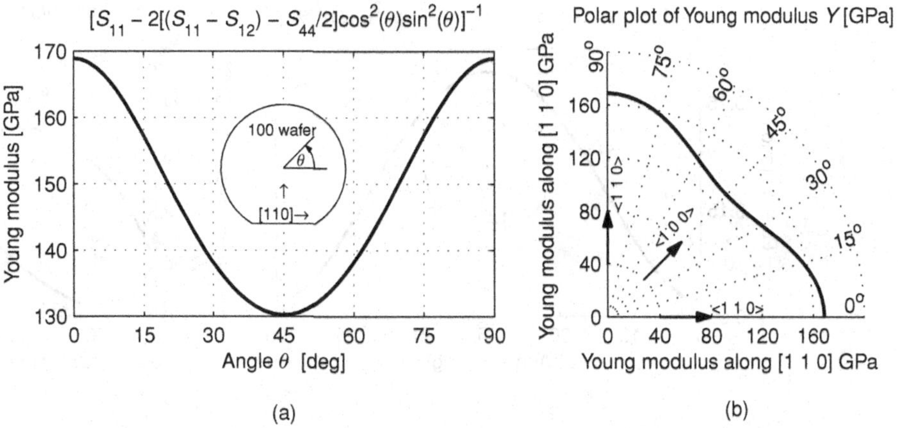

Figure D.15 Variation of Young modulus as a function of angle with respect to the flat of the wafer. (a) Young modulus plotted versus angle for (100) wafer. (b) Polar plot, based on [10], depicting variation of Young modulus with respect to the crystallographic axes.

For Si, the most important directions are (100), (110), and (111). For (100), only one direction cosine is non-zero, viz. $l = m = 0$ and $n = 1$, so

$$Y_{100} = \frac{1}{S_{11}} = 130 \text{ GPa}. \tag{D.98}$$

For (110), $l = 0$, $m = n = \cos(\pi/4) = 1/\sqrt{2}$, so

$$Y_{110} = \left[S_{11} - \frac{1}{2}\left[(S_{11} - S_{12}) - \frac{1}{2}S_{44}\right]\right]^{-1} = 169 \text{ GPa}. \tag{D.99}$$

Finally, for (111), $l = m = n = 1/\sqrt{3}$, so

$$Y_{111} = \left[S_{11} - \frac{2}{3}\left[(S_{11} - S_{12}) - \frac{1}{2}S_{44}\right]\right]^{-1} = 189 \text{ GPa}. \tag{D.100}$$

Since Si beams are fabricated in the wafer plane, their Young moduli vary continuously with the angle with respect of the flat of the wafer, as depicted in Fig. D.15. For this case $l = 0$, $m = \cos(\theta)$, and $n = \sin(\theta)$, so Eq. (97) simplifies to

$$Y = \left[S_{11} - 2\left[(S_{11} - S_{12}) - \frac{1}{2}S_{44}\right](\cos^2(\theta)\sin^2(\theta))\right]^{-1}. \tag{D.101}$$

Figure D.15a explicitly shows the dependence of Young modulus on the angle θ. The polar plot of Fig. D.15b introduced by Wortman and Evans [10] is more general because it does not depend on a particular wafer type.

The Poisson ratio is defined with respect to two mutually perpendicular planes:

$$\nu_{ij} = -\frac{S_{12} + \left(S_{11} - S_{12} - \frac{1}{2}S_{44}\right)(l_i^2 l_j^2 + m_i^2 m_j^2 + n_i^2 n_j^2)}{S_{11} - 2\left[(S_{11} - S_{12} - \frac{1}{2}S_{44}\right](l_i^2 m_i^2 + m_i^2 n_i^2 + l_i^2 n_i^2)}, \tag{D.102}$$

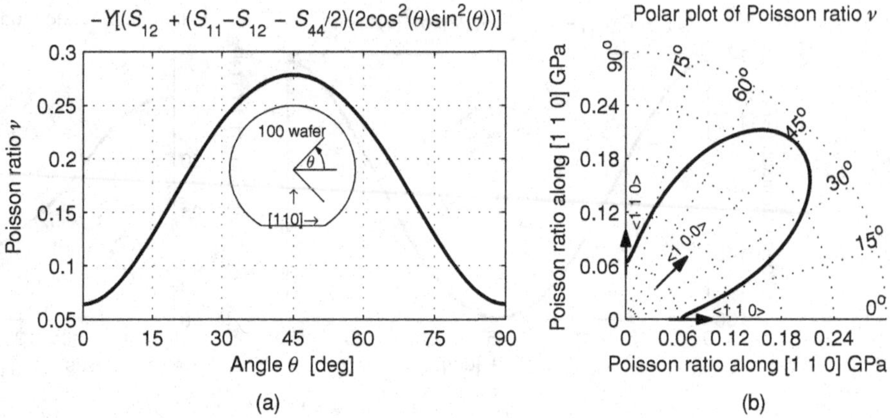

Figure D.16 Variation of Poisson ratio as a function of angle with respect to the flat of the wafer. (a) Poisson ratio plotted vs. angle for (100) wafer. (b) Polar plot, based on [10].

where (l_i, m_i, n_i) and (l_j, m_j, n_j) are the direction cosines of the two planes with respect to the cubic axes. For the two perpendicular directions in the plane of a (100) wafer, the two direction cosines are closely related $(l_i, m_i, n_i) = (0, \cos(\theta), \sin(\theta))$, and $(l_j, m_j, n_j) = (0, \sin(\theta), -\cos(\theta))$, as shown in the inset of Fig. D.16a. For this case Eq. (D.102) simplifies to

$$\nu = -Y \left[S_{12} + \left(S_{11} - S_{12} - \frac{1}{2} S_{44} \right) (2 \cos^2(\theta) \sin^2(\theta)) \right]^{-1}, \tag{D.103}$$

where Y is the Young modulus from Eq. (D.101). The corresponding variation of ν in the (100) wafer plane is plotted in Fig. D.16a. The variation of ν with respect to the crystallographic axes is shown in Fig. D.16b.

Figures D.15 and D.16 reveal that the elastic constants of Si are strong functions of direction. The mechanical design of MEMS must account for this anisotropy to assure desired performance specifications.

The interested reader is referred to Hopcroft *et al.* [12] for more details on the anisotropy of the Young modulus and Poisson ratio in the plane of (100) Si wafers. Because it depends on the crystallographic orientation of a Si wafer, anisotropy is a major consideration in MEMS processing and integration. In fact, Kim *et al.* have argued that (111) wafers are the best choice for MEMS structures because the elastic material properties in this plane are isotropic [13].

References

1. K.-J. Bathe, *Finite Element Procedures* (Englewood Cliffs, NJ: Prentice Hall, 1996).
2. P. C. Chou and N. J. Pagano, *Elasticity: Tensor, Dyadic, and Engineering Approaches* (New York: Dover Publications, 1992).

3. J. M. Gere and S. Timoshenko, *Mechanics of Materials*, 4th edn (Boston: PWS Pub Co., 1997).
4. E. Popov, *Engineering Mechanics of Solids* (Englewood Cliffs, NJ: Prentice-Hall, 1990).
5. R. R. Craig, *Structural Dynamics: An Introduction to Computer Methods* (New York: Wiley, 1981).
6. S. Timoshenko, Analysis of bi-metal thermostats. *Journal of the Optical Society of America*, 11 (1925), 233–255.
7. W.-H. Chu, M. Mehregany and R. L. Mullen, Analysis of tip deflection and force of a bimetallic cantilever microactuator. *Journal of Micromechanics and Microengineering*, 3 (1993), 4.
8. A. S. Saada, *Elasticity Theory and Applications*, 2nd edn (Malabar, FL: Krieger, 1993).
9. S. Timoshenko and S. Woinowsky-Krieger, *Theory of Plates and Shells*, 2nd edn (New York: McGraw-Hill, 1959).
10. J. J. Wortman and R. A. Evans, Young's modulus, shear modulus, and Poisson's ratio in silicon and germanium. *Journal of Applied Physics*, 36 (1965), 153–156.
11. J. F. Nye, *Physical Properties of Crystals* (Oxford: Clarendon Press, 1957).
12. M. A. Hopcroft, W. D. Nix, and T. W. Kenny, What is Young's modulus of silicon? *Journal of Microelectromechanical Systems*, 19 (2010), 229–238.
13. J. Kim, D. Cho, and R. S. Muller, Why is (111) silicon a better mechanical material for MEMS? In *Digest, Transducers 01*, Munich, Germany, 2001, pp. 662–665.

Index